INTERNATIONAL SERIES OF MONOGRAPHS ON PHYSICS

International Series of Monographs on Physics

161. V.Z. Kresin, H. Morawitz, S.A. Wolf: *Superconducting state—mechanisms and properties*
160. C. Barrabès, P.A. Hogan: *Advanced general relativity—gravity waves, spinning particles, and black holes*
159. W. Barford: *Electronic and optical properties of conjugated polymers, Second edition*
158. F. Strocchi: *An introduction to non-perturbative foundations of quantum field theory*
157. K.H. Bennemann, J.B. Ketterson: *Novel superfluids, Volume 2*
156. K.H. Bennemann, J.B. Ketterson: *Novel superfluids, Volume 1*
155. C. Kiefer: *Quantum gravity, Third edition*
154. L. Mestel: *Stellar magnetism, Second edition*
153. R.A. Klemm: *Layered superconductors, Volume 1*
152. E.L. Wolf: *Principles of electron tunneling spectroscopy, Second edition*
151. R. Blinc: *Advanced ferroelectricity*
150. L. Berthier, G. Biroli, J.-P. Bouchaud, W. van Saarloos, L. Cipelletti: *Dynamical heterogeneities in glasses, colloids, and granular media*
149. J. Wesson: *Tokamaks, Fourth edition*
148. H. Asada, T. Futamase, P. Hogan: *Equations of motion in general relativity*
147. A. Yaouanc, P. Dalmas de Réotier: *Muon spin rotation, relaxation, and resonance*
146. B. McCoy: *Advanced statistical mechanics*
145. M. Bordag, G.L. Klimchitskaya, U. Mohideen, V.M. Mostepanenko: *Advances in the Casimir effect*
144. T.R. Field: *Electromagnetic scattering from random media*
143. W. Götze: *Complexdynamics of glass-forming liquids—a mode-coupling theory*
142. V.M. Agranovich: *Excitations in organic solids*
141. W.T. Grandy: *Entropy and the time evolution of macroscopic systems*
140. M. Alcubierre: *Introduction to 3+1 numerical relativity*
139. A.L. Ivanov, S.G. Tikhodeev: *Problems of condensed matter physics—quantum coherence phenomena in electron-hole and coupled matter-light systems*
138. I.M. Vardavas, F.W. Taylor: *Radiation and climate*
137. A.F. Borghesani: *Ions and electrons in liquid helium*
135. V. Fortov, I. Iakubov, A. Khrapak: *Physics of strongly coupled plasma*
134. G. Fredrickson: *The equilibrium theory of inhomogeneous polymers*
133. H. Suhl: *Relaxation processes in micromagnetics*
132. J. Terning: *Modern supersymmetry*
131. M. Mariño: *Chern-Simons theory, matrix models, and topological strings*
130. V. Gantmakher: *Electrons and disorder in solids*
129. W. Barford: *Electronic and optical properties of conjugated polymers*
128. R.E. Raab, O.L. de Lange: *Multipole theory in electromagnetism*
127. A. Larkin, A. Varlamov: *Theory of fluctuations in superconductors*
126. P. Goldbart, N. Goldenfeld, D. Sherrington: *Stealing the gold*
125. S. Atzeni, J. Meyer-ter-Vehn: *The physics of inertial fusion*
123. T. Fujimoto: *Plasma spectroscopy*
122. K. Fujikawa, H. Suzuki: *Path integrals and quantum anomalies*
121. T. Giamarchi: *Quantum physics in one dimension*
120. M. Warner, E. Terentjev: *Liquid crystal elastomers*
119. L. Jacak, P. Sitko, K. Wieczorek, A. Wojs: *Quantum Hall systems*
117. G. Volovik: *The Universe in a helium droplet*
116. L. Pitaevskii, S. Stringari: *Bose-Einstein condensation*
115. G. Dissertori, I.G. Knowles, M. Schmelling: *Quantum chromodynamics*
114. B. DeWitt: *The global approach to quantum field theory*
113. J. Zinn-Justin: *Quantum field theory and critical phenomena, Fourth edition*
112. R.M. Mazo: *Brownian motion—fluctuations, dynamics, and applications*
111. H. Nishimori: *Statistical physics of spin glasses and information processing—an introduction*
110. N.B. Kopnin: *Theory of nonequilibrium superconductivity*
109. A. Aharoni: *Introduction to the theory of ferromagnetism, Second edition*
108. R. Dobbs: *Helium three*
107. R. Wigmans: *Calorimetry*
106. J. Kübler: *Theory of itinerant electron magnetism*
105. Y. Kuramoto, Y. Kitaoka: *Dynamics of heavy electrons*
104. D. Bardin, G. Passarino: *The Standard Model in the making*
103. G.C. Branco, L. Lavoura, J.P. Silva: *CP Violation*
102. T.C. Choy: *Effective medium theory*
101. H. Araki: *Mathematical theory of quantum fields*
100. L.M. Pismen: *Vortices in nonlinear fields*
99. L. Mestel: *Stellar magnetism*
98. K.H. Bennemann: *Nonlinear optics in metals*
94. S. Chikazumi: *Physics of ferromagnetism*
91. R.A. Bertlmann: *Anomalies in quantum field theory*
90. P.K. Gosh: *Ion traps*
87. P.S. Joshi: *Global aspects in gravitation and cosmology*
86. E.R. Pike, S. Sarkar: *The quantum theory of radiation*
83. P.G. de Gennes, J. Prost: *The physics of liquid crystals*
73. M. Doi, S.F. Edwards: *The theory of polymer dynamics*
69. S. Chandrasekhar: *The mathematical theory of black holes*
51. C. Møller: *The theory of relativity*
46. H.E. Stanley: *Introduction to phase transitions and critical phenomena*
32. A. Abragam: *Principles of nuclear magnetism*
27. P.A.M. Dirac: *Principles of quantum mechanics*
23. R.E. Peierls: *Quantum theory of solids*

Superconducting State

Mechanisms and Materials

Vladimir Z. Kresin
Lawrence Berkeley Laboratory

Sergei G. Ovchinnikov
Kirensky Institute of Physics,
Federal Research Center "Krasnoyarsk Science Centre",
Siberian Branch of the Russian Academy of Science"
and Siberian Federal University

Stuart A. Wolf
University of Virginia

OXFORD
UNIVERSITY PRESS

OXFORD
UNIVERSITY PRESS

Great Clarendon Street, Oxford, OX2 6DP,
United Kingdom

Oxford University Press is a department of the University of Oxford.
It furthers the University's objective of excellence in research, scholarship,
and education by publishing worldwide. Oxford is a registered trade mark of
Oxford University Press in the UK and in certain other countries

First Edition published in 2021

Impression: 2

Published in the United States of America by Oxford University Press
198 Madison Avenue, New York, NY 10016, United States of America

British Library Cataloguing in Publication Data
Data available

Library of Congress Control Number: 2021931165

ISBN 978–0–19–884533–1

DOI: 10.1093/oso/9780198845331.001.0001

Printed and bound by
CPI Group (UK) Ltd, Croydon, CR0 4YY

This book is dedicated in memory of
Boris Geilikman
Eugene Kuz'min
Bernard Serin

Preface

During the last several years, the field of superconductivity has undergone intensive developments. Several new families of superconducting materials have been discovered and are under active investigation. These include high-T_c hydrides, which have record values for T_c under high pressure, topological superconductors, graphene-like systems, and others. Superconductivity at room temperature has been achieved under high pressure in 2020.

At no other time have so many different types of superconducting materials been investigated. This positions the field of superconductivity at the intersection of physics, chemistry, and materials science.

Acknowledging that it would be unrealistic to try to describe the full range of superconducting materials, in writing this book, our goal has been to demonstrate the richness of the field by focusing on certain selected systems, and to identify general features that are shared by the various materials families. For example, the novel superconductors are complex compounds, many of which contain mixed-valence ions. This feature plays a key role in the properties of high-T_c cuprates, of manganites, and of high-T_c hydrides.

This book does not aim to be a textbook or a complete review of the field. Its focus is on the mechanisms of superconductivity, their related spectroscopies, and a description of various superconducting materials. Admittedly, the selection of topics has been influenced by the authors' own scientific interests, and some important subjects have not been included, such as transport properties (thermal conductivity) and ac response.

We assume that the reader is familiar with the fundamentals of solid state physics, for example, the work by Ashcroft and Mermin (1976). The theoretical section also assumes a knowledge of statistical mechanics and many-body techniques at the level of the works by Abrikosov *et al.* (1975), Lifshitz and Pitaevskii (2002), and Mahan (1993). Some background information is presented in appendix E.

At the same time, by providing qualitative explanations of the main concepts and results, we hope that the book will be accessible, without a detailed focus on the calculations, making it useful and interesting for a broad audience.

VZK wrote sections 1.1, 2.2–2.5, 3.2, 3.5, 5.1, 5.3, 7.1, and 7.2, chapters 8–10, subsections 1.2.1–1.2.3, 2.1.1–2.16, 3.1.1, 3.1.2, 3.1.4, 3.1.5, 5.4.1–5.4.4, 7.4.3, and 7.4.4, and appendices A–C and E. SGO wrote sections 2.6, 3.3, 3.6, 7.3, and 7.5, subsections 1.2.4, 2.1.7, 3.1.3, 5.4.5, 5.4.6, 7.4.1, and 7.4.2, and appendix D. SAW wrote chapter 4. Chapter 6 was written jointly by VZK and SAW. Sections 3.4 and 5.2 were written jointly by VZK and SGO.

Acknowledgements

Over the last several years, we have benefited from many fruitful discussions. VZK is indebted to Defang Duan, Lev Gor'kov, Vitaly Kresin, Alex Mueller, and Yury Ovchinnikov for many discussions and collaborations. SGO is thankful to Igor Sandalov, Sergey Nikolaev, Vladimir Gavrichkov, Maxim Korshunov, and Elena Shneyder for many helpful discussions and for the pleasure of working together. SAW would like to thank Dr Jiwei Lu and Dr Ale Lukaszew for many interesting discussions.

We are very grateful to Oxford University Press for making this book possible. We would like to thank especially Sonke Adlung, Francesca McMahon, Sumintra Gaur for all their help in setting up and carrying out the production of this book.

We thank Changgyoo Park (Berkeley) and Tamara Ovchinnikova (Krasnoyarsk) for their careful assistance in preparing the manuscript.

This book, as well as our other scientific endeavours, would not have been possible without the forbearance of Lilia Kresin, Tamara Ovchinnikova, and Iris Wolf.

Contents

1

Introduction

1.1 Historical Perspective

The story of superconductivity began early in the twentieth century. The timing was not an accident: quantum physics also was born at the turn of the century. More specifically, its foundation was formed by studies of radiation, by Max Planck, and the photoelectric effect, by Albert Einstein. The first important preliminary step towards the discovery of superconductivity was the liquefication of helium. This was first achieved on 10 July 1908, in Kamerlingh Onnes's laboratory (in Leiden) and that event can be considered to be the beginning of low-temperature physics. Initially, developments in quantum physics and low-temperature studies were independent, but it later became clear that these two fields are interconnected. It was first thought that quantum physics only describes phenomena occurring at the atomic scale. In contrast, many-body systems, like solids, are normally described by classical physics, because thermal motion spreads the quantum behaviour of atoms (or nuclei). However, transitioning to low temperatures allows one to observe macroscopic quantum phenomena, such as superconductivity.

The unique feature of helium is that it stays liquid (at ambient pressure) even if temperature goes down to $T = 0$ K. This behaviour seems strange, because thermal motion is absent at $T = 0$ K and the interatomic interaction should lead to the formation of a solid structure; each ion is 'frozen' near its equilibrium position, which corresponds to the minimum of energy. However, according to quantum mechanics, this is not the case. Each ion even at absolute zero is moving (zero-order vibrations) and, because for helium the interaction is weak (He is an inert atom) and the amplitude of the vibrations is rather large (He is one of the lightest atoms), a solid cannot be formed and helium stays liquid up to $T = 0$ K. The existence of such a liquid allows us to study properties of various materials in the low-temperature region (sometimes liquid helium is called a 'quantum liquid').

As mentioned above, liquid helium was obtained on 10 July 1908. To follow this important result, Kamerlingh Onnes decided to study the temperature dependence of the resistance at such low temperatures. In 1911, Kamerlingh Onnes and his student Gilles Kolst discovered that, at temperatures close to 4 K, the electrical resistance of mercury vanished (Fig. 1.1). According to Kamerlingh Onnes: 'The resistance disappeared ...

Superconducting State: Mechanisms and Materials. Vladimir Z. Kresin, Sergei G. Ovchinnikov, and Stuart A. Wolf, Oxford University Press (2021). © Kresin, Ovchinnikov, Wolf. DOI: 10.1093/oso/9780198845331.003.0001

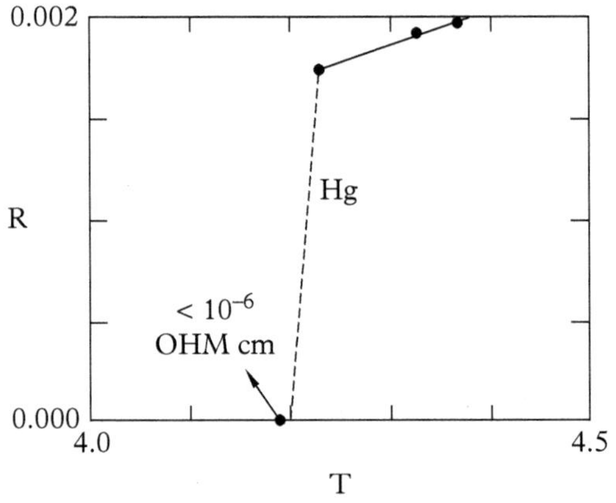

Figure 1.1 *Resistance of mercury vs temperature. (From Kamerlingh Onnes, 1911; the discovery of superconductivity)*

Thus the mercury at 4.2 K has entered a new state which, owing to its particular electric properties, can be called a state of superconductivity' (Onnes, 1911).

Recently, scientists celebrated the one-hundred-year anniversary of this discovery (see e.g. Delft and Kes, 2010). There has been significant progress during this century, so that at present we are dealing with a rather complex and strongly developed field.

The discovery of superconductivity created new area of research with several directions. At first, it was established that absence of resistance is a remarkable phenomenon, but, nevertheless, this is just one of many features of superconductors. In reality we are dealing with a new state of matter, which is characterised by magnetic properties, transport properties, optical properties, and so on, which are entirely different from those for usual metals. There were many important discoveries during these years, and the first of them was the Meissner effect (Meissner and Ochsenfeld, 1933) whereby magnetic flux is excluded from the inside of a superconducting body. Subsequently, a serious challenge appeared to develop the theory of the phenomenon. In addition to finding the origin of superconductivity, it was necessary to explain why this new state is not universal. Indeed, some metals (e.g. Al, Zn, Pb, Nb) are superconductors, whereas other metals (e.g. Cu, Ag, Au) are not.

And, finally, shortly after the discovery room temperature superconductivity was envisioned. Driven by this goal, the search for new superconducting materials with higher values for the critical temperature has started. Mercury was the first superconductor discovered ($T_c \simeq 4.2$ K) At present, we know about twenty pure elements, and Nb has the highest value of T_c among them ($T_c \simeq 9.2$ K). Since 1911 many new superconducting materials were discovered and the list now includes more than 6000. For the first fifty years, the maximum transition temperature rose slowly (about 3 K per decade), reaching about 18 K in the late 1950s in both the B1 structure NbCN and the A15 structure Nb_3Sn.

When discussing theoretical studies, one should mention several important developments preceding the microscopic theory. The two-fluid model (Gorter and Casimir, 1934) provided a useful phenomenological picture, and the London equations (1935) provided a description of the Meissner effect.

An important theoretical advance came with the development of the Ginzburg–Landau theory (1950), which contains a phenomenological description of the superconducting state. It was the solution to these equations that provided the description of the magnetic vortices and the two fundamentally different types of superconductors distinguished by their very different behaviours in a magnetic field. These two kinds of superconductors are type I, which exhibits only the Meissner state, and type II, which exhibits a mixed state where flux is not fully excluded but is incorporated into the superconductor in quantised flux entities called vortices (Abrikosov, 1957).

The main challenge was to explain the phenomenon of superconductivity. This problem was worked on by the most prominent physicists of the twentieth century. A. Einstein thought about peculiar molecular currents (1922). It is interesting that, according to his picture, the sample containing a contact of two independent superconducting pieces should not carry a supercurrent. He convinced Meissner and Holm to perform the experiment (see Huebener and Luebbig, 2008) but, to his disappointment, such a current did flow from one piece to the other. It is interesting that, in reality, Meissner and Holm observed the Josephson effect, which was rediscovered and appreciated twenty years later, after the creation of the microscopic theory.

Many other prominent scientists were trying to resolve the puzzle, among them Landau, Bloch, Heisenberg, and so on (see Sommerfeld and Bethe, 1933).

A major discovery, namely, the observation of the isotope shift in T_c, was made in 1950 (Fröhlich, 1950; Maxwell, 1950; Reynolds *et al.*, 1950). This discovery provided an important clue for the microscopic mechanism since it showed that that the lattice and, hence, the phonons were strongly involved in the superconductivity.

In 1957, a microscopic description of the superconducting state, based on the concept of pairing opposite-momentum, opposite-spin electrons on the Fermi surface, was developed and published in the landmark paper by Bardeen, Cooper, and Schrieffer (BCS; 1957). This concept had been introduced earlier by Cooper (1956). The BCS paper explains the phenomenon of superconductivity. It also contains many predictions about the behaviour of this unique paired electron state that were almost immediately verified and provided the rapid acceptance of this theory.

In the early 1960s, superconducting junctions were shown to exhibit some additional very unusual properties. Both single-electron excitations (quasiparticles) and pairs could tunnel across a thin insulating barrier separating two superconducting electrodes. Quasiparticle tunnelling was discovered and understood by Giaever (1960), whereas Josephson (1962) predicted the tunnelling of pairs and the very unusual behaviour of such junctions. In addition, the tunnel junctions provided the ultimate test of the generalised strong-coupling theory as well as providing a basis for determining the underlying mechanism (McMillan and Rowell, 1965).

W. Little (1964) described a model of high-T_c organic polymers. This paper marked the beginning of the search for high-T_c superconductivity and introduced the non-phonon (electronic) mechanism as a key ingredient for such a search. The paper also predicted the phenomenon of organic superconductivity. The first organic superconductor (TMTSF-PF_6; $T_c = 1.5$ K) was discovered by Jerome *et al.* (1980). Organic superconductors have now been synthesised with transition temperatures close to 40 K.

The discovery of a new family, high-T_c oxides (cuprates), came in 1986—seventy-five years after the initial discovery of superconductivity! The La(Ba,Sr)CuO compounds were found to be superconducting by Bednorz and Mueller (1986) and have a maximum T_c of ~40 K. The cuprates themselves have been very rich in new superconductors and there are presently more than fifty such compounds. Wu *et al.* (1987) reported a YBaCuO compound with a transition temperature of about 90 K, which is well above the temperature of liquid nitrogen. The cuprate with the highest transition temperature is an HgBaCaCuO compound with a transition temperature of $T_c \approx 133$ K (Schilling *et al.*, 1993). Under pressure, the value of T_c was raised to 150 K (Gao *et al.*, 1994). The discovery of the copper oxides (cuprates) moved the field of superconductivity to

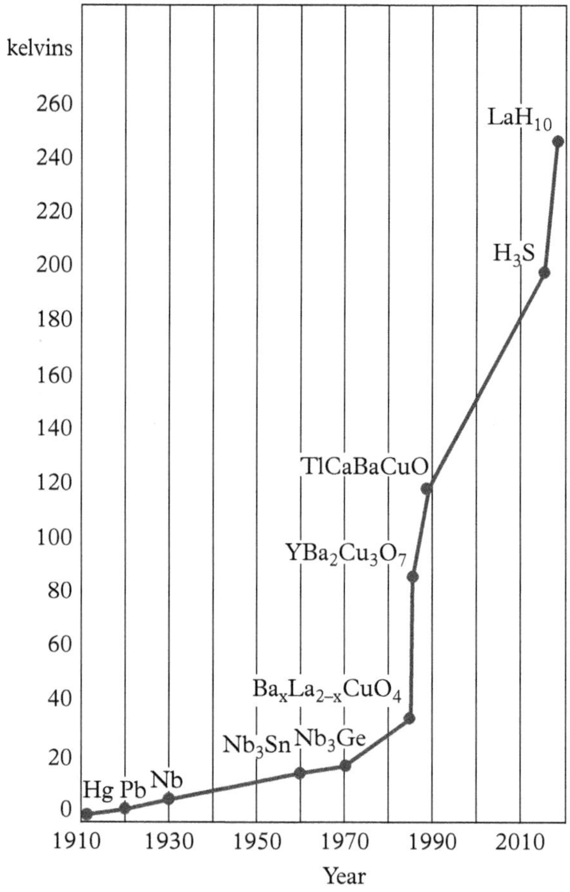

Figure 1.2 *Increase in maximum value of T_c with time.*

an entirely different and much higher energy scale. This energy scale corresponds to temperatures, whose values are much closer to room temperature (Fig. 1.2).

During the last twenty years, many new superconducting systems have been discovered—among them, fullerenes, MgB_2 compounds, pnictides, and so on. We will discuss the properties of these new systems in chapters 5 and 6. As a whole, we are witnessing a unique time in the history of superconductivity, as never before have so many novel superconducting systems been studied at the same time.

Recently, we have been witnessing remarkable progress towards higher values of T_c. This progress will be described in more detail in several chapters in this book. The breakthrough made the goal of reaching room temperature superconductivity perfectly realistic. This was confirmed in October 2020 when there appeared a report on the first room temperature superconductor: a three-component hydride with $T_c = 288$ K under high pressure (Snider *et al.*, 2020). The compound contains hydrogen, sulphur, and carbon ions (see section 7.2).

Thus, superconductivity is a very rich field, and one can expect many surprises in the future.

1.2 Excitations in Solids

1.2.1 Phonons and the Electron–Phonon Interaction

The adiabatic approach. Adiabatic approximation is the foundation of the theory of solids and molecular systems. It makes it possible to describe electronic states and introduces phonons, classifies energy spectra, and also introduces the electron–phonon interaction, which plays a key role in transport properties and also in superconductivity.

The starting point of the theory of metals is the approximation introduced by Born and Oppenheimer (1927; see e.g. the reviews by Born and Huang (1954), Ziman (1960), and Grimvall (1981)). This approximation is based on the fact that the masses of the ions and the conduction electrons are very different. The electrons adiabatically follow the motion of the ions. However, the description based on the adiabatic approximation is not exact. The residual non-adiabaticity is a measure of the 'friction' between the electrons and the lattice, representing what we refer to as the electron–phonon interaction.

The full Hamiltonian of the electron–ion system is

$$\widehat{H} = \widehat{T}_{\boldsymbol{r}} + \widehat{T}_{\boldsymbol{R}} + V(\boldsymbol{r}, \boldsymbol{R}) \tag{1.1}$$

Here $\{\boldsymbol{r}, \boldsymbol{R}\}$ are the sets of electronic and ionic coordinates, respectively, $\widehat{T}_r$ and $\widehat{T}_R$ are the kinetic energy operators for electronic and ionic motions, respectively, and $V(\boldsymbol{r}, \boldsymbol{R}) = V_1(\boldsymbol{r}) + V_2(\boldsymbol{R}) + V_3(\boldsymbol{r}, \boldsymbol{R})$ is the total potential energy (V_1 and V_2 correspond to the electron–electron and ion–ion interactions; V_3 describes the electron–ion interaction).

Our goal is to describe the interaction between the conduction electrons and the crystal lattice in a metal. At first glance, it may seem that it should be possible to treat this interaction in a manner analogous to that used in the theory of radiation, where the full Hamiltonian contains free-electron and photon fields, plus a small interaction term. However, the present problem turns out to be quite delicate, the reason being that the very procedure of introducing the phonon field requires great care. In addition, as it will be shown below, the term describing the electron–phonon coupling, unlike the electron–photon interaction, appears to be not small.

To proceed, we need to introduce the phonon subsystem. Furthermore, a rigorous description of the electron–phonon interaction requires that the Hamiltonian be written in the form

$$\widehat{H} = \widehat{H}_0 + \widehat{H}' \tag{1.2}$$

where

$$\widehat{H}_0 = \widehat{H}_0^{el}(\boldsymbol{r}) + \widehat{H}_0^{ph}(\boldsymbol{R}) \tag{1.3}$$

and

$$\widehat{H}' \equiv U(\boldsymbol{r}, \boldsymbol{R}) \tag{1.4}$$

Here $\widehat{H}_0^{el}(\boldsymbol{r})$ and $\widehat{H}_0^{ph}(\boldsymbol{R})$ describe the free-electron and phonon fields, so that

$$\widehat{H}_0^{el}(\boldsymbol{r}) = \widehat{T}_{\boldsymbol{r}} + U^{el}(\boldsymbol{r}) \tag{1.3$'$}$$

$$\widehat{H}_0^{ph}(\boldsymbol{R}) = \widehat{T}_{\boldsymbol{R}} + U^{ph}(\boldsymbol{R}) \tag{1.3$''$}$$

U^{el} and U^{ph} are the effective potential energies.

A rigorous theory has to be based on the representation (1.2), with the operator (1.4) corresponding to the electron–phonon interaction. The effects of this interaction can be calculated by perturbation theory (if $\widehat{H}'$ is small), or with the help of the diagrammatic technique. However, it turns out that the problem of presenting the Hamiltonian (1.1) in the form (1.2), and, therefore, correctly introducing $\widehat{H}'$ is quite non-trivial.

The next section contains the descriptions of two different methods: the Born–Oppenheimer approach and the 'crude' approach. These methods are effectively used, but nevertheless they do not have the structure corresponding to eqns (1.2) and (1.3). Still, these descriptions are useful, since they allow us to obtain a number of useful relations. In addition, they form a foundation for developing the approach, satisfying conditions (1.2) and (1.3). This approach will be presented below (see e.g. eqn (1.15)). Indeed, it will be shown that the term $\widehat{H}'$ (electron–phonon interaction) is not small relative to $\widehat{H}_o$. More specifically, it is even large relative to the phonon energy.

Adiabatic approximation and non-adiabaticity: The Born–Oppenheimer approach and the crude approach. The total wave function of the system $\Psi(\boldsymbol{r}, \boldsymbol{R})$ is the solution of the Schrödinger equation

$$\widehat{H}\Psi(\boldsymbol{r},\boldsymbol{R}) = E\Psi(\boldsymbol{r},\boldsymbol{R}) \tag{1.5}$$

The total Hamiltonian is defined by eqn (1.1). Let us neglect, in the zeroth-order approximation, the term corresponding to the kinetic energy of a heavy ionic system. Then we can introduce a complete orthogonal set of electronic wave functions, which are the solutions of

$$\widehat{H}_r\psi_n(\boldsymbol{r},\boldsymbol{R}) = \varepsilon_n(\boldsymbol{R})\psi_n(\boldsymbol{r},\boldsymbol{R}) \tag{1.6}$$

where $\widehat{H}_r = \widehat{T}_r + V(\boldsymbol{r},\boldsymbol{R})$.

The functions $\psi_n(\boldsymbol{r},\boldsymbol{R})$ and their eigenvalues $\varepsilon_n(\boldsymbol{R})$ (electronic terms) depend parametrically on the nuclear (ionic) positions. The full wave functions $\psi_n(\boldsymbol{r},\boldsymbol{R})$ associated with the nth electronic state can be sought in the form of an expansion in the complete orthogonal set of the solutions of the electron Schrödinger equation (1.6), that is,

$$\Psi_n(\boldsymbol{r},\boldsymbol{R}) = \sum_m \phi_{nm}(\boldsymbol{R})\psi_m(\boldsymbol{r},\boldsymbol{R}) \tag{1.7}$$

Substituting expansion (1.7) into eqn (1.5), multiplying by $\psi_\alpha(\boldsymbol{r},\boldsymbol{R})$, and integrating over **r** yields the following equation:

$$\left[\widehat{T}_{\boldsymbol{R}} + \varepsilon_\alpha(\boldsymbol{R})\right]\phi_{n\alpha}(\boldsymbol{R}) = E\phi_{n\alpha}(\boldsymbol{R}) \tag{1.8}$$

We neglect the sum $\sum_m C_{am}\phi_{nm}(\boldsymbol{R})$, where $C_{\alpha m}$ contains derivatives of ψ_n with respect to the nuclear coordinates.

According to eqn (1.8), $\Phi_{n\alpha}(\boldsymbol{R})$ can be treated as a nuclear wave function; as for the electronic term $\varepsilon_\alpha(\boldsymbol{R})$, it appears as the potential energy for the motion of the heavy particles. If we also neglect off-diagonal contributions to expansion (1.7), we arrive at the Born–Oppenheimer expression for the total wave function:

$$\Psi_{nv}(\boldsymbol{r},\boldsymbol{R}) = \psi_n(\boldsymbol{r},\boldsymbol{R})\phi_{nv}(\boldsymbol{R}) \tag{1.9}$$

v denotes the set of quantum numbers describing nuclear motion. Equations (1.6), (1.8), and (1.9) define the Born–Oppenheimer adiabatic approximation. It is obvious from the derivation that expression (1.9) is not an exact solution of the total Schrödinger equation.

Let us evaluate the non-adiabaticity neglected in the course of deriving approximation (1.9). The non-adiabaticity operator can be found in the following way (see e.g. Kresin and Lester, 1984). The zeroth-order operator, which has function (1.9) as its eigenfunction, has the form

$$\widehat{H}_0 = \widehat{T}_{\boldsymbol{r}} + V(\boldsymbol{r},\boldsymbol{R}) + \widehat{H}^0_{\boldsymbol{R}} \tag{1.10}$$

where the operator $\widehat{H}^0_{\boldsymbol{R}}$ is defined by the relation

$$\widehat{H}^0_{\boldsymbol{R}}\psi_n(\boldsymbol{r},\boldsymbol{R})\phi_{nv}(\boldsymbol{R}) = \psi_n(\boldsymbol{r},\boldsymbol{R})\widehat{T}_{\boldsymbol{R}}\phi_{nv}(\boldsymbol{R}) \tag{1.10'}$$

Indeed, one can see by direct calculation that

$$\begin{aligned}
\widehat{H}_0\Psi_{\rm BO}(\boldsymbol{r},\boldsymbol{R}) &= \left[\widehat{T}_{\boldsymbol{r}} + V(\boldsymbol{r},\boldsymbol{R})\right]\psi_n(\boldsymbol{r},\boldsymbol{R})\phi_{nv} + \widehat{H}^0_{\boldsymbol{R}}\psi_n(\boldsymbol{r},\boldsymbol{R})\phi_{nv}(\boldsymbol{R}) \\
&= \varepsilon_n(\boldsymbol{R})\psi_n(\boldsymbol{r},\boldsymbol{R})\phi_{nv}(\boldsymbol{R}) + \psi_n(\boldsymbol{r},\boldsymbol{R})\widehat{T}_{\boldsymbol{R}}\phi_n(\boldsymbol{R}) \\
&= \psi_n(\boldsymbol{r},\boldsymbol{R})\left[\widehat{T}_{\boldsymbol{R}} + \varepsilon_n(\boldsymbol{R})\right]\phi_{nv}(\boldsymbol{R}) \\
&= E_{nv}\Psi_{\rm BO}(\boldsymbol{r},\boldsymbol{R})
\end{aligned}$$

where BO stands for Born–Oppenheimer. Based on eqns (1.1), (1.2), and (1.10), one can now determine the non-adiabaticity operator $\widehat{H}'$:

$$\widehat{H}' = \widehat{T}_{\boldsymbol{R}} - \widehat{H}^0_{\boldsymbol{R}} \tag{1.11}$$

Hence

$$\widehat{H}'\Psi_{\rm BO}(\boldsymbol{r},\boldsymbol{R}) = \widehat{T}_{\boldsymbol{R}}\psi_n(\boldsymbol{r},\boldsymbol{R})\phi_{nv}(\boldsymbol{R}) - \psi_n(\boldsymbol{r},\boldsymbol{R})\widehat{T}_{\boldsymbol{R}}\phi_{nv}(\boldsymbol{R}) \tag{1.12}$$

or, in matrix representation,

$$\begin{aligned}
\left\langle mv'|\widehat{H}'|nv\right\rangle &= \int \psi^*_m(\boldsymbol{r},\boldsymbol{R})\phi^*_{mv}(\boldsymbol{R})\widehat{T}_{\boldsymbol{R}}\phi_{nv}(\boldsymbol{R})d\boldsymbol{R}dt \\
&\quad - \delta_{mn}\int \phi^*_{mv}(\boldsymbol{R})\widehat{T}_{\boldsymbol{R}}\psi_n(\boldsymbol{r},\boldsymbol{R})\phi_{nv}(\boldsymbol{R})d\boldsymbol{R}
\end{aligned} \tag{1.12'}$$

Let us estimate the accuracy of the Born–Oppenheimer approximation. The characteristic scale of the nuclear wave function $\phi_{nv}(\boldsymbol{R})$ is of the order of the vibrational amplitude a, whereas that of the electronic wave function is the lattice period L. As a result, we obtain

$$\begin{aligned}
\widehat{H}'\Psi_{\rm BO}(\boldsymbol{r},\boldsymbol{R}) &\sim -\frac{\hbar^2}{M}\frac{\partial\psi}{\partial R}\frac{\partial\phi}{\partial R} \\
&\sim -\frac{\hbar^2}{M}\frac{\psi}{L}\frac{\phi}{a} \sim \frac{\hbar^2}{Ma^2}\frac{\mathrm{a}}{L}\Psi_{\rm BO} \sim \frac{\mathrm{a}}{L}\hbar\omega\Psi_{\rm BO}
\end{aligned}$$

that is,

$$\widehat{H}' \sim \kappa = \frac{\mathrm{a}}{L} \ll 1 \tag{1.13}$$

Hence non-adiabaticity is proportional to the small parameter $k = a/L$ (we have used the relation $a = [\hbar/M\omega]^{1/2}$). Moreover, since $\omega = (K_n/M)^{1/2}$, where K_n is the force constant due to the Coulomb internuclear interaction, we can write $a^2 = \hbar/(K_n/M)^{1/2}$. Similarly, $L^2 \cong \hbar/(K_e m)^{1/2}$. Since $K_e = K_n$ (the Coulomb interaction does not depend on the mass factor), we arrive at $k \cong (m/M)^{1/4}$.

It is useful also to present the following classical estimate (Migdal, 2000). Equation (1.12) can be written in the form

$$\widehat{H}'\Psi_{\mathrm{BO}}(\boldsymbol{r},\boldsymbol{R}) = \left[\widehat{T}_{\boldsymbol{R}}, \psi_n\right]\phi_{nv}(\boldsymbol{R}) \tag{1.12''}$$

If we assume that the motion of heavy particles is classical, then $i\left[\widehat{T}_{\boldsymbol{R}}, \psi_n\right] \to (\partial\psi_n/\partial R)v$, where v is the velocity of the heavy particles. Thus, the qualitative aspect of the adiabatic approximation is indeed connected with the small velocity of the ions relative to the much greater velocity of the electrons. Averaging the total potential energy $V(\boldsymbol{r}, \boldsymbol{R})$ over the fast electronic motion, we obtain the effective potential $\varepsilon(\boldsymbol{R})$ for the nuclear motion.

The adiabatic method described above is a very elegant and powerful approach, which provides a foundation for the theory of solids.

Nevertheless, if we want to develop a rigorous and complete approach, we should state that this method has a definite shortcoming. Indeed, one can see from eqns (1.10) and (1.10′) that the zeroth-order Hamiltonian does not have the desired form (1.2).

Another conventional form of adiabatic theory (the 'crude' or 'clamped' approximation) is based on the fact that the ions are located near their equilibrium positions. In this version of the adiabatic theory, the expansion of the total wave function uses the electronic wave functions $\psi_n(\boldsymbol{r},\boldsymbol{R}) \cong \psi_n(\boldsymbol{r},\boldsymbol{R}_0)$, where $\boldsymbol{R}_0$ corresponds to some equilibrium configuration, that is,

$$\Psi_n(\boldsymbol{r},\boldsymbol{R}) = \sum_m \tilde{\phi}_{nm}(\boldsymbol{R})\psi_m(\boldsymbol{r},\boldsymbol{R}_0) \tag{1.14}$$

The total wave function in the zeroth-order approximation is $\Psi_n(\boldsymbol{r},\boldsymbol{R}) = \Psi_n(\boldsymbol{r},\boldsymbol{R}_0)\,\tilde{\phi}_{nv}(\boldsymbol{R})$. Strictly speaking, the nuclear wave functions ϕ_{nv} (see eqn (1.8)) and $\tilde{\phi}_{nv}$ are different. The non-adiabaticity operator in the 'crude' approximation can be written in a relatively simple form:

$$H_1 = \sum_{i,a} \frac{\partial \mathrm{V}}{\partial \boldsymbol{R}_{i\alpha|0}} \Delta \boldsymbol{R}_{i\alpha} \tag{1.14'}$$

The 'crude' approach has the same weakness as the Born–Oppenheimer approximation: the zeroth-order Hamiltonian does not have the form (1.2). As a result, neither approach is adaptable to the usual perturbation theory treatment. It is difficult to evaluate higher-order corrections or to consider the case of strong non-adiabaticity.

Now we turn to a different approach to adiabatic theory, which will allow us to overcome these problems.

Electron–phonon coupling Rigorous expansion. Let us introduce the functions (Geilikman, 1971, 1975):

$$\Psi^0(\boldsymbol{r},\boldsymbol{R}) = \psi_n(\boldsymbol{r},\boldsymbol{R}_0)\phi_{nv}(\boldsymbol{R}) \tag{1.15}$$

which form a complete basis set. The functions $\psi_n(\boldsymbol{r},\boldsymbol{R}_0)$ and $\phi_{nv}(\boldsymbol{R})$ are defined by the equation

$$\left[\widehat{T}_{\boldsymbol{r}} + V(\boldsymbol{r},\boldsymbol{R})\right]\psi_n(\boldsymbol{r},\boldsymbol{R}) = \varepsilon_n(\boldsymbol{R})\psi_n(\boldsymbol{r},\boldsymbol{R})_{|\mathbf{R}=\mathbf{R}_0} \tag{1.16}$$

and by eqn (1.8).

Here n is the set of quantum numbers for the electronic states, ν is the vibrational quantum number, and $\boldsymbol{R}_0$ is the equilibrium configuration. Strictly speaking, $\boldsymbol{R}_0$ depends on n but, for levels close to the ground state, this dependence may be neglected.

The functions (1.15) form the basis set in the present method. It is interesting that the total function (1.15) is a combination of the Born–Oppenheimer and the 'crude' functions. Indeed, the electron wave function is that of the 'crude' approach, whereas $\phi_{nv}(\boldsymbol{R})$ is a solution of the nuclear Schrödinger equation of the Born–Oppenheimer version. Equation (1.8) together with the wave functions $\phi_{nv}(\boldsymbol{R})$ serves to define phonons in the adiabatic theory.

We can write down the Hamiltonian in the zeroth-order approximation (the functions (1.15) are its eigenfunctions; see eqn (1.19′)). Namely,

$$\widehat{H}_0 = \widehat{H}_{0\boldsymbol{r}} + \widehat{H}_{0\boldsymbol{R}} \tag{1.17}$$

where

$$\widehat{H}_{0\boldsymbol{R}} = \widehat{T}_{\boldsymbol{r}} + V(\boldsymbol{r},\boldsymbol{R}_0) \tag{1.17′}$$

$$\widehat{H}_{0\boldsymbol{R}} = \widehat{T}_{\boldsymbol{R}} + \tilde{A}(\boldsymbol{R}) \tag{1.17″}$$

The operators $\widehat{T}_{\boldsymbol{r}}$ and $\widehat{T}_{\boldsymbol{R}}$ are defined according to (1.1). The operator $\widehat{A}$ is defined by the following matrix form:

$$\widehat{A}_{nv}^{n'v'} = \delta_{nn'}\int \phi_{n'v'}(\boldsymbol{R})\left[\varepsilon_n(\boldsymbol{R}) - \varepsilon_n(\boldsymbol{R}_0)\right]\phi_{nv}(\boldsymbol{R})d\boldsymbol{R} \tag{1.18}$$

It is easy to show that the operator $\widehat{H}_0$ is diagonal in the space of the functions (1.15). Indeed,

$$\widehat{H}_{0;nv}^{n'v'} = \int \psi_n^*(\boldsymbol{r},\boldsymbol{R}_0)\phi_{n'v'}(\boldsymbol{R})\left[\widehat{T}_{\boldsymbol{r}} + V(\boldsymbol{r},\boldsymbol{R}_0) + \widehat{T}_{\boldsymbol{R}} + \widehat{A}(\boldsymbol{R})\right] \times \psi_n(\boldsymbol{r},\boldsymbol{R}_0)\phi_{nv}(\boldsymbol{R})d\boldsymbol{r}d\boldsymbol{R}$$
$$= \delta_{nn'}\int \phi_{n'v'}(\boldsymbol{R})\left[\widehat{T}_{\boldsymbol{R}} + \varepsilon_n(\boldsymbol{R})\right]\phi_{nv}(\boldsymbol{R})d\boldsymbol{R}$$
$$= \delta_{nn'}\delta_{\nu\nu'}E_n \tag{1.19'}$$

Note the important fact that the operator $\widehat{H}_0$ represents a sum of two terms, one which depends on $\boldsymbol{r}$, and the other which depends on $\boldsymbol{R}$. These terms correspond to the electron and the phonon fields, respectively. The electron–phonon interaction is given by the following non-adiabatic term in the full Hamiltonian:

$$\widehat{H}' = \widehat{H} - \widehat{H}_0 \tag{1.20}$$

where $\widehat{H}$ and $\widehat{H}_0$ are defined by eqns (1.1) and (1.17). As a result, we find

$$\widehat{H}' = V(\boldsymbol{r},\boldsymbol{R}) - V(\boldsymbol{r},\boldsymbol{R}_0) - \widehat{A}(\boldsymbol{R}) \tag{1.20'}$$

or, in matrix representation ($\widehat{H}_{na} \equiv \widehat{H}'$),

$$H_{na;nv}^{n'v'} = [V(\mathbf{r},\mathbf{R}) - V(\mathbf{r},\mathbf{R}_0)]\,\Big|_{nv}^{n'v'} - \widehat{A}(\mathbf{R})\Big|_{nv}^{n'v'} \tag{1.20''}$$

The last term is defined by eqn (1.18).

The non-adiabatic Hamiltonian can be written as an expansion about the equilibrium coordinates:

$$H_{na;nv}^{n'v'} = \frac{\partial V}{\partial \mathbf{R}_{|0}}\delta\mathbf{R}\Bigg|_{nv}^{n'v'} - \frac{\partial \varepsilon_n}{\partial \mathbf{R}_{|0}}\delta\mathbf{R}\Bigg|_{nv}^{n'v'} + \frac{1}{2}\frac{\partial^2 V}{\partial \mathbf{R}_i \partial \mathbf{R}_{k|0}}\delta\mathbf{R}_i\delta\mathbf{R}_k\Bigg|_{nv}^{n'v'}$$
$$- \frac{1}{2}\frac{\partial^2 \varepsilon_n}{\partial \mathbf{R}_i \partial \mathbf{R}_{k|0}}\delta\mathbf{R}_i\delta\mathbf{R}_k\Bigg|_{nv}^{n'v'} + \frac{1}{6}\frac{\partial^2 \varepsilon_n}{\partial \mathbf{R}_i \partial \mathbf{R}_k \partial \mathbf{R}_{l|0}}\delta\mathbf{R}_i\delta\mathbf{R}_k\delta\mathbf{R}_l\Bigg|_{nv}^{n'v'} + \cdots \tag{1.21}$$

The terms $\partial\varepsilon_n(\boldsymbol{R})/\partial\boldsymbol{R}|_0$ vanish. The first term in (1.21) corresponds to the 'crude' approximation (see eqn (1.14)).

Note that the result (1.21) represents a power-series expansion in terms of the parameter (1.13). Indeed, in eqn (1.21) the deviation from the equilibrium position is $\delta R \cong a$, while the equilibrium distance is $R_0 \cong L$. It is convenient to rewrite eqn (1.21) in the following way:

$$H_{na}\Big|_{nv}^{n'v'} = H_1\Big|_{nv}^{n'v'} + H_2\Big|_{nv}^{n'v'} + \cdots \tag{1.22}$$

where

$$H_1\Big|_{nv}^{n'v'} = \frac{\partial V}{\partial \mathbf{R}_{|0}}\delta\mathbf{R}\Big|_{nv}^{n'v'} \tag{1.22$'$}$$

$$H_2\Big|_{nv}^{n'v'} = \left[\frac{1}{2}\frac{\partial^2 V}{\partial \mathbf{R}_i \partial \mathbf{R}_{k|0}}\partial\mathbf{R}_i\partial\mathbf{R}_k - \frac{1}{2}\frac{\partial^2 \varepsilon_n}{\partial \mathbf{R}_i \partial \mathbf{R}_{k|0}}\partial\mathbf{R}_i\partial\mathbf{R}_k\right]\Big|_{nv}^{n'v'} \tag{1.22$''$}$$

Let us estimate the magnitude of the terms $\widehat{H}_1$, $\widehat{H}_2$, … . Starting with $\widehat{H}_2$, we have $H_2 \cong \hbar\omega$. Note that each additional factor $\delta\boldsymbol{R}_{i\alpha}$ brings in the parameter $k = a/L \ll 1$. Hence

$$H_1 \sim \hbar\omega/k \tag{1.23}$$

$$H_2 \sim \hbar\omega \tag{1.23$'$}$$

$$H_3 \sim k\hbar\omega \tag{1.23$''$}$$

and so on. One can see from eqns (1.17), (1.20), and (1.22) that the total Hamiltonian can be written in the form (1.2), that is, as a sum of the terms describing free-electron and phonon fields and an additional term corresponding to the electron–phonon interaction.

It is interesting that the first term in (1.22) (see eqn (1.23)) can make the contribution larger than the phonon energy. This is a peculiar feature of the theory.

One should stress the very important fact that the first non-adiabatic terms ((1.23) and (1.23′)) do not contain the smallness $\propto \kappa$. Moreover, the first term is proportional to κ^{-1}. This is the reason why one must be careful when analysing the effects of electron–phonon coupling.

In addition, because we are dealing with operators, the corresponding contribution depends on the type of the transitions (diagonal vs non-diagonal). For example, one finds that the electron–phonon interaction makes only a very small contribution to the total energy of the system. At the same time, some other effects, such as mass renormalisation, turn out to be very significant. The phenomenon of superconductivity is another example of a gigantic non-adiabatic effect (see section 2.1.2).

Electron–phonon interaction and renormalisation of normal parameters. The electron–phonon interaction plays an extremely important role in the physics of metals. For example, this interaction is one of the principal relaxation mechanisms defining electrical conductivity. In this case, we are dealing with electron scattering by real thermal phonons. But, even in the absence of thermal phonons, for example, at $T = 0$ K, the polarizability of the lattice (i.e. changes in the regime of zero-point vibrations of the ions) often plays a significant role. Superconductivity represents probably the most spectacular manifestation of this type of interaction. However, it affects the normal state properties as well. For example, the effective mass m* measured in certain experiments (e.g. cyclotron resonance or the de Haas–van Alphen effect) comes out different from the so-called band value m^b corresponding to a frozen lattice.

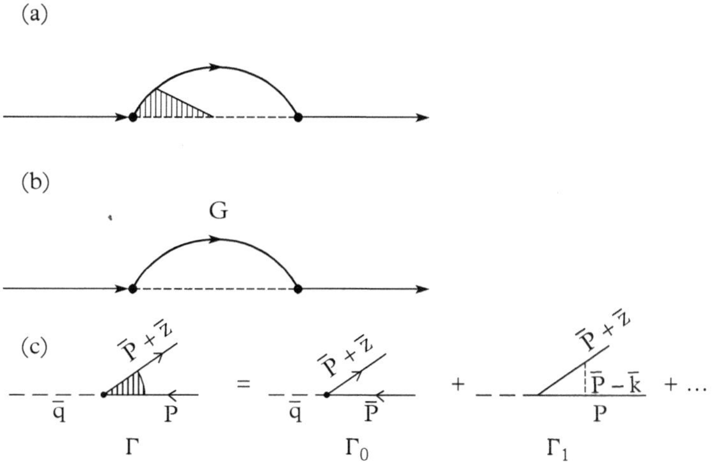

Figure 1.3 *Lattice polarisation. (a,b) Self-energy part. (c) Electron–phonon vertex.*

Qualitatively speaking, as a result of lattice polarisation, electrons find themselves 'dressed' in ionic coats. This mass renormalisation effect turns out to be quite significant for a number of metals. Namely, it is found that $m^*(0) = m^b(1+\lambda)$, where λ is the coupling constant describing the electron–phonon interaction ($m^*(0)$ is the value of the effective mass at $T = 0$ K). For example, in lead, $\lambda \cong 1.4$, so that renormalisation leads to the electron becoming about 2.5 times heavier than what would be expected for a frozen lattice. A similar relationship holds for the Sommerfeld constant describing the electronic heat capacity: $\gamma(0) = \gamma^b(1+\lambda)$.

In quantum language, renormalisation effects are described by the diagram shown in Fig. 1.3. We are dealing with emission and subsequent absorption of a virtual phonon.

Analytically, the equation shown in Fig. 1.3a can be written in the form

$$\sum_1(\boldsymbol{k},\omega_n) = \pi T \sum_{\omega_{n'}} \int d\boldsymbol{k}' \zeta^2_{\boldsymbol{q},\boldsymbol{\kappa}'} D(\boldsymbol{q},\omega_n - \omega_{n'}) G(\boldsymbol{k}',\omega_{n'}) \tag{1.24}$$

Here Σ is the self-energy part describing the scattering, D and G are the phonon propagator and the one-particle Green's function, respectively, $\boldsymbol{q} = \boldsymbol{\kappa} - \boldsymbol{\kappa}'$, $\omega_n = (2n+1)\pi T$, and $\zeta_{\boldsymbol{q},\boldsymbol{\kappa}'}$ is the matric element. We employ the thermodynamic Green's functions formalism (see e.g. Abrikosov *et al.*, 1975, and appendix E). A usual equation containing time dependence can be obtained by an analytical continuation to the upper plane. Transforming the integration over the phonon momentum **q** and then phonon frequency, we obtain

$$\sum_1(\boldsymbol{k},\omega_n) = \pi T \sum_{\omega_{n'}} \int d\Omega \frac{g(\Omega)}{\Omega} D(\Omega,\omega_n - \omega_{n'}) G(\boldsymbol{k}',\omega_{n'}) \tag{1.24$'$}$$

$$k \cong k' \cong k_F$$

Here we introduced the important function

$$g(\Omega) = \alpha^2(\Omega)\mathrm{F}(\Omega) \tag{1.25}$$

($\alpha^2(\Omega)$ describes the electron–phonon interaction, and $\mathrm{F}(\Omega)$ is the phonon density of states); it is defined by the expression (see e.g. Scalapino, 1969, and McMillan and Rowell, 1969)

$$\alpha^2(\Omega)F(\Omega) = \frac{\int \frac{dS_{k'}}{|v_{k'}|} \int \frac{dS_{k'}}{|v_k|} \frac{1}{(2\pi)^3 \hbar} \sum_\lambda |\zeta_{k',k,\lambda}|^2 \delta\left[\Omega - \omega_{\lambda,k'-k}\right]}{\int \frac{dS_k}{|v_k|}} \tag{1.25$'$}$$

For many materials one can introduce the effective average phonon frequency

$$\tilde{\Omega} = \left\langle \Omega^2 \right\rangle^{1/2} \tag{1.26}$$

$$\langle \Omega^n \rangle = (2/\lambda) \int g(\Omega)\Omega^{n-1} d\Omega \tag{1.26$'$}$$

Then equation (1.24′) can be written in the form

$$\sum_1 = \pi T \lambda \sum_{n' \geq 0} D(\tilde{\Omega}, \omega_n - \omega_{n'}) G(\omega_{n'}) \tag{1.24$''$}$$

Here

$$\lambda = 2 \int \frac{d\Omega}{\Omega} \alpha^2(\Omega) F(\Omega) \tag{1.27}$$

is the electron–phonon-coupling constant, which is rigorously defined by eqn (1.27).

It should be pointed out that some phenomena do not involve mass renormalisation (e.g. high-frequency effects, the Pauli spin susceptibility). A detailed review can be found in the book by Grimvall (1981).

Let us come back to the renormalisation of the Sommerfeld constant (Grimvall, 1969, 1981; Kresin and Zaitsev, 1978). The entropy of the electron system can be written as (Eliashberg, 1963)

$$S_c = \frac{v_0}{T^2} \int_0^\infty \frac{d\varepsilon \varepsilon}{\cosh^2(\varepsilon/2T)} \left[\varepsilon - f(\varepsilon)\right] \tag{1.28}$$

Here v_0 is the unrenormalised density of states at the Fermi level, and $f(\varepsilon)$ is the odd part of the self-energy function, which describes the electron–phonon interaction (see Fig. 1.3a).

If we neglect the function $f(\varepsilon)$ on the right-hand side of eqn (1.27), we recover the usual expressions for the entropy and, consequently, for the heat capacity of a free-electron gas. Renormalisation effects are contained in the function $f(\varepsilon)$.

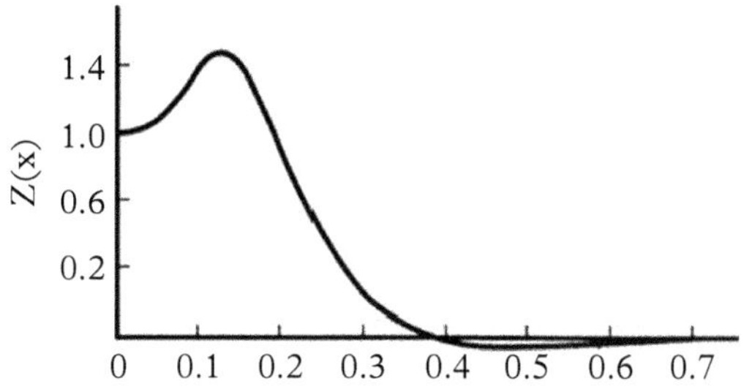

Figure 1.4 *Universal function Z(x).*

A calculation of the function $f(\varepsilon)$ leads to the following result for the electronic heat capacity:

$$C_e(T) = \gamma(T)T \tag{1.28$'$}$$

where

$$\gamma(T) = \gamma^0 \left[1 + 2\int_0^\infty \frac{d\Omega}{\Omega} g(\Omega)\mathrm{Z}(\mathrm{T}/\Omega)\right]$$

In this expression, $g(\Omega)$ is defined by eqn (1.25), and $Z(x)$ is a universal function (Fig. 1.4).

For $T = 0$ K, we find

$$\gamma(0) = \gamma^b(1 + \lambda) \tag{1.28$''$}$$

where λ is the electron–phonon coupling constant (eqn (1.27)).

As can be seen from eqn (1.28), the electron–phonon interaction leads to a deviation from the linear temperature dependence of the electronic heat capacity. This effect is due to the fact that the state of the lattice changes with temperature as thermal phonons appear.

Result (1.28$'$) can be written in a form which contains only experimentally measured quantities:

$$\gamma(T) = \gamma(0)\left[1 + \rho\left(\frac{\kappa(T)}{\kappa(0)} - 1\right)\right] \tag{1.28$'''$}$$

$$\kappa(T) = 2\int d\Omega\ \mathrm{g}(\Omega)\Omega^{-1}Z(T/\Omega)$$

where $(\gamma(0) = m^* \mathrm{p_F}/3, m^* = m^b(1 + \lambda)$ is the renormalised effective mass, $\rho = \lambda(1 + \lambda)^{-1}$ is the renormalised coupling constant, and λ is defined by eqn (1.27). For $T \to 0$ K,

$$[\gamma(T)/\gamma(0) - 1] \sim T^2 \ln(\Omega_D/T)$$

This dependence was derived by Eliashberg (1963). It is interesting that, at sufficiently low temperatures, the additional contribution to the electronic heat capacity given, according to this dependence, by $\Delta C \propto T^2 \ln(\Omega_D/T)$ exceeds the lattice heat capacity ($\propto T^3$).

As the temperature increases, $\gamma(T)$ goes through a maximum; upon a further rise in temperature, the function $Z(T/\Omega)$ decreases and approaches zero. It becomes small for $x \cong 0.3$, whereby $\gamma(T)$ acquires the unrenormalised value γ^b. Qualitatively, this is explained by the fact that intense thermal motion washed out the ionic 'coat', so that the electrons find themselves 'undressed' at high temperatures.

One can also calculate the temperature dependence of the effective mass, which is described by the relation

$$\frac{m^*(T)}{m^*(0)} = 1 + 2\int \frac{d\Omega}{\Omega} g(\Omega) G(T/\Omega) \tag{1.29}$$

where $G(x) = (2\pi x)^2 \sum_{n=0}^{\infty} (2n+1)/\{1 + [(2n+1)nx]^2\}^2$ is the so-called Grimvall function (Grimvall, 1969; see also Allen and Cohen, 1970). The calculated dependence $m^*(T)$ is in good agreement with the cyclotron resonance measurements by Sabo for Zn (1969) and by Krasnopolin and Khaikin for Pb (1973).

So far, we have been discussing corrections to the electronic parameters induced by the electron–phonon interaction. As we have seen, non-adiabaticity leads to significant changes in the effective mass, electronic heat capacity, the Fermi velocity $v_F = p_F/m^*$, and so on. Let us now look at the effect of this interaction on the phonon spectrum. The results here turn out to be completely different (Brovman and Kagan, 1967; Geilikman, 1971). The correction to the phonon frequency $\Delta\Omega$ can be evaluated as the variation of the non-adiabatic contribution to the total energy. A calculation of this correction leads to $\Delta\Omega \propto \kappa^2$ (κ is defined by eqn (1.13)); thus, the shift in the phonon frequency is relatively small.

This is an important result, which has direct bearing on the problem of lattice instability. It also has an interesting history. The fact of the matter is that, if one directly uses the Fröhlich model, in which the full Hamiltonian is made up of an electronic term, a phonon term with an acoustic dispersion law ($\omega = uq, q \to 0$), and an interaction term given by H_1 (see eqn (1.14′)), then one obtains the result that $\Omega = \Omega_0(1-2\lambda)^{1/2}$. In other words, one finds a significant shift in the phonon frequency. Furthermore, for $\lambda > \lambda_{\max} = 0.5$, the frequency is imaginary, that is, the lattice becomes unstable. Thus, in the early years following the BCS work, it was assumed that λ cannot exceed $\lambda_{\max}$, which then leads to an upper limit on the superconducting critical temperature T_c, $T_{c,\max} \cong \Omega_D \exp(-1/\lambda_{\max}) \cong 0.1\Omega_D$ achievable by the phonon mechanism. This result is in contradiction with subsequent experimental observations; at present, we know many superconductors with $\lambda > 0.5$ (see e.g. Wolf, 2012).

The problem with the above analysis is that the Fröhlich Hamiltonian is not a valid tool. If we start with the rigorous adiabatic theory, the interaction is given not by H_1 but by the full expression H_{na} (see eqn (1.22)). The use of this expression leads to the aforementioned weak renormalisation of the phonon frequency $\sim\kappa^2$. A rigorous systematic application of the adiabatic theory leads to only weak corrections due to non-adiabaticity.

The 'Migdal' theorem. In the previous section, we took into account the lowest-order terms in the electron–phonon interaction. This can be seen directly from Fig. 1.3c Indeed, electron–phonon scattering is described by a factor containing just the matrix element of the electron–phonon interaction (the 'single vertex', Γ_0), and higher-order terms, such as Γ', are neglected. According to Migdal (1958), this approximation is justified, because higher-order corrections contain an additional small adiabatic parameter, $\tilde{\Omega}/E_F$.

Indeed, consider the term Γ', which can be analytically written in the following form (see Fig. 1.3b):

$$\Gamma'_{(p,q)} = \zeta^3 \int d\boldsymbol{k}d\omega D(\boldsymbol{p}-\boldsymbol{k})G(\boldsymbol{k})G(\boldsymbol{k}+\boldsymbol{q})$$

where $k = \{\boldsymbol{k}, \omega\}$

The presence of the D-function provides the cut-off in the integration at $\omega \cong \tilde{\Omega}$; indeed, at $\omega \gg \tilde{\Omega}$, this function decreases rapidly ($\sim \tilde{\Omega}^{-2}$). On the other hand, the characteristic scale for the electronic Green's function is $\sim E_F{}^{-1}$. As a result, the quantity Γ' can be estimated as $\propto \lambda^2(\frac{\tilde{\Omega}}{E_F})\Gamma_0$, where $\lambda^2 \propto \zeta\nu_0/\tilde{\Omega}$, and Γ' contains the small adiabatic parameter $\tilde{\Omega}/E_F$. A more detailed derivation and some special cases (e.g. when $u \gg v_F$, where u is the sound velocity) are described by Scalapino (1969) and Grimvall (1981).

The 'Migdal' theorem was one of the first applications of quantum field theory to condensed matter physics. The analysis was based on the pioneering work by Galitski and Migdal (1958). This theorem is essential for the analysis of the superconducting state (see section 2.2).

1.2.2 Diabatic Representation and Polaronic States

Diabatic representation. According to the usual adiabatic theory (the Born–Oppenheimer approximation; see section 1.2.1), the electronic Hamiltonian $\widehat{H}_e = \widehat{T}_r + V(\boldsymbol{r}, \boldsymbol{R})$ is diagonal in the representation defined by eqn (1.9). The analysis is based on the usual stationary Schrödinger equation (1.5). Equation (1.6) determines the electronic terms $\varepsilon_m(\boldsymbol{R})$. In quantum chemistry, the dependence $\varepsilon_m(\boldsymbol{R})$ is 'visualised' as the potential energy surface.

Assume that the energy terms have the structure shown in Fig. 1.5a (solid lines). The lowest term contains two regions. One of them (L) corresponds to the bound ionic state, and the second region (R) to the continuum spectrum.

Another example is the case when the potential has two close minima (double-well structure); see Fig. 1.5b. The problem of propagation $L \rightarrow R$ can be solved with use of the usual method with matching at the boundaries, and so on. However, one can employ a different method, the so-called diabatic representation (see e.g. O'Malley, 1967, and Kresin and Lester, 1984).

Let us introduce a new potential $\tilde{V}(\boldsymbol{r}, \boldsymbol{R})$, so that the effect of the substitution $V \rightarrow \tilde{V}$ in the electronic equation (1.6) is to change the adiabatic terms to diabatic, that is, to the picture of crossing terms $\tilde{\varepsilon_1}$ and $\tilde{\varepsilon_2}$. and the wave functions $\tilde{\psi}_1(\boldsymbol{r}, \boldsymbol{R})$ and $\tilde{\psi}_2(\boldsymbol{r}, \boldsymbol{R})$ In addition, the analysis in the diabatic representation is based not on the stationary but on the time-dependent Schrödinger equation. In other words, the $L \rightarrow R$ transfer can be described as a quantum transition between the terms 1 and 2.

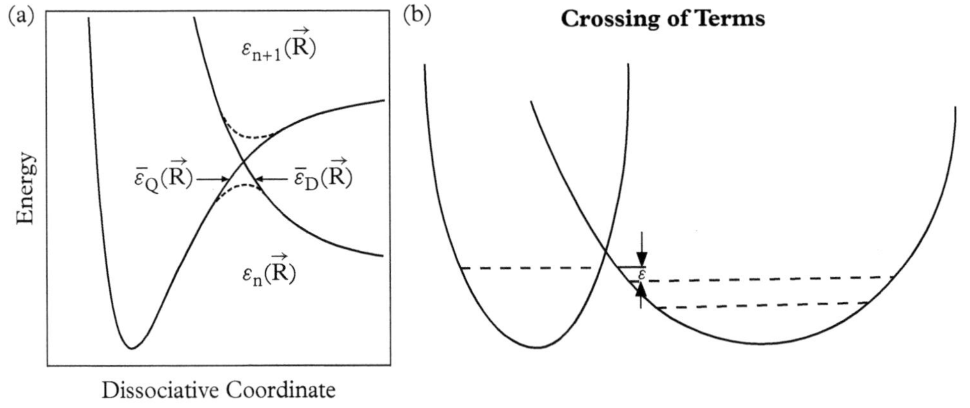

Figure 1.5 *(a) Adiabatic potential energy surfaces $\varepsilon_n(R)$ and $\varepsilon_{n+1}(R)$ (from the diagonalisation of the electronic Hamiltonian), and diabatic potential energy surfaces $\tilde{\varepsilon}_D(R)$ and $\tilde{\varepsilon}_Q(R)$, which cross (b) electronic terms (the diabatic representation).*

The total wave functions are sought in the form

$$\psi(\boldsymbol{r},\boldsymbol{R},t) = a(t)\tilde{\psi}_1(\boldsymbol{r},\boldsymbol{R}) + b(t)\tilde{\psi}_2(\boldsymbol{r},\boldsymbol{R}) \tag{1.30}$$

where

$$\begin{aligned}\tilde{\psi}_1(\boldsymbol{r},\boldsymbol{R}) &= \tilde{\psi}_1(\boldsymbol{r},\boldsymbol{R})\phi_{1,v_1}(\boldsymbol{R})\\ \tilde{\psi}_2(\boldsymbol{r},\boldsymbol{R}) &= \tilde{\psi}_2(\boldsymbol{r},\boldsymbol{R})\phi_{2,v_2}(\boldsymbol{R})\end{aligned} \tag{1.30$'$}$$

It is essential that the functions $\phi_{1,v_1}(\boldsymbol{R})$ and $\phi_{2,v_2}(\boldsymbol{R})$ are not orthogonal, since they belong to different energy terms.

Assume that $a(0) = 1$ and $b(0) = 0$. Then the probability of the 1–2 transfer is equal to the following (Landau and Lifshitz, 1977):

$$|b(t)|^2 = \tilde{b}^2\left\{1 - \cos 2\left[(\varepsilon^2/4) + (\widehat{H}_{el|12})^2\right]^{1/2} t\right\} \tag{1.31}$$

where

$$\tilde{b}^2 = \frac{(\widehat{H}_{el|12})^2}{2\left[(\frac{\varepsilon^2}{4}) + (\widehat{H}_{el|12})^2\right]}$$

$$\widehat{H}_{el|12} = \int d\boldsymbol{R}\cdot L(\boldsymbol{R})\phi_2^*(\boldsymbol{R})\phi_1(\boldsymbol{R})$$

$$L(\boldsymbol{R}) = \int d\boldsymbol{r}\cdot \psi_2^*(\boldsymbol{r},\boldsymbol{R})\widehat{H}_{el}\psi_1(\boldsymbol{r},\boldsymbol{R})$$

In eqn (1.31) ε is the energy difference (see Fig. 1.5b), and $\widehat{H}_{el|12}$ is the matrix element, describing the transition between the terms (in the diabatic representation). The function $L(\boldsymbol{R})$ slowly depends on $\boldsymbol{R}$. The main contribution to the matrix element comes from the region of the overlap of nuclear wave functions. As a result, one can write

$$H_{el;1,2} \simeq L_0 F \tag{1.32}$$

where $L_0 = L(\boldsymbol{R}_0)$, R_0 is the crossing point, and

$$F = \int \phi_2^*(\boldsymbol{R})\phi_1(\boldsymbol{R})d\boldsymbol{R} \tag{1.32'}$$

is the Franck–Condon factor. One can write

$$\tilde{b}^2 = \frac{(L_0 F)^2}{2\left[(\frac{\varepsilon^2}{4}) + (L_0 F)^2\right]} \tag{1.33}$$

As was noted above, the nuclear wave functions $\tilde{\phi}_2$ and $\tilde{\phi}_1$ are not orthogonal. The scale of the electronic wave function $\tilde{\psi}_l(\boldsymbol{r}, \boldsymbol{R})$ is much larger than that for the nuclear function $\tilde{\phi}_i(\boldsymbol{R})$, which is localised in the region $\delta R \sim a$, where a is the vibrational amplitude.

The probability $[b(t)]^2$ oscillates according to eqn (1.31). Its average value, which corresponds to the equilibrium-doped state, is equal to $\tilde{b}^2$ (see eqn (1.33)).

It is essential that, unlike the case with the usual Born–Oppenheimer approximation, the operator $\widehat{H}_{el}$ has non-diagonal matrix elements. Such an element describes the transition (in our case, tunnelling) between the terms.

The described approach is similar to that employed by Bardeen (1961) to describe the tunnelling effect (see also e.g. Kane, 1969, and Reittu, 1995). Instead of the usual method based on the use of the stationary Schrödinger equation, Bardeen analysed tunnelling as a time-dependent phenomenon, and, more specifically, as a quantum transition between two (initial and final) states. Such transitions are caused by specific Hamiltonians (see e.g. Landau and Lifshitz, 1977). Correspondingly, the tunnelling Hamiltonian formalism (Harrison, 1961; Cohen *et al.*, 1962) appears to be a powerful tool in solid state physics, and, especially, in physics of superconductivity. Such a treatment corresponds to the dynamic Jahn–Teller effect (see appendix A).

Diabatic representation is a convenient tool in molecular spectroscopy and solid state physics. It is employed to describe the polaronic state when each ion is characterised by two close minima (see section 2.3.1), the isotope effect (see sections 3.5.7, 5.43, and 9.53), and the quasi-resonant states of nanoclusters (see section 8.3.1).

Polarons: Concept. A strong electron–lattice interaction could lead to the formation of polarons. The concept of polarons was introduced and studied by Landau (1933); see Landau (1965) and Landau and Pekar (1946). A polaron can be created if an electron is added to the crystal with a small carrier concentration (see e.g. Ashcroft and Mermin, 1976). Because of strong local electron–ion interactions, the electron appears to be trapped and can be viewed as being dressed in a 'heavy' ionic 'coat'. In reality, we are dealing with a strong (non-linear) manifestation of the electron–lattice interaction.

The concept of polarons is an essential ingredient of the physics of high-T_c superconductivity. In fact, the concept of a Jahn–Teller polaronic state (see e.g. Hock *et al.*, 1983) was a main motivation for Bednorz and Mueller to search for superconductivity in the oxides, and this led to their breakthrough discovery (1986, see chapter 5).

The formation of polarons is a strong non-adiabatic phenomenon. As we know, the usual adiabatic approximation (see above and sections 2.1.1–2.1.3) allows us to separate electronic and ionic motions. Indeed, this approximation is based on the fact that the ionic motion is much slower than the motion of electrons. As a first step, one can neglect the kinetic energy of the ions and study the electronic structure for a 'frozen' lattice. The electronic energy (electronic terms) appears to be function of the ionic positions ($\varepsilon_{el} = \varepsilon_n(\boldsymbol{R})$). Next, one can consider the ionic dynamics; it turns out that the electronic terms $\varepsilon_n(\boldsymbol{R})$ form the potential for the ionic motion. The total wave function Ψ can be written as a product: $\Psi = \Psi_{el}\Phi_{ionic}$ (see eqns (1.9) and (1.15)). However, such a separation of electronic and ionic terms is impossible for polaronic states: the ions actively participate in their formation.

In order to study the case of small carrier concentration, it is convenient to use the so-called Holstein Hamiltonian (1959):

$$\widehat{H} = J\sum_{i,\delta} a^{+}_{j+\delta}a_j + \sum_{\boldsymbol{q}} \omega_{\boldsymbol{q}} b^{+}_{\boldsymbol{q}} b_{\boldsymbol{q}} + \sum_{j,\boldsymbol{q}} M_{\boldsymbol{q}} a^{+}_{j} a_j e^{i\boldsymbol{q}\boldsymbol{R}}(b_{\boldsymbol{q}} + b^{+}_{-\boldsymbol{q}}) \tag{1.34}$$

which was introduced to describe small polarons in semiconductors. The first term describes the kinetic energy (hopping) of the particles; the second term is the phonon energy, and the last term corresponds to the interaction. By direct comparison of eqns (1.1), (1.6), (1.8), and (1.34), one can see that the polaronic state described by the Hamiltonian (1.34) corresponds to the energy scales opposite to those in the usual adiabatic theory, that is, to the case when the electronic energy is small relative to the vibrational scale. The Hamiltonian (1.29) was derived by Holstein explicitly for such a case.

Dynamic polaron. Consider what can be dubbed the 'dynamic polaron' state. In this case, ions do not reside, as usual, in a definite local minimum of the potential, but are instead presented with two closely spaced minima (a 'double-well' structure; see Fig. 1.5b). Such a situation indeed occurs in practice, and is especially interesting for layered structures (such as the cuprates; see chapter 5).

Note that the 'double-well' structure is a result of the crossing of electronic terms. The ionic configuration at this crossing corresponds to a degeneracy of the electronic states, which is a key ingredient of the Jahn–Teller effect (1937; see e.g. Landau and Lifschitz, 1977; Bersuker, 2006).

In order to describe such a case, it is convenient to use a 'diabatic' representation (see above). In this representation, we are dealing directly with the crossing of electronic terms. The operator $\widehat{H}_{el} = \widehat{T}_{\boldsymbol{r}} + V(\boldsymbol{r}, \boldsymbol{R})$, where $\widehat{T}_{\boldsymbol{r}}$ is a kinetic energy operator, $V(\boldsymbol{r}, \boldsymbol{R})$ is the total potential energy, and $\boldsymbol{r}$ and $\mathbf{R}$ are the electronic and nuclear coordinates, respectively) has non-diagonal terms (unlike the usual adiabatic picture when $\widehat{H}_{el}$ is diagonal). The charge transfer in this picture is accompanied by the transition to another electronic term. Such a process is analogous to the Landau–Zener effect (see Landau and Lifshitz, 1977).

The total wave function can be written in the form (1.30) (see above); $\tilde{\psi}_i(\boldsymbol{r},\boldsymbol{R}), \Phi_i(\boldsymbol{R})$, $i = \{1,2\}$, are the electronic and vibrational wave functions that correspond to two different electronic terms (see Fig. 1.5).

Let us stress at first that it is impossible to separate the electronic and vibrational motion; one can see this directly from eqn (1.30). Indeed, unlike the usual adiabatic picture (cf. eqns (1.9) and (1.15)), the wave function (1.30) is not a product of the electronic and nuclear wave functions.

If terms are asymmetric ($\varepsilon \neq 0$), the quantity b^2 depends on the Franck–Condon factor. Therefore, the presence of the Franck–Condon factor is essential; it leads to a peculiar isotope effect in the high-T_c cuprates and manganites (see sections 5.4.3 and 9.5.3).

The described dynamic polaronic effect is the manifestation of the dynamic Jahn–Teller effect (see appendix A and e.g. Bersuker, 2006, and Salem, 1966). It is essential for studying the superconducting state in layered materials and, especially, in the high-T_c cuprates (see chapter 5).

1.2.3 Electronic Excitations

Plasmons. Usual plasmons in three-dimensional metals are described in many textbooks (see e.g. Ashcroft and Mermin, 1976, and Mahan, 1993). As we know, plasmons describe the collective oscillations of electrons with respect to a positive background. Recall that the usual electronic quasiparticle excitations correspond to the poles of the one-particle Green's function. Correspondingly, the spectrum of plasma oscillations which represent the collective motion is given by the poles of the two-particle Green's function (Fig. 1.6) or by poles of the vertex Γ.

Note that, if we imagine the system of non-interacting electrons, then the one-particle spectrum does not disappear, but is reduced to the kinetic energy of a free electron. As for collective excitations, such as plasmons, their existence is caused explicitly by the interaction only. That is why the calculation is directly related to the analysis of the Coulomb vertex Γ.

Usually the value of the plasma frequency for an isotropic three-dimensional system is quite high (~5–10 eV). The concept of plasmons as collective excitations was introduced by Goldman (1947) and by Bohm and Pines (1953). However, when dealing with more complicated band structures, one encounters additional low-lying plasmon branches. Such low-lying branches could be also observed in low-dimensional systems, even in the simplest, one-band case.

In this section, we focus on layered conductors. The reason for such a focus is twofold. Firstly, many novel superconducting systems contain conducting layers. In addition,

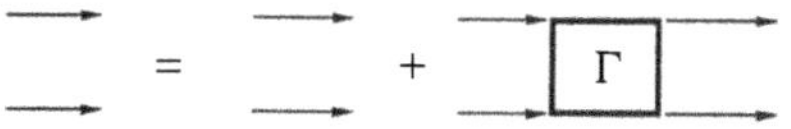

Figure 1.6 *The two-particle Green's function and the total vertex* Γ.

the plasmon spectrum for such conductors drastically differs from that for usual three-dimensional metals. It is different also from the spectrum for two-dimensional systems (Ando *et al.*, 1982). One can demonstrate (see below) that the plasmon spectrum of layered conductors contains an acoustic branch which is similar to acoustic phonons. As a consequence, the presence of such a low-lying plasmon branch leads to its noticeable contribution to the interaction and formation of many-layered superconducting systems (see section 2.6). Moreover, it turns out that, for the family of superconducting nitrides, the exchange by acoustic plasmons plays a key role (see section 7.4.3).

Plasmons in layered systems, dispersion law, and 'electronic sound'. As was noted above, the situation in the layered conductors is very different from that in bulk three-dimensional materials. Unlike isotropic metals, these materials have considerable anisotropy in their conductivities along and perpendicular to the layers. As a consequence, the screening properties of such a system are quite different from the isotropic case. The energy spectrum has been analysed by Vissher and Falicov (1971), Fetter (1974), and Kresin and Morawitz (1990).

The term 'layered electron gas' (LEG) describes the basic physical concept, namely an infinite set of two-dimensional layers of carriers (two-dimensional electron gas), separated by electronically inactive (insulating) spacer layers.

We assumed that we are dealing with the in-plane motion of electrons only; the interlayer hopping, in a first approximation, is neglected. At the same time, the Coulomb interaction is, as usual, three-dimensional. In other words, in the LEG model, the Coulomb interaction between charge carriers on different layers is included exactly, while the electronic collective motion is calculated for a single layer with two-dimensional plane-wave wave functions for the carriers. In addition, the dielectric response of the intervening, non-metallic region is included by using the low-frequency limit of its dielectric constant (e.g. $\varepsilon \approx 3$–6 for the cuprates) to screen the interaction between conducting layers. The layering leads to a drastic change in the plasmon spectrum.

A direct consequence of the layering of the carriers is that the spatial periodicity along the normal to the planes allows the definition of a corresponding wave vector $q_z = \pi/L$, L is the interlayer spacing.

Unlike the one-particle excitations, the spectrum of plasma oscillations, which represents the collective motion, corresponds to the poles of the two-particle Green's function. Since the existence of plasmons is caused by interaction only, the calculation is directly related to the analysis of the Coulomb vertex $\Gamma \equiv \Gamma_c$ (Fig. 1.6).

We employ random-phase approximation (RPA); this method allows us to obtain the complete solution. In the RPA method, the total vortex Γ_c is reduced to the 'chain' diagram (Fig. 1.7). Therefore, $\Gamma_c = V_c + V_c \sqcap \Gamma_c$, and

$$\Gamma_c = \frac{V_c}{1 - V_c \sqcap} \tag{1.35}$$

Here $V_c \equiv V_c(q)$ is the Fourier component of the unscreened Coulomb potential, and $\sqcap$ is the polarisation (see Fig. 1.7). RPA provides a rigorous result for $\boldsymbol{r}_s \ll 1$ (the high-density limit; $r_s = r_0/r_B$; see e.g. Mahan (1993)), r_B is the Bohr radius, and

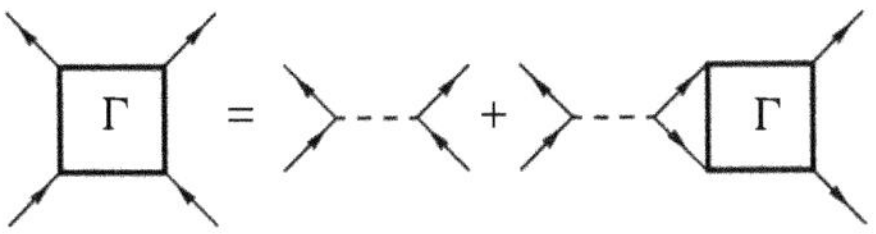

Figure 1.7 *Equation for the total vertex (random-phase approximation).*

$r_0 \approx 0.6\ n{-}^{1/3}$, and n is the carrier concentration). For real metals, $5.5 \gtrsim r_s \gtrsim 2$. Based on RPA, one can obtain a good qualitative description of the screening and the spectrum (plasmons). Γ_c can be written in the form

$$\Gamma_c(\mathbf{q};|n-m|)\frac{V_c(\mathbf{q})}{\varepsilon(\mathbf{q},|n-m|)} \tag{1.35$'$}$$

The Coulomb term is written in its most general form as the ratio of the bare Coulomb interaction and the dielectric function. The dielectric function $\varepsilon(\mathbf{q},\omega_n)$ which describes the electronic screening of the Coulomb interaction and also enters eqn (1.35′) can be written in the form (see eqn (1.35))

$$\varepsilon(\mathbf{q},\omega_n-\omega_m)=1-V_c(\mathbf{q})\sqcap(\mathbf{q},\omega_n-\omega_m)$$

Both the Coulomb interaction and the dielectric function have to be calculated for a layered structure. The Coulomb potential $V_c(\mathbf{q})$ is the Fourier transform of the three-dimensional Coulomb interaction

$$V_c(\mathrm{r})=e^2/\varepsilon_M|r| \tag{1.36}$$

(ε_M is the dielectric constant of the spacers).

To perform more rigorous evaluation, one should calculate the quantity (matrix element) $V_c(\boldsymbol{q})$ directly for the layered structure.

One can write the expression for $V_c(r)$ in the form

$$V_c=\frac{1}{\varepsilon_M q_{\|}}\exp(-q_{\|}|l-l'|) \tag{1.36$'$}$$

where l and l' are the labels of the layers. After summation over $m=l-l'$, one can obtain

$$V_c(\boldsymbol{q})=\frac{\lambda_c}{\nu}\frac{R(\tilde{q};q_z)}{\tilde{q}} \tag{1.37}$$

with

$$R(\tilde{q},q_z)=\frac{\sinh(\kappa_F\tilde{q})}{\cosh(\kappa_F\tilde{q})-\cos(q_zL)}(1-\delta_{\mathbf{q},0}) \tag{1.37$'$}$$

We introduce dimensionless quantities $\tilde{q} = q_{\parallel}/2\kappa_F$, $\kappa_F = 2k_F L$, ($q_{\parallel}$ is the in-plane momentum, L is the interlayer distance, and k_F is the in-plane Fermi wave vector) as well as $\nu = m_b/2\pi\hbar^2$, the two-dimensional electronic density of states (m_b is the band mass). Note that $R(\tilde{q}; q_z)$ contains the factor $(1 - \delta_{q,0})$ reflecting the presence of the neutralising positive ion counter charges. Equation (1.37) contains the dimensionless Coulomb interaction constant defined by

$$\lambda_c = \frac{e^2}{2\varepsilon_M \hbar v_F} \tag{1.37''}$$

v_F is the Fermi velocity. Note that $\lambda_c \sim r_s$, where $r_s = \frac{\sqrt{2}}{k_F r_B}$ ($r_B = \hbar^2 \varepsilon_M / me^2$ is the Bohr radius) is the well-known dimensionless electron density radius defined here for a LEG. Note that the presence of the dielectric constant $\varepsilon_M (\varepsilon_M > 1)$ makes the applicability of the RPA approximation more favourable. The plasmons are defined by the poles of Γ ($\Gamma \equiv \Gamma_c(q,\omega)$ is the analytical continuation of the function $\Gamma_c(q,\omega_n)$) or, equivalently, by the zeros of the dielectric function.

Let us introduce the dimensionless parameter $\alpha = \omega/q_{\parallel} v_F$. One has to distinguish two cases: $\alpha < 1$ and $\alpha > 1$. Plasmons appear in the region of $\alpha > 1$. Let us evaluate the polarisation parameter describing the two-dimensional electronic system:

$$\sqcap(\mathbf{q},\omega) = \sum_{\mathbf{\kappa}} \frac{n_{\mathbf{\kappa}+\frac{\mathbf{q}}{2}} - n_{\mathbf{\kappa}-\frac{\mathbf{q}}{2}}}{\omega + \varepsilon_{\mathbf{\kappa}+\frac{\mathbf{q}}{2}} - \varepsilon_{\mathbf{\kappa}-\frac{\mathbf{q}}{2}}} \tag{1.38}$$

Here $n_{\mathbf{\kappa}}$ is the Fermi distribution function of the carriers with the dispersion relation $(\mathbf{\kappa}) = \frac{\kappa^2}{2m^*}$, and m^* is the effective mass. We assume also that $q_{\parallel} \ll 2p_F$. Then we obtain

$$\sqcap = -\frac{m^*}{2\pi^2} \int_0^{2\pi} d\varphi [\alpha - \cos\varphi + i\delta\cos\varphi]^{-1} \tag{1.38'}$$

If $\alpha > 1$ (this region corresponds to plasmons), we obtain

$$\sqcap = -\left(\frac{m^*}{\pi}\right)\left[(1-\alpha^{-2})^{-\frac{1}{2}} - 1\right] \tag{1.38''}$$

For simplicity, consider the case $\alpha >> 1$. Then the polarisation operator is given by

$$\sqcap(q_{\parallel},\omega) = v(v_{Fq_{\parallel}}/\omega)^2 \tag{1.39}$$

where v is the density of states.

Let us demonstrate that the layering leads to a dispersion relation which is entirely different from that for three-dimensional bulk metals (then $\omega_{pl} \cong \omega_0 + aq^2$; usually, the value of ω_0 is high; $\omega_0 \cong 5-10$ eV); we use here the semi-quantitative treatment. The expression for the Coulomb potential can be written in the form

$$V_c(\mathbf{q}) = \frac{e^2}{(q_{\|}^2 + q_z^2)} \tag{1.39'}$$

This expression is valid for the usual three-dimensional system. The quantisation can be approximately taken into account by considering the quantised values of $q_z = \pi n/L$, where L is the interlayer distance.

Let us remember that we are considering the two-dimensional in-plane electronic motion, but the three-dimensional Coulomb interaction. Then one can see from eqns (1.35″) and (1.39′), that the dependence $\omega(q_{\|})$ is determined by the value of q_z. For example, for $q_z = 0$, we obtain the usual three-dimensional value of the plasmon frequency, so that $\omega = \omega_0$ for $q_{\|} = 0$. However, for $q_z = \pi/L$, we obtain $\omega \propto q_{\|}$, that is, the acoustic dispersion law. Therefore, layering leads to a drastic change in the plasmon spectrum.

To perform a more rigorous evaluation, one should calculate the quantity (matrix element) $V_c(\mathbf{q})$ directly for the layered structure. This quantity is the Fourier transform of the three-dimensional Coulomb interaction, eqn (1.36).

With the use of eqns (1.35) and (1.38), one can obtain the following general expression for the plasmon spectrum:

$$\omega = q_{\|} v_F \left[1 + (\nu V_c)^2 (0.25 + \nu V_c)^{-1}\right]^{1/2} \tag{1.40}$$

where $V_c \equiv V_c(\tilde{q}, q_z)$ is the Coulomb interaction defined in eqn (1.37). If $\nu V_c \gg 1$, we obtain the dispersion law: $\omega = \hbar q_{\|} v_F \sqrt{1 + \nu V_c}$; this corresponds to the hydrodynamic approximation for small $q_{\|}$ (see Fetter, 1974). For $\omega \gg \hbar q_{\|} \nu_F$, eqn (1.40) reduces to the expression $\omega \cong \hbar q_{\|} v_F \sqrt{\nu V_c}$, which at $q_z = 0$ leads to the usual 'optical' plasmon with $\Omega_{pl}^2 = \omega^2(q_{\|} = 0, q_z = 0) = 4e^2 \varepsilon_F / \epsilon_M L$. For $q_z = \pi/L$, on the other hand, we obtain the acoustic dispersion law.

One can see directly from eqns (1.37) and (1.40) that the plasmon spectrum of the layered conductor is described not by the single branch $\omega = \omega_0 + aq^2$ but by the plasmon band with branches corresponding to $0 < q_z < \pi/L$ (Fig. 1.8). The plasmon band $\omega = \omega(q_{\|}, q_z)$ is confined between the upper branch at $q_z = 0$ (the in-phase motion of the charge carriers) and the lower branch at $q_z = \pi/L$ (the out-of-phase motion of the carriers). It is also possible to include a small hopping term t in the c-axis direction; one now finds a small gap at $q_{\|} = 0$ of order $4t$.

One can show that the density of states for this plasmon band peaks near its boundaries, that is, near $q_z = \pi/L$ and $q_z = 0$ (Morawitz *et al.*, 1993). The indication for the peak structure can be seen from eqn (1.40). Indeed, the derivative $\frac{\partial \omega}{\partial q_z} = 0$ for the values $q_z = \frac{\pi}{L}$ and $q_z = 0$. Because of the peaked structure, the plasmon spectrum can be visualised as a set of two branches (Fig. 1.9). The upper branch ($q_z = 0$) is similar to the usual plasmon branch for an isotropic three-dimensional metal; it corresponds to the in-phase motion of neighbouring layers.

The most important feature of the plasmon spectrum of a layered metal is an appearance of the lower branch ($q_z = \pi/L$), which has an acoustic nature. This branch

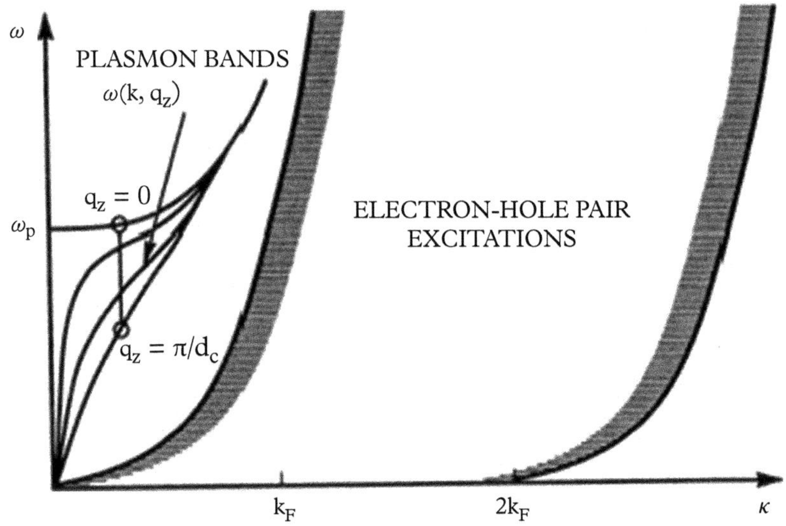

Figure 1.8 *The plasmon band in a layer conductor.*

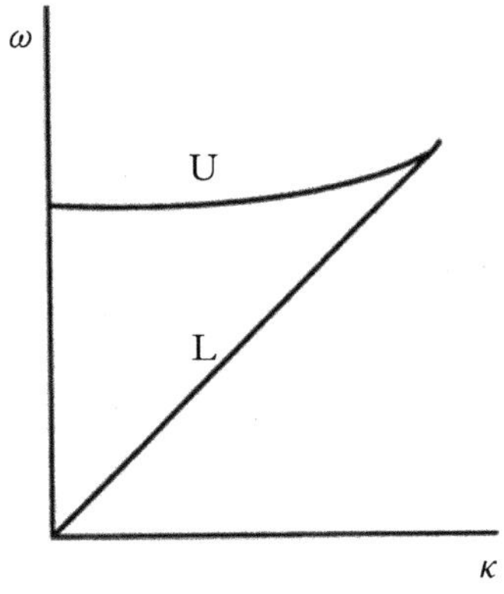

Figure 1.9 *Upper (U) and lower (L) plasmon branches in a layer conductor.*

corresponds to the out-of-phase motion of neighbouring layers, that is, they oscillate relative each over. This mode can be called an 'electronic sound'.

There are several novel consequences from the splitting of the collective excitation spectrum into layer plasmon branches. These arise specifically because of the existence of a low-frequency electronic response. The first of these is the possibility of pairing the carriers by exchange of acoustic plasmons (see section 2.6).

Manifestation of low-frequency plasmons: Observables. Let us discuss the question of how to observe the existence of plasmons and, especially, low-frequency electronic collective excitations ('electronic sound') in the layered conductors. The usual method is to measure the energy lost by the electron moving in the crystal (see e.g. Hofer *et al.*, 2016, and Larkin, 1960). Such an electron can radiate (or absorb) plasmons (this is similar to Cherenkov radiation), and this phenomenon can be described by the so-called loss function.

The plasmon spectrum of a layered system contains two branches (Fig. 1.9): an upper branch (U), which is similar to that in usual three-dimensional metals, and the low-frequency acoustic branch (L).

The loss function γ contains two terms: $\gamma = \gamma_U + \gamma_L$, where the terms describe the contributions of the upper and lower branches, respectively. The Hamiltonian describing the interaction between the particle and the material carriers has the form

$$\widehat{H} = \sum_{\boldsymbol{\kappa},\boldsymbol{\kappa}',\mathbf{p},\mathbf{q}} V_{\boldsymbol{\kappa},\mathbf{p}}^{\boldsymbol{\kappa}',\mathbf{p}-\mathbf{q}} a_{\boldsymbol{\kappa}'}^{+} a_{\mathbf{p}-\mathbf{q}}^{+} a_{\boldsymbol{\kappa}} a_{\mathbf{p}} \tag{1.41}$$

Here $\mathbf{p}$ is the momentum of the charged particle, $\boldsymbol{\kappa}$ and $\boldsymbol{\kappa}'$ are the initial and final momenta, respectively, of the material carriers, and $\mathbf{q}$ is the transferred momentum:

$$V_{\boldsymbol{\kappa},\mathbf{p}}^{\boldsymbol{\kappa}',\mathbf{p}-\mathbf{q}} = \int e^{-(\mathbf{p}-\mathbf{q})\mathbf{r}_2} \phi_{\boldsymbol{\kappa}'}^{*}(\mathbf{r}_2) \frac{e^2}{|\mathbf{r}_1 - \mathbf{r}_2|} e^{i\mathbf{p}\mathbf{r}_2} \phi_{\boldsymbol{\kappa}}(\mathbf{r}_1) d\mathbf{r}_1 d\mathbf{r}_2 \tag{1.41$'$}$$

is the matrix element of the Coulomb interaction. The wave function of the passing particle is taken as a plane wave. It is assumed that its movement is fast ($\frac{e^2}{\hbar\nu} \ll 1$), so that the interaction can be treated in the main approximation. The function $\phi_{\boldsymbol{\kappa}}(\mathbf{r})$ can be written in the form

$$\phi_{\boldsymbol{\kappa}}(\mathbf{r}) = e^{i\boldsymbol{\kappa}\boldsymbol{\rho}} \chi_{\boldsymbol{\kappa}_z}(z)$$

Passage of a charged particle through a layered conductor is accompanied by the energy loss

$$\frac{\partial E}{\partial t} = (2\pi)^{-3} \int (\varepsilon_{\mathbf{p}} - \varepsilon_{\mathbf{p}-\mathbf{q}}) W_{\mathbf{q}} d^3\mathbf{q} \tag{1.42}$$

($\mathbf{q}$ is a momentum transfer, and $W_{\mathbf{q}}$ is a total probability), which is a temperature dependent (Kresin and Morawitz, 1991). The probability $W_{\mathbf{q}}$ has the form

$$W_{\mathbf{q}} = 2\pi |V_{\boldsymbol{\kappa},\mathbf{q}_z}|^2 \phi_{\mathbf{q}}(\varepsilon_{\mathbf{p}} - \varepsilon_{\mathbf{p}-\mathbf{q}}) \tag{1.42$'$}$$

where $\phi_q(\omega) = \sum_{m,n} e^{\beta(\Omega - \mu N_m - E_m)} \langle m | \sum a_{k+q}^{+} a_k | n \rangle \delta(E_n - E_m - \omega)$

The exponential factor describes the statistical averaging (density matrix; Ω is the thermodynamic potential, $\beta = 1/T$). The key factor is the appearance of temperature in the expression for the loss function. Since $\omega_v \gg \Gamma$ (ω_v is the frequency corresponding to the upper branch) the term γ_v is temperature independent. As for the term γ_L, the situation is different. An increase in temperature leads to an increase in the number of acoustic plasmons, in accordance with the statistic (it is similar to the temperature dependence for acoustic phonons). As a result, one can obtain

$$\gamma_L \sim T^3 \tag{1.43}$$

Such a dependence ($\sim T^3$) could be observed and is a direct manifestation of a peculiar plasmon spectrum in layered conductor.

Another inelastic process, namely, resonant inelastic X-ray scattering can be also used to detect the presence of low-frequency plasmon modes (see chapter 4). This 'photon in–photon out' spectroscopy has even greater accuracy than electron energy loss spectroscopy for low-frequency collective charge excitations (see e.g. the review by Ament *et al.* (2011)). As was noted above, the plasmon spectrum of layered conductors combines the almost two-dimensional nature of the carrier dynamics (with very small interlayer hopping) with the three-dimensional nature of the Coulomb interaction (intralayer and interlayer components). As a result, the plasmon spectrum has a rather peculiar dispersion law with its dependence on the in-plane momentum q_{II} and the out-plane momentum q_z. It is impossible to measure the dispersion law with the use of usual light, because it has small value of the momentum. The situation is different with X-ray quanta. The large value of their momenta allows to use the conservation laws for momentum and energy and recover the dispersion law for plasmons. In addition, the change in the polarisation of the radiation allows us to separate the charge excitations, such as plasmons, from magnons. Indeed, the change in the polarisation does not affect the plasmons, but it leads, because of the spin conversation, to a change in the electronic spin; this affects the magnetic excitations.

Recent studies (Hepting *et al.*, 2018) made it possible to determine the presence of acoustic plasmons in cuprates. We will discuss it later on, in section 5.4, describing properties and mechanisms of superconductivity in the high-T_c cuprates.

Therefore, the plasmon spectrum in layered conductors such as cuprates represents the plasmon band (Fig. 1.8). The density of states peaks near the upper ($q_z = 0$) and lower ($q_z = q/L$) boundaries. Qualitatively, one visualises the plasmon band as a set of two branches (Fig. 1.9); the upper (U) branch is similar to that in usual metals. A very important feature of the layered metals is the appearance of the lower (L) branch, which has an acoustic dispersion law and can be called 'electronic' sound.

The three-dimensional case: 'Demons'. The value of the plasmon frequency for a one-band three-dimensional metal is very high. As a result, the contribution of such plasmons to the pairing is not essential, because the corresponding coupling constant is negligibly small. However, when dealing with more complicated band structures, one encounters additional low-lying plasmon branches. Below, we describe such a case.

Consider a metal with two overlapping energy bands, one of which ('*a*') contains light carriers while the other ('*b*') is narrow and contains heavy carriers. It turns out that its energy spectrum is characterised by the presence of an acoustic plasmon branch ('demons'). This mode corresponds to the collective motion of the light carriers with respect to the heavy ones. It turns out that this mode is described by the acoustic dispersion law; this acoustic branch is similar to phonons. Indeed, the system is described by the following equations (see e.g. Geilikman, 1966):

$$\Gamma_{aa} = (V_{aa} + \Pi_{bb} R)\, S^{-1}$$
$$S = 1 + V_{aa}\Pi_a + V_{bb}\Pi_b + \Pi_a \Pi_b R, \qquad R = V_{ab}^2 - V_{aa}V_{bb} \tag{1.44}$$

where V_{aa}, V_{bb}, and V_{ab} are the Coulomb matrix elements. In this case, Γ_{aa} is defined by two parameters: $\alpha_a = \omega/qv_F^a$ and $\alpha_b = \omega/qv_F^b$, where the indices a and b refer to the two bands. Expression (1.44) is a generalisation of the equation describing usual three-dimensional plasmons (see e.g. Abrikosov *et al.*, 1975) for the case of two bands.

Let us assume that $m_b \gg m_a$ and $v_F^a \gg v_F^b$, and consider the case of $\alpha_a \ll 1$ and $\alpha_b \gg 1$. Then

$$\Pi_{a;0} = -\frac{m_a k_F}{\pi^2}, \qquad \Pi_{b;\alpha_b \gg 1} = \left(\frac{v_{Fb} q}{\omega}\right)^2 \tag{1.44$'$}$$

In this case, the light carriers provide screening, while the presence of the heavy one results in the appearance of an additional plasmon branch. Indeed, from eqns (1.44) and (1.44′) and the condition $S(\mathbf{q}, \omega) = 0$, we obtain the dependence $\omega \propto q$, that is, the new plasmon branch has an acoustic character.

These so-called demons, acoustic plasmons which are due to the presence of two groups of carriers of different masses, are similar to phonons. A similar picture can also arise in multivalley semiconductors, semimetals, and so on.

The possibility of superconducting pairing due to exchange of 'demons' has been considered in a number of papers (Garland, 1963; Geilikman, 1966; Fröhlich, 1968; Ihm *et al.*, 1981); see also the review by Ruvalds (1981).

1.2.4 Magnetic Excitations

The low-energy magnetic excitations in ferromagnets and antiferromagnets are known as spin waves (magnons in quantum theory). Similar to phonons describing the periodical oscillations of the ionic density, and plasmons related to the periodical fluctuations of the charge density, magnons result from the periodical fluctuations of the spin density. Their description is different in metals with itinerant electrons and in insulators with localised electrons.

Spin waves in the band theory of magnetism. Many years ago, Bloch (1929) showed that electrons with low density may become ferromagnetic due to the exchange matrix elements of the Coulomb interaction. The mean field band theory of ferromagnets was developed by Stoner (1938); according to this theory, the electronic band in ferromagnets is split by spin projection on spin up and spin down sub-bands:

$$\varepsilon_{k\uparrow} = \varepsilon_k - JM, \quad \varepsilon_{k\downarrow} = \varepsilon_k + JM \tag{1.45}$$

Here ε_k is the band energy of an electron in the paramagnetic phase, $J > 0$ is the exchange interaction, and $M = (N_\uparrow - N_\downarrow)/N$ is the magnetisation given by the difference of the spin up and spin down electronic concentrations. Magnons in this model describe the collective oscillations of the electron's spin density, similar to plasmons describing the collective oscillations of the electron's charge density. The magnon with the wave number q creation operator is given by $b_q^+ = \sum_k c_{k\downarrow}^+ c_{k-q,\uparrow}$. The magnon spectrum and thermodynamics can be found from the two-particle Green's function, given by the sum of the bubble diagrams within the RPA. Formally, it looks similar to the plasmon oscillations; the only difference is that one of the Green's functions in the bubble has spin up, and the other one has spin down. A very important difference is that the Coulomb interaction for plasmons is a long-range one, while the exchange interaction for magnons is a short-range one. That is why in three-dimensional materials the plasmon has large energy at $q = 0$, the plasma frequency, while the magnon energy is zero at $q = 0$.

For three-dimensional isotropic ferromagnets, the magnon dispersion at small q is given by a quadratic dependence $\omega_q = Dq^2$, while, for three-dimensional isotropic antiferromagnets, the magnon dispersion is linear. The exchange interaction is isotropic, and the gapless magnon dispersion is a manifestation of the Goldstone theorem: the long-range ferromagnet and antiferromagnet states have lower symmetry then the Hamiltonian, so massless bosons should appear. In two-dimensional isotropic systems, the long-range ferromagnet and antiferromagnet orders at finite temperature are unstable due to the Mermin–Wagner (1966) theorem. The short-range magnetic order in layered superconductors like cuprates and Fe pnictides is essential, and its effect on superconducting pairing will be discussed below. In general, the electron–magnon interaction results in the spin-dependent pairing of two electrons, which is essential to superconductivity; this effect will be discussed in section 2.7.

Spin waves in magnetic insulators: The Heisenberg model. The Heisenberg model describes the exchange interaction of spins localised in the lattice sites $\boldsymbol{R}_i$. This model is valid for magnetic insulators when the kinetic energy of electrons is negligibly small and may be neglected. The model Hamiltonian is given by

$$H_J = -\frac{1}{2}\sum_{i,j} J_{i,j} \boldsymbol{S}_i \boldsymbol{S}_j \tag{1.46}$$

Usually, the main contribution results from the nearest neighbours interaction $\mathcal{J}$, and the ground state is ferromagnet for $\mathcal{J} > 0$, and antiferromagnet for $\mathcal{J} < 0$. In the ferromagnet ground state, all spins are parallel to each other with maximal total magnetisation with the ground state wave function

$$|\Psi_0\rangle = |\uparrow\uparrow\uparrow\uparrow\uparrow\uparrow\uparrow\rangle$$

Due to the isotropy of the Heisenberg Hamiltonian, the direction of magnetisation is determined by some external anisotropic perturbation; it may be an external magnetic field or weak magnetic anisotropy. Let us choose the z-axis along this direction; then $S^z = S$ for each ion in the ground state. An individual spin-flip at one site $\boldsymbol{R}_i$ is an elementary excitation with the wave function $|\Psi_i\rangle = S_i^- |\Psi_0\rangle$. Due to the interatomic exchange interaction $\mathcal{J}_{ij}$, this excitation may move from the site $\boldsymbol{R}_i$ to the site $\boldsymbol{R}_j$. Coherent movement of this excitation is the spin wave (magnon) described by the wave function

$$|\Psi_q\rangle = \frac{1}{\sqrt{N}} \sum_{R_i} e^{iqR_i} S_i^- |\Psi_0\rangle$$

with the wavenumber $\boldsymbol{q}$ and the energy $\omega(q) = S(J(o) - J(q))$.

In the limit $q \to 0$, the magnon energy in a ferromagnet is equal to $\omega(q) = Dq^2$. The magnon creation operator is given by the spin operator S_q^-, so magnons are not bosons. In the limit of infinite spin value, their statistics tend to the Bose–Einstein one. The spin operators may be represented by Bose-type operators in several ways, for example by non-linear Holstein and Primakoff (1940) transformation. Thermodynamics of a ferromagnet at low temperatures $T \ll T_K$, where T_K is the Curie temperature, can be described by the gas of non-interacting magnons. Their contribution to the decreasing magnetisation is given by the Bloch (1930) law

$$\frac{M(0) - M(T)}{M(0)} \sim \left(\frac{T}{T_K}\right)^{3/2}$$

More complicated magnetic structures are described by several magnetic sublattices with equal and oppositely directed sublattices in antiferromagnets, and non-equivalent sublattices in ferrimagnets. Nevertheless, the concept of magnons as low-energy excitations is valid.

2

Mechanisms

2.1 Pair Correlation

2.1.1 The General Picture and Main Challenges

The fundamental concept of the present theory is that there is an attractive interaction between electrons in superconducting materials. This feature distinguishes this type of electronic system from the usual electron gas. At first glance, it seems strange because we are used to seeing electrons repel each other. However, the electrons are located inside a crystal, and the presence of such a medium can change the sign of the interaction. It could be understood as a negative sign of the dielectric constant. Qualitatively, this additional electron–electron interaction mediated by an ionic system can be visualised as follows: an electron attracts the positive ion located in its proximity, and the subsequent shift in the ion's position affects the state of another electron. The first electron is 'dressed' by a positive ionic coat and, if the positive charge exceeds that of an electron, then this electron and its 'coat' form a positive unit that is attracted to the second electron (Fig. 2.1).

In quantum theory, this additional interaction mediated by an ionic subsystem can be described as phonon radiation and its subsequent adsorption by the second electron (exchange by phonons). The discovery of the isotope effect, which preceded the creation of microscopic theory, and tunnelling spectroscopy (see section 2.3.4) directly shows that the pairing can be provided by an electron–lattice interaction. The attraction is caused by the electron–lattice interaction, and its strength must be sufficient to overcome the Coulomb repulsion forces. The latter are actually somewhat weakened by a peculiar logarithmic factor (see section 2.2.2). Still, the inter-electron attraction mediated by phonon exchange must be sufficiently strong: its energy must be of the order of ε_F, since the Coulomb repulsion corresponds to this energy scale.

There are two fundamental questions that should be addressed before introducing the main equations of the microscopic theory and evaluating the major parameters. The first question is formulated above and concerns the scale of the electron–phonon interaction relative to the Coulomb repulsion.

Superconducting State: Mechanisms and Materials. Vladimir Z. Kresin, Sergei G. Ovchinnikov, and Stuart A. Wolf, Oxford University Press (2021). © Kresin, Ovchinnikov, Wolf. DOI: 10.1093/oso/9780198845331.003.0002

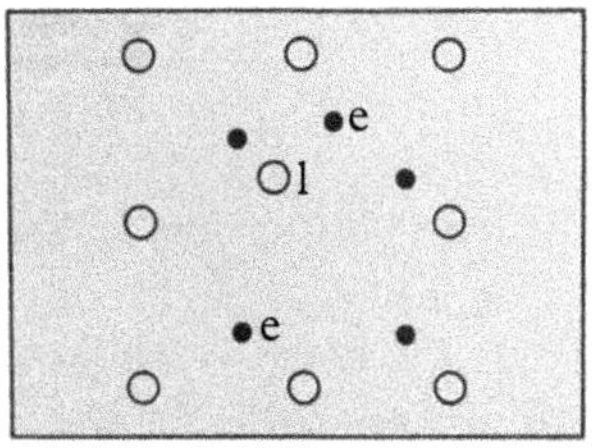

Figure 2.1 *Inter-electron attraction. The displacement of the ion I, caused by its interaction with the electron, affects the state of the other electrons.*

For superconductors, the interplay between the electron–lattice and Coulomb interactions leads to the resulting electron–electron attraction and the formation of bound electron pairs. Indeed, analysis of the electron–electron attraction shows that, for some electrons, the strongest attraction will be with another electron, the state of which is characterised by an opposite momentum and spin. As a result, the electronic system in a superconductor can be described as a set of electron pairs. The concept of pairing was introduced by Cooper (1956), and the electron pairs are often called 'Cooper pairs'. The concept of pairing was the key step allowing Bardeen, Cooper, and Schrieffer to create the complete BCS theory (1957).

The Cooper pair forms the bound state. It is important to stress that the attraction is important but, according to quantum mechanics, is not a sufficient condition for forming the bound state. Indeed, generally speaking, the strength of the attraction should exceed some value in order to obtain the negative electronic energy level, which corresponds to the stationary bound state. However, as was proven by Cooper (the 'Cooper theorem', 1956), any small attraction of electrons in metal leads to the formation of the bound state. The essence of this theorem will be discussed later on in this chapter. The pairing occurs for any small electron–electron attraction and also will be explained. Qualitatively, this is due to the fact that the pairing occurs for the electrons moving on the Fermi surface. The presence of the Fermi surface leads to the two-dimensional picture. As for the two-dimensional scenario, we know (see e.g. Landau and Lifshitz, 1977) that any weak attraction leads to the formation of the bound state.

2.1.2 Superconductivity as a 'Giant' Non-Adiabatic Phenomenon

Here we focus on the scale of the electron–phonon interaction. Indeed, one may wonder whether the electron–phonon interaction is capable of producing such a large effect ($\sim \varepsilon_F$). It turns out that the answer is yes, and the reason has to do with the presence of strong non-adiabaticity.

Consider electrons near the Fermi level and their transitions between energy states in the region $\sim \tilde{\Omega}(\tilde{\Omega} \cong \Omega_D)$ near the Fermi level, so that

$$\Delta\varepsilon < \tilde{\Omega}$$

Transitions of this kind are non-adiabatic, since the change in the electronic energy is smaller than the characteristic phonon energy. As we shall see, these non-adiabatic transitions are responsible for the observed large effect.

Let us estimate the strength of the phonon-mediated electron–electron interaction. The matrix elements of interest, M_{fi}, can be written in the form

$$M_{fi} = \frac{|\widehat{H}_1|_{fk}|\widehat{H}_1|_{ki}}{\varepsilon_{\boldsymbol{p}_1} - \varepsilon_{\boldsymbol{p}_3} - \hbar\Omega_{\boldsymbol{q}}} + \frac{|\widehat{H}_1|_{fl}|\widehat{H}_1|_{li}}{\varepsilon_{\boldsymbol{p}_2} - \varepsilon_{\boldsymbol{p}_4} - \hbar\Omega_{\boldsymbol{q}}} \tag{2.1}$$

Here $i \equiv \{\boldsymbol{p}_1, \boldsymbol{p}_2\}$ and $f \equiv \{\boldsymbol{p}_3, \boldsymbol{p}_4\}$ are the initial and final states, respectively; *k*, *l* are the virtual (intermediate) states, and $\boldsymbol{p}_i$ are the corresponding momenta. The following conservation rules hold: $\boldsymbol{p}_3 - \boldsymbol{p}_1 = \boldsymbol{p}_2 - \boldsymbol{p}_4 = \boldsymbol{q}$, and $\varepsilon_{\boldsymbol{p}_1} + \varepsilon_{\boldsymbol{p}_2} = \varepsilon_{\boldsymbol{p}_3} + \varepsilon_{\boldsymbol{p}_4}$, where $\boldsymbol{q}$ is the phonon momentum. The electron–phonon interaction is described by the operator $\widehat{H}_i$ (see eqns (1.22′) and (1.23)), that is, by the first term in the total Hamiltonian $\widehat{H}_{na}$.

Note that expression (2.1) contains terms non-diagonal in the operator $\widehat{H}_1$. These non-diagonal terms are finite. Since we are considering virtual phonon-exchange processes, the conservation of energy for the elementary act of electron–phonon scattering does not need to be held. The main contribution to the interaction comes from the non-adiabatic region. Unlike the case of small molecules, the existence of such a region is guaranteed by the continuous nature of the electron and phonon spectra.

Let us estimate the matrix element (2.1) in this region. Using eqn (1.23), we obtain

$$M_{fi} \approx \frac{\left(\tilde{\Omega}/k\right)^2}{\tilde{\Omega}} = \frac{\tilde{\Omega}}{k^2} = \frac{\tilde{\Omega}}{\left(\tilde{\Omega}/\varepsilon_F\right)} \approx \varepsilon_F \tag{2.2}$$

Hence, the phonon-exchange interaction is very strong $(\sim \varepsilon_F)$. This is due to the large contribution of $\widehat{H}'$, eqn (1.23), which comes from the non-adiabatic domain $\Delta\varepsilon_{el.} \lesssim \Omega_D$. Therefore, indeed, we are dealing with a giant non-adiabatic effect.

The semi-qualitative arguments presented above explain how a strong attraction can arise and overcome the Coulomb repulsion. A detailed quantitative analysis must be based on the methods of many-body theory, since effects of this magnitude cannot be rigorously treated by perturbation theory.

2.1.3 The ‘Cooper’ Theorem

According to the Cooper theorem (Cooper, 1956), the electron–electron attraction in a metallic system leads to the formation of the bound state (‘Cooper pairs’) for any weak interaction. This is not the case for the usual three-dimensional Fermi gas. It turns out that the formation of the bound state for any small interaction is caused by the Pauli principle and, subsequently, by the presence of the Fermi surface. This key factor makes the situation seem similar to the two-dimensional case.

In order to clarify this question, which is fundamental for understanding the nature of the superconducting state, let us first consider the quantum mechanical problem regarding the electronic state in the two-dimensional potential well $U(\rho)$, where ρ is the polar coordinate (see Landau and Lifshitz, 1977, and chapter 6). The Schrödinger equation

$$\widehat{T}\psi = [E - U(\rho)]\,\psi \tag{2.3}$$

($\widehat{T}$ is the kinetic energy operator) can be written, with a good accuracy, in the form

$$\rho^{-1}\frac{\partial}{\partial\rho}\left(\rho\frac{\partial\psi}{\partial\rho}\right) = 2mU \tag{2.3$'$}$$

We are looking for the solution $|E| \ll |U|$. Then one can neglect E in the region $\rho < a$, where a is the width of the well. In addition, ψ slowly depends on ρ, and one can assume that $\psi = \text{const} = 1$. One can see from eqn (2.3) that, outside the potential well, the Schrödinger equation for the free motion has the following form:

$$\rho^{-1}\frac{\partial}{\partial\rho}\left(\rho\frac{\partial\psi}{\partial\rho}\right) + 2mE\psi = 0 \tag{2.3$''$}$$

Equating the logarithmic derivatives inside and outside of the well, one can evaluate the value of the energy level with its peculiar dependence on the strength of the potential:

$$E \propto \left(\frac{\hbar^2}{ma^2}\right) e^{-\frac{1}{\lambda^*}} \tag{2.4}$$

where $\lambda^* = \left(\frac{m}{\hbar^2}\right)\int_0^\infty U(\rho)\,\rho d\rho$. Indeed, one can see that the energy level, that is, the bound state, exists for any two-dimensional potential and is exponentially small relative to the depth of the well.

Next, let us discuss the 'Cooper theorem' (Cooper, 1956; see also e.g. Ashcroft and Mermin, 1976). Consider the degenerate electronic gas and two electronic states at the Fermi surface. These electrons interact via a weak attractive force, which is caused by the exchange by virtual phonons. The Schrödinger equation in the momentum representation has the following form:

$$(E - 2\varepsilon_{\boldsymbol{k}})\,\varphi_{\boldsymbol{k}} = \int \frac{d\boldsymbol{k}'}{(2\pi)^3} U_{\boldsymbol{k},\boldsymbol{k}'},\varphi_{\boldsymbol{k}'} \tag{2.5}$$

Here, $\varepsilon_{\boldsymbol{k}} = \frac{k^2}{2m}$ is the energy of a free electron, and U is the attractive potential. The presence of the Fermi surface is manifested in the condition that the pairing wave function $\varphi_{\boldsymbol{k}} = 0$ at $k < k_F$. In addition, we assume that $U_{\boldsymbol{k},\boldsymbol{k}'} = U_0 = $ constant in the small region and $\varepsilon_F < \varepsilon_k < \varepsilon_F + \tilde{\Omega}$, where $\tilde{\Omega}$ is the characteristic phonon energy, $\tilde{\Omega} \sim \Omega_D$, Ω_D is the Debye frequency, and $V = 0$ outside of this region. One can also consider $\varphi_{\overline{K}} \simeq$ constant inside of this small region.

Then we obtain the following simple equation:

$$1 = \lambda \int_{\varepsilon_F}^{\varepsilon_F + \Omega} d\varepsilon (2\varepsilon - E)^{-1} \tag{2.6}$$

We also assume that the electronic density of states $\nu(\varepsilon) \simeq \nu(\varepsilon_F)$, so that $\lambda = V\nu_F/2$. After simple calculation, we obtain

$$E = 2\varepsilon_F - \delta \tag{2.7}$$

where

$$\delta = 2\tilde{\Omega} e^{-\frac{1}{\lambda}} \tag{2.7$'$}$$

Therefore, two electrons form the bound state with an exponentially small energy δ. Because of the presence of the Fermi surface, the problem becomes two dimensional (in the momentum space). As a result, we are dealing with the formation of the bound state with an exponential dependence of its energy on the strength of the interaction (cf. eqns (2.4) and (2.7′)).

It is important to stress that the analysis described above by no means represents a rigorous treatment of the superconducting state. Superconductivity is a collective phenomenon, and such a treatment should be based on the many-body theory. At the same time, the theorem proven by Cooper is very important, since it demonstrates the basic physics underlying the phenomenon. The rigorous analysis (see the next section) provides the validity of the fundamental concept of Cooper pairing and contains the analysis of many aspects of the superconducting state.

2.1.4 The BCS Model

The BCS theory (Bardeen *et al.*, 1957) was developed within the so-called weak coupling approximation. Indeed, it is based on the Hamiltonian

$$H_{int} = \lambda \sum_{\boldsymbol{k},\boldsymbol{k}'} a^{+}_{\boldsymbol{k}',1/2} a^{+}_{-\boldsymbol{k}',-1/2} a_{\boldsymbol{k},-1/2} a_{-\boldsymbol{k},1/2} \tag{2.8}$$

which describes the effective electron–electron attraction, where λ is the coupling constant. The Hamiltonian (2.8) does not contain the field of bosons (phonons form such a field for conventional superconductors), which provide the attraction. Such an exclusion of the field can only be justified for a weak interaction ($\lambda \ll 1$). This is analogous to exclusion of the electromagnetic field (photons) from the Hamiltonian function for the system of charges in classical electrodynamics (see Landau and Lifshitz, 1975).

Note also that the BCS theory contains many universal relations, for example

$$2\varepsilon(0) = 3.52T_c \tag{2.9}$$

$$\frac{\Delta C}{C_{n_{IT_c}}} = 1.43 \tag{2.9'}$$

Here, $\varepsilon(0)$ is the energy gap at $T = 0$ K and $\Delta C = C_s - C_n$ is a jump in heat capacity at T_c $(C_n = \gamma T_c)$. This universality is not surprising because the starting Hamiltonian (2.8) contains only a single parameter, the coupling constant λ. Instead of λ, one can select another parameter that can be measured experimentally, namely, the value of the critical temperature, T_c. As a result, all quantities can be expressed in terms of T_c, and this leads to the universality. We will discuss these relations in section 2.2.3. Note also that one can derive the expression for the critical temperature, which for the BCS model has the following form:

$$T_c = a\tilde{\Omega}e^{-\frac{1}{\lambda}}; a_{BCS} = 1.14 \tag{2.10}$$

where $\tilde{\Omega}$ is the characteristic phonon frequency ($\tilde{\Omega} \sim \Omega_D, \Omega_D$ is the Debay frequency). One should also include the term describing the Coulomb repulsion: $\lambda \to \lambda - \mu^*$, where $\mu^* \simeq 0.1$–0.15 (see below and section 2.2.2). We will discuss expression (2.10) in more detail below and also in section 2.2. Note that the dependence T_c upon the coupling constant is rather peculiar ($\propto e^{-\frac{1}{\lambda}}$). Such a dependence is a fundamental feature of the theory of superconductivity for a weak coupling case ($\lambda \ll 1$). Indeed, one might think that, for a weak electron–phonon interaction, one can use the quantum mechanical perturbation theory. However, one can see directly from eqn (2.10) that this is not the case. The perturbation theory is based on an expansion in a series of small parameters (in this case, the value of the coupling constant is such a parameter). But the function $e^{-\frac{1}{x}}$ cannot be expanded in a series of small x; the reader can try and see it directly. It is surprising that the solution of the problem of superconductivity required the creation of a new method that is different from the perturbative approach. This is true even in a weak coupling case because, in reality, the phenomenon of superconductivity is non-adiabatic by its nature (see section 1.2) and the interaction inside the non-adiabatic shell ($\sim \tilde{\Omega}$) leads to a large effect on the scale of the electronic energy.

In addition to the exponential term, please note that the expression (2.10) contains a pre-exponential factor, which is proportional to the characteristic phonon energy, and also the numerical factor $a_{\text{BCS}} = 1.14$. It turns out that this factor is incorrect. Therefore, calculations based on eqn (2.10) lead to incorrect results. For example, consider the superconducting *Al*. Its main parameters are $\tilde{\Omega} \simeq 300$ K and $\lambda \simeq 0.38$ (see Wolf, 2012). With the use of eqn (2.10), we obtain $T_c \simeq 4.5$ K, which totally contradicts the experimental value $T_c \simeq 1.2$ K.

As noted above, this incorrect result is caused by the incorrect numerical factor a. Because of this, it is known that the BCS expression (eqn (2.10)) has a pre-exponential accuracy. In order to obtain the correct value (see eqns (2.21) and (2.23)), it is necessary to go beyond the BCS model and calculate T_c based on a more general approach, which is valid for any strength of the electron–phonon interaction, including the case of weak coupling. We will describe this approach in section 2.2.2.

2.1.5 Microscopic Theories and the Gor'kov Method

The BCS model is based on the Hamiltonian (2.8) and the Cooper theorem, that is, this model describes the scenario when the many-electron system with weak attractions is unstable with respect to the formation of the correlated bound state with the main channel of the Cooper pairing; the pair is formed by electrons with opposite momenta and spins. The article by Bardeen *et al.* (1957) is rather long and contains cumbersome calculations with the introduction of a special trail function, Hartree–Fock approximations, and so on. It contains a rigorous evaluation of the energy spectrum, critical temperature, electromagnetic and transport properties, and so on. However, other theoretical methods were developed later. One of them is the Bogoluybov approach (1958; see also Bogoluybov *et al.*, 1959 and reviews by Abrikosov and Khalatnikov, 1959 and Lifshitz and Pitaevskii, 2002). The most popular method, which mainly has been used during the past 60 years, was developed by Gor'kov (1958) (see also Abrikosov *et al.*, 1975, Lifshitz and Pitaevskii, 2002, and Mahan, 1993).

The starting point for both the Bogoluybov and Gor'kov methods is the reduced BCS Hamiltonian (2.8). Such a choice is justified because the main attraction occurs between electrons with opposite momenta and spins (see e.g. Lifshitz and Pitaevskii, 2002).

The Bogoluybov method is based on the canonical transformation

$$\alpha_{\mathbf{k}_0} = u_{\mathbf{k}} a_{\mathbf{k},-1/2} - v_{\mathbf{k}} a^{+}_{-\mathbf{k},-1/2}$$
$$\alpha_{\mathbf{k}_1} = u_{\mathbf{k}} a_{-\mathbf{k},-1/2} + v_{\mathbf{k}} a^{+}_{\mathbf{k},1/2} \tag{2.11}$$

where the new quasiparticles $\alpha_{\mathbf{k}_0}, \alpha_{\mathbf{k}_1}$ are superpositions of electrons and holes.

The requirement for the energy to be at a minimum allows one to evaluate the amplitudes u_k, v_k and the energy spectrum, described by the well-known expression

$$\varepsilon_{\mathbf{k}} = \sqrt{\xi^2_{\mathbf{k}} + \Delta^2} \tag{2.11$'$}$$

where $\xi_{\mathbf{k}}$ is the energy of an electron in a normal metal, referred to as the chemical potential, and Δ is the energy gap. The Bogoluybov method is very convenient for studying the thermodynamic and transport properties, especially thermal conductivity (Geilikman, 1958).

As mentioned above, the method developed by Gor'kov (1958) has been used in the great majority of papers concerned with various aspects of the theory of superconductivity. It is based on the Green's function methods, which play a key role in modern many-body physics. This book focuses on the mechanisms of superconductivity and the properties of various materials. The description of the methods of many-body physics and, correspondingly, Gor'kov's approach, is outside of the scope of this work (see appendix E, where the major concepts are formulated along with some references).

As also mentioned above, the main quantity used by Gor'kov is the Green's function $G(x,x') = <T\psi_\alpha(x)\psi_\beta^+(x')>$, where $x = \{\vec{r},t\}$; $\psi_\alpha(x), \psi_\beta^+(x')$ are the operators in the Heisenberg representation, and α, β are the spin indexes (see appendix E). The Green's function provides uniquely rich information about many-body systems. It is directly related to the density matrix and, therefore, to the statistical description; moreover, it allows us to evaluate the energy spectrum and also to study various properties (e.g. electrodynamics) of the system of interest. The related method of the thermodynamic Green's function provides the description of finite temperatures, including evaluation of the critical temperature.

For Gor'kov's method, one should stress the following points: first, the superconducting state is characterised by the presence of the condensate of the Cooper pairs. The electronic transition into the condensate is described by the so-called pairing Green's function, introduced by Gor'kov:

$$F(x,x') =< N|T\psi_\alpha(x)\psi_\beta(x')|N+2>$$

and, correspondingly,

$$F^+(x,x') =< N+2|T\psi_\alpha^+(x)\psi_\beta^+(x')|N>$$

Note that the quantity $F(x,x)$ can be understood as the wave function of two electrons forming the Cooper pair.

Second, Gor'kov's method allows the properties of superconductors to be studied in real space. This feature is key if we are interested in the behaviour of superconductors in an external field, in the impact of impurities, and so on.

The equations for the usual and the pairing Green's functions can be derived from the equations for the operators $\psi_\alpha^+(x)$, $\psi_\beta(x)$ and the BCS Hamiltonian, which can be written in the following form:

$$\widehat{H} = \int \left[-\left(\psi^+ \frac{\nabla^2}{2m} \psi \right) + \frac{\lambda}{2} \left(\psi^+ \left(\psi^+ \psi \right) \psi \right] d\vec{r}$$

Consider, as an example, the superconducting system, which is homogeneous; then $G(x,x') \equiv G(x-x')$. Let us also ignore the dependence on the spin variables. Then Gor'kov's equations have the following form:

$$\begin{aligned} &\left(i\frac{\partial}{\partial t} - \frac{\nabla^2}{2m} \right) G(x-x') - i\lambda F(O+) F(x-x') = \delta(x-x') \\ &\left(i\frac{\partial}{\partial t} - \frac{\nabla^2}{2m} - 2\mu \right) F^+(x-x') + i\lambda F(O+) G(x-x') = 0 \end{aligned} \tag{2.12}$$

One can transform the momentum representation so that the following can be obtained:

$$G(p) = \frac{u_p^2}{\omega - E(\boldsymbol{p}) + i\delta} + \frac{v_p^2}{\omega + E(\boldsymbol{p}) - i\delta} \tag{2.13'}$$

$$F^{+}(p) = -i\lambda \frac{F^{+}(O+)}{[\omega - E(\boldsymbol{p}) + i\delta]\,[\omega + E(\boldsymbol{p}) - i\delta]}$$

$$u_p^2 = 0.5\left[1 + \frac{\xi_{\boldsymbol{p}}}{E(\boldsymbol{p})}\right];\ v_p^2 = 0.5\left[1 - \frac{\xi_{\boldsymbol{p}}}{E(\boldsymbol{p})}\right]$$

$$E(\boldsymbol{p}) = \left(\xi_{\boldsymbol{p}}^2 + \Delta^2\right)^{1/2}$$

(see eqn (2.11)). The poles of the Green's function determine the energy spectrum $E(\boldsymbol{k})$, which, as noted above, contains an energy gap.

Everything written in this chapter is related to the BCS model, which is valid for a weak coupling approximation. We are not focusing here on derivations of various properties of superconductors. The BCS model was generalised for a more general treatment and strong coupling theory. This theory will be described in the following section (section 2.2) and valid expressions for materials with weak coupling ($\lambda \ll 1$) will follow from the more general results.

As mentioned above, also note that Gor'kov's method allows the superconducting phenomenon to be studied in real space. As a result, one can study the impact of impurities, the behaviour of superconductors in an arbitrary strong magnetic field, the appearance of the vortex structure, and so on. The corresponding phenomena will be discussed later on, in section 3.3.

2.1.6 The Energy Gap and the Coherence Length

Based on the Hamiltonian (2.8), one can evaluate the energy spectrum in the superconducting state. We will discuss the derivation for the general case (without the restriction $\lambda \ll 1$) in section 2.2.1. Here, we write out the result as $E = \left(\xi^2 + \Delta^2\right)^{\frac{1}{2}}$ (see eqn (2.11′)), where $\xi = \varepsilon(\boldsymbol{k}) - \varepsilon_F$ is the electronic energy referred to as the Fermi level and $\boldsymbol{k}$ is the electronic momentum; in the simplest case of the isotropic system and effective mass approximation, $\varepsilon(\boldsymbol{k}) = \frac{k^2}{2m^*}$. The parameter Δ is the so-called energy gap, which is the minimum value of the electronic energy. It can be visualised as the energy required to break up the Cooper pair. The value of the energy gap depends on temperature; therefore, its maximum value corresponds to $T = 0\ K$. In addition, it also depends on the direction of the momentum **p**. In other words, the energy gap is anisotropic, although this anisotropy is rather small for conventional materials (we will discuss the reason for weak anisotropy in section 3.2).

Another important parameter is the coherence length, ξ_0. Qualitatively speaking, the coherence length can be visualised as the size of the Cooper pair. Rigorously speaking, one should remember that the pairing occurs not in real space but in momentum space. As for the coherence length, it can be understood as the scale of the pairing wave function. More specifically, the pairing is described in many-body theory by the pairing Green's function $F(\boldsymbol{r},\boldsymbol{r}')$ (Gor'kov, 1958; see appendix E), and the coherence length describes the spatial extent of this function.

One can estimate the value of the coherence length in the following way. The transition into the superconducting state corresponds to the instability of the usual Fermi distribution towards the formation of the Cooper pairs and the appearance of the energy gap Δ near the Fermi surface (see eqn (2.11′)). As a result, the value of the Fermi energy is smeared near E_F, and the value of Δ is the scale of this smearing. Since $E_F = \frac{{p_F}^2}{2m^*}$, this smearing corresponds to the uncertainty of the Fermi momentum so that $p_F\delta\frac{p_F}{2m^*} \sim \Delta$. Therefore, $\delta p_F \sim \Delta/v_F$. Using the uncertainty relation $\delta r \delta p \simeq \hbar$, we obtain $\xi_0 \simeq \delta r \simeq \hbar/\delta k_F \simeq \hbar v_F/\Delta$. Thus,

$$\xi_0 = \frac{\hbar v_F}{\Delta} \tag{2.13}$$

The value of the coherence length for conventional superconductors is rather large. Indeed, $\Delta \sim T_c$ (more exactly, $\Delta = 1.76\ T_c$; see eqn (2.9)). Using the values $T_c \sim 5$ K and $V_F \sim 1.5 \times 10^8$ sm/s, we obtain $\xi_0 \simeq 10^{-4}$ cm. The lattice period is $d \sim 10^{-8}$ cm. Therefore, the coherence length is on the scale of $\sim 10^4$ lattice periods. Qualitatively speaking, this means that we are dealing with the correlated motion of two electrons separated by a large distance. There are many electrons between these two correlated electrons. Qualitatively, one can imagine the following picture: the electrons in the superconductors have come together for a gala ball and are performing a dance. The distance between the partners is rather large, so there are many dancing 'electrons' between them; however, their movement is correlated. In addition, the 'dancers' are constantly changing their partners.

The novel high-T_c superconductors, unlike conventional ones, are characterised by a short coherence length. This is due to the high values of T_c and the small values of v_F. For example, for high-T_c cuprates, $\xi_0 \simeq$ 15–20 Å.

The large scale of the coherence length relative to the average inter-electron distance reflects the fact that we are dealing with the coherent state of a whole electronic system; thus, the energy gap is the special feature of all systems. This is the minimum value of energy required for the transition of a whole electronic system into the excited state.

In usual normal metals, electron scattering, for example electron–ion collisions, leads to the transition of the energy of the moving electrons into the thermal energy of the chaotic motion. However, if the sample is in the superconducting state, then the presence of an energy gap means that such a transition is forbidden until the value of the current is below some threshold. Therefore, below some value, we are dealing with the lossless superconducting current.

There is a key difference between the energy gaps, that is, the spectral features, for the superconducting state and for the band structure. If we are dealing with a filled valence band and an empty conducting band (insulator), then the weak electric field will not create an electric current. For the superconducting state, the picture is different. The energy gap appears at the Fermi surface and separates the pairing electron condensate from the excited states. In other words, the presence of a gap for an insulator reflects its zero conductance, whereas, for the superconducting state, it corresponds to the opposite case, namely, to its infinite conductance.

The stationary current in a normal metal is provided by an external electric field; its presence allows electrons to overcome the resistance caused by various collisions, and the charge transfer (current) is accompanied by the appearance of thermal motion (losses). As for the superconducting state, the electric field is absent for the stationary case, and the dissipationless current is related to the magnetic field, which does not perform any work. As a result, the electrodynamic properties are described by different equations.

2.1.7 Pairing and Orbital Momenta

The general equation for the gap parameter (see Fig. 2.2) has the following form:

$$\Delta(\boldsymbol{k}) = \pi T \sum_{\omega_n} \int d\boldsymbol{k}' \Gamma(\boldsymbol{k}, \boldsymbol{k}) \frac{\Delta(\boldsymbol{k}')}{\omega_n^2 + \xi^2(\boldsymbol{k}') + \Delta^2(\boldsymbol{k}')} \tag{2.14}$$

Here $\Delta(\boldsymbol{k})$ is the energy gap, $\xi(\boldsymbol{k}') = \varepsilon(\boldsymbol{k}') - \varepsilon_F$ is the electron energy referred to the Fermi level, and Γ is the so-called total vertex describing the pairing interaction. For the weak coupling case, Γ contains the corresponding matrix element. In the general case of strong electron–phonon interaction, Γ also contains the so-called phonon propagator (see section 2.1). We assume, in accordance with the Cooper theorem, that the pairing occurs on the Fermi surface. It allows us to put $|\boldsymbol{k}| = |\boldsymbol{k}'| = k_F$. Then the effective pairing interaction depends on the angle θ between $\boldsymbol{k}$ and $\boldsymbol{k}'$. The quantity $\Gamma(\boldsymbol{k}, \boldsymbol{k}')_{k_F}$ can be expanded in the series of the Legendre polynomials:

$$\Gamma = \sum_{l=0}^{\infty} (2l+1) \Gamma_l P_l(\cos\theta) \tag{2.14'}$$

Correspondingly, the gap parameter $\Delta(\boldsymbol{k})$ is also angle dependent. Note that we are assuming that the system is isotropic; then, in accordance with the general principles of quantum mechanics, one classifies the states with orbital momenta l. A positive value of some amplitude Γ_l, which corresponds to attraction between electrons, leads to the formation of Cooper pairs. Indeed, according to the Cooper theorem, any small attraction leads to the instability of the normal state and to the formation of electronic pairs.

For usual conventional superconductors, the pairing is dominated by the component with $l = 0$ (*s*-wave pairing; the classification is similar to that for the atomic orbitals). Then the energy gap is almost isotropic, so that the state is gapped in all directions. Note that the usual superconductors are characterised by singlet pairing, that is, the paired electrons have opposite spins. This is consistent with the symmetry requirements.

Indeed, the total wave function describing the pair of electrons should be antisymmetric, and the '*s*' state ($l = 0$) can be combined with the spin wave function with opposite spins.

The next singlet state can be combined with $l = 2$ (*d*-wave pairing). For the important case of layered superconductors, and the presence of the layered structure is an important feature of many novel superconductors (see chapters 5 and 6), the structure of the gap parameters is rather peculiar. We are dealing with alternating signs and, as a result, the order parameters have nodes in specific directions. This leads to the different temperature dependences of some parameters (e.g. the penetration depth). The penetration depth is described by the exponential dependence as $T \to 0$ for the *s*-wave scenario, whereas this dependence is linear for *d*-wave pairing. Note that the *d*-wave picture could be energetically favourable if the coherence length is small. Indeed, a large coherence length (this is the typical case for conventional superconductors) makes it possible to overcome the Coulomb repulsion, which is screened at the distances of the order of the lattice constant. As for the novel high-T_c superconductors, they are characterised by a short coherence length comparable to a lattice period (see e.g. section 5.3). Then the *d*-wave symmetry helps to overcome the Coulomb repulsion, since the negative total effective interaction (if the Coulomb repulsion prevails) can be compensated by the negative sign of the gap parameter (in some direction); this is favourable for the pairing picture.

The case of triplet pairing, when the electrons forming the pair have spins aligned in the same direction, is also realistic but, in this case, the orbital momentum should be odd, for example $l = 1$ (*p*-wave pairing). More complicated triplet pairing with $l = 3$ (*f*-wave pairing) is also discussed, for example in the heavy fermion system (see chapter 5 and section 6.4).

It is important to stress that the *s*, *d*, *p*, and *f* states introduced above describe the superconducting order parameter in momentum space, not real space.

2.2 The Phonon Mechanism

2.2.1 Main Equations

As was mentioned above, the BCS model describes the Fermi gas with weak attraction. The attraction is provided by exchange by bosons between the electrons (e.g. by phonons). Strictly speaking, the nature of this attraction is not essential in the BCS model and enters the theory only as the corresponding energy scale; it determines the cut-off. Nevertheless, specific mechanisms of superconductivity are defined by the type of the bosons. This and following sections (sections 2.2–2.7) are devoted to description of various mechanisms.

This section is concerned with the phonon mechanism. The purpose of the analysis is twofold. Firstly, it will be demonstrated that, for weak electron–phonon interaction, we obtain the same result as in the BCS theory. Secondly, and this is very important, the analysis will allow us to go beyond the weak coupling approximation. Indeed, there are many superconducting materials where the electron–phonon interaction responsible for pairing is not weak, and their properties should be described properly.

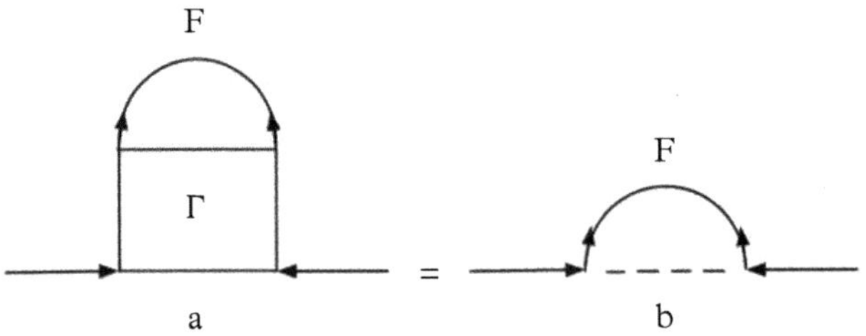

Figure 2.2 *(a) The pairing self-energy part. (b) The phonon mechanism* $\Gamma \equiv D$.

As was noted above, the BCS theory contains many universal relations, like eqns (2.9) and (2.9′). All properties could be described by a single parameter, namely, by the value of the critical temperature, T_c. For the strong coupling case, the situation is different. The universality is lost, because, in addition to depending on T_c, the properties depend also on the phonon spectrum and, more specifically (see below), on the value of the characteristic phonon frequency.

As a first step, let us formulate the main equations describing the phonon mechanism. We employ the thermodynamic Green's function formalism (see the appendix and also e.g. Abrikosov *et al.*, 1975 and Lifshitz and Pitaevskii, 2002), because our special focus will be on the evaluation of the critical temperature. The pairing is described by the diagrammatic equation for the so-called pairing self-energy part (Fig. 2.2).

F is the pairing Green's function introduced by Gor'kov (1958), and Γ is the total vertex describing the pairing interaction. For the phonon mechanism $\Gamma \cong D$, D is the phonon propagator; it has the form

$$D(\Omega, \omega_n - \omega_{n'}) = \frac{\Omega^2}{\Omega^2 + (\omega_n - \omega_{n'})^2} \tag{2.15}$$

Ω is the phonon frequency, $\omega_n = (2n+1)\pi T$.

One should also consider the effect described above(section 1.2.1). An electron moving through the crystal lattice polarises the latter. Of course, this polarisation also acts back on the electron and affects its motion. In quantum language, polarisation can be described as the emission and subsequent absorption of virtual phonons. The self-energy part $\sum_1$, depicted in the diagram in Fig. 1.3 describes this process.

The Green's function G and the pairing function F are given by the diagrammatic equations (Fig. 2.3).

The pairing is described by the following analytical equation (Eliashberg, 1960, 1961):

$$\Delta(\omega_n, T) = T \sum_{\omega_{n'}} \int d\mathbf{k}' \left|\tilde{\zeta}\right|^2 D(\omega_n - \omega_{n'}; \Omega) \frac{\Delta(\omega_n, T)}{\omega_{n'}^2 + \xi_{\mathbf{k}'}^2 + \Delta^2(\omega_n, T)} \tag{2.16}$$

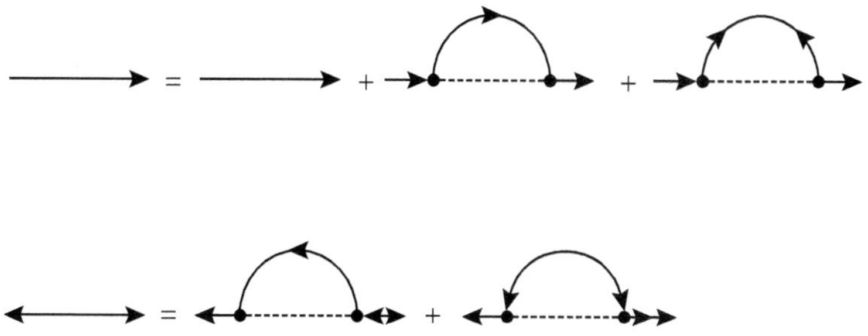

Figure 2.3 *Diagrammatic equations for (upper panel) the Green's function G and (lower panel) the pairing function F^+.*

Here the D-function is defined by eqn (2.15), $\tilde{\zeta} = \tilde{\zeta}_{\boldsymbol{k},\boldsymbol{k}'}$ is the matrix element (cf. eqn (1.24)), $\xi^2_{\boldsymbol{k}'}$ is the electron energy referred to the Fermi level, and $\Delta(\omega_n, T)$ is the order parameter. We can make a transformation to the integration over phonon momentum $\mathbf{q} = \mathbf{k}' - \mathbf{k}$ and then to phonon frequency to obtain

$$\Delta(\omega_n) Z = \pi T \sum_{n'} \int d\Omega \frac{g(\Omega)}{\Omega} D(\omega_n - \omega_{n'}; \Omega) \frac{\Delta(\omega_{n'})}{\left[\omega^2_{n'} + \Delta^2(\omega_{n'})\right]^{\frac{1}{2}}} \tag{2.17}$$

$$Z = \pi T \sum \int d\Omega \frac{g(\Omega)}{\Omega} D\left(\omega_n - \omega_{n'}; \tilde{\Omega}\right) \frac{\omega_{n'}}{\left[\omega^2_{n'} + \Delta^2(\omega_{n'})\right]^{\frac{1}{2}}} \tag{2.17$'$}$$

Z is the so-called renormalisation function describing the electron–phonon scattering. Here $\Delta(\omega_n)$ is the so-called thermodynamic pairing order parameter, and the function $g(\Omega)$ is defined by eqn (1.25). At $T = T_c$, we can put $\Delta(\omega_n) = 0$ in the denominator of the integrand to obtain

$$\Delta(\omega_n) Z = \pi T \sum_{\omega_{n'}} \int d\Omega \frac{g(\Omega)}{\Omega} D(\omega_n - \omega_{n'}, \Omega) \frac{\Delta(\omega_{n'})}{|\omega_{n'}|} \tag{2.18}$$

The expression for Z at $T = T_c$ has the form

$$Z = 1 + \pi T \sum_{\omega_{n'}} \int d\Omega \frac{g(\Omega)}{\Omega} D(\omega_n - \omega_{n'}, \Omega) \frac{\omega_{n'}}{|\omega_{n'}|} \tag{2.18$'$}$$

Equations (2.18) and (2.18′) contain the very important function $g(\Omega) = \alpha^2(\Omega) F(\Omega)$; see eqn (1.25). One can also introduce the coupling constant λ (see eqn 1.27). Note that the coupling constant does not directly appear in eqns (2.18) and (2.18′). It arises explicitly, for example, if we consider a model with a single phonon frequency, for example a delta-function-like peak with $g(\Omega) = \lambda \Omega \delta\left(\Omega - \tilde{\Omega}\right)$.

In real life, one deals with complicated phonon spectra, and the superconducting parameters correlate directly with the function $\alpha^2(\Omega)F(\Omega)$. Note that eqns (2.18) and (2.18′) can be solved numerically without involving the coupling constant concept. Such calculations have been performed, for example, by Errea *et al.* (2015) and Flores-Livas *et al.* (2016). Strictly speaking, it is not possible to rewrite eqns (2.18) and (2.18′) in a form which would explicitly contain the coupling constant. Nevertheless, the coupling constant concept remains very useful, since it makes it possible to obtain an analytical expression for T_c and study various dependences. The possibility of introducing the coupling constant is due to the differences in the behaviour of the phonon Green's function (2.15) and of the function $g(\Omega) = \alpha^2(\Omega)F(\Omega)$; they enter as the factor into the integrand in eqns (2.18) and (2.18′). The Green's function D is a relatively smooth function of frequency. The behaviour is usually quite different for the function $g(\Omega)$ and especially for the phonon density of states factor $F(\Omega)$. The latter typically contains a number of rather sharp peaks whose origin is as follows.

The phonon spectrum of a solid consists of several branches corresponding to longitudinal and transverse acoustic and optical phonons. At small wave vectors q, the acoustic branches have simple linear dispersion laws, for example $\Omega_{tr} = u_{tr}q$. For large q, on the other hand, there are deviations from the linear behaviour that give rise to regions with a high density of states ($q^2 dq/d\Omega$ is large). In these regions, the frequency varies only weakly with the momentum, with the result that the density of states becomes very high, as we have stated. These regions, where the function $g(\Omega) = \alpha^2(\Omega)F(\Omega)$ contains peaks, make the strongest contribution to electron pairing.

As for the optical phonons, their dispersion is quite flat, so that the frequency does not vanish in the long-wavelength limit. As a result, their presence leads to the appearance of strong peaks in the phonon density of states in the broad range of momenta, so that their contribution to the pairing is very substantial. As we know, the existence of optical modes is caused by the presence of two or more ions in the unit cell (see e.g. Aschroft and Mermin, 1976). Especially, their impact on the superconducting state is strong for hydrides (see section 7.2), the presence of the light hydrogen ions leads to an appearance of high-frequency phonon modes, and this is beneficial for the pairing.

Let us now consider the factor $g(\Omega)D(\Omega)/\Omega$. Breaking up the integration over frequencies into separate intervals containing different peaks, and making use of the smoothness of the phonon Green's function, we can write the right-hand side of eqn (2.18) as a sum of terms of the form $\lambda_i\, D(\Omega_i)$, where the coupling constants are defined by expressions of the type (1.27), each referring to a different peak. Note that the presence of any additional phonon mode, even that with low frequency, increases the value of T_c (Bergmann and Rainer, 1973).

For many materials, one can introduce an effective average phonon frequency, thereby casting eqn (2.18) into a form corresponding to a single characteristic frequency $\tilde{\Omega}$ and a single coupling constant. The value $\tilde{\Omega}$ can be taken as $\tilde{\Omega} = <\Omega^2>^{1/2}$; see eqn (1.26). The average $<f>$ is defined by

$$<f> = \frac{2}{\lambda} \int d\Omega\, f(\Omega) \frac{\alpha^2(\Omega)\mathrm{F}(\Omega)}{\Omega}$$

Sometimes the close value $\tilde{\Omega} = <\Omega_{log}>$ is selected, so that $f = \log\Omega$; however, we think that such a choice is not satisfactory, since under the 'log' sign there should be always the dimensionless quantity. Below, we use the value $\tilde{\Omega} = <\Omega^2>^{1/2}$, so that $f(\Omega) = \Omega^2$.

As was shown by Louie and Cohen (1977), the function

$$K_{n-n'} = 2\int d\Omega\, g(\Omega)\,\Omega \times \left[\Omega^2 + (n-n')^2(2\pi T)^2\right]^{-1}$$

can be replaced by $K^0_{n-n'} \approx \lambda \mathrm{D}\left(\tilde{\Omega}, \omega_n - \omega_{n'}\right)$, where the D-function and $\tilde{\Omega}$ are defined by eqns (2.15) and (1.26), respectively. With high accuracy, $K_n = K^0_n(1-r)$, with $r \ll 1$ ($r \propto \delta = \frac{\tilde{\Omega}}{<\Omega>} - 1$). As a result, eqn (2.18) can be written in the form

$$\Delta(\omega_n) Z = 1 + \pi T\lambda \sum_{\omega_{n'} \geq 0} D\left(\omega_n - \omega_n; \tilde{\Omega}\right) \frac{\Delta|\omega_{n'}|}{|\omega_{n'}|} \tag{2.19}$$

The phonon propagator is defined by eqn (2.15). Equation (2.18′) for the renormalisation function Z can be written in a similar way

$$Z = 1 + \pi T\lambda \sum_{\omega_{n'} \geq 0} D\left(\omega_n - \omega_{n'}, \tilde{\Omega}\right) \frac{\omega_{n'}}{|\omega_{n'}|} \tag{2.19′}$$

Equations (2.17) and (2.18), and, correspondingly, eqns (2.19) and (2.19′), are the main equations of the strong coupling theory. They are valid for any value of the coupling constant and, therefore, for any relation between T_c and $\tilde{\Omega}$. Note, however, that, in agreement with the 'Migdal' theorem (section 1.2.1), we neglected the higher-order corrections of the vortex; such approximation is justified if $\tilde{\Omega} \ll E_F$ (or $\lambda \ll 1$). This is the only condition of the applicability of the theory.

The function D guarantees the convergence of the summation at the upper limit by effectively cutting off this summation at $\tilde{\Omega}$.

2.2.2 Critical Temperature

Equations (2.19) and (2.19′) are valid for any strength of the electron–phonon interaction. Based on these equations, let us evaluate the main parameter of the theory, the critical temperature.

The value of T_c. is determined by several quantities: $T_c = T_c(\lambda, \mu^*, \tilde{\Omega})$, where λ describes the strength of the electron–phonon coupling, $\tilde{\Omega}$ stands for the characteristic phonon frequency and sets the energy scale (in a rough approximation, $\tilde{\Omega} \sim \Omega_D$, where Ω_D is the Debye temperature), and μ^* describes the Coulomb repulsion (see below). Curiously, it will turn out that the specific forms of the analytic expressions for T_c are different for different coupling strengths.

Weak coupling. The limit of weak coupling deserves special attention, both for its own sake and in order to establish contact with the usual formalism of the BCS theory. This limit corresponds to $\lambda \ll 1$. As can be shown self-consistently from the results of weak coupling theory, this means that $\pi T_c. \ll \tilde{\Omega}$, where $\tilde{\Omega}$ corresponds to the important short-wavelength part of the phonon spectrum. In the weak coupling approximation ($\lambda \ll 1$), we can neglect the dependence of the phonon Green's function on ω_n. Then $\Delta(\omega_n)$ also becomes independent of ω_n, and we arrive at the equation

$$1 = \left(\frac{\lambda}{Z}\right) \pi T_c \sum_{\omega_n} \frac{\tilde{\Omega}^2}{\tilde{\Omega}^2 + \omega_n^2} \cdot \frac{1}{|\omega_n|}_{1T=T_c} \tag{2.20}$$

As for the renormalisation function Z, one might assume by looking at eqn (1.19′) that, for the weak coupling case ($\lambda \ll 1$), one can put $Z \simeq 1$. In addition, one can approximate $D \simeq 1$ and terminate the summation at $\tilde{\Omega}$. Then, after straightforward calculation of the sum in eqn (2.20), one obtains $1 = \ln\left[\left(\frac{\pi}{\gamma}\right)\left(\frac{T_c}{\tilde{\Omega}}\right)\right]$, where $\gamma \simeq 2.75$ is the Euler constant), and we arrive at eqn (2.11) with $a \simeq 1.14$. However, this approximation appears to be incorrect if we are concerned not only with the exponential dependence but also with the pre-exponential factor. Qualitatively, this can be seen from eqn (2.20), which has the form at the usual BCS equation for T_c with the replacement $\lambda \to \lambda^* = \lambda(1+\lambda)^{-1}$ (we used the expression $Z \simeq 1 + \lambda$, which can serve as an approximation, following from eqn (1.19′) for the case $\pi T_c \ll \tilde{\Omega}$). If we use the BCS expression, eqn (2.11), and replace $\lambda \to \lambda^* = \lambda(1+\lambda)^{-1}$ in the exponential form, we can see directly that the replacement leads to large decrease in the pre-exponential factor ($a_{BCS} = 1.14 \to a_{BCS}/e$). Therefore, the electron–phonon scattering term (renormalisation factor)Z plays an important role even in a weak coupling case.

To obtain a more accurate expression for T_c in the weak coupling case, one should perform more exact calculation using the general expression for Z, eqn (2.19′). Such calculation, performed analytically for the more general case $\lambda \tilde{<} 1$ (see the next section) leads to the following expression for T_c for $\lambda \ll 1$ (weak coupling):

$$T_c = a\tilde{\Omega} e^{-\frac{1}{\lambda - \mu^*}}; \quad a \simeq 0.25 \tag{2.21}$$

$\mu^* \simeq 0.1 - 0.15$ is the Coulomb pseudopotential (see section 2.2.2).The expression (2.21) provides good description of experimental data.

Note that the problem of pre-exponential accuracy did arise shortly after the creation of the BCS theory. It was studied in a number of papers (Karakosov *et al.*, 1976). The explicit calculation discussed by Wang and Chiubukov (2013) leads also to numerical factor 0.25 (see eqn (2.21)).

Intermediate coupling ($\lambda \cong 1.5$). In the preceding section, we discussed the case of weak coupling ($\lambda \ll 1$). If the superconductor is characterised by a larger coupling constant ($\lambda \gtrsim 1$), the approximations employed there are not valid. In particular, we can no longer neglect the dependence of Δ on ω_n. In addition, as was noted above (section 2.1.1), it is necessary to take into account the presence of the renormalisation factor Z. As a first step, one can just replace $\lambda \to \lambda^* = \lambda(1+\lambda)^{-1}$, since $Z \simeq 1+\lambda$ (Nakajima and Watabe, 1963). Then one can obtain the expression for T_c:

$$T_c = 1.14\tilde{\Omega} e^{-\frac{(1+\lambda)}{\lambda}}$$

which can be used for semi-quantitative analysis. A more detailed study was carried out by McMillan (1968a). McMillan made use of eqn (2.19) and computed the critical temperature for a phonon spectrum of the type found in Nb. This phonon spectrum (or, more precisely, the phonon density of states $F(\Omega)$) is known from neutron spectroscopy data, and has a structure typical of many metals. This numerically calculated T_c can be analytically fitted by the following formula:

$$T_c = \frac{\Theta_D}{1.45} \exp\left(-\frac{1.04}{\rho - \frac{\mu^*}{1+\lambda} - 0.62\mu^* \rho} \right) \tag{2.22}$$

Here μ^* is the Coulomb pseudopotential (see the next section), and $\rho = \lambda(\lambda+1)^{-1}$.

Dynes (1972) modified the McMillan equation by introducing the average phonon frequency $\tilde{\Omega}$ (see eqn (1.26)). The McMillan–Dynes expression is highly cited and has the form

$$T_c = \frac{\tilde{\Omega}}{1.2} \exp\left(-\frac{1.04(1+\lambda)}{\lambda - \mu^*(1+0.62\lambda)} \right) \tag{2.23}$$

This equation is valid for $\lambda \lesssim 1.5$. One can use the expression, which is close to eqn (2.23) and also valid for an intermediate strength of the electron–phonon interaction (then $\left(\frac{\pi T_c}{\tilde{\Omega}}\right)^2 \ll 1$). This expression was obtained analytically (Geilikman *et al.*, 1975) and has the following form (if $\mu^* = 0$):

$$T_c = 1.14 e^{-\frac{1+1.5\lambda}{\lambda}} \tag{2.23$'$}$$

This expression can be written in the form (2.21), and the coefficient $a \simeq 0.25$. Note that the McMillan–Dynes equation can be approximated by expression (2.21) with the coefficient $a \simeq 0.3$, which is close to that in eqn (2.21). We have already mentioned that the McMillan equation was obtained as a fit to the numerical solution for a Nb-like phonon spectrum: since this spectrum is rather typical, this equation has been applied with great success to many superconductors (see e.g. Allen and Dynes, 1975; Wolf, 2012).

The paper by McMillan (1968) also contains a useful expression for the electron–phonon coupling constant:

$$\lambda = \frac{\eta}{\tilde{\Omega}^2} \tag{2.24}$$

where

$$\eta = \frac{<I>\nu}{M} \tag{2.24'}$$

Here $<I>$ is the average matrix element of the electron–phonon interaction, ν_F is the density of electronic states, and M is the ion mass; $\tilde{\Omega}$ is defined by expression (1.26). It is important that the dependence of T_c. on the phonon frequency is not determined solely by the pre-exponential factor: in principle, a strong dependence is also contained in the exponent itself. Note that these two dependences have opposite effects on T_c. This fact has led McMillan to propose a softening mechanism for increasing T_c. The previously accepted notion, based on the original BCS theory, was that a rise in the phonon frequency serves only to increase the pre-exponential factor and, consequently, increases T_c. The dependence (2.24) shows that a rise in the phonon frequency may, generally speaking, lead to the opposite result as well.

The softening mechanism, that is, the decrease in the effective phonon frequency, can lead to an increase in the value of $A\ell_{1-x}Si_x$ the critical temperature. This has been observed for a number of compounds. For example, the solid solution obtained by substituting Si for Al(Sluchanko *et al.*, 1995) is characterised by an increase in T_c (1.2K $\rightarrow$ 11K). Such a drastic increase is caused by phonon softening (Chevrier *et al.*, 1988).

The case when the Fermi surface has a flat region leads to an interesting manifestation of the softening mechanism. Indeed, as we know, the electron–phonon interaction leads, in the presence of such a region, to an appearance of the so-called charge density waves (CDWs). More specifically, at some specific temperature, T_{CDW}, one can observe the transition into the CDW state, which is characterised by the opening of the energy gap near the flat region of the Fermi surface. This transition is preceded by the logarithmic instability and corresponding softening of the phonon spectrum (see e.g. Afanas'ev and Kagan (1963) and the book edited by Gor'kov and Gruner (1989). Therefore, the electron–phonon interaction may lead to the superconducting transition (at T_c) as well as to the CDW transition (at T_{CDW}). The most interesting case is when $T_c > T_{CDW}$. In such a case, the softening proceeding the CDW transition leads to an increase in the electron–phonon coupling constant λ (see eqn (2.24)), and this increase leads to a noticeable increase in the value of T_c. Such a scenario plays a key role in softening the optical phonons in LaH_{10}. This compound is characterised by the low value of the optical phonon frequency and, correspondingly, by a very large electron–phonon coupling constant; this leads to a record value of T_c. We will discuss this case in section 7.2.

As was mentioned above, the McMillan equation is valid for $\lambda \lesssim 1.5$. In connection with this fact, it is interesting to note that from eqns (2.2.2) and (2.24) it appears that there exists an upper limit on the attainable value of T_c. This was pointed out in the original McMillan paper. Indeed, neglecting μ^* for simplicity and calculating $\partial T_c/\partial\tilde{\Omega}$, we find, with the use of eqns (2.23) and (2.24), the maximum value, which corresponds to $\lambda \simeq 2$.

This conclusion, however, has to be taken with a great deal of caution. The fact of the matter is that the value $\lambda = 2$ is outside the range where the McMillan equation (2.22) and, correspondingly, the McMillan–Dynes equation (2.23), are applicable; this is even more so for higher values of λ. We will come back to this point later when we discuss the expression for T_c, which holds for larger values of λ.

The Coulomb interaction. In addition to the attraction mediated by phonon exchange, we must keep in mind the presence of Coulomb repulsion. It turns out that an important aspect of this problem is the phenomenon of logarithmic weakening of the repulsion. This factor has to do with the difference in the energy scales of the attraction and repulsion effects. The attraction is important in an energy interval $\tilde{\Omega} \cong \Omega_D$. The repulsion, on the other hand, is characterised by the electronic energy scale $\sim\varepsilon_F$.

Let us look at this problem in more detail (Tyablikov and Tolmachev, 1958; Bogoluybov *et al.*, 1959; Abrikosov and Khalatnikov, 1959; Morel and Anderson, 1962; McMillan, 1968; Geilikman *et al.*, 1975). The equation for the order parameter $\Delta(\omega_n)$ in the presence of Coulomb repulsion can be written as follows (at $T = Tc$):

$$\Delta(\omega_n) = \frac{1}{1 - f^n/\omega_n}\pi T \sum_{\omega_{n'}} \left[\lambda D\left(\omega_n - \omega_{n'}, \tilde{\Omega}\right) - \frac{1}{2}V_c\ \Theta(\varepsilon_0 - |\omega_{n'}|)\right] \frac{\Delta(\omega_{n'})}{|\omega_{n'}|} \tag{2.25}$$

The second term in the square brackets describes the Coulomb force ($\varepsilon_0 \cong \varepsilon_F$). Recall that the phonon Green's function D provides a cut-off at energies $\sim\tilde{\Omega}$. We can seek the solution in the following form:

$$\Delta(\omega_n) = \Delta_0 \frac{\tilde{\Omega}^2}{\tilde{\Omega}^2 + {\omega_n}^2} + \Delta_\infty \frac{{\omega_n}^2}{\tilde{\Omega}^2 + {\omega_n}^2} \tag{2.25$'$}$$

The second term reflects the presence of the repulsion. It results in the order parameter continuing outside the region $\omega_n \simeq \tilde{\Omega}$.

The quantities Δ_0 and Δ_∞ are defined by

$$\Delta_0 = \rho\pi T \sum_{\omega_{n'}} \frac{\tilde{\Omega}^2}{\tilde{\Omega}^2 + {\omega_{n'}}^2} \frac{\Delta(\omega_{n'})}{|\omega_{n'}|} - \frac{V_c}{1+\lambda}\pi T \sum_{\omega_{n'}}^{\varepsilon_0} \frac{\Delta(\omega_{n'})}{|\omega_{n'}|};$$

$$\Delta_\infty = -V_c \pi T \sum_{\omega_{n'}} \frac{\Delta(\omega_{n'})}{|\omega_{n'}|} \tag{2.25$''$}$$

Substituting eqn (2.25′) and eliminating the constants Δ_0 and Δ_∞, we arrive at the following result for T_c:

$$T_c = 1.14\tilde{\Omega}\exp\left(-\frac{1 + 0.5\rho - 0.35\rho^2 + 0.8\rho\left[\mu^*/(1+\lambda)\right] + 0.4\mu^*\rho^2}{\rho - \left[\mu^*/(1+\lambda)\right] - 0.5\rho\mu^* - 1.5\mu^*\rho^2}\right) \tag{2.26}$$

Here

$$\mu^* = \frac{V_c}{1 + V_c ln\left(\varepsilon_0/\tilde{\Omega}\right)} \tag{2.27}$$

is the so-called Coulomb pseudopotential. It contains the large logarithmic factor $ln\left(\varepsilon_0/\tilde{\Omega}\right)$, which reduces the contribution of the Coulomb repulsion. Usually, $\mu^* \cong 0.1 - 0.15$, although its value might be different.

For the weak coupling case ($\lambda \ll 1$), we obtain an expression close to eqn (2.21). For the intermediate case, eqn (2.26) leads to an expression which is close to eqn (2.23). The presence of the large logarithmic factor and the corresponding effective weakening of the Coulomb repulsion is very beneficial for appearance of superconductivity. It reflects the difference in the energy scales for the attraction (characteristic phonon frequency) and the repulsion (electronic energy $\sim\varepsilon_F$).

Note that, if the superconductor contains, in addition to the phonon subsystem, some high-energy electronic excitations which provide an additional attraction, this leads to an effective decrease in μ^*. We will discuss this question in more detail in section 2.5.

Very strong coupling. We now consider the case of very strong coupling ($\lambda \gg 1$). We shall see that this regime corresponds to the criterion $2\pi T_C \gg \tilde{\Omega}$. This case was first analysed by Allen and Dynes (1975) by means of numerical calculations (for $\mu^* = 0$). An analytical treatment based on a direct solution of eqn (2.19) was developed by Kresin *et al.* (1984).

One can see directly from the main equation (2.19) that, for this case, the dependence T_c on the coupling constant λ drastically differs from the exponential dependence, eqn (2.23). The term $m = n$ is cancelled. Because of it and the inequality $2\pi T_C \gg \tilde{\Omega}$, one can neglect the phonon frequency $\tilde{\Omega}$ in the denominator of the phonon propagator D, eqn (2.15). Then it is clear from the simple consideration of dimensionality that $Tc \propto \lambda^{\frac{1}{2}}\tilde{\Omega}$. The calculation of the corresponding coefficient will be described below, in this section.

It is convenient to use the dimensionless form of eqns (2.19) and (2.19′). It turns out that, for $\lambda \gg 1$, the function $\Delta(n)$ decays rapidly. In this case, it is very useful to employ the matrix method (Owen and Scalapino, 1971). The equation which determines T_c can be written in the following form:

$$\phi_n = \sum_{m\geq 0} \tilde{\mathrm{K}}_{\mathrm{nm}}\phi_{\mathrm{m}} \tag{2.28}$$

$$\tilde{\mathrm{K}}_{\mathrm{nm}} = \frac{\alpha}{(2n+1)^{1/2}(2m+1)^{1/2}} F_{n,m,v} \tag{2.28′}$$

Here

$$F_{n,m,v} = \frac{1}{v^2 + (n-m)^2} + \frac{1}{v^2 + (n+m+1)^2} - \delta_{nm}\sum_{i=0}^{2n}\frac{1}{v^2 + (n-1)^2} \tag{2.28″}$$

where $\phi_n = \Delta_n/(2n+1)^{1/2}$, $\nu = \tilde{\Omega}/2\pi T_C$, $\tilde{\Omega} = <\Omega^2>^{1/2}$, $\alpha = \lambda\nu^2$, and the coupling constant λ is defined, as usual, by eqn (1.26). It is important that eqn (2.28) is valid for any λ.

It should be pointed out that the presence of the renormalisation function Z (corresponding to the last term on the right-hand side of eqn (2.28″)) plays an important role: it results in a cancellation of the diagonal term $n = m$.

Therefore, in the zero-order approximation, the quantity ν in the denominator of the expression (2.28″) can be neglected, and only $\varphi_0 \neq 0$. Then eqn (2.28) gives $\tilde{K}_{00} = 1$. On the other hand, according to eqns (2.28′) and (2.28″), $\tilde{K}_{00} = \alpha$. We then obtain

$$T_{c0} = (2\pi)^{-1}\lambda^{1/2}\tilde{\Omega} \tag{2.29}$$

In the next approximation, we keep more terms. Specifically, we solve the equation $det\widehat{M} = 0$, where the matrix $\widehat{M}$ is defined by $\widehat{M} = \widehat{K} - \widehat{1}$. Then $\widehat{M}_{00} = \alpha - 1$, $\widehat{M}_{01} = \widehat{M}_{10} = 0.72\alpha$, and $\widehat{M}_{11} = -0.63\alpha - 1$. After a simple calculation, we obtain $\alpha = 0.785$ and, based on eqn (2.28), we finally arrive at the following expression:

$$T_{c0} = 0.18\ \lambda^{1/2}\tilde{\Omega} \tag{2.30}$$

The next iteration changes the coefficient in eqn (2.30) by only a negligible amount. It is remarkable that even the zeroth-order approximation describes T_c with good accuracy ($\cong$13%). One can see from eqn (2.30) that the dependence T_c on λ in this case is drastically different from that for the weak coupling limit (eqn (2.21)). Equation (2.30) is valid for $\lambda \tilde{>} 5$.

It is interesting to consider the effect of the Coulomb interaction on T_c in the limit of strong coupling. The equation can be reduced to the form (cf. eqn (2.28))

$$\phi_n = \sum_{m \geq 0} \tilde{\mathrm{K}}^{\mathrm{c}}_{\mathrm{nm}} \phi_{\mathrm{m}} \tag{2.31}$$

$\tilde{\mathrm{K}}^{\mathrm{c}}_{\mathrm{nm}}$ is described by eqn (2.28′), but contains

$$F^c{}_{n,m,v} = F_{n,m,v} - 2\mu^*$$

Solving eqn (2.31) by analogy with the case $\mu^* = 0$, we obtain

$$T_c = 0.18\ \lambda_{eff}{}^{1/2}\tilde{\Omega} \tag{2.32}$$

$$\lambda_{eff} \cong \lambda(1 + 2.6\mu^*)^{-1} \tag{2.32′}$$

Since $\mu^* \ll 1$, we can restrict ourselves to the linear term.

We see from eqn (2.32′) that the Coulomb term decreases the effective constant ($\lambda_{\text{eff}} < \lambda$), but this decrease differs in a striking way from that in the weak coupling approximation. In the latter case, the effective constant has the well-known form $\lambda_{eff} = \lambda - \mu^*$. In the strong coupling limit, the decrease is not given by a difference but rather is described by the ratio (2.32′). This gives a stronger dependence on μ^*. If the effect of the Coulomb interaction were described by the difference $\lambda - \mu^*$, as in the case of weak coupling, this effect would be negligibly small in the limit of $\lambda \gg 1$. We see that an increase in the electron–phonon coupling is accompanied by a transition to the dependence (2.32′), which is more drastic in the $\lambda \gg 1$ limit than a simple subtraction. As a result, the Coulomb term still makes a noticeable contribution despite the fact that $\lambda \gg 1$.

This change in the dependence of T_c on μ^* is connected with the features of the kernel of eqn (2.31). In the weak coupling approximation, the temperature dependence of the phonon Green's function is negligibly small; this smallness is due to the presence of the small parameter $(T_c/\Omega_D)^2$. In this case, the kernel contains the difference $\lambda - \mu^*$. In the limit $\lambda \gg 1$, on the other hand, the phonon Green's function depends strongly upon T, and this dependence results in a different relation between λ and μ^*.

Therefore, in the limit of very strong coupling, T_c is described by expressions (2.30) and (2.32), which are entirely different from the exponential dependences of the weak coupling and intermediate cases (see eqns (2.21) and (2.23)).

Arbitrary coupling strength. In previous sections, we described several special cases applying to various strengths of electron–phonon coupling. We have seen the interesting fact that the analytical expressions for T_c are very different for different coupling strengths. The weak and intermediate coupling cases are described by the exponential dependences (2.21) and (2.23), whereas the super-strong coupling case is characterised by the dependence $T_c \propto \lambda^{1/2}$, eqn (2.31).

Clearly, it is attractive to try to obtain an expression valid for an arbitrary value of the coupling strength. Furthermore, from the practical point of view, it is important to be able to analyse the case of $1.5 < \lambda < 5$, which is outside the limits of applicability of both the McMillan equation ($\lambda < 1.5$) and the very strong coupling limit ($\lambda > 5$), eqn (2.31). Such a universal equation has been derived by Kresin (1987a). Below, we describe the derivation and discuss its implications.

We will make use of the matrix representation (2.28). The most interesting case corresponds to $\nu^2 = \left(\tilde{\Omega}/2\pi T_C\right)^2 \lesssim 1$, with $\tilde{\Omega} = <\Omega^2>^{1/2}$. The region of $\nu^2 \ll 1$ can be treated analytically and is described by expression (2.30).

The problem reduces to that of solving the matrix equation: $\det|\tilde{K} - \tilde{I}| = 0$. In the region $\nu^2 \lesssim 1$ the convergence is relatively rapid and, to a 1% accuracy, it is sufficient to stop at $m = 5$. The calculation can be performed for a range of values of ν and the corresponding values of α determined. As a result, we obtain a dependence of $T_C/\tilde{\Omega}$ upon λ as shown in Fig. 2.4.

With high accuracy, the dependence $T_C(\lambda)$ can be described by the following analytical expression:

$$T_C = \frac{0.25\tilde{\Omega}}{\left(e^{2/\lambda} - 1\right)^{1/2}} \tag{2.33}$$

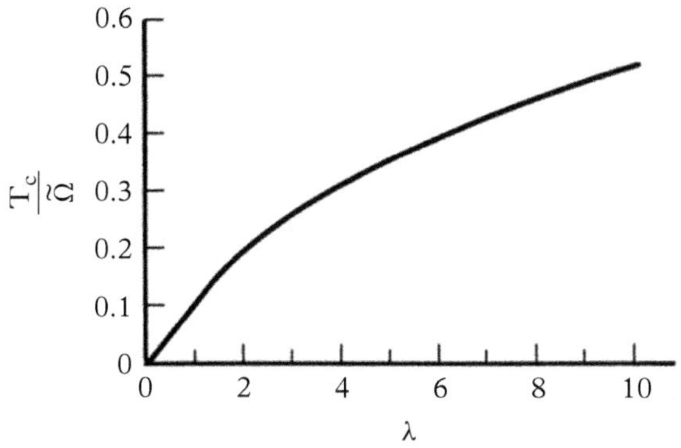

Figure 2.4 *The dependence* $T_c(\lambda)$.

The dependence (2.33) is the solution of the general equation (2.19), or (2.28). Indeed, it was used as a trial function, and then it was demonstrated that it satisfied eqn (2.19) with a high degree of precision.

Let us now consider the impact of the Coulomb interaction on T_c. This question can be analysed by analogy with the derivation of eqn (2.33). We need to solve the equation $\det|\tilde{K}_c - \tilde{I}| = 0$; see eqn (2.31). Usually $\mu^* \cong 0.1$, that is, $\mu^* \ll 1$. The calculation could be carried out for $\mu^* \lesssim 0.2$. The dependence obtained can be described by the following expression:

$$T_C = 0.25\tilde{\Omega}\left(e^{2/\lambda_{\text{eff}}} - 1\right)^{-1/2} \tag{2.34}$$

where

$$\lambda_{\text{eff}} = (\lambda - \mu^*)\left[1 + 2\mu^* + \lambda\mu^* t(\lambda)\right]^{-1} \tag{2.34$'$}$$

The function $t(\lambda)$ decreases exponentially with increasing λ and can be written in the universal form proposed by Tewari and Gumber (1990): $t(\lambda) = 1.5\, e^{-0.28\lambda}$. Thus, T_C is described by the analytical expression (2.34), which has the same form as eqn (2.33), with the replacement $\lambda \to \lambda_{\text{eff}}$.

Equation (2.34) explicitly describes T_C for an arbitrary value of the coupling constant λ. One can see from Table 2.1 that there is a good agreement between eqn (2.34) and experimental data.

Let us analyse various cases. If $\lambda \gg 1$, the exponential factor can be expanded in powers of $1/\lambda$ and we find $T_c = 0.18\ \lambda^{1/2}\tilde{\Omega}$, in accordance with eqn (2.30). In the opposite limit of weak coupling, we obtain eqn (2.21) with $a = 0.25$. For intermediate coupling, we recover an expression essentially identical to eqn (2.23).

As was noted above (see the discussion below eqns (2.24) and (2.24′)), based on the McMillan–Dynes equation (2.23), one can obtain the maximum value of T_c. However, this value corresponds to $\lambda = 2$ and therefore lies outside of the limit of applicability of

Table 2.1 *The ratio* $T_c/\tilde{\Omega}$.

Superconductor	$\left[T_c/\tilde{\Omega}\right]_{th}$	$\left[T_c/\tilde{\Omega}\right]_{\exp}$
In	0.04	0.04
$Pb_{0.4}Tl_{0.6}$	0.07	0.07
In_2Bi	0.08	0.09
Hg	0.09	0.11
Pn	0.10	0.105
$Pb_{0.9}Bi_{0.1}$	0.125	0.13
$Pb_{0.45}Bi_{0.55}$	0.16	0.15

the McMillan–Dynes equation ($\lambda \lesssim 1.5$). The question about the upper limit of T_c and its behaviour with an increase in value of the coupling constant λ can be analysed with use of the general equation (2.33), which is valid for larger values of λ (we neglect μ^* similarly to the case of the McMillan–Dynes equation). Then, based on eqns (2.33) and (2.24), we find that $\partial T_c/\partial\tilde{\Omega} < 0$. The phonon softening results in an increase in T_c. Note that this increase slows down at large λ, and one can see from eqns (2.30), (2.24), and (2.24′) that T_c saturates at the value $T_c = 0.18\eta^{1/2}$.

One can also use the following expression (Allen and Dynes, 1975):

$$T_c = \frac{f_1 f_2}{1.2}\left[\frac{1.04(1+\lambda)}{\lambda - \mu^* - 0.62\lambda\mu^*}\right] \tag{2.35}$$

which is similar to eqn (2.23) but the pre-exponential factor is multiplied by the product $f_1 f_2$; the functions f_1 and f_2 are numerically fit for the solution valid at large λ and have the following form:

$$f_1 = \left[1 + \left(\frac{\lambda}{\Lambda_1}\right)^{\frac{3}{2}}\right]^{\frac{1}{3}}$$

$$f_2 = 1 + \lambda^2\left(\frac{\tilde{\Omega}}{\Omega_{log}} - 1\right)\left(\lambda^2 + \Lambda_2^2\right)^{-1}$$

$$\Lambda_1 = 2.46(1 + 3.8\mu^*)$$

$$\Lambda_2 = 1.82(1 + 6.3\mu^*)\left(\frac{\tilde{\Omega}}{\Omega_{log}}\right)$$

Two coupling constants method. As was noted above (section 2.1.2), the main equation describing the electron–phonon mechanism (eqn (2.17)) does not explicitly contain the coupling constant. Nevertheless, the coupling constant concept is very useful, because it makes it possible to derive useful analytical expressions for the critical temperature, such as eqns (2.21), (2.23), and (2.34), and to study various dependences. At the same time, as was stressed in section 2.1, introducing the electron–phonon coupling constant λ is not a straightforward task. This is an approximation, which is based on a comparison of the characteristic scales of the phonon density of states $F(\Omega)$ and the phonon propagator D. The quantity λ can be introduced if the phonon function D changes slowly on the scale of the peak structure of the function $F(\Omega)$. The late condition is satisfied for many usual superconductors (see the discussion in section 2.2.1). However, for systems characterised by broad and complex phonon spectra, especially systems containing, in addition to acoustic phonon modes, high-frequency optical branches, an approximation with a single coupling constant and a single characteristic frequency is too crude. This is especially true for compounds which have light and heavy ions in the unit cell, for example, hydrides (see section 7.2). For example, later on, in chapter 7, Fig. 7.4 shows that the superconducting material H_3S contains well-separated groups of the low-frequency acoustic phonon modes and high-frequency optical phonons.

The more general approach (Gor'kov and Kresin, 2016, 2018) is based on the use of two coupling constants, λ_{opt} and λ_{ac}, and two characteristic frequencies, $\tilde{\Omega}_{opt}$ and $\tilde{\Omega}_{ac}$.

The phonon spectrum contains two well-separated frequency intervals for the optical and acoustic branches. Let us introduce the coupling constants λ_{opt} and λ_{ac} for each of these regions, and the corresponding average frequencies $\tilde{\Omega}_{opt}$ and $\tilde{\Omega}_{ac}$. Then the equation for the order parameter at $T = T_c$ takes the following form (cf. eqn (2.19)):

$$\Delta(\omega_n) Z = \pi T_c \sum_m \int d\Omega \left[(\lambda_{opt} - \mu^*) D\left(\tilde{\Omega}_{opt}, \omega_n - \omega_m\right) + \lambda_{ac} D\left(\tilde{\Omega}_{ac}, \omega_n - \omega_m\right)\right] \frac{\Delta(\omega_m)}{|\omega_m|} \tag{2.36}$$

Here we separate the integration over the optical and acoustical frequencies to introduce two coupling constants (i = opt, ac):

$$\lambda_i = \int d\Omega_i \frac{\alpha^2(\Omega) F(\Omega)}{\Omega}, \quad \tilde{\Omega}_i = <\Omega^2>_i^{1/2} \tag{2.37}$$

$$<\Omega^2>_i = \frac{2}{\lambda_i} \int d\Omega_i \Omega \alpha^2(\Omega) F(\Omega) \tag{2.37'}$$

The critical temperature must be calculated with the use of eqn (2.36). This equation is the generalisation of eqn (2.19) for the presence of two phonon groups: the acoustic and the optical modes.

Consider initially the weak coupling case ($T_c \ll \tilde{\Omega}_{ac} \ll \tilde{\Omega}_{opt}$). Then one can evaluate T_c within the usual BCS logarithmic approximation while adding the renormalisation function $Z \approx 1 + \lambda_T$ ($\lambda_T = \lambda_{OPT} + \lambda_{ac}$) into the exponent (Grimvall, 1981):

$$T_c \approx 0.25 \left(\tilde{\Omega}_{opt}\right)^{\frac{\lambda_{opt}}{\lambda_T}} \left(\tilde{\Omega}_{ac}\right)^{\frac{\lambda_{ac}}{\lambda_T}} e^{-\frac{1+\lambda_T}{\lambda_T - \mu^*}} \tag{2.38}$$

If $\lambda_{opt} = 0$, we obtain the expression for usual weakly coupled superconductors (eqn (2.21)).

Consider now the different case when $\lambda_{opt} \gg \lambda_{ac}$, that is, the situation when the main contribution to the pairing is provided by the optical phonons. Let us write T_c as

$$T_c = T_c^0 + \Delta T_c^{ac},\ T_c^0 \equiv T_c^{opt}$$

and assume that $\Delta T_c^{ac} \ll T_c^0$. Based on eqn (2.36), with the use of these assumptions, one can obtain the following analytical expression for the critical temperature of the high-T_c phase:

$$T_c = \left[1 + 2\frac{\lambda_{ac}}{\lambda_{opt} - \mu^*} \frac{1}{1 + \left(\pi T_c^0/\tilde{\Omega}_{ac}\right)^2}\right] T_c^0 \tag{2.39}$$

For $\lambda_{opt} \lesssim 1.5$, one can use the McMillan–Dynes expression (eqn (2.23)):

$$T_c^0 = \left(\frac{\tilde{\Omega}_{opt}}{1.2}\right) e^{-\frac{1.04(1+\lambda_{opt})}{\lambda_{opt} - \mu^*(1+0.62\lambda_{opt})}} \tag{2.40}$$

or eqn (2.23) with $\lambda \equiv \lambda_{opt}$. For larger strengths of the electron–phonon interaction $\left(\lambda > 1.5\right)$, one should use expression (2.33):

$$T_c^0 = \frac{0.25\tilde{\Omega}_{opt}}{\left(e^{\frac{2}{\lambda_{eff}}} - 1\right)^{\frac{1}{2}}}$$
$$\lambda_{eff} = (\lambda_{opt} - \mu^*)\left[1 + 2\mu^* + \lambda_{opt}\mu^* t(\lambda_{opt})\right]^{-1}$$
$$t(x) = 1.5e^{-0.28x} \tag{2.41}$$

The method described in this section will be used later on for analysis of the high-T_c hydrides (section 7.2).

2.2.3 Properties of Superconductors with Strong Coupling

As has been stressed earlier, the weak coupling approximation (the BCS model) is remarkable for the universality of its results. The best-known one relates the energy gap at $T = 0$ K and T_c:

$$2\varepsilon(0) = 3.52T_c$$

The model yields many other universal relations, such as $\varepsilon(T)/T_c = 3.06\left[1 - T/T_c\right]^{1/2}_{T \to T_c}$, $\Delta C/C_n(T_c) = 1.4$, and so on; see eqns (2.9) and (2.9′). The origin of this universality is in the fact that all superconducting properties in the BCS model are determined by the single parameter λ, the coupling constant (or, more rigorously, by $-\mu^*$).

Instead of λ, one can equally well select T_c as such a parameter; indeed, it is usually much better to deal with an experimentally measured quantity. As a result, all the other parameters (energy gap, heat capacity, critical field, etc.) are related in a universal way to T_c.

Strong coupling effects destroy this universality. This is caused by the fact that strong coupling theory contains a direct dependence not only on λ but also on the structure of the phonon spectrum.

It turns out (see below) that an increase in the coupling strength leads to the ratio $\eta = 2\varepsilon(0)/T_c$ growing beyond the BCS value of 3.52. Since the ratio η can be measured experimentally, this provides a relatively simple test of the magnitude of the coupling constant. One has to keep in mind, though, that the above argument is valid only if we are dealing with only one gap (and consequently only one coupling constant). This is the case for practically all conventional superconductors. The case for a multiband structure requires a separate analysis (see section 3.2).

Let us look at the relationship between $\varepsilon(0)$ and T_c. We start with eqn (2.17). The non-linear equation can be simplified by introducing the coupling constant and characteristic phonon frequency $\tilde{\Omega}$. We will assume that $T_c \ll \tilde{\Omega}$; nevertheless, our goal is to calculate corrections $\sim(T_c/\tilde{\Omega})^2$.

The solution can be found by quasi-linearising eqn (2.17). The idea of this method is as follows (Zubarev, 1960). The equation for $\Delta(\omega)$ at $T = 0$ K can be written in the form

$$V(\omega) = \frac{\tilde{\Omega}^2}{\tilde{\Omega}^2 + \omega^2} + Z^{-1} \int_{-\infty}^{\infty} d\omega' R(\Omega, \omega, \omega') \frac{V(\omega')}{\sqrt{\omega'^2 + \Delta^2(0) V^2(\omega')}} \tag{2.42}$$

where $\omega' \cong \tilde{\Omega}, V(\omega) = \Delta(\omega)/\Delta(0)$, and the kernel is

$$R(\omega, \omega') = \frac{\tilde{\Omega}^2}{\tilde{\Omega}^2 + (\omega - \omega')^2} - \frac{\tilde{\Omega}^2}{\tilde{\Omega}^2 + \omega^2} \frac{\tilde{\Omega}^2}{\tilde{\Omega}^2 + \omega'^2} \tag{2.42$'$}$$

It is important that $R(\omega, 0, \Omega) = 0$, and $\boldsymbol{R}(\omega, \omega, \Omega) \sim \omega^2$ for small ω. It follows that the main contribution to the integral over ω' comes from the region $\omega' \cong \tilde{\Omega}$ and, in the zeroth-order approximation, we can neglect the factor $\Delta^2(0)V^2(\omega')$ in the denominator on the right-hand side of eqn (2.42). As a result, we end up with a linear equation for the zeroth-order function $V_0(\omega)$. The solution of the full equation (2.42$'$) can be sought in the form $V(\omega) = V_0(\omega) + V'(\omega)$, where $V'(\omega) \propto \Delta^2(0)/\tilde{\Omega}^2$. It is also possible to derive a linear integral equation for the function $V'(\omega)$. Thus, the quasi-linearisation method reduces the solution of the non-linear equation to that of a system of linear equations.

For $T = T_c$, eqn (2.17) can be transformed with the aid of the Poisson summation formula

$$\sum_{\omega_n > 0} f(\omega_n) = (1/2\pi T) \int_0^{\infty} f(z) dz + (1/\pi T) \sum_{s=1}^{\infty} (-1)^s \cos(sz/T) dz$$

By comparing the equations for $= \Delta(\omega)$ at $T = 0$ K and $T = T_c$, we can derive an expression relating the energy gap $\varepsilon(0)$ at $T = 0$ K and the critical temperature T_c (Geilikman and Kresin, 1966, 1972; Geilikman *et al.*, 1975):

$$\frac{2\varepsilon(0)}{T_c} = 3.52\left[1 + a\left(\frac{T_c}{\tilde{\Omega}}\right)^2 \ln\frac{\tilde{\Omega}}{T_c}\right] \tag{2.43}$$

where a is a numerical factor ($a \cong 5.3$).

Several comments should be made concerning eqn (2.43). Clearly, this expression generalises eqn (2.9), which was obtained in the weak coupling approximation: we recover the latter if the second term in the brackets is neglected. This second term, which reflects the effects of strong coupling, has a peculiar form. In addition to the ordinary quadratic term, it contains the large logarithmic factor $\ln\frac{\tilde{\Omega}}{T_c}$.

We see that taking account of the effects of strong coupling destroys the universality of the relation between $\varepsilon(0)$ and T_c. The expression we obtained contains a characteristic phonon frequency, which varies from one superconductor to another.

Furthermore, the relationship (2.43) is no longer that of direct proportionality. This statement can be checked experimentally by studying the dependence of the energy gap and T_c on pressure. Such an experiment (Zavaritski *et al.*, 1971), which utilised a tunnel junction under pressure, has shown that, indeed, in lead the energy gap and the critical temperature do not vary in an identical manner. This is a manifestation of strong coupling effects. The experimental method employed in this work also made it possible to control the shift of the characteristic phonon frequency . The results were in very good agreement with eqn (2.43).

Instead of $\tilde{\Omega}$, we could choose other characteristic frequencies, such as with the averaging carried out with respect to the function $\alpha^2(\Omega)F(\Omega)$: $<\Omega>= \int d\Omega\alpha^2(\Omega)F(\Omega)$, $<\Omega_{log}>$, and so on (see Carbotte, 1990; Wolf, 2012). The comparison between theory and the experimental data depends, of course, on the choice of the characteristic frequency $\tilde{\Omega}$, although the difference is not dramatic. The general qualitative features described above remain valid.

Equation (2.43) shows that one of the effects of strong coupling is to make the ratio $\eta = 2\varepsilon(0)/T_c$ exceed the value $\eta_{\mathrm{BCS}} = 3.52$ (the extra term in parentheses is always positive). There is a one-to-one correlation between η and the magnitude of the coupling constant λ. An increase in λ leads to an increase in T_c and, as result, to an increase in the ratio η.

It is important to keep in mind, however, that the result (2.43) was obtained under the assumption that $T_c \ll \tilde{\Omega}$. As λ, and consequently T_c, increase, we come to a point (at $\lambda \cong 2-3$) when the above inequality does not hold anymore. The following question arises then: what happens to the ratio η in the regime of super-strong coupling where the critical temperature is described by eqns (2.30) and (2.33)?

It can be shown in general (Kresin, 1987c) that the ratio $2\varepsilon(0)/T_c$ saturates with increasing λ. The energy gap is defined as the root of the equation $\varepsilon = \Delta(-i\varepsilon)$. We are interested in the behaviour of the function $\Delta(\omega)$ for high frequencies, where it is determined from eqn (2.17):

$$\Delta(\omega) \simeq \frac{\lambda\tilde{\Omega}^2}{\omega^2}\int d\omega'\Delta(\omega')\left|\omega'^2 + \Delta^2(\omega')\right|^{-1/2} \tag{2.42''}$$

As a result, the following expression is obtained for the gap $\varepsilon(0)$ in the regime $\lambda \gg 1$: $\varepsilon(0) = C\sqrt{\lambda}$. This dependence is identical with that of T_c (see eqn (2.30)) in the region of large λ. Consequently, the ratio $\eta = 2\varepsilon(0)/T_c$ becomes a universal constant as λ becomes large: $\eta_{|\lambda \gg 1} \cong 13.5$. Note that, according to some studies of organics and the cuprates, one can observe values close to η_{max} (see Gray, 1987).

Hence, the ratio η has universal values in both the region of weak coupling and that of strong coupling, and $\eta_{BCS} = \eta_{|\lambda \ll 1} = 3.52 \ll \eta_{|\lambda \gg 1}$. The transition from weak electron–phonon coupling to strong coupling is accompanied by an increase in η, but the dependence $\eta(\lambda)$ saturates for large λ, even though both T_c and $\varepsilon(0)$ continue to grow.

The temperature dependence of the energy gap is also different from that in superconductors with weak coupling. It has the following form (Geilikman and Kresin, 1968b):

$$\varepsilon(T) = a\left[1 - \left(\frac{T}{T_c}\right)\right]^{\frac{1}{2}}_{|T \to T_c},$$

$$a = 3.06\left[1 + 8.8\left(T_c^2/\tilde{\Omega}^2\right)\ln\left(\tilde{\Omega}/T_c\right)\right] \tag{2.44}$$

For example, for lead $a \approxeq 4$. The dependence $\varepsilon(T)$ influences many properties of superconductors, such as heat capacity, thermal conductivity, ultrasound attenuation, and so on. For example, the jump in the heat capacity β is directly related to the change in entropy $(S^s - S^n)(T_c) \propto a^2$. A more detailed evaluation based on the general expression for the thermodynamic potential leads to the following result (Kresin and Parchomenko, 1975):

$$\beta = 1.43\left[1 + b\left(\frac{T_c}{\tilde{\Omega}}\right)^2\left(\ln\frac{\tilde{\Omega}}{T_c} + \frac{1}{2}\right)\right], \quad b \simeq 18 \tag{2.44'}$$

For example, for lead, $\beta = 2.4$ (cf. $\beta = 1.4$ in the BCS theory, see eqn (2.9′)) and, for gallium, $\beta = 2.3$. Ultrasound attenuation and electronic thermal conductivity decrease much more sharply with decreasing temperature below T_c than in weakly coupled superconductors. The properties of strongly coupled superconductors can be analysed in detail by tunnelling spectroscopy (Scalapino *et al.*, 1966; Scalapino, 1969; Wolf, 2012). This important method will be discussed in sections 3.3.4 and 4.1.

2.3 The Electron–Lattice Interaction: Special Aspects

2.3.1 The Polaronic Effect and its Impact on T_c

The double-well structure. We described above (section 1.2.2) the dynamic polaronic state, which is a manifestation of the dynamic Jahn–Teller effect. Here we discuss its impact on the superconducting state and, especially, on T_c. More specifically, we address the following question: why is the presence of polaronic states beneficial for superconductivity? As is known, the polaronic effect provided the main motivation for the original search for high T_c. in the cuprates (Bednorz and Mueller, 1986; see chapter 5).

Consider a complex compound where one ionic subsystem is such that an ion is characterised by two close equilibrium positions (the double-well potential; see Fig. 1.5b).

Such a strongly anharmonic potential leads to a peculiar non-adiabatic polaronic effect; indeed, in this case, the electronic and local lattice degrees of freedom turn out to be inseparable. Especially interesting are the cases of layered systems, and here we focus on such a case. For example, oxygen ions in the cuprates do form such a subsystem; the double-well structure has been observed experimentally (see section 5.4.3).

One can show that the presence of such a structure leads to a noticeable increase in the effective strength of electron–lattice coupling relative to the usual case of a single potential minimum for the ionic coordinate (Kresin, 2010). The thing is that the pairing interaction is described by an equation which contains a matrix element connecting the initial and the virtual states. One can show that, even in the absence of direct interaction with the double-well structure, the resulting coupling constant will increase. This is due to the increased phase space for virtual transitions or, in other words, due to an increased number of these transitions.

Consider the special case when the electronic term for some ionic subsystem (e.g. for oxygen ions in the cuprates) contains two close minima positions. As an example, we focus on the case when the double-well potential corresponds to some direction; for layered systems, this direction is perpendicular to the layers ($OZ||c$). As for the dependence of the ionic motion on X,Y, it is described by the usual harmonic dynamics with single-minima equilibrium positions. Therefore, $\varepsilon_n(\boldsymbol{R}) \equiv \varepsilon_n(\boldsymbol{\rho}, Z)$ with the double-well structure in the Z direction, where $\boldsymbol{\rho}$ denotes the in-plane ionic position.

We employ the tight-binding approximation. As a first step, we should write down the local wave function. In our case, the ion is affected by the double-well potential (the Z direction; Fig. 2.5). At this stage, it is very convenient to employ the diabatic representation (see section1.2.2). As a result of the transformation, the double-well potential is replaced by two crossing harmonic energy terms (Fig. 2.5).

Each of these terms also contains its vibrational manifold. Therefore, the local state is described by two groups of terms ('a' and 'b'), so that

$$\Psi_{loc.} = c_a \Psi_{\rm a}(\boldsymbol{r}, \boldsymbol{R}) + c_b \Psi_{\rm b}(\boldsymbol{r}, \boldsymbol{R}) \tag{2.45}$$

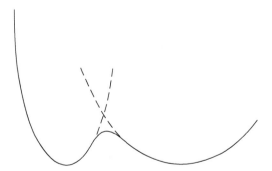

Figure 2.5 *Transformation to the diabatic representation: solid line, initial term (double-well structure); broken lines, crossing terms.*

where

$$\Psi_{a(b)}(\boldsymbol{r},\boldsymbol{R}) = \Psi_{a(b)}\left(\boldsymbol{r},\boldsymbol{R}_0^{a(b)}\right)\chi_{a(b)}\left(z - z_0^{a(b)}\right)$$

Note that $\Psi_a = \psi_s\chi_{sv}$, and $\Psi_b = \psi_{s'}\chi_{s'v'}$. Here $\{s, s'\}$ and $\{n, n'\}$ are the electronic and vibrational quantum numbers, respectively. If both terms are equivalent, then $c_a^2 = c_b^2 = 0.5$. Because of the inter-term tunnelling, the energy levels, which correspond to isolated terms, are split into symmetric and antisymmetric; the scale of the splitting is of order of ε_{ab}, where (see section 1.2.2)

$$\varepsilon_{ab} = \int d\boldsymbol{R}\chi_a \widehat{H}_{\boldsymbol{r};ab}\chi_b$$

$$\widehat{H}_{\boldsymbol{r};ab} = \int d\boldsymbol{r}\psi_b^*\widehat{H}_{\boldsymbol{r}}\psi_a. \tag{2.45$'$}$$

It is essential that, unlike the usual adiabatic picture, the operator $\widehat{H}_{\boldsymbol{r}} = \widehat{T}_{\boldsymbol{r}} + V(\boldsymbol{r},\boldsymbol{R})$ has non-diagonal terms in the diabatic representation. One can see directly from eqn (2.45) that the wave function $\Psi_{loc.}$ cannot be written as a product of electronic and ionic wave functions. In other words, the electronic and local ionic motions cannot be separated as in the usual adiabatic theory. Such a polaronic effect is a strong non-adiabatic phenomenon.

In the tight-binding approximation, each local term, including those formed by the transformation (Fig. 2.5), is broadened into the energy band. Note at first that the local electronic function (see eqn (2.45)) $\psi_i\left(\boldsymbol{r},\boldsymbol{R}_0^i\right)$ ($i = a, b$) can be written in the form $\psi_i\left(\boldsymbol{r}-\boldsymbol{R}_0^n\right),\boldsymbol{R}_0^n$, corresponding to the crossing point for the nth ion ($i = a, b$). Indeed, because the electronic wave function has a scale of the length of the bond that greatly exceeds the amplitude of vibrations and the distance 'δ' between the minima, one can neglect the difference between $\boldsymbol{R}_0^n$ and $\boldsymbol{R}_0^i$.

The total wave function also contains the vibrational part. Firstly, let us separate the local vibrational mode that corresponds initially to the double-well structure or, in the diabatic approximation, to two crossing terms (see above). As for the dependence of ionic motion on X and Y, it can be described by a set of normal modes. Then the total wave function has the form

$$\Psi(\boldsymbol{r},\boldsymbol{R}) = u_{\boldsymbol{k}}(\boldsymbol{r},\boldsymbol{R})\,e^{i\boldsymbol{k}\boldsymbol{r}}\Phi_{vib.}$$

$$u_{\boldsymbol{k}}(\boldsymbol{r},\boldsymbol{R}) = \sum_n \Psi_{loc.}(\boldsymbol{r},\boldsymbol{R})\,e^{i\boldsymbol{k}\left(\boldsymbol{R}_0^n-\boldsymbol{r}\right)} \tag{2.45$''$}$$

$\Psi_{loc.}$ is determined by eqn (2.45). The local vibrational wave functions are centred at z_0^i and at $z_0^i - \delta$. If the terms are similar, then $\chi^b(z) = \chi^a(z-\delta)$.

One should stress a key difference between the ionic motions, which is described by the function

$$\Phi_{vib}(R_x,R_y) = \Pi\phi(\Omega_m)$$

(Ω_m are the frequencies of the usual normal modes) and the function χ (R_z). There is no in-plane ionic hopping between various ionic sites. However, oxygen can tunnel between the two minima, and this is reflected in a noticeable overlap of vibrational functions χ^{i} corresponding to the two crossing terms (Fig. 2.5).

The superconducting state. Let us consider the case when the main contribution to the pairing is coming from the interaction with usual normal (harmonic) modes Q_m (one can easily describe a more general case). Then the interaction Hamiltonian can be written in the usual form (see eqn (1.21)):

$$H = \frac{\partial V}{\partial Q_m} \delta Q_m \tag{1.21'}$$

Summation over m is implied. The equation for the pairing amplitude has the following form (see eqn (2.16)):

$$\Delta(\omega_n) = T \sum_{\omega_n{}'} \int d\xi' |\zeta|^2 \nu_F D\left(\omega_n - \omega_{n'}, \tilde{\Omega}\right) F^+(\omega_{n'}) \tag{2.16'}$$

Here D is the phonon propagator defined by eqn (2.15), $\omega_n = (2n+1)\pi T, \tilde{\boldsymbol{\Omega}}$ is the characteristic phonon frequency, $F^+(\omega_n) = \Delta(\omega_n) / \left[\omega_n{}^2 + \xi_{nv}^2 + \Delta^2(\omega_n)\right]$ is the Gor'kov's pairing function, ξ_{nv} is the excitation energy relative to the Fermi energy, and ν_F is the density of states. It is essential that ξ_{nv} is the energy of virtual transitions, which contribute to pairing. Note that the integrand contains the matrix element ζ, which has the form

$$\zeta = \int \Psi^{*f} \widehat{H} \Psi^i d\boldsymbol{r} d\boldsymbol{R} \tag{2.46}$$

where $\widehat{H}$ is defined by eqn (1.21′), i and f denote the initial and the virtual states, respectively, and the total function has the form (2.45″).

Let us introduce some characteristic frequency $\tilde{\Omega}$ and a single mode Q_0, which provides one of the main contributions to the pairing (see eqn (1.21′)). Then the product $\Pi\phi_{\nu m}(\Omega_m)$ can be replaced by the function $\phi_\nu(\,\mathrm{Q})$. Correspondingly, $H = \frac{\partial V}{\partial Q_0} \delta Q_0$. In other words, we model the phonon spectrum as consisting of two modes. One of them, the harmonic mode $\tilde{Q}$, provides a major contribution to the coupling interaction, and the second mode which is characterised by a double-well structure (Fig. 2.5), provides an additional number of the virtual states. This model can be easily generalised. One can see from eqn (2.46) that the average value of the coupling matrix element is a sum:

$$\zeta_{\mathrm{av}} = \zeta_0 + \zeta_1 \tag{2.46'}$$

where

$$\zeta = \int d\boldsymbol{r} d\tilde{Q} \sum_n \psi_s(\boldsymbol{r} - \boldsymbol{R}_0^n) \psi_{s'}(\boldsymbol{r} - \boldsymbol{R}_0^n) \left(\frac{\partial V}{\partial \tilde{Q}}\right)_0 \delta \tilde{Q}_{av}$$

$$\zeta_1 \approx \zeta_0 \bar{F}^{ab} \tag{2.46''}$$

where

$$\bar{F}^{ab} = 2\sum_{s,v_1} \chi_{sv}\left(z - z_0^a\right)\chi_{s_1v'}\left(z - z_0^b\right) \qquad (2.46''')$$

As usual (see e.g. Anselm, 1982), the displacement $\delta\tilde{Q}_{av}$ can be expanded in a Fourier series. It is essential that the Franck–Condon factor $\bar{F}^{ab} \neq 0$. Indeed, eqn (2.16′) assumes summation over virtual states, that is, integration over ξ and summation over ν.The vibrational wave functions for the same term are orthogonal. However, this is not the case for different crossing terms. The additional contribution ζ_1 corresponds to virtual transitions, which are accompanied by the change of electronic terms upon ionic tunnelling between the two minima. In other words, ζ_1 corresponds to transitions to states with various ν' (for given ν). It is important to note that the contribution comes only from the terms with the same symmetry, that is, from terms with the same sign of the coefficient 'c_b' (see eqn (2.45)). Therefore, there is an additional contribution (relative to the usual case), caused by the transitions between the local vibrational levels. Such an increase in phase space for virtual transitions leads to an increase in the value of the coupling matrix element ζ.

Consider in more detail the Franck–Condon factor $\bar{F}^{ab}$. Assume for concreteness that the initial state corresponds to $\nu_a = 0$. Equation (2.16′) contains a sum $\sum_{\nu'} \bar{F}^{ab}{}_{\nu'0}$, which represents the contributions of virtual transitions to the manifold of the coherence states. In the diabatic representation, a single non-harmonic term is replaced by two crossing terms, and each of them can be treated in harmonic approximation. The Franck–Condon factor is equal (see e.g. Landau and Lifshitz, 1977) $\bar{F}^{ab} = \left(\frac{\beta^\nu}{\nu!}\right)\exp(-\beta)$, where $\beta = (\delta/a)^2$ (a is the amplitude of vibrations, and δ is the distance between the minima). One can see that $\bar{F}^{ab}{}_{\nu'0}$ contains a small factor, $\exp\left(-(\delta/a)^2\right)$, which describes the probability of the ionic tunnelling between two minima. But this smallness is compensated by summation over all virtual transitions. If, for example, $\delta/a = 2$, then $\zeta_1 \approx 0.7\zeta_0$. If $\zeta_0 \approx 1$, then $\zeta = \zeta_0 + \zeta_1 \approx 1.7$. The value of T_c is determined by the coupling constant $\lambda \propto \zeta^2$, and we obtain $\lambda \approx 3$. Therefore, the polaronic effect leads to a large increase in the effective strength of the electron–lattice coupling. As was stressed above, this enhancement arises due to the increase in phase space for virtual transitions, which promotes the pairing.

2.3.2 The Van Hove Scenario

Properties of low-dimensional systems can be strongly affected by the presence of the Van Hove singularity (1953). For the one-dimensional structure, this singularity in the electronic density of states has the form $(E - E_s)^{-1/2}$. As for the two-dimensional structure (layered systems), the divergence has a logarithmic dependence.

This issue is important for the description of the A15 compounds A_3B (e.g. V_3Si, Nb_3Sn). These materials have the highest values of T_c among conventional superconductors. According to Labbe and Friedel (1966) and Labbe *et al.* (1967), it can be explained by the fact that A-ions form quasi-one-dimensional structures; this leads to structure instability, lattice softening, and an increase in T_c (see section 7.1.2).

The Van Hove scenario for high-T_c cuprates was discussed by Barisic *et al.* (1987); also see the review by Bok and Bouvier (2012). For the case when the half-width of the peak, $D/2$, is of the same order as the characteristic phonon frequency $\tilde{\Omega}$ but still $D > \tilde{\Omega}$, one can use the following expression for T_c (in a weak coupling approximation):

$$T_c = D \exp\left(-\frac{1}{\sqrt{\lambda}}\right) \tag{2.47}$$

Since $\lambda \ll 1$, the dependence containing $\lambda^{-1/2}$ leads to larger values of T_c relative to the BCS expression (2.21). The isotopic dependence of T_c is also different, because the pre-exponential factor is different from that in eqn (2.21).

2.3.3 Bipolarons, and BEC versus BCS

The bipolaronic scenario ('local' pairs) was proposed as an explanation of superconductivity even before the BCS theory (Schafroth, 1955). A more rigorous concept of a bipolaron, which is a bound state of two polarons, was introduced by Vinetskii (1961) and Eagles (1969). A more detailed model of bipolaronic superconductivity, namely, the picture that bosons (bipolarons) formed on a lattice could form a superconducting system, was proposed by Alexandrov and Ranninger (1981). The quantitative picture of bipolaronic superconductivity is rather elegant and is very different from the conventional BCS concept. The main difference is the nature of the normal state. As we know, the starting point of the BCS picture is that, in the normal state (above T_c or above the critical field), we are dealing with the usual fermions (delocalised electrons) and, correspondingly, with a Fermi surface. According to the bipolaronic picture, the normal state represents a Bose system formed by pairs of polarons: pairing occurs in real space. As a result, the nature of the phase transition at T_c is entirely different. According to the bipolaronic scenario, we are dealing with the Bose–Einstein condensation (BEC) of bosons, whereas the formation of pairs (Cooper pairs) in usual superconductors occurs at T_c. The Cooper pair is formed by two electrons with opposite momenta, so that the pairs are formed in momentum, not real space.

There is an interesting question of bipolaronic instability. One can expect that, for large values of the electron–phonon coupling constant λ, the usual Fermi system is unstable with respect to formation of bipolarons. One should also note that an analysis of bipolarons is often based on the Holstein Hamiltonian (1.34). However, as was discussed above (section 1.2.2), this Hamiltonian is applicable if the electronic frequency $E_F \ll \tilde{\Omega}$. Such a situation, indeed, is favourable, because, in this case, the size of a bipolaron could be less than that of the inter-electron spacing, and this factor diminishes the overlap of neighbouring wave functions.

As was noted above, the bipolaronic picture requires that the carriers have a bosonic nature. This factor is important, since it contradicts the existence of the Fermi surface. A more general picture was described by Mueller (2007). According to this approach, the compound may contain two components: bipolarons and free fermions. The presence of free fermions explains the presence of the Fermi surface.

As was mentioned above, bipolaronic superconductivity is caused by BEC, that is, that picture is similar to that for liquid helium and its superfluidity. In connection with this, one can study a general question about the scenario that is intermediate between BEC and Cooper pairing (the BCS case). Such a generalisation was considered initially by Leggett (1980) and later by Nozieres and Schmitt-Rink (1985) and Nozieres (1995). The properties of a Fermi gas with an attractive potential have been studied as a function of the coupling strength. The BEC and BCS cases correspond to two limits (strong and weak coupling). The analysis shows the importance of one-particle excitations for Cooper pairing versus collective excitations for the BEC case.

2.3.4 Manifestations of the Phonon Mechanism, and a Proposed Experiment

The question of determining the phonon mechanism is not a trivial one. Indeed, many superconductor properties are not sensitive to the specific mechanism of the pairing. For example, for weakly coupled superconductors, which are described by the BCS theory, the main quantities, such as heat capacity, electromagnetic properties, and transport properties, can be expressed in a single parameter, namely the critical temperature (see section 2.1.4), regardless of the mechanism of pairing in the material. There is just a limited number of methods allowing to determine whether or not the superconducting state is provided by the phonon mechanism. We described above the main features of the phonon mechanism of superconductivity.

Below, we will discuss other mechanisms (electronic, magnetic), which also can provide the pairing of the carriers. Correspondingly, one can raise the following question: how can we distinguish these mechanisms? More specifically, if we are concerned with some specific superconducting material, how can we determine which mechanism is dominant and which kind of interaction provides the pairing for the sample of interest? We used the word 'dominant' because it is perfectly realistic that the superconducting state for some systems is supported by, for example, two mechanisms—phononic and electronic—and we will be dealing with such combinations. If the pairing is caused by combination of various mechanisms and, in such a case, it is interesting to determine the degree of participation for each of them. Below we describe these methods.

The isotope effect. Let us discuss here the manifestations of the phonon mechanism. Firstly, one should mention the isotope effect. Indeed, as was indicated in the introduction (section 1.1), the discovery of the isotope effect directly preceded the creation of the BCS theory. The dependence of the critical temperature on the isotopic mass ($T_c \propto M^{-\alpha}, M$ is the ionic mass, α is the isotope coefficient; for the ideal case, $\alpha_{id} = 0.5$) demonstrates that the lattice is involved in the formation of the superconducting state. The impact of the isotope substitution looks straightforward to understand but, in reality, this is rather complex phenomenon (see section 3.5). There are a number of factors affecting the value of the isotope coefficient α. As a whole, one can state that, if the value of the isotope coefficient is large, relatively close to ideal $\alpha_{max} = 0.5$, then the phonon mechanism is dominant. For hydrides (see section 7.2), the substitution of deuterium for hydrogen leads, for example, for H_3S compounds, to a shift in T_c, corresponding with $\alpha \simeq 0.35$. For the LaH_{10} compound, the value of α is even larger.

Tunnelling spectroscopy. The most rigorous evidence for the participation of phonons in the pairing is provided by tunnelling spectroscopy, more specifically, by the special method, developed by McMillan and Rowell (1965, 1969). Tunnelling spectroscopy will be discussed in detail in section 4.1.1. Here we describe only the idea of the method. The main equation describing the phonon mechanism, eqn (2.17), contains the spectral function $\alpha^2(\Omega)F(\Omega)$, $F(\Omega)$ is the phonon density of states, and $\alpha^2(\Omega)$ contains the matrix element for the electron–phonon interaction. The McMillan–Rowell method allows one to use the tunnelling $I-V$ characteristics (I is the tunnelling current, and V is the voltage) and to perform a so-called inversion of the Eliashberg equation in order to obtain the spectral function. The spectral function is characterised by the presence of the peak structure; its origin is related to the corresponding dependence of the phonon density of states $F(\Omega)$; see section 2.2.1. One should stress that the function $\alpha^2(\Omega)F(\Omega)$ is obtained by the 'inversion' of eqn (2.17), that is, by a method which is based on the assumption that the pairing is caused by the phonon mechanism. Correspondingly, the structure of the spectral function, including the positions of the peaks in the phonon density of states $F(\Omega)$, is directly related to this assumption. At the same time, the phonon density of states $F(\Omega)$ can be measured directly, without relation to either the phonon mechanism of the pairing or even to the phenomenon of superconductivity. Indeed, it can be obtained with the use of neutron scattering. One can compare the positions of the $F(\Omega)$ peaks obtained by these two methods. The coincidence of these positions confirms the validity of the assumption that the superconducting state is caused by the electron–phonon interaction. It is important also that the McMillan–Rowell method makes it possible to determine the value of the Coulomb pseudopotential μ^*. Moreover, knowledge of the spectral function allows us to calculate the value of the electron–phonon coupling constant λ and the characteristic phonon frequency $\tilde{\Omega}$ (see eqn (1.27)), and then with the use, for example, of the McMillan–Dynes expression (eqn (2.23)) to evaluate the value of T_c, which should coincide with the measured value.

This method represents the most rigorous way to determine whether or not the superconducting state is provided by the electron–phonon interaction.

Note that recent progress in the development of ab initio calculations makes it possible to calculate the phonon spectrum and, correspondingly, the phonon density of states for specific class of materials, and then the spectral function of interest (see e.g. Li *et al.*, 2014, and Duan *et al.*, 2014) without invoking tunnelling measurements. Such an approach appears to be very successful for studying the mechanism of high-T_c superconductivity in the hydrides under pressure (see section 7.2). These materials display a record high value for the critical temperature, one that is close to room temperature. The ab initio calculations, along with measured large values of the isotope coefficient, provide rigorous evidence that the electron–phonon interaction is the dominant mechanism for high-T_c superconductivity in hydrides.

Relaxation, and a proposed experiment. The experiments described in the previous section provide strong support for phonon-mediated superconductivity in the cuprates. One more interesting experiment aiming at determination of the mechanism can also be proposed (Ovchinnikov and Kresin, 1998; a detailed description can be found in the review by Kresin and Wolf, 2009).

The method is based on the technique of using Josephson junctions for the generation of phonons (Eisenmerger and Dayem, 1967; Eisenmerger, 1969; Dynes *et al.*, 1971; Dynes and Narayanamurti, 1973). One can modify this technique for any boson contributing to the pairing. A non-equilibrium superconducting state is formed by incoming radiation. The creation of excited quasiparticles is followed by a relaxation process. By the end of this process, a noticeable number of quasiparticles are concentrated at or very near the energy gap edge, $E \approx \varepsilon$, where ε is the pairing gap. The final stage of relaxation is the recombination of Cooper pairs. For conventional superconductors, this stage is accompanied by radiation of phonons.

In a classic experiment (Eisenmenger and Dayem, 1967; Eisenmenger, 1969), the generation and detection of phonons propagating through a sapphire substrate was demonstrated using two Josephson junctions located on opposite sides of a cylindrical sapphire block. The study was aimed at the generation of almost monochromatic phonons. We now look at such experiments from a different point of view. Indeed, these experiments were possible only because phonons were responsible for pairing in the electrodes of the emitting junction and are thus emitted when quasiparticle excitations relax to the gap edge and recombine to form pairs. In other words, one can observe the recombination of electrons with energies near the gap edge; these electrons can form Cooper pairs, and this process is accompanied by radiation of phonons with $\hbar\Omega \approx 2\varepsilon$.

One can raise the following question: why are other excitations not radiated, only phonons? The answer is obvious and directly reflects the fact that phonons form the glue for pairing. In fact, radiation of phonons created by recombination is an additional support for the phonon mechanism of pairing in conventional superconductors. If pairing is provided, for example, by magnetic excitations, the recombination would be accompanied by the radiation of magnons.

One can propose the following experiment. On one side of a high-quality sapphire (or other nearly defect-free single-crystal substrate), one can prepare an Nb or NbN tunnel junction as a detector of phonons. This detector will be most sensitive to phonons that are above the gap energy 2ε of the electrodes. On the other side of the substrate, we prepare a junction containing the superconductor of interest. After the current or light has been removed, the generated quasiparticles will relax very rapidly to the gap edge and, as they recombine to form pairs, they will emit phonons. The observation of a signal from the superconductor, similar in magnitude and temporal behaviour to that of the control junction, would be extremely strong evidence that phonons were the primary excitation from the recombination of excited quasiparticles.

If we now assume that spin fluctuations are the primary pairing excitation, then we would replace the substrate that was a very good phonon propagator with a substrate that would not support the propagation of high-energy phonons but was magnetic and would be an excellent propagator of magnons. Perhaps single-crystal yttrium iron garnet with appropriate impurities could be prepared into such a substrate. The same (e.g. NbN) junction would be placed on one side of this substrate and the same two experiments should be performed. The conventional emitter should give a very small signal, whereas the signal in the studied material should be much larger, an indicator that magnons are the primary recombination excitation.

2.4 Is There an Upper Limit for T_c?

The question of whether or not there exists some upper limit to T_c for the phonon mechanism has an interesting history. Based on the Fröhlich Hamiltonian, which is the sum of the electronic term, the phonon term with an experimentally measured phonon frequency, and the electron–phonon interaction, the following expression can be obtained: $\Omega = \Omega_0(1-2\lambda)^{1/2}$ (Migdal, 1960). Based on this expression, it can be concluded that the value of the coupling constant λ cannot exceed $\lambda_{max} = 0.5$, implying that $T_c \approx 0.1\Omega_0(\Omega_0 \approx \Omega_D)$ is the upper limit of T_c. Indeed, such a point of view was almost generally accepted after the appearance of the BCS theory. However, very soon it became clear that something is wrong with this criterion, since many superconductors were discovered with $\lambda > 0.5$ (e.g. Sn, Pb, and Hg). The problem was clarified later by Brovman and Kagan (1967) and, especially, by Geilikman (1971, 1975).

The expression $\Omega = \Omega_0(1-2\lambda)^{1/2}$ was derived from the Fröhlich model. However, the electron–ion interaction is also crucial for the formation of a realistic phonon spectrum. In other words, in this model, there is double counting. The analysis of the electron–phonon interaction and, correspondingly, the lattice stability has to be carried out with considerable care. This has been done and should be based directly on the adiabatic approximation. The conclusion is that the electron–phonon interaction does not lead to the dependence when the phonon frequency becomes equal to zero at some value of the coupling constant. Rigorous analysis shows that this interaction leads only to the formation of the observed acoustic dispersion law.

The Fröhlich Hamiltonian and the subsequent conclusion about instability is valid for the non-adiabatic case, when $\tilde{\Omega} \gg E_F$; it is irrelevant for the majority of systems. The same is true for the possibility of bipolaronic instability at $\lambda \simeq 1$, which also corresponds to the non-adiabatic case. Indeed, the Holdstein model (eqn (1.31)) was developed for semiconductors with a small concentration.

Another faulty restriction on T_c, based on the McMillan equation (2.22), was proposed later. Indeed, this equation taken at face value leads to an upper limit of T_c. As noted above (section 2.2.2), the maximum value of T_c corresponds to $\lambda = 2$. However, the McMillan equation is valid only for $\lambda \lesssim 1.5$. Therefore, the value $\lambda = 2$ is outside the range of its applicability.

The treatment based on eqns (2.33) and (2.34) leads to a very different conclusion. As noted in the previous section (see Table 2.1), experimentally large values of λ have been determined. for several superconductors. For example, $\lambda \approx 2.1$ for $Pb_{0.65}Bi$, and $\lambda \approx 2.6$ for Am-$Pb_{0.45}Bi_{0.55}$ (Allen and Dynes, 1975; see also the review by Wolf, 2012). And if the material is characterised by relatively large values of both λ and $\tilde{\Omega}$, it might have a very high T_c value.

Recently, high values of T_c have been observed for a novel family of superconducting hydrides under high pressure (see section 7.2). The observed T_c values are close to room temperature ($T_c \simeq 250$ K for La-H hydrides). Because of the large value of the isotope coefficient and based on the ab initio calculations, it is generally accepted that the superconductivity in high-T_c hydrides is caused by the electron–phonon interaction.

As for predicting the limiting value of the T_c, the history of superconductivity has shown that several breakthroughs were contrary to the predictions. In reality, the lattice instability probably exists at some large values of the coupling constants, but this is still an open question. The recent discovery of high-T_c hydrides, with values not distant from room temperature, clearly demonstrates that the strong coupling theory works for coupling strength values of $\lambda \approx 2$–3, which is sufficient for achieving superconductivity at room temperature. We can look forward to this event as there are reasons to expect it in the near future.

2.5 Electronic Mechanisms and the Little Model

Up to this point, we have focused on the phonon mechanism of superconductivity. However, electron pairing can be created not only by phonon exchange but also by the exchange of other excitations. In this and the next chapters, we will discuss the electronic mechanism of superconductivity. The most common (but not unique) instance of such a mechanism arises in the presence of two groups of electronic states. Excitations within one of these groups serve as 'agents' giving rise to pairing in the other group.

The study of electronic mechanisms of superconductivity began with the work of Little (1964). This work also served to initiate the general discussion of the relationship between dimensionality and the superconducting transition. Below, we shall describe some relevant models. At first, we describe the Little model and its development. Other electronic mechanisms will be also described.

2.5.1 Two Electronic Groups, and High T_c

Little (1964) considered the case of a one-dimensional organic polymer. Remarkably, at that time, no superconducting polymers existed; there also were no organic superconductors. The first superconducting polymer was synthesised in 1975 with Little's participation (Gill *et al.*, 1975; Greene *et al.*, 1975). As for organic superconductivity, this phenomenon was discovered in 1980 (Jerome *et al.*, 1980) and at present represents one of the most actively developing areas of research (see e.g. Ishiguro, 2001, and Jerome, 2012; also see section 8.1). Therefore, the paper by Little (1964), in addition to introducing a novel mechanism of pairing, also contains the prediction of superconductivity in organic materials.

In the Little model, the conduction electrons in the primary chain (see Fig. 2.6) are paired thanks to their interaction with electronic excitations in the side branches. The critical temperature is given by a BCS-like expression (2.10). but, in the present case, the pre-exponential factor corresponds to the electronic, rather than ionic, energy scale, so that

$$T_c = \Delta E_{el} \exp\left(-\lambda_{el}\right) \tag{2.48}$$

The fact that $\Delta E_{el.} \gg \tilde{\Omega}$ is responsible for the hope that high critical temperatures may be achieved.

Here it is worth making a brief historical digression. The goal of attaining high-temperature superconductivity culminating in an ideal case of observing the

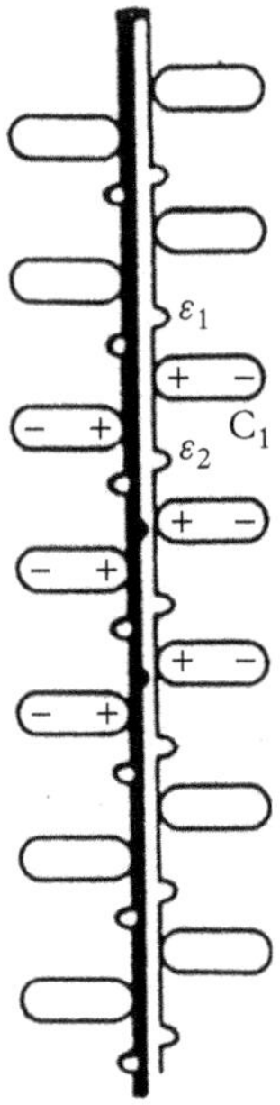

Figure 2.6 *Little's model.*

phenomenon at room temperature was born a long time ago, even before the appearance of the BCS theory. One of the first questions this theory was confronted was that of the upper limit of the critical temperature. In other words, it was essential to understand whether superconductivity is a low-temperature phenomenon or is not subject to such a restriction. The initial answer was, unfortunately, not optimistic. Indeed, as was noted above (section 2.8), in the early 1960s, the common understanding was that the value of the electron–phonon coupling could not exceed $\lambda_{\max}.= 0.5$ and therefore, according to expression (2.21), the critical temperature was not expected to exceed some optimum value $T_{c;max} \approx 0.1\tilde{\Omega}, \tilde{\Omega} \approx \Omega_D$. As a result, the dream of high-temperature superconductivity was thought to be unrealistic. As we have said above, this conclusion was erroneous, and it became clear that, even with phonons, one can reach a high T_c. However, in the early 1960s, this was not thought to be the case. Little's work (published in 1964) revived the dream of high T_c and initiated a search for this phenomenon. This milestone paper contained the following key components:

- The new mechanism of the pairing different from the electron–phonon interaction was prepared. As a result, the search for high-temperature superconductivity had started.
- The paper had resulted in intensive studies of low-dimensioned structures.
- The phenomenon of organic superconductivity was predicted.

It is hard to name another scientific paper which was so rich in its content and made a similar impact (see also Gutfreund, 2016; Little, 2016).

The idea to search for an electronic mechanism is very fruitful, but its realisation is quite non-trivial. Clearly, the pre-exponential factor acts to raise T_c. However, the coupling constant λ also depends sensitively on $\Delta E_{el.}$, and it is essential that this constant not turn out to be too small.

Furthermore, we are really dealing with the difference $\lambda - \mu^*$. Because of the inequality $\Delta E_{el} \gg \tilde{\Omega}$, the logarithmic weakening of the Coulomb repulsion is not as strong as in the phonon case. This factor also decreases the exponent. Finally, certain fundamental problems concerning the superconducting transition arise in low-dimensional cases.

The problem of the intensity of the pairing interaction has been discussed by Gutfreund and Little (1979). According to quantum mechanical calculations, in the ordinary model of the type shown in Fig. 2.6, the average spacing between the conducting chain and the charge localised on the side branches is quite large (~4.5 A). As a result, the value of λ is quite low. As a consequence, a different approach was developed by Little (1983). Here pairing takes place in a chain containing a transition-metal atom chain (see Fig. 2.7). The conduction electrons move in the $d_{z^2} - p_z$ orbital band. Strong coupling occurs due to the overlap of the d_{xz} and d_{yz} orbitals with those of the ligands. In this situation, it is critical that the energy levels of the conduction electron orbitals be close to those of the ligands' electrons. It is also necessary to have very dense packing of the polarisable ligands. All this works to strengthen the coupling and to optimise the delicate balance of factors controlling the electron–electron interaction.

Another important problem has to do with low dimensionality and fluctuations (Ferrell, 1964; Rice, 1965; Hohenberg, 1967). The fact of the matter is that fluctuations of the order parameter destroy long-range order: the correlation function $K(x_1 - x_2)$ vanishes for $|x_1 - x_2| \to \infty$ at finite temperatures. This difficulty can be overcome by creating 'quasi-one-dimensional' systems made up of bundles of superconducting threads (Dzyaloshinski and Katz, 1968). The probability of jumps between the threads is small but sufficient to stabilise the system with respect to fluctuations.

In the one-dimensional case, the Fermi 'surface' is made up of two points at $\pm k_F$. This imposes severe restrictions on the allowed values of transferred momenta. The

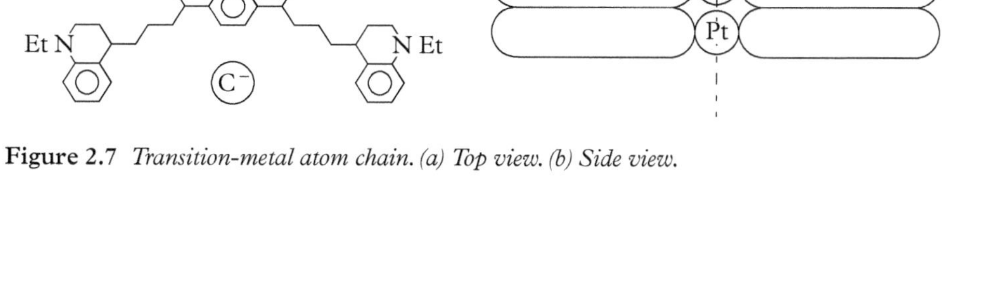

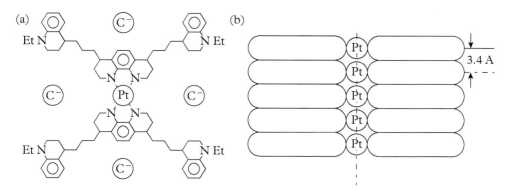

Figure 2.7 *Transition-metal atom chain. (a) Top view. (b) Side view.*

different possible types of instabilities in the one-dimensional case form the subject of *g*-ology (Gutfreund and Little, 1979). One should distinguish various coupling constants: λ_{SS}, λ_{TS}, λ_{CDW}, and λ_{SDW}, where SS and TS correspond to singlet and triplet superconducting states, and CDW and SDW denote transitions with the appearance of CDWs and spin-density waves (SDWs), respectively. A rigorous treatment of various instabilities has been carried out by Bychkov *et al.* (1966) and by Dzyaloshinski and Larkin (1972).

2.5.2 The 'Sandwich' Excitonic Mechanism

As described above, the main feature of the Little model was the presence of two groups of electronic states. Based on this idea, Ginzburg (1965; see also Ginzburg and Kirzhnits, 1982) analysed the case of two-dimensional conducting layers with coupling due to the exchange of excitons in an adjacent dielectric film. The idea is to create a heterogeneous layered structure with alternating metallic and dielectric planes.

A similar geometry was considered by Allender *et al.* (1973), but with semiconducting rather than dielectric layers. Clearly, the effectiveness of such a system is highly dependent on the thickness of the metallic film, which must be on the order of the screening length, that is, it may not exceed 10–15 A. In this model, an important parameter is the penetration depth D of an electron tunnelling from the metallic film into the semiconducting covering. This depth can be shown to be equal to

$$D = \frac{\pi}{3} \frac{\hbar \kappa_F}{2m\Delta} \tag{2.49}$$

Substituting $\Delta \approx 1$ eV and $k_F \approx 1.5$ A^{-1}, we find $D \approx 5$ A. Thus, we are dealing with quite substantial penetration depths.

2.5.3 Three-Dimensional Systems

A system with two different coexisting electron groups can be realised in three dimensions as well. This case was first studied by B. Geilikman (1965, 1971, 1973). Here one has to consider two special cases: (1) one where the group providing the attraction is localised, and (2) one where both groups are delocalised. Let us briefly discuss these two cases.

Speaking of the first case, we can consider, for example, a metal with non-metallic impurities. The interaction of the conduction electrons with the electrons in the non-metal atoms leads to an additional effective interaction between the conduction electrons. This effective interaction has the form

$$\Gamma_{p1,p2;p3,p4} \approx \Gamma'_c + \Gamma''_{p1,p2;p3,p4} \tag{2.50}$$

where

$$\Gamma''_{p_1p_2;p_3p_4} \approx \sum_{n,\lambda,\lambda'} V_{p_1,n\lambda;p_3,n\lambda'} V_{n\lambda',p_2;n\lambda,p_4} \prod_{\lambda\lambda'}$$

$$\prod_{\lambda\lambda'} = \frac{n_\lambda - n_{\lambda'}}{E_\lambda - E_{\lambda'} + \hbar\omega}$$

$$\hbar\omega = \xi_3 - \xi_1$$

and $\boldsymbol{p}_i$ are the momenta of the conduction electrons. The first term in (2.50) corresponds to the usual screened Coulomb repulsion, and the second, to the attractive interaction. Here $V_{p_1,n\lambda;p_3,n\lambda'}$ is the matrix element of the interaction between the delocalised conduction electrons and the localised electrons, λ is the set of quantum numbers of the electron in the non-metal atom at the lattice site n, E_λ is the energy of this electron, and $\prod_{\lambda\lambda'}$ is the polarisation operator. Because of the summation over n, $\Gamma \propto N$, N being the concentration of the non-metal atoms.

Superconducting pairing can arise if the attraction outweighs the Coulomb repulsion. Note that a number of factors serve to decrease the value of the effective coupling constant. These include the exchange interaction, the smallness of the ratio a/L (where a is the radius of the orbit of the localised electrons, and L is the lattice parameter), and so on. In addition, the spacing $\Delta E = E_{loc}$ should be sufficiently small. On the other hand, this smallness leads to a decrease in T_c; see eqn (2.48). Therefore, there exists an optimum value for ΔE. It appears that ΔE should be of the order of 0.1-0.5 eV. Values of such a scale are possible, for example, in the case of fine-structure levels.

It should be pointed out that not only electronic but also vibrational states can be localised. For example, a molecule can be placed on the surface of a thin film or introduced into the sample. Then the interaction of the carriers with molecular vibrations can provide additional inter-electron attraction (Kresin, 1974). As a result, the presence of such molecules leads to an increase in T_c. This effect has been observed (Gamble and McConnell, 1968). Inclusion of organic molecules (tetracyanoguinodimethane) raised the value of the critical temperature of Al films up to $T_c = 5.24$ K.

A special case is that of a superconducting molecular crystal. Systems of this type have lattices made up of molecules with their own internal electronic and vibrational levels.

It is interesting to consider the case of pairing provided both by the ordinary phonon mechanism and by an additional interaction with localised electronic states. Making use of the step method (cf. section 2.2.2), one finds the following in the weak coupling approximation:

$$T_c \cong \tilde{\Omega} e^{\frac{1}{g_0}}$$

where

$$g_0 = g_{ph} - (g_e - g_c) \Big/ \left[1 - (g_e - g_c) \ln\left(\Delta E_a / \hbar\tilde{\Omega}\right)\right] \tag{2.51}$$

Note that the difference $g_e - g_c$ enters the expression for g_0 with a minus sign. This means that the additional attraction is logarithmically enhanced.

The case when both electronic groups are delocalised is realised, for example, in a metal with two overlapping bands ('*a*' and '*b*' electronic states) of very different widths, for example, an *s*-band and a *p*-band. Virtual electronic transitions in the narrow band caused by the Coulomb interaction of the '*a*' and '*b*' electrons, can lead to pairing in the wide band. The calculations (Geilikman, 1965, 1973) were performed within the random-phase approximation; they are at least of qualitative interest when applied to real metals. One can consider also multi-valley semiconductors, where the large value of the dielectric constant ε makes RPA more applicable (Geilikman and Kresin, 1968a). The calculation leads to the following expression for the effective interaction of the '*a*' electrons:

$$\Gamma_{aa} = \left(V_{aa} + \prod_{aa} R\right) S^{-1}, \qquad \Gamma_{bb} = \left(V_{bb} + \prod_{bb} R\right) S^{-1}$$

$$S = 1 + V_{aa}\prod_a + V_{bb}\prod_b + \prod_a\prod_b \boldsymbol{R}, \quad \boldsymbol{R} = V_{ab}^2 - V_{aa}V_{bb} \tag{2.52}$$

Here $\prod_{aa}$ and $\prod_{bb}$ are the polarisation operators for the '*a*' and '*b*' electrons, and V_{ik} are the Coulomb matrix elements. Conditions under which the effective interaction becomes attractive turn out to be rather non-trivial.

Systems with overlapping bands can give rise to low-lying collective branches (acoustic plasmons). The exchange of these collective excitations also can lead to pairing.

2.6 Plasmons in Layered Conductors

The plasmon mechanism implies that pairing occurs via the exchange of plasmons; in other words, plasmons play a role similar to that of phonons. In principle, the pairing can be provided by contributions of both channels, that is, by phonons and plasmons.

Let us make the following general remark. Lately, we have been witnessing the discovery of many new superconducting materials: high-temperature cuprates, fullerides, borocarbides, ruthenates, MgB_2, metal-intercalated halide nitrides, intercalated Na_xCoO_2, and so on. Systems such as organics, heavy fermions, and nanoparticles have also been intensively studied. Many of these systems belong to the family of layered (quasi-two-dimensional) conductors and characterised, for example, by strongly anisotropic transport properties. An interesting question raised by the observation of superconductivity in all the systems is the following: why is layering a favourable factor for superconductivity? In connection with this interesting question, one should stress that layering leads to peculiar dynamic screening of the Coulomb interaction; this is important for the description of the superconducting state in layered conductors. As was discussed above (section 1.2.3), layered conductors have a plasmon spectrum that differs fundamentally from three-dimensional metals. In addition to having a high-energy 'optical' collective mode, the spectrum also contains an important low-frequency part (see chapter 1, Fig. 1.9). The screening of the Coulomb interaction is incomplete, and the *dynamic* nature of the Coulomb interaction becomes important. The contribution of the plasmons *in conjunction* with the phonon mechanism may lead to high values of T_c.

As was noted above (section 1.2.3), we consider a layered system consisting of a stack of conducting sheets along the z-axis, where the sheets are separated by dielectric spacers. Because of the large anisotropy of the conductivity, it is a good approximation to neglect transport between the layers. On the other hand, the Coulomb interaction between charge carriers is effective both *within* and *between* the sheets. To ensure charge neutrality, we further introduce positive countercharges spread out homogeneously over the sheets. In other words, the description of layered conductors can be made by neglecting the small interlayer hopping, in a first approximation. On the other hand, it is essential to take into account the screened interlayer Coulomb interaction, which has an important dynamic part. Indeed, it is known that, for usual three-dimensional materials, this interaction can be considered in the static limit, since electronic collective modes are very high in energy. The optical plasmon energies are of the order of 5–30 eV in metals (see e.g. Ashcroft and Mermin, 1976). Therefore, Coulomb repulsion enters the theory of superconductivity as a single constant pseudopotential μ^*. The situation is very different in layered conductors: incomplete screening of the Coulomb interaction results from the layering. The response to a charge fluctuation is time dependent, and the frequency dependence of the screened Coulomb interaction becomes important. This leads to the presence of *low*-energy electronic collective modes: the acoustic plasmons. It is this particular feature of layered materials that brings about an additional contribution to the pairing interaction between electrons.

In order to calculate the superconducting critical temperature T_c, one can use the following equations for the superconducting order parameter and the renormalisation function, respectively:

$$\phi_n(\mathbf{k}) = T\sum_{m=-\infty}^{\infty}\int\frac{dk}{(2\pi)^3}\Gamma\left(\mathbf{k},\mathbf{k}';\omega_n-\omega_m\right)\frac{\phi_m(\mathbf{k}')}{\omega_m^2+\xi_{k'}^2}\bigg|_{T_c} \tag{2.53}$$

$$\omega_n(\mathbf{k}) - \omega_n = T\sum_{m=-\infty}^{\infty}\int\frac{d^3\mathbf{k}'}{(2\pi)^3}\times\Gamma\left(\mathbf{k},\mathbf{k}';\omega_n-\omega_m\right)\frac{\omega_m(\mathbf{k}')}{\omega_m^2(\mathbf{k}')+\xi_m^2}\bigg|_{T_c} \tag{2.53$'$}$$

Here $\phi_n(\boldsymbol{k}) = \Delta_n(\mathbf{k})Z, \omega_n(\boldsymbol{k}) = \omega_n Z, \Delta_n \equiv \Delta(\omega_n)$; we shall not write out the expression for Z. The interaction kernel can be written as a sum of electron–phonon and Coulomb interactions. The Coulomb term contains the plasmon excitations (see eqns (1.35) and (1.37)).

We focus on the case where the interaction kernel is a sum of electron–phonon and Coulomb interactions. Then, the total kernel $\Gamma \equiv \Gamma(\mathbf{q},\omega_n-\omega_m)$, with $\mathbf{q} = \mathbf{k}-\mathbf{k}'$, is written in the form

$$\Gamma = \Gamma_{ph} + \Gamma_c \tag{2.54}$$

with

$$\Gamma_{ph}(\mathbf{q}; |n-m|) = |g_v(\mathbf{q})|^2 D(\mathbf{q}, |n-m|) = |g_v(\mathbf{q})|^2 \frac{\Omega_v^2(\mathbf{q})}{(\omega_n - \omega_m)^2 + \Omega_v^2(\mathbf{q})}$$

and

$$\Gamma_c(\mathbf{q}; |n-m|) = \frac{V_c(\mathbf{q})}{\epsilon(\mathbf{q}, |n-m|)}$$

$D(\mathbf{q}, |n-m|)$ is the phonon Green's function, and $\Omega_v^2(\mathbf{q})$ is the phonon dispersion relation; summation over phonon branches v is assumed. The Coulomb term Γ_c is written in its most general form as the ratio of the bare Coulomb interaction $V_c(\mathbf{q})$ and the dielectric function $\epsilon(\mathbf{q}, |n-m|)$ (see eqns (1.35′) and (1.37)). The plasmons appear in the region: $\omega > q_{\|} V_F$ (region I). As for the region II ($\omega < q_{\|} V_F$), it corresponds to the static response. Then the total vertex Γ contains two parts: Γ_c^{I} and Γ_c^{II}. We focus on the term Γ_c^{I}, which is dynamic since it is determined by the contribution of plasmons. Therefore, the contribution of Γ_{pH} and Γ_c^{II} can be treated by analogy with an analysis of conventional superconductors and corresponding evaluation of two parameters: λ (the electron–phonon coupling constant) and μ^* (the Coulomb pseudopotential). In other words, we assume that the strength of the electron–phonon interaction is sufficient to overcome the Coulomb repulsion. As for the term Γ^{I}, it provides an additional and sometimes (see below) very essential attractive contribution leading to a noticeable increase in T_c.

It is worth noting that such an approach is different from studies where the authors try to calculate the difference between the attractive (Γ_c^{I}) and the repulsive (Γ_c^{II}) terms to decide whether or not the materials make the transition into the superconducting state (see e.g. Takada, 1993). However, such calculations are unrealistic. Indeed, as noted above (section 2.1.1), the criterion of superconductivity is determined by small difference ($\sim T_c$) of two large quantities; each of them is of order of ε_F. However, both terms can be calculated with use of approximations (like RPA), and their accuracies are not sufficient for such calculation.

The approach, presented above, is based on the assumption that the Coulomb repulsion (the static part, Γ_c^{II}) is overcomed by the electron–phonon coupling. It makes it possible to consider the term Γ_c^{I} (the contribution of the plasmons) as an additional attraction and estimate its contribution to the superconducting state of various systems. In other words, we assume that the phonons are sufficient for the occurrence of the superconducting state, which makes it possible to use the conventional approach for the static term. The contribution of the low-frequency plasmons can be studied for various layered systems (see sections 5.4.4 and 7.4.3) with the use of normal-state parameters (e.g. v_F and ϵ_M).

Therefore, layered materials have distinctive *low-energy* electronic collective excitations that provide exchange bosons for the pairing between electrons. These acoustic plasmons lead to enhancement of the superconducting T_c. Based on eqn (2.53) and numerical calculations for the phonon–plasmon model, an increasing influence of the electronic pairing mechanism for the three classes of layered superconductors can be demonstrated (Bill *et al.*, 2003). In metal-intercalated halide nitrides, the contribution arising from low-energy electronic collective modes is dominant. Thus, these materials are unique: the exchange bosons are made of the same particles (the electrons) as those that bind into pairs below the T_c. In the case of organic layered materials, the electronic and phonon energies as well as the structure of the conducting layers and insulating spacers lead to a situation in which the contribution of the phonons and acoustic plasmons is of the same order. Finally, in the case of high-temperature superconductors, the contribution of low-energy plasmons is significant, but not dominant. All of these systems will be described later on (chapters 5–8).

An essential conclusion is that the physics of layered superconductors cannot be reduced to the study of one conducting layer (or the layers belonging to one unit cell, as in some high-temperature superconductors). Such simplification relies on the observation of 'quasi-two-dimensional' transport. However, it does not account for the screening properties of the electron–electron Coulomb interaction. As discussed above, the screening is very different in layered materials than in two-dimensional and three-dimensional isotropic metals. The particular screening properties are essential for the behaviour of layered conductors. The size of the effect of the screened Coulomb interaction depends very much on the specific features of the material. For example, the covalency within the conducting layers, the structure of the spacers, and the presence of van der Waals gaps will determine its contribution to both the normal- and the superconducting-state properties. Therefore, the study of screening properties in layered conductors is a promising direction to take to better understand the similarities and differences between different classes of materials and will serve as a bridge to analyse the properties of two-dimensional and three-dimensional systems.

We discussed above (section 1.2.3) the appearance of acoustic plasmons ('demons') for three-dimensional systems with very different overlapping bands.

The possibility of superconducting pairing due to the exchange of 'demons' has been considered in a number of papers (Garland, 1963; Geilikman, 1966; Fröhlich, 1968; Ihm *et al.*, 1981); see also the review by Ruvalds (1981).

2.7 Magnetic Mechanisms

2.7.1 Introduction

The origin of the phonon mechanism of pairing is the Coulomb interaction between an itinerant electron and a localised ion. The magnetic mechanism of pairing also results from the part of Coulomb interaction that depends on spins, and is known as an exchange interaction. Via this interaction, one electron violates the spin polarisation of surrounding spins, and another electron detects this violation. In quantum field theoretical language, both interactions are given by a second-order Feynman diagram for two electrons interacting via the emission and absorption of bosons, phonons in the former case, and magnons in the latter case. The basic idea that elementary excitations arising from magnetic degrees of freedom (magnons) may affect superconducting pairing between electrons was discussed by Akhiezer and Galitsky (1959) and Vonsovskii and Izyumov (1962) soon after the publication of the BCS theory. They showed that exchange by magnons in ferromagnetic metals results in *p*-type pairing. The effect of magnon-mediated pairing was considered also in papers by Berk and Schrieffer (1966) and Doniach and Engelsberg (1966), who showed that incipient ferromagnetic order could suppress phonon-induced *s*-wave pairing. Pairing in ^{3}He associated with nearly ferromagnetic spin fluctuations in the *p*-wave channel was proposed by Anderson and Brinkman (1973). For the heavy fermion materials, *d*-wave pairing due to nearly antiferromagnetic spin-fluctuations was proposed by Scalapino *et al.* (1986–7) and Miyake *et al.* (1986). The review of early papers on magnetic correlations and magnetic pairing in cuprates had been done by Kampf (1994).

We note at the outset that pairing interactions based on the repulsive ($U > 0$) single-band Hubbard model lead to *d*-wave pairing (Scalapino *et al.*, 1986). It is naturally not surprising that, after the superconducting cuprates were found to be high-temperature superconductors (Bednorz and Mueller, 1986) with parent compounds showing antiferromagnetic order, various extensions of this result were applied to them and have led to a breathtaking advance in understanding the manifestations of magnetic interactions in highly correlated materials. In the following, we present a very cursory overview of this huge field, and refer the interested reader to a few of the original publications in this field as well as to existing summaries specifically aimed at spin-fluctuation-induced superconductivity.

The majority of papers discussing the magnetic mechanism of pairing use the Hubbard model as a starting Hamiltonian, and then use diagrammatic, perturbative, or numerical techniques to study the physical consequences, such as various instabilities (SDWs, superconductivity, energy gap formation, etc.), one- and two-particle Green's functions, and their relationship to experiments. Such an approach can demonstrate the importance of magnetic mechanism of pairing and give a very general picture of unusual properties of cuprates. Nevertheless, a more detailed understanding of unusual normal and superconducting properties of cuprates without doubt requires a more realistic description of its electronic structure and a combination of phonon, magnetic, and other possible mechanisms of pairing. This will be discussed later on in section 5.3.

2.7.2 Localised versus Itinerant Aspects of Electrons in Transition Metals and Compounds

The conventional wisdom of the band theory of solids is that band electrons are somehow renormalised almost-free electrons, like almost-free particles with modified dispersion law $\varepsilon(k)$. This single-particle approach works rather well in metals and semiconductors. For $3d$- and $4f$-electrons in transition metals and compounds, the situation is more complicated. These materials form a class of so-called narrow-band materials. The bandwidth is a measure of the kinetic energy; a small bandwidth means that the kinetic energy is rather small. While, in conventional metals, the kinetic energy of electrons is much higher than their potential energy from the Coulomb interaction, and the interaction can be considered via a perturbation theory, in narrow-band materials, kinetic energy may be of the same order of magnitude or even smaller than the Coulomb one. In this situation, full or partial localisation of electrons may occur, resulting in the duality for $3d$-electrons that may be partially localised and partially itinerant. The problem of the duality of the d-electrons in the Fe-group $3d$-metals has been known for a long time. Herring (1960) discussed how some of the properties of these ferromagnets can be considered from the itinerant point of view, and others from the localised one. In $3d$-metal ionic compounds like oxides, where the $3d$-electrons are more separated by anions, the tendency for electron localisation is even more pronounced. The simplest model incorporated itinerant and localised properties of d-electrons is the Hubbard (1963) model. This model is a particular version of a single-band tight-binding model where only one Coulomb matrix element for two electrons which have opposite spins but are located in the same ion is considered. We will discuss this model in different regimes below. The Hamiltonian of the Hubbard model,

$$H = \sum_{i\sigma} \left\{ (\varepsilon - \mu) n_{i\sigma} + \frac{U}{2} n_{i\sigma} n_{i\bar{\sigma}} \right\} + \sum_{i \neq j, \sigma} t_{ij} a_{i\sigma}^{+} a_{j\sigma} \tag{2.55}$$

describes a system of electrons in orbital non-degenerates states (s-like) at each site i of the crystal lattice with single-electron energy ε and chemical potential μ and which contains two parameters: the interatomic hopping t (kinetic energy with the bandwidth $2W$, in a simplest version of hopping t to z nearest neighbour, with the half-bandwidth $W = zt$) and the intra-atomic Coulomb interaction U (potential energy). Three different regimes may be outlined: (1) the band limit $W \gg U$, (2) the atomic limit $W \ll U$, and (3) the intermediate limit $W \sim U$.

In the band limit, the model describes conventional metal with electron concentration $0 \leq n \leq 2$ and weakly interacting electrons. The only difference is that neglecting the long-range Coulomb interaction results in the incorrect description of the charge density fluctuations: instead, the plasmon excitations with plasma frequency at $q = 0$ in the metal in the Hubbard model have zero sound excitations with zero energy at $q = 0$. Nevertheless, this drawback of the Hubbard model is not important for the magnetic pairing in the Hubbard model. The conventional perturbation theory can be applied and may result in the SDW state (Slater, 1961; Overhauser, 1962). If the Fermi surface has a special nesting symmetry providing the electron–hole symmetry, the SDW at half-filling with concentration $n = 1$ occurs for arbitrary small repulsion U and results in the metal–insulator transition (Kozlov and Maksimov, 1965). Deviation of the electron concentration from half-filling (doping) may result in a coexistence of SDW and superconductivity (Kopaev, 1982). Detailed calculations of the spin fluctuation exchange interaction with full momentum and frequency dependence were carried out within FLEX approximation to the Hubbard model (Bickers *et al.*, 1989). Beyond the repulsive spin fluctuation exchange interaction, there is a higher-order attractive contribution considered under the so-called spin-bag concept by Schrieffer *et al.* (1989). The magnetic pairing in the band limit will be discussed below in the next section.

In the opposite atomic limit for the half-filled model, electrons are localised, and the antiferromagnetic Heisenberg exchange model is a more appropriate starting approach (Anderson, 1963). The ground state for the alternant lattices (which can be separated into two equivalent sublattices, A and B, so that all atoms from A are surrounded by atoms from B, and vice versa, examples being a two-dimensional square lattice and a three-dimensional simple cubic lattice) is the antiferromagnetic Mott–Hubbard insulator. Deviations of electron concentration from $n = 1$ by electron $n = 1 + x$ or hole $n = 1 - x$ doping produce some extra carriers that may hop from site to site and provide some conductivity. Since $U/W \gg 1$, the conventional perturbation theory and diagram technique cannot be used, and theoretical description of this strongly correlated matter is a very complicated and not fully resolved problem. Some perturbation approaches using the small parameter t/U will be discussed in section 2.7.4.

In the intermediate case $W \sim U$, there is no small parameter, and this is the most difficult situation for a theory. The Mott–Hubbard insulator–metal transition is expected in this case (Mott, 1949; Hubbard, 1965; Mott, 1974). Only numerical approaches can be used to interpolate between weak and strong correlations, which will also be discussed in section 2.7.4, as well as the exact solution of the one-dimensional Hubbard chain by Lieb and Wu (1968).

2.7.3 Magnetic Pairing in the Band Limit

Pairing from spin fluctuation exchange. Just one year after the Bednorz and Muller (1986) discovery of superconductivity in cuprates, Bickers *et al.* (1987) suggested a singlet $d_{x^2-y^2}$ pairing induced by antiferromagnetic spin fluctuations. This idea was based on the previous study of superconductivity in heavy fermion materials by Scalapino *et al.* (1986–7) and Miyake *et al.* (1986). In the band limit, we start from a system of free electrons with a tight-binding dispersion $t(k)$ given by a Fourier transform of the interatomic hopping. The Hamiltonian (2.55) with $U = 0$ after the Fourier transform

$$a_{\boldsymbol{k}\sigma} = \frac{1}{\sqrt{N}} \sum_{\boldsymbol{f}} a_{\boldsymbol{f}\sigma} \exp\left(i\boldsymbol{k}\boldsymbol{r}_{\boldsymbol{f}}\right)$$

becomes diagonal:

$$H_0 = \sum_{\boldsymbol{k}\sigma} \left(\varepsilon\left(\boldsymbol{k}\right) - \mu\right) a^{+}_{\boldsymbol{k}\sigma} a_{\boldsymbol{k}\sigma}) \tag{2.56}$$

with the electron band energy $\varepsilon\left(\boldsymbol{k}\right) = \varepsilon + t\left(\boldsymbol{k}\right)$, where

$$t\left(\boldsymbol{k}\right) = \sum_{\boldsymbol{g}} t_{\boldsymbol{f}\boldsymbol{g}} \exp(i\boldsymbol{k}\left(\boldsymbol{r}_{\boldsymbol{f}} - \boldsymbol{r}_{\boldsymbol{g}}\right) \tag{2.57}$$

In the normal paramagnetic phase, there are spin fluctuations (paramagnons); electrons interacting with the spin fluctuations results in the exchange interaction considered by Berk and Schrieffer (Fig. 2.8).

At small doping away from half-filling, the bare susceptibility $\chi_0(\boldsymbol{q},\omega)$ given by an electron–hole bubble peaks at an incommensurate wavenumber $\boldsymbol{Q}$ near (π, π) and the effective particle–particle interaction $V(\boldsymbol{k},\boldsymbol{k}')$ is repulsive (Kampf, 1994). To understand how repulsive interaction may result in pairing, it is necessary to examine the spatial Fourier transform $V(\mathbf{l})$ of the effective interaction (Fig. 2.9).

In spite of the total repulsion, it is clear from Fig. 2.9 that the anisotropic attraction takes place along the x and y directions. It results in the superconducting d-pairing with the gap $\Delta_{\mathbf{k}} = 0.5\Delta_0(\cos k_x - \cos k_y)$. This gap satisfies the BSC-type equation

$$\Delta_{\boldsymbol{k}} = -1/N\sum_{\boldsymbol{p}} V(\boldsymbol{k},\boldsymbol{p})\,\Delta_{\boldsymbol{p}}/2E_{\boldsymbol{p}}, E_{\boldsymbol{k}} = \sqrt{(\epsilon_{\boldsymbol{k}}^2 + \Delta_{\boldsymbol{k}}^2)}, \tag{2.58}$$

Due to the repulsive interaction $V(\boldsymbol{k},\boldsymbol{p}) > 0$, a solution of this equation requires that the gap change sign on the Fermi surface. The gap is zero in the nodal direction $kx = ky$.

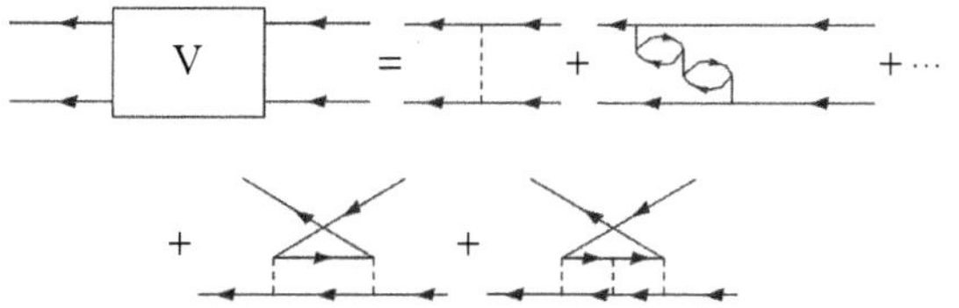

Figure 2.8 *The Berk–Schrieffer diagrams for the spin fluctuation exchange interaction. The dashed line denotes the Coulomb repulsion U. (From Kampf, 1994)*

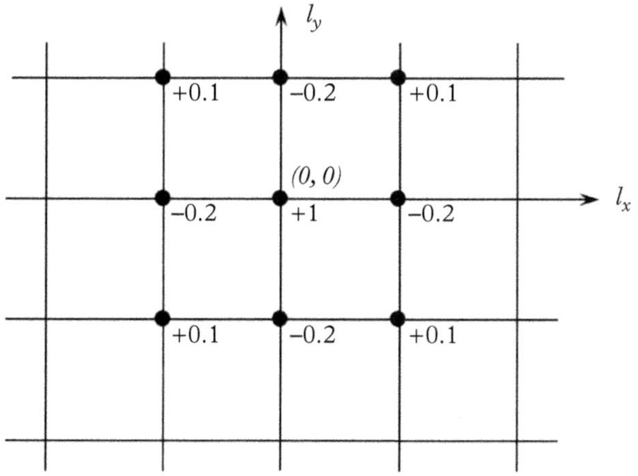

Figure 2.9 *Spatial variation $V(l_x,l_y)$ for the square lattice. (From Kampf, 1994)*

The spin-bag model. If we consider electrons in the SDW state below the Néel temperature T_N, their band structure differs from the paramagnetic phase, and the pairing interaction may be also different. This situation has been discussed by Schrieffer *et al.* (1989) in the spin-bag model. The magnetic polaron physics is captured in this model, in which one carrier locally disturbs the antiferromagnetic order by its own spin, forming a spin bag. Another electron within the coherence length ξ_{AFM} of the local distortion (where AFM denotes antiferromagnetic) experiences an attractive interaction by sharing the bag. For weak interaction, $U \ll W$, quite good approximation is given by the Hartree–Fock solution with the decoupling of the Coulomb interaction:

$$U n_{f\uparrow} n_{f\downarrow} \to U n_{f\uparrow} \langle n_{f\downarrow} \rangle + U \langle n_{f\uparrow} \rangle n_{f\downarrow}$$

The Hartree–Fock electron energy depends on the electron occupation numbers and spin projection

$$\varepsilon_{HF}(\boldsymbol{k},\sigma) = \epsilon(k) + U \langle n_{-\sigma} \rangle \tag{2.59}$$

In the antiferromagnetic phase sublattice magnetisation $\langle S^z \rangle$ for half-filled bands, the static SDW results in the insulator with dispersion of electrons at the bottom of the empty conductivity band and at the top of the filled valence band given by

$$E^{\pm}(\boldsymbol{k}) = \varepsilon \pm \left(t^2(\boldsymbol{k}) + U^2 \langle S^z \rangle^2 \right)^{1/2} \tag{2.60}$$

The effective particle–particle interaction in the spin-bag model is given by a set of diagrams similar to those in Fig. 2.8, with the difference being that required spin susceptibilities have to be calculated in the broken symmetry SDW state with the electronic band structure given by eqn (2.60). Since, in the spin-bag model, the reduced SDW amplitude in the vicinity of the hole is important for the pairing, Schrieffer *et al.* (1989) have discussed the exchange interaction induced by longitudinal χ^{zz} spin fluctuations. The exchange by transverse χ^{+-} spin fluctuations is equally important for a complete picture of the effective interaction in the SDW state (Frenkel and Hanke, 1990). The detailed analysis of this interaction made by Schrieffer *et al.* (1989), Vignale and Singwi (1989), Frenkel and Hanke (1990), and in the review paper by Kampf (1994) has revealed that transverse interaction $V^{zz}(\boldsymbol{k},\boldsymbol{p})$ provides the attraction for small momentum transfer $\boldsymbol{k} - \boldsymbol{p} = \boldsymbol{q} \sim 0$, as was assumed in the spin-bag idea. At large $\boldsymbol{q} \cong \boldsymbol{Q}$, the interaction is repulsive, like the paramagnon contribution in the paramagnetic phase. The antiperiodicity of the pairing interaction $V^{zz}(\boldsymbol{k},\boldsymbol{p}) = -V^{zz}(\boldsymbol{k}+\boldsymbol{Q},\boldsymbol{p}) = -V^{zz}(\boldsymbol{k},\boldsymbol{p}+\boldsymbol{Q})$ implies that the d-wave gap satisfies the gap equation, like eqn (4.4). The spatial Fourier transform of the transverse interaction $V^{+-}(\boldsymbol{k},\boldsymbol{p})$ shows that it does not lead to pairing in the singlet channel.

Similar spin-bag analysis can be extended to the short-order antiferromagnetic correlations that exist also in the paramagnetic metal phase of the doped two-dimensional Hubbard model. Here the pairing mechanism originates from the same effects that

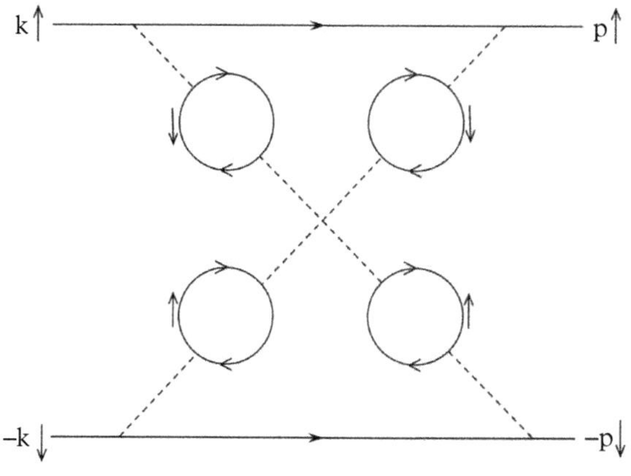

Figure 2.10 *Lowest-order spin-bag contribution to the irreducible pairing interaction in the paramagnetic phase of the Hubbard model. (From Kampf and Schrieffer, 1990)*

result in the formation of the pseudogap in the single-particle density of states. The effective interaction appears in the sixth order perturbation represented by the crossed-line diagram (Fig. 2.10).

This diagram contains an attractive contribution for small momentum transfer (Kampf and Schrieffer, 1990).

The self-consistent theory of d-wave superconductivity from the band limit. In the single-band Hubbard model, the same electrons participate in the formation of the antiferromagnetic fluctuations and provide the electron–electron coupling considered above. That is why the electronic self-energy and the magnetic susceptibility have to be calculated self-consistently. Several groups have realised this programme by a perturbation approach from the band limit (Tewordt, 1979; Bickers *et al.*, 1989; Lenek *et al.*, 1990; Bickers and White, 1991; Wermbter and Tewordt, 1991; Lenek *et al.*, 1994; Monthoux and Scalapino, 1994; Pao and Bickers, 1994). For electronic self-energy, the effective interaction has been written in the RPA. In the superconducting phase, the electronic self-energy in the Nambu representation with the Pauli matrices τ_α ($\alpha = 0, 1, 2, 3$) is given by the following equation (Lenek *et al.*, 1994):

$$\sum(\boldsymbol{k}, i\omega_n) = i\omega_n \,[1 - Z(\boldsymbol{k}, i\omega_n)]\,\tau_0 + \xi(\boldsymbol{k}, i\omega_n)\,\tau_3 + \Phi(\boldsymbol{k}, i\omega_n)\,\tau_1 \qquad (2.61)$$

Here Z is the renormalisation factor for the frequency $\omega(\boldsymbol{k}, i\omega_n) = \omega_n \; Z(\boldsymbol{k}, i\omega_n)$, $\xi(\boldsymbol{k}, i\omega_n)$ determines the energy shift, and the function $\Phi(\boldsymbol{k}, i\omega_n)$ gives the superconducting gap

$$\Delta(\boldsymbol{k}, i\omega_n) = \Phi(\boldsymbol{k}, i\omega_n)\,/Z(\boldsymbol{k}, i\omega_n) \qquad (2.62)$$

Three non-linear equations for the functions $\omega(\boldsymbol{k}, i\omega_n), \xi(\boldsymbol{k}, i\omega_n)$, and $\Phi(\boldsymbol{k}, i\omega_n)$, together with the RPA equations with dynamical susceptibilities and the chemical potential equation, describe the superconducting phase for a given set of parameters t, U, electronic concentration n, and temperature T. The numerical solution by Lenek *et al.* (1994) for the parameter set $t = 0.1$ eV, $U = 4.28t$, $n = 0.84$ (2.63) has proved the existence of d-wave superconductivity with a momentum-dependent gap $\sim \cos k_x - \cos k_y$ below $T_c = 0.016t$. At $T = 0.6\ T_c$, the maximal value of the gap is equal to $2\Delta_{max}/T_c = 7.7$. The very important difference from the BCS theory is the importance of all wave *numbers* for magnetic pairing, not only the narrow region close to the Fermi level.

The analytical continuation of the Matsubara magnetic susceptibility $\chi(\boldsymbol{q}, i\omega_m)$ on the real frequency axis results in the magnetic response being determined like

$$\lim_{\omega \to 0} \frac{\chi(\boldsymbol{q}, i\omega_m = \omega + i\delta)}{\omega} \tag{2.64}$$

For the parameters set (2.63) at $T = 0.91\ T_c$, this response function has four sharp peaks near the wave number (π, π), indicating strong antiferromagnetic fluctuations. Indeed, for $U = 4.30t$, the antiferromagnetic phase stabilises (Lenek *et al.*, 1994). A similar conclusion was obtained by Pao and Bickers (1994). The general conclusion from all of the papers cited above is that the self-consistent description of the electronic spectrum and the magnetic susceptibility in the band limit $U \ll W \sim zt$ makes it possible to describe d-wave superconductivity with T_c up to 60 K; nevertheless, this solution appears for the parameters $U \sim zt$, at the border of applicability of the band limit.

Conclusions. Nowadays, the magnetic mechanism of pairing is more often involved in discussions of superconductivity in cuprates. Nevertheless, we want to emphasise here that it is a rather general phenomenon that has been used for liquid ^{3}He, heavy fermions, Sr_2RuO_4, and Fe-pnictides. Some other examples are related to astrophysics (neutron stars) and nuclear physics. With respect to cuprates, the relevant regime for superconductivity starts at small values of the carrier concentration n_c, near the half-filled band, which characterises the antiferromagnetic starting materials La_2CuO_4 and $YBa_2Cu_3O_6$. The central idea of the models we have described is that the motion of the carriers introduced by doping is dominated by the proximity to antiferromagnetic order in the CuO_2 planes. Superconductivity within the band limit of the two-dimensional Hubbard model occurs via the exchange of paramagnons, and leads to a d-wave order parameter for the superconducting gap function. Nevertheless, the dome-like doping dependence of the superconducting critical temperature T_c, as well as many other unusual features of the rich phase diagram of hole-doped and electron-doped cuprates, was not reproduced within perturbation theory approaches.

The intra-atomic Coulomb repulsion U between two holes on the same site is assumed to remain the dominant energy scale ($U > 4t$). Indeed, for $3d$ elements, the atomic value for U is 15–20 eV. Due to the screening effect in a crystal, it may be significantly reduced. Typical values are $U = 4$ eV and $t = 0.4$ eV, with $U/t = 10$ (Hybertsen *et al.*, 1992). That is why the band limit is hardly able to describe physics of cuprates.

Nevertheless, the principal demonstration of possible superconducting pairing with a magnetic mechanism has been done in this limit. In the next section, we will discuss the magnetic mechanism of pairing in the regime of strong electron correlations, the atomic limit $U \gg W$, and the intermediate case $U \sim W$.

2.7.4 Magnetic Pairing in the Hubbard Model in the Regime of Strong Electron Correlations

Electronic properties of the Mott–Hubbard insulator. In the absence of hopping, $t_{fg} = 0$, the model describes a system of individual sites, each of which can be in one of the following eigenstates:

$$|0\rangle = \text{vacuum}; \ |\sigma\rangle = a^{+}_{f\sigma}|0\rangle, \sigma = \pm 1/2; \ |2\rangle = a^{+}_{f\uparrow}a^{+}_{f\downarrow}|0\rangle \tag{2.65}$$

with energies $\varepsilon_0 = 0, \varepsilon_1 = \varepsilon - \mu, \varepsilon_2 = 2\varepsilon_1 + U$. The number of electrons per site is equal to

$$n = \frac{1}{N}\sum_{f}\langle\widehat{N_f}\rangle = \frac{1}{N}\sum_{f\sigma}\langle a^{+}_{f\sigma}a_{f\sigma}\rangle = N_+ + N_- + 2N_2 \tag{2.66}$$

Here N_+, N_-, and N_2 are the occupation numbers of states (2.65) with spin up, spin down, and two electrons, respectively. The total energy is given by $E = \sum_{f\sigma}\varepsilon_1 N_{f\sigma} + \sum_f \varepsilon_2 N_{f2}$. In general, some sites are occupied by electrons with spin up or down (paramagnetic sites), others are in diamagnetic singlet states $|0\rangle$ and $|2\rangle$, and the whole system may have different magnetic properties. The half-filled band with $n = 1$ is a special case because each site has spin ½. The ground state is highly degenerate, and this degeneracy may be eliminated by weak intersite hopping. To obtain the quasiparticle spectrum, one should calculate the single-electron Green's function

$$\begin{aligned} G_\sigma(\boldsymbol{k}, E) &= \langle\langle a_{\boldsymbol{k}\sigma}|a^{+}_{\boldsymbol{k}\sigma}\rangle\rangle_E \\ &= \sum_{(\boldsymbol{R}_f - \boldsymbol{R}_{f'})} exp(-i\boldsymbol{k}\cdot(\boldsymbol{R}_f - \boldsymbol{R}_{f'}))\langle\langle a_{\boldsymbol{f}\sigma}|a^{+}_{\boldsymbol{f}'\sigma}\rangle\rangle_E \end{aligned} \tag{2.67}$$

An exact equation of motion for the Green's function is given by

$$\begin{aligned} (E - \varepsilon + \mu)\langle\langle a_{\boldsymbol{f}\sigma}|a^{+}_{\boldsymbol{f}'}\rangle\rangle &= \delta_{\boldsymbol{f}\boldsymbol{f}'} + \sum_{\boldsymbol{h}}\langle\langle a_{\boldsymbol{f}+\boldsymbol{h},\sigma}|a^{+}_{\boldsymbol{f}'}\rangle\rangle\, t(\boldsymbol{h}) + UF_\sigma(\boldsymbol{f} - \boldsymbol{f}', E) \\ F_\sigma(\boldsymbol{f} - \boldsymbol{f}', E) &= \langle\langle a_{\boldsymbol{f},\sigma}n_{\boldsymbol{f},-\sigma}|a^{+}_{\boldsymbol{f}'}\rangle\rangle \end{aligned} \tag{2.68}$$

and involves a higher-order Green's function F. The exact equation of motion for the F function has the form

$$(E-\varepsilon+\mu)\langle\langle \mathrm{a}_{\mathbf{f},\sigma}\mathrm{n}_{\mathbf{f},-\sigma}|\mathrm{a}^{+}_{\mathbf{f}',\sigma}\rangle\rangle \tag{2.69}$$
$$=\delta_{\mathbf{ff}'}\langle \mathrm{n}_{\mathbf{f},-\sigma}\rangle+U\langle\langle \mathrm{a}_{\mathbf{f},\sigma}\mathrm{n}_{\mathbf{f},-\sigma}|\mathrm{a}^{+}_{\mathbf{f}',\sigma}\rangle\rangle+\sum_{\mathbf{h}} t(\boldsymbol{h})(\langle\langle \mathrm{a}_{\mathbf{f}+\mathbf{h},\sigma}\mathrm{n}_{\mathbf{f},-\sigma}|\mathrm{a}^{+}_{\mathbf{f}',\sigma}\rangle\rangle$$
$$+\langle\langle \mathrm{a}^{+}_{\mathbf{f},-\sigma}\mathrm{a}_{\mathbf{f}+\mathbf{h},-\sigma}\mathrm{a}_{\mathbf{f},\sigma}|\mathrm{a}^{+}_{\mathbf{f}',\sigma}\rangle\rangle+\langle\langle \mathrm{a}_{\mathbf{f}-\mathbf{h},-\sigma}\mathrm{a}_{\mathbf{f},\sigma}\mathrm{a}_{\mathbf{f},-\sigma}|\mathrm{a}^{+}_{\mathbf{f}',\sigma}\rangle\rangle)$$

The simplest decoupling in the last three terms of eqn (2.69) is known as the Hubbard I decoupling (Hubbard, 1963):

$$\langle\langle \mathrm{a}_{\mathbf{f}+\mathbf{h},\sigma}\mathrm{n}_{\mathbf{f},-\sigma}|\mathrm{a}^{+}_{\mathbf{g},\sigma}\rangle\rangle\rightarrow\langle \mathrm{n}_{\mathbf{f},-\sigma}\rangle\,\langle\langle \mathrm{a}_{\mathbf{f}+\mathbf{h},\sigma}|\mathrm{a}^{+}_{\mathbf{g},\sigma}\rangle\rangle$$

In the Hubbard I approximation, the system of equations for functions G and F is closed and, after the Fourier transformation, it may be written as

$$\begin{aligned}(E-\varepsilon(\boldsymbol{k})+\mu)G_\sigma(\boldsymbol{k},E)&=1+UF_\sigma(\boldsymbol{k}.E)\\(E-\varepsilon-U+\mu)F_\sigma(\boldsymbol{k},E)&=\langle n_{-\sigma}\rangle+\langle n_{-\sigma}\rangle t(\boldsymbol{k})G_\sigma(\boldsymbol{k},E)\end{aligned} \tag{2.70}$$

Here the space uniform solution $\langle \mathrm{n}_{\mathbf{f},-\sigma}\rangle=\langle \mathrm{n}_{-\sigma}\rangle$ is assumed. The single-electron Green's function (2.67) is given by

$$\begin{aligned}G_\sigma(\boldsymbol{k}.E)&=\frac{E-\varepsilon-U(1-\langle n_{-\sigma}\rangle)+\mu}{(E-E_1(\boldsymbol{k}))(E-E_2(\boldsymbol{k}))}\\E(\boldsymbol{k})&=\varepsilon+\frac{U+t(\boldsymbol{k})}{2}\pm\left[\left(\frac{U-t(\boldsymbol{k})}{2}\right)^2+Ut(\boldsymbol{k})\langle n_{-\sigma}\rangle\right]^{1/2}-\mu.\end{aligned} \tag{2.71}$$

In the paramagnetic half-filled case $\langle n_{-\sigma}\rangle=1/2, \mu=\varepsilon+U/2$, the quasiparticle spectrum has the form

$$E_{1,2}(\boldsymbol{k})=\frac{1}{2}\left(t(\boldsymbol{k})\pm\left(U^2+t^2(\boldsymbol{k})^{1/2}\right)\right) \tag{2.72}$$

The subbands $E_1(\boldsymbol{k})$ and $E_2(\boldsymbol{k})$ are called the Hubbard subbands. The Hubbard I solution has been subjected to criticism in the literature because it contains a gap between the $E_1(\boldsymbol{k})$ and $E_2(\boldsymbol{k})$ subbands for all values of U while the metal–insulator transition and the disappearance of the gap are intuitively expected at $U\sim W$. Nevertheless, in the atomic limit $U\gg W$, the Hubbard I solution is a good starting approximation. In this limit, one can write the energies of the Hubbard subbands as (neglecting second-order t^2/U contribution)

$$E_1(\boldsymbol{k})=\frac{1}{2}(t(\boldsymbol{k})+U),\;\;E_2(\boldsymbol{k})=\frac{1}{2}(t(\boldsymbol{k})-U) \tag{2.73}$$

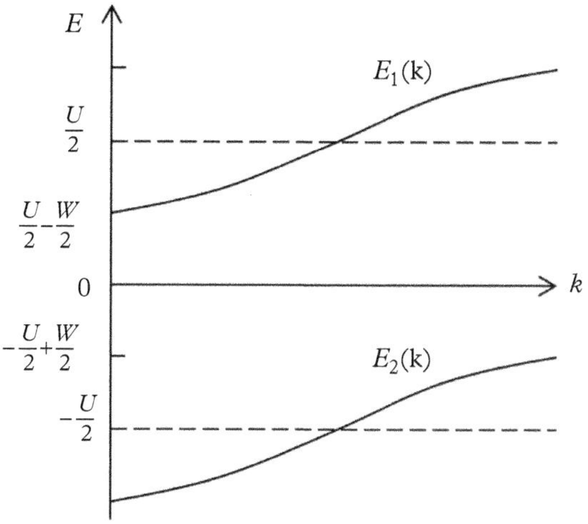

Figure 2.11 *Scheme of the Mott–Hubbard insulator band structure in the atomic limit $U \gg W$ for the half-filled Hubbard model.*

The direct gap for optical excitations is U, while the insulator gap $E_g = U - W$ (Fig. 2.11). Note also that the bandwidth of the Hubbard bands is W, two times smaller than the free-electron bandwidth $2W$.

In Fig. 2.12 we show a scheme of the electronic density of states $N(E)$ for the half-filled Hubbard model in the weak correlation band limit $U \ll W$, the intermediate case $U \sim W$, and in the strong correlation atomic limit $U \gg W$. In the band limit, one has a half-filled metal; in the atomic limit, the Mott–Hubbard insulator occurs with a gap that goes to zero when W tends to U. Of course, this is not a solid theoretical result, just a hint that corresponds to the expected transition from insulator to metal. Several approximations have been developed to obtain the metal–insulator transition: Hubbard III (1964), the Gutzwiller approximation (1963, 1964, 1965), coherent potential approximation (Velicky *et al.*, 1968), single-loop approximation in the X-operator diagram technique (Zaitsev, 1976), and dynamical mean field approximation in the infinite dimension (Metzner and Vollhardt, 1989; Georges *et al.*, 1996), cluster dynamical mean field theory (cluster-DMFT; Stanescu and Kotliar, 2006), and different numerical methods that will be discussed later on in section 2.7.4. Still, the properties of the Hubbard model at $W \sim U$ are far from being clear. Experimentally, such transitions are known for many materials (Mott, 1974); the Coulomb parameter U is the intra-atomic value that is supposed to be constant, while the half-bandwidth $W = zt$ depends on the interatomic distance and may be increased by external pressure or chemical pressure due to ionic substitution.

One important distinction between the Hubbard subbands and the free-electron bands is that, for the former, the number of states in a given subband depends on the electron occupation numbers, while, for conventional single-electron bands, the total

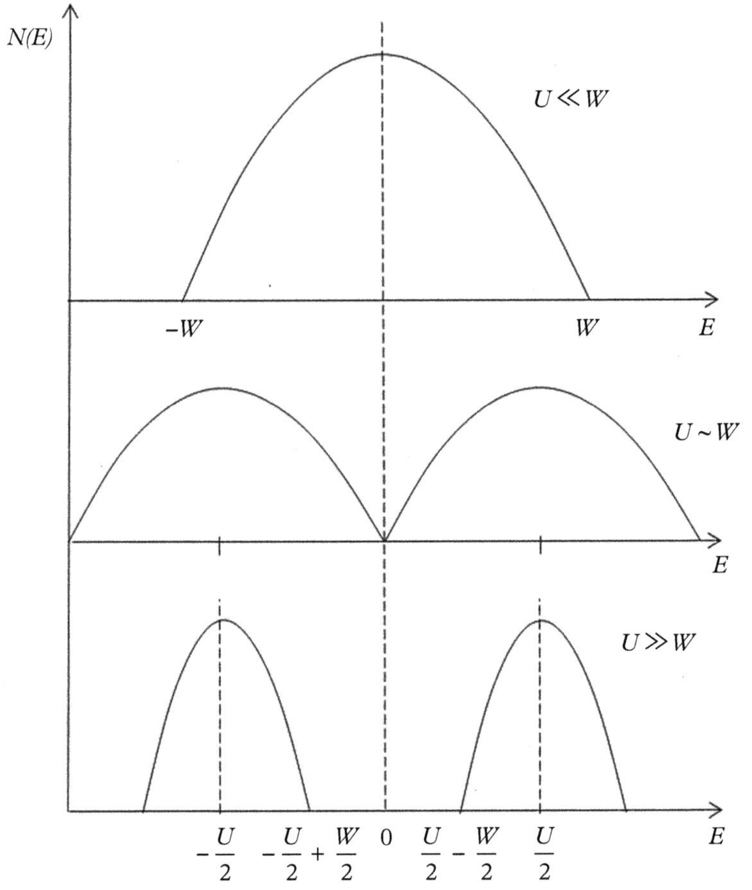

Figure 2.12 *Scheme of the electronic density of states: metal in the band limit, insulator in the atomic limit, and interpolation of the atomic limit to the intermediate case.*

number of states in each band is equal to 2, taking the spin into account. To prove this peculiarity, we introduce the single-electron spectral density function

$$A_\sigma(\boldsymbol{k}, E) = (-1/\pi)\, ImG_\sigma(\boldsymbol{k}, E + i\delta)$$

which, in the Hubbard I approximation with the Green's function (2.61), is given by the sum of the upper and lower Hubbard band contributions:

$$\begin{aligned} A_\sigma(\boldsymbol{k}.E) &= \frac{E_1(\boldsymbol{k}) - \varepsilon - U(1 - \langle n_{-\sigma}\rangle) + \mu}{E_1(\boldsymbol{k}) - E_2(\boldsymbol{k})}\delta(E - E_1(\boldsymbol{k})) \\ &+ \frac{U(1 - \langle n_{-\sigma}\rangle) + \varepsilon - E_2(\boldsymbol{k}) - \mu}{E_1(\boldsymbol{k}) - E_2(\boldsymbol{k})}\delta(E - E_2(\boldsymbol{k})) \\ &\equiv A_{1\sigma}(\boldsymbol{k}.E)\,\delta(E - E_1(\boldsymbol{k})) + A_{2\sigma}(\boldsymbol{k}.E)\,\delta(E - E_2(\boldsymbol{k})) \end{aligned} \tag{2.74}$$

It is clear from eqn (2.74) that $A_{1\sigma} + A_{2\sigma} = 1$, that is, only the sum from both bands and over spin projection will be equal to two. For the separate bands, the density of states is equal to

$$N_{1\sigma,2\sigma}(E) = \frac{1}{N}\sum_{\boldsymbol{k}} A_{1\sigma,2\sigma}(\boldsymbol{k}, E)$$

Integration over energy results in the following equation (Fulde, 1991):

$$N_{1\sigma} = \int N_{1\sigma}(E)dE = \langle n_{-\sigma}\rangle, \quad N_{2\sigma} = \int N_{2\sigma}(E)dE = 1 - \langle n_{-\sigma}\rangle$$

For the half-filled case $N_{1\sigma} = N_{2\sigma} = 1/2$, with spin taken into account, the total number of states in each Hubbard band is one. This fact is often used to justify the claim that, in the limit $U \to \infty$, an electron becomes a spinless particle. However, this is not true, because there is a twofold spin degeneracy in the paramagnetic phase, and the Zeeman splitting takes place in the external magnetic field. The situation with the quasiparticle statistics is more complicated.

The Hubbard model in the one-dimension and infinite-dimension cases. For the one-dimensional case $D = 1$, there is an exact solution by Lieb and Wu (1968). The ground state energy is given for the half-filled case $n_e = 1$ by

$$\frac{E_0}{N} = -4|t| \int_0^{\infty} \frac{dx\, J_0(x) J_1(x)}{x\left[1 + exp\left(xU/|t|\right)\right]} \tag{2.75}$$

where $\mathcal{J}_n(x)$ is the Bessel function. The ground state energy in the band limit $U \ll t$ is

$$\frac{E_0}{N} = -4\frac{|t|}{\pi} + \frac{U}{4} - 0.017\frac{U^2}{|t|}$$

In the atomic limit $U \gg t$, it is

$$\frac{E_0}{N} = -\frac{4|t|^2}{U} ln2$$

The last expression is exactly the energy of the $D = 1$, Heisenberg $S = 1/2$ antiferromagnetic chain with the intersite exchange interaction $\mathcal{J} = 4t^2/U$. The ground state for $n_e = 1$ is an insulator for all U/t values. It is proved by the fact that the chemical potential required to add one electron μ_+ is larger than the potential to remove one electron μ_-. Relation $\mu_+ > \mu_-$ is typical for an insulator with an energy gap between the filled and the empty bands.

The exact wave function of the ground state is given by the Bethe ansatz and has a rather complicated structure that may be simplified in the limit $U \to \infty$ when all two-electron states at one site can be excluded. It results in the following wave function (Ogata and Shiba, 1990):

$$\Psi\left(x_1, \cdots, x_N\right) = det\left|exp\left(ik_j, x_j\right)\right| \Phi\left(y_1, \cdots, y_M\right) \tag{2.76}$$

where $x_1, \ldots, x_M$ are the spin-up electron sites, and $x_{M+1}, \ldots, x_N$ are the spin-down electron sites. In the right side of eqn (2.76), the Slater determinant describes the non-interacting spinless fermions with moment $k_1, \ldots, k_N$, and the function Φ in eqn (1.28) is the exact wave function of the Heisenberg $D = 1$ chain with $S = 1/2$, where $y_1, \ldots, y_M$ are the 'pseudocoordinates' of the spin-down particles.

Factorisation of the wave function (2.76) into the product of two factors, with one corresponding only to the charge degree of freedom and the other corresponding only to the spin variables, is a very important result. Sometimes this is claimed to be the general property of strongly correlated electronic systems and is used in two dimensions. However, it is doubtful that this result is true for a lattice with dimension $D > 1$. Even having a small transverse hopping $t_\perp$ between two parallel chains does not allow one to reproduce the exact solution (2.76) as the zero approximation with some perturbative corrections by $t_\perp$, as no exact solution is known for such a model, which is called a two-leg ladder.

The other non-trivial solution of the Hubbard model in the infinite dimension $D \to \infty$ gives no indication for the charge-spin separation at the level of one electron. The $D = 3$ systems seems to be appropriate to start with, using the $D \to \infty$ limit and the $1/D$ expansion. The most unclear situation is for $D = 2$ systems.

The low-energy excitations in the $D = 1$ Hubbard model have been studied in detail. Instead of producing the Fermi liquid, the strong correlations result in the Luttinger liquid (Tomonaga, 1950; Luttinger, 1968; Solyom, 1979; Haldane, 1981). For the Luttinger liquid, the distribution function $n_k = \langle a_k^+ a_k \rangle$ has no jump at the Fermi level:

$$n_k - n_{kF} \sim sign\left(k - k_F\right) \left|k - k_F\right|^\alpha$$

where α is a non-universal constant and the value of α is determined by the interaction constant. In the limit $U \to \infty$, $\alpha = 1/8$ (Parolla and Sorella, 1990). Instead of having the delta function (as in the Fermi liquid), the spectral density in the Luttinger liquid near the Fermi level has the form $A(\omega) \sim |\omega - \varepsilon_F|^\alpha$. Zero spectral density at ε_F in the one-dimensional Hubbard model was the main reason for abandoning the description of the electron as a particle with charge e and spin 1/2 in strongly correlated electronic systems (Anderson, 1990). Meanwhile, quantum Monte Carlo (QMC) calculations (Preuss *et al.*, 1994) have revealed the quasiparticle peak in $A(k, \omega)$ typical for the Fermi liquid. In the limit $U = \infty$, the spectral density $A(k,\omega)$ has been calculated (Penc *et al.*, 1995) using the exact wave function (2.76). The multi-electron effects decrease the Van Hove singular contributions to $A(k, \omega)$ at the band edge, and a new singularity, $A(\omega) \sim |\omega - \varepsilon_F|^{-3/8}$, opposite to the Luttinger liquid expectations (for electron concentration $n_e < 1$) has been

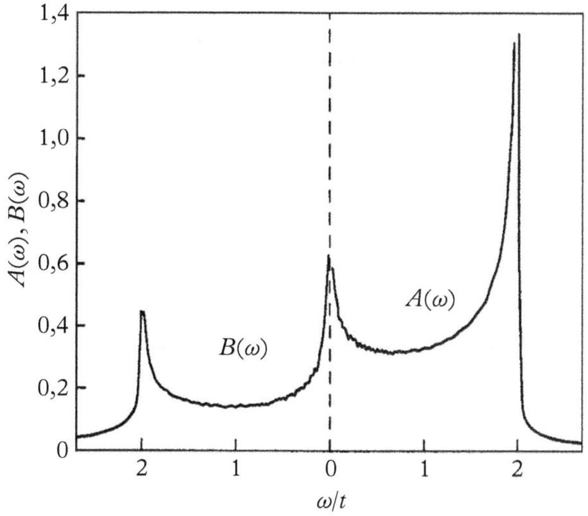

Figure 2.13 *Density of states of the lower Hubbard band for $n_e = 1/2$ and $U = \infty$, for the one-dimensional Hubbard model. (From Penc et al., 1995)*

found (Fig. 2.13). The non-quasiparticle tails in the spectral density above the bandwidth are also the result of strong electron correlations.

The main contributions to $A(\boldsymbol{k}, \omega)$ (more than 99.99%) are given by the sum of the single-particle excitations and the three-particle excitations involving the electron–hole pairs. The applicability of the Luttinger liquid ideas has also been analysed by Penc *et al.* (1995) and they have found $\alpha = 1/8$ in the region $|\omega - \varepsilon_F| < \frac{t^2}{U}$; outside this region, $\alpha = -3/8$. When $U \to \infty$, the region of the Luttinger liquid tends to zero; so, at large but finite U, the spectral density $A(\omega)$ has its peak at $\omega \sim \frac{t^2}{U}$ because of the singularity with $\alpha = -3/8$. This result explains the peak in $A(\omega)$ obtained previously by QMC (Preuss *et al.*, 1994). We want to emphasise that, even in the limit $U \to \infty$ the spectral weight will be zero exactly at the Fermi level, while a minimal energy shift from the Fermi level would provide the peak in the spectral weight and destroy the Luttinger liquid.

In the other limit, $D \to \infty$, the non-trivial solutions for the Hubbard model have been found by Metzner and Vollhardt (1989). In this limit, the self-energy becomes local (Muller-Hartmann, 1989):

$$\sum_{fg} (\omega) = \sum (\omega)\ \delta_{fg} \tag{2.77}$$

This makes it possible to construct a non-trivial dynamical mean field theory (DMFT) (which is exact in the $D \to \infty$ limit) that covers the main results expected in $D = 3$: the metal–insulator transition at $U \sim W$, the antiferromagnetic order with the local magnetic moments in the half-filled Mott–Hubbard insulator at $U \gg W$, and the Fermi-liquid behaviour in the metal phase at $U \ll W$. The large difference from Fig. 2.12 at $W \sim U$ is that the DMFT has predicted a narrow Kondo-like peak at the Fermi level, with the electron effective mass increasing near the border of localisation. There are several reviews devoted to the $D = \infty$ limit (Izyumov, 1995; Georges *et al.*, 1996; Gebhard, 1997). Much of the literature was sceptical, considering the Kondo peak to be an artefact of DMFT, one caused by neglecting the wavenumber dependence of the self-energy. Indeed, cluster-DMFT, which takes into account nearest-neighbour spatial correlations, has not revealed the Kondo peak (Kyung *et al.*, 2006). More complicated evolution of the density of states occurs with increasing U/t ratio (Fig. 2.14).

At $U = 4t$, the paramagnetic phase is metal, while, in the antiferromagnetic state, there is an SDW gap. At $U = 5t$, a small (compared to the SDW gap) dip appears at the Fermi level of the paramagnetic phase, but the paramagnetic phase is still metal. At $U = 6t$, an insulator gap appears for both phases, and two sharp bands begin to develop inside the large gap between two Hubbard bands centred around $+U/2$ and $-U/2$. Comparison of the paramagnetic and antiferromagnetic densities of states at $U = 6t$ and $U = 8t$ has revealed that strong local spin correlations within 2×2 clusters results in a paramagnetic band structure that is very similar to the antiferromagnetic one. That is why more simple calculation of the antiferromagnetic band structure within the Hubbard I approximation gives the splitting of each paramagnetic Hubbard band into two antiferromagnetic one (Ovchinnikov and Shneyder, 2003), as in the more complicated cluster-DMFT theory.

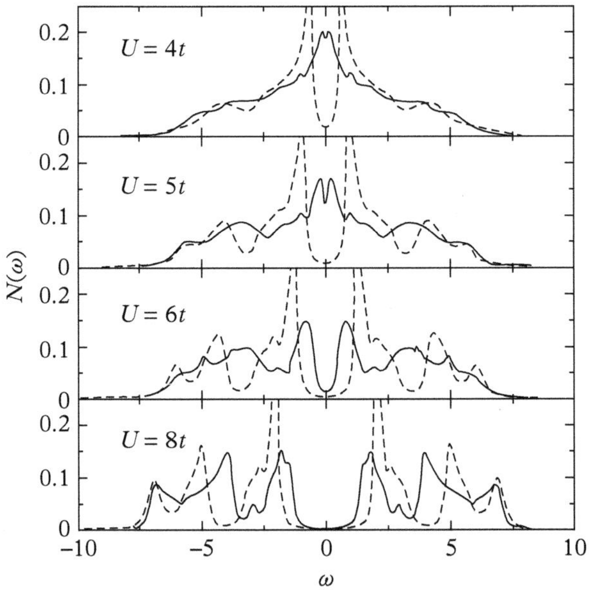

Figure 2.14 *Density of states for the two-dimensional half-filled Hubbard model for several values of U: solid line, paramagnetic solution; dashed line, antiferromagnetic solution. (From Kyung et al., 2006)*

The $t-\mathcal{J}$ model. This model was derived from the Hubbard model in the limit $t/U \ll 1$. Using the canonical transformation to exclude the local two-electron states, we get the following Hamiltonian (Bulayevsky *et al.*, 1968; Chao *et al.*, 1977; Hirsch, 1985; Gros *et al.*, 1987):

$$H_{t-J^*} = -t \sum_{\langle i,j \rangle \sigma} \left(c_{i\sigma}^{+} c_{j\sigma} + c_{j\sigma}^{+} c_{i\sigma} \right) + \frac{t^2}{U} \sum_{j,\delta,\delta'} \left(c_{j+\delta\uparrow}^{+} c_{j\downarrow}^{+} c_{j\downarrow} c_{j+\delta'\uparrow} + c_{j\uparrow}^{+} c_{j+\delta\downarrow}^{+} c_{j+\delta'\downarrow} c_{j\uparrow} + c_{j+\delta\uparrow}^{+} c_{j\downarrow}^{+} c_{j+\delta'\downarrow} c_{j\uparrow} + c_{j\uparrow}^{+} c_{j+\delta\downarrow}^{+} c_{j\downarrow} c_{j+\delta'\uparrow} \right) \tag{2.78}$$

where $\mathbf{j}+\boldsymbol{\delta}'$ are the nearest neighbours to the site $\mathbf{j}$. Neglecting all three-site terms and taking $\boldsymbol{\delta}' = \boldsymbol{\delta}$, we obtain the $t-\mathcal{J}$ model Hamiltonian with $J = 4t^2/U$:

$$H_{t-j} = -t \sum_{(i,j)\sigma} \left(c_{i\sigma}^{+} c_{j\sigma} + h.c. \right) + \sum_{(i,j)} J \left(\boldsymbol{S}_i \boldsymbol{S}_j - n_i n_j/4 \right) \tag{2.79}$$

Here the Fermi operators are defined in a restricted Hilbert space without double-electron occupations at one site:

$$c_{i\sigma} \to c_{i\sigma} \left(1 - n_{i,-\sigma} \right)$$

The three-site term in eqn (2.78) provides its contribution only in the case of non-zero numbers of electrons (or holes). It may be rewritten as follows:

$$H_j^{(3)} = -\frac{t^2}{U} \sum_{j,\sigma} \sum_{\delta \neq \delta'} \left(c_{j+\delta,\sigma}^{+} n_{j,-\sigma} c_{j+\delta',\sigma} - c_{j+\delta,\sigma}^{+} c_{j,-\sigma}^{+} c_{j,\sigma} c_{j+\delta',-\sigma} \right) \tag{2.80}$$

where the first term corresponds to the intrasublattice hopping that does not affect the antiferromagnetic order, and the second term describes the hopping between the next-nearest neighbours accompanied by the spin fluctuation. Both of these hoppings bring forth the so-called $t - t' - t'' - \mathcal{J}$ model (Belinicher *et al.*, 1996). We want to emphasise that, in the limit $t \ll U$, the effective model for the Hubbard model is the $t-\mathcal{J}^*$ model, which includes both the conventional $t-\mathcal{J}$ Hamiltonian (2.79) and the three-site correlated hopping (2.80).

To estimate the role of the three-site terms in eqn (2.78) and the accuracy of the $t-\mathcal{J}$ model, the energy of the ground state was computed by the exact diagonalisation method for the $D = 1$ chain with ten atoms, and the $D = 2$ lattice cluster with ten atoms, in the framework of a few models: the Hubbard model, the $t-\mathcal{J}$ model without carriers, the $t-\mathcal{J}$ model, the $t-\mathcal{J}^*$ model, and the Hubbard model with one carrier (von Szczepanski *et al.*, 1990). It turns out that, at $U/t = 20$, all models give the same value of energy, with differences less than 1%. With U/t decreasing, for example, at $U/t = 8$, the difference in the $D = 2$ cluster is more than 20%, which is a clear indication to take into account correlated hopping terms (2.80) in the effective Hamiltonian. Another comparison of the $t - J$ model, the Hubbard model, and the $t - J^*$ model was carried within the cluster perturbation theory (CPT), where the two-dimensional square lattice was covered by a set of 2×2 clusters, with exact diagonalisation within each cluster, and perturbation treatment of the intercluster hopping (Kuzmin *et al.*, 2014). It is clear from Fig. 2.15 that the central and the right columns are practically identical in dispersion, with a small difference in the spectral intensity near the M point far from the Fermi level, while the difference with the $t - J$ model (left column) is more pronounced, both in the dispersion and in the spectral intensity. Of course, the difference is only quantitative, not qualitative. Similar conclusions were obtained later with 4×4 clusters by Wang (2015).

Strictly speaking, the $t - J$ model is deduced from the Hubbard model at $J \ll t$. Nevertheless, due to the non-trivial properties of the $t - J$ model, one often considers arbitrary values of J/t. Some justification for a broader consideration of the t/J model has been provided by Eder *et al.* (1996), who showed that the additional Coulomb matrix elements V_{ij} can result in instability with respect to the CDW state. A system close to this instability may be a realisation of the $t - J$ model with $J \sim t$.

The quasiparticles in the $t - J$ model and in the atomic limit of the Hubbard model have been intensively discussed in the literature; see, for example, the reviews by Fulde

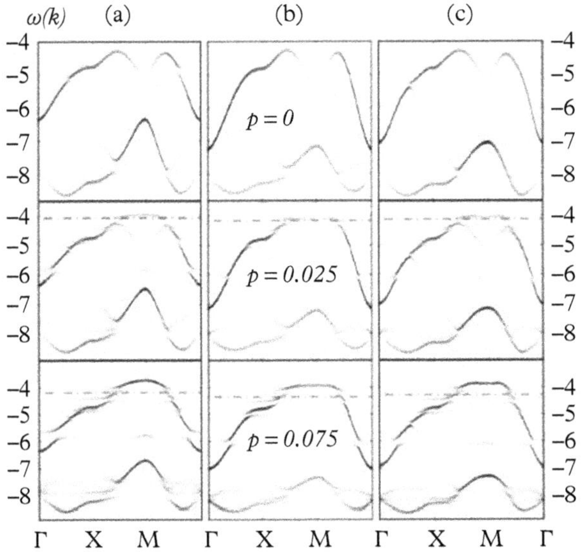

Figure 2.15 *Comparison of the lower Hubbard band dispersion for doping levels* $p = 0, p = 0.025,$ *and* $p = 0.075$ *within the cluster perturbation theory at* $U = 12, t = 1$ *in (a) the* $t - J$ *model, (b) the Hubbard model, and (c) the* $t - J^*$ *model. The dash-dotted line denotes the chemical potential;* $\Gamma = (0, 0), X = (\pi, 0),$ *and* $M = (\pi, \pi)$*. The intensity of the lines corresponds to the quasiparticle spectral density. (From Kuzmin et al., 2014)*

(1991), Dagotto (1994), Izyumov (1995), Kampf (1994), Brenig (1995), Ovchinnikov (1997), and Maier *et al.* (2005a), and the papers by Gull and Millis (2015) and by Braganca *et al.* (2018). In most approaches, the quasiparticle is a spin polaron: it is an electron dressed by the spin fluctuations. The distinction between the free electron and the spin polaron has been demonstrated by Eder and Ohta (1994), who calculated the spectral density $A(\boldsymbol{k}, \omega)$ for the $t - \mathcal{J}$ model via the exact diagonalisation method. The annihilation operator of a polaron destruction can be written in the following form:

$$c_{k,\uparrow} = \frac{1}{\sqrt{N}} \sum_{j} \sum_{\lambda=0}^{3} e^{i\mathbf{k}\cdot\mathbf{R}_j} \alpha_\lambda(\mathbf{k}) A_{j,\lambda} \tag{2.81}$$

$$A_{j,0} = c_{j,\uparrow}, \quad A_{j,1} = \sum_{h\in N(j)} S_j^- c_{h,\downarrow}, \qquad A_{j,2} = \sum_{h\in N(j)} \sum_{l\in N(h)} S_j^- S_h^+ c_{l,\uparrow}$$

$$A_{j,3} = \sum_{h\in N(j)} \sum_{l\in N(h)} \sum_{m\in N(l)} S_j^- S_h^+ S_l^- c_{m,\downarrow}$$

Here $N(j)$ denotes a set of nearest neighbours to $\boldsymbol{R}_j$, and $\alpha_\lambda(\boldsymbol{k})$ are the variational parameters. For the free electron, the spectral density has a wide spectrum for almost all $\boldsymbol{k}$ (Fig. 2.16) of non-coherent origin; this means that the free-electron state is not the

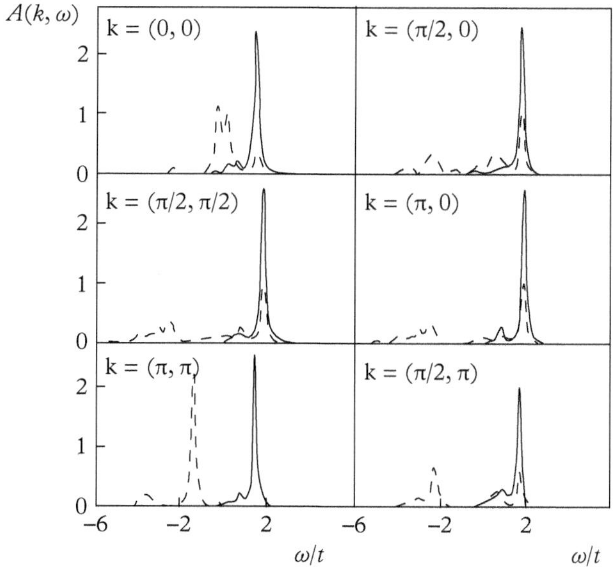

Figure 2.16 *The spectral functions of a free electron (dashed line) and a spin polaron (solid line) for the $t - \mathcal{J}$ model with the electron concentration $n_e = 1$ obtained by the exact diagonalisation of a 4×4 cluster. (From Eder and Ohta, 1994)*

proper state of the system. Conversely the narrow peaks in $A(\boldsymbol{k}, \omega)$ for the spin polaron mean that it is a well-defined quasiparticle for the $t - \mathcal{J}$ model.

The Hubbard model in an X-operator representation. In the atomic limit, it is convenient to use the X operators (Hubbard, 1965). It turns out that the X-operator representation is also useful for a broader range of different models (Ovchinnikov and Val'kov, 2004), so we will define X operators in a more general form then was done by Hubbard. Let us consider any local Hamiltonian $H_{\mathbf{f}}$, which may describe a unit cell centred at a site **f** with several atoms per cell.

Let $\{|p\rangle\} = \{|1\rangle, |2\rangle, \cdots |n\rangle\}$ be a complete set of orthogonal and normalised eigenstates of H_f with the eigenenergy E_n. This set forms a local basis at the cell **f** of the Hilbert subspace H_n, and all physical operators corresponding to the given cell are the operators in H_n. The total number of linear independent operators in H_n is equal to n^2. We take these operators in the form of the Hubbard X operators

$$X_{\mathrm{f}}^{p,q} = |p\rangle\langle q| \tag{2.82}$$

where the indices p and q denote the corresponding eigenstates of H_f ($1 \le p \le n, \le 1 \le q \le n$). By the action of $\mathbf{X}_{\mathbf{f}}^{\mathbf{p},\mathbf{q}}$ on the state $|l\rangle$, we have

$$X^{p,q}|l\rangle = |p\rangle\langle q|l\rangle = \delta_{ql}|p\rangle \tag{2.83}$$

From eqn (2.83), it is obvious that $X_{\boldsymbol{f}}^{p,q}$ is a transition operator that transforms the initial state $|q\rangle$ into the final state $|p\rangle$. The X-operators algebra is given by the following rules:

- the multiplication rule:

$$X_{\boldsymbol{f}}^{p_1,q_1} X_{\boldsymbol{f}}^{p_2,q_2} = |p_1\rangle\langle q_1||p_2\rangle\langle q_2| = \delta_{q_1 p_2} X_f^{p_2,q_2}$$

- the commutation rule:

$$\left[X_{\boldsymbol{f}}^{p_1,q_1} X_{\boldsymbol{g}}^{p_2,q_2}\right]_{\pm} = \delta_{\boldsymbol{fg}}\left(\delta_{q_1 p_2} X_f^{p_1,q_2} \pm \delta_{p_1 q_2} X_f^{p_2,q_1}\right)$$

- the sum rule (completeness of the basis set):

$$\sum_{\mathrm{p}} X_{\boldsymbol{f}}^{p,p} = \sum_{p} |p\rangle\langle p| = 1$$

The diagonal operators $X_{\boldsymbol{f}}^{n,n}$ are the projection operators

$$(X^{n,n})^2 = X^{n.n} X^{n,n} = X^{n,n} \tag{2.84}$$

while the non-diagonal operators have no projection properties as $(X^{p,q})^2 = 0$.

One very useful property of X operators is that any local operator $\widehat{Q}_f$ in H_n can be rewritten in the X-operator representation by using the identity

$$\widehat{Q} = \widehat{1} \cdot \widehat{Q} \cdot \widehat{1} = \sum_{pq} |p\rangle\langle p|\widehat{Q}|q\rangle\langle q| = \sum_{pq} \langle p|\widehat{Q}|q\rangle X^{p,q} \tag{2.85}$$

where the matrix element $\langle p|\widehat{Q}|q\rangle$ gives the probability of the system under the action of the operator $\widehat{Q}$ going from the initial state $|q\rangle$ to the final state $|p\rangle$. It is not convenient to describe each X operator by the pair of indices (p, q). Therefore, to simplify the notation, we introduce an ordered set of n^2 elements (p, q), for example, as follows:

$$(11), (12), \ldots (1\mathrm{n}), (21), (22), \ldots, (2n), \ldots, (n1), (n2), \ldots, (nn)$$

We make a one-to-one correspondence of this set to the other set of n^2 elements,

$$\lambda_1, \lambda_2, \ldots, \lambda_{n^2}$$

Using this notation, we can rewrite eqn (2.85) in the form

$$Q = \sum_{\lambda} \gamma(\lambda) X^{\lambda},\ \gamma(\lambda) = \langle p|\widehat{Q}|q\rangle \tag{2.86}$$

The coefficients $\gamma(\lambda)$ are the parameters of the X representation for the Q operator. The single index form (2.86) is more convenient than eqn (2.85) for two reasons. Firstly, it gives a simpler formulation of the Wick theorem for the X operators, which is more complicated than to the Wick theorem for Fermi/Bose operators. Each pairing of two X operators provides a new X operator similar to a spin operator. Nevertheless, a modified Wick theorem has been formulated by Westwanski and Pawlikovski (1973). The diagram technique (Zaitsev, 1975, 1976) is more complicated and rather awkward. Further development of the diagram technique, including general rules for arbitrary order diagrams and the generalised Dyson equation was given by Ovchinnikov and Val'kov (2004). Secondly, the interaction matrix $V_{\lambda 1, \lambda 2}$ turns out to be a product of two factors, $f(\lambda 1)^* \cdot f(\lambda 2)$, which is very important for solving the system of equations for the Green's functions.

All X operators can be divided into two classes: quasi-Bose operators and quasi-Fermi operators, depending on the difference of the fermion numbers n_p and n_q in the initial state $|q\rangle$ and the final state $|p\rangle$. If $n_p - n_q$ is even, then the $X^{p,q}$ operator is of the Bose type and, when $n_p - n_q$ is odd, the $X^{p,q}$ operator is of the Fermi type. It is obvious that diagonal operators are of the Bose type.

Coming back to the Hubbard model where the atomic basis consists of four states $|0\rangle, |\sigma\rangle|, |2\rangle$, eqn (2.65), the X operator $X^{p,q}$ can be represented by a 4×4 matrix with all elements equal to zero except one, which is equal to 1, on the intersection of the p line and the q column. There are eight operators of the Fermi type, $X^{0,\sigma}$, $X^{\sigma,0}$, $X^{2,\sigma}$, $X^{\sigma,2}$; four non-diagonal Bose-type operators, $X^{+,-}, X^{-,+}, X^{2,0}, X^{0,2}$; and four diagonal Bose-type operators, $X^{0,0}$, $X^{+,+}$, $X^{-,-}$, $X^{2,2}$. For the Hubbard model, the X operators can be expressed via the Fermi operators $a_{f\sigma}$:

$$\begin{aligned}
X_f^{0,0} &= (1-n_{f\uparrow})(1-n_{f\downarrow}), \quad X_f^{\sigma,\sigma} = n_{f,\sigma}(1-n_{f,-\sigma}), \quad X_f^{2,2} = n_{f\uparrow}n_{f\downarrow} \\
X_f^{\sigma,0} &= a_{f,\sigma}^{+}(1-n_{f,-\sigma}), \quad X_f^{2,\sigma} = 2\sigma a_{f,-\sigma}^{+} n_{f,\sigma} \\
X_f^{\sigma,-\sigma} &= a_{f,\sigma}^{+} a_{f,-\sigma}, \quad X_f^{2,0} = 2\sigma a_{f,-\sigma}^{+} a_{f,\sigma}^{+}
\end{aligned}$$

Conversely, the Fermi operators according to eqn (2.85) may be written in the X-operator representation as

$$a_{f,\sigma}^{+} = X_f^{\sigma,0} + 2\sigma X_f^{2,-\sigma} \tag{2.87}$$

The number operator $\widehat{N}$ of electrons is given by

$$\widehat{N} = \sum_{f,\sigma} a_{f\sigma}^{+} a_{f,\sigma} = \sum_{f} \left(0 \cdot X_f^{0,0} + 1 \cdot \left(X_f^{+,+} + X_f^{-,-}\right) + 2 \cdot X_f^{2,2}\right)$$

The Hubbard model Hamiltonian in the X representation has the form

$$\begin{aligned}
H &= H_0 + H_{int} \qquad (2.88) \\
H_0 &= \sum_f \left\{ (\varepsilon - \mu) \sum_\sigma X_f^{\sigma,\sigma} + (2\varepsilon + U - 2\mu) X_f^{2,2} \right\} \\
H_{int} &= \sum_{\mathbf{f},\mathbf{g}} t_{\mathbf{f},\mathbf{g}}^{p_1 q_1, p_2 q_2} X_f^{p_1,q_1} X_g^{q_2,p_2} \\
&= \sum_{\mathrm{f,g}} \mathrm{t_{f,g}} \left\{ \left(\mathrm{X_f^{+,0}} + \mathrm{X_f^{2,-}}\right)\left(\mathrm{X_g^{0,+}} + \mathrm{X_g^{-,2}}\right) + \left(\mathrm{X_f^{-,0}} - \mathrm{X_f^{2,+}}\right)\left(\mathrm{X_g^{0,-}} - \mathrm{X_g^{+,2}}\right) \right\}
\end{aligned}$$

The Coulomb term in the Hamiltonian that is a product of four Fermi operators becomes linear in the X operators while the hopping term is a quadratic form in X operators. The representation of the Fermi operator (2.87) as a sum of two X operators corresponds to the splitting of the free-electron band onto two Hubbard subbands. To prove this statement, we write down the exact equations of motion for operators in the Heisenberg representation

$$
\begin{aligned}
i\dot{X}_{f}^{0,\sigma} &= \left[X_{f}^{0,\sigma}, H\right]_{-} \\
&= \varepsilon_1 X_{f}^{0,\sigma} + \sum_{g} t_{f,g} \left\{ \left(X_{f}^{0,0} + X_{f}^{\sigma,\sigma}\right)\left(X_{g}^{0,\sigma} + 2\sigma X_{g}^{-\sigma,2}\right) \right. \\
&\quad \left. + X_{f}^{-\sigma,\sigma}\left(X_{g}^{0,-\sigma} - 2\sigma X_{g}^{\sigma,2}\right) + 2\sigma X_{f}^{0,2}\left(X_{g}^{-\sigma,0} - 2\sigma X_{g}^{2,\sigma}\right)\right\} \\
2i\sigma\dot{X}_{f}^{-\sigma,2} &= \left(\varepsilon_1 + U\right) 2\sigma X_{f}^{-\sigma,2} \\
&\quad + \sum_{g} t_{f,g} \left\{ \left(X_{f}^{-\sigma,-\sigma} + X_{f}^{2,2}\right)\left(X_{g}^{0,\sigma} + 2\sigma X_{g}^{-\sigma,2}\right) \right. \\
&\quad \left. - X_{f}^{-\sigma,\sigma}\left(X_{g}^{0,-\sigma} - 2\sigma X_{g}^{\sigma,2}\right) - 2\sigma X_{f}^{0,2}\left(X_{g}^{-\sigma,0} - 2\sigma X_{g}^{2,\sigma}\right)\right\}
\end{aligned}
$$

For the Green's functions

$$
G_{11}(\boldsymbol{f}, \boldsymbol{f}'E) = \langle\langle X_{f}^{0,\sigma} | X_{f'}^{0,\sigma} \rangle\rangle_E, G_{21}(\boldsymbol{f}, \boldsymbol{f}'E) = \langle\langle 2\sigma X_{f}^{-\sigma,2} | X_{f'}^{0,\sigma} \rangle\rangle_E
$$

in the Hubbard I approximation, we have the following system of equations

$$
\begin{aligned}
&(E - \varepsilon_1 - F_{0,\sigma} t(\boldsymbol{k})) G_{11}(\boldsymbol{k}, E) - F_{0,\sigma} t(\boldsymbol{k}) G_{21}(\boldsymbol{k}, E) = F_{0,\sigma} \\
&- F_{-\sigma,2} t(\boldsymbol{k}) G_{11}(\boldsymbol{k}, E) + (E - \varepsilon_1 - U - F_{-\sigma,0} t(\boldsymbol{k})) G_{21}(\boldsymbol{k}, E) = 0
\end{aligned} \tag{2.89}
$$

Here $F_{0,\sigma} = \langle X^{0,0} + X^{\sigma,\sigma} \rangle, F_{-\sigma,2} = \langle X^{-\sigma,-\sigma} + X^{2,2} \rangle$ are the filling factors that depend on the occupation numbers of the involved local states. From the sum rule, we have $F_{0,\sigma} + F_{-\sigma,2} = 1$. It is apparent that the quasiparticle spectrum given by the solution of eqn (2.89) is described by eqn (2.71). Moreover, in the limit $U \gg t$, when the intersubband hopping in eqn (2.88) is of minor importance, we can write the solution of eqn (2.89) for the lower Hubbard band as

$$
G_{11}^{(0)}(\boldsymbol{k}, E) = \frac{F_{0,\sigma}}{E - \varepsilon_1 - F_{0,\sigma} t(\boldsymbol{k})} \tag{2.90}
$$

and for the upper Hubbard band as

$$
G_{22}^{(0)}(\boldsymbol{k}, E) = \langle\langle X_{k}^{-\sigma,2} | X_{k}^{2,-\sigma} \rangle\rangle_E = \frac{F_{-\sigma,2}}{(E - \varepsilon_1 - U - F_{-\sigma,2} t(\boldsymbol{k}))} \tag{2.91}
$$

From eqns (2.90) and (2.91), it is obvious that the operators $X^{\sigma,0}$ and $X^{2,-\sigma}$ describe the creation of a Fermi-type quasiparticle with spin projection σ in the lower Hubbard band and the upper Hubbard band, respectively. As will be shown later on in this chapter, the Hubbard I decoupling scheme is equivalent to the Hartree–Fock approximation in the X-operator diagram technique (Zaitsev, 1976; Ovchinnikov and Val'kov, 2004). The half-filled Hubbard model with $n_e = 1$ in the limit $U \gg W$ describes the Mott–Hubbard insulator with an empty upper Hubbard band and an occupied lower Hubbard band. With doping by electrons ($n_e = 1 + x, x > 0$) or by holes ($n_e = 1 - x,$

$x > 0$), the Fermi level moves into the upper Hubbard band or the lower Hubbard band, respectively. Beyond the Hubbard I approximation, one can include the quasiparticle high-order scattering processes and the interaction of the Fermi-type quasiparticle with the Bose-type spin and charge fluctuations. Nevertheless, even in the simple eqns (2.90) and (2.91), one can see the following effects of strong electron correlations on the quasiparticle bands:

(a) *The correlation narrowing of the quasiparticle bandwidth that is determined self-consistently by the filling factors.* The effective bandwidth becomes dependent on the temperature, the magnetic field, and the electron concentration.

(b) *The spectral weight decreasing for each Hubbard subband caused by the filling factor in the numerator of the Green's functions (2.90) and (2.91).* Due to the sum rule, only the total number of states for both the lower Hubbard band and the upper Hubbard band is one for each spin projection, and two for both spin projections. In other words, the Fermi-type quasiparticles in the X-operator representation have both the electrical charge ($n_p - n_q = \pm 1$) and the spin $S = 1/2$. The principal difference from the free particles is the reduced spectral weight that is calculated self-consistently and depends on the temperature, the magnetic field, and the electron concentration. Due to the sum rule, the spectral weight of one quasiparticle depends on the spectral weight of the other quasiparticles.

Since the Hubbard I approximation contains these essential correlation effects, it is more convenient to use it as a zero approximation for the description of the strongly correlated system than the weak coupling approach, which starts from the band limit. Several methods on how to proceed beyond the Hubbard I approximation are given in the literature, and the diagram technique is one of them. Not only for the Hubbard model but also for the general approach, starting with some local set of multi-electron eigenstates, one can write the electron annihilation/creation operator at a site f with the orbital index λ and the spin projection σ, following eqn (2.86), as given by a linear combination of Hubbard fermions:

$$a_{f\lambda\sigma} = \sum_{pq} \gamma_{\lambda\sigma}(p,q) X_{\boldsymbol{f}}^{p,q} = \sum_{m} \gamma_{\lambda\sigma}(m) X_{\boldsymbol{f}}^{m}, \; a_{\boldsymbol{f}\lambda\sigma}^{+} = \sum_{m} \gamma_{\lambda\sigma}^{*}(m) X_{\boldsymbol{f}}^{m*} \quad (2.92)$$

The single-particle electronic Green's function $G_\sigma(\boldsymbol{k},\omega) = \langle\langle a_{\boldsymbol{k}\lambda\sigma} | a_{\boldsymbol{k}\lambda'\sigma}^{+} \rangle\rangle$ is given by a linear combination of the Green's functions

$$D_{\boldsymbol{k}\omega}^{mn} = \langle\langle X_{\boldsymbol{k}}^{m} | X_{\boldsymbol{k}}^{n*} \rangle\rangle_{\omega}$$
$$G_\sigma(\boldsymbol{k},\omega) = \sum_{mn} \gamma_{\lambda\sigma}(m) \gamma_{\lambda'\sigma}^{*}(n) D_{\boldsymbol{k}\omega}^{mn}. \quad (2.93)$$

The analysis of the diagram expansion results in the following exact generalised Dyson equation (Ovchinnikov and Val'kov, 2004) for a paramagnetic phase in the matrix form

$$\widehat{D}(\boldsymbol{k},\omega)=\left[\widehat{G}_0^{-1}(\omega)-\widehat{P}(\boldsymbol{k},\omega)\widehat{t}(\boldsymbol{k})+\widehat{\Sigma}(\boldsymbol{k},\omega)\right]^{-1}\widehat{P}(\boldsymbol{k},\omega) \tag{2.94}$$

Here $\widehat{G}_0^{-1}(\omega)$ is the exact local Green's function with the matrix elements $G_0^{m,n} = \delta_{m,n}/[\omega-(E_m-E_n)]$, where $\widehat{\Sigma}(\boldsymbol{k},\omega)$ and $\widehat{P}(\boldsymbol{k},\omega)$ are the self-energy and the strength operators, respectively. The matrix $\widehat{t}(\boldsymbol{k})$ describes the Fourier transform of the interatomic hopping $t_{\boldsymbol{fg}}^{m,n}=\sum_{\lambda,\lambda'}\gamma_{\lambda\sigma}(m)\gamma_{\lambda'\sigma}^{*}(n)t_{\boldsymbol{fg}}^{\lambda,\lambda'}$. The presence of the strength operator results in the spectral weight redistribution between the Hubbard subbands that is the intrinsic feature of strongly correlated systems. The term 'strength' is related to the oscillator strength of the excitation between the initial and the final local states p and q. It should be stressed that the self-energy in the X-operator representation differs from the conventional self-energy in the Fermi/Bose diagram techniques. A Dyson equation that is similar was derived earlier in the diagram technique for the spin operators by Bar'yakhtar *et al.* (1984).

The Hubbard I approximation is obtained from the exact eqn (2.94) by neglecting the self-energy term $\Sigma=0$ (that is why it is the Hartree–Fock approximation in the X-operator perturbation theory), and the strength operator $P^{m,n}=\delta_{m,n}F_n$, where, for the quasiparticle band index $n=(p,q)$, the value $F_n=\langle X^{p,p}+X^{q,q}\rangle$. We call this value the occupation factor; it is this factor that determines the spectral weight of the Hubbard subbands in eqns (2.90) and (2.91). One important difference from the free electron is that the Hubbard subband corresponding to the excitation from the empty initial state $|q\rangle$ to the empty final state $|p\rangle$ with the energy E_p-E_q will have zero spectral weight. For non-zero spectral weight, at least one of the eigenstates involved has to be occupied at least partially.

Beyond the Hubbard I approximation, one should take into account the interaction of the electron with spin and charge fluctuations. The effect of the long-range antiferromagnetic order can be easily incorporated into the Hubbard I approximation (Ovchinnikov and Shneyder, 2003); see the discussion of Fig. 2.14 above. A more complicated problem concerns the effect of a short-range spin and charge fluctuation on the electronic properties. Within the X-operator approach, the self-energy for the Hubbard model has been calculated in the non-crossing approximation (Plakida and Oudovenko, 2007) by neglecting a vertex renormalisation. The self-consistent system of equations for the Green's function and the self-energy is similar to the system of equations obtained by the composite operator method (Manchini and Avella, 2004; Krivenko *et al.*, 2005). There are two contributions to the self-energy determined by the two Hubbard subbands, while, in the $t-\mathcal{J}$ model, only one subband is considered (Plakida and Oudovenko, 1999; Val'kov and Dzebisashvili, 2005). However, depending on the doping value and the position of the chemical potential, a substantial contribution to the self-energy comes only from the subband that is close to the Fermi level.

The typical electronic dispersion and the spectral function for the underdoped $x=0.05$ and the overdoped $x=0.3$ cases are shown in Figs. 2.17 and 2.18, respectively. The Fermi surfaces for three doping values are shown in Fig. 2.19.

A similar evolution of the Fermi surface with doping has been obtained for the $t-\mathcal{J}^*$ model also in the non-crossing approximation by Korshunov and Ovchinnikov (2007).

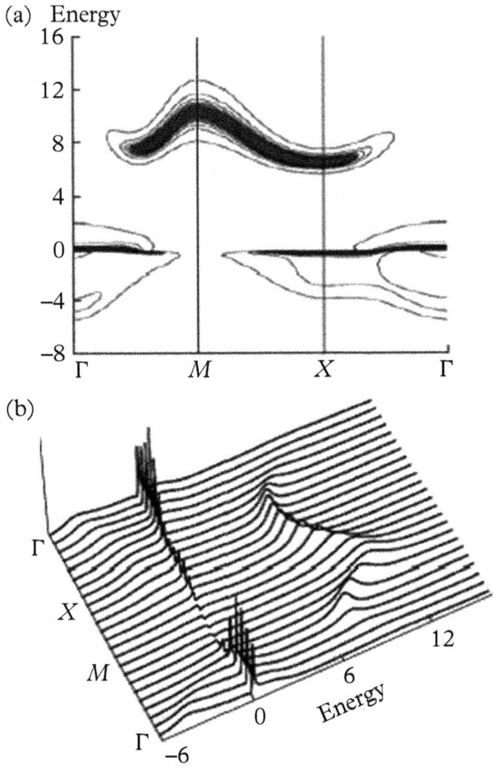

Figure 2.17 *Dispersion curves (a) and spectral functions (b) for $U = 8$ in units of t along the symmetry directions $\Gamma(0, 0) - M(\pi, \pi) - X(\pi, 0)$ for hole doping at $x = 0.05$. (From Plakida and Oudovenko, 2007)*

In agreement with the exact Dyson equation (2.94), the electronic Green's function is equal to

$$\langle\langle X_k^{0,\sigma} | X_k^{\sigma,0} \rangle\rangle_E = \frac{(1+x)/2}{E - (\varepsilon - \mu) - \frac{1+x}{2} t_k - \frac{1-x^2}{4} \frac{\tilde{t}_k^2}{E_{ct}} + \sum(k)} \tag{2.95}$$

where the self-energy $\Sigma(\boldsymbol{k})$ is considered in the static limit and is given by

$$\Sigma(\boldsymbol{k}) = \frac{2}{1+x} \frac{1}{N} \sum_{\boldsymbol{q}} \left[Y_1(\boldsymbol{k}, \boldsymbol{q}) K_{\boldsymbol{q}} - Y_2(\boldsymbol{k}, \boldsymbol{q}) \frac{3}{2} C_{\boldsymbol{q}} \right]$$

$$Y_1(\boldsymbol{k}, \boldsymbol{q}) = t\boldsymbol{q} - \frac{1-x}{2} J_{\boldsymbol{k}-\boldsymbol{q}} - x \frac{\tilde{t}_{\boldsymbol{q}}^2}{E_{ct}} - \frac{1+x}{2} \frac{2\tilde{t}_{\boldsymbol{k}} \tilde{t}_{\boldsymbol{q}}}{E_{ct}}$$

$$Y_2(\boldsymbol{k}, \boldsymbol{q}) = t_{\boldsymbol{k}-\boldsymbol{q}} - \frac{1-x}{2} \left(J_{\boldsymbol{q}} - \frac{\tilde{t}_{\boldsymbol{k}-\boldsymbol{q}}^2}{E_{ct}} \right) - \frac{1+x}{2} \frac{2\tilde{t}_{\boldsymbol{k}} \tilde{t}_{\boldsymbol{k}-\boldsymbol{q}}}{E_{ct}}$$

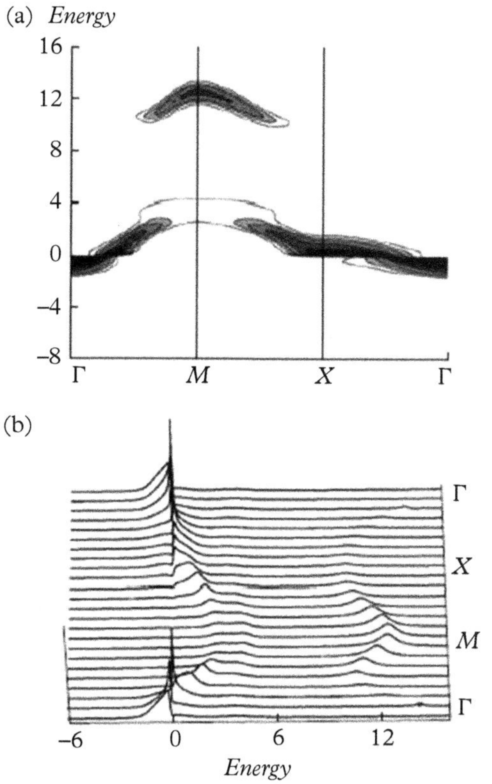

Figure 2.18 *Dispersion curves (a) and spectral functions (b) for $U = 8$ in units of t along the symmetry directions $\Gamma(0,0)-M(\pi,\pi)-X(\pi,0)$ for hole doping at $x = 0.3$. (From Plakida and Oudovenko, 2007)*

Here E_{ct} is the charge transfer gap of the pristine multiband Hubbard-type model that was mapped onto the effective low-energy $t - \mathcal{J}^*$ model by the LDA+GTB approach (Korshunov *et al.*, 2005) with all of the parameters calculated ab initio for $La_{2-x}Sr_xCuO_4$, where t and $\tilde{t}$ are the intraband and interband hopping parameters and their Fourier transforms, respectively, and where the exchange interaction is determined by the interband hopping $J = 2\tilde{t}^2/E_{ct}$, where K_q and C_q are the kinematic and spin correlation functions. The spin system is considered to be the isotropic spin liquid with zero average value for each spin projection and non-zero interatomic correlation functions describing the short-range antiferromagnetic order (Shimahara and Takada, 1991; Barabanov and Berezovsky, 1992). This correlation functions have been calculated up to the ninth neighbours following the work by Val'kov and Dzebisashvili (2005). The doping evolution of the electronic structure is accompanied with three quantum phase transitions: Lifshitz transitions with the change of the Fermi surface (Fig. 2.20).

The first Lifshitz transition occurs at $x_{c1} = 0.151$ from four small hole pockets centred near $(\pi/2, \pi/2)$ to two large hole pockets around (π, π). With further doping, the

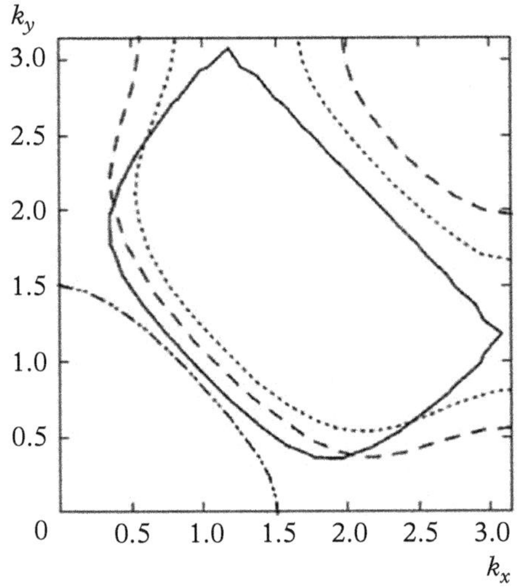

Figure 2.19 *Doping dependence of the Fermi surface for* $x = 0.1$ *(solid line at temperature* $T = 0.03$, *and dotted line at* $T = 0.3$*),* $x = 0.2$ *(dashed line), and* $x = 0.3$ *(dot-dashed line) for* $U = 4$. *(From Plakida and Oudovenko, 2007)*

smaller pockets disappear at $x_{c2} = 0.246$ and, above x_{c2}, there is one large hole pocket. The third transition between the large hole surface centred around M(π, π) and the large electronic surface around the $\Gamma(0, 0)$ point takes place in the strongly overdoped regime at $x_{c3} = 0.53$ (Ovchinnikov *et al.*, 2009). While, in three-dimensional materials, both the Lifshitz transition with the topology change as for x_{c1} and the Lifshitz transition with the disappearance of one surface as for x_{c2} have the same square root peculiarities in the electronic density of states (Lifshitz, 1960), in the two-dimensional system considered here, the first transition is accompanied by the logarithmic divergence of the density of states, and the second one by the step singularity. The step singularity at x_{c2} separates the Fermi-liquid regime at $x > x_{c2}$ from the pseudogap regime with the depleted kinetic energy of electrons at $x < x_{c2}$ (Ovchinnikov *et al.*, 2011). The doping-dependent spin correlation functions proved that, above x_{c2}, only the nearest-neighbour correlations occur, while, below x_{c2}, the short-range spin fluctuations increase with decreasing doping. This approach has emphasised the potential 'quantum critical' nature of the Lifshitz transition at x_{c2}. The idea is that, at a critical doping, antiferromagnetic fluctuations disappear and the pseudogap temperature T^* goes to zero (Laughlin *et al.*, 2001). We would like to stress that both logarithmic and step density-of-state singularities are in perfect agreement with the general properties of the van Hole singularities for two-dimensional electrons (Ziman, 1964; Nedorezov, 1966). Unlike three-dimensional systems with the 5/2 singularity for the thermodynamical potential $\delta\Omega \sim (x - x_c)^{2.5}$, the thermodynamical potential for two-dimensional electrons has the singular contribution $\delta\Omega \sim (x - x_{c2})^2$ for the step singularity, and $\delta\Omega \sim (x - x_{c1})^2 ln|x - x_{c1}|$ for the logarithmic singularity. Thus, the quantum phase transition at xc_2 is the second-order transition, while the singularity at

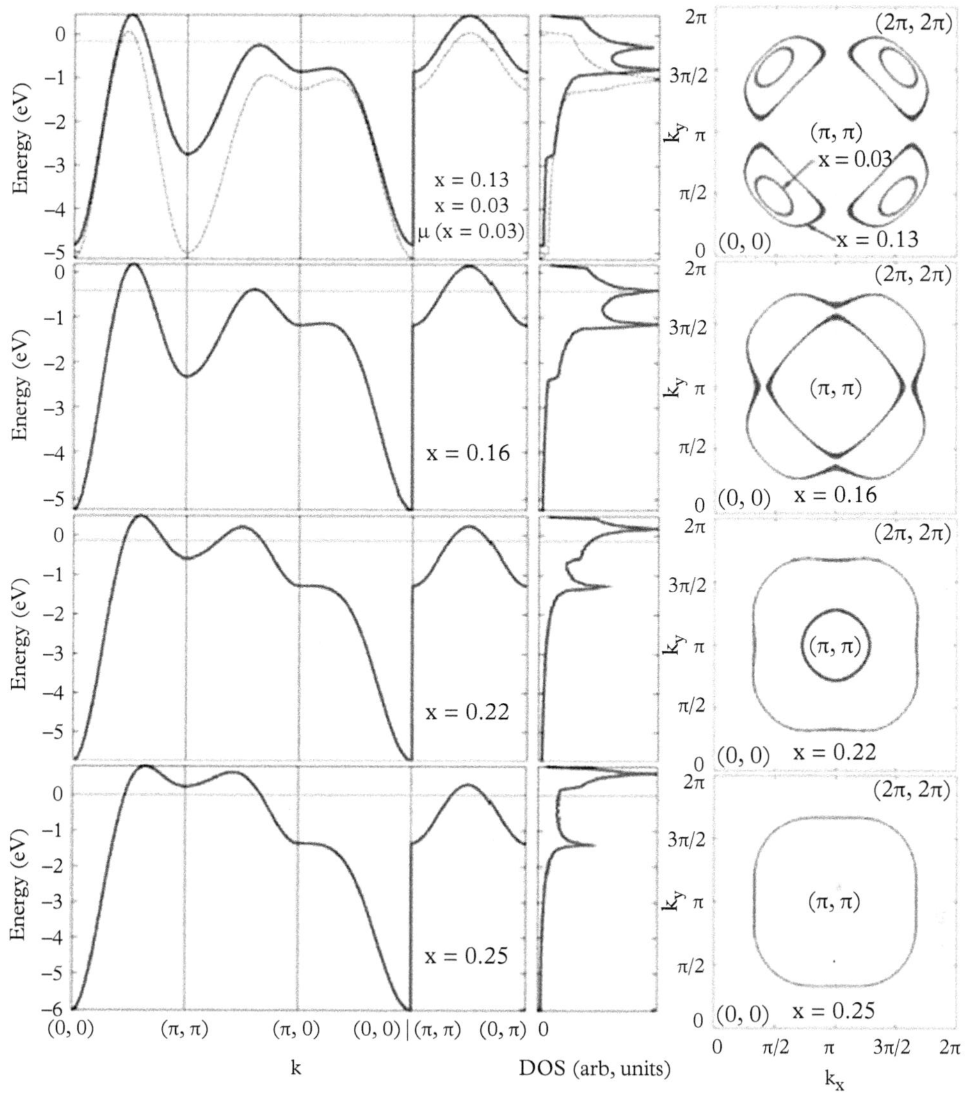

Figure 2.20 *The evolution of the electronic structure and the Fermi surface (a) and the spin correlation functions $C_n = \langle S_0^+ S_n^- \rangle$ (b) with hole doping within the $t - J^*$ model. (From Korshunov and Ovchinnikov, 2007)*

x_{c1} is stronger. It immediately follows that the Sommerfeld parameter γ in the electronic heat capacity $\gamma = C_e/T$ has also a singular step contribution at x_{c2}, and $\delta\gamma \sim ln|x - x_{c1}|$ near x_{c1}. Similar divergence in the specific heat was found within the dynamical cluster approximation (DCA) for the Hubbard model (Mikelsons *et al.*, 2009).

The Fermi surfaces shown in Figs. 2.19 and 2.20 reflect only the real part of the quasiparticle energy. The electronic scattering on spin fluctuations (Barabanov *et al.*,

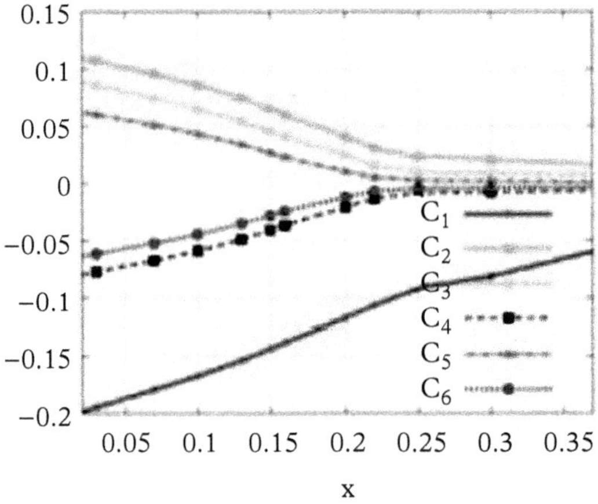

Figure 2.20 *Continued (b)*

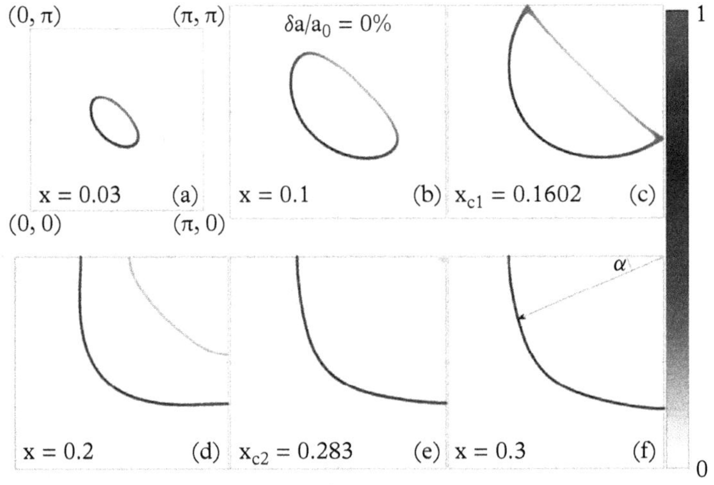

Figure 2.21 *The formation of a Fermi arc instead of a Fermi circle in the pseudogap state. Here* $\frac{\delta a}{a_0} = 0$ *means the perfect square lattice with the parameters of the Hubbard model obtained from the ab initio calculations for* La_2CuO_4. *(From Makarov et al., 2018)*

2001; Plakida and Oudovenko, 2007) or the interband scattering between the lower Hubbard band and the upper Hubbard band (Makarov *et al.*, 2018) provides the non-uniform distribution of the spectral weight along the Fermi contour (Fig. 2.21) and the transformation of the continuous Fermi contour in the arc with depleted spectral weight in the antinodal direction X–M in the pseudogap state for $x < x_{c2}$. Note that the critical concentrations for the Hubbard model are slightly different that for the $t - \mathcal{J}^*$ model in Fig. 2.20.

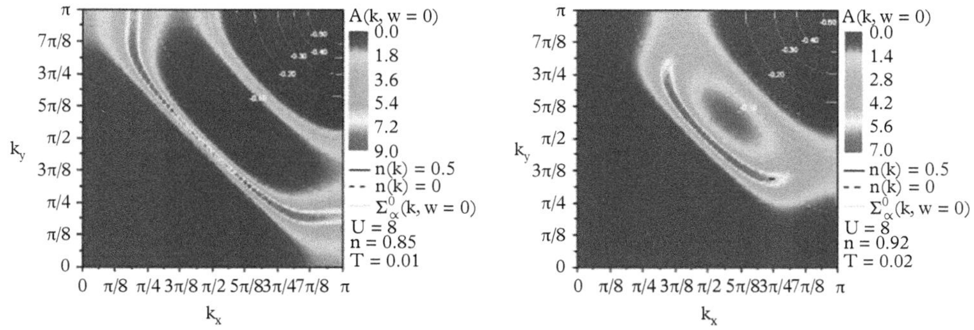

Figure 2.22 *Spectral function of the Hubbard model at the chemical potential as a function of momentum for* $U = 8$ *and hole doping* $x = 0.15$ *(left) and* $x = 0.08$ *(right). (From Avella, 2014)*

A similar evolution of the Fermi surface with arc formation was obtained for the two-dimensional Hubbard model on the square lattice within the composite operator method by Avella (2014); see Fig. 2.22, where the electron concentration $n = 1 - x$.

CPT. CPT is an intermediate approach that treats the short-range interaction and correlations exactly while the long-range correlations are considered by perturbation theory. Cluster models can be formulated in a general functional framework (Maier *et al.*, 2005a; Potthoff, 2005; Kotliar, 2006). The first version of CPT (Senechal *et al.*, 2000) consists of (i) dividing the lattice into identical N-site clusters, (ii) evaluating by exact diagonalisation the single-particle Green's function $G_0(\omega)$ for a cluster with open boundary conditions, and (iii) treating the intercluster hopping t in perturbation theory and recovering the crystal Green's function $G(\boldsymbol{k},\omega)$. The perturbation theory is given by the Hubbard I approximation. The Green's function calculated by CPT contains a set of discrete poles, as in ordinary exact diagonalisations, except that (i) more poles have substantial weight and (ii) they disperse continuously with the wave vector, allowing for clear momentum distribution curves. The results presented here in Figs 2.23 and 2.24 (Senechal and Tremblay, 2004) have been calculated on twelve-site rectangular clusters. The resulting Green's function is averaged over the (3×4) and (4×3) clusters to recover the original symmetry of the lattice. The main features are the same when using clusters of different shapes. A similar result was obtained later on by different groups (Kohno, 2012, 2014; Nikolaev *et al.*, 2014). In the Hubbard I approximation, there is no imaginary part of the quasiparticle energy so, to calculate the spectral function, usually each delta function is broadened by the Lorentzian with some width δ, typically $\delta \sim 0.1t$ corresponds to the experimental resolution of the line width via angle-resolved photoemission spectroscopy (ARPES).

Figure 2.23 illustrates the effect of increasing interaction strength on a 17% hole-doped system. Clearly, there is a range of frequencies with zero spectral weight for all wave vectors. This is the Mott–Hubbard gap. At finite doping, it always opens up away from the zero-energy Fermi level when U is sufficiently large. A new important

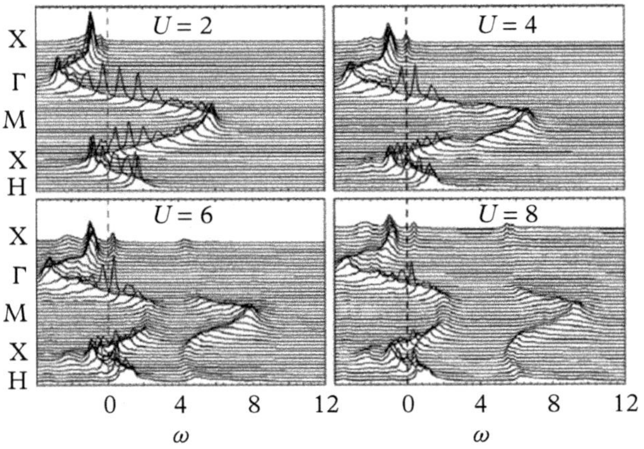

Figure 2.23 *Single-particle spectral weight as a function of energy for wave vectors along the high-symmetry directions for hole concentration* $x = 0.17$ *and several U values obtained by cluster perturbation theory calculations on a* 3×4 *cluster, with Lorentzian broadening* $\delta = 0.12t$. *Here* $H = (\pi/2, \pi/2)$. *(From Senechal and Tremblay, 2004)*

phenomenon found and confirmed by many experiments in high-T_c superconductors is a pseudogap in the normal state that has been the focus of much attention in the field. Contrary to one of the central tenets of Fermi-liquid theory, sharp zero-energy excitations do not enclose a definite volume in the Brillouin zone. Certain directions are almost completely gapped while others are not. This is illustrated in Fig. 2.24 (Senechal and Tremblay, 2004).

Usually, the exact diagonalisation of an individual cluster Hamiltonian in the CPT approach is carried out with Lanczos method, which takes into account only the ground state and a few excited states, ignoring all others. The other version of CPT, based on a full set of the exact eigenstates of the cluster $\{|p\rangle\}$ and a construction of the cluster Hubbard operators $X_f^{pq} = |p\rangle \langle q|$ that is similar to eqn (2.82) was proposed by Nikolaev and Ovchinnikov (2010). Within this approach, the intercluster hopping can be easily written in a form similar to that for eqn (2.88), as for the standard Hubbard model; the only difference is the matrix dimension for the Hubbard operator: it is a 4×4 matrix in the conventional Hubbard model, while it is $N \times N$ in CPT. For a minimal 2×2 cluster, the number of eigenstates $N = 256$. From one side, it is a more complicated approach. But, from another side, it makes it possible to take into account the short-range correlations and formulate the exact sum rule based on the Fermi operator anticommutator; this sum rule requires that the total single-electron spectral weight be as follows:

$$\int A(\mathbf{k}, \omega)\, d\omega = 1 \tag{2.96}$$

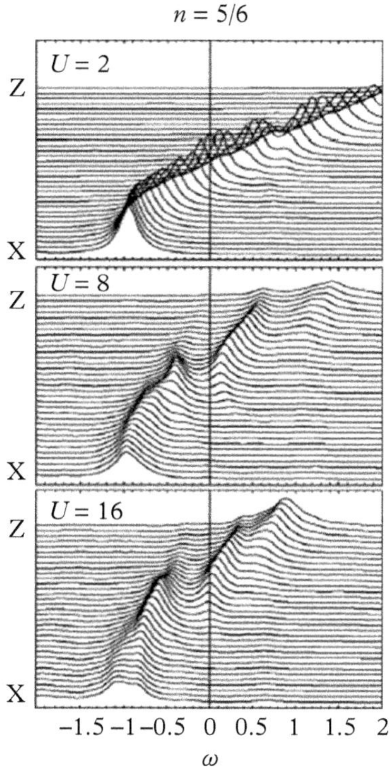

Figure 2.24 *Spectral function for a hole-doped system along the direction* $X(\pi, 0)$ *to* $Z = (\pi, \pi/2)$. *At* $U = 2$, *a depression in the spectral function is visible slightly away from the Fermi level* $\omega = 0$, *while the pseudogap is fully opened at* $U = 8$. *(From Senechal and Tremblay, 2004)*

As we discussed in the previous chapter, quasiparticles with excitation from the empty initial states to the empty final states have zero spectral weight. Also, some matrix elements of the Fermi operator $\langle p|a_{f\sigma}|q\rangle$ are zero, for example when a spin value of the initial and the final states differs by one or more. So, a norm-conserving version of CPT, NC-CPT, was proposed by Nikolaev and Ovchinnikov (2010); in this version, it is possible to decrease the number of local eigenstates and to check how close the total spectral weight is to 1. For example, in the undoped case, it is enough to choose 15–20 eigenstates to get the integral value 0.999 in eqn (2.96). This approach gives some more deep understanding of how the second Lifshitz transition from the pseudogap state to the normal Fermi liquid occurs at the doping x_{c2} shown in Figs 2.20 and 2.21. In the pseudogap state beside the poles of the Green's function, a line of zeros appears; see Fig. 2.25 (Ovchinnikov and Nikolaev, 2011). A similar line of zeros where the spectral weight is zero was obtained earlier via cellular DMFT plus the exact diagonalisation method (Stanescu and Kotliar, 2006; Sakai *et al.*, 2009).

The critical concentration of the Lifshitz transition for a 2×2 cluster is different from those shown in Figs 2.20 and 2.21 but, for larger clusters, such as 3×3 and

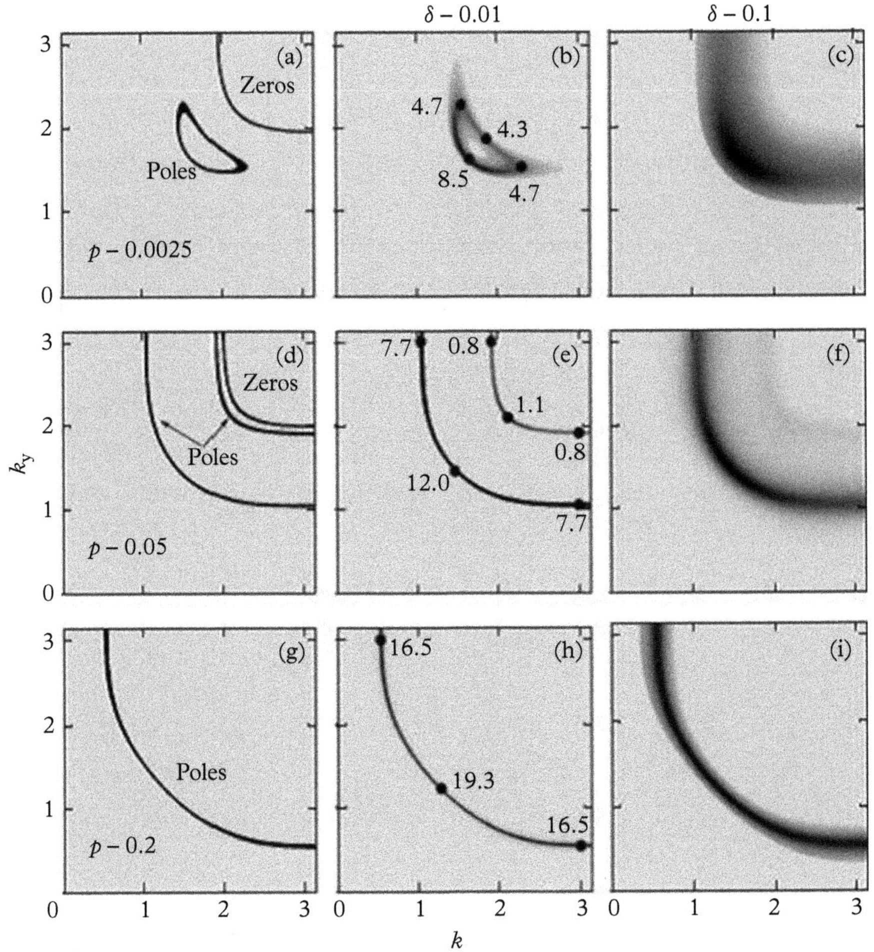

Figure 2.25 *The Fermi surfaces for three doping concentrations p: from poles and zeros of the Green's function (a) and a spectral weight map with high* $\delta = 0.01$ *(b) and low* $\delta = 0.1$ *(c) resolution for* $U = 4.3$ *at doping* $p = 0.0025$*, the same in (d), (e), and (f) at* $p = 0.05$*, and in (g), (h), and (i) at* $p = 0.20$*. The numbers at the different k points in (b), (e), and (h) indicate the spectral weight at these points. (From Ovchinnikov and Nikolaev, 2011)*

3×4 clusters, we get x_{c1} and x_{c2} close to the values in Figs 2.20 and 2.21. Nevertheless, the qualitative picture is the same: below x_{c1} we found a close circle near $H = (\pi/2, \pi/2)$, with a smaller weight for k points closer to the line of zeros; between x_{c1} and x_{c2}, the inner circle in (d) moves to the line of zeros and its spectra weight is strongly suppressed (e); at the Lifshitz transition x_{c2}, the inner circle and the line of zeros mutually annihilate; and, above x_{c2}, the large Fermi surface corresponding to the Fermi liquid occurs. Nowadays, CPT calculations with nine, twelve, or 16 atoms per cluster are available (Senechal and

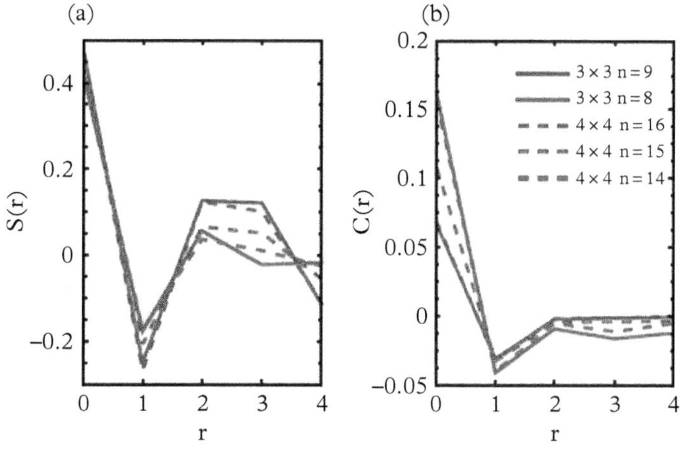

Figure 2.26 *Spin (a) and charge (b) correlation functions calculated within clusters with the specified size and number of electrons. (From Kuz'min et al., 2018)*

Tremblay, 2004; Kohno, 2012, 2014). To get reliable results, the size of the cluster should be larger than the correlation length of the spin and charge fluctuations shown in Fig. 2.26 (Kuz'min *et al.*, 2018).

It is clear from Fig. 2.26 that short-range spin correlations exist up to the fifth neighbour in the undoped case and decrease with doping, while the charge correlation is almost negligible at the second neighbour and above. That is why (3×3)- and (4×4)-cluster CPT can reproduce such fine structures of the electronic dispersion as kinks and waterfalls (Kohno, 2012, 2014).

Superconductivity in the $t - J$ model. A first indication of hole–hole attraction in the $t - J$ model was revealed from the formation of a bound state between the holes in the two-hole ground state (Bonca *et al.*, 1989; Hasegawa and Poilblanc, 1989; Riera and Young, 1989; Dagotto *et al.*, 1990; Boninsegni and Manousakis, 1993; Poilblanc *et al.*, 1994). The bound state has $d_{x^2-y^2}$ symmetry. The origin of binding is clear: two holes added to a quantum antiferromagnet with $S = 1/2$ on a square lattice at each ion will minimise the exchange energy by occupying nearest-neighbour sites. When two holes move independently, eight antiferromagnetic bonds are broken, while, for the pair, only seven bonds are destroyed. This mechanism is similar to the spin-bag one discussed in section 2.7.3 for the weak correlation limit.

For the $t - J$ model in the atomic limit, the theory of singlet superconducting pairing of magnetic polarons was developed in the self-consistent Born approximation by Plakida *et al.* (1997). In this approach, the electronic quasiparticles form a narrow energy band with doping-dependent width and with the Fermi surface consisting of the four circles centred in the $(\pi/2, \pi/2)$ points, as shown in Fig. 2.20 at small doping. In the simplest mean field approximation, neglecting the self-energy renormalisation, the equation for T_c in the $t - J$ model looks similar to the BCS equation (Plakida, 2001):

$$\Delta(\boldsymbol{k}) = \frac{1}{N}\sum_{\boldsymbol{q}} J_{\boldsymbol{k}-\boldsymbol{q}} \frac{\Delta(\boldsymbol{q})}{2E_{\boldsymbol{q}}}\, tanh \frac{E_{\boldsymbol{q}}}{2T_c}, \quad E_{\boldsymbol{q}} = \sqrt{\varepsilon^2(\boldsymbol{q}) + |\Delta(\boldsymbol{q})|^2} \tag{2.97}$$

Here $J_{\boldsymbol{k}-\boldsymbol{q}}$ is the Fourier transform of the interatomic exchange interaction, and the superconducting gap has $d_{x^2-y^2}$ symmetry; $\Delta(\boldsymbol{q}) = \frac{cosq_x - cosq_y}{2}$. The effect of the three-site correlated hopping (2.80) on superconductivity is the renormalisation of the coupling parameter, $J_{\boldsymbol{k}-\boldsymbol{q}} \to \frac{n}{2} J_{\boldsymbol{k}-\boldsymbol{q}}$ (Val'kov *et al.*, 2002), where n is the electron concentration, and $n = 1 - x$ for hole doping.

To take into account the retardation effects, within the self-consistent Born approximation, the equations for the normal Σ_{hh} and the abnormal Σ_{hf} parts of the self-energy may be written as the Eliashberg equations

$$\Sigma_{hh}(\boldsymbol{k}, i\omega_n) = -T\sum_{\boldsymbol{q}m} G_{hh}(\boldsymbol{q}, i\omega_m)\, \lambda^{11}_{\boldsymbol{k},\boldsymbol{k}-\boldsymbol{q}}(i\omega_n - i\omega_m) \tag{2.98}$$

$$\Sigma_{hf}(\boldsymbol{k}, i\omega_n) = -T\sum_{\boldsymbol{q}m} G_{hf}(\boldsymbol{q}, i\omega_m)\, \lambda^{12}_{\boldsymbol{k},\boldsymbol{k}-\boldsymbol{q}}(i\omega_n - i\omega_m)$$

where the following notations have been introduced for the linear combinations of the magnon Green's function $D(\boldsymbol{q}, -i\omega_n)$:

$$\lambda^{11}_{\boldsymbol{k},\boldsymbol{q}}(i\omega_n) = g(\boldsymbol{k},\boldsymbol{q})^2\, D(\boldsymbol{q}, -i\omega_n) + g(\boldsymbol{q}-\boldsymbol{k},\boldsymbol{q})^2\, D(-\boldsymbol{q}, i\omega_n)$$
$$\lambda^{12}_{\boldsymbol{k},\boldsymbol{q}}(i\omega_n) = g(\boldsymbol{k},\boldsymbol{q})\, g(\boldsymbol{q}-\boldsymbol{k},\boldsymbol{q})\, [D(\boldsymbol{q}, -i\omega_n) + D(-\boldsymbol{q}, i\omega_n)]$$

The magnon Green's function can be expressed over the electronic Green's function. In the simplest case, it is given by a bare loop with the magnon dispersion

$$\omega(\boldsymbol{k}) = zJS(1-\delta)^2 \sqrt{1 - (cosk_x + cosk_y)^2/4}$$

where $S = 1/2$ and δ is the concentration of doped holes. Close to T_c, the Eliashberg equations (2.98) may be linearised and solved numerically. The superconducting gap has d-wave symmetry, and the doping dependence of T_c is shown in Fig. 2.27.

The effect of three-site correlated hopping, eqn (2.80), on T_c has been considered by Val'kov *et al.* (2002). The amplitude of this hopping, $\sim\frac{t^2}{U}$, is small in comparison to the single-electron hopping t, so its contribution to the electron energy spectrum is small. In contrast, its contribution to the exchange coupling J is of the same order. Indeed, the three-site correlated hopping results in a doping-dependent decrease in the effective coupling for d-wave superconductivity, $J \to \frac{1-x}{2} J$, that reduces T_c more than ten times. The next weak point of Fig. 2.27 is that the maximal critical temperature of superconductivity occurs for doping 20% and more, where long-range antiferromagnetic order does not exist.

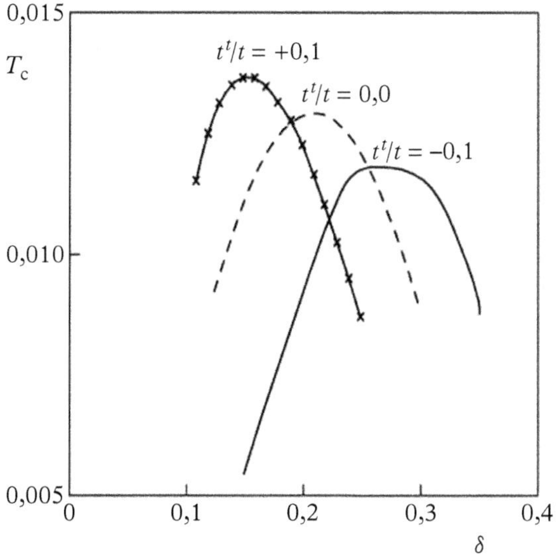

Figure 2.27 *The critical temperature T_c in units of t in the $t - \mathcal{J}$ model as a function of doping for several values of the second-neighbour hopping t'. (From Plakida et al., 1997)*

For the paramagnetic phase, Plakida and Oudovenko (1999) considered the effect of short-range antiferromagnetic correlations on the electronic structure and superconductivity. The problem was, how to calculate the spin susceptibility, as the unknown vertex corrections in the atomic limit did not allow for the accurate solution of the problem similar to the band limit. So the authors used non-self-consistent expressions for the spin correlation functions. Later, a more accurate calculation of the dynamical spin susceptibility was obtained via the relaxation function theory by Vladimirov *et al.* (2009). The magnetic short-range effects on the electronic structure and critical temperature in the $t - \mathcal{J}^*$ model have also been considered by Val'kov and Dzebisashvili (2005). They wrote the electronic self-energy in a non-crossing approximation similar to eqn (2.95), both in the normal and in the superconducting phases. The short-range spin correlations were calculated in the spin-liquid state with spherical symmetry. Depending on the model parameters set, the authors obtained d-wave superconductivity with $T_c(x)$ with one or two maxima. The shape of the $T_c(x)$ with the maximum at the optimal doping is related to the Lifshitz quantum phase transition at $x_{c1} = 0.151$ (as shown above in Fig. 2.20) that results in the density of states singularity (see Fig. 2.28). As usual, close to the Lifshitz transition, the density of states has the regular contribution and the singular one from the change of the Fermi surface topology (Lifshitz, 1960).

The description of superconductivity in the self-consistent Born approximation for the Hubbard model with a non-zero concentration of local pairs $N_2 \sim t^2/U^2$ within the Hubbard operator approach requires some extra effort to define the anomalous average for the d-wave pairing (Plakida *et al.*, 2003). In the Hubbard model, there is no prepared

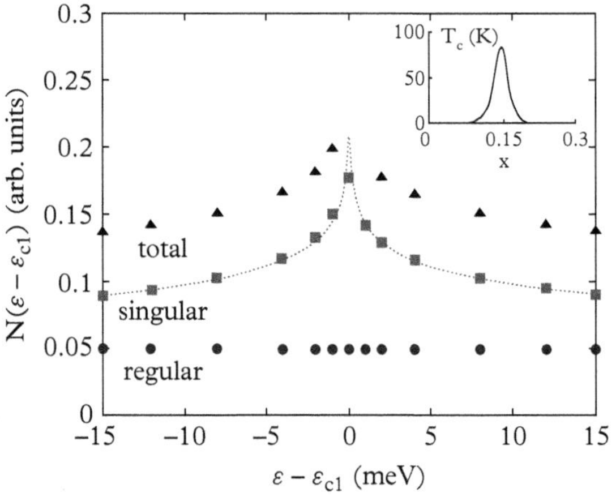

Figure 2.28 *The effect of logarithmic singularity at* $x_{c1} = 0.151$ *induced by the Lifshitz quantum phase transition on the critical temperature shown in the inset. (From Ovchinnikov et al., 2011)*

exchange interaction as in the $t - \mathcal{J}$ model and, to introduce the d-wave order parameter $\Delta^{22}_{ij\sigma}$ for the holes doped in the upper Hubbard band (the holes at the top of the electronic valence band), Plakida *et al.* (2003) introduced the following anomalous average:

$$\langle X_i^{0,2} N_j \rangle = -\frac{1}{U} \sum_{m \neq i,\sigma} \sigma t_{im} \langle X_i^{\sigma,2} X_m^{-\sigma,2} N_j \rangle \cong -\frac{4t_{ij}}{U} \sigma \langle X_i^{\sigma,2} X_j^{-\sigma,2} \rangle$$

Here the last expression is generated by the term in summation with $m = j$. The superconducting gap is

$$\Delta^{22}_{ij\sigma} = -\sigma t_{ij} \langle X_i^{0,2} N_j \rangle / F_2 = J_{ij} \langle X_i^{\sigma,2} X_j^{-\sigma,2} \rangle / F_2$$

where $F_2 = \langle X^{\sigma\sigma} + X^{22} \rangle = (1 + x)/2$ is the filling factor for the upper Hubbard band with the doped hole concentration x. This equation is equivalent to the gap equation of the $t - \mathcal{J}$ model with the exchange interaction J_{ij}. Due to the retardation effect, a new contribution to d-wave pairing has been found by Plakida (2014), the kinematical spin fluctuation mechanism, which is related to the dynamical spin susceptibility $\chi_{sf}(q,\omega)$. The coupling parameter in the equation for T_c is renormalised as $J_{\boldsymbol{k}-\boldsymbol{q}} \to J_{\boldsymbol{k}-\boldsymbol{q}} - t^2(\mathbf{q})\chi_{sf}(\boldsymbol{k} - \boldsymbol{q}, \omega)$. Due to the antiferromagnetic spin correlations, the spin susceptibility is negative, and the spin fluctuation term results in T_c increasing.

Superconductivity in the Hubbard model. Soon after the Bednorz and Muller (1986) discovery of the superconductivity in cuprates, Anderson (1987) proposed the resonating valence bond (RVB) conception based on the strongly correlated limit of the two-dimensional Hubbard model. He assumed that quantum fluctuations (due to the low spin $S = 1/2$ of the Cu d^9 ion and the quasi-two dimensionality of the crystal structure)

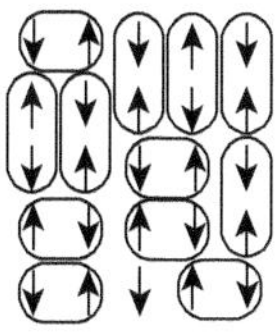

Figure 2.29 *The spin-singlet liquid of the resonating valence bond state, with unpaired spins denoted as spinons. (From Norman, 2014)*

would melt the Néel order typically associated with Mott insulators. He proposed that the resulting state would be a liquid of spin singlets. As a given singlet involves coupling two of the copper ions, and each copper ion is surrounded by four other copper ions, then each bond can be either part of a singlet or not. Therefore, each bond can 'resonate' between being part of a singlet or not, hence the term 'resonating valence bond'. Now imagine doping a hole into such a configuration. Since each copper ion in the undoped case participates in a singlet, then one singlet is broken. This leaves a free chargeless spin (denoted a 'spinon') and a spinless charged hole (denoted a 'holon'); see Fig. 2.29. This implies the presence of spin-charge separation.

This idea met with a lot of criticism in the literature. Firstly, the long-range order in the undoped Hubbard model in the two-dimensional square lattice does not melt. Its stability has been demonstrated by many QMC calculations for 4×4, 8×8, and 12×12 clusters (Hirsch and Tang, 1989; Scalapino, 1991; Moreo, 1993), as well as by the exact diagonalisation method; see the review by Dagotto (1994). The Néel state remains the ground state also in the $t - \mathcal{J}$ model (Bonesteel and Wilkins, 1991; Dagotto *et al.*, 1992a) and the extended model with next-nearest interaction (Dagotto, 1991). Secondly, the notions of holons and spinons had appeared in the exact solution of the one-dimensional Hubbard chain at $U = \infty$. For large but finite U, and for a two-dimensional plane or even the model closest to a one-dimensional chain model, that of a two-leg chain, there is no proof for either the spinless charged holon or the chargeless spinon. The dynamics of the hole in the antiferromagnetic background has been studied by the exact diagonalisation of the Hubbard model (Dagotto *et al.*, 1992b; Ohta *et al.*, 1992) and the $t - \mathcal{J}$ model (Stephan and Horsch, 1991; Otaki *et al.*, 2015) and is described by spin polaron physics. Nevertheless, the RVB model gives rise to a large number of investigations of the electronic spectra and possible phase formation in the field of strongly correlated electrons. This great progress has been achieved due to the use of cluster extensions of the DMFT; see the recent review by Rohringer *et al.* (2018). This will be discussed in the next section.

The interpolation for $U \sim W$ usually may be obtained by numerical non-perturbative methods like exact diagonalisation (see the review by Dagotto, 1994), QMC (see the recent review by Becca and Sorella, 2017), and by hybrid methods combining exact treatment of local correlations with perturbation theory, like CPT, DMFT, and DCA. The interpolation solutions may also be obtained within the strong coupling diagram technique and the variational Gutzwiller approach.

The strong coupling diagram technique. The electronic structure of the two-dimensional Hubbard model in the regime of strong correlations has been also studied by the strong coupling diagram technique by Sherman (2018). This approach is based on a regular series expansion of Green's functions in powers of the hopping parameter. Previously it was shown (Vladimir and Moskalenko, 1990; Sherman, 2006) that the two lowest-order diagrams in the expansion of the irreducible part are already enough to describe the Mott metal–insulator transition. The effect of spin and charge fluctuations on electronic spectra in this theory is described by diagrams containing ladder inserts in the irreducible part. The electronic spectra have been self-consistently found for the ranges of Hubbard repulsions $2t < U < 10t$, temperatures $0.2t < T < t$, and electron concentrations $0.7 < n < 1$. For undoped cases, this approach yields the temperature-dependent band structure and the density of states (Fig. 2.30).

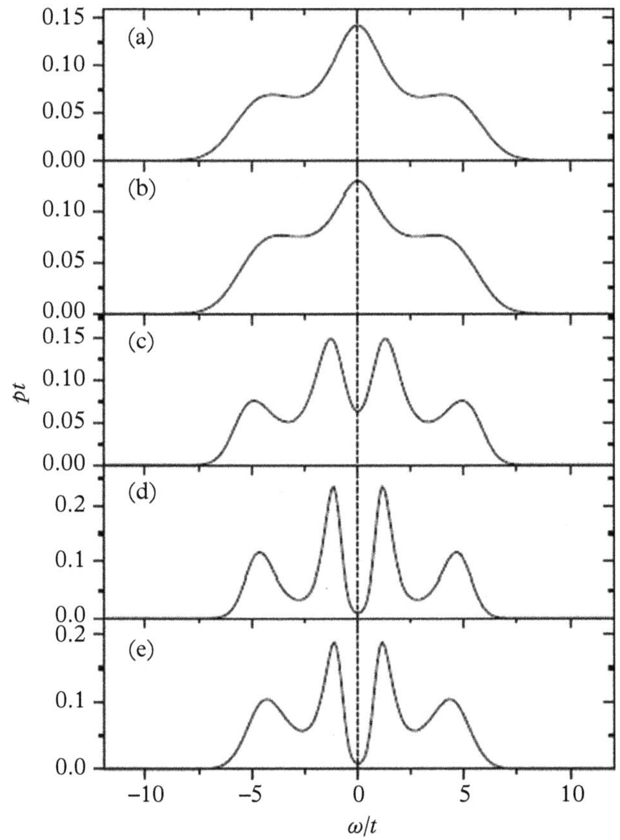

Figure 2.30 *Temperature-dependent density of states for the half-filled Hubbard model at* $U = 5.1t$, *and temperature* $T/t = 1.03$ *(a), 0.62 (b), 0.49 (c), 0.35 (d), and 0.29 (e). (From Sherman, 2018)*

The transition from metal at high temperature to insulator at low temperature, shown in Fig. 2.30, is similar to the transition from metal to insulator at zero temperature with increasing the coupling parameter U/t in Fig. 2.14. The Fermi surface obtained by Sherman (2018) for the underdoped Hubbard model reveals the maximum at the nodal direction, and the minimal in the antinodal one, providing a pseudogap similar to the ones in Figs 2.21, 2.22, and 2.25.

The variational Gutzwiller solution. When working with a variational Gutzwiller wave function, the main task is the calculation of the expectation value of the Hamiltonian with respect to this trial wave function. It yields the ground state properties, including the superconducting gap as a function of doping. The statistically consistent Gutzwiller approach (Jedrak and Spalek, 2011; Spalek, 2015) is equivalent, with a renormalised mean field theory that describes superconductivity in the $t - \mathcal{J}$ model (Zhang *et al.*, 1988; Edegger *et al.*, 2007; Guertler *et al.*, 2009) and can be related directly to the slave-boson method (Ruckenstein *et al.*, 1987; Kotliar and Liu, 1988; Comjayi *et al.*, 2007); also see the review paper by Lee *et al.* (2006). However, within the Gutzwiller-type approach, no extra Bose fields are required, as the electronic correlations are evaluated directly.

Solutions of the system of self-consistent equations for the superconducting order parameter (Paramekanti *et al.*, 2001) and the amplitude of the d-wave gap (Paramekanti *et al.*, 2004) obtained within the Gutzwiller approximation and variational Monte Carlo calculations are shown in Fig. 2.31.

Recent diagrammatic expansion of the Gutzwiller approach starts with the renormalised mean field theory as a zero approximation and takes into account non-local charge and spin correlations; it was used for the description of the d-wave superconductivity in the Hubbard model (Kaczmarczyk *et al.*, 2013) and in the $t - \mathcal{J}$ model (Kaczmarczyk *et al.*, 2014). For the case of the Hubbard model, a narrow stability range of the coexistent superconducting–nematic phase appears between the pure

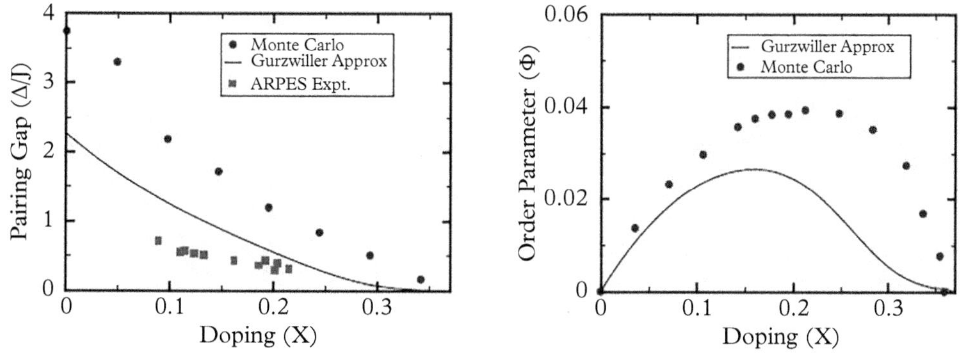

Figure 2.31 *Superconductivity in the ground state of the hole-doped Hubbard model from the Gutzwiller approximation and variational Monte Carlo calculations. The pairing amplitude in units of $\mathcal{J}$ (a) and the superconducting order parameter (b) as function of doping are shown. (From Randeria et al., 2011)*

superconducting phase and the pair density wave (PDW) + CDW phases. In this phase, the rotational symmetry is broken; however, the translational symmetry is conserved. Such a phase can be considered a precursor state for the formation of the CDW + PDW phase with decreasing doping (Zegrodnik and Spalek, 2018).

The exact diagonalisation approach. The exact diagonalisation method is based on the numerical computation of the ground state wave function and the expectation values of the Hamiltonian and various correlation functions for finite size clusters; see the review by Dagotto (1994). The strong advantage of the exact diagonalisation method is its non-perturbative character; this is especially useful in the region of a model parameter where all perturbation theories failed, as in the intermediate regime $U \sim W$ in the Hubbard model. The main disadvantage of the exact diagonalisation method is the restriction of the cluster size L, as there is no way to obtain results in the thermodynamic limit $L \to \infty$. In order to do exact diagonalisation in a cluster with N sites, one has to diagonalise numerically the matrix $N_s \times N_s$, where N_s is the number of states in the Hilbert space. For the $t - \mathcal{J}$ model, $N_s = 3^N$ and, for the Hubbard model, $N_s = 4^N$. The standard diagonalisation technique, which requires full storage of the matrix elements, is of no practical use, due to the limitation of the memory size in present computers. The Lanczos (1950) method avoids the problem of calculating and saving huge matrices in the computer memory by constructing a basis that directly leads to a tridiagonal matrix. It is an example of a family of projection techniques known as the Krylov subspace methods. That is why the Lanczos algorithm is widely used for obtaining the ground state and a few excited states (Maekawa *et al.*, 2004). The application of this algorithm to finite temperature has also been realised (Jaclic and Prelovsek, 2000), and various quantities have been calculated for the $t - \mathcal{J}$ and Hubbard models. The maximum sizes of square lattices examined up until now are $N = 20$ for the Hubbard model, and $N = 32$ for the $t - \mathcal{J}$ model. For example, the exact diagonalisation studies of the $t - \mathcal{J}$ model by Otaki *et al.* (2015) is restricted by number of atoms $N = 16$ for the one-dimensional model and $N = 16$ (4×4 cluster) for the square lattice in two dimensions.

For the $t - \mathcal{J}$ model, the superconducting ground state with $d_{x^2-y^2}$ symmetry was found by the exact diagonalisation method by Dagotto and Riera (1993), and Dagotto (1994); see Fig. 2.32.

It is necessary to mention that the $t - \mathcal{J}$ model results from the Hubbard model in the perturbation approach, so the region $\mathcal{J} \ll t$ is physically relevant. There are no clear indications for superconductivity in this parameter region from the exact diagonalisation method. Nevertheless, exact diagonalisation is now used quite often as a zero approximation in various hybrid approaches, such as exact diagonalisation in direct perturbation theory, and exact diagonalisation DMFT.

QMC. QMC simulations have yielded conflicting results, with some indicating an enhancement of pairing, while others not. The issue, of course, is the inability of accessing very low temperatures because of the fermion sign problem. Nevertheless, in the past two decades, there has been a substantial development in Monte Carlo methods for the optimisation of variational wave functions and the definition of stable projection

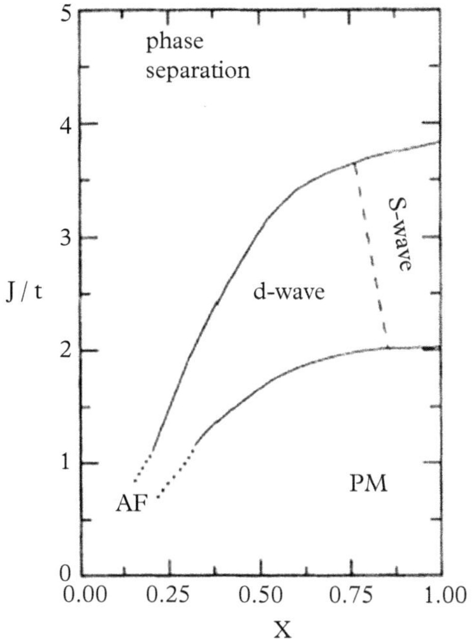

Figure 2.32 *Schematic phase diagram of the two-dimensional* $t - \mathcal{J}$ *model at zero temperature; x is the hole density. Here AF indicates the antiferromagnetic phase near half-filling, and PM indicates a weakly interacting gas of electrons in the paramagnetic state in the overdoped region. The superconducting area has two possible states, with d-wave and s-wave order parameters. (From Dagotto, 1994)*

techniques, especially for fermionic models. Even though a final solution of correlated models in more than one spatial dimension is still elusive, variational wave functions have been proven to provide very accurate approximations of ground states, when compared to exact diagonalisation of small clusters or in particular regimes where exact solutions are known (e.g. when the sign problem is not present or for particularly designed cases where the ground state can be exactly evaluated). In particular, a deep investigation has been dedicated to frustrated Heisenberg models on various lattices (e.g. on square, triangular, honeycomb, and kagome lattices), to pursue the possibility of stabilising so-called quantum spin liquids (Becca and Sorella, 2017).

The QMC simulations by Scalapino *et al.* (1992) for the Hubbard model for $U/t = 4$ and down to temperatures $T = t/10$ have revealed that the doped Hubbard model has the metallic but not the superconducting state. Later it was shown by Sorella *et al.* (2002) that pairing is a robust property of hole-doped antiferromagnetic insulators. In one dimension and for two-leg ladder systems, a BCS-like variational wave function with long-bond spin singlets and a Jastrow factor provides an accurate representation of the ground state of the $t - \mathcal{J}$ model, even though strong quantum fluctuations destroy the off-diagonal superconducting long-range order in this case. However, in two dimensions, it has been argued, and numerically confirmed using several techniques, especially

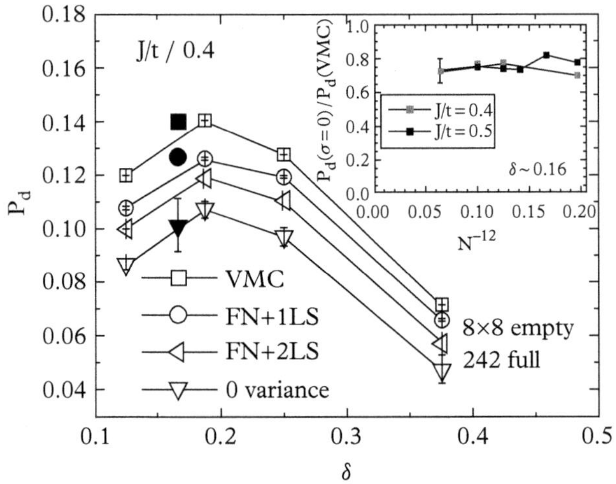

Figure 2.33 *Superconducting order parameter P_d as function of hole concentration δ at $\mathcal{J}/t = 0.4$ for clusters of 64 sites at several densities and 242 sites close to optimal doping. The inset shows the size scaling of P_d for $\delta \sim 0.16$. (From Sorella et al., 2002)*

QMC, that quantum fluctuations are not strong enough to suppress superconductivity (Fig. 2.33).

The possible coexistence of antiferromagnetism and superconductivity at low doping has been also found by Sorella *et al.* (2002). This result illustrates the flexibility of the QMC approach.

DMFT and its cluster extensions. DMFT was discussed in section 2.7.2. The space correlations neglected in DMFT are very important in two dimensions; for example, the short-range antiferromagnetic correlations in cuprates are known in a wide range of doping. That is why cluster extensions are important.

In the CPT discussed above, a cluster is embedded in its environment by Dyson's equation only. There are other methods currently or recently developed that simulate the cluster's environment, that is, the rest of the lattice, by a bath of fermionic levels chosen in a self-consistent way. These are DCA (Hettler *et al.*, 1998; Maier *et al.*, 2000) and cellular DMFT (CDMFT; Lichtenstein and Katsnelson, 2000; Kotliar *et al.*, 2001). They are both offshoots of DMFT, in which a lattice model maps onto the single-site Anderson impurity model (Georges and Kotliar, 1992), which can be solved numerically exactly by QMC simulations (Jarrel, 1992) For a review and additional references, see the work by Georges *et al.* (1996), Vollhardt *et al.* (2011), and Potthoff (2011). Common to both DCA and CDMFT is the necessity to couple the Hubbard model defined on a cluster with an Anderson impurity model (the bath). The whole system (bath + cluster) is then solved, for instance by QMC or exact diagonalisation, which implies in the latter case that the physical size of the cluster has to be small (e.g. 4–6 sites) to leave space for the bath. DCA and CDMFT differ in that the former is a momentum-space technique using periodic boundary conditions on the cluster, whereas the latter is a real-space technique using open boundary conditions.

Impressive successes of the DCA and CDMFT approaches include the description of the pseudogap physics and unconventional superconductivity in the Hubbard model; for more results and larger clusters, see the work by Gull *et al.* (2013), Harland *et al.* (2016), Sordi *et al.* (2011), and Sakai *et al.* (2012). In practice, numerical limitations due to the exponential growth of the cluster Hilbert space restrict the cluster size to about 10×10 sites. While correlations are included non-perturbatively, they remain short-ranged even in two dimensions and for a single orbital. A recent review of the effects of non-local correlations is given by Rohringer *et al.* (2018), where the diagrammatic extensions of DMFT are discussed. In this approach, self-energy corrections to DMFT are computed through Feynman diagrams, which allows one to reach significantly larger lattice sizes. Pertinent steps have also been taken towards understanding high-temperature superconductivity. Kitatani *et al.* (2015) and Otsuki *et al.* (2014) have studied superconducting instabilities, and Gunnarsson *et al.* (2015) have considered the effect of fluctuations on the pseudogap.

In Fig. 2.34, the phase diagram of the half-filled three-dimensional Hubbard model with nearest-neighbour hopping is shown. Being exact at infinite dimensions, DMFT is expected to work better in a three-dimensional lattice than in a two-dimensional lattice. Indeed, the diagram in Fig. 2.34 demonstrates to which extent the different methods can be applied. In the limit $U \gg t$, localised spins described by the Heisenberg model with an antiferromagnetic phase below the Néel temperature T_N, and a paramagnetic insulator above T_N are expected. In the opposite weak correlation limit $U \ll t$, an antiferromagnetic SDW below the Néel temperature T_N, and a paramagnetic metal above T_N, are expected. At high T, a crossover from the paramagnetic metal to the paramagnetic insulator occurs for $U \sim 10t$; see the work by Georges *et al.* (1996) and Kent *et al.* (2005) for the DMFT phase diagram. For the simple cubic lattice, the bandwidth $2W = 12t$, so the crossover takes place for the expected parameters when the Coulomb and kinetic energies are of the same order. At weak coupling, DMFT, the dual fermion approach (DF), and dynamical vertex approximation (DΓA) provide quite similar Néel temperature dependences on the U/t ratio, with smaller values given by DF and DΓA than by DMFT, due to the effect of non-local fluctuations. In the strong coupling limit, the T_N from DΓA correctly approaches the behaviour of the Heisenberg model, while the T_N from DMFT tends to the Weiss limit. In the most interesting regime of intermediate coupling, $8 < U/t < 12$, rather similar results are given by DΓA, DF, DCA, and diagrammatic determinant Monte Carlo. It is quite noticeable that DMFT is equivalent to the Weiss mean field in spite of the fluctuations in time considered by DMFT. The space fluctuations are important even for a three-dimensional lattice. For the two-dimensional Hubbard model that is of the most interest for high-T_c superconductivity, the space fluctuations are even more important.

Confirmation of *d*-wave superconductivity in the two-dimensional Hubbard model has been obtained by several methods, among them cluster-DMFT and DCA (Lichtenstein and Katsnelson, 2000; Maier *et al.*, 2005b; Capone and Kotliar, 2006; Haule and Kotliar, 2007; Sordi *et al.*, 2012; Gull *et al.*, 2013). The transition temperature is found

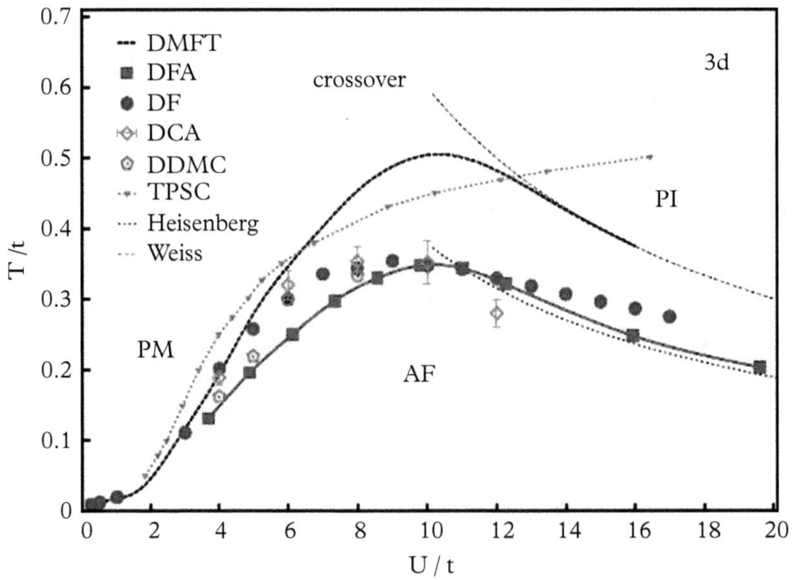

Figure 2.34 *Phase diagram of the three-dimensional Hubbard model, as obtained by various methods. Here the following notations are used for different modifications of the theory: dynamical vertex approximation (DΓA), the dual fermion approach (DF), dynamical cluster approximation (DCA), diagrammatic determinant Monte Carlo (DDMC), the two-particle self-consistent theory (TPSC), the solution of the Heisenberg model from the strong coupling limit (Heisenberg), and the Weiss mean field theory of the Heisenberg model (Weiss). (From Rohringer et al., 2018)*

by computing the leading eigenvalue of a linearised Eliashberg-like equation (Rohringer *et al.*, 2018):

$$\sum_{\boldsymbol{k}\omega} \Gamma^{\nu\nu'\omega=0}_{pp,r,kk'\boldsymbol{q}=0} G_{(-\boldsymbol{k}')(-\omega')} G_{\boldsymbol{k}'\omega} \varphi_{\boldsymbol{k}'\omega'} = \varphi_{\boldsymbol{k}\omega} \tag{2.99}$$

where $r = s, t$ stands for the singlet and triplet channels, respectively. Results classified according to the different symmetries have been obtained by DF and offer a way to perform diagrammatic expansion around DMFT (Otsuki *et al.*, 2014); these are shown in Fig. 2.35.

It should be noted that DF cannot reproduce the dome structure of $T_c(x)$ dependence with a downturn towards half-filling. In contrast, such a superconducting dome is found in some other diagrammatic extensions of DMFT such as DΓA (Kitatani *et al.*, 2019), FLEX + DMFT (Kitatani *et al.*, 2015), EDMFT + GW, and TRILEX (Vucicevic *et al.*, 2017) in agreement with DCA (Maier *et al.*, 2005a) and the two-particle self-consistent approach (Kyung *et al.*, 2003). As discussed by Kitatani *et al.* (2019), the local

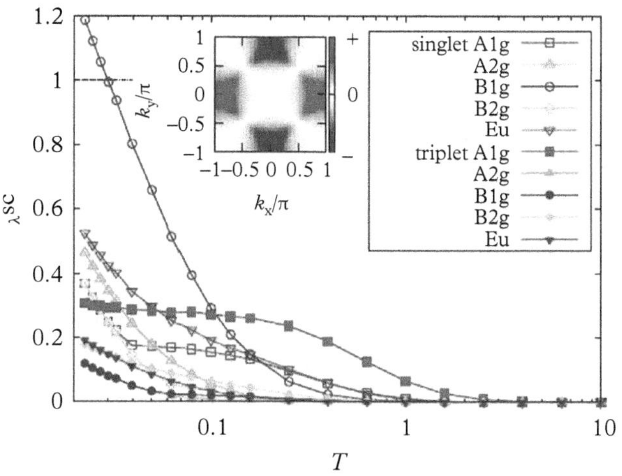

Figure 2.35 *Temperature dependence of the leading eigenvalue in the particle–particle channel for the hole-doped two-dimensional Hubbard model at* $U/t = 8$ *and doping* $x = 0.14$*, separated into contributions for the given symmetry. Only the eigenvalue for B1g symmetry crosses 1, implying the transition to the superconducting state. The inset shows symmetry of the gap. (From Otsuki et al., 2014)*

particle–particle diagrams significantly screen the bare interaction at low frequencies and hence the pairing interaction. Thus, dynamical vertex correction is one of main reasons for T_c suppression from the mean field value.

Conclusions. As we have discussed above, the antiferromagnetic exchange interaction between two electrons (or holes) on an antiferromagnetic background results in their singlet pairing. In the strong correlation regime, this pair is rather local. The strong intra-atomic Coulomb interaction suppresses the probability of two electrons being at one site; hence, the *s*-pairing cannot be realised and the gap in the two-dimensional Hubbard model on a square lattice has the *d* symmetry.

The magnetic mechanism of pairing is assumed to be responsible for high-temperature superconductivity in cuprates and Fe-based pnictides and chalcogenides. Nevertheless, for any compound, the electronic structure contains many bands, and usually more complicated multiband models are required to discuss its electronic, magnetic, and superconducting properties. Only with some assumptions and approximations, a Hubbard model that may be used for cuprates can be obtained from more general multiband $p - d$ model as the effective low-energy model. As for the Fe-based pnictides and chalcogenides, they have to be discussed within the multiband Hubbard-like model. The other important ingredient of the theory of superconductivity in such systems is the electron–phonon pairing, which should be considered to be on equal footing with magnetic and other electronic mechanisms of pairing. All these aspects will be discussed later on.

3

Properties: Spectroscopy

3.1 Macroscopic Quantisation

Many phenomena observed on a macroscopic scale are based on quantum physics, including the linear dependence of electronic heat capacity, ferromagnetism, many transport phenomena, and so on. The same is true about the basic concepts of superconductivity such as, for example, the 'Cooper theorem'. However, in this chapter, more specific concepts that are usually manifested on an atomic scale, such as the quantisation of energy levels and the phase of the wave function, will be discussed. This manifestation occurs on the macroscopic scale in superconductors; this is possible because a highly correlated state of electrons is present. The strongest manifestation of this correlation is the existence of the wave function for a whole superconducting sample:

$$\psi = fe^{i\theta} \tag{3.1}$$

The description of macroscopic quantum phenomena begins with the quantisation of the magnetic flux.

3.1.1 Flux Quantisation

Considering the metallic ring shown in Fig. 3.1, its thickness 'd' greatly exceeds the penetration depth $\delta : d \gg \delta$. Placing the ring in an external magnetic field that is parallel to the ring's axis causes the magnetic field to penetrate the ring. It is in the normal state $(T > T_c)$. If the temperature is decreased below the T_c, then the ring makes a transition into the superconducting state, and this transition is accompanied by the expulsion of the magnetic field from the ring. If the external magnetic field is turned off, then its absence is also expected inside the ring. Nevertheless, the magnetic field inside of the ring stays 'frozen', that is, it continues to be different from zero. This 'strange' behaviour follows directly from the absence of an electric field E inside of the ring; otherwise, in the absence of resistance, this field will accelerate the electrons. The constant value of the magnetic flux $\Phi = \oint \boldsymbol{H} d\boldsymbol{S}$ (integration is taken over the area of the ring) follows directly from the Maxwell equation $\frac{\partial \boldsymbol{H}}{\partial t} = \text{rot}\boldsymbol{E}$. Note that the presence of the 'frozen' magnetic field (Fig. 3.1) is accompanied by the superconducting current so that there

Superconducting State: Mechanisms and Materials. Vladimir Z. Kresin, Sergei G. Ovchinnikov, and Stuart A. Wolf, Oxford University Press (2021). © Kresin, Ovchinnikov, Wolf. DOI: 10.1093/oso/9780198845331.003.0003

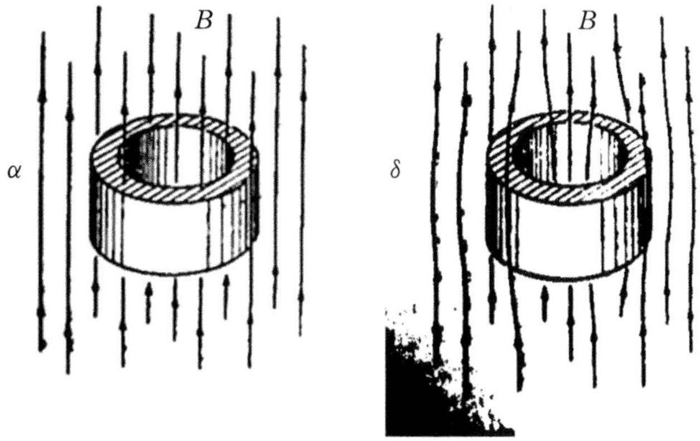

Figure 3.1 *Metallic ring, and magnetic field. (Left) Above T_c. (Right) Below T_c; an external magnetic field is switched off.*

is a self-consistent picture of the dissipationless current, supporting the presence of the magnetic field. This current exists in the thin layer (~δ) near the surface of the ring (remember that its thickness $d \gg \delta$ so that the current inside of the ring is absent).

In order to obtain even more impressive results, the Bohr quantisation condition (the gigantic Bohr orbit) should be used:

$$\oint \boldsymbol{p_s} \boldsymbol{dl} = N\hbar \tag{3.2}$$

where $\boldsymbol{p}_s$ is the momentum of the electronic pair moving in an external magnetic field: $\boldsymbol{p_s} = 2(m\boldsymbol{v_s} + e\boldsymbol{A}/c)$, and $\boldsymbol{A}$ is the vector potential. Since the current $\boldsymbol{j_s} = ne\boldsymbol{v_s}$, one obtains

$$\frac{mc}{ne^2} \oint \boldsymbol{j_s dl} + \Phi = N\frac{\hbar c}{2e} \tag{3.2'}$$

Choosing the line for integration to be in the middle of the ring, the thickness $d \gg \delta$, and the current $j_s = 0$, the following important result is determined:

$$\Phi = N\Phi_0; \Phi_0 = \frac{\hbar c}{2e} \tag{3.3}$$

Therefore, the value of the flux is quantised ($N = 0, 1, 2, \ldots$) and $\Phi_0 = 2 \times 10^{-7}\mathrm{G} \cdot \mathrm{sm}^2$. The permitted values of the flux should form a multisequence. The value of the magnetic flux inside the ring cannot be equal to an arbitrary value but is quantised, according to eqn (3.3).

In order to estimate the scale of this quantisation, a cylinder with an inner diameter of 0.1 mm should be imagined. If the magnetic field inside of this cylinder is equal to

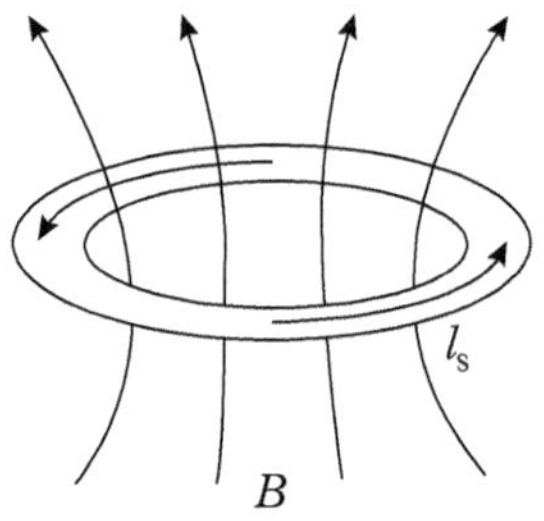

Figure 3.2 *Flux quantisation.*

quantum Φ_0, then it would be equal to approximately 1% of Earth's magnetic field. This value is not small. Therefore, the macroscopic value of the magnetic field and its quantum are involved.

Flux quantisation was observed experimentally in 1961 (Deaver and Fairbanks, 1961; Doll and Nabauer, 1961). Deaver and Fairbanks used a tin pipe with a small inner diameter ($\sim 1.5 \times 10^{-3}$cm (!)). Because of such a small diameter, the magnetic field required to create one magnetic flux quantum (fluxoid) was rather large (0.1 G). The value of the magnetic field was determined by measuring the voltage induced by the tube motion in coils placed near its ends. Indeed, it was observed that the trapped flux has only discrete values (Fig. 3.2).

Note also that the flux quantisation was predicted by F. London in 1951, that is, before the creation of the BCS theory. As mentioned above, the prediction was verified experimentally, but the measured value of the flux quantum was two times lower than that predicted by F. London. This means that the characteristic charge $e^* = 2e$ (e is the electronic charge). This difference is caused by the pairing. According to the BCS theory, the charge in superconductors is transferred by bound pairs of electrons (Cooper pairs). A similar characteristic charge e^* enters the Ginzburg–Landau theory (see section 3.1.3).

3.1.2 The Josephson Effect

In 1962, W. Josephson predicted this effect, which was named after him. The prediction was based on the BCS theory and, especially, on the concept of the phase coherence in the superconducting state. In fact, there are two Josephson effects: (1) the stationary Josephson effect (the 'dc' effect), where the superconducting current can flow through a tunnel junction without any applied voltage, and (2) the non-stationary effect (the 'ac' effect), where the applied constant voltage leads to the appearance of an alternating current flowing across the junction.

The Josephson predictions were verified shortly after the appearance of his paper (Anderson and Rowell, 1963; Yanson *et al.*, 1965). Josephson's initial paper studied the fundamentals of the theory of superconductivity, but very soon the effects formed the basics for superconducting electronics and, correspondingly, for endless applications (see the work by Barone and Paterno, 1982 and Van Duzer and Turner, 1999 and a recent review by Braginski, 2019). As a whole, it is difficult to name another example of

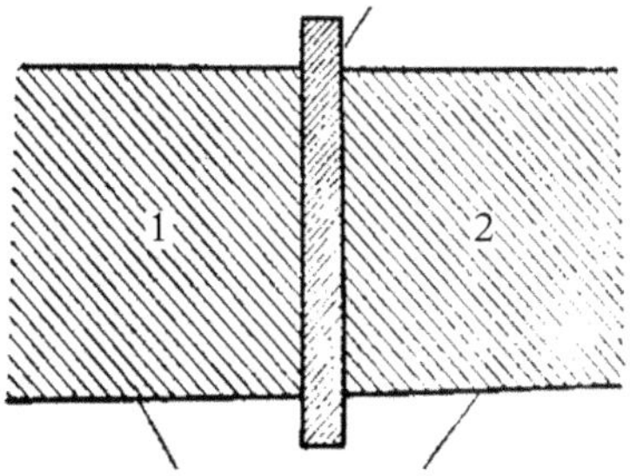

Figure 3.3 *A Josephson contact.*

such a short period of time between the abstract analysis of a fundamental concept and the development of its applications.

Below, both the stationary and non-stationary Josephson effects are discussed.

The stationary Josephson effect. Starting from the well-known quantum mechanical expression for the current density, $\boldsymbol{J} = (ie\hbar/2m)(\psi\nabla\psi^* - \psi^*\nabla\psi)$, and considering a single electron with its wave function $\psi = |\psi| \exp(i\varphi)$ (φ is the phase), then the charge transfer is determined by the phase factor, namely $j \propto \nabla\varphi$. However, such a charge transfer does not provide the macroscopic current because of the chaotic distribution of the phases for the many-electron system.

The situation in superconductors is entirely different, and this difference is caused by the phase correlation (see eqn (3.1)). Considering two superconductors separated by a thin tunnelling barrier (Fig. 3.3), each of these superconductors is described by its wave function; the phase factors are θ_1 and θ_2 (see eqn (3.1)):

$$\psi_1 = n_1^{1/2} \exp(i\theta_1) \tag{3.4}$$

$$\psi_2 = n_2^{1/2} \exp(i\theta_2) \tag{3.4$'$}$$

where n_1 and n_2 are the electron concentrations.

Using the elegant derivation by Fainman (1963), the following general expressions can be written:

$$i\frac{\partial\psi_1}{\partial t} = U_1\psi_1 + K\psi_2 \tag{3.5}$$

$$i\frac{\partial\psi_2}{\partial t} = U_2\psi_2 + K\psi_1 \tag{3.5$'$}$$

The quantity K describes the tunnelling channel connecting the superconductors. Substituting the expressions (3.4) and (3.4′) into (3.5) and (3.5′), respectively, and equalising the real and imaginary parts, the following equations are obtained:

$$\frac{\partial n_1}{\partial t} = 2(n_1 n_2)^{1/2} K \sin\delta \tag{3.6}$$

$$\frac{\partial n_2}{\partial t} = -2(n_1 n_2)^{1/2} K \sin\delta \tag{3.6'}$$

$$-\frac{\partial \delta}{\partial t} = (U_2 - U_1) \tag{3.7}$$

Focusing on eqns (3.6) and (3.6′) (eqn (3.7) will be discussed in the next section), if the tunnelling is negligibly small or absent ($K = 0$), then $n_1 =$ constant. Therefore, this derivative, which is proportional to the tunnelling probability, directly represents the tunnelling current:

$$J = J_{\max} \sin\delta \tag{3.8}$$

where $\delta = \theta_2 - \theta_1$ is the phase difference across the barrier and $J_{\max} = 2(n_1 n_2)^{1/2} K$ is the amplitude of the current.

Therefore, it can be seen from eqn (3.8) that the superconducting current flows through the tunnelling barrier, and the presence of such a current does not require any external voltage. The current is caused by the phase difference. As mentioned above, the stationary Josephson effect was experimentally observed shortly after the appearance of the theoretical paper. It is remarkable that the phase of the wave function, which is a fundamental quantum mechanical quantity, is manifested in the macroscopic experiment, and the Josephson current can be observed.

The Josephson current can be imagined as the flow of Cooper pairs. However, it should be stressed that the probability of tunnelling is the same as that for a single electron. The correlation (pairing) does not affect the probability of tunnelling.

The current–voltage characteristic curve is plotted in Fig. 3.4. The value of the dc Josephson current increases at zero voltage (line ‘a’), up to its maximum value, J_m. Then the switch to the usual one-particle tunnelling dependence (line ‘c’) can be observed. If the external current in the source is fixed, then the transition occurs along the line ‘b’.

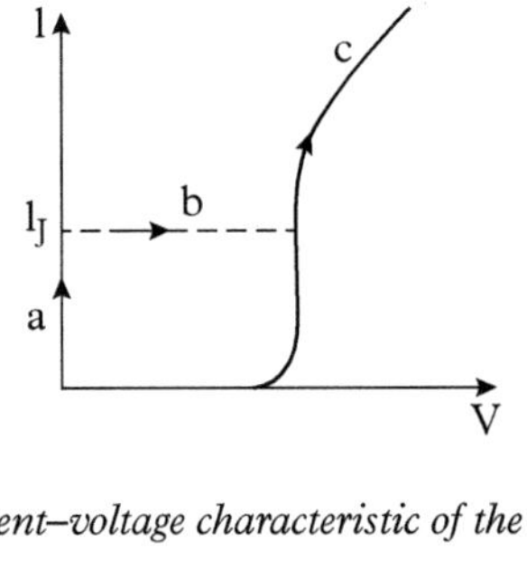

Figure 3.4 *The current–voltage characteristic of the dc Josephson junction.*

A more rigorous evaluation can be performed with the use of the many-body theory (see Kulik and Yanson, 1972). More specifically, the tunnelling Hamiltonian can be used:

$$\hat{H}_T = \sum_{\boldsymbol{p},\boldsymbol{q}} T_{\boldsymbol{p},\boldsymbol{q}} a_{\boldsymbol{p}}^{+} b_{\boldsymbol{q}}$$

where the operators $a_{\boldsymbol{p}}^{+}$ and $b_{\boldsymbol{q}}$ describe the electronic states in the first and second superconductors, respectively (see Fig. 3.3), and $T_{\boldsymbol{p},\boldsymbol{q}}$ is the so-called tunnelling matrix element. The tunnelling Hamiltonian method was developed after the appearance of the seminal paper by Bardeen (1961) (see also Kane, 1969). It is well known that quantum mechanical tunnelling is usually treated with the stationary Schrödinger equation with the corresponding matching solutions inside of the barrier. Bardeen's approach is based on the time-dependent Schrödinger equation, and the tunnelling is a quantum transition between electrodes 1 and 2 (it is described by the 'Golden Rule'; see Landau and Lifshitz, 1977). Such a transition should be provided by a Hamiltonian, and the tunnelling Hamiltonian plays this role.

The current is described by the relation: $I = e\langle \dot{n}_1 \rangle$ (cf. eqn (3.6)), and this relation can be written as follows:

$$I = -ie\langle [n_1, H_T] \rangle\,; \text{where n}_1 \text{ is the operator } n_1 = \sum_p a_p^{+} a_p$$

With the use of the total amplitude of the transition

$$S(t) = T\exp\left(-i\int_{-\infty}^{\infty} H_T(t')\,dt'\right)$$

the expression for the tunnelling current (eqn (3.8)) can be reduced to the following form:

$$I_{\max} = \frac{T}{neR}\sum_{\omega_n}\int d\xi_{\boldsymbol{p}}\int d\xi_{\boldsymbol{q}} F_1(\boldsymbol{p},\omega_n)\, F_2(\boldsymbol{q},-\omega_n)$$

where R is the so-called tunnelling resistance, $R^{-1} = 4\pi e^2 \nu_1 \nu_2 \langle |T_{tun}^2| \rangle$, ν_1 and ν_2 are the electronic densities of states near the Fermi level, $\langle |T_{tun}^2| \rangle$ is the average value of T_{tun}^2, T_{tun} is the tunnelling matrix element, and F_1 and F_2 are the pairing Green's functions (see eqn (2.13′) for superconductors 1 and 2, respectively. It is convenient to use the thermodynamic Green's function ($\omega_n = (2n+1)\pi T$) (see e.g. Abrikosov *et al.*, 1975 and appendix E).

Therefore, the temperature dependence of the stationary Josephson current can be studied. Substituting the expression

$$F(\mathbf{p},\omega_n) = \frac{\Delta}{\omega_n^2 + \Delta^2 + \xi_p^2}$$

into the equation for the amplitude of the current $I_{\max}$, one obtains:

$$I_{\max} = \frac{\pi}{eR} T \sum_{\omega_n} \tilde{F}(\Delta_1)\tilde{F}(\Delta_2)$$

where $\tilde{F}(\Delta) = \Delta\left(\Delta^2 + \omega_n^2\right)^{-1}$.

Considering the case of similar superconductors, then $\Delta_1 = \Delta_2 \equiv \Delta(T)$, and the summation can be easily performed. As a result, the following expression is obtained:

$$I_{\max} = \frac{\pi}{2}\frac{\Delta(T)}{eR}\,\mathrm{th}\frac{\Delta(T)}{2T} \tag{3.8$'$}$$

The amplitude of the current has a maximum value at $T = 0$ and then decreases as $T \to T_c$. The temperature dependence of the stationary Josephson current is close to that for the energy gap.

Therefore, the tunnelling junction (Fig. 3.1) below T_c is capable of transferring the superconducting current, which flows in the absence of an external voltage; its value is determined by the phase difference δ (eqn (3.8)) with the amplitude, which is temperature dependent, according to eqn (3.8′).

The value of the direct current depends strongly on the phase difference. Note that, in reality, the value of the current is determined by an internal source, and the phase difference is adjusted correspondingly so that eqn (3.8) is satisfied.

Note also that the dc Josephson current can be visualised as the flow of Cooper pairs but, as mentioned above, the probability of Josephson tunnelling is the same as that for the tunnelling of a single electron. This can be seen directly from the expression for the amplitude of the current $I_{\max}$ with its dependence on the tunnelling resistance: $I_{\max} \sim R^{-1}$. The expression for the tunnelling resistance contains the tunnelling matrix element $R^{-1} \propto \langle |T_{tun}^2| \rangle$. As for the tunnelling matrix element, it determines the Josephson tunnelling at $T < T_c$ as well as the one-electron tunnelling in the normal state ($T > T_c$). In both cases, the corresponding expressions contain a quadratic power of the tunnelling matrix element, that is, their probabilities are described by the lowest order of the expansion of the tunnelling amplitude in the powers of T_{tun}.

In order to complete the description of the fundamentals of the stationary Josephson effect, the tunnelling junction (Fig. 3.1) should be considered in an external magnetic field. Assuming that the field is parallel to the junction, and, more specifically, to one of its edges, for example, $\vec{B} \| OY$ (the axis OZ is chosen to be perpendicular to the junction), then the presence of the magnetic field affects the phase of the junction by adding the value $\oint \mathbf{A}d\mathbf{l}$, where $\mathbf{A}$ is the vector potential; the contour of the integration consists of two lines inside of electrodes 1 and 2. They are parallel to the axis OX and two short lines along the axis OZ so that a closed loop can be formed. The additional phase factor $\oint \boldsymbol{A}dl = \int \boldsymbol{H}ds = \Phi$ is equal to the total flux inside the tunnelling barrier. The total current can be calculated as follows:

$$\mathrm{I}_{\max} = \int \mathrm{j}_{\mathrm{m}} \sin\delta \, ds = \int \mathrm{j}_{\mathrm{m}} \mathrm{I}_{\mathrm{m}} e^{i\delta} d\delta$$

The integration is taken over the surface of the junction, leading to the following dependence:

$$I_{\max} = I^0_{\max} \left| \frac{\sin(\pi\Phi/\Phi_0)}{\pi\Phi/\Phi_0} \right| \tag{3.9}$$

where $I^0_{\max} = I_{\max}$ $(H = 0)$, and Φ_0 is the flux quantum (eqn (3.3)). Therefore, the stationary Josephson current oscillates as a function of the applied magnetic field (Fig. 3.5). The period of oscillations is directly related to the value of the flux quantum Φ_0. The tunnelling current is absent if the total flux is a multiple of Φ_0 (see the work by Langenberg *et al.*, 1966 and Solymar, 1972). It is remarkable that the dc Josephson effect demonstrates both of the fundamental quantum mechanical concepts: the phase of the wave function and the quantisation of the spectrum.

The non-stationary Josephson effect. As noted at the beginning of this chapter, the non-stationary ('ac') Josephson effect is created by the constant voltage V, which is applied

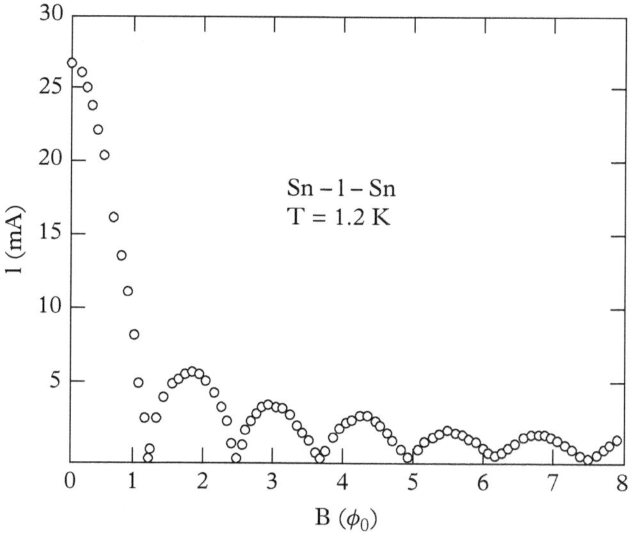

Figure 3.5 *The Josephson current as a function of the magnetic field. (After Landenberg et al., 1966)*

to the tunnelling junction. In usual metals, the existence of such a voltage leads to the appearance of the direct current. For superconductors, the situation is different. The direct current flows through the junction without any applied voltage. If the voltage V is applied, it produces work equivalent to 2 eV. Since this additional energy is not needed for the current flow, it is radiated by the electrons as a quantum $\varepsilon = \hbar\omega$ so that $\hbar\omega = 2$ eV. As a result, the presence of the alternating current is the source of the observed radiation. Therefore, the applied fixed voltage V creates the ac tunnelling. This is the qualitative picture of the non-stationary Josephson effect.

A more detailed consideration follows from eqn (3.7). In the absence of an applied voltage, $U_1 = U_2$ and the phase difference $\delta =$ constant. The voltage applied across the junction creates the relation $U_1 = U_1 = 2$ eV. Then the phase difference becomes time dependent, and δ can be written in the form

$$\delta = \delta_0 + 2e \int_0^t V(t')\, dt'$$

The Josephson current is described by eqn (3.8), but the value of the phase difference depends on time. If the applied voltage $V =$ constant, then the following equation can be obtained with the use of eqn (3.8):

$$I = I_c \sin(\delta_0 + 2eVt)$$

Indeed, an alternating current with a frequency $\omega = 2$ eV can be observed. The radiation caused by this current was directly detected by Yanson *et al.* (1965).

Considering the case when an ac field is applied across a junction in addition to the dc voltage V_0, the total voltage $V(t)$ is expressed as follows:

$$V = V_0 + V_1 \cos \Omega t \tag{3.10}$$

Then, based on eqn (3.7),

$$I = I_{\max} \sin\left(\delta_0 + 2eV_0 t + (2eV_1/\Omega) \sin \Omega t + \tilde{\delta}\right)$$

As a next step, the following identity can be used:

$$\exp(if \sin x) = \sum_{n=-\infty}^{\infty} J_n(f) \exp(inx)$$

where $J_n(x)$ is the Bessel function, and the current can be expressed in the form

$$I = I_{\max} \sum_{n=-\infty}^{\infty} (-1)^n J_n\left(\frac{2eV_1}{\hbar\Omega}\right) \sin\left[\left(\frac{2eV_0}{\hbar} - n\Omega\right) t - n\delta_0 + \tilde{\delta}\right]$$

The most interesting feature of this expression is that the dc can be observed if the external voltage satisfies the condition:

$$2eV_0 = n\hbar\Omega \tag{3.10'}$$

Indeed, the total tunnelling current does not depend on time and, as a result, jumps corresponding to various values '*n*' ('Shapiro steps') can be observed. These steps were first observed by Shapiro (1963) and then by Grimes and Shapiro (1968) (see Fig. 3.6). In addition, D. Tilley and J. Tilley (1986) described an interesting manifestation of the non-stationary Josephson effect.

The properties at an isolated Josephson junction are discussed above. Furthermore, the properties of the network containing many junctions can be studied. The analysis of such networks is interesting because it may lead to many potential applications. Each junction placed in an external voltage V_0 will radiate with the frequency $\omega = 2eV_0/\hbar$. Considering a network consisting of N interconnected tunnelling junctions with an external voltage V_0 applied to the ends of the network, each junction would radiate. Therefore, there is the interesting question of the synchronisation of a whole network (see Fig. 3.7). If all junctions are similar, then each of them is affected by an applied

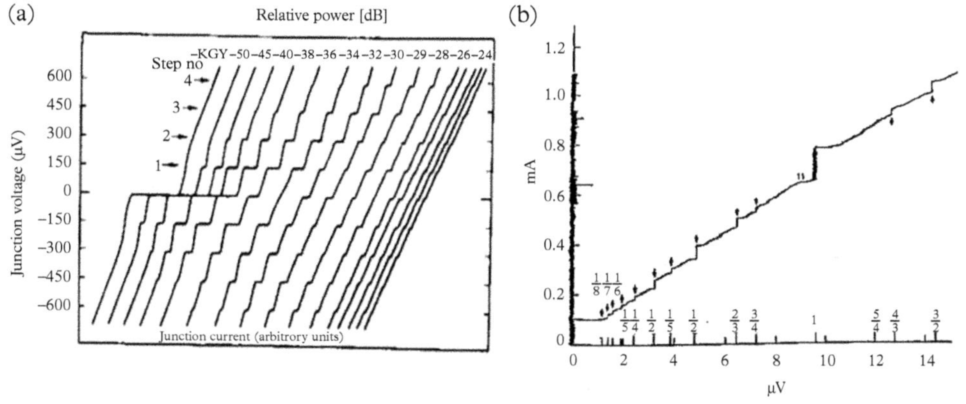

Figure 3.6 *Shapiro steps. From D. Tilley and J. Tilley, (1986)*

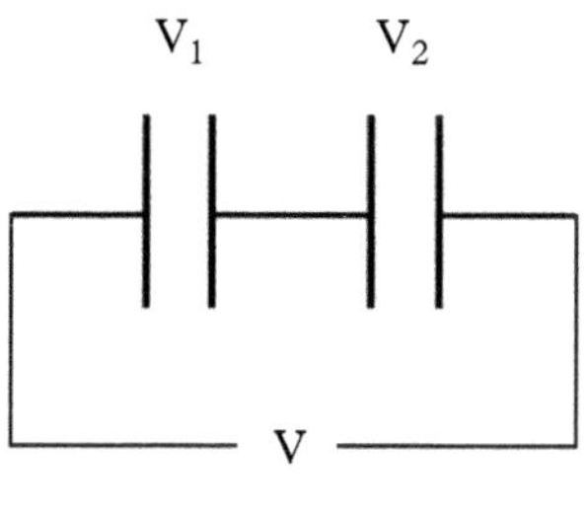

Figure 3.7 *A Josephson network.*

voltage $\delta V = V_0/N$, and synchronised radiation is present so that $\hbar\omega = 2eV_0/N$. The measurements were mainly focused on the building of such networks with equivalent junctions (e.g. Cawthorne *et al.*, 1999). However, synchronisation and even the Schapiro steps for a network containing non-equivalent junctions can be observed (Kresin and Ovchinnikov, 2013). It is hypothesised that, in such a case, the voltages applied to different junctions are different and a chaotic distribution of frequencies should result. However, this is not the case; the synchronisation occurs, thanks to charge conservation, that is, the equality of the currents flowing through the neighbouring junctions:

$$j^{s}_{max}\left(2e\int_0^t V_s dt + \theta_s\right) = j^{s+1}_{max}\left(2e\int_0^t V_{s+1} dt + \theta_{s+1}\right)$$

This condition should be satisfied along with the relation

$$\sum_{s=1}^{N} V_s = V_0 = \text{constant}$$

Indeed, the analysis shows that, for equivalent junctions, all of them radiate with the frequency $\omega = 2eV_0/\hbar N$. For example, in the simplest case of two equivalent junctions, each of them is affected by the voltage $V_0/2$. As for different junctions, the solution is less trivial. In this case, the distribution of voltages applied to such junctions is rather peculiar: it is time dependent, whereas the total voltage at the end of the networks is fixed. For the case of two interconnected junctions with different tunnelling resistances R_1 and R_2 and, correspondingly, two different amplitudes $j^{(1)}_{\max}$ and $j^{(2)}_{\max}$, the voltages are $V_1 \equiv V_1(t)$ and $V_2 = V_0 - V_1(t)$. They are time dependent, but the radiation appears to be synchronised, with frequencies $\omega = 2eV_0/\hbar$, $4eV_0/\hbar$; they are multiples of the total applied voltage.

The Shapiro steps also can be observed. Indeed, such steps were observed for the network containing two different $YBa_2Cu_3O_{7-\delta}$ junctions under the additional 12 GHz microwave radiation (Chen *et al.*, 2005).

3.1.3 The Ginzburg–Landau theory, and Vortices

The Ginzburg–Landau theory (1950) was the first quantum theory of superconductivity. The idea of Cooper pairs was not known in 1950. The Ginzburg–Landau theory was a phenomenological one based on the general ideas of the Landau (1937) theory of phase transitions. Its success demonstrates the power of a general symmetry-based approach joined with physical intuition. The initial version of the Ginzburg–Landau theory corresponds to the local London and London (1935) equations. Later Bardeen (1954) showed that it may also be formulated in agreement with a non-local Pippard (1953) type version. Microscopic derivation of the Ginzburg–Landau equation from the BCS theory has been done by Gor'kov (1959).

Superconductivity as an example of the Landau theory of second-order phase transitions. In the Landau theory of second-order phase transitions, the free energy of the ordered phase close to the transition temperature may be written as the power series of the order parameter. For superconductors, the electronic wave function $\Psi(\mathbf{r})$ is taken as the order parameter with the condition that $|\Psi|^2 = n_s/2$, where n_s is the density of superconducting electrons, In the simplest case of a uniform superconductor without an external magnetic field, Ψ does not depend on $\mathbf{r}$. The free energy of the superconducting phase close to T_c may be written as

$$F_{s0} = F_n + \alpha|\Psi|^2 + \frac{\beta}{2}|\Psi|^4 \tag{3.11}$$

Here F_n is the normal phase free energy, and α and β are some phenomenological temperature dependent parameters of a material. Standard minimisation of the free energy results in

$$|\Psi_0|^2 = -\alpha/\beta \tag{3.12}$$

and the free energies difference

$$F_n - F_{s0} = {}^{\alpha^2}\!/_{2\beta} \tag{3.13}$$

Landau has assumed the parameters may be written as

$$\alpha(T) = \tilde{\alpha}(T - T_c) \tag{3.14}$$

while $\beta > 0$ and is temperature independent. With this choice of parameters, the superconducting phase exists below T_c, and the normal phase is stable above T_c. Moreover, the critical value of the uniform external magnetic field ${}^{H_c^2}\!/_{8\pi} = F_n - F_{s0}$ depends on temperature as $H_c \sim T_c - T$, in agreement with experiment.

The Ginzburg–Landau functional and equations. Ginzburg and Landau (1950) generalised the ideas discussed above to a spatially non-uniform superconductor in an external magnetic field. They wrote the Gibbs free energy density close to T_c as

$$g_{sH} = g_n + \alpha|\Psi|^2 + \frac{\beta}{2}|\Psi|^4 + \frac{1}{2m^*}\left|-i\hbar\nabla\Psi - \frac{e^*}{c}\mathbf{A}\Psi\right|^2 + \frac{H^2}{8\pi} - \frac{\mathbf{H}\mathbf{H}_0}{4\pi} \tag{3.15}$$

Here $\mathbf{H}_0$ is the uniform external magnetic field, and $\mathbf{H}(\mathbf{r})$ is the microscopic non-uniform magnetic field. The gradient term is a kinetic energy of superconducting electrons. Indeed, in quantum mechanics, the kinetic energy of a particle with mass m is equal to

$$\frac{1}{2m}|-i\hbar\nabla\Psi|^2$$

When a particle has electric charge *e* and moves in the field given by a vector potential ***A***, its momentum is given by

$$-i\hbar\nabla \to -i\hbar\nabla - \frac{e}{c}\boldsymbol{A} = m\boldsymbol{v} \tag{3.15'}$$

The Ginzburg–Landau theory contains the effective charge and mass of superconducting electrons; after the development of the BCS theory, it become clear that $e^* = 2e$, $m^* = 2m$. The total free energy of a superconductor in the Ginzburg–Landau theory is equal to

$$G_{sH} = G_n + \int \left[\alpha|\Psi|^2 + \frac{\beta}{2}|\Psi|^4 + \frac{1}{4m}\left|-i\hbar\nabla\Psi - \frac{2e}{c}\mathbf{A}\Psi\right|^2 + \frac{(\mathrm{rot}\mathbf{A})^2}{8\pi} - \frac{\mathrm{rot}\mathbf{A}\cdot\mathbf{H}_0}{4\pi}\right] dV \tag{3.16}$$

One has to obtain such equations for unknown functions $\Psi(\boldsymbol{r})$ and $\boldsymbol{A}(\boldsymbol{r})$ to minimise the Ginzburg–Landau free energy when solutions of these equations are used to calculate the integral in (3.16). To solve this standard variation problem, we start with the variation over $\Psi^*(\boldsymbol{r})$, assuming $\Psi(\boldsymbol{r})$ and $\boldsymbol{A}(\boldsymbol{r})$ fixed:

$$\delta_{\Psi^*} G_{sH} = 0 \tag{3.17}$$

Separating the volume and surface integrals in the left side of eqn (3.17), one finds the first Ginzburg–Landau equation and the broader condition

$$\alpha\Psi + \beta\Psi|\Psi|^2 + \frac{1}{4m}\left(i\hbar\nabla + \frac{2e}{c}\mathbf{A}\right)^2\Psi = 0 \tag{3.18}$$

$$\left(i\hbar\nabla\Psi + \frac{2e}{c}\mathbf{A}\Psi\right)\cdot\mathbf{n} = 0 \tag{3.19}$$

where **n** is the unit vector orthogonal to the superconductor surface. The variation over the vector potential $\boldsymbol{A}(\boldsymbol{r})$ results in the second Ginzburg–Landau equation,

$$\boldsymbol{j}_s = -i\frac{\hbar e}{2m}(\Psi^*\nabla\Psi - \Psi\nabla\Psi^*) - \frac{2e^2}{mc}|\Psi|^2\mathbf{A} \tag{3.20}$$

where the current density j_s in superconductor is given by

$$j_s = \frac{c}{4\pi} rot\ rot\mathbf{A}, \quad \mathbf{H} = rot\mathbf{A}$$

Introducing the dimensionless wave function $\psi(\mathbf{r}) = |\psi| \, e^{i\theta} = \frac{\Psi(\mathbf{r})}{\Psi_0}, |\Psi_0|^2 = \frac{n_s}{2} = \frac{|\alpha|}{\beta}$ and the notations

$$\xi^2 = \frac{\hbar^2}{4m|\alpha|}, \lambda^2 = \frac{mc^2}{4\pi n_s e^2} = \frac{mc^2\beta}{8\pi e^2|\alpha|} \tag{3.21}$$

we may write down the Ginzburg–Landau equations in a more convenient and compact form:

$$\xi^2\left(i\nabla + \frac{2\pi}{\Phi_0}\mathbf{A}\right)^2\psi - \psi + \psi|\psi|^2 = 0 \tag{3.22}$$

$$rot\ rot\mathbf{A} = \frac{|\psi|^2}{\lambda^2}\left(\frac{\Phi_0}{2\pi}\nabla\theta - \mathbf{A}\right) \tag{3.23}$$

Here $\Phi_0 = \pi\hbar c/e$ is the flux quantum. The parameters introduced above are the coherence length ξ and the penetration length of the magnetic field λ. These parameters are temperature dependent, close to T_c:

$$\xi \sim (T_c - T)^{-0.5}, \lambda \sim (T_c - T)^{-0.5} \tag{3.24}$$

These parameters determine the Ginzburg–Landau parameter $\varkappa$:

$$\varkappa = \lambda/\xi \tag{3.25}$$

This parameter makes it possible to categorise all superconductors under an external magnetic field into two groups: those with $\varkappa < 1/\sqrt{2}$, and those with $\varkappa > 1/\sqrt{2}$.

Now we consider the gauge invariance of the Ginzburg–Landau theory. The vector potential $\boldsymbol{A}(\boldsymbol{r})$ in the Ginzburg–Landau equations can be changed by the gauge transformation

$$\boldsymbol{A} = \boldsymbol{A}' + \nabla\varphi \tag{3.26}$$

where $\varphi(\mathbf{r})$ is an arbitrary scalar function. It is easy to check that the Ginzburg–Landau equations are gauge invariant if the transformation (3.26) for the potential is combined with the transformation of the wave function

$$\psi = \psi' e^{i\frac{2\pi}{\Phi_0}\varphi(\mathbf{r})} \tag{3.27}$$

The Ginzburg–Landau equations are widely used to describe non-uniform superconductivity, including the proximity effect, the critical field, and the critical current of superconducting thin films, fluctuations near T_c, Abrikosov vortices in superconductors under an external magnetic field with $\varkappa > 1/\sqrt{2}$, and so on.

Microscopic derivation via the Gor'kov approach. In the initial version of the Ginzburg–Landau theory, the temperature was assumed to be close to T_c, to have small values for the superconducting gap and its space variation. Its microscopic derivation from the BCS theory by Gor'kov (1959) was also restricted by temperatures near T_c. The most general version of the Ginzburg–Landau theory valid for arbitrary gap values was derived by Eilenberger (1966). Gor'kov (1959) used an integral equation for the Green's functions. In the dirty limit $l \ll \xi_0$, the integral equations may be reformulated as differential equations (de Gennes, 1964; Maki, 1966). Differential equations describe local effects. Determining the non-local electrodynamics of superconductors requires integral equations, which was considered by Eilenberger (1967) in the limit of small order parameters.

We shall start from the Gor'kov equations for the thermodynamic Green's functions $G(r,\tau;r',\tau'), F^{+}(r,\tau;r',\tau')$. These equations are in a magnetic field for the Fourier transforms $G_\omega(r,r'), F^{+}_\omega(r,r')$ and, over $u=\tau-\tau'$, have the form

$$\left\{i\omega_n + \frac{1}{2m}(\boldsymbol{\nabla} - \mathrm{ie}\boldsymbol{A}(\boldsymbol{r}))^2 + \mu\right\} G_\omega(\boldsymbol{r},\boldsymbol{r}') + \Delta(\mathbf{r}) F^{+}_\omega(\boldsymbol{r},\boldsymbol{r}') = \delta(\boldsymbol{r}-\boldsymbol{r}')$$

$$\left\{-i\omega_n + \frac{1}{2m}(\boldsymbol{\nabla} + \mathrm{ie}\boldsymbol{A}(\boldsymbol{r}))^2 + \mu\right\} F^{+}_\omega(\boldsymbol{r},\boldsymbol{r}') - \Delta^{*}(\mathbf{r}) G_\omega(\boldsymbol{r},\boldsymbol{r}') = 0 \tag{3.28}$$

where $\Delta^{*}(\mathbf{r}) = \mathrm{T}\sum_{\mathrm{n}} F^{+}_\omega(\boldsymbol{r},\boldsymbol{r})$. Introducing the Green's function $G^{(0)}_\omega(\boldsymbol{r},\boldsymbol{r}')$ of an electron in a normal metal in the same magnetic field, eqns (3.28) can be written in the integral form

$$G_\omega(\boldsymbol{r},\boldsymbol{r}') = G^{(0)}_\omega(\boldsymbol{r},\boldsymbol{r}') - \int G^{(0)}_\omega(\boldsymbol{r},\boldsymbol{l})\ \Delta(\mathbf{l}) F^{+}_\omega(\boldsymbol{l},\boldsymbol{r}')\, d^3\boldsymbol{l}$$

$$F^{+}_\omega(\boldsymbol{r},\boldsymbol{r}') = \int G^{(0)}_{-\omega}(\boldsymbol{l},\boldsymbol{r})\ \Delta^{*}(\boldsymbol{l}) G_\omega(\boldsymbol{l},\boldsymbol{r}')\ \boldsymbol{l} \tag{3.29}$$

Near the transition temperature, $\Delta(\mathbf{r})$ (as well as $F^{+}_\omega(\boldsymbol{r},\boldsymbol{r}')$) is small, and eqns (3.29) may be expanded in powers of Δ expressing $F^{+}_\omega(\boldsymbol{r},\boldsymbol{r}')$ up to terms of the fourth order, and $G_\omega(\boldsymbol{r},\boldsymbol{r}')$ up to terms of the second order. This procedure results in an equation connecting $\Delta(\mathbf{r})$ and potential $\boldsymbol{A}(\boldsymbol{r})$:

$$\begin{aligned}\Delta^{*}(\mathbf{r}) = {} & \mathrm{gT}\sum_{\mathrm{n}} \int G^{(0)}_\omega(\boldsymbol{r},\boldsymbol{r}') G^{(0)}_{-\omega}(\boldsymbol{r},\boldsymbol{r}') \Delta^{*}(\mathbf{r}')\, \mathrm{d}^3\mathbf{r}' \\ & - \mathrm{gT}\sum_{\mathrm{n}} \int G^{(0)}_\omega(\boldsymbol{s},\boldsymbol{r}) G^{(0)}_{-\omega}(\boldsymbol{s},\boldsymbol{l}) G^{(0)}_\omega G^{(0)}_{-\omega}\ \Delta(\mathbf{s}) \Delta^{*}(\mathbf{l}) \Delta^{*}(\mathbf{m})\, d^3\boldsymbol{s}\, d^3\boldsymbol{l} d^3\boldsymbol{m}\end{aligned} \tag{3.30}$$

The important distances in eqn (3.30) are of the order $\xi_0 = \frac{\hbar v}{T_c}$; at the same time, the changes in the gap $\Delta(\mathbf{r})$ and the potential $\boldsymbol{A}(\boldsymbol{r})$ take place at distances of the order of the penetration depth λ, which is much larger than ξ_0 near T_c. Using these facts, it is possible to perform the integration in eqn (3.30) and obtain the following equations (Gor'kov, 1959):

$$\left\{\frac{1}{2m}(\nabla + \mathrm{ie}\boldsymbol{A}(\boldsymbol{r}))^2 + \frac{1}{\lambda}\left[\frac{T_c - T}{T_c} - \frac{3\lambda}{2E_F}|\Delta(\mathbf{r})|^2\right]\right\}\Delta^*(\mathbf{r}) = 0 \tag{3.31}$$

$$\boldsymbol{j}(\boldsymbol{r}) = \left\{\frac{ie}{m}(\Delta\nabla\Delta^* - \Delta^*\nabla\Delta) - \frac{4e^2|\Delta|^2}{mc}\boldsymbol{A}\right\}\frac{7\zeta(3)n_e}{16(\pi T)^2} \tag{3.32}$$

Here $\zeta(x)$ is Riemann's zeta function, and n_e is the electronic concentration. Introducing the 'wave function'

$$\Psi(r) = \Delta(\mathbf{r})\sqrt{7\zeta(3)n_e \Big/ 4\pi T_c}$$

one can notice that eqns (3.31) and (3.32) are completely analogous to the Ginzburg–Landau equations with the charge $e^* = 2e$.

Abrikosov vortices. One very important prediction derived from the Ginzburg–Landau theory was obtained by Abrikosov (1957), who has developed the theory of type II superconductivity and explained its magnetic properties. The most unusual was the intermediate Shubnikov phase, where superconductivity can exist in a magnetic field without the Meissner effect (Shubnikov *et al.*, 1937). The typical magnetisation curve for a type II superconductor is shown in Fig. 3.8.

When the external field H_0 is smaller the first critical field H_{c1}, the internal field inside the superconductor $B = 0$. When $H_{c1} < H_0 < H_{c2}$, the non-zero field B appears smaller than H_0, and superconductivity coexists with the external field. Finally, when $H_0 = H_{c2}$, superconductivity disappears in the sample volume. Abrikosov (1957) has shown that the magnetic field in the intermediate Shubnikov phase at $H_{c1} < H_0 < H_{c2}$ penetrates into the sample as a set of vortices. Each vortex has a thin cylindrical core in the normal state, with a zero superconducting order parameter $\Psi(r) = 0$, where the core radius is of the order of the coherence length ξ. The superconducting current circulates without damping around the core, providing an inner magnetic field H parallel to the external one. The supercurrent flow takes place in the region with the radius of penetration length λ and, in type II superconductors, usually $\lambda \gg \xi$.

One vortex provides one quantum of the magnetic flux $\Phi_0 = \pi\hbar c/e$. Previously for superfluid ^{4}He, Onsager (1949) and later, independently, Feynman (1955) showed that vorticity enters by quantised vortex lines. When $H_0 > H_{c1}$, vortices inside the superconductor form the triangular lattice with the distance between them of the order of λ. With further increase of the external magnetic field H_0, the density of vortices increases and the lattice period decreases. At $H_0 = H_{c2}$, the lattice parameter is of the of the coherence length ξ. The neighbour vortices touch each other by the normal core, and the order parameter $\Psi(r) = 0$ throughout the whole volume. The second-order phase transition on the normal state occurs. There are many experimental confirmations of the

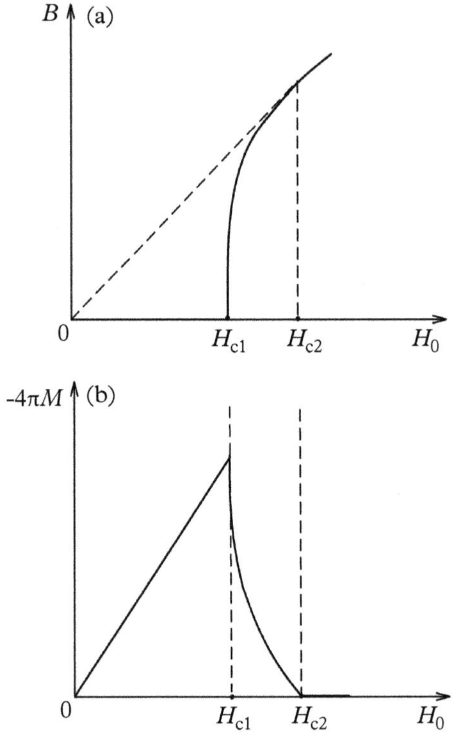

Figure 3.8 *Magnetic induction B (a) and magnetisation M (b) as a function of the external magnetic field H_0 in the type II superconductor cylinder with a magnetic field parallel to the cylinder axes.*

vortices lattice, by neutron diffraction (Cribier *et al.*, 1964) and by electron microscopy (Essmann and Trauble, 1967).

Let us consider a single vortex in an infinite superconductor with the Ginzburg–Landau parameter $\varkappa \gg 1$ and $\lambda \gg \xi$. At a distance from the centre of the core $r \gg \xi$, one has $|\Psi(r)|^2 = 1$. In this area, the Ginzburg–Landau eqn (3.23) for the vector potential looks like

$$rot\ rot\mathbf{A} = \frac{1}{\lambda^2}\left(\frac{\Phi_0}{2\pi}\nabla\theta - \mathbf{A}\right)$$

Calculating *rot* from both parts of this equation, we may write

$$\boldsymbol{H} + \lambda^2 rot\ rot\mathbf{H} = \frac{\Phi_0}{2\pi}\mathrm{rot}\nabla\theta \tag{3.33}$$

For any $r \neq 0$, we have $rot\nabla\theta = 0$ but, in the centre of the core, there is a peculiarity: $|\nabla\theta| \to \infty$. The whole rotation around the core line results in a 2π phase change, so we may write (here e_v is an unit vector along the vortex)

$$\mathrm{rot}\nabla\theta = 2\pi\delta(\boldsymbol{r})\boldsymbol{e}_v$$

Equation (3.33) now may be written as

$$\boldsymbol{H} + \lambda^2 rot\ rot\mathbf{H} = \Phi_0\ \delta(\boldsymbol{r})\boldsymbol{e}_v \tag{3.34}$$

with the boundary condition $H(\infty) = 0$. The solution of this equation is given by

$$H(r) = \frac{\Phi_0}{2\pi\lambda^2}\mathrm{K}_0\left(\frac{\mathrm{r}}{\lambda}\right) \tag{3.35}$$

where $\mathrm{K}_0(\mathrm{z})$ is the MacDonald function with asymptote

$$\mathrm{K}_0(\mathrm{z}) \sim \begin{cases} \ln(1/z) & at\ z \ll 1, \\ e^{-z}/z^{1/2} & at\ z \gg 1 \end{cases} \tag{3.36}$$

This solution results in the infinite magnetic field in the core centre. This artefact indicates that solution (3.35) is not valid at $r \sim \xi$. With logarithmic precision, the value of $H(0)$ may be obtained by numerical solution of the Ginzburg–Landau equations (Hu, 1972):

$$H(0) = \frac{\Phi_0}{2\pi\lambda^2}(\ln \varkappa - 0.28) \tag{3.37}$$

The space distribution of the order parameter and the magnetic field of a separate vortex is shown in Fig. 3.9.

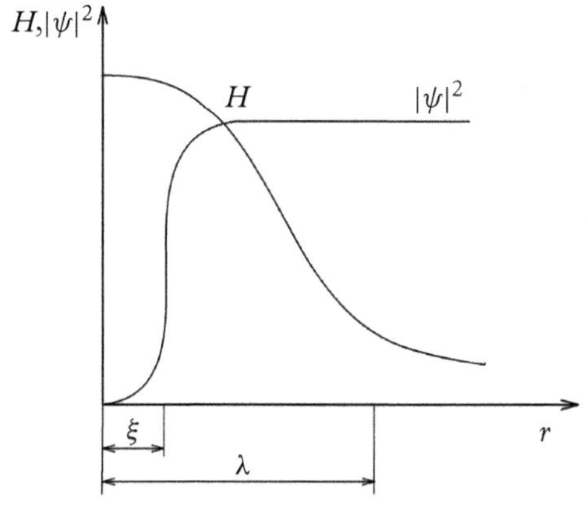

Figure 3.9 *Distribution of the order parameter and the field in a single vortex.*

3.1.4 The Little–Parks Effect

The Little–Parks effect was discovered and described by Little and Parks (1964) (see also the reviews by de Gennes (1996), Tinkham (1996), and Waldram (1996)) and represents a remarkable manifestation of macroscopic quantisation.

Considering a thin cylindrical film placed in an external magnetic field H (Fig. 3.10), the magnetic field is parallel to the cylindrical axis. In addition, $d \ll R$, where d is the thickness of the film and R is the radius of the cylinder. It can be observed that the T_c value oscillates (!) as a function of the external magnetic field. This is the essence of the Little–Parks effect.

Next, the origin of the effect will be discussed. The order parameter $\Psi(r)$ can be written in the form

$$\Psi = |\Psi| \exp(i\theta(\boldsymbol{r})) \tag{3.38}$$

where $|\Psi|$ is the constant inside the film, and θ is the phase.

Because of the selected orientation of the magnetic field, the velocity $\boldsymbol{v}_s$ is tangential to the cylinder. The integral $\oint \boldsymbol{v}_s d\boldsymbol{l}$, where the integration is taken over the circle with the radius R, can be written as follows:

$$\oint \boldsymbol{v}_s d\boldsymbol{l} = 2\pi R v \tag{3.39}$$

With the use of eqns (3.39) and (3.2)–(3.3), the following expression is derived:

$$v_s = \frac{\hbar}{mR}\left(n - \frac{\Phi}{\Phi_0}\right) \tag{3.40}$$

where Φ is the total magnetic flux ($\Phi = \pi R^2 H$) and Φ_0 is the flux quantum. As can be seen from eqn (3.40), the velocity has discrete values. At the same time, according

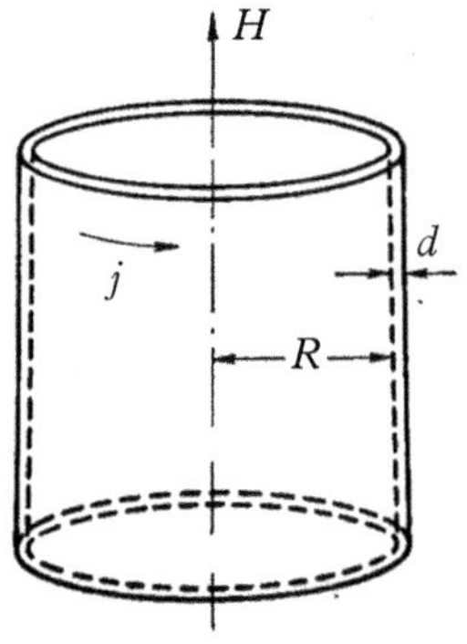

Figure 3.10 *The Little–Parks experiment.*

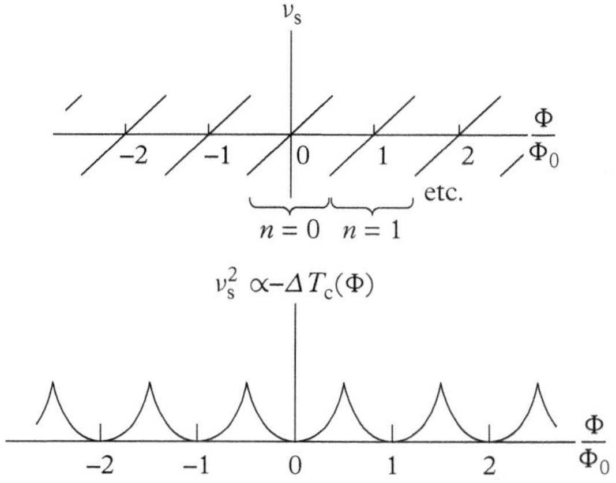

Figure 3.11 *The Little–Parks effect. (Upper panel) Supercurrent velocity as a periodic function of* $\frac{\Phi}{\Phi_0}$. *(Lower panel)* T_c *oscillates as a function of the field.*

to the Ginzburg–Landau equation (eqn (3.15′)), the value of the velocity should have the minimum value; then it will correspond to the minimum value of the free energy. Therefore, the velocity is determined by the minimum values of the right side of eqn (3.40); as a result, the velocity is the periodic function of the field H. According to eqn (3.40) and the relation $\Phi = \pi R^2 H$, the period is equal to $\tau = \Phi_0/\pi R^2$ (Fig. 3.11, top). Little and Parks used tin films evaporated around organic filaments of ~1 μm in diameter. The radius of the film $R \simeq 7 \times 10^3$Å. Correspondingly, the periodicity was set by the field $M \simeq 14$G. The change in velocity affects the T_c value. Indeed, the requirement for the free energy to be at a minimum value leads to a change (at a given v_s) of the value of the parameter 'α' entering the Ginzburg–Landau expression (3.15). According to microscopic theory (see eqn (3.31)), the value of α depends on T_c, and the T_c value is the periodic function of the external field H. Experimentally, the effect is manifested as the oscillations in resistance as the function of the field (Fig. 3.11).

3.1.5 The Search for the Lossless Current State

The most remarkable manifestation of the superconducting state is the absence of resistivity, that is, the dissipationless charge transfer. The appearance of a lossless current is provided by the Cooper pairing and, consequently, by the gap in the electronic energy spectrum. The electric field is absent inside a superconductor in its stationary state. Indeed, in the absence of resistance, the electric field would accelerate electrons so that the transfer of the direct current would be impossible. The macroscopic direct superconducting current creates a permanent magnetic field, thus forming a self-consistent picture; the value of the current is determined by the strength of the field, which can affect the charge's movement without producing any work.

Contrary to this picture, lossless charge transfer at the atomic scale is well known. The electrons in atoms of molecules are not paired, but they are moving without losses. Therefore, an interesting question can be raised: is it possible to build a macroscopic system that is capable of transferring the lossless current by analogy with that occurring at the atomic scale? This question has been discussed in depth by Mannhart (2018, 2020) and by Mannhart and Braak (2019).

The existence of such a lossless current would mean that the ground state of our system is the current state. At first sight, the creation of such a system is unrealistic. Indeed, the properties of the macroscopic system are described by statistical physics, that is, the distribution function. This function depends only on the absolute value of the velocity of the particle. Therefore, the average value of the current, which is the integral containing the one-particle currents and the distribution function, is equal to zero. The fact that the distribution function, which describes the chaotic electronic system, does not depend on the direction of the velocity but only on its absolute value reflects the time reversal symmetry ($t \to -t$). However, the creation of a crystal without such a symmetry can be envisioned. Such an attempt was made by Tavger (1986), who had earlier developed the concept of magnetic symmetry (Tavger and Zaitsev, 1956; see also Landau and Lifshitz, 1963).

There are solids with so-called magnetic structure. For such crystals, the change $t \to -t$ and the corresponding change of sign of the local current $\boldsymbol{j}$ leads to a change in the state. The usual group theory can be generalised by including the new element R, changing $\boldsymbol{j} \to -\boldsymbol{j}$ and, especially, its combination with other elements (rotations, translations, etc.; see the discussion in Landau and Lifshitz, 1963). In principle, crystals with magnetic symmetry can display 'spontaneous conductivity' (Tavger, 1986), which is a phenomenon similar to superconductivity.

As indicated above, a detailed study was carried out by Mannhart (2018, 2020) and Mannhart and Braak (2019). They proposed the model and, more specifically, the quantum device that can provide the lossless charge transfer. The concept of this device is based on the Rashba effect (Rashba, 1960; Bychkov and Rashba, 1984). This effect is described by the Hamiltonian

$$\hat{H} = \alpha (\boldsymbol{\sigma} \times \boldsymbol{p}) \cdot \hat{z} \tag{3.41}$$

where $\boldsymbol{p}$ is the momentum, $\boldsymbol{\sigma}$ is the electronic spin matrix, and α is the so-called Rashba coupling; this Hamiltonian determines the scale of the spin–momentum interaction. It is especially important for studying low-dimensional systems. Qualitatively, this effect can be described as follows. If the electric field is perpendicular to the surface, then the interaction $V = -Ez$ (the z-axis is perpendicular to the surface), and the electron will be affected by the magnetic field $H = -(\boldsymbol{v} \times \boldsymbol{E})/c^2$ (the relativistic effect). Its interaction with the electronic spin leads to the effective Hamiltonian (eqn (3.41)).

Mannhart introduced the quantum loop, which is depicted in Fig. 3.12. The left conductor, which is in an external electric field, displays the Rashba effect, whereas the right conductor is the casual one. As a result, the charge transfers for parts 1 and 2 are different, and the lossless charge transfer can be observed. The scales of the electronic

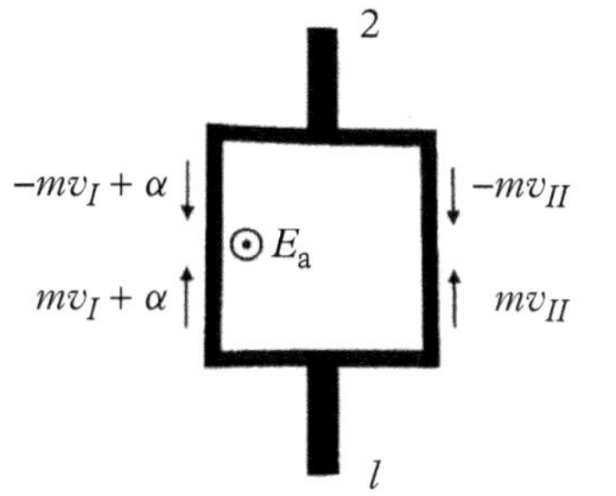

Figure 3.12 *A quantum loop with non-reciprocal transport. (After Mannhart and Braak, 2019)*

wave functions in these parts are different; therefore, some peculiar non-linear quantum mechanics are involved.

Importantly, strictly speaking, the non-stationary Schrödinger equation does not have time-reverse symmetry. The symmetry requirement also necessitates the change $\Psi \to \Psi^*$. According to Landau and Lifshitz (1977), this feature probably leads to the asymmetrical nature of the second law of thermodynamics. In any case, the search for a system displaying the lossless current ground state is a very interesting direction invoking a number of fundamental concepts.

3.2 Multigap Superconductivity

3.2.1 Multigap Superconductivity: The General Picture

Consider a superconductor containing several different groups of electrons occupying distinct quantum states. The most typical example is a material with several overlapping energy bands (Fig. 3.13).

One can expect that each band will possess its own energy gap. Let us clarify this point and introduce the necessary definition. Of course, if the energy gap were defined as the smallest quantum of energy which can be absorbed by the material, then only the smallest gap of the system would satisfy this definition; it would be impossible for the system to have several energy gaps. To avoid misunderstanding, when talking about the multigap structure of a spectrum, we will mean explicitly that the density of states of the superconducting electrons contains several peaks. The magnitudes of these gaps will differ according to the variations in the densities of states and the coupling constants across the bands.

The two-gap model was first considered by Suhl *et al.* (1958) and independently by Moskalenko (1959). A general treatment based on quantum field-theoretical techniques, together with an analysis of various thermodynamic, magnetic, and transport properties, was given by Geilikman *et al.* (1967).

Each band contains its own set of Cooper pairs. Since, generally speaking, $p_{F;i} \neq p_{F;l}$ (here $p_{F;i}$ and $p_{F;l}$ are the Fermi momenta for different bands), there is negligible pairing

Figure 3.13 *Overlapping bands.*

of electrons belonging to different energy bands. Indeed, only electrons with equal and opposite momenta form the pairs. This does not mean, however, that, within each band, the pairing is completely insensitive to the presence of the other. On the contrary, a peculiar interband interaction and the appearance of non-diagonal (interband) coupling constants are fundamental properties of the multigap model.

Indeed, consider two electrons belonging to band i. They exchange phonons and, as a result, form a pair. There exist two pairing scenarios. In one of them, the first electron emits a virtual phonon and makes a transition into a state within the same energy band. The second electron from the same band absorbs the phonon and also remains in the band, forming a bound pair with the first one. This is the usual pairing picture of the BCS theory, described by a coupling constant λ_i. However, the presence of the other energy band gives rise to an additional channel. Namely, the first electron, originally located in the i band, can emit a virtual phonon and make a transition into the l band. The phonon is absorbed by the second i-band electron, which also is scattered into the l band, where it pairs up with the first electron. Thus, the initial state had two electrons in the i band, while the final state finds a pair in the l band. Of course, the initial and final states must obey conservation of energy (both must be located at the Fermi level) and conservation of momentum (total momentum must be equal to zero) laws. Charge-transfer processes like this are described by non-diagonal coupling constants λ_{il} and, because of them, the system is characterised by a single critical temperature (see below). Otherwise, each set of electrons would have its own T_c.

The Hamiltonian describing the pairing interaction has the following form (in a weak coupling approximation):

$$\hat{H}_{int} = \sum_{\substack{i,l \\ \boldsymbol{\kappa},\mathbf{p}}} g_{il} a^{+}_{i;\boldsymbol{\kappa},+} a^{+}_{i;-\boldsymbol{\kappa},-} a_{l;-\mathbf{p},-} a_{l;\mathbf{p},+} \tag{3.42}$$

The terms with $i = l$ (then $g_{il} = g_{li} = g_i$) correspond to the intraband transition, whereas the terms with $i \neq l$ describe the interband transitions; the indices '$+$' and '$-$' correspond to various spin's projections.

3.2.2 Critical Temperature

We introduce self-energy parts $\boldsymbol{\Delta}_i(\mathbf{p}, \omega_n)$ which describe electron pairing in the i-th band. They satisfy the system of equations shown in Fig. 3.14, or, in analytical form,

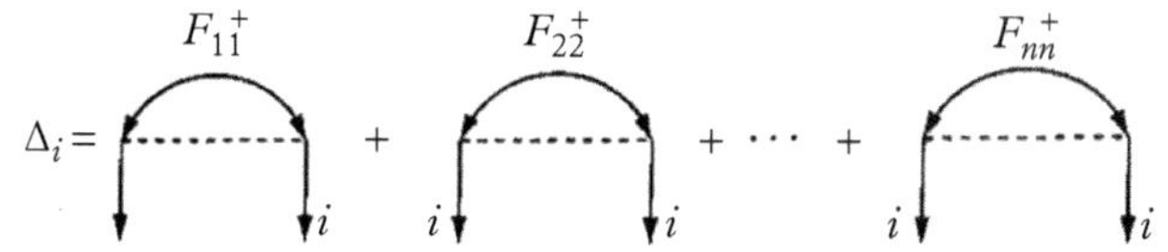

Figure 3.14 *Diagrammatic equations for the order parameter.*

$$\Delta_i(\mathbf{p},\omega_n) = \sum_{l=1}^{n}\sum_{\omega_{n'}} \int \lambda_{il} D(\mathbf{p}-\mathbf{p}',\omega_n-\omega_{n'}) F_{ll}^{+}(\mathbf{p}',\omega_{n'})$$

where

$$F_{ll}^{+}(\mathbf{p}',\omega_{n'}) = \frac{\Delta_l}{\omega_n^2+\xi^2+\Delta_l^2} \tag{3.43}$$

D is the phonon Green's function; λ_{il} describes the transition of an electron pair from the i-th to the l-th band accompanied by phonon absorption or emission, Δ_i is the order parameter, and Z_i is the renormalisation function such that $\lambda_{il} = |g_{il}|^2 v_e/Z_i$. Equation (3.43) is the generalisation of eqn (2.18) for the multigap case.

Let us look at the case of weak coupling. Summing over ω_n', we reduce eqn (3.43) to the form

$$\varepsilon_i = \sum_{l=1}^{n} \lambda_{il}\varepsilon_l \int d\xi_l \, \frac{\tanh(E_l/2T)}{E_l}$$
$$E_l = \left(\xi_l^2 + |\varepsilon_l|^2\right)^{1/2} \tag{3.44}$$

Then $D \simeq 1$ and $\Delta_i \simeq \varepsilon_i$ (ε_i is the energy gap). One can see from eqn (3.44) that the gaps ε_i in different bands appear at the same temperature T_c (see also Fig. 3.15). In the most interesting two-gap case, the critical temperature turns out to be

$$T_c = a\tilde{\Omega}e^{-1/\tilde{\lambda}}$$
$$\tilde{\lambda} = \frac{1}{2}\left[\lambda_1+\lambda_2+\sqrt{(\lambda_2-\lambda_1)^2+4\lambda_{12}\lambda_{21}}\right] \tag{3.45}$$

This result can be easily obtained from eqn (3.44) ($i,l = 1.2$) taken at $T = T_c$. The expression (3.45) is a generalisation of eqns (2.10) and (2.21) for the two-gap case; $a \simeq 0.25$.

One can see directly from eqn (3.45) that the presence of the second gap is a favourable factor for T_c, that is, its value is larger than that for the one-gap case. This statement is valid even if $\lambda_2 = 0$ (see section 3.2.6), that is, for the case when the pairing

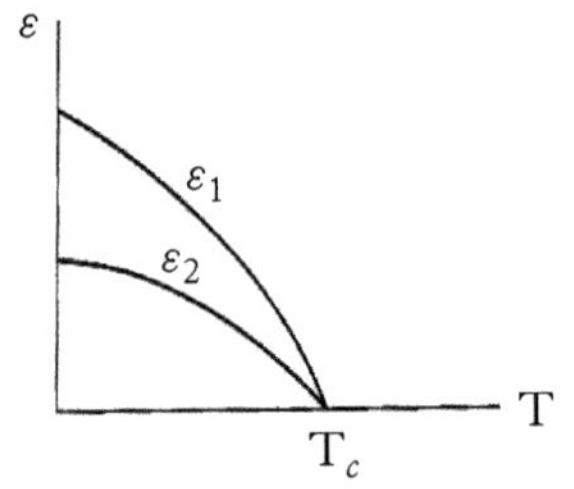

Figure 3.15 *The dependences* $\varepsilon_1(T)$ *and* $\varepsilon_2(T)$.

in the second band is not intrinsic, but is induced by the interband transition. One can say that this feature is caused by an increase in a phase space for virtual transitions.

The presence of overlapping bands is typical for many superconductors. Nevertheless, it was not possible to observe multigap structure in conventional superconductors. This is due to their large coherence lengths: the inequality $\xi \gg l$ (l is the mean free path), which holds for most conventional superconductors, leads to averaging because of interband scattering. As a result, the multigap structure is washed out and the one-gap model is applicable.

The situation in novel superconductors, for example, the high-T_c oxides, and MgB_2, is entirely different. The coherence length is small, and this leads to the observation of a multigap structure (see section 7.1.3).

A two-gap spectrum was first observed in the exotic system, Nb-doped $SrTiO_3$ compound (Binnig *et al.*, 1980). The STM technique (see section 4.1.2) was employed, and the two-gap structure of the spectrum was observed. The structure appears as a result of doping and the corresponding filling of the second band. As was noted above, this was the first observation of the two-gap structure. According to a theoretical analysis by Fernandes *et al.* (2013), the two-gap structure can also be observed for Nb-doped $SrTiO_3$ films and their interface, that is, for the quasi-two-dimensional case.

3.2.3 The Energy Spectrum

We focus here mainly on the two-gap case, because this case is relatively simple and the most important one. The temperature dependences of the energy gaps can be evaluated from eqn (3.44) and, generally speaking, are very different from that for the single-band BCS case; the typical dependences can be seen in Fig. 3.15.

The energy spectrum and the properties of two-gap superconductors are described by three coupling constants, $\lambda_1, \lambda_2, \lambda_{12}$, which enter the Hamiltonian (3.42).

Let us express the energy gaps ε_i in terms of these coupling constants. One can employ a method similar to that used by Pokrovsky (1961) to study the anisotropy of the energy gap. The solution of eqn (3.44) at $T = 0$ K can be sought in the form

$$\varepsilon_i = 2\tilde{\Omega}\phi_i \exp(-\alpha) \tag{3.46}$$

Then, $\phi_i = QX_i$, where X_i is the normalised solution of the following system of linear equations:

$$\phi_i = \alpha \sum_l \lambda_{il} \phi_l \tag{3.46'}$$

corresponding to the smallest eigenvalue α. The calculation leads to the following expression for coefficient Q:

$$Q = \exp\left[-\sum_i \nu_i |X_i|^2 ln|X_i| \Big/ \sum_i \nu_i |X_i|^2 \right] \tag{3.46''}$$

for the two-gap case

$$X_1 = \lambda_{12} S^{-1}, X_2 = \left(\tilde{\lambda} - \lambda_1\right) S^{-1}$$
$$S = \left[\lambda_{12}^2 + (\tilde{\lambda} - \lambda_1)^2\right]^{1/2}$$

where $\tilde{\lambda}$ is defined by eqn (3.45). From these results, one can show (Kresin and Wolf, 1990b) that one of the gaps always exceeds the BCS value, while the other is less than this value. Indeed, we can write

$$\frac{2\varepsilon_1(0)}{T_c} = 3.52 e^{A_1}, \frac{2\varepsilon_2(0)}{T_c} = 3.52 e^{A_2} \tag{3.47}$$

where

$$A_1 = \nu_2 \tau^2 R \ln\left(\tau^{-1}\right), \quad A_2 = -\nu_1 R \ln\left(\tau^{-1}\right)$$
$$R = \left(\nu_1 + \nu_2 \tau^2\right)^{-1}, \qquad \tau = \left(\tilde{\lambda} - \lambda_1\right)/\lambda_{12}$$

or

$$\tau = \frac{1}{2}\left(\lambda_2 - \lambda_1 + \left[(\lambda_2 - \lambda_1)^2 + 4\lambda_{12}\lambda_{21}\right]^{1/2}\right)\lambda_{12}^{-1}$$

One can see that, regardless of the value of τ, we always have $A_1 A_2 < 0$. Therefore (see eqn (3.47)), one can state the following theorem: one of the gaps in a two-gap system is always smaller than the BCS value, whereas the other gap is greater than this value.

Note that, qualitatively, this result is related to the aforementioned (section 2.2) impact of impurities on the multigap structure: the impurity scattering leads to averaging of the gaps and to the one-gap picture which is observed for conventional superconductors. Such an averaging would lead to the BCS relation $2\varepsilon_i = 3.52 T_c$ if, indeed, the value of one gap exceeds the BCS value and, for the second gap, $2\varepsilon(0)_{min} < 3.52 T_c$. Otherwise, the average value would be larger (or smaller) than that, following from the BCS theory.

Therefore, the inequality $2\varepsilon(0)/T_c > 3.52$ can be caused not only by strong coupling effects (see section 2.2.3), but by the presence of the multigap structure. That is why an analysis of the experimental data when $2\varepsilon(0)/T_c > 3.52$ should be performed with considerable care (see e.g. Santiago and de Menezes, 1989).

Temperature dependences of the gaps for some specific two-gap superconductors are determined by the values of the coupling constants $\lambda_1, \lambda_2, \lambda_{12}$. Nevertheless, one can formulate the general feature, namely that the ratios $\eta = \varepsilon_2/\varepsilon_1$ are equal at $T = 0$ K and near $T = T_c$, that is, $\eta(0) = \eta(T_c)$ (Kresin, 1973). Indeed, the gaps at $T = 0$ K and $T \to T_c$ are the solutions of the following set of equations, which follow directly from (3.44) $(i, k = 1, 2)$:

$$T = 0\ \mathrm{K} : \varepsilon_i = \sum_{i,k} \lambda_{ik} \ln \frac{2\tilde{\Omega}}{|\varepsilon_k|} \tag{3.44$'$}$$

$$T \to T_c : \varepsilon_i = \ln \frac{\tilde{\Omega}}{T_c} \sum_{i,k} \lambda_{ik} \varepsilon_k \tag{3.44$''$}$$

As was noted above, the solution of eqn (3.44′) can be sought in the form $\varepsilon_i = 2\tilde{\Omega}\varphi_i \exp(-\alpha)$. As a result, we obtain eqn (3.46′), allowing us to determine $\eta(0)$.

The ratio $\eta(T_c)$ can be determined from eqn (3.44″) and is the solution of the equation $\lambda_{12}\eta^2 + (\lambda_1 - \lambda_2)\eta - \lambda_{12} = 0$. This equation permits us to express the ratio $\eta = \eta(T_c)$ in terms of the coupling constants $\lambda_1, \lambda_2, \lambda_{12} = \lambda_{21}$. As for the ratio $\eta(0)$, it can be obtained from eqns (3.46) and (3.46′). One can see that $\eta(0)$ and $\eta(T_c)$ are described by similar equations and, therefore, indeed, the ratios of the gaps at $T = 0$ K and $T = T_c$ are equal.

Let us stress again that the deviations from the BCS behaviour and the inequality $2\varepsilon_i(0) \neq 3.52T_c$ here are caused not by strong coupling (cf. section 2.3), but by the presence of a multigap structure.

3.2.4 Properties of Two-Gap Superconductors

Let us, at first, stress one more important feature of the two-gap picture. In the usual theory of superconductivity, in the weak coupling approximation, the energy gap is related to T_c by the universal relation $2\varepsilon(0) = 3.52T_c$ (see section 2.1.6). This is due to the fact that, for such a case, the system is characterised by a single parameter, namely the coupling constant λ. All other parameters can be expressed in terms of λ. Instead of λ, one can select the critical temperature T_c as the parameter. As a result, all other characteristic quantities, including the energy gap at $T = 0$ K, $\varepsilon(0)$, can be expressed in terms of T_c by universal relations. In a multiband model with a large number of coupling constants, there is no such universal relation between the energy gaps $\varepsilon(0)$, and T_c.

For two-gap superconductors, one can select as the independent and experimentally measured parameters the value T_c and the values of the energy gaps $\varepsilon_1(0)$ and $\varepsilon_2(0)$.

The temperature dependences of the thermodynamic, electromagnetic, and transport parameters differ from those obtained in the usual one-gap BCS theory. For example, for

$T \to 0$, the heat capacity is the sum of two exponents; the band with the smaller superconducting gap makes the dominant contribution and, as a result, $C_{el.|T\to 0}$ decreases more slowly than in the one-band case.

Penetration depth and surface resistance. The superconducting current for the multigap superconductor can be written in the form

$$j(\boldsymbol{q}) = \sum_i Q_i(\boldsymbol{q})\boldsymbol{A}(\boldsymbol{q}) \tag{3.48}$$

Here $\boldsymbol{A}$ is a vector potential, summation is over all overlapping energy bands, and the kernel $Q_i(\boldsymbol{q})$ is (see e.g. Abrikosov *et al.*, 1975)

$$Q_i(\boldsymbol{q}) = \frac{3\pi}{4}\frac{N_i e^2}{m_i} T \sum_n \int_{-1}^{1} \frac{(1-\mu^2)\,d\mu}{\left(\omega_n^2 + |\varepsilon_i|^2\right)^{1/2}} \frac{\varepsilon_i^2}{\left(\omega_n^2 + {\varepsilon_i}^2 + \left(\frac{v_i q \mu}{2}\right)^2\right)} \tag{3.48$'$}$$

Here N_i and m_i are the electron concentration and the effective mass, respectively, for the i-th band, and ε_i and v_i are the corresponding energy gap and the Fermi velocity, $\omega_n = (2n+1)\pi T$.

The penetration depth λ can be calculated from the following expression (see e.g. Abrikosov and Khalatnikov, 1959; we consider the case of diffuse reflection):

$$\lambda = \pi\left(\int_0^\infty dq \ln\left[1 + \frac{Q(q)}{q^2}\right]\right)^{-1} \tag{3.49}$$

One can consider various limiting cases. If $v_i q \ll T_c, \varepsilon_i(0)$ (the London case), then

$$Q_i(q) = 4\pi e^2 \sum_i n_{i;s}/m_i \tag{3.49$'$}$$

where $n_{i;s} = \pi n_i \varepsilon_i^2 T \sum_{\omega_n} \left(\omega_n^2 + \varepsilon_i^2\right)^{-3/2}$ is the concentration of 'paired' electrons in the i-th band.

Let us consider the low-temperature region. Based on eqns (3.49) and (3.49′), one can obtain for the two-gap picture (this is the most interesting case) that the quantity $\Delta\lambda = \lambda(T) - \lambda(0)_{|T\to 0}$ is a sum of two exponentional terms, that is,

$$\Delta\lambda = a e^{-\frac{\varepsilon_1(0)}{T}} + b e^{-\frac{\varepsilon_2(0)}{T}} \tag{3.50}$$

This expression is valid for the usual *s*-wave pairing. According to eqn (3.50), at $T \to 0$, the temperature dependence of $\Delta\lambda$ is determined by the smaller gap, that is, the penetration depth decreases more slowly than in the BCS one-band case.

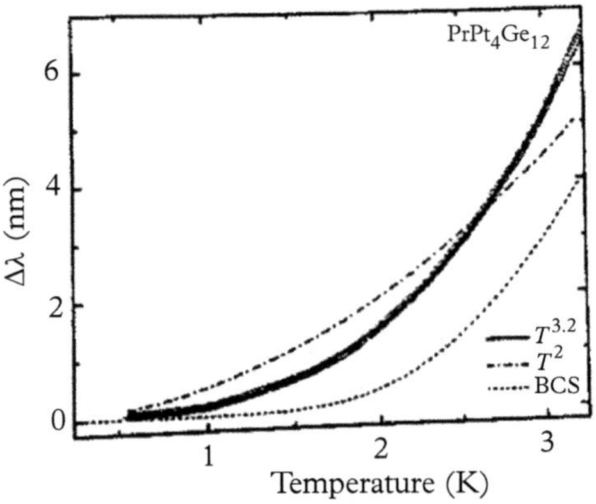

Figure 3.16 *Temperature dependence of the penetration depth $\Delta\delta \sim T^2$ (dash-dotted line) and the single-gap Bardeen, Cooper, and Schrieffer (BCS) model (dotted line); see the discussion in the text. (From Zhang et al., 2013)*

Using eqns (3.49) and (3.49′), one can evaluate the dependence of $\Delta\lambda(T)$ for a whole temperature range up to $T = T_c$ (see Fig. 3.16 and the discussion below; also see section 2.5.3). One can also study the opposite limiting case, $v_i q \gg T_c, \varepsilon_i(0)$ (the Pippard case).

One can have peculiar situations in which $\xi_2 \ll \delta \ll \xi_1$ (δ is the penetration depth, ξ_1, ξ_2 are the coherence lengths in the two bands). In this case, a single sample will contain two groups of electrons, one of which is Pippard-like, while the other is London-like (Geilkman *et al.*, 1967b, 1968).

The surface resistance at $T \to 0$ can also be written as a sum of two exponents.

Heat capacity. In the low-temperature region ($T \to 0$) the heat capacity is a sum of two exponents; the band with the smaller energy gap makes the dominant contribution, and, as a result, $C_{\text{el.}|T\to 0}$ decreases more slowly than in the one-gap case. The dependence is similar to eqn (3.50).

In order to evaluate the jump in heat capacity at $T = T_c$, one can use the expression for the entropy at $T \to T_c$ (see e.g. Lifshitz and Pitaevskii, 2002):

$$S = \frac{2\pi^2}{3} T \sum_i \nu_i - T \sum_i \nu_i \left(\frac{\varepsilon_i}{T}\right)^2 \tag{3.51}$$

Based on eqn (3.53), one can obtain

$$\frac{\Delta C}{C_n(T_c)} = \left(\frac{\Delta C}{C_n(T_c)}\right)_{BCS} \cdot \rho \tag{3.51′}$$

where $\left(\frac{\Delta C}{C_n(T_c)}\right)_{BCS} \approx 1.4$ and $\rho = 1 - \frac{\sum_{i,k} \nu_i v_k \left(|X_i|^2 - |X_k|^2\right)^2}{2\sum_{i,k} \nu_i v_k |X_i|^4}$ (see eqns (3.46) and (3.46″)). It is essential that $\rho < 1$ and, therefore, the multigap situation leads to the jump in heat capacity to be smaller than for the usual one-gap case. The impact is similar to that for the energy gap anisotropy (Pokrovsky, 1961). For some superconductors, the jump in heat capacity is larger than that in the BCS theory. This phenomenon is caused by the strong coupling effects (see section 2.2.3).

Experimental observations. As was stressed in the beginning of this chapter, the term 'two-gap superconductivity' refers to the situation when the electronic density of states displays two distinct peaks. As a consequence, the two-gap structure can be detected by tunnelling spectroscopy. In this way, a two-gap structure in the tunnelling spectrum of the high-T_c compound YBCO (see section 5.3.3) was observed by Geerk *et al.* (1988).

As was noted in the previous section, the presence of the two-gap structure greatly affects the temperature dependences of various quantities. In chapter 5 and sections 7.1.3, 7.2.3, and 7.3, we will discuss such manifestations for high-T_c cuprates, MgB_2, and other compounds. Let us discuss here, as an example, the data obtained for a relatively new superconducting material, $PrPt_4Ge_{12}$, which belongs to the family of so-called filled-skutterudite compounds. It has a relatively high value of T_c ($T_c \approx 8$K, see Gumeniuk *et al.*, 2008). Various manifestations of the multigap structure of this superconductor were observed by Zhang *et al.* (2013). They studied a high-quality single crystal of the compound, so that the mean free path ($l \approx 10^2$ nm) greatly exceeded the value of the coherence length ($\xi \approx 13.5$ nm). Because $l \gg \xi$, there was no averaging of the gaps (see above), and it was natural to expect the presence of several gaps for such a complex material.

Figure 3.16 shows the measured dependence of $\Delta\lambda = \lambda(T) - \lambda(0)$ on temperature (solid line) for $PrPt_4Ge_{12}$. The observed dependence ($\propto T^{3.2}$) cannot be fitted by the usual BCS exponential curve or by the power law ($\propto T^2$) valid in the presence of the nodes. One can provide an excellent fit with the use of the two-gap model (see eqn (3.50)) with the values of the energy gaps $\varepsilon_1(0) = 0.8T_c$ and $\varepsilon_2(0) = 2T_c$. Note that, according to Zhang *et al.* (2013), these values are in agreement with criterion (3.47). Indeed, $(\varepsilon_1(0)/T_c) < (\varepsilon(0)/T_c)_{BCS}$, whereas $(\varepsilon_2(0)/T_c) > (\varepsilon(0)/T_c)_{BCS}$. The two-gap picture with these values of the gaps allowed Zhang *et al.* (2013) to describe with great accuracy the data on heat capacity.

3.2.5 The Strong Magnetic Field and the Ginzburg–Landau Equations for a Multigap Superconductor

One can also study the behaviour of a multigap superconductor in an arbitrary magnetic field near T_c. In other words, we are talking about the generalisation of the Ginzburg–Landau equations for the multigap case (Geilikman *et al.*, 1967). For the usual one-gap case, the Ginzburg–Landau equations were derived from the microscopic theory by Gor'kov (1959); see also the review by Abrikosov *et al.* (1975). For the multigap case, one can use the expansion near T_c and the expression for the current:

$$\Delta_i^*(\boldsymbol{r}) = \ln\left(\frac{2\gamma\tilde{\Omega}}{\pi T}\right)\sum_k \lambda_{ik}\Delta_k^*(\boldsymbol{r})$$
$$+\frac{7\zeta(3)}{48(\pi T_c)^2}\left[\sum_k \lambda_{ik}v_k\left(\frac{\partial}{\partial r} - 2ie\boldsymbol{A}\right)^2\Delta_k^*(\boldsymbol{r}) - 6\sum_k \lambda_{ik}\Delta_k^*(\boldsymbol{r})|\Delta_k^*(\boldsymbol{r})|^2\right] \tag{3.52}$$

$$\boldsymbol{j}(\boldsymbol{r}) = \sum_k\left[\frac{ie}{m_k}\left(\Delta_k\frac{\partial\Delta_k^*}{\partial\boldsymbol{r}} - \Delta_k^*\frac{\partial\Delta_k}{\partial\boldsymbol{r}}\right) - \frac{4e^2}{m_k}|\Delta_k|\boldsymbol{A}\right]\frac{7\zeta(3)n_k}{16(\pi T_c)^2}$$

Using the method proposed by Gor'kov and Melik-Barkhudarov (1963), one can seek the solution of eqn (3.51) in the form

$$\Delta_i(\boldsymbol{r}) = X_i Q^*(\boldsymbol{r}) + \varphi_i^*(T) \tag{3.52'}$$

where $Q(\boldsymbol{r}) \propto \left[\frac{T_c - T}{T_c}\right]^{\frac{1}{2}}$ and slowly depends on coordinates, φ_i is a small correction, $\varphi_i \propto \left[\frac{T_c - T}{T_c}\right]^{\frac{3}{2}}$, and X_i are defined by eqns (3.46″). Substituting eqn (3.52′) in eqn (3.52), one can arrive at the following equations:

$$a\left(\frac{\partial}{\partial\boldsymbol{r}} - 2ie\boldsymbol{A}\right)^2 Q^*(\boldsymbol{r}) + b\left[\frac{T_c - T}{T_c}\right]Q^*(\boldsymbol{r}) - cQ^*(\boldsymbol{r})|Q(\boldsymbol{r})|^2 = 0$$
$$j = \left[ie\left(Q(\boldsymbol{r})\frac{\partial Q^*(\boldsymbol{r})}{\partial\boldsymbol{r}}Q^*(\boldsymbol{r})\frac{\partial Q(\boldsymbol{r})}{\partial\boldsymbol{r}}\right) - 4e^2|Q(\boldsymbol{r})|^2\boldsymbol{A}(\boldsymbol{r})\right]d \tag{3.53}$$

where

$$a = \rho\sum_i v_i|X_i|^2;\ b = \sum_i \nu_i|X_i|^2$$
$$c = 6\rho\sum_i \nu_i|X_i|^4$$
$$d = 3\rho\sum_i \frac{n_i}{M_i}|X_i|^2;\ \rho = \frac{7\zeta(3)}{48(\pi T_c)^2}$$

One can see from eqn (3.53) that the Ginzburg–Landau theory can be used for the multigap superconductor; the function $Q(\boldsymbol{r})$ plays the role of the order parameter, and the values of coefficients are different from those for the usual one-gap case. It is essential that the Ginzburg–Landau theory contain a single order parameter, $Q(\boldsymbol{r})$, even for the multigap case.

3.2.6 Induced Two-Band Superconductivity

Let us consider the two-band model and the special case of $\lambda_2 = 0$, that is, let us suppose that one of the bands is intrinsically normal. Below the critical temperature, both bands are in the superconducting state. In other words, superconductivity in the second band is induced by the interband transitions. The critical temperature is given by (3.45) with

$$\tilde{\lambda} = \frac{1}{2}\left[\lambda_1 + \left(\lambda_1^2 + 4\lambda_{12}\lambda_{21}\right)^{1/2}\right] \tag{3.45$'$}$$

In the present case, the bands are not spatially separated. Instead, near the Fermi surface, the states belonging to different bands are separated in momentum space. The induced superconducting state arises thanks to phonon exchange. Note that other excitations such as excitons, plasmons, and so on are also capable of providing interband transitions.

It is important to note that $\tilde{\lambda} > \lambda_1$, which is due to an effective increase in the phonon phase space. As was noted above, this means that the presence of the second band is favourable for superconductivity. As a result of charge transfer, the second band acquires an induced energy gap which depends on the constant λ_{12}.

As remarked above, the ratios $2\varepsilon_1(0)/T_c$ and $2\varepsilon_2(0)/T_c$ differ from the value $2\varepsilon(0)/T_c = 3.52$ of the BCS theory. Namely, $2\varepsilon_1(0)/T_c > 3.52$, and $2\varepsilon_2(0)/T_c < 3.52$. In the case of induced two-gap superconductivity, the ratio λ_{21}/λ_1 could be very small. As a result, the values of the energy gaps may differ very noticeably from each other.

Now consider the case of strong coupling. We can write

$$\Delta_1(\omega_n) Z_1 \cong \lambda_1 \sum_{\omega_{n'}} k_{\omega_n,\omega_{n'}} \Delta_1(\omega_{n'}) \tag{3.54}$$

$$\Delta_2(\omega_n) Z_1 \cong \lambda_{21} \sum_{\omega_{n'}} k_{\omega_n,\omega_{n'}} \Delta_1(\omega_{n'}) \tag{3.54$'$}$$

where

$$k_{\omega_n,\omega_{n'}} = \tilde{\Omega}^2\left[\tilde{\Omega}^2 + (\omega_n - \omega_{n'})^2\right]^{-1}\left(\omega_{n'}^2 + \Delta_1^2\right)^{-1/2}$$

The energy gap is the solution of the equation $\omega = \Delta(-i\omega)$, where $\Delta(z)$ is the analytical continuation of the function $\Delta(\omega_n)$.

We obtain the following for the ratio of the gaps:

$$\tau = \frac{\varepsilon_2(0)}{\varepsilon_1(0)} \cong \frac{\lambda_{21}(1+\lambda_1)}{\lambda_1} \tag{3.55}$$

The presence of strong coupling leads to the appearance of the factor $(1+\lambda_1)$. We see that strong coupling tends to decrease the relative difference in the values of the gaps.

3.2.7 Symmetry of the Order Parameter and the Multiband Superconductor

Consider the case when the Fermi surface, in addition to the main ('large') part, contains small pockets. Such complex structure is not uncommon for many novel systems, such as heavy fermion materials, organics, borocarbides, high-T_c cuprates, and hydrides. If, in addition, the d-wave component of the order parameter dominates, then the value of the induced order parameter for the pocket may end up being close to zero (Gor'kov, 2012). In other words, we are dealing with the coexistence of the d-wave superconductivity in one band and the 'unpaired' carriers in the pocket. Indeed, the induced order parameter is determined by eqn (3.45) with $\lambda_{22} = 0$ (see eqn (3.45′)). Because of the d-wave symmetry, the order parameter in the 'large' band has parts with opposite signs, and their compensations lead to the absence of pairing amplitude in the pocket.

Two-band induced superconductivity is caused by phonon-mediated exchange, but does not require a spatial separation of the intrinsically superconducting and normal subsystems. Another type of induced superconductivity, the proximity effect, is based on such a separation and will be considered in section 3.4.

3.3 Impurity Scattering and Pair Breaking

3.3.1 Pair Breaking

The effect of impurities on superconductivity is stronger than in normal metals. Non-magnetic impurities at small concentration in a normal metal interact with the charge of an itinerant electron, resulting in its scattering and providing a finite mean free path l and a finite rest resistivity ρ_0 at $T = 0$. In a superconductor with s-pairing, the charge scattering influences both electrons forming a Cooper pair, and the pair survives such scattering, so the influence of normal impurities on superconductivity is very weak. Its effect is of the order $a/l \ll 1$, where a is an atomic size and l is the mean free path. In contrast, paramagnetic impurities result in a spin-dependent scattering accompanied with a spin flip. If one electron in the singlet Cooper pair reversed its spin, the pair with parallel spins may be destroyed. Thus, spin-dependent scattering by magnetic impurities results in pair breaking (Abrikosov and Gor'kov, 1961). Its effect is determined by the ratio ξ/l_s, where ξ is the coherence length and l_s is the mean free path. Because of the spin-flip scattering, the lifetime of a pair state is finite and this results in a rapidly decreasing transition temperature.

The Abrikosov and Gor'kov (1961) theory is based on the Gor'kov (1958) Green's function version of the BCS theory with the normal Green's function $G(\boldsymbol{p}, \omega)$ and the pairing function $F(\boldsymbol{p}, \omega)$. The paramagnetic impurities give rise to the perturbation

$$H^{imp}(\boldsymbol{r}) = \sum_i \{v_1(\boldsymbol{r} - \boldsymbol{R}_i) + v_2(\boldsymbol{r} - \boldsymbol{R}_i)\boldsymbol{S_i} \cdot \boldsymbol{s}\} \tag{3.56}$$

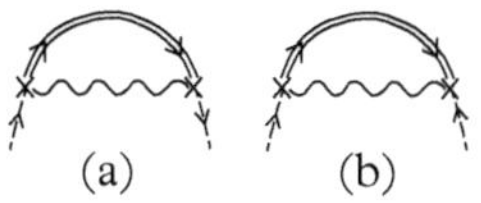

Figure 3.17 *The self-energy corrections to the normal (a) and pairing (b) Green's functions. The double line with arrows in the same direction represents the diagonal part of the propagator. The double line with arrows opposed is the off-diagonal part of the propagator. The crosses denote a (double) interaction with the same impurity. (From Skalski et al., 1964)*

where $\boldsymbol{R}_i$ denotes the position of the impurity i with the spin $\boldsymbol{S}_i$, and $\boldsymbol{s}$ is the electronic spin operator. The positions of the impurities are random, and the impurity spins are assumed to be uncorrelated with each other due to small concentrations of impurities. The gap equation used to find the critical temperature T_c should be averaged over the random impurity distribution. To provide this averaging, the special 'cross' diagram technique has been invented (Abrikosov and Gor'kov, 1961). Below we use clearer points of view developed later by Phillips (1963), and Skalski *et al.* (1964); see Fig. 3.17.

In the absence of impurities, the poles of the Green's function G give the usual quasiparticle energies

$$\omega = \pm\left(\xi^2 + \Delta^2\right)^{1/2}$$

where $\xi = v_0\,(p - p_0)\,, v_0$ is the Fermi velocity. In the presence of impurity scattering, ω and Δ are altered to

$$\tilde{\omega} = \omega + \left(\frac{1}{2\tau_1}\right) u/\left(1 - u^2\right)^{1/2} \tag{3.57}$$

$$\tilde{\Delta} = \Delta + \left(\frac{1}{2\tau_2}\right) 1/\left(1 - u^2\right)^{1/2} \tag{3.58}$$

Here u is an implicit function of ω determined by $u = \tilde{\omega}/\tilde{\Delta}$. The total scattering rate is $1/\,\tau_1$, and the spin-flip scattering rate is proportional to

$$\frac{1}{\tau_s} = \frac{1}{2\tau_1} - \frac{1}{2\tau_2} \tag{3.59}$$

From eqns (3.57), (3.58), and (3.59) one finds a consistency equation which determines the contribution of each state $\boldsymbol{p}$ to

$$\frac{\omega}{\Delta} = u\left[1 - \left(\frac{1}{\Delta\tau_s}\right) 1/\left(1 - u^2\right)^{1/2}\right] \tag{3.60}$$

As long as the solutions of eqn (3.60) are real at $T = 0$, the quasiparticle states associated with Cooper pairing do not decay. For $\Delta\tau_s > 1$, all quasiparticle states $\boldsymbol{p}$ and all frequencies ω decay (Abrikosov and Gor'kov, 1961). To see what is happening to the gap, we consider several special cases (Phillips, 1963):

(a) No magnetic scattering: $\tau_1 = \tau_2 = \tau$. Then $\tilde{\omega} = \eta\omega$ and $\tilde{\Delta} = \eta\Delta$ with

$$\eta = 1 + i/2\tau\left(\omega^2 - \Delta^2\right)^{1/2}$$

Because $\tilde{\omega}$ and $\tilde{\Delta}$ are multiplied by the same factor, the shape of the quasiparticle density of states near the Fermi energy and $\omega = \Delta$ is unchanged.

(b) All scattering flips spins: $1/\tau_2 = 0$. Then $\tilde{\Delta} = \Delta$, and

$$\left(1 - u^2\right)^{1/2} = i\,\xi/\Delta,$$
$$\tilde{\omega} = \omega - i\left(\Delta^2 + \xi^2\right)^{1/2}/\xi\tau_s$$

According to the last equation, the spin-flip scattering gives all quasiparticle states a complex frequency $\tilde{\omega} = \omega + i\Gamma$, where the decay rate $\Gamma \sim 1/\xi$ as $\xi \to 0$. The states nearest the gap (which is sharp in the absence of magnetic scattering) are the ones that suffer the greatest lifetime broadening in the presence of magnetic scattering. For $\Delta\tau_s = M > 1$, the states $0 \le \xi \le \Delta/M$ will spread into the gap. The density of states for magnetic scattering only and M = 2 is shown in Fig. 3.18.

From eqn (3.60), it is clear that no decay takes place for $\Delta\tau_s > 1$ and

$$\omega < \omega_0 = \Delta\left[1 - (\Delta\tau_s)^{-2/3}\right]^{3/2} \tag{3.61}$$

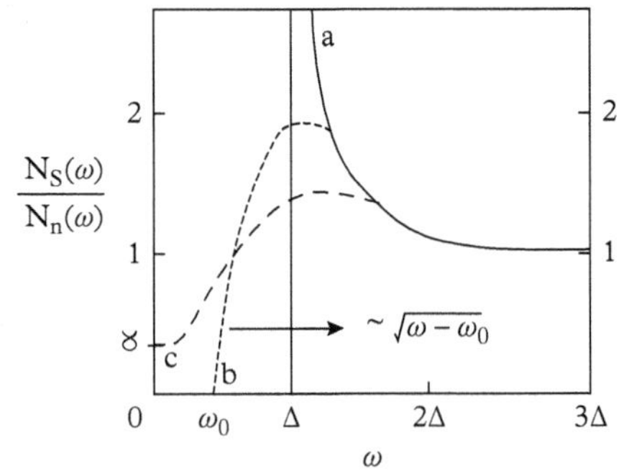

Figure 3.18 *The density of states, which is symmetrical about $\omega = 0$; (a) A pure superconductor or a superconductor with non-magnetic scattering: $1/\Delta\tau_S = 0$; (b) $\Delta\tau_s = 2$; (c) $\Delta\tau_s$ slightly less than one. (From Phillips, 1963)*

Thus, ω_0 plays the role of the band gap. Although imaginary frequencies would imply Lorentzian broadened quasiparticle states, eqn (3.61) shows that each Lorentzian is cut off at ω_0. This means that, even in the presence of completely magnetic scattering (when all quasiparticles decay), it is possible to form wave packets of stationary states at $\omega > \omega_0$. This conclusion is similar to the ones discussed by Anderson (1959) for non-magnetic scattering.

The pair-breaking effect of magnetic impurities on the critical temperature T_c is given by (Abrikosov and Gor'kov, 1961)

$$ln\left(\frac{T_c^0}{T_c}\right) = \psi(0.5+\gamma_s) - \psi(0.5) \tag{3.62}$$

where T_c^0 is the critical temperature without impurities, ψ is the digamma function, and $\gamma_s = 1/2\pi T_c \tau_s$. At small concentrations of magnetic impurities, the value γ_s is small, so $\psi(0.5+\gamma_s) \approx \psi(0.5) + \pi^2\gamma_s/2$, and one obtain linear decreasing:

$$T_c = T_c^0 - \pi/4\tau_s \tag{3.63}$$

With greater concentrations, $T_c \to 0$ and the parameter γ_s is large. For large x,

$$\psi(x) = \ln x - (2x)^{-1} - \left(12x^2\right)^{-1}$$

Finally, the critical concentration n_c when $T_c \to 0$ is given by

$$1/\tau_{sc} = \pi T_c^0/3.56 = \Delta_{00}/2 \tag{3.64}$$

Here Δ_{00} is the superconductor order parameter at $T = 0$ without impurities. Close to the critical concentration, T_c is equal to

$$T_c \approx \sqrt{6}/\pi(\tau_s - \tau_{sc})^{1/2}\tau_{sc}{}^{-3/2}$$

The critical value of the 'magnetic' mean free pass $l_{sc} = v_0\tau_{sc} \sim 10^{-4}$ cm. It should be noted that the magnetic interaction v_2 in the impurity perturbation (3.56) is much weaker than the charge one v_1, the typical ratio being 10^{-2}. So $l_{sc} \sim 10^{-4}$ cm corresponds to the mean free pass l that determines the conductivity of the normal metal to be $l \sim 10^{-6}$ cm and the critical concentration of magnetic impurities is of the order 1%.

3.3.2 Gapless Superconductivity

From eqn (3.61) for the gap and eqn (3.63) for the critical temperature, it is clear that the value of the energy gap is suppressed by magnetic impurities stronger than T_c. When $\Delta\tau_s = 1$, the gap tends to zero, while the order parameter at zero temperature $\Delta_0 = e^{-\pi/4}\Delta_{00}$. For $\Delta_0\tau_s < 1$, there is a concentration range $n' < n < n_c, n' = 2e^{-\pi/4}n_c \approx 0.91n_c$, where the superconcting order still exists but the gap is zero. This state was called gapless superconductivity (Abrikosov and Gor'kov, 1961). As can be seen from line (c) in Fig. 3.18, the electronic density of states in this state has a non-zero value $\gamma = [1-(\Delta_0\tau_s)^2]^{1/2}$ that determines the electronic heat capacity $C \sim \gamma T$ as in a normal metal but with a suppressed Sommerfeld parameter. Nevertheless, the superconductivity still exists; the peak in the density of states at $\omega \sim \Delta$ in Fig. 3.18 indicates that the non-Ohmic tunnelling characteristic can be measured, although it will be much weaker than for non-magnetic scattering.

This result is of fundamental importance. It illustrates the fact that the presence of the energy gap is an important feature but not a condition necessary for the existence of superconductivity. The order parameter may exist even in the absence of the energy gap.

The gapless state can be visualised qualitatively as being described by the 'two-liquid' model (Kresin *et al.*, 2014). According to this model, the superconductor contains two 'components'. The presence of a 'normal' component leads to the absence of the energy gap, whereas the 'superfluid' component is responsible for zero resistance, the Meissner effect, and so on.

3.3.3 Symmetry of the Order Parameter and Pair Breaking

Spin triplet p- and f-wave pairing. Here we will discuss the impurity effect on unconventional superconductors with *p*, *d*, and *f* -types of pairing. Spin triplet superconductors may have $l = 1$ (*p*-pairing) or $l = 3$ (*f*-pairing), as was discussed in section 3.1. This type of pairing is known for different phases of ^3He (Leggett, 1975; Wheatley, 1975; Mineev, 1983), for Sr_2RuO_4 (Rice and Sigrist, 1995; Ovchinnikov, 2003), and for some heavy fermion compounds (Ott *et al.*, 1983; Pfleiderer, 2009). The pair-wave function for $S = 1$ can be written as (Mineev and Samokhin, 1998)

$$\psi^t_{pair} = (\boldsymbol{d}(\boldsymbol{k})\boldsymbol{\sigma})\, i\sigma_y \tag{3.65}$$

Here $\boldsymbol{\sigma} = (\sigma_x, \sigma_y, \sigma_z)$ are the Pauli matrices. The space inversion for the triplet wave function requires the order parameter $\boldsymbol{d}(\boldsymbol{k}) = -\boldsymbol{d}(-\boldsymbol{k})$. This inversion is crucial for pair breaking in triplet superconductors. The non-magnetic impurity scattering results in T_c suppression, which is similar to the effect of magnetic impurities on *s*-wave superconductivity (Balian and Werthammer, 1963). Similar scattering of the triplet Cooper pair from the surface or the grain border suppresses the T_c value. That is why, for all known triplet superconductors, T_c is rather small.

For example, in the first report of triplet superconductivity in Sr_2RuO_4, the value $T_c = 0.93$ K was obtained (Maeno *et al.*, 1994). Later it was found that T_c strongly depends on the concentration of non-magnetic impurities and lattice defects (Mao *et al.*, 1999). The high-quality single crystals grown at Kyoto University (Mao *et al.*, 2000) are 4 cm large with the mean free path of electrons $l \sim 2000$ nm and $l/\xi = 30$ so the clean limit is attained. The maximal value $T_c = 1.489$ K was obtained. Figure 3.19 shows the T_c dependence on the residual resistivity ρ_0 found from the temperature dependence of the resistivity $\rho = \rho_0 + AT^2$ in the temperature interval 4.2 K–25 K.

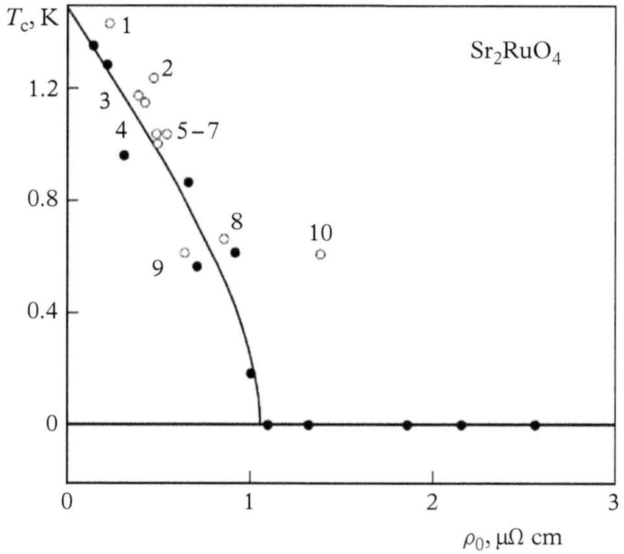

Figure 3.19 *The critical temperature T_c plotted as a function of the residual resistivity ρ_0 in Sr_2RuO_4; the numbers correspond to the different samples from Mao et al. (1999), and the solid curve correspond to the results that follow from the modified Abrikosov–Gor'kov theory.*

The magnitude of ρ_0 is determined by the lattice defect concentration (open circles with numbers from 1 to 10) and the Al impurity (black circles; data from Mackenzie *et al.*, 1998). The solid curve represents the fit by Mackenzie *et al.* (1998) to the Abrikosov–Gor'kov theory for pair destruction by magnetic impurities. The modified Abrikosov–Gor'kov theory for non-magnetic impurities effect in unconventional superconductivity is discussed by Mineev and Samokhin (1998).

Singlet d-wave pairing. For superconductors with d-pairing, the effect of impurities is also peculiar. The most studied d-wave superconductors are cuprates, where, in the square lattice, the d-wave gap is given by

$$\Delta(\mathbf{k}) = \Delta_0/2\,(\cos \mathrm{k_x a} - \cos \mathrm{k_y a}) \tag{3.66}$$

This gap is zero in the nodal direction [1, 1] and is maximal in the antinodal directions [1, 0] and [0, 1]. Up to now, there has been no generally accepted microscopic theory for superconductivity in cuprates, as strong electron correlations and short-range antiferromagnetic fluctuations prevent the description of electrons as a normal Fermi liquid and the superconducting state within the BCS theory. So, we start our discussion with a review of the experimental data. Phase diagram asymmetry for hole- and electron-doped cuprates is closely related to the impact that non-magnetic and magnetic impurities replacing copper sites has on superconducting properties. In electron-doped systems (n-type), the situation is analogous to that for conventional superconductors: non-magnetic impurities weakly suppress T_c, while magnetic ones cause a collapse of superconductivity for the impurity concentration of ~1%, which is in close agreement with the Abrikosov–Gor'kov theory. These results follow from studies of both polycrystalline samples (Tarascon *et al.*, 1990; Jayaram *et al.*, 1995) and $Pr_{2-x}Ce_xCu_{1-y}MyO_{4+z}$ single crystals with M = Ni, Co (Brinkmann, 1996).

Unlike n-type cuprates, their hole-doped counterparts show a different behaviour. Notice, however, that, to compare effects of different types of impurities on T_c, it is more convenient to study the lanthanum-based system $La_{2-x}Sr_xCu_{1-y}M_yO_4$ (with y being the amount of M, Fe, Co, Ni, Zn, Al, and Ga), where all impurities are located in the CuO_2 layer, for which the following results were obtained by Xiao *et al.* (1990): both the magnetic impurity Co and the non-magnetic impurities Zn, Al, and Ga result in almost the same $T_c(y)$ dependence. At the same time, Fe causes the most rapid suppression of T_c, while Ni gives the slowest decrease in it, although both should be magnetic due to their atomic structure. The static susceptibility measurements by Xiao *et al.* (1990) and Ting *et al.* (1992) have revealed the presence of an effective magnetic moment at the impurity site in all systems studied. Moreover, it becomes clear that the rate of T_c suppression has a weak correlation with the impurity valence. Furthermore, the magnitude of the moment strongly correlates with the critical impurity concentration at which T_c vanishes. This argues in favour of the magnetic mechanism of pair breaking and against pair breaking originating from the change in the hole doping level. A qualitative explanation (Xiao *et al.*, 1990) for the concentration dependence of $T_c(y)$ and for the magnetic properties of impurities is based on indications that all impurities with zero spin (non-magnetic Zn, Al, and Ga, as well as Co^{3+} in the low-spin state) induce an effective magnetic moment in close proximity to the Cu^{2+} moment. That is, one has to consider the copper ion spin to be removed by the impurity. For an impurity with an open d-shell (Fe, Co, Ni), it is necessary to consider not only the removed copper spin but also its own impurity moment. The experimental value of the Fe^{3+}-ion effective moment suggests that it resides in a high-spin state with $S = 5/2$ in the lanthanum system and generates an effective moment significantly larger than the Cu^{2+} moment. On the other hand, the anomalously small experimental value of the Ni^{2+} effective moment suggests that the spin should be no more than 0.32, instead of the expected $S = 1$. Xiao *et al.* (1990) explain this by the significant delocalisation of the Ni spin state, in contrast to the strong localisation of the Fe state.

Besides providing a qualitative explanation of the anomalous result of Cu with Ni replacement, proper treatment of multi-electron effects in a correlated band structure leads to a quantitative description of the $T_c(y)$ dependence (Ovchinnikov, 1995). With the diamagnetic replacement of Cu with Zn, the fraction of ions in configuration d^{10} (Zn^{2+}) is equal to y. The model for such systems is the antiferromagnetic lattice of $S=1/2$ spins with one empty site that behaves as one paramagnetic centre due to the uncompensation of sublattices. That is why the effect of the formally non-magnetic impurity Zn in an antiferromagnetic background is similar to that of a paramagnetic impurity in a normal metal. As for the replacement of Cu with Ni, the ion Ni^{2+}, which formally should be in the d^8 state with the spin $S = 1$, due to strong intra-atomic Coulomb repulsion and strong covalence, acquires an intermediate valence. In the limit $U_d \to \infty$ the probability of the d^8 state is zero. The probability of it being in the non-magnetic d^{10} state with spin $S = 0$ is equal to $u_0^2 = (1 - \Delta_0/\nu_0)/2$, where $\Delta_0 = \varepsilon_p - \varepsilon_d - V_{pd}$, $\nu_0^2 = \Delta_0^2 + 8t_{pd}^2$. These expressions are obtained within the p – d model of cuprates (Emery, 1987; Varma *et al.*, 1987); here ε_p and ε_d are the energies of a hole on oxygen and copper sites, respectively, V_{pd} is their Coulomb interaction, and t_{pd} is the amplitude of a hole hopping. The probability of the d^9 state with $S = 1/2$ is $v_0^2 = 1 - u_0^2$. For typical parameters $\Delta_0 = 1.4$ eV and $t_{pd} = 1$ eV, $u_0^2 = 0.28$ and $v_0^2 = 0.72$. So, the effective spin at the Ni^{2+} impurity is $S_{Ni} = 0.5v_0^2 = 0.36$ agrees with the nuclear magnetic resonance (NMR) data (Alloul *et al.*, 1995). For the finite and large $U_d = 10$ eV, the high-spin $S = 2$ contribution in the d^8 state is still negligibly small, of the order of 1%.

Therefore, with the substitution of Ni for Cu, the amount of impurity ions in the d^{10} state (the same state as for Zn) is equal to $u_0^2 y$. The probability of ions being in the d^9 state is $v_0^2 y$, and since their magnetic and charge characteristics are almost the same as those of Cu, such ions should not suppress superconductivity. The resulting ratio of $T_c(y)$ slopes for Ni and Zn impurities is $u_0^2 = 0.28$, which is close to the experimental value of 0.38 for $La_{2-x}Sr_xCu_{1-y}M_yO_4$, M = Zn, Ni (Xiao, 1990). Summarising the discussion, we notice that, while the microscopic theory of d-wave superconductivity in cuprates is still far from being complete, the final understanding of the impurity scattering effects is lacking. We direct the curious reader to some reviews, like those by Ovchinnikov (1997), Hirschfeld and Atkinson (2002), Balatsky *et al.* (2006), Alloul *et al.* (2009), and Pogorelov *et al.* (2011).

Pair breaking in multiband superconductors. Here we will discuss specific effect of impurities in a superconductor with a multi-valley structure for the Fermi surface, such as the layered Fe-based compounds (Korshunov *et al.*, 2016). Superconducting Fe-based materials can be divided into two subclasses: pnictides and chalcogenides. The square lattice of Fe is the basic element of all the construction. Fe is surrounded by As or P pnictogens situated in the tetrahedral positions within the first subclass, and by Se, Te, or S chalcogens within the second subclass. All five orbitals, d_{x2-y2}, d_{3z2-r2}, d_{xy}, d_{xz}, and d_{yz}, are at or near the Fermi level. This results in a significantly multiorbital and multiband low-energy electronic structure, which could not be described within the single-band model. For example, within the five-orbital model (Graser *et al.*, 2009) correctly reproducing density functional theory band structure (Cao *et al.*, 2008), the Fermi surface comprises four sheets: two hole pockets around the point (0, 0), and two electron pockets around the points (π, 0) and (0, π). The presence of a few pockets and the multiorbital band character significantly affect superconducting pairing. Each band participates in the superconductor pairing due to intraband and interband interactions.

The symmetry class of Fe-based superconductors is the generalised A_{1g} (extended s-wave symmetry), probably involving a sign change of the order parameter between Fermi surface sheets or its parts (Hirschfeld *et al.*, 2011). Gaps with the same symmetry may have different structures. Figure 3.20 illustrates this for s-wave symmetry. Fully gapped s-states without nodes on the Fermi surface differ only by a relative gap sign between the hole and the electron pockets, which is positive in the s++ state and negative

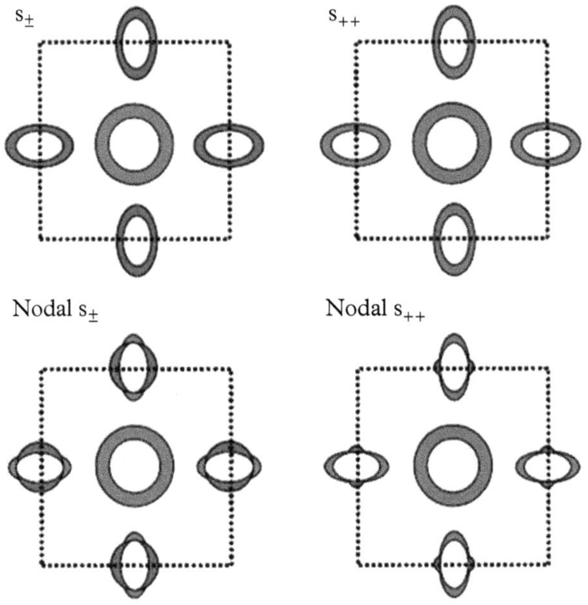

Figure 3.20 *Four types of order parameter structures with the s-wave symmetry in the two-dimensional Brillouin zone (the dotted squares) corresponding to one Fe atom per unit cell. Different colours stand for different signs of the order parameter. (From Korshunov et al., 2016)*

in the $s\pm$ state. In addition to a global change of sign, which is equivalent to a gauge transformation, we can have individual rotations on single pockets and still preserve symmetry. These states are called 'nodal $s_\pm$ ' ('nodal s_{++} ') and are characterised by the opposite (same) averaged signs of the order parameter on the hole and electron pockets. Nodes of this type are sometimes described as 'accidental', since their existence is not dictated by symmetry, unlike the symmetry nodes of the d-wave gap. In general, the extended s-wave gap may be written as (here $\alpha = a, b$, denotes Fermi surfaces centred at $(0, 0)$ and $(\pi, 0)/(0, \pi)$)

$$\Delta^{\alpha}(\mathbf{k}) = \Delta_0^{\alpha} + \frac{1}{2}\Delta_1^{\alpha}(\cos \mathrm{k_x a} + \cos \mathrm{k_y a}) \tag{3.67}$$

where the first term is the spherical s-gap and the second one is the anisotropic first harmonic. For equal signs of Δ_0^{α} and Δ_1^{α}, we have the s_{++} state if $\Delta_1^{\alpha} < \Delta_0^{\alpha}$, and the nodal s_{++} if $\Delta_1^{\alpha} > \Delta_0^{\alpha}$. Similarly for opposite signs of Δ_0^{α} and Δ_1^{α}, we have the $s_\pm$ state if $\Delta_1^{\alpha} < \Delta_0^{\alpha}$, and the nodal $s_\pm$ if $\Delta_1^{\alpha} > \Delta_0^{\alpha}$.

Superconducting states with different symmetries and structures of order parameters respond differently when subject to disorder. In unconventional superconductors, suppression of the critical temperature as a function of parameter Γ characterising impurity scattering may follow a complicated law. Several experiments on Fe-based systems show that T_c suppression is much weaker than expected in the framework of

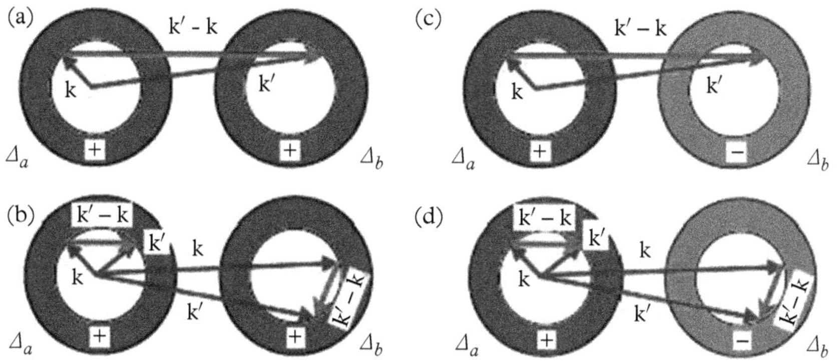

Figure 3.21 *Diagram of two Fermi surfaces with the superconducting gaps Δ_a and Δ_b, with the same signs (a, b) and with the opposite signs (c, d). Interband impurity scattering (a, c) mixes states with Δ_a and Δ_b, while the intraband scattering (b, d) involves states within the boundaries of each Fermi surface. (From Korshunov et al., 2016)*

the Abrikosov–Gor'kov theory for both non-magnetic (Karkin *et al.*, 2009; Kim *et al.*, 2014) and magnetic disorder (Cheng *et al.*, 2010; Grinenko *et al.*, 2011).

Non-magnetic impurities in a conventional two-band superconductor with two isotropic gaps lead to scattering of quasiparticles between bands or within each band. For impurity scattering within only one band ('intraband impurities'), the pair-breaking effects for different bands are not coupled. Therefore, all conclusions made for the single-band case above are also true here for each band in the presence of both non-magnetic and magnetic impurities. Interband processes, as shown in Fig. 3.21, result in an averaging of the gaps and, therefore, in the initial suppression of T_c, after which T_c saturates and stays constant until localisation effects become important (Sadovskii, 1997). For the interband scattering in the s_{++} state with $\Delta_a = \Delta_b$, T_c is independent of disorder. Therefore, there is no impurity effect on the multiband isotropic *s*-wave superconducting state.

Interband scattering in a two-band system with the sign-changing order parameter leads to a much more complicated behaviour (Preosti and Muzikar, 1996; Golubov and Mazin, 1997). In this case, non-magnetic impurities with the interband component of the scattering potential destroy the superconductivity even for equal magnitudes of gaps and densities of states (the so-called symmetric model). The reason is quite simple: interband scattering results in averaging the gaps in two different bands and, since Δ_a and Δ_b have opposite signs in the $s_{\pm}$ state, their average goes to zero. The critical temperature T_c would then vanish for the finite critical impurity concentration, as in the theory of magnetic impurity scattering in a single-band *s*-wave superconductor (Abrikosov and Gor'kov, 1961).

Such a characteristic feature of Fe-based superconductors was quickly noticed by different groups of researchers (Chubukov *et al.*, 2008; Mazin *et al.*, 2008; Senga and Kontani, 2008). In contrast, magnetic impurities with interband scattering do not affect the $s_{\pm}$ state with equal absolute values of gaps $\Delta_a = -\Delta_b$.

As for the effect of non-magnetic and magnetic disorder on a multiband anisotropic superconductor, the simple (and naive!) qualitative rule of thumb is as follows: when a non-magnetic impurity scatters a pair from one point on the Fermi surface to another point such that the order parameter does not change sign, scattering is not pair breaking; if the order parameter flips its sign, it is pair breaking. For a magnetic impurity, the opposite is true: scattering with an order parameter sign change is not pair breaking; otherwise, it is. As follows from calculations for some particular cases, however, such a naive qualitative rule collapses, and quite unexpected results appear. More complicated behaviour results from different types of impurity scattering, the intraband and interband ones. If non-magnetic impurity scattering occurs *only* between the bands with different signs of the gaps, than the critical temperature T_c for the $s_\pm$ state is suppressed, as in the case of magnetic impurity scattering in a single-band BCS superconductor.

Even assuming isotropic gaps on two different Fermi surface sheets and non-magnetic scattering, we find the suppression of superconductivity for a system with mainly intraband scattering to be slower than expected. Anderson's theorem is applicable in the limit of pure *intraband* scattering: the system is 'insensitive' to the signs of the gaps, and T_c is not suppressed. Therefore, the T_c suppression rate depends on the ratio between intra- and interband scattering rates, and making conclusions about the superconducting state on the basis of systematic disorder studies is harder than in the single-band case.

As concerns magnetic impurities and their effect on the properties of the $s_\pm$ and s_{++} states, there are few cases when the critical temperature T_c saturates and stays finite in contrast to T_c, following from the prediction of the Abrikosov–Gor'kov theory (Korshunov *et al.*, 2014). The critical temperature T_c in the $s_\pm$ state becomes insensitive to impurities for pure interband scattering. This partially confirms qualitative arguments that the $s_\pm$ state with magnetic impurities behaves like the s_{++} state with non-magnetic disorder (Golubov and Mazin, 1997), and agrees with theoretical calculations in the Born limit (Li and Wang, 2009). For the initially unequal absolute value of the gaps, $\Delta_a \neq -\Delta_b$, there is an initial decrease in T_c for a small scattering parameter Γ until the renormalised gaps become equal and then T_c saturates, since the analogue of Anderson's theorem is achieved. For the finite intraband and interband scattering, the intraband scattering on the magnetic disorder averages gaps up to zero and thus suppresses T_c.

A very unusual effect of non-magnetic impurities on the order parameter and T_c in the two-band model has been predicted by Efremov *et al.* (2011). Within the T-matrix approximation, they have shown that, for a finite non-magnetic impurity scattering rate, the transition from the $s_\pm$ state to s_{++} occurs, that is, one of the two gaps changes sign when going through zero. At the same time, T_c stays finite and almost independent of the impurity scattering rate, which is proportional to the impurity concentration and magnitude of the scattering potential. The experimental measurements of the inverse London penetration depth λ_L^{-2} in proton-irradiated $Ba(Fe_{1-x}Rh_x)_2As_2$ have confirmed such transition (Ghigo *et al.*, 2018).

This transition is shown in Fig. 3.22 and occurs with the change of the temperature and the impurity scattering rate (Shestakov *et al.*, 2018a) and for some magnitude of the scattering potential, the sequence of transitions $s_\pm \rightarrow s_{++} \rightarrow s_\pm$ is found. Such a transition is accompanied by a sharp change of the smaller gap Δ_b; nevertheless, the T_c remains a smooth function of the scattering rate (Shestakov *et al.*, 2018b).

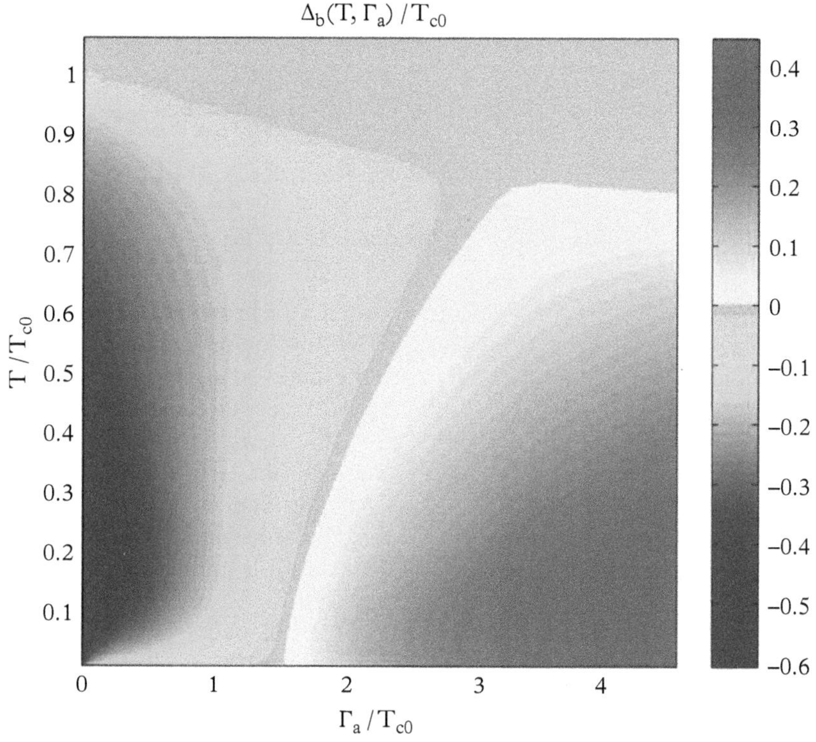

Figure 3.22 *The temperature magnitude of the scattering potential map of the smaller gap Δ_b demonstrates the transition from the $s_{\pm}$ state to the s_{++} one. (From Shestakov et al., 2018a)*

3.4 Induced Superconductivity: The Proximity Effect

3.4.1 The Proximity 'Sandwich'

The proximity effect was discovered by Meissner (1960) and involves spatially separated normal and superconducting subsystems. The simplest type of proximity system is shown schematically in Fig. 3.23.

A superconducting state is induced in the normal film N affected by the neighbouring superconducting film S. Experimentally, one finds that film N begins to exhibit the Meissner effect (see e.g. Simon and Chaikin, 1981). Qualitatively, such an effect means that two electrons from the N film can tunnel into the S film, forming the Cooper pair: this leads to an appearance of the pairing function F_N^+ (see Figs 2.2 and 2.3). One can also introduce the order parameter Δ_N (see below).

An important parameter of the proximity system is the coherence length in the normal film, ξ_N. It describes the scale of penetration of the superconducting state into N film. This quantity, introduced by Clarke (1969), is temperature dependent and equals

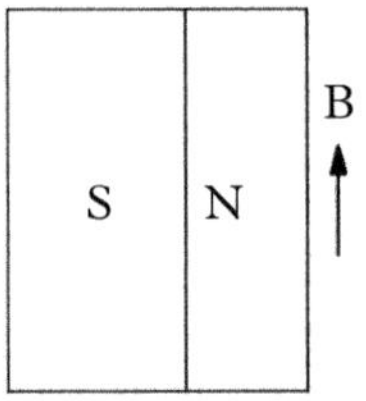

Figure 3.23 *The proximity S–N system.*

$\xi_N = \hbar v_f/kT$. Thus, lowering the temperature benefits the proximity effect. For example, it has been observed (Mota *et al.*, 1989) that, for NbTe–Cu(Ag) systems, large scales (up to $2 \cdot 10^5$ A) of induced superconductivity could be achieved at ultra-low temperatures.

There exist different methods of studying the proximity effect; among them, two are particularly prominent. One of them is based on the Ginzburg–Landau theory and makes use of the boundary conditions at the *S–N* interface (the 'Orsay School'; see e.g. the work by de Gennes (1964, 1966) and Deutscher and de Gennes (1969)). This approach directly takes into account the spatial variation of the order parameter $\Delta(\boldsymbol{r})$. Since this description is based on the Ginzburg–Landau theory, it is applicable near T_c only (or near H_{c_2}).

The other approach is the tunnelling model developed by McMillan (1968b). The superconducting state in the N film arises thanks to Cooper pair tunnelling from the superconducting film into the normal one. In the tunnelling model, it is assumed that the order parameters in the S and N films are constant throughout the films. This model can be used, if the thickness of the normal film is much smaller than the coherence length ξ_N, that is, $L_N \ll \xi_N$. An attractive feature of this model, in addition to its elegance and simplicity, is that it is applicable not only near T_c but at any temperature.

Usually, the approach based on use of the boundary conditions is applicable for good-quality contacts. Otherwise, the tunnelling model is more suitable.

3.4.2 Critical Temperature

Let us calculate the critical temperature of an *S–N* proximity sandwich within the framework of the tunnelling model. We will employ the method of thermodynamic Green's functions. Let us introduce the self-energy parts $\sum_2^\alpha$ and $\sum_2^\beta$ ($\alpha(\beta) \equiv S(N)$) describing pairing in the films; they satisfy the following equations:

$$\sum\nolimits_2^\alpha = \sum\nolimits_{2,ph}^\alpha + \tilde{T}^2 \int d\mathbf{p}' F^\beta(\mathbf{p}', \omega_n)$$

$$\sum\nolimits_2^\beta = \tilde{T}^2 \int d\mathbf{p}' F^\alpha(\mathbf{p}', \omega_n)$$

$$\sum\nolimits_{2,ph}^\alpha = T \sum_{\omega n'} \int d\mathbf{p}' \zeta_\alpha^2(\mathbf{p}, \mathbf{p}') D(\omega_n - \omega_{n'}, \Omega(q)) F^\alpha(\mathbf{p}', \omega_{n'}) \qquad (3.68)$$

Here D is the phonon Green's function (see chapter 2, eqn (2.15)), and F^{α} and F^{β} are the pairing Green's functions, for example,

$$F^{\alpha}(\omega_n,\mathbf{p}) = -\frac{\sum_2^{\alpha}(\omega_n,\mathbf{p})}{\left[\omega_n^2 Z_{\alpha}^2 + \xi_{\alpha}^2(\mathbf{p}) + \sum_2^{(\alpha)^2}(\omega_n,\mathbf{p})\right]} \tag{3.68$'$}$$

A similar expression can be written for F^{β}. Here ξ_{α} is the energy of an ordinary electron referred to the Fermi level, Z is the renormalisation function, and $\tilde{T}$ is the averaged value of the tunnelling matrix element. Equations (3.68) are written for the case when the existence of the pair condensate in the β film is totally due to the proximity effect.

In the weak coupling approximation, the renormalisation functions at $T = T_c$ are equal to

$$Z_i(\omega_n) = 1 + \frac{\Gamma^{ik}}{|\omega_n|}; \Gamma^{ik} = \pi\tilde{T}^2 \nu_k V_k;\ i,k = \alpha,\beta;\ i \neq k$$

where ν_k and V_k are the density of states and volume, respectively. The parameters Γ^{ik} were introduced by McMillan (1968). For example, $\Gamma^{\beta\alpha}$ can be written in the following form:

$$\Gamma^{\beta\alpha} = \frac{v_{F_\perp}\sigma}{2BL_{\beta}}$$

where $v_{F_\perp}$ is the Fermi velocity, σ is the barrier penetration probability, and B is a function of the ratio of the mean free path to the film thickness. If $T = T_c$, we should put $\sum_i = 0$ in the denominator of (3.68′). Making use of eqns (3.68) and (3.68′), we arrive, after some manipulations, at an equation for T_c (Kresin, 1982):

$$\begin{aligned} \ln\frac{T_c}{T_c^{\alpha}} &= -\frac{\Gamma_{\alpha\beta}}{\Gamma}\frac{1}{\lambda_{\alpha}}\int d\Omega g(\Omega) \\ &\times\left\{\left[\psi\left(\frac{1}{2}+\frac{\Gamma}{2\pi T_c}\right)-\psi\left(\frac{1}{2}\right)\right]\frac{\Omega^2}{\Omega^2+\Gamma^2}+\frac{\Gamma^2}{\Omega^2+\Gamma^2}\ln\frac{2\Omega}{\pi T_c}\right\} \end{aligned} \tag{3.69}$$

Here T_c^{α} is the critical temperature in an isolated α film, $\Gamma = \Gamma_{\alpha\beta} + \Gamma_{\beta\alpha}$, and ψ is the digamma function. It should be noted that, in general, $g(\Omega) \neq g^a(\Omega)$; in other words, the presence of the interface may alter the function $g(\Omega)$. Equation (3.69) allows us to evaluate T_c for any proximity system and is valid for any relation between Γ and T_c^{α}.

Let us consider some special cases. Assume that $\Gamma \gg T_c$. Making use of the asymptotic form of the digamma function, we find, after simple substitutions, that

$$T_c = T_c^{\alpha}\left(\frac{\pi T_c^{\alpha}}{2 < u >_{\gamma}}\right)^{\rho} \tag{3.70}$$

where

$$\rho = \Gamma_{\alpha\beta}/\Gamma_{\beta\alpha}; \Gamma = \Gamma_{\beta\alpha}(1+\rho)$$
$$u = \Gamma^{1-\delta}\Omega^{\delta}; \delta = \Gamma^2\left(\Omega^2 + \Gamma^2\right)^{-1} \tag{3.70'}$$

or

$$\rho = (\nu_\beta/\nu_\alpha)(L_\beta/L_\alpha) \tag{3.70''}$$

Recall that it is assumed that $L_i \ll \xi_i\,(i \equiv \alpha, B)$. The mean value is to be understood in the following sense:

$$\ln\frac{\langle u(\Omega)\rangle}{T_c} = \frac{1}{\lambda_\alpha}\int d\Omega g_\alpha(\Omega)\ln\frac{u(\Omega)}{T_c}$$

Equation (3.70) is valid for any relation between Γ and $<\Omega>$ (keeping in mind that $\Gamma \gg T_c$). If $\Gamma \ll <\Omega>$, we arrive at the expression obtained by McMillan (1968):

$$T_c = T_c^\alpha\left(\frac{\pi T_c^\alpha}{2\,\Gamma\gamma}\right)^\rho \tag{3.71}$$

In the opposite limit, ($\Gamma \gg \tilde{\Omega}$) (Cooper, 1961), we obtain the result

$$T_c = T_c^\alpha\left(\frac{\pi T_c^\alpha}{2\langle\Omega\rangle\gamma}\right)^\rho \tag{3.71'}$$

Finally, if $\Gamma \lesssim T_c$, $\tilde{\Omega} \gg T_c$, and $\rho \ll 1$, we find $T_c = T_c^\alpha\ \exp(-F)$, where

$$F = (\Gamma_{\alpha\beta}/\Gamma)\left[\psi\left(\frac{1}{2}+\frac{\Gamma}{2\pi T_c}\right)-\psi\left(\frac{1}{2}\right)\right]$$

Another interesting quantity to determine is the induced energy gap $\varepsilon_\beta(0)$. It can be found from eqn (3.68) and is given by the simple expression (McMillan, 1968b)

$$\varepsilon_\beta(0) = \Gamma^{\alpha\beta} \tag{3.68''}$$

Figure 3.24 displays the density of states of the superconducting electrons. The density of states is characterised by two distinct peaks, corresponding to two gaps, $\varepsilon_\alpha(0)$ and $\varepsilon_\beta(0)$.

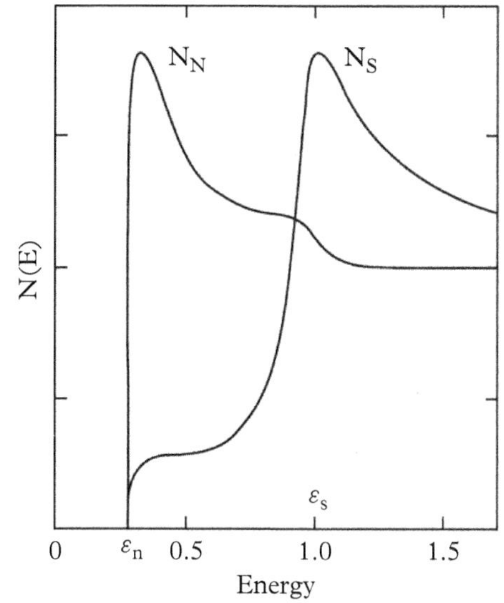

Figure 3.24 *The two-gap structure in the densities of states for an S–N system.*

Clearly, the behaviour of an isolated metallic film is strongly affected by the boundary conditions. Indeed, electrons are scattered by the surface and, for example, one has to distinguish the case of specular or diffuse scattering. However, in the case of a proximity system, for example an N–S sandwich, the scattering picture becomes quite peculiar. Indeed, two N electrons can tunnel into the S film (we assume that $E_N < \varepsilon_s$), making a transition into the Cooper pair condensate. That is to say, the presence of the N–S boundary leads to the 'disappearance' of two N electrons. At the same time, we know that the disappearance of an electron is equivalent to the appearance of a hole. As a result, the transfer of two N electrons into a Cooper pair on the S side can be described as the disappearance of one electron with the simultaneous creation of a hole (envisioning this creation instead of the disappearance of the second electron). Therefore, in terms of the scattering picture, the process at the N–S boundary can be described as the disappearance of an electron and the creation of a hole (Andreev scattering (1964); see also the work by Blonder *et al.* (1982) and the reviews by Tinkham (1996), Waldram (1996), and Deutscher (2005)). Generally speaking, the whole picture corresponds to the Cooper diagrams in Figs 2.2 and 2.3; the pairing can be described as an electron $\rightarrow$ hole transformation.

Here, we focused on S–N proximity systems. One can also study the contact of two superconductors $S_\alpha - S_\beta$ with different values of the critical temperature. In this case, the proximity sandwich has a T_c for which the value is intermediate between $T_{c;\alpha}$ and $T_{c;\beta}$.

The physics of proximity systems is an interesting and broad area. For example, the study of Josephson junctions S–N–I–S and S–N–S is concerned with Josephson

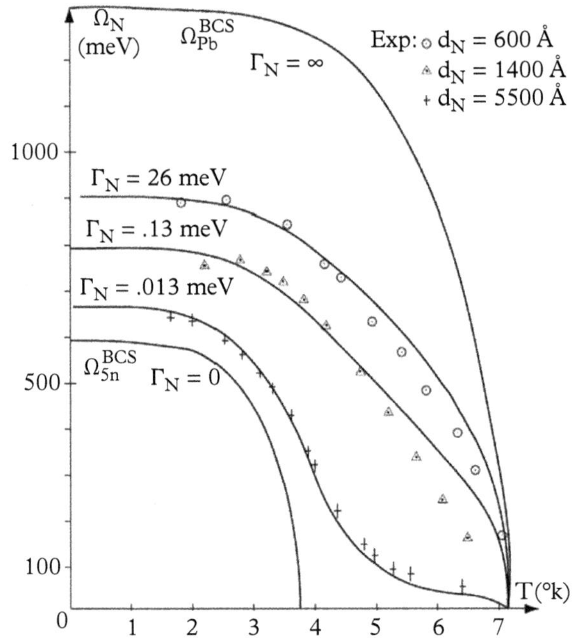

Figure 3.25 *Temperature variation in the energy gap of tin films backed by 5000 Å of lead. Experimental curves are compared with the theoretical curves from the MacMillan tunnelling model: $\Gamma_N = \Gamma^{ab}$. (Figure reproduced from Gilabert et al. (1971). @1971, with permission from Elsevier)*

tunnelling in proximity systems. Other active areas include the electrodynamics of proximity systems, as well as the properties of $S_\alpha - S_\beta$ contacts ($T_c^\alpha \neq T_c^\beta$). If, for example, $T_c^\alpha > T_c^\beta$, then, at T > T_c^β, the superconducting state penetrates into the β film, but the decay of the pairing function is slower than for the *SN* contact (Covaci and Marsiglio, 2006).

Note that, for the $S_\alpha - S_\beta$ system at $T_c^\alpha > T > T_c^\beta$, the McMillan tunnelling model is applicable, and equations similar to eqn (3.69) can be used to calculate, for example, the temperature dependence of the energy gaps. For example, the data by Gilabert *et al.* (1971) for the Pb–Sn system (Fig. 3.25) are in a good agreement with the tunnelling model.

As was noted above, the tunnelling model is valid if the thickness of the normal (*N*) film is smaller than the coherence length ξ_N. If we are concerned with the spatial dependence of the order parameter, we should use a more general approach, for example the approach based on the Ginzburg–Landau theory with corresponding boundary conditions (see e.g. Deutcher and de Gennes, 1969). A very efficient method based on the Green's functions, averaged over energy (Eilenberger, 1968; Larkin and Ovchinnikov, 1969, 1986) was developed by Usadel (1970); see, for example, the work by Gueron *et al.*

(1996). In this section, we will discuss the Usadel equation, following his paper (Usadel, 1970).

To simplify the original Gor'kov (1959; 1960) equations, Eilenberger (1968) derived transport-like equations for type II superconductors. He introduced the functions $g(\omega, \boldsymbol{r}, \boldsymbol{v})$ and $f(\omega, \boldsymbol{r}, \boldsymbol{v})$, which are closely related to the Green's functions for a superconductor. Indeed, these quantities are the impurity-averaged Green's functions integrated over the energy variable. They are solutions of the following transport-like equation:

$$\left\{2\omega + \boldsymbol{v}\cdot\hat{\boldsymbol{\partial}}\right\} f(\omega, \boldsymbol{r}, \boldsymbol{v}) = \mathbf{2}\Delta$$
$$g(\omega, \boldsymbol{r}, \boldsymbol{v}) + nv \oint d\Omega_{v'} \frac{d\sigma(\boldsymbol{v}, \boldsymbol{v}')}{d\Omega} \{g(\omega, \boldsymbol{r}, \boldsymbol{v}) f(\omega, \boldsymbol{r}, \boldsymbol{v}') - f(\omega, \boldsymbol{r}, \boldsymbol{v}) g(\omega, \boldsymbol{r}, \boldsymbol{v}')\}. \quad (3.72)$$

Here $\boldsymbol{v}'$ denotes the Fermi velocity, $\omega = (2n+1)\pi T$, and $\hat{\boldsymbol{\partial}}$ is defined by $\hat{\boldsymbol{\partial}} = \boldsymbol{\partial} + 2ie\boldsymbol{A}(\boldsymbol{r})$, with $\boldsymbol{A(r)}$ a vector potential of the magnetic field. The functions f and g are connected by

$$g(\omega, \boldsymbol{r}, \boldsymbol{v}) = [1 - f(\omega, \boldsymbol{r}, \boldsymbol{v}) f^*(\omega, -\boldsymbol{r}, \boldsymbol{v})]^{1/2}$$

These equations are completed by the self-consistency conditions

$$\Delta(\boldsymbol{r})\, ln\frac{T}{T_c} + 2\pi T \sum_{\omega>0} \left[\frac{\Delta(r)}{\omega} - \oint \frac{d\Omega_v}{4\pi} f(\omega, \boldsymbol{r}, \boldsymbol{v})\right] = 0$$
$$\boldsymbol{j}_s(\boldsymbol{r}) = \frac{1}{4\pi} \boldsymbol{\nabla} \times [\boldsymbol{B}(\boldsymbol{r}) - \boldsymbol{B}_e(\boldsymbol{r})] = 2ieN(0) 2\,\pi T \sum_{\omega>0} \oint \frac{d\Omega_v}{4\pi} \boldsymbol{\nu} g(\omega, r, \boldsymbol{\nu})$$

where T_c denotes the transition temperature of the bulk superconductor without a magnetic field. For a dirty superconductor, one expects the motion of electrons to be nearly isotropic. The assumption that the functions $f(\omega, \boldsymbol{r}, \boldsymbol{v})$ and $g(\omega, \boldsymbol{r}, \boldsymbol{v})$ are nearly isotropic with respect to ν led Luders (1967) in the limit $\Delta \to 0$ to the diffusion equation for the de Gennes' (1964) kernel of the linearised self-consistency equation. We apply Luders's method, expanding $f(\omega, \boldsymbol{r}, \boldsymbol{v})$:

$$f(\omega, \boldsymbol{r}, \boldsymbol{v}) = F(\omega, \boldsymbol{r}) + \hat{\boldsymbol{\nu}}\cdot\boldsymbol{F}(\omega, \boldsymbol{r}) \quad (3.73)$$

Here $\hat{\boldsymbol{\nu}} = \boldsymbol{\nu}/\nu$. In the lowest order for the g function, we get

$$g(\omega, \boldsymbol{r}, \boldsymbol{\nu}) = G(\omega, \boldsymbol{r}) + \hat{\boldsymbol{\nu}}\cdot\boldsymbol{G}(\omega, \boldsymbol{r})$$

where $G(\omega, r)$ is defined by

$$G(\omega,\boldsymbol{r}) = \left[1 - \left|F\left(\omega, r\right)\right|^2\right]^{1/2}$$

and $\boldsymbol{G}(\omega, \boldsymbol{r})$ is defined by

$$\boldsymbol{G}(\omega,\boldsymbol{r}) = \frac{F(\omega,\boldsymbol{r})\boldsymbol{F}^*(\omega,\boldsymbol{r}) - F^*(\omega,\boldsymbol{r})\boldsymbol{F}(\omega,\boldsymbol{r})}{2G(\omega,\boldsymbol{r})} \tag{3.74}$$

The scalar and vector parts of the functions f and g are the 'densities' $F(\omega,\boldsymbol{r}), G(\omega,\boldsymbol{r})$ and the 'currents' $\boldsymbol{F}(\omega,\boldsymbol{r}), \boldsymbol{G}(\omega,\boldsymbol{r})$. Integrating eqn (3.72) over all directions of the Fermi velocity, we get

$$2\omega F(\omega,\boldsymbol{r}) + \frac{1}{3}\boldsymbol{\nu}\hat{\boldsymbol{\partial}}\boldsymbol{F}(\omega,\boldsymbol{r}) = 2\Delta G(\omega,\boldsymbol{r}) \tag{3.75}$$

and, by first multiplying eqn (3.72) by $\hat{\boldsymbol{\nu}}$ and then integrating over $\hat{\boldsymbol{\nu}}$, we find

$$2\omega\boldsymbol{F}(\omega,\boldsymbol{r}) + \nu\hat{\boldsymbol{\partial}}F(\omega,\boldsymbol{r}) = 2\Delta\boldsymbol{G}(\omega,\boldsymbol{r}) + \tau_{tr}^{-1}\left[\boldsymbol{G}(\omega,\boldsymbol{r})F(\omega,\boldsymbol{r}) - \boldsymbol{F}(\omega,\boldsymbol{r})G(\omega,\boldsymbol{r})\right] \tag{3.76}$$

where τ_{tr} is the transport lifetime. With the help of eqn (3.74), we can eliminate the 'currents' $\boldsymbol{F}$ and $\boldsymbol{G}$ from eqns (3.75) and (3.76) and obtain an equation for the density F (below, we have defined $D = \tau_{tr}\nu^2/3$):

$$2\omega F(\omega,\boldsymbol{r}) - D\hat{\boldsymbol{\partial}}\left[G(\omega,\boldsymbol{r})\hat{\boldsymbol{\partial}}\boldsymbol{F}(\omega,\boldsymbol{r}) + \frac{1}{2}\frac{F(\omega,\boldsymbol{r})}{G(\omega,\boldsymbol{r})}\boldsymbol{\partial}\left|F\left(\omega,\boldsymbol{r}\right)\right|^2\right] = \boldsymbol{2}\Delta(\boldsymbol{r})G(\omega,\boldsymbol{r}) \tag{3.77}$$

This is the Usadel equation for the abnormal Usadel function $F(\omega,r)$. It is a generalisation of de Gennes' (1964) diffusion equation and is valid only in the dirty limit $2\pi T_c\tau_{tr} \ll 1$ under the following assumptions:

$$G \gg 2\tau_{tr}\omega, F \gg 2\tau_{tr}\Delta, |\, F \,| \gg |\, \hat{\boldsymbol{\nu}}\cdot\boldsymbol{F} \,|$$

The expansion of the Usadel equation in terms of Δ up to the term Δ^3 has been given by Maki (1964). The self-consistency conditions in terms of the Usadel function looks like

$$\Delta(\boldsymbol{r})\, ln\frac{T}{T_c} + 2\pi T\sum_{\omega>0}\left[\frac{\Delta(\boldsymbol{r})}{\omega} - F(\omega,\boldsymbol{r})\right] = 0$$

for the order parameter and, for the current density,

$$\boldsymbol{j}_s(\boldsymbol{r}) = 2ieN(0)2\pi TD\sum_{\omega>0}\left[F^*(\omega,\boldsymbol{r})\,\hat{\boldsymbol{\partial}}F(\omega,\boldsymbol{r}) - F(\omega,\boldsymbol{r})\left(\hat{\boldsymbol{\partial}}F(\omega,\boldsymbol{r})\right)^*\right]$$

At the surfaces and interfaces, the Usadel equation has to be completed by boundary conditions. In the linear case, these conditions at the *S–N* contact reads

$$\boldsymbol{n}\cdot\hat{\boldsymbol{\partial}}F(\omega,\boldsymbol{r}) = 0$$

In general, the Usadel equation is a quasi-classical approximation of the Gor'kov equations.

3.4.3 Proximity Effects in Ferromagnetic–Superconductor Heterostructures

One more interesting and actively developing area of *S–N* proximity systems is the competition between superconductivity and magnetism in ferromagnet–superconductor (*F–S*) heterostructures. The mutual influence of superconductivity and magnetism in *F–S* systems is different for *F* metal or insulator layers. For systems with ferromagnetic metal (*FM*) layers, the superconducting state of an *FM–S* system is a superposition of two pairing mechanisms, the BCS mechanism in the *S* layers and the Larkin–Ovchinnikov–Fulde–Ferrell (LOFF) mechanism in the *FM* layers. The competition between ferromagnetic and antiferromagnetic spontaneous moment orientations in *FM* layers is explored for 0- and π-phase superconductivity in *FM–S* systems. For *FI–S* structures, where *FI* is a ferromagnetic insulator, the exchange interaction between magnetic ions forms by the direct exchange inside the *FI* layers, as well as the indirect Ruderman–Kittel–Kasuya–Yosida (RKKY) exchange between localised spins via *S*-layer conduction electrons. Within this framework, possible mutual accommodation scenarios for superconducting and magnetic order parameters are possible.

Superconductivity and ferromagnetism are antagonistic phenomena, and their coexistence in uniform materials requires that special, difficultly realisable conditions be fulfilled. This antagonism manifests itself first in the relation of these phenomena to a magnetic field. A superconductor tends to expel a magnetic field (Meissner effect), whereas a ferromagnet concentrates the force lines of the field inside its volume (the effect of magnetic induction). The first explanation of the suppression of superconductivity via ferromagnetic ordering in transition metals was given by Ginzburg (1957), who indicated that, in these metals, magnetic induction exceeds the critical field. This antagonism is also understandable from the viewpoint of microscopic theory: attraction between electrons creates Cooper pairs in a singlet state, whereas exchange interaction, which produces ferromagnetism, tends to arrange electron spins in parallel to one another. Therefore, when the Zeeman energy of the electrons of a pair in an exchange field I exceeds the superconducting gap Δ, the superconducting state is destroyed. The corresponding critical field is $I_c \sim \Delta/\mu_B$, where μ_B is the Bohr magneton. Unlike the way the critical field H_c acts on the orbital states of the electrons of a pair, the critical field I_c acts on electron spins; therefore, the destruction of superconductivity due to this field is called the paramagnetic effect.

For the above reasons, the coexistence of the superconducting and ferromagnetic order parameters is unlikely in a uniform system, although it is easily achievable in an artificially prepared layered F–S system. Owing to the proximity effect, a superconducting order parameter can be induced in the F layer; on the other hand, the neighbouring pair of F layers can interact with one another via the S layer. Such systems exhibit rich physics, which can be controlled by varying the thicknesses of the $F(d_f)$ and $S(d_s)$ layers or by placing the F–S structure in an external magnetic field. Numerous experiments on F–S structures revealed non-trivial dependences of the temperature of the superconducting transition T_c on the thickness of the ferromagnetic layer. The magnitude of T_c depends to a significant extent on the transparency of the N–S interface (Khusainov, 1991), on the relation between the thicknesses of the metal layers and the coherence length, and on the relation between the parameters of the electron structure and the electron interaction of the contacting metals (de Gennes, 1964). The experimental and theoretical investigations of the proximity effect in various N–S systems have been comprehensively reflected in the review by Jin and Ketterson (1989) and for F–S systems by Izyumov *et al.* (2002)). It may seem that the greater the thickness of the ferromagnetic metallic layer d_f, the greater should be the effect of suppression of superconductivity in such a system. However, in experiments, a non-monotonic and even oscillating dependence of T_c on d_f was frequently observed.

The interaction between itinerant electrons (s) spin $\boldsymbol{\sigma}$ and localised electrons (d) spin $\boldsymbol{S}$ in a ferromagnetic metal is described by the sd-exchange Hamiltonian

$$H_{sd}(\boldsymbol{r}) = \sum_j J_{sd}(\boldsymbol{r} - \boldsymbol{R}_j)\,\boldsymbol{S}_j \cdot \boldsymbol{\sigma} \tag{3.78}$$

In the paramagnetic phase the sd-exchange interaction leads to s-electron scattering by localised spins and, in the F phase, it results in the effective exchange field I, acting on s-electron spin:

$$H_s d = I\sigma^z, \; I = \sum_j J_{sd}(\boldsymbol{r} - \boldsymbol{R}_j)\langle S_j^z \rangle$$

This effective field I leads to a splitting s-electron energies for spins up and down similar to the Zeeman splitting by external magnetic fields. The suppression of T_c by this field is described by the Baltensperger–Sarma equation (Baltensperger, 1958; Sarma, 1963)

$$ln\frac{T_c}{T_{cs}} = \Psi\left(\frac{1}{2}\right) - \Psi\left(\frac{1}{2} + i\frac{I}{2\pi T_c}\right) \tag{3.79}$$

where T_c and T_{cs} is the critical temperature of the superconducting phase in the presence and absence, respectively, of the sd-exchange. The T_c falls off rapidly and vanishes at

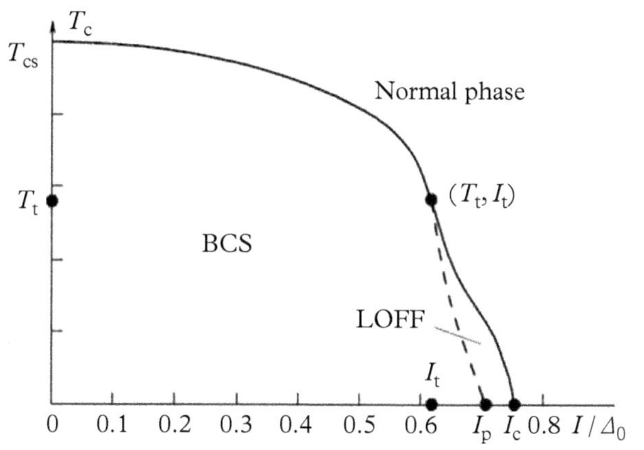

Figure 3.26 *The critical temperature T_c as a function of the exchange field I in a ferromagnetic metal: solid line, second-order phase transitions; dashed line, first-order phase transitions. (From Izyumov, Yu., Proshin, Yu., Khusainov, M. (2002), Physics-Uspekhi 45, 109.)*

$I \sim T_c$; the numerical solution of eqn (3.79) is shown in Fig. 3.26 by a solid line in the interval between T_{cs} and T_t, corresponding to second-order phase transitions. In the interval of I from $I = 0.5\Delta_0$ to $I_t = 0.62\Delta_0$, where $dT_c/dI = \infty$, the T_{cs} (I) curve is two-valued, which indicates the instability of the system and the possibility of a first-order phase transition. In this case, a more general equation than eqn (3.79) is required. Sarma (1963) has shown that, beginning from the field I_t corresponding to the tricritical point $T_t = 0.56T_{cs}$, the phase transition is first order. The corresponding (dashed) line terminates at the point $I_p = \Delta_0/\sqrt{2}$, which is called the Chandrasekhar–Clogston paramagnetic limit (Chandrasekhar, 1962; Clogston, 1962). Above this field, an F metal at $T = 0$ cannot remain a superconductor with a uniform order parameter.

In the narrow range of fields (from I_t to $I_c \cong 0.76\Delta_0$) exceeding the paramagnetic limit, a new, LOFF (Larkin and Ovchinnikov, 1964; Fulde and Ferell, 1964), spatially non-uniform superconducting state occurs. In this state, pairing of electrons that belong to the exchange-split Fermi surface occurs; therefore, pairs are forming from the $|\boldsymbol{p}\uparrow\rangle$, $|-\boldsymbol{p}+\boldsymbol{k}\downarrow\rangle$ states, where $k \sim {}^{I}\!/_{v_F}$. In the simplest case, the gap in the LOFF state depends on the coordinate as follows:

$$\Delta(\boldsymbol{r}) = \Delta_0 \exp(i\boldsymbol{kr}) \tag{3.80}$$

Since the LOFF state is formed at $\sim \Delta_0$, the wave vector of modulation

$$k \sim {}^{\Delta_0}\!\Big/_{v_F} \sim {}^{1}\!\Big/_{\xi_{s0}}$$

is determined by the inverse correlation length of the superconductor. The spin density in the LOFF state is modulated with the same wave vector. In the phase diagram shown in Fig. 3.26, the LOFF state is separated from the BCS state by the line of the first-order transition, whereas the transition from the LOFF phase into the normal phase occurs along the line of the second-order phase transition (Saint-James *et al.*, 1969). The LOFF phase plays a key role in the formation of the superconducting state in *FM–S* structures. A spatial modulation of the gap in eqn (3.80) indicates that the gap in the *FM–S* heterostructure may change its phase from layer to layer. When the phase is changed by π, and the sign of the superconducting order parameter is reversed, this type of superconductivity is called π-phase superconductivity (Bulaevskii *et al.*, 1977). When the *FM* thickness d_f changes, the transition from 0-phase to π-phase superconductivity can occur (Radovic *et al.*, 1991); in this case, the oscillations of T_c versus d_f takes place. However, oscillations were also observed in three-layer structures such as Fe/Nb/Fe (Muhge *et al.*, 1997) and Fe/Pb/Fe (Lazar *et al.*, 2000), in which π-phase superconductivity is impossible. To explain this fact, further development of the theory was required.

The boundary-value problem and calculations of T_c versus d_f have been carried out by Radovic *et al.* (1991) in the dirty limit, assuming the high transparency of the *FM–S* interface. A more general theory based on the Usadel equation with the arbitrary value of transparency and the complex nature of the diffusion coefficient results in the other types of T_c (d_f) behaviour, including re-entrant and periodically re-entrant superconductivity (Proshin and Khusainov, 1997; Tagirov, 1998; Izyumov *et al.*, 2000). This theory leads to the conclusion that the superconductivity in *FM–S* systems is a combination of BCS pairing in *S* layers and LOFF pairing in *FM* layers. The oscillating T_c (d_f) dependence takes place in the case of the high transparency of the junction, while, at a low or moderate transparency, the T_c (d_f) curve may be smoothed and monotonic.

Other interesting aspects of the problem of coexistence and mutual adjustment of superconductivity and ferromagnetism arise in *FI–S* structures (see e.g. the review by Tedrow and Meservey, 1994). In particular, no clarity exists so far as to the nature of internal fields that lead to splitting of the BCS peak in the density of states of aluminium quasiparticles in tunnel junctions such as EuO/Al/Al_2O_3/Al (Tedrow *et al.*, 1986), EuS/Al/Al_2O_3/Ag (Hao *et al.*, 1991), and Au/EuS/Al (Moodera *et al.*, 1988), where EuO and EuS are ferromagnetic insulators. A specific feature of *FI–S* systems (as compared to *FM–S* structures) is that the *FI* layers are non-transparent for the conduction electrons of the *S* layers. Therefore, the superconducting layers are subjected to the action of only the exchange field of localised spins located at the *FI–S* interfaces. From physical considerations, the presence of an internal field that provides the splitting of the BCS peak in the density of states of *FI–S* junctions, and its saturation in a magnetic field can be described by a non-uniform magnetic ordering which is induced in the ferromagnetic film by the superconducting substrate. Therefore, the problem of the mechanisms of intralayer and interlayer exchange coupling in *FI–S* superlattices and multilayers is a key point for the understanding of effects related to the coexistence and mutual accommodation of two competing types of long-range order. Such a mechanism, which ensures the long-range interaction between localised spins belonging to the same *FI–S* boundary, as well as between localised spins of neighbouring *FI–S* boundaries in superlattices, may serve the RKKY indirect exchange via the conduction electrons of superconducting interlayers (Khusainov, 1996). For further discussion of *FI–S* systems, we recommended the review by Izyumov *et al.* (2002).

3.4.4 The Proximity Effect versus the Two-Gap Model

At this point, we may compare the results found for the proximity system with those of the two-gap model. There is a profound analogy between these two situations. In both cases, we are dealing with induced superconductivity (we are assuming that, in the second band, as in the N film, there is no intrinsic pairing). In both cases, an induced energy gap appears. However, the mechanisms giving rise to induced superconductivity are very different. In the two-band model, the system's bands are 'separated' in momentum space, and the second band acquires an order parameter thanks to phonon exchange. The phase space is effectively increased. In the proximity effect, on the other hand, the systems are spatially separated, and superconductivity is induced by the tunnelling of Cooper pairs.

In both cases, we end up with an induced energy gap, but the physical differences manifest themselves in the behaviour of the critical temperature. The two-band picture turns out to be favourable for T_c, which becomes higher than in the single-band case (eqn (3.45)). In contrast, the proximity effect depresses T_c. The critical temperature of a proximity sandwich is lower than that of an isolated S film. This can be seen directly from eqns (3.71) and (3.71′).

3.5 The Isotope Effect

3.5.1 General Remarks

The isotope effect occupies a special place in the physics of superconductivity. Its discovery was the major step preceding the BCS theory by demonstrating that a lattice is involved in the formation of the superconducting state. The effect was studied theoretically by Fröhlich (1950) and was discovered by measuring the critical temperature of various isotopes of mercury (Maxwell, 1950; Reynolds *et al.*, 1950). As the mass M was varied between 199.5 and 203.4, the value of T_c changed from 4.185 K to 4.140 K. It was established that the following relation holds:

$$T_c = (\text{const}) \cdot M^{-\alpha}; \alpha \approx 0.5 \tag{3.81}$$

The dependence shown in eqn (3.81), and value $\alpha \approx 0.5$, are in total agreement with the BCS expression (2.10) for T_c, since the vibrational frequency $\tilde{\Omega} \propto M^{-1/2}$. This relation, as well as eqn (3.81), defines the main quantity, the isotope coefficient α.

At first sight, the isotope effect looks like a rather straightforward phenomenon, but such an impression is misleading. In reality, it is a complex effect controlled by many factors. In many cases, one can observe large derivations from the canonical value $\alpha = 0.5$. As a result, the presence of the isotopic dependence means that the lattice is involved in the pairing, but it is difficult to provide a quantitative estimation of the degree of this involvement.

Below we discuss various factors affecting the value of the isotope coefficient.

3.5.2 The Coulomb Pseudopotential

The BCS expression (2.10) contains the Coulomb pseudopotential μ^*, which is defined by eqn (2.27). It can be seen directly from eqn (2.27) that the value of μ^* depends on the phonon frequency, and its change upon the isotope substitution also affects the value of T_c and, correspondingly, the value of the isotope coefficient α.

Starting from eqn (3.81), defining the isotope coefficient, the following general expression can be obtained:

$$\alpha = -(M/T_c)(\partial T_c/\partial M) \tag{3.82}$$

Then,

$$\alpha = -(M/T_c)\ (\partial T_c/\partial \Omega)\ (\partial \Omega/\partial M) \tag{3.83}$$

Based on eqns (2.27), (2.10), and (3.83),

$$\alpha = \alpha_{|\mu^*=0}\left[1 - \left(\frac{\mu^*}{\lambda - \mu^*}\right)^2\right] \tag{3.84}$$

It can be seen that the presence of the Coulomb factor leads to a decrease in the value of the isotope coefficient. For example, if $\lambda = 0.5$ and $\mu^* = 0.15$, then $\alpha = 0.4$. For $\lambda = 0.4$ and $\mu^* = 0.15$, the decrease is even more noticeable: $\alpha \approx 0.25$.

Equation (2.10) is valid for the case of weak coupling. It can be seen from eqns (2.23) and (2.32) that, for intermediate and strong coupling (see also eqn (2.34)), the presence of the Coulomb pseudopotential also leads to a noticeable impact on the isotope effect.

3.5.3 Multicomponent Lattices and Two Coupling Constants

As shown above, the Coulomb pseudopotential noticeably affects the value of the isotope coefficient. Still, for a mono-atomic lattice, the picture looks rather straightforward. However, it becomes more complicated for a multicomponent lattice, which is the case when there are different ions in the unit cell. Indeed, in addition to the acoustic phonon modes, the presence of various ions in the unit cell leads to the appearance of optical modes. Of course, this addition is favourable for the pairing, but the description of the isotopic dependence of T_c caused by replacement of one of the ions by its isotope becomes more complicated.

To demonstrate the main point, let us consider the simple case of a two-component linear system. It is known (see Maradudin *et al.*, 1971; Ashcroft and Mermin, 1976) that the phonon spectrum of such a system is described by two modes with the following dispersion relations:

$$\Omega_1^2 = \gamma\left(\frac{1}{M_1}+\frac{1}{M_2}\right) - \gamma\left[\left(\frac{1}{M_1}-\frac{1}{M_2}\right)^2 + \frac{4}{M_1 M_2}\cos^2 qd\right]^{\frac{1}{2}}$$
$$\Omega_2^2 = \gamma\left(\frac{1}{M_1}+\frac{1}{M_2}\right) + \gamma\left[\left(\frac{1}{M_1}-\frac{1}{M_2}\right)^2 + \frac{4}{M_1 M_2}\cos^2 qd\right]^{\frac{1}{2}} \quad (3.85)$$

where M_1 and M_2 are the masses of the ions, γ is the elastic constant, d is the lattice period, and q is the phonon momentum.

Consider the most interesting case when the masses of the two ions in the unit cell are very different, that is, one of them is very light relative to the mass of the other ion; assume, for concreteness, that $M_1 \gg M_2$. For example, this case corresponds to the family of two-component hydrides (see section 7. 2), which contain H ions and some heavy ions (e.g. H_3S, LaH_{10}). These materials display one of the highest known values of T_c. Qualitatively, it is clear that the optical phonon modes, corresponding to the relative motion of the ions in the unit cell, represent such a case with the vibrational motion of the light ion with a mass M_2. The frequency, expressed as $\Omega_2 \propto (\gamma/M_2)^{1/2}$, is expected to be rather high.

Indeed, from eqn (3.85), one can obtain

$$\Omega_1^2 = \gamma\left(\frac{1}{M_1}+\frac{1}{M_2}\right) - \gamma\left(\frac{1}{M_1}+\frac{1}{M_2}\right)\left[1 - \frac{4M_1M_2\sin^2 qd}{(M_1+M_2)}\right]^{1/2}$$

Since $M_1 \gg M_2$, we obtain (at $q \to 0$)

$$\Omega_1 \equiv \Omega_{ac} \propto \left(\frac{\gamma}{M_1}\right)^{1/2} q \quad (3.86)$$

and, with use of eqn (3.85),

$$\Omega_2 \equiv \Omega_{opt} \propto \sqrt{\frac{\gamma}{M_2}} \quad (3.87)$$

Therefore, the optical modes are due to the motion of the lightest ion, whereas, for the acoustic modes, the participation of the heavy ion prevails. As for the superconducting state, both branches (optical and acoustic) make their contribution to the pairing. Correspondingly, one can introduce two coupling constants, λ_{opt} and λ_{ac} (see section 2.2.2).

In principle, the isotope substitution can be performed independently for both ions. The most interesting case corresponds to substitution of just the light ion, for example, the H $\to$ D substitution for superconducting hydrides.

The value of the isotope coefficient depends on the relation between the coupling constants λ_{opt} and λ_{ac}. It also depends on the intensity of the electron–phonon interaction (weak vs. strong coupling), since this intensity affects the value of T_c (see section 2.2.2). If the coupling of both the optical and acoustic branches is weak ($\lambda_{opt}, \lambda_{ac} \ll 1$), then one should use eqn (2.38) for T_c. Then, with the use of the expression

$$\alpha = 0.5 \frac{\Omega_{opt}}{T_c} \frac{\partial T_c}{\partial \Omega_{opt}} \tag{3.88}$$

which follows directly from eqn (3.83) within the harmonic approximation, we obtain

$$\alpha = 0.5\,(\lambda_{opt}/\lambda_T); \; \lambda_T = \lambda_{opt} + \lambda_{ac} \tag{3.89}$$

In the more interesting case, when $\lambda_{ac} \ll \lambda_{opt}$ and $\lambda_{opt} \gtrsim 1$, then the superconducting state is mainly determined by the strong coupling to high-frequency optical phonons; this is the case for hydrides with a high T_c (see section 7.2), and one should use a different expression for T_c, eqn (2.39). Then the following result for the isotopic coefficient is obtained:

$$\alpha = 0.5 \left[1 - 4 \frac{\lambda_{ac}}{\lambda_{opt}} \cdot \frac{\rho^2}{(\rho^2 + 1)^2} \right] \tag{3.90}$$

where $\rho = \tilde{\Omega}_{ac}/\pi T_c^0$, and T_c^0 is the value of T_c for $\lambda_{ac} = 0$; see eqn (2.40).

It can be seen that the value of the isotope coefficient α is different from $\alpha_{max} = 0.5$, and the deviation is determined by the relative contributions of the optical and acoustic phonon branches. Therefore, the experimentally measured values of the isotope coefficient for substitution of the light ion allow description of the interplay and relative contributions of the optical and acoustic branches into the T_c value of the superconducting transition.

3.5.4 Anharmonicity

Anharmonicity of the lattice can strongly affect the value of the isotope coefficient. A well-known superconductor displaying such an effect is the Pd–H compound. The isotope substitution H $\rightarrow$ D leads to an increase in T_c; in other words, the isotope coefficient α (see eqn (3.81)) appears to be negative. The idea that such an unusual behaviour of the isotopic dependence is caused by anharmonicity was proposed by Ganzuly (1973). A detailed analysis, based on the strong coupling theory (see section 2.2.2), was carried out by Klein and Cohen (1992). They demonstrated that anharmonicity of the lattice leads to the dependence of T_c on mass, which is drastically different from that determined by the law $\tilde{\Omega} \propto M^{-1/2}$. The calculation for the Pd–H system performed up to the sixth order of the ionic displacement shows that the coupling constant λ (see eqn (3.23)) increases upon the isotope substitution. This increase, caused by the lattice dynamics, overcomes the decrease in the pre-exponential factor. The value of T_c was determined from eqn (2.34). The value of μ^* for the Pd–H system was chosen to be $\mu^* = 0.115$, since it provides the best agreement with the data (Schirber and Northrop, 1974). The value of the isotope coefficient obtained by Klein and Cohen (1992) appears to be negative, with agreement with the experimental observation.

A similar phenomenon has been observed for organic superconductors, namely BEDT-TTF-based systems; see section 8.1.2 (Whangbo *et al.*, 1987; Schlueter *et al.*, 2001). Indeed, in the BCS model, the isotope effect is caused by the pre-exponential factor only, and it leads to the relation $\alpha = 0.5$ (since, in the harmonic approximation, $\tilde{\Omega} \sim M^{-1/2}$). The coupling constant λ in this approximation defined by eqn (3.23) does not depend on the ionic mass M. Anharmonicity leads to an additional dependence of λ on M. This correction is rather small, since the anharmonicity, similar to a whole Born–Oppenheimer adiabatic picture, is described by the small parameter a/L (a is the amplitude of vibration, L is the lattice period). But. as a whole, this might be a noticeable effect, because of the exponential dependence of T_c on the coupling strength.

Therefore, anharmonicity can drastically change the isotopic dependence of T_c, and it may even lead to a negative sign of the isotope coefficient.

3.5.5 The Isotope Effect in Proximity Systems

In this and the following sections, factors that are not directly related to the pairing mechanism but nevertheless affect the isotopic dependence are considered. It turns out that the value of T_c can be changed by some external factors ($T_{c0} \to T_c$, where T_{c0} is the intrinsic value of the critical temperature). These factors do not affect the lattice dynamics, but their presence leads to a shift in the value of the isotope coefficient ($\alpha_0 \to \alpha$).

The most interesting situation occurs when the relation between T_{c0} and T_c is not linear. More specifically, consider the isotope effect in a proximity system S–N (where S and N are superconducting and normal films, respectively; see section 3.4 and Fig. 3.23). It can be seen that the value of α depends on the relative thicknesses of the films. Indeed, assuming that the thickness $L_N \ll \xi_N$, where $\xi_N = hv_{F;N}/2\pi T$ is the coherence length for the film N, then the well-known McMillan tunnelling model (1968) can be used (see section 3.4.2). According to this model, the proximity effect is described by the parameter $\Gamma = \Gamma_{SN} + \Gamma_{NS}$, where $\Gamma_{ik} = \tilde{T}^2_{ik}\nu_k\nu_k$, $\tilde{T}_{ik}$ is the tunnelling matrix element, υ_k is the density of states (per unit of volume V), and i, $k = \{S, N\}$, $i \neq k$. Assuming that $\Gamma \ll \tilde{\Omega}$, it is obvious that the relation between T_c and T_{c0} is non-linear. The T_c of the whole system differs from T_{c0} (T_{c0} is the critical temperature of the isolated S film) and is described by eqn (3.71).

If the isotope substitution $M \to M^*$for the isolated S film is made, the shift in T_{c0} can be measured and the isotope coefficient α_0, which is described by eqn (3.82), can be determined. The presence of the N film leads to a change in T_c and, consequently, the value of the isotope coefficient $\alpha = -(M/\Delta M)(\Delta T_c/T_c)$. It can be seen directly from eqns (3.71) and (3.71′) that the relative shift in T_c differs from ΔT_{c0} / T_{c0}. Indeed, the value of T_c for the whole sandwich is determined by eqn (3.71). Then the value of the shift ΔT_c is determined not only by ΔT_{c0} but also by the value of the parameter ρ, which is defined by eqn (3.70′); it reflects the presence of the N film. More specifically, on the basis of eqn (3.70), we obtain

$$\Delta T_c/T_c = (\Delta T_{c0}/T_{c0})\,(1+\rho)$$

As for the value of the isotope coefficient, we obtain

$$\alpha = \alpha_0\,[1 + (\nu_N L_N/\nu_S L_S)] \tag{3.91}$$

Indeed, it can be seen that the value of the isotope coefficient is modified by the proximity effect. Moreover, $\alpha > \alpha_0$. Therefore, a decrease in T_c, which is a well-known feature of the proximity effect and is described by eqn (3.70), is accompanied by an increase in the isotope coefficient. It is interesting that the value of α can be modified by changing the thicknesses of the films. The increase in the thickness of a normal film L_N leads to a decrease in T_c, but the value of α increases. For example, if $\alpha_0 = 0.2$, $\nu_N/\nu_S = 0.8$, and $L_N/L_S = 0.5$, then $\alpha = 0.28$. If the thickness of the normal film is increased so that $L_N = L_S$, then $\alpha = 0.36$.

Note that the increase of the isotope coefficient discussed in this section is not related to lattice dynamics; as a result, the value of α can, in principle, exceed the value of $\alpha_{0;\mathrm{ph.,max}} = 0.5$.

The presence of a normal film affects the state of the condensate of Cooper pairs ('number of pairs') and acts like a pair breaker (see section 3.3.1). It affects the value of T_c without any change in the pairing mechanism and its strength. Another more well-known example of pair breaking occurs in the presence of magnetic impurities (see section 3.3). Their impact on the isotope effect will be described in the next section.

3.5.6 Magnetic Impurities and the Isotope Effect

In this section, we focus on another isotope effect that is not related to lattice dynamics: we consider a superconductor that contains magnetic impurities.

The presence of magnetic impurities leads to a decrease in the critical temperature, T_c, relative to the intrinsic value, T_{c0}, because of the pair-breaking effect described in section 3.3.1. This depression of T_c is described by the equation

$$ln\,(T_{c0}/T_c) = \Psi\,[0.5 + \gamma_s] - \Psi(0.5); \gamma_s = \left(2\pi T_c^0 \tau_s\right)^{-1} \tag{3.62}$$

where Ψ is the digamma function. In this case (cf. eqn (3.71)), there is a non-linear relation between T_{c0} and T_c.

The isotope substitution $M \to M^*$for the sample without magnetic impurities allows the shift in T_{c0} to be observed and the isotope coefficient to be measured. It can be seen directly from eqn (3.82) that the relative shift in T_c and the value of the isotope coefficient α in the presence of magnetic impurities differ from ΔT_{c0} / T_{c0} and the value of α_0, respectively.

With the use of eqns (3.82) and (3.62), one can arrive at the following expression (Carbotte *et al.*, 1991; Kresin *et al.*, 1997):

$$\alpha = \alpha_0 [1 - \Psi' (0.5 + \gamma_s) \gamma_s]^{-1} \tag{3.92}$$

Note that an increase in the concentration of magnetic impurities n_M leads to an increase in γ_s. Equation (3.92) is valid in a broad range except for a very small region near n_{cr} (then T_c is close to $T = 0$ K, and the condition $\Delta T_c / T_c \ll 1$ is not satisfied).

The presence of magnetic impurities leads to an increase of the isotope coefficient ($\alpha > \alpha_0$). It follows directly from eqn (3.92) and the inequality $\Psi' > 0$. The validity of this inequality can be seen directly from the expression

$$\Psi' (0.5 + x) = \sum_{k=0}^{\infty} (k + 0.5 + x)^{-2}$$

Thus, the dependence of α on the concentration of magnetic impurities can be studied. For small γ_s (small values of n_M), $\Delta\alpha \propto n_M$. Therefore, near T_{c0}, the critical temperature displays a linear decrease with an increase in n_M, whereas the isotope coefficient increases linearly as a function of n_M. In the region $\gamma_s \gg 1$, an asymptotic expression for the digamma function can be used, and $\Delta\alpha \propto (n_M)^2$ is obtained, that is, the dependence becomes strongly non-linear.

It can be seen directly from eqns (3.91) and (3.92) that the impact of the proximity effect as well as magnetic scattering on the value of the isotope coefficient is caused by the non-linear relation between T_c and T_{c0}. Qualitatively, both factors provide pair breaking, and this leads to an effective increase in the normal electronic component.

3.5.7 The Polaronic Effect and Isotope Substitution

Some novel materials, including high-T_c cuprates, display a dynamic polaronic effect (see sections 1.2.2 and 2.3.1). These materials contain subgroups of ions (e.g. oxygen ions in the cuprates), and each of the ions is characterised by not one but two close equilibrium positions. In addition, the carriers in the materials of interest (see chapters 5 and 9) are provided by doping. For the doped materials, the value of T_c depends strongly on n (n is the carrier concentration). The doping is the charge-transfer process. If two close ionic equilibrium positions are present, then the ionic movement includes not only the vibration near its equilibrium position but also the tunnelling between their minima. Therefore, the ion participates in the charge-transfer process (doping; see below). The isotope substitution changes the ionic mass and affects the probability of the tunnelling. As a result, the doping and, consequently, the carrier concentration n are also affected. As noted above, since the value of T_c strongly depends on n, a novel isotope affect can be observed (Kresin and Wolf, 1994). Later on, we will discuss all the implications of such an isotope effect for the physics of cuprates (section 5.4.3) and manganites (section 9.5.3) with a high T_c value; here we describe the polaronic isotope effect in more detail.

In general, if the charge transfer occurs in the framework of the usual adiabatic picture (section 1.2.1) so that only the carrier motion is involved, then the isotope substitution does not affect the Coulomb forces and therefore does not change the charge transfer dynamics. However, the situation of strong non-adiabaticity (polaronic effect; see section 1.2.2) is different and does not allow the separation of electronic and nuclear motions. In this case, charge transfer appears to be a more complex phenomenon that involves nuclear motion, and this leads to a dependence of the doping on the isotopic mass.

Let us consider the case when the lattice configuration, that is, the positions of some definite ions (e.g. oxygen for YBCO; see section 5.4.3), corresponds to the crossing of electronic terms (degenerate or near-degenerate electronic states; see chapter 1, Fig. 1.5). Then the ion has two close equilibrium positions.

The charge transfer in this picture is accompanied by the transition to another electronic term. Such a process is analogous to the Landau–Zener effect (see Landau and Lifshitz, 1977). The charge transfer corresponds to the transitions between the first and second terms.

The total wave function (see section 1.2.2, eqn (1.30)) can be written in the form

$$\Psi(\boldsymbol{r},\boldsymbol{R},t) = a(t)\Psi_1(\boldsymbol{r},\boldsymbol{R}) + b(t)\Psi_2(\boldsymbol{r},\boldsymbol{R})$$

where $\Psi_i(\boldsymbol{r},\boldsymbol{R}) = \Psi_i(\boldsymbol{r},\boldsymbol{R})\Phi_i(\boldsymbol{R}), i = \{1,2\}$ and $\Psi_i(\mathbf{r},\mathbf{R})$ and $\Phi_i(\mathbf{R})$ are the electronic and vibrational wave functions, respectively, that correspond to two different electronic terms (see chapter 1, Fig. 1.5).

Qualitatively, the charge transfer for such a non-adiabatic scenario can be visualised as a multistep process. First, the carrier makes a transition from the reservoir site to the ion. Then the ion transfers to another term, and this is finally followed by the second transition of the carriers (e.g. for the cuprates, this is the transition to the Cu–O plane; see chapter 5). It is important that the second step (ion transfer) is affected by the isotope substitution.

Let us separate the z-coordinate of the ion (the z-axis has been chosen to be parallel to the ionic motion between the minima) so that $\Phi_i(\boldsymbol{R}) = \Phi_i(\boldsymbol{\rho}_i)\varphi_i(Z)$, where $\boldsymbol{\rho}_i$ corresponds to the other degrees of freedom. It cannot be assumed that the electronic terms are similar; they differ in the energy level spacing. This leads to some splitting between the vibrational levels (see chapter 1, Fig. 1.5) and the transition of the ion is not a resonant one. In the harmonic approximation,

$$\varphi_1(Z) = (\pi a_1)^{-1/2} \exp\left(-(Z - Z_{10})^2/2a_1^2\right) \tag{3.93}$$

$$\varphi_2(Z) = (\pi a_2)^{-1/2} \exp\left[-(Z - Z_{20})^2/2a_2^2\right] H_\nu\left[(Z - Z_{20})a_2^{-1}\right] \tag{3.93'}$$

where $a_i = (M\Omega_i)^{-1}$ is the vibrational amplitude, H_ν is the Hermite polynomial, M is the ionic mass, and Z_{i0} is the equilibrium position. Assume that the ionic motion corresponds to the zeroth vibrational state of the first term, and the v-th level for the second term. Assume also that initially the electron occupies the first electronic state so that the amplitude $b(0) = 0$. The time dependence $b(t)$ describes the dynamics of the charge transfer $(1 \to 2)$.

As was noted earlier (section 1.2.2), it is convenient to employ the diabatic representation; then, for the average value, $\tilde{b}^2$, we obtain expression (1.33). The asymmetry of the potential and, correspondingly, the inequality $\varepsilon \neq 0$ play key roles. For a large asymmetry (see chapter 1, Fig. 1.5), $| L_0 F_{12} | \ll \varepsilon$. Then the carrier concentration is

$$n_{pl} \propto \overline{b^2} = 2L_0^2 |F_{12}|^2 / \varepsilon^2 \tag{3.94}$$

Since the Franck–Condon factor F_{12} and the value of ε depend directly on the mass M, the carrier concentration also depends on M. This dependence reflects the fact that the electronic and vibrational states are not separate (the polaronic effect).

Let us focus now on the isotope coefficient $\alpha = (-M/T_c)(\partial T_c/\partial M)$. The two contributions should be distinguished so that $\alpha = \alpha_0 + \alpha_{na}$, where $\alpha_0 = -(M/T_c)(dT_c/d\Omega)(d\Omega/d\mathrm{M})$ describes the usual isotope effect caused by a change in the phonon spectrum (Ω is a characteristic phonon frequency for that mass). If the corresponding mode in the polyatomic system does not contribute noticeably to the pairing, then the coefficient α_0 is small.

The term α_{na} corresponds to a different isotope effect, which is due to the dependence $n(M)$ described above and also the dependence $T_c(n)$. Namely,

$$\alpha_{na} = -\frac{M}{T_c}\frac{\partial T_c}{\partial n}\frac{\partial n}{\partial M} \tag{3.95}$$

According to eqns (3.94) and (3.95), the dependence $n(M)$ is determined mainly by the Franck–Condon factor. Based on eqns (3.93) to (3.95) and the dependence $\varepsilon^2 \propto M^{-1}(n \propto Me^{-sM^{0.5}}, s = k^{0.5}d^2/2\hbar, k^{-0.5} = k_1^{-0.5} + k_2^{-0.5}$, where k_i is the elastic constant and d is the distance between the minima), the following expression for the isotope coefficient is obtained:

$$\alpha_{na} = \gamma \frac{n}{T_c}\frac{\partial T_c}{\partial n} \tag{3.96}$$

where $\gamma \cong$ constant (it has a weak logarithmic dependence on M). Expression (3.96) can be compared directly with experimental data (see section 5.4.3).

Equation (3.96) contains the derivative $\partial T_c/\partial n$. Therefore, if $T_c = T_{c,\max}$, the isotope coefficient α_{na} is equal to zero. For cuprates, T_c has a maximum at some value $n = n_{max}$. Therefore, the described mechanism leads to the situation when the maximum value of T_c corresponds to a minimum in the value of the isotope coefficient. This phenomenon, indeed, has been observed experimentally (see section 5.2.2).

Notably, continuous doping leads not only to an increase in the concentration n but to a change in the electronic terms, and the structure of the terms becomes less asymmetric. For example, for cuprates, this is reflected in the movement of the average position of the axial oxygen towards the plane. This motion has been observed experimentally

(Jorgensen *et al.*, 1990). Such an evolution leads to an additional decrease in the value of α.

The second comment is related to the overdoped region. The analysis described above is applicable to a single-phase system, that is, for the underdoped region of the phase diagram. Structural transitions can drastically modify this picture. If the overdoped region corresponds to the same phase as the underdoped region, which is likely the case for BiPbCaSrCuO, then, in the region $n > n_{max}$, the isotope coefficient should become negative; indeed, such an effect has been observed by Bornemann *et al.* (1991). However, if, in the region $n > n_{max}$, there is a multiphase compound (as is the case for $La_{2-x}Sr_xCuO_4$; see Bozin *et al.*, 2000), then the picture described above can be used only for $n < n_{max}$. Finally, it should be mentioned that, according to the described model, the values of α are not limited to a maximum of 0.5 and may exceed this value.

Note also that experimental studies of the isotope effect should be carried out with considerable care. It is important to perform the measurements with the same sample (before and after the isotope substitution) and not use various samples with different isotopes. This is essential because different samples can have different T_c values, and the measurements performed with such samples could be misleading.

Therefore, a strong non-adiabaticity of the ions in doped superconductors, such as high-T_c oxides or manganites, leads to the dependence of the carrier concentration on M, and this factor affects the T_c value.

3.5.8 Penetration Depth and Isotopic Dependence

The polaronic isotope effect described above is a consequence of the dependence of the carrier concentration on the ionic mass M: $n = n(M)$. Since properties other than T_c are also affected by the value of the carrier concentration, the isotopic dependence of various quantities is expected. As an important example, let us discuss the isotopic dependence of the penetration depth δ. This dependence was theoretically introduced by Kresin and Wolf (1995) and then analysed in detail by Bill *et al.* (1998). The effect has been observed experimentally for oxides with high T_c values and is discussed below (see section 5.4.3).

By analogy with T_c, a new isotope coefficient β can be defined by the dependence

$$\delta \propto M^{-\beta} \tag{3.97}$$

As a result, β is determined by the expression (c.f. eqn (3.82))

$$\beta = -(M/\delta)\ (\partial\delta/\partial M) \tag{3.98}$$

describing the change in the value of the penetration depth induced by the isotope substitution $M \to M^* = M + \Delta M$.

In the London limit, the penetration depth is given by the well-known relation

$$\delta^2 = \frac{mc^2}{4\pi n_s e^2} = \frac{mc^2}{4\pi n\varphi(T/T_c)\, e^2} \tag{3.99}$$

where m is the effective mass and n_s is the superconducting density of the charge carriers, which is related to the normal density n through the relation $n_s = n\varphi(T/T_c)$. The function $\varphi(T/T_c)$ is a universal function of T/T_c. For example, for conventional superconductors, $\varphi \simeq 1 - (T/T_c)^4$ near T_c, whereas $\varphi \simeq 1$ near $T = 0$. The isotope coefficient β of the penetration depth can then be determined from the relation that follows from eqns (3.98) and (3.99):

$$\beta \equiv -\frac{M}{\delta}\frac{\partial \delta}{\partial n_s}\frac{\partial n_s}{\partial M} = \frac{M}{2n_s}\frac{\partial n_s}{\partial M}. \tag{3.100}$$

Because of the relation $n_s = n\varphi(T)$, one has to distinguish the two contributions to β. There is the usual (BCS) contribution, β_{ph}, arising from the fact that $\varphi(T/T_c)$ depends on ionic mass through the dependency of T_c on the characteristic phonon frequency. Indeed, isotopic substitution leads to a shift in T_c and thus in β, which might be noticeable near T_c.

The presence of the factor n (see eqn (3.99)) leads to a peculiar isotope effect for the penetration depth, with $\beta \equiv \beta_{na}$ (β_{na} corresponds to the non-adiabatic contribution).

In the region near $T = 0$ K,

$$\beta_{na} \cong (M/2n)\cdot(dn/dM) \tag{3.101}$$

Comparing eqns (3.95), (3.96), and (3.101), it can be inferred that $\beta_{na} = -\frac{\gamma}{2}$; thus, a relation between the non-adiabatic isotope coefficient of T_c, α_{na} (see above), and β_{na} is established:

$$\alpha_{na} = -2\beta_{na}\frac{n}{T_c}\frac{\partial T_c}{\partial n} \tag{3.102}$$

The equation contains only measurable quantities; therefore, it can be verified experimentally. It is interesting to note that β_{na} and α_{na} have opposite signs ($\partial T_c/\partial n > 0$ in the underdoped region in cuprates with a high T_c value).

Let us consider another case when the penetration depth is affected by isotopic substitution but the dependence is not related directly to the lattice dynamics (cf. section 5.4.3). Namely, consider the *S–N* proximity system (Bill *et al.*, 1998), where S is a superconductor and N is a metal or a semiconductor. Assume that $\delta < L_N \ll \xi_N$ (this condition is certainly satisfied in the low-temperature regime, since $\xi_N = \hbar v_F/2\pi T$), where L_N and ξ_N are the thickness and the coherence length of the normal film, respectively. In this case, the penetration depth is described by the expression

$$\delta^{-3} = a_N \Phi \tag{3.103}$$

where $a_N =$ constant, its value depends only on the material properties of the normal film (it is independent of the ionic mass) and

$$\Phi = \pi T \sum_{n \geq 0} \frac{1}{x_n^2 p_n^2 + 1}, p_n = 1 + t\sqrt{x_n^2 + 1} \qquad (3.103')$$

where $x_n = \omega_n / \varepsilon_s(T)$, $\omega_n = (2n+1)\pi\mathrm{T}$, and $\varepsilon_s(T)$ is the superconducting energy gap of the S film. In the weak coupling limit considered here, $\varepsilon_s(0) = rT_{c0}$ with $r = r_{BCS} = 1.72$; t is a dimensionless parameter: $t = r(L_N/L_S)(T_{c;S}/\Gamma_0)$ ($\Gamma_0 \equiv \Gamma^{\alpha\beta} \propto L_s^{-1}$ is the McMillan parameter; see section 3.4.2); and L_N and L_S are the thicknesses of normal and superconducting films, respectively.

Since δ depends non-linearly on T_{c0} (through the proximity parameter t), the penetration depth will display an isotope shift due to the proximity effect. From eqn (3.103), the isotope coefficient β_{prox} of the penetration depth can be calculated as follows:

$$\beta_{\mathrm{prox}} = -\frac{2\alpha_0}{3\Phi} \sum_{n>0} \frac{x_n^2 p_n^2}{x_n^2 p_n^2 + 1} \left(1 - \frac{t}{p_n\sqrt{x_n^2 + 1}}\right) \qquad (3.104)$$

where α_0 is the isotope coefficient of T_{c0} for the isolated superconducting film S.

Note that the isotope coefficients β_{prox} and α_0 generally have opposite signs. In addition, the isotope coefficient depends on the proximity parameter t, that is, on the thickness ratio $l = L_N/L_S$ of the normal and superconducting films and on the MacMillan tunnelling parameter Γ_0. Note also that the isotopic shift of the penetration depth is temperature dependent: $|\beta_{prox}|$ increases with increasing temperature. Therefore, the presence of a normal layer on a superconductor induces an isotope shift of the penetration depth.

Like the proximity effect, the isotopic dependence of the penetration depth is also affected by another pair-breaking effect: magnetic scattering (Bill *et al.*, 1998). However, a detailed description of this effect is beyond of the scope of this book.

3.6 Fluctuations in Superconductors

Both the Ginzburg–Landau theory and the BCS theory are formulated in the mean field approximation. Usually, this approximation describes physics only qualitatively but, for superconductivity, the situation is better, as was demonstrated by Ginzburg (1960), who showed that fluctuations in clean superconductors give an essential contribution only in a very narrow temperature region ($\sim 10^{-12}$ K) around the critical temperature T_c. Later, Aslamazov and Larkin (1968) showed that, in dirty superconductor films, the fluctuation range of temperatures may be much larger, and fluctuations occur not only for thermodynamical properties, but also in transport properties. They predicted a paraconductivity phenomenon, when the conductor resistivity drops in the normal phase above T_c. This phenomenon was subsequently found experimentally by Glover (1967). Since that time, many fluctuation phenomena have been studied in superconductors, some of which will be discussed in this chapter. A detailed description of fluctuations in superconductivity can be found in the book by Varlamov and Larkin (2005).

3.6.1 The Ginzburg–Levanyuk Parameter

We start with a question: where can the Ginzburg–Landau theory be applied? By construction, it cannot be valid too far from T_c, where the size of the Cooper pair $\xi(T) \sim \xi_0$. In this temperature range, short-wavelength fluctuations are important, and restriction by a square of the order parameter gradient in the Ginzburg–Landau functional (3.16) becomes incorrect. Thus, restriction for temperature $T_c - T \ll T_c$ or $|\tau| \ll 1, \tau = (T - T_c)/T_c$ is evident. On the other hand, close to the critical temperature, fluctuation contributions to the thermodynamics become important. The area of strong fluctuations was estimated by Levanyuk (1959) within the general phenomenological Landau theory of second-order phase transitions. For superconductivity, Ginzburg (1960) estimated the fluctuation contribution to the specific heat; it becomes comparable with the specific heat jump at T_c when

$$|\tau| \sim Gi_{(3)} \sim \left(\frac{T_c}{E_F}\right)^4 \sim 10^{-12} \div 10^{-14} \tag{3.105}$$

A dimensionless width of the critical area in the theory of superconducting fluctuations is called the Ginzburg–Levanyuk parameter (number) $Gi_{(D)}$. It strongly depends on the effective sample dimension D. Within the microscopic BSC theory, the role of superconducting fluctuations in the normal phase has been discussed by Aslamazov and Larkin (1968), Maki (1968), and Thompson (1970). This microscopic approach confirms the Ginzburg estimation of the critical area for a clean superconductor, as given above. Moreover, fluctuations appear to be much stronger in dirty superconducting films and whiskers. That is why the Ginzburg–Levanyuk parameter strongly depends on the effective sample dimension D. Even with an extremely small value of $Gi_{(3)}$, the fluctuation effects for small granules with effective dimension $D = 0$ are quite strong. The microscopic theory makes it possible to calculate the Ginzburg–Levanyuk parameter $Gi_{(D)}$ for different dimensions D both in the clean and in the dirty limit (Table 3.1). In a dirty superconductor where the mean free pass $l \ll \xi_0$, the coherence length is reduced: $\xi_{0d} \ll \xi_0$.

Table 3.1 shows the Ginzburg–Levanyuk parameters for different superconductors in the clean (c) and dirty (d) limits; $D = 0$ means a granule with volume V with T_{c0}, and ξ_0 corresponds to the bulk of the same material. For $D = 2$, the thickness of the film is d. For $D = 1$, S is the cross-section area, and τ_{tr} is the transport relaxation time (Varlamov and Larkin, 2005).

It worth noting the relation between the Ginzburg–Levanyuk parameter for the dirty limit with a $D = 2$ film and the dimensionless conductance $G_{\square} = h/e^2 R_{\square}$, where $R_{\square}$ is the resistivity for the unit area of the film. For a dirty film, $Gi_{(2)}^{(d)} \approx 1/23 G_{\square}$. The Ginzburg–

Table 3.1 *The Ginzburg–Levanyuk parameter for clean (c) and dirty (d) superconductors with different shape and dimensions D. (From Varlamov and Larkin, 2005.)*

$Gi_{(3)}$	$Gi_{(2)}$	$Gi_{(1)}$	$Gi_{(0)}$
$80\left(\frac{T_c}{E_F}\right)^4$, (c)	$\left(\frac{\mathbf{T_c}}{\mathbf{E_F}}\right)$, **(d)**	0.5, *(d)*	$13{,}3\frac{T_{c0}}{E_F}\sqrt{\frac{\xi_0^3}{\nu}}$
$\frac{1.6}{(p_F l)^3}\left(\frac{T_c}{E_F}\right)$, *(d)*	$\frac{0.27}{p_F l}$, (d)	$1.3\left(p_F^2 S\right)^{-2/3}(\mathrm{T_c}\tau_{tr})^{-1/3}$, (d) wire	
	$\frac{1.3}{p_F{}^2 l d}$, (d) film	$2.3\left(p_F^2 S\right)^{-2/3}$, (d) whisker	

Levanyuk parameter may be introduced in different ways. For example, one can compare the fluctuation energy $k_B T_c$ with the magnetic energy of the volume ξ^3 for the critical value of the magnetic field H_{c2}^2 (Blatter *et al.*, 1994) or calculate the temperature value when the Aslamazov and Larkin contribution to conductivity becomes equal to its normal value (Larkin and Ovchinnikov, 2001). All definitions give parameters that are equal by order of magnitude.

3.6.2 The Effect of Fluctuations on Specific Heat

A specific heat anomaly is a universal feature of all second-order phase transitions. According to Landau theory, a finite jump of $C(T)$ at T_c is expected. According to the modern theory of second-order phase transitions, the critical behaviour is determined only by symmetry of the order parameter and the dimension of space (Kadanoff, 1966; Patashinskii and Pokrovsky, 1966; Vaks and Larkin, 1966; Halperin and Hohenberg, 1969; Polyakov, 1969; Wilson, 1971). For singlet superconductors and superfluid ^{4}He, the complex single-component order parameters are similar; nevertheless, for ^{4}He, the specific heat reveals the logarithmic divergence $C(T) \sim -ln\mid 1-T/T_c\mid$ instead of the finite jump. The reason for this is the large value of the Ginzburg–Levanyuk parameter for ^{4}He, $Gi \sim 1$. In bulk superconductors, Gi is very small, so usually the logarithmic singularity cannot be observed.

Outside the fluctuation range of temperatures where $\mid 1-T/T_c\mid \geq Gi$, fluctuation contributions to the specific heat are small:

$$\frac{\delta C(T)}{C_0} \sim \begin{cases} \left(\frac{Gi_{(3)}}{|1-T/T_c|}\right)^{\frac{1}{2}} & at\ D=3, \\ \frac{Gi_{(2)}}{|1-T/T_c|} & at\ D=2, \end{cases} \tag{3.106}$$

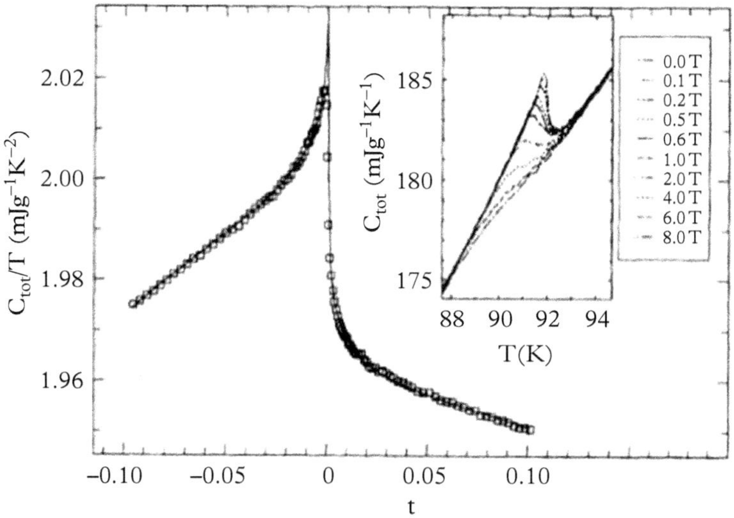

Figure 3.27 *The specific heat singularity in the YBaCuO single crystal; here* t $= (T - T_c)/T_c$. *(From Overend et al., 1994)*

where $C_0 = C_s - C_n$ is the specific heat jump at T_c. In $D = 2$ systems, the fluctuation contributions are larger but, for a thin film, it is not easy to measure them, due to the small sample volume. In a quasi-$D = 2$ system like the high-T_c cuprates with weak interlayer bonding, it is possible to measure increases in the specific heat, which are similar to the λ-singularity in ^{4}He (Fig. 3.27).

The interaction of fluctuations in the critical area $\tau \leq Gi$ changes the temperature dependence not only for the specific heat; the correlation length divergence is given by $\xi(T) \sim \tau^{-\nu}$, and the penetration depth looks like $\lambda(T) \sim \tau^{-\nu/2}$, where the critical index $\nu \approx 2/3$ in the $D = 3$ system with a complex order parameter (Patashinskii and Pokrovsky, 1966).

3.6.3 Magnetic Susceptibility

The diamagnetic properties of a superconductor are characterised by the magnetic susceptibility $\chi(T)$. In normal metals, χ is very small, typically $\chi \leq 10^{-5}$. In superconductors, $\chi = -1/4\pi$. Thus, at the transition, a jump that is several orders of magnitude in size occurs. One may expect a noticeable fluctuation contribution to magnetic susceptibility even outside the critical region at $|1 - T/T_c| \geq Gi$.

The magnetic moment $M = -\partial G/\partial H_0$, where the free energy G of a superconductor in the external magnetic field H_0 can be found following the general rules of statistical physics:

$$\exp\left(-G/K_BT\right) = \int \mathrm{D}\Psi(r)\exp\left(-G\left[\Psi(r)\right]/K_BT\right) \tag{3.107}$$

Here D means integration over all possible configurations of the order parameter $\Psi(\boldsymbol{r})$ with the statistical weight determined by the Ginzburg–Landau functional $[\Psi(\boldsymbol{r})]$, eqn (3.16). For $|1 - T/T_c| \geq Gi$, the term $|\Psi(\boldsymbol{r})|^4$ in the Ginzburg–Landau functional may be neglected, and the functional integral (3.107) is given by a product of independent integrals:

$$\exp\left(-G/K_BT\right) = \prod_{\mu} da_\mu d\tilde{a}_\mu \exp\left(-E_\mu |a_\mu|^2\right) \tag{3.108}$$

Here we use a decomposition of the order parameter $\Psi(\boldsymbol{r}) = \sum a_\mu \psi_\mu$ over the eigenfunctions ψ_μ of the Hamiltonian (where $\alpha = \tilde{\alpha}(T - T_c), \tilde{\alpha} = const$):

$$H_2 = \frac{1}{4m}\left(-i\hbar\nabla - \frac{2e}{c}\boldsymbol{A}\right)^2 + \alpha$$

with the eigenenergies E_μ. The integration over each pair $a_\mu, \tilde{a}_\mu$ gives a multiplier $\pi k_BT/E_\mu$. The magnetic moment is written as follows:

$$M = -\frac{\partial G}{\partial H_0} = -K_BT\sum_{\mu}\frac{\partial \ln E_\mu}{\partial H_0} \tag{3.109}$$

We calculate the diamagnetic susceptibility for the simplest case of a spherical sample with the radius $R \ll \xi(T)$, where only the lowest-energy eigenstate E_0 is important in eqn (3.109). The corresponding eigenfunction is the constant $\psi_\mu = \left(4\pi R^2/3\right)^{-1/2}$. The energy E_0 is given by the mean value of the operator H_2:

$$E_0 = \alpha + \psi_0^2 \int_{r<R} d^3r \frac{e^2}{mc^2} A^2(r) = \alpha + \frac{e^2 H_0^2 R^2}{10mc^2} \tag{3.110}$$

The vector potential in eqn (3.110) is perpendicular to the sphere radius and given by $\boldsymbol{A} = [\boldsymbol{H}_0\boldsymbol{r}]/2$. Finally, the magnetic moment is equal to

$$M(H_0) = -\frac{E^2 T_c R^2 H_0}{5m^2c^2\left(\tilde{\alpha}(T - T_c) + \frac{e^2 H_0^2 R^2}{10mc^2}\right)} \tag{3.111}$$

and the linear part of the fluctuation diamagnetic susceptibility per unit volume looks like

$$\chi_{fluct}(T) = \frac{3e^2 T_c}{20\pi mc^2 R\tilde{\alpha}(T - T_c)} \tag{3.112}$$

The normal metal susceptibility is $\chi_0 \sim {}^{e^2 k_F}\!/\!_{mc^2}$. Using eqn (3.21), we can express parameter $\alpha = \tilde{\alpha}(T - T_c)$ from the Ginzburg–Landau functional over ξ_0 and T_c. The ratio of the fluctuation contribution to the normal one is given by

$$\frac{\chi_{fluct}}{\chi_0} \sim \frac{\min(\xi_0, l)}{R} \frac{T_c}{T - T_c} \tag{3.113}$$

It is clear from eqn (3.113) that the fluctuation contribution to susceptibility is not determined by the Ginzburg–Levanyuk parameter Gi. Moreover, it decreases when the coherence length ξ_0 or (in the dirty limit) the mean free path l becomes smaller and the parameter Gi increases.We have discussed above the small spherical particle. For a bulk superconductor, the ratio $\chi_{fluct}/\chi_0 \sim (T_c/T - T_c)^{1/2}$ does not depend on material parameters at all (Lifshitz and Pitaevskii, 1978). The difference between the fluctuation contribution to susceptibility and heat capacity and conductivity (see below) results from the non-locality in space of the diamagnetic response to the external magnetic field. Indeed, the diamagnetic response is determined by currents over the surface. That is why the sphere radius R appears in eqn (3.113).

3.6.4 Paraconductivity

In the microscopic theory of superconductivity, three main mechanisms of fluctuation conductivity are known (Varlamov and Larkin, 2005). The most general relation between current density $\boldsymbol{j}(\boldsymbol{r}, t)$ and the vector potential $\boldsymbol{A}(\boldsymbol{r}', t')$ is given by the operator of the electromagnetic response $Q_{\alpha\beta}$ (Abrikosov *et al.*, 1963):

$$j_\alpha(\boldsymbol{r}, t) = -\int Q_{\alpha\beta}(\boldsymbol{r}, \boldsymbol{r}', t, t') A_\beta(\boldsymbol{r}', t') d\boldsymbol{r}' dt' \tag{3.114}$$

The operator $Q_{\alpha\beta}$ can be represented by a correlator of two single-electron Green's functions averaged over the impurity positions and including electron–electron interactions in the Cooper channel. It is given by ten diagrams in first-order perturbation theory (Varlamov and Larkin, 2005). We restrict ourselves here to a qualitative discussion of the main mechanisms of fluctuation conductivity. The first one is given by the fluctuation of Cooper pairs above T_c, which provides an additional conductivity channel and reduces the resistivity (Aslamazov and Larkin, 1968). This is the direct fluctuation contribution to the conductivity; its mechanism can be clarified within the conventional Drude expression for the conductivity of a normal metal with impurities:

$$\sigma = {}^{e^2 n_e \tau_{tr}}\!\big/\!_{m} \tag{3.115}$$

Here n_e is the density of electrons, and τ_{tr} is the transport relaxation time. The density of fluctuating Cooper pairs with momentum $\boldsymbol{p}$ above T_c may be given by $(p) = \langle \Psi_p \Psi_{-p} \rangle$, where the product $n_e \tau_{tr}$ in eqn (3.115) should be replaced by $\int dp\, n(p) \tau(p)$. To estimate the pair density, we can use the quadratic part of the Ginzburg–Landau functional similar to eqn (3.108), where the eigenfunctions are the plane waves e^{ipr} and the eigenenergies are equal to $E_p = \hbar^2 p^2/4m + \alpha$. This gives

$$n(p) = \frac{k_B T}{\hbar^2 p^2/4m + \alpha} \tag{3.116}$$

For a fluctuating pair, the relaxation time $\tau(p)$ is determined not by the impurity scattering but by the pair breaking into two separate electrons. This time may be estimated by using a time-dependent Ginzburg–Landau equation that contains the kinetic coefficient γ: $\tau(p) = \gamma / \left(\hbar^2 p^2/4m + \alpha\right)$. Finally, we obtain the Aslamazov–Larkin fluctuation conductivity (paraconductivity) as

$$\sigma_{AL} \sim \frac{e^2}{m} \sum_p \frac{k_B T \gamma}{\left(\hbar^2 p^2/4m + \alpha\right)^2} \tag{3.117}$$

This conductivity is divergent at T_c; for a superconductor sample with dimension $D = 1, 2, 3$, the divergence is given by

$$\sigma_{AL}^{(D)} \sim \left(\frac{T_c}{T - T_c}\right)^{(4-D)/2} \tag{3.118}$$

The second mechanism is indirect and is related to the fluctuation renormalisation of a single-electron density of states; it is called the density of states contribution. Due to the formation of the fluctuating pairs, the density of single electrons close to the Fermi energy is reduced so that it looks like a fluctuation pseudogap near the Fermi level. The Drude conductivity in the normal state decreases. So, the density of states contribution has a negative sign, unlike the Aslamazov–Larkin one and is less singular in parameter $(T - T_c)^{-1}$. Nevertheless, there are several examples when the Aslamazov–Larkin contribution is suppressed and the density of states contribution is the main one (Aslamazov and Larkin, 1974; Altshuler *et al.*, 1983; Gray and Kim, 1993).

The third mechanism is also indirect and is a quantum effect due to coherent elastic scattering of both electrons forming the Cooper pair on the same impurities. It is called the Maki–Thompson contribution (Maki, 1968; Thompson, 1970) and appears only in transport properties. The Maki–Thompson contribution may be interpreted as the Andreev reflection of an electron on the fluctuating Cooper pairs. The temperature dependence of the Maki–Thompson contribution is similar to the paraconductivity, and its magnitude is very sensitive to the orbital symmetry of pairing. The maximal Maki–Thompson contribution occurs for *s*-pairing, while *d*-pairing strongly suppress it in high-T_c cuprates (Yip, 1990).

3.6.5 NMR relaxation

Here we will discuss fluctuation contributions to spin susceptibility and the NMR relaxation rate. These contributions may result in pseudogap phenomena typical for high-T_c superconductors in the normal phase. The dynamic spin susceptibility is determined by a finite-temperature transverse-spin Green's function:

$$\chi_{\pm}^{(R)}(\boldsymbol{k},\omega) = \chi_{\pm}\left(\boldsymbol{k}, i\omega_\nu \to \omega + i0^+\right)$$

where

$$\chi_{\pm}(\boldsymbol{k}, i\omega_\nu) = \int_0^{1/T} d\tau e^{i\omega\tau} \langle\langle \hat{T}_\tau (S_+(\boldsymbol{k},\tau) S_-(-\boldsymbol{k},0)\rangle\rangle \tag{3.119}$$

The static spin susceptibility $\chi_s = \chi_{\pm}^{(R)}(\boldsymbol{k}\to 0, \omega = 0)$ and the dynamic NMR relaxation rate $1/T_1$ is given by (A is a constant)

$$\frac{1}{T_1 T} = \lim_{\omega\to 0} \frac{A}{\omega} \int \frac{d^3k}{(2\pi)^3} Im\chi_{\pm}^{(R)}(\boldsymbol{k},\omega) \tag{3.120}$$

For free electrons at low temperature, $T \ll E_F$, the theory of metals gives the Pauli susceptibility $\chi_S^0 = \nu$ (where ν is the density of states at the Fermi level), and the Korringa law $(1/T_1T)^0 = A\pi\nu$. The fluctuation contributions below will be normalised to these values. The dynamic spin susceptibility is given by a set of diagrams for the polarisation operator that are similar to those for the electromagnetic response discussed above (eqn (3.114)). The main difference is that two electrons interacting with the external field $S_\pm$ should have the opposite spin projections, so it excludes the Aslamazov–Larkin contribution. Below we will consider the layered superconductor. Both for the clean and dirty limit, the fluctuation contribution to the longitudinal susceptibility is equal to (Randeira and Varlamov, 1994)

$$\frac{\chi_s^{||}}{\chi_s^{(0)}} = -2Gi_{(2)} ln\left(\frac{2}{\sqrt{\tau}+\sqrt{\tau+r}}\right) \tag{3.121}$$

Here $\tau = (T - T_c)/T_c$, and r is the anisotropy parameter that was introduced by Lawrence and Doniach (1971) and plays an important role in the theory of layered superconductors: $r = (2\xi_z(T_c)/s)^2$, where $\xi_z(T)$ is the correlation length perpendicular to the layers, and s is the distance between the layers. The calculation of the NMR relaxation rate is much more complicated. Besides the finite transport relaxation time of the electron, τ_{tr}, an important role is played by the decoherence time τ_φ, the time it takes for an electron to lose its phase. The NMR relaxation rate has been analysed in different limits by Kuboki and Fukuyama (1989), Heym (1992), and Randeira and Varlamov (1994). In the ultraclean limit, the density of states contributions and the

Maki–Thompson contributions cancel each other out, and the fluctuation contribution is absent. For long-living pairs $(1/\tau_\varphi \ll T_c)$, the NMR relaxation rate increases when T tends to T_c from the normal phase. For strong decoherence, the abnormal Maki–Thompson contribution is suppressed, and the NMR relaxation rate is determined by less singular density of states contributions and regular Maki–Thompson contributions. It results in the change of the sign of the fluctuation contribution to the NMR relaxation rate and decreasing $1/T_1$ when T tends to T_c.

4 Experimental Methods

This chapter explores the experimental techniques that have led to many of the advances in the understanding of superconductivity and complement the theoretical advances described in other chapters in this book. Section 4.1 describes the tools that have been used to probe the key properties of the superconducting state, that is, the energy gap and the nature of the pairing interaction. Understanding these properties was key to the search for high T_c and provided the key elements for turning superconductivity into a viable technology.

Section 4.2 describes the techniques responsible for discovering the highest T_c hydrides. This technique has provided proof that room-temperature superconductivity not only is theoretically possible but is a reality. The hope is that fully understanding these materials will lead to discovering materials with comparable T_c's but without the need for high pressure.

Section 4.3 describes the various techniques for preparing high-quality thin films of the most significant superconductors. These techniques have made possible much of superconducting technology that continues to provide impetus for future research on superconducting devices.

4.1 Spectroscopy

In this section, we describe several of the most important experimental techniques that have been used to probe the most fundamental properties of the superconducting state: the energy gap and the pairing interaction. The techniques that will be described are all spectroscopic: they involve the tunnelling of quasiparticles through an insulating barrier or through a point contact, the interaction of electromagnetic waves or photons with a superconducting film or surface, the attenuation of ultrasonic sound waves, and the relaxation and or resonance of muons interacting with a superconducting compound.

4.1.1 Tunnelling Spectroscopy

The BCS theory was very successful in explaining many of the fundamental properties of elemental superconductors. However, the mechanism of pairing in many of the more

Superconducting State: Mechanisms and Materials. Vladimir Z. Kresin, Sergei G. Ovchinnikov, and Stuart A. Wolf, Oxford University Press (2021). © Kresin, Ovchinnikov, Wolf. DOI: 10.1093/oso/9780198845331.003.0004

exotic materials was uncertain. Tunnelling spectroscopy has proved to be a very powerful tool for both probing the nature of the interaction and verifying the validity of the strong coupling theory of Eliashberg (1960). In this section, we will describe the experimental techniques that are used to obtain the tunnelling spectra, present the relevant equations for the tunnelling density of states, and then describe the beautiful inversion method developed by McMillan and Rowell (1969).

Experimental methods. There are several experimental methods that have been used to generate tunnelling spectra. The most widely used are methods based on the deposition of a barrier and a counter electrode. The simplest manifestation of this method requires the deposition of the superconducting electrode; the formation of the barrier, either by oxidation of the superconductor or by a deposited insulating layer; and the final deposition of another metallic or superconducting electrode at right angles to the original film. This simple cross-stripe structure can be made using contact masks and does not require sophisticated lithography. An example of such a structure is illustrated in Fig. 4.1. Subsequently, junctions have been made via a trilayer process where the electrodes and the barrier are deposited sequentially in the same deposition run and the junction is defined photolithographically afterwards. This method allows the preparation of small and very uniform tunnel structures. Point contacts have also been used very successfully to make tunnelling measurements. In this case, a sharpened and sometimes oxidised needle with a small voltage applied to it is slowly moved into the proximity of a superconducting surface. The needle position is adjusted until the desired current starts to flow. Current–voltage characteristics can then be measured. This technique has been reasonably successful in probing oxide superconductors, since it has been very difficult to prepare thin insulating barriers by conventional deposition methods.

The important quantities that need to be measured are the direct current–voltage characteristic, I–V; the derivative of the I–V, dI/dV; and the second derivative, d^2I/dV^2.

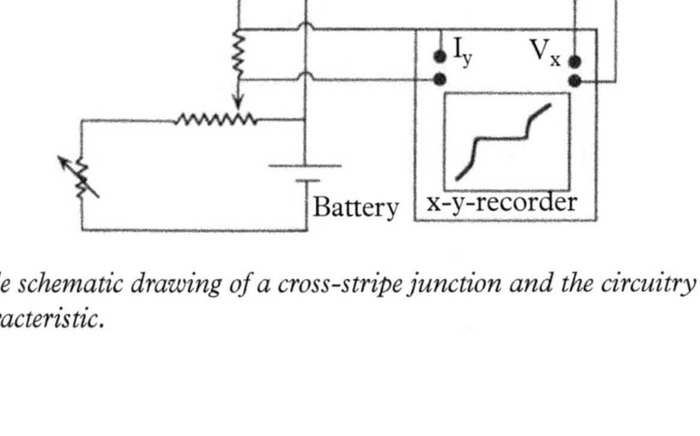

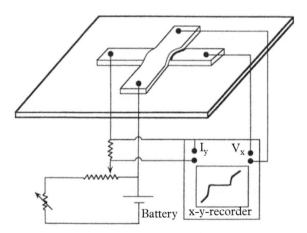

Figure 4.1 *A simple schematic drawing of a cross-stripe junction and the circuitry necessary to measure an I-V characteristic.*

This data must be taken with the sample in both the normal and the superconducting state and with the Josephson or pair tunnelling contribution quenched by a small magnetic field. These measurements are somewhat tricky and require ac lock-in techniques (see McMillan and Rowell, 1969). The I–V characteristic and its first derivative, dI/dV, are essential for determining the tunnelling density of states $N_T(E)$ and is therefore very important in making the connection to the theory. The first-derivative data must be taken to very high precision because only the deviations from the 'BCS' (or weak coupling) density of states are significant, and these deviations should be measured to about 1%. Indeed, from these measurements, one can find the energy gap as well as detailed information about the phonon modes (or other excitations) responsible for the attractive pairing interaction. Just as importantly, this technique may be unique in determining whether the excitations responsible for the superconductivity are not phononic. Details of these analyses will be described below.

The energy gap and the transition temperature. In any complete analysis of tunnelling data, it is important to know the T_c for the junction. Since the tunnelling process involves a very thin region at the surface of the superconductor, the T_c of the junction may be slightly different then the 'bulk' T_c for the rest of the film. Since T_c is an important parameter in the rest of the analysis, it is important to know it quite accurately. The simplest method for determining the T_c of the junctions is to plot the ratio of the conductance in the superconducting and the normal states for various temperatures and then extrapolate to a ratio of 1. This is the best estimate of the transition temperature of the superconducting material.

Determining the gap from the conductance is simple in principle but, in practice, is complicated by the non-idealities of real junctions.

Metal–insulator–semiconductor junctions. If the tunnel junction counter electrode is a normal metal, then the normalised conductance of the junction is given simply by

$$\sigma = (\mathrm{dJ/dV})_\mathrm{s}(\mathrm{dJ/dV})_\mathrm{n} = \int \mathrm{N_T}(\omega')\left[-\mathrm{d}f(\omega' + \mathrm{eV})/\mathrm{d})\,\omega''\right]\mathrm{d}\omega' \tag{4.1}$$

where $\mathrm{N_T}(\omega) = \mathrm{Re}\left[|\omega|/(\omega^2 - \Delta^2)^{1/2}\right]$, f is sharply peaked and positive at $\mathrm{E}' = -\mathrm{eV}$ with a half width of order kT, and Δ is the energy gap. Thus, at $\mathrm{T} = 0$, $\sigma = \mathrm{N_T}(\omega)$. Direct information about the gap is contained in the normalised conductance. In Fig. 4.2 we show he current–voltage characteristics of Al–Pb junctions, taken by Nicol, Shapiro, and Smith (Nicol *et al.*, 1960) above the transition temperature of aluminium, are shown in Fig. 4.2. These were the first data where a quantitative comparison between the experiment and the BCS theory was made.

In order to determine the gap from arbitrary tunnelling data, a simple method is to measure the normalised conductance σ at zero bias and compare with the tabulated data of σ versus Δ/T (Bermon, 1960). Naturally, the complete conductance curve can be fitted to the functional form of eqn (6.1) in chapter 6 or one that has been fully calculated and tabulated by Bermon. Of course, these results are in the weak coupling limit and they will lose some accuracy as the coupling gets stronger.

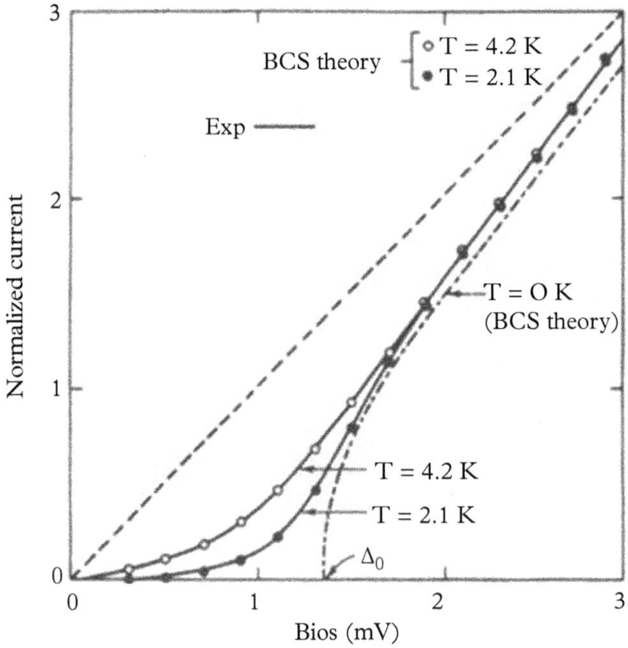

Figure 4.2 *Comparison between experimental and theoretical I–V characteristics of an Al–Pb junction. This figure demonstrates excellent agreement with the Bardeen, Cooper, and Schrieffer (BCS) theory. (Figure reproduced from Nicol et al., 1960)*

Semiconductor–insulator–semiconductorjunctions. For junctions with two superconducting electrodes, the determination of the gap is more straightforward. If both electrodes are equivalent, then, for $T \ll T_c$, there should be a sharp discontinuity in the conductance at 2Δ. Of course, for non-ideal (real) junctions, the rise is not abrupt but is typically broadened by lifetime effects or by a distribution of gaps. In this case, the point of maximum slope, dI/dV, should be used. In the case of a nearly linear rise, the midpoint should be used (McMillan and Rowell, 1969).

Semiconductor$_1$–insulator–semiconductor$_2$junctions. For junctions with inequivalent superconducting electrodes, there is a cusp in the conductance at $\Delta_2 - \Delta_1$ and an abrupt rise in the conductance at $\Delta_1 + \Delta_2$. Thus, by clearly locating the position of the cusp and using the same technique for the steep rise that was described above for the semiconductor–insulator–semiconductor junctions, one can determine the gaps for both superconducting electrodes.

If $T \ll T_c$, then the value of the junction conductivity is an exponentially decreasing function of the smaller gap divided by temperature and, in fact, obeys the relation (see Giaever and Mergerle, 1961)

$$\sigma(V = 0) = (2\pi\Delta_1/kT)^{1/2} \exp(-\Delta_1/kT) \tag{4.2}$$

Thus, the value of the gap for the lower T_c electrode can be estimated from the temperature dependence of the conductivity at very low bias.

Inversion of the gap equation and $\alpha^2F(\omega)$. The Eliashberg gap equations (1960, 1961) describe the direct influence of the phonon density of states, $F(\omega)$ on the energy-dependent gap function $\Delta(\omega)$. In turn, the gap function modifies the electronic density of states that can be directly determined by tunnelling conductance measurements. Just how the phonon spectrum influences $\Delta(\omega)$ and the normalised conductance was shown by Scalapino, Schrieffer, and Wilkins (1966). They calculated the effect of a single peak in $F(\omega)$ on $\Delta(\omega)/\Delta_o(\omega)$ and on $N_T(\omega)/N(0)$. The results are shown in Fig. 4.3. It is clear from this calculation that strong peaks in the phonon density of states show up as dips in the normalised tunnelling conductance.

All the information about the phonon density of states and the strength of the electron–phonon interaction is contained in the tunnelling measurements. The method by which the information is extracted is by a numerical inversion of the gap equations.

Thus, inverting the Eliashberg equations to determine the coupled phonon density of states $\alpha^2F(\omega)$ and μ^* (the Coulomb pseudopotential) is an extremely powerful

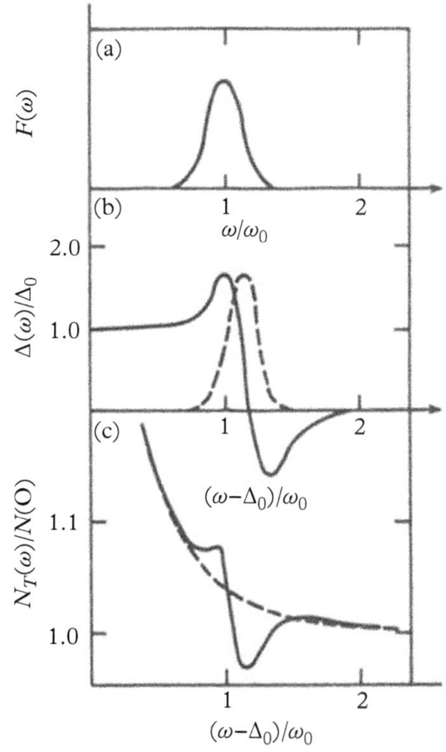

Figure 4.3 *The effect of a single peak in the phonon spectrum (a) on the gap function (b) and on the tunnelling density of states. (Figure reproduced from Scalapino et al., 1966)*

method for determining the degree to which any superconductor can be described in the framework of BCS superconductivity and, of course, was initially used to determine the accuracy of the Eliashberg theory itself, which was found to be accurate to within several per cent. The method developed by McMillan and Rowell (1965) involves numerically solving the integral Eliashberg equations for a given set of parameters, calculating the density of states, comparing the calculated values to the measured values, adjusting the input parameters, and iterating the procedure until the calculated density of states matches the measured one. A detailed description of the procedure and its application to Pb can be found in McMillan and Rowell (1969) but the highlights of this treatment are included here.

The integral equations for the normal and pairing self-energies for a dirty superconductor are

$$\xi(\omega) = [1 - Z(\omega)]\,\omega = \int^{\infty}_{\Delta_o} d\omega' \operatorname{Re}\left[\omega'/(\omega^2 - \Delta^2)^{1/2}\right]$$

$$\int d\omega_q\, \alpha^2(\omega_q)\, F(\omega_q)\,[Dq(\omega' + \omega) - D_q(\omega' - \omega)] \tag{4.3}$$

$$\varphi(\omega) = \int^{\omega^c}_{\Delta_o} d\omega' \operatorname{Re}\left[\Delta'/\left(\omega'^2 - \Delta'^2\right)^{1/2}\right]$$

$$\int d\omega_q\, \alpha^2(\omega_q)\, F(\omega_q)\,[D_q(\omega' + \omega) - D_q(\omega' - \omega) - \mu^*] \tag{4.4}$$

where $D_q(\omega) = (\omega + \omega_q + i0^+)^{-1}$, $\Delta(\omega) = \varphi(\omega)/Z(\omega)$, and $\Delta_o = \Delta(\Delta_o)$. $F(\omega)$ is the phonon density of states and $\alpha^2(\omega)$ is an effective electron–phonon coupling function for phonons of energy ω.

For an isotropic superconductor, the energy-dependent normalised tunnelling conductance (see above) gives the superconducting density of states

$$N(\omega) = \sigma(\omega) = \operatorname{Re}\left[\omega/\left(\omega^2 - \Delta(\omega)^2\right)^{1/2}\right] \tag{4.5}$$

Equations (4.3) and (4.4) are solved by iterating them both together. We start with a guess for $\alpha^2(\omega)F(\omega)$ (based on the measured differential conductivity) and μ^* and a zeroth-order guess for the gap function, that is, $\Delta^{(0)} = \Delta_o$ for $\omega < \omega_o$, and $\Delta^{(0)} = 0$ for $\omega > \omega_o$, where Δ_o is the measured energy gap (see above) and ω_o is the maximum phonon frequency. Using these parameters, we find $\xi^{(1)}$, $\phi^{(1)}$, $Z^{(1)}$, and hence $\Delta^{(1)}$. The iteration is continued until $\Delta^{(n)}$ converges to three decimal places. At this point, $N(\omega)$ is computed and compared to the experimental result.

Now the task is to adjust the zeroth-order $\alpha^2F(\omega)$ and μ^* to provide convergence of the calculated density of states to the measured density of states. To do this, it is necessary to calculate the linear response of the density of states to a small change in $\alpha^2F(\omega)$. This amounts to calculating the derivative $\delta N_c(\omega)/\delta\alpha^2F(\omega)$. This allows us to estimate the change in the zeroth-order guess for $\alpha^2F(\omega)$ necessary for the first-order iteration. The change is given by

$$\delta\alpha^2F(\omega) = \int d\omega' \left[\delta N_c(\omega)/\delta\alpha^2F(\omega)\right]^{-1}\left[N_c(\omega') - N_c(o)(\omega')\right] \tag{4.6}$$

and thus

$$\left[\alpha^2F(\omega)\right]^{(1)} = \left[\alpha^2F(\omega)\right]^{(o)} + \delta\alpha^2F(\omega) \tag{4.7}$$

Since the gap equations are not linear, the iteration needs to be continued until the calculated density of states reproduces the measured density of states. During the iteration process, μ^* is adjusted so that the calculated energy gap agrees with the measured energy gap. For many conventional superconductors, μ^* is found to be approximately 0.1; however, by virtue of the role it plays in the inversion procedure, it can be zero or even negative. If some spectral weight is missed in the tunnelling data because it occurs at energies beyond the voltage range of the measurements, then the inversion procedure reflects this in a reduced value of μ^*. A recent confirmation of this fact was provided by Schneider *et al.* (1987) for the cluster compound YB_6.

This procedure has been used to show that most of the conventional superconductors are indeed very well described by the Eliashberg equations. For example, the best results have been obtained for Pb (see McMillan and Rowell, 1969) where the Eliashberg theory has been fully tested and found to be very accurate.

Figure 4.4 shows the normalised density of states from the tunnelling conductance measurements, and Figure 4.5 shows the calculated $\alpha^2F(\omega)$ resulting from the inversion procedure previously described. (McMillan and Rowell, 1969).

The electron–phonon coupling parameter λ. The Eliashberg equations can be solved for T_c, and the solution can be simplified with the use of the parameter λ, which is simply related to $\alpha^2F(\omega)$ by the following equation:

$$x\lambda = \int d\omega 2\alpha^2F(\omega)/\omega \tag{4.8}$$

One can show (see chapter 3, eqn (3.35)) that, for an isotropic superconductor for any value of λ,

$$T_c = 0.25\,\tilde{\Omega}/\left(e2/\lambda_{eff} - 1\right)1/2 \tag{4.9}$$

where $\lambda_{eff} = (\lambda - \mu^*)\left[1 + 2\mu^* + \lambda\mu^* t(\lambda)\right]^{-1}$ and $\tilde{\Omega}$ is the average coupled phonon energy. The parameter λ is considered to be the most direct measure of the strength of the electron–phonon interaction, which strongly affects the nature of the superconductivity, that is, if $\lambda \ll 1$, then we have weak coupling or 'BCS' superconductivity whereas, if $\lambda \gg 1$, we have strongly coupled superconductors.

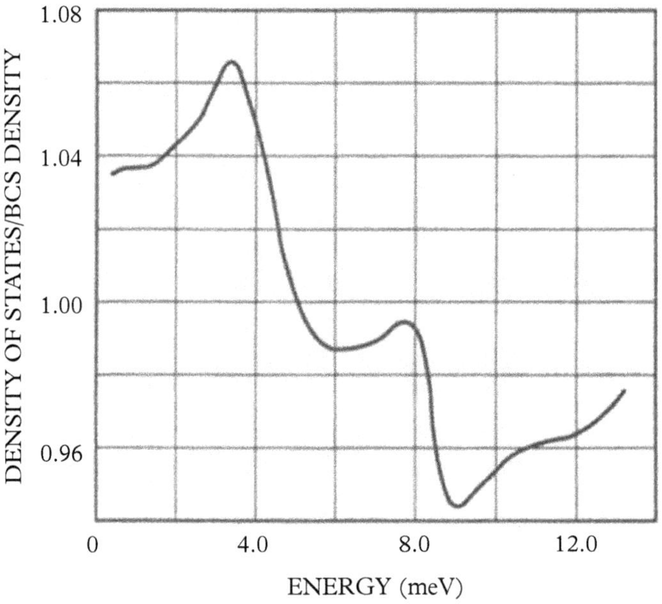

Figure 4.4 *The electronic density of states of Pb divided by the Bardeen, Cooper, and Schrieffer (BCS) density of states versus* $E - \Delta_0$. *(Figure reproduced from McMillian and Rowell, 1969)*

It is also important to note that normal electrons near the Fermi surface are dressed with a cloud of 'virtual' phonons. This dressing of the electron properties shows up as an enhancement of the cyclotron mass, the Fermi velocity, and the electronic heat capacity. The enhancement over the band value is given by exactly the same parameter λ that is implicit in the Eliashberg gap equations. Thus, detailed Fermiology provides an independent method for determining the strength of the electron–phonon interaction, as well as the accuracy of the inversion.

In fact, the degree to which phonons are responsible for the superconductivity can be tested by detailed comparisons of calculated specific heat and measured specific heat. The method depends on measurements of the temperature dependence of the specific heat, determinations of $\alpha^2F(\omega)$ from tunnelling spectroscopy, and values of $F(\omega)$ from neutron-scattering experiments. The main idea of this procedure is to use the manner in which the phonon dressing of the electronic part of the Sommerfeld constant is un-renormalised at high temperatures. By comparing the high-temperature and low-temperature values of the Sommerfeld constant, one can determine λ, the electron–phonon coupling strength. The electronic specific heat is separated from the total specific heat by integrating the measured $F(\omega)$ to determine the lattice specific heat and then subtracting it. Tunnelling spectroscopy, through its determination of $\alpha^2F(\omega)$ is also a measure of λ, provided that the phonons are fully responsible for the superconductivity. If some other boson contributes strongly to the pairing, than the $\alpha^2F(\omega)$ determined by tunnelling will not have the correct magnitude, since the

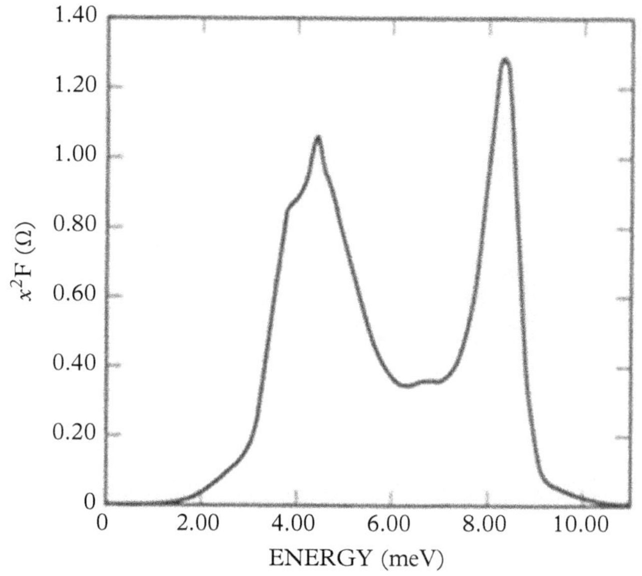

Figure 4.5 *The function* $\alpha^2F(\Omega)$ *for Pb found by fitting the data from Fig. 3.4. (After McMillian and Rowell, 1965)*

inversion procedure, which assumes that phonons are fully responsible, will improperly weight the phonon peaks. A simple integration of the tunnelling $\alpha^2F(\omega)$ will give a value of λ which can be compared to the value determined by the analysis of the specific heat. If they agree, then phonons account for the superconductivity; if they disagree, then some other pairing interaction is present. This procedure was carried out by Kihlstrom *et al.* (1988) for several A-15 structure superconductors. It was determined that V_3Si was completely phononic whereas Nb_3Ge was not! It would be very useful to carry out this procedure for the cuprate superconductors. Unfortunately, the state of the art in tunnelling spectroscopy at the present time for these compounds is not adequate to have a very reliable determination of $\alpha^2F(\omega)$. Some of the more interesting data on tunnelling into the cuprate superconductors will be presented in chapter 5, which treats the cuprates in great detail.

4.1.2 Scanning Tunnelling Microscopy and Spectroscopy

Scanning tunnelling microscopy and spectroscopy has become an extremely valuable tool for the understanding of the properties of superconducting compounds, especially the cuprates and the Fe-based compounds. The scanning tunnelling microscope was developed by Binnig and Rohrer (1982) and it revolutionised the study of surfaces as it was able to image single atoms on the surface (see Fig. 4.6). It was soon realised that this tool could also locally measure the electronic structure of the surface with atomic

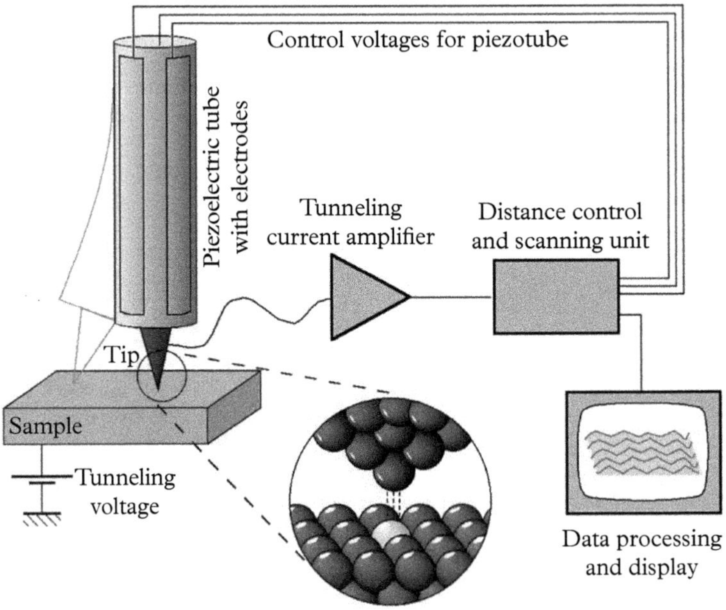

Figure 4.6 *A schematic drawing of a scanning tunnelling microscope.*

resolution by varying the tip voltage and measuring the current at fixed distance from the surface.

The derivative of the I–V curve, namely dI/dV, can be easily shown to be proportional to the density of electronic states at the surface being explored (Tersoff and Hamann, 1985). By scanning the tip over a surface, a very high-resolution map of the density of states and hence, if the surface is superconducting, a map of the energy gap can be obtained. Since the spatial resolution of the microscope is much higher than any of the relevant superconducting lengths, these maps can provide information on the effect of local perturbations like magnetic or non-magnetic impurities, as was first demonstrated by Ji *et al.* (2008) and is illustrated in Figure 4.7. What is clearly observed in this figure is the effect of a magnetic impurity (Mn or Cr) on the density of states in the superconducting gap of Pb, indicating the local destruction of superconductivity by the magnetic impurity. This destruction is very localised around the impurity and covers a region that is approximately the coherence length of Pb. This ability to look at the spectrum reflecting the local density of states for the cuprates in the pseudogap region is crucial to understanding the nature of this region, as will be described and illustrated in detail in chapter 5.

4.1.3 Other Spectroscopic Techniques

Infrared spectroscopy. The study of the absorption of electromagnetic energy has played a very important role in the history of superconductivity. It provided some very early confirmation of the BCS theory and provides a simple method for determining the energy gap, especially for thin films where the transmission, the reflection, and the absorption can be estimated. Mattis and Bardeen (1958) and, independently, Abrikosov, Gor'kov, and Khalatnikov (1959) derived the general equations for complex conductivity. For the convention where the incoming radiation is in a plane wave $\exp[\mathrm{i}(\omega t - \mathbf{qr})]$, $\sigma = \sigma_1 - \mathrm{i}\sigma_2$. The results of the Mattis–Bardeen theory for σ_1/σ_n and σ_2/σ_n are given by the following two equations (Ginsberg and Hebel, 1969):

$$\frac{\sigma_1}{\sigma_2} = \frac{2}{\hbar\omega}\int_{\Delta}^{\infty}[f(E) - f(E+\hbar\omega)]\,g(E)dE + \frac{1}{\hbar\omega}\int_{\Delta-\hbar\omega}^{-\Delta}[1 - 2f(E+\hbar\omega)]\,g(E)dE \tag{4.10}$$

$$\frac{\sigma_1}{\sigma_2} = \frac{1}{\hbar\omega}\int_{\Delta-\hbar\omega,-\Delta}^{\Delta}\frac{[1-2f(E+\hbar\omega)]\left(E^2+\Delta^2+\hbar\omega E\right)}{(\Delta^2-E^2)^{1/2}\left[(E+\hbar\omega)^2-\Delta^2\right]^{1/2}}dE \tag{4.11}$$

If $\hbar\omega > 2\Delta$, then the second term in eqn (4.10) is included and the lower limit of integration in eqn (6.11) should be $-\Delta$ instead of $\Delta - \hbar\omega$.

Measuring any two of the rates energy propagation, reflection, and transmission or absorption for a plane wave normal to the surface of a thin film whose thickness is less than the penetration depth and the coherence length makes it possible to obtain a direct determination of σ_1/σ_n and σ_2/σ_n. These results can then be directly compared to the theoretical expressions, eqns (4.10) and (4.11).

The earliest experiments by Glover and Tinkham (1957), however, only measured the transmitted energy so that unambiguous values of σ_1/σ_n and σ_2/σ_n could not be obtained. However, they could estimate these quantities using the Kramers–Kronig transform relation between σ_1 and σ_2, as well as the sum rule, which requires that

$$\int \sigma_{1s}(\omega)\,d\omega = \int \sigma_{1n}(\omega)\,d\omega \tag{4.12}$$

Equations (6.10) and (6.11) predict that, for $T = 0$, σ_1/σ_n is zero for frequencies up to the gap frequency $2\Delta/h$ and finite above this frequency whereas, for finite temperatures, σ_1/σ_n will be finite but vanishingly small if $(-\Delta/kT) \ll 1$. The Kramers–Kronig transformation requires that σ_2/σ_n also be small near the gap frequency, which means that both the reflection and the absorption will be small, and the result is that the transmission has a sharp peak at the gap frequency. Thus, a rather unambiguous determination of the gap was made by these early infrared measurements.

Later measurements were made of both transmission and reflection. In these cases, rather accurate determinations of σ_1/σ_n and σ_2/σ_n could be made and compared to the theory (eqns (4.10) and (4.11)). An example of this type of result is shown in Fig. 4.7, which shows both the measured curves of σ_1/σ_n and the theoretical curve based on the Mattis–Bardeen theory.

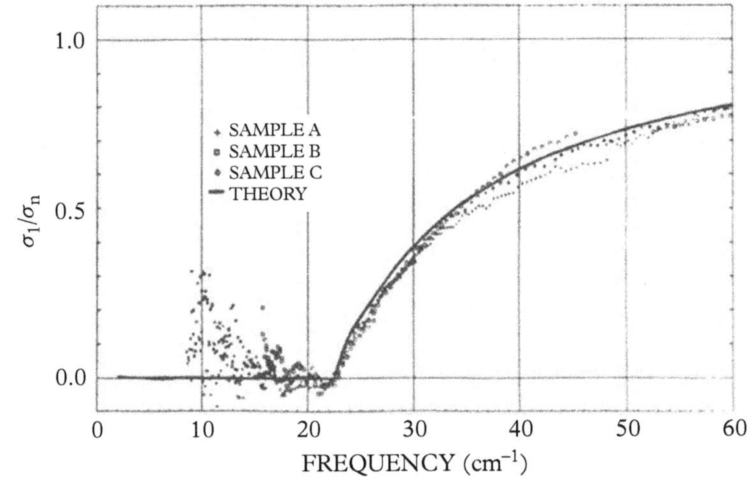

Figure 4.7 *Measured and calculated values of the conductivity ratio for three Pb films as a function of the phonon frequency. (Reproduced from Palmer, 1966)*

Direct measurements of the absorption of electromagnetic energy as a function of frequency can be used to determine the energy gap accurately. If the temperature is low enough so that $\exp[-\Delta/kT] \ll 1$, then there is a well-defined absorption edge at 2Δ. Measuring the absorption is typically done by bolometric methods, where the sample is thermally isolated from its surroundings and the absorption is measured by the temperature rise.

Precise infrared spectroscopy can also be used to reconstruct the function $\alpha^2(\Omega)F(\Omega)$. This method was proposed by Little and collaborators (Holcomb *et al.*, 1996; Little *et al.*, 1999) and is based on the thermal-difference–reflectance spectroscopy technique. This technique enables the determination of $\alpha^2(\Omega)F(\Omega)$ for an energy interval that is much larger than could be determined using the tunnelling method (see section 4.1). The reflectivity of the sample is measured with a high degree of precision at different temperatures, and the ratio of the difference relative to the sum is determined. Using a theoretical method developed by Shaw and Swihart (1968), an inversion of the data is performed and the function $\alpha^2(\Omega)F(\Omega)$ can be obtained. The large range of energies allowed using this method enables higher-energy electronic modes to be revealed if they are present. This method has been successfully applied to the cuprates (see chapter 5).

Ultrasonic attenuation. Historically, the first method used to measure the energy gap and check the validity of the temperature dependence of the BCS gap was ultrasonic attenuation. The experimental method involves mounting transducers on opposite side of a crystal of the superconducting material, launching a sound wave into the sample, and measuring the ratio of the attenuation below the transition temperature to its value just above the transition. The simplest case to consider is for the propagation of longitudinal

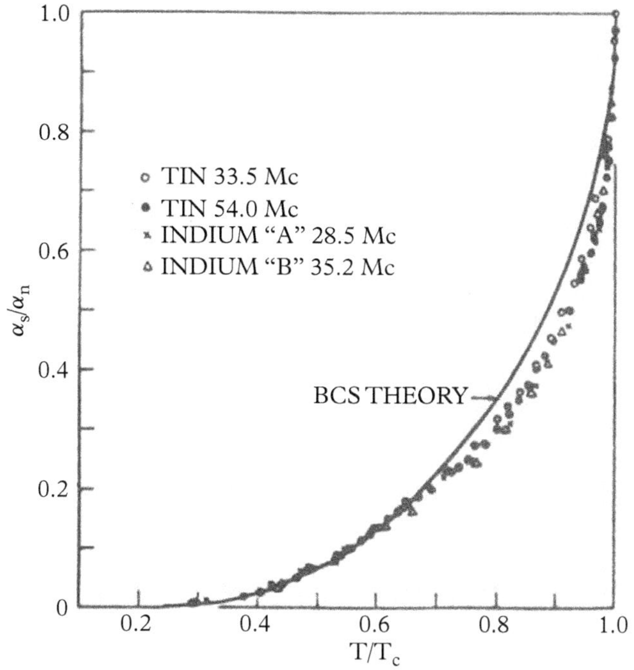

Figure 4.8 *The attenuation ratio as a function of reduced temperature compared to the Bardeen, Cooper, and Schrieffer (BCS) variation. (Reproduced from Morse and Bohm, 1957)*

sound waves where the interaction of the sound wave with the crystal is due to the change in the crystal potential which accompanies the wave. If the phonons associated with the sound wave have energies $\hbar(\omega)$ less than the gap energy 2Δ, then the attenuation at very low temperatures ($T \ll T_c$) should be proportional to $\exp[-\Delta/kT]$, since only the thermally excited quasiparticles can absorb the phonons.

In their landmark paper, Bardeen, Cooper, and Schrieffer (1957) calculated the ratio of the ultrasonic attenuation in the superconducting state α_s divided by the attenuation in the normal state α_n for the case when $q_l \gg 1$, where q is the wavenumber of the sound wave, and l is the electron mean free path, and $\hbar\omega < 2\Delta$. In this case, the ratio α_s/α_n is equal to $2f(\Delta(T), T)$ where f is the Fermi function. This expression is valid even for $q_l \ll 1$. Thus, by fitting $\alpha s/\alpha n$ as a function of temperature to the Fermi function, one can determine the value of the energy gap and its temperature dependence. Figure 4.8 shows the result from the paper by Morse and Bohm (1957) where the measurements on tin and indium are compared with the BCS theory.

Angle-resolved photoemission. Angle-resolved photoemission has emerged as one of the best tools for experimentally determining the electronic structure of highly correlated superconductors, especially superconducting oxides like the cuprates and ruthenates. By measuring the energy and angular distribution of electrons that are photoemitted

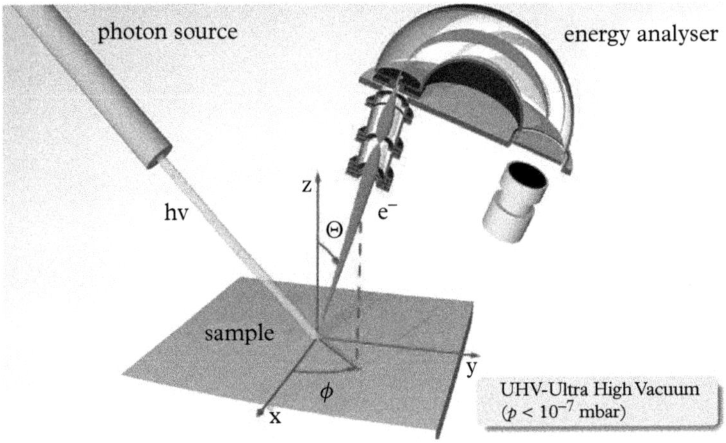

Figure 4.9 *A highly focused photon source from a synchrotron impinges on the sample, and the electrons emitted are analysed using a double-hemispherical analyser that can measure the dispersion (E vs k) along an angular cut in k-space.*

from the material in response to a beam of high-energy photons, one can determine much about the energy and the wave vector (momentum) of the electrons that are moving in the material, particularly those materials that cannot be explained using an independent particle approximation. Over the last score of years, this technique has reached extraordinary resolutions in energy and angular distribution, better than 2 meV in energy and 0.2 degrees of angular resolution. For a detailed review of the state of the art for ARPES, see the work by Damascelli (2004). ARPES can provide information on the band structure and band dispersion, determine the complicated Fermi surface, and provide information on the strength and nature of the many-body correlations that are important for the single-electron excitation spectrum, and how they are responsible for many of the important physical properties. A key feature of the cuprates in the pseudogap (mainly underdoped) region are Fermi arcs that are incomplete Fermi contours that change in length with temperature and doping. These structures reflect the unusual properties of the pseudogap region that may reflect the underlying inhomogeneity of these compounds. Fermi arcs were first discovered using ARPES and they provide important information about the unusual electronic structure of the cuprates, especially in the underdoped pseudogap part of the phase diagram.

A typical ARPES system resides at the end of a synchrotron beamline, and such a system is very schematically shown in Fig. 4.9.

In addition to finding dispersion curves and constructing the band structure and the Fermi surface, ARPES can determine the k dependence of the superconducting energy gap. The symmetry of the gap can be determined, as well as strong anisotropies.

Muon spin relaxation or resonance. Muon spin relaxation or resonance (μSR) has become an important technique (see the reviews by Keller, 1989 and Schenck, 1985) for studying the magnetic properties of novel materials and, in particular, the magnetic behaviour of novel superconductors, including many that will be described in chapters 5 (cuprates) and 6 (novel superconducting materials). A muon is an unstable particle with either positive charge or negative charge (μ^+orμ^-) that lives for about 2 microseconds and then spontaneously decays into a positron (electron) and a neutrino–antineutrino pair. Muons are produced in an accelerator that uses high-energy protons that are accelerated to more than 500 MeV and then collide with a light-element target such as beryllium or carbon. This nuclear reaction produces positively or negatively charged pions. The pions are very unstable and decay in about 25 nanoseconds into muons and muon neutrinos or antineutrinos. The muons of interest to the studies of most solids are the positively charged muons (μ^+), and beams of these muons are produced from pions decaying at rest in the surface layer of the primary light-element target; these muons are called surface muons. Negative-muon beams cannot be produced in this manner, but they are not of significant interest for studies of the solid state, for reasons that will not be discussed here. The surface-muon (μ^+) beam has a momentum of 2.98 MEV/c (in the rest frame of the pion) and an energy of 4.119 MeV. This beam emerges from the target isotropically from the decaying pion at rest.

Serendipitously, the decay of the pion exhibits maximal parity violation for this weak interaction and thus the surface muons are perfectly spin polarised *opposite* to their momentum; thus, when the beam hits the material being studied, the muons arrive nearly 100% spin polarised. This makes this technique, in many respects, superior to other magnetic resonance probes that rely on obtaining a high-spin polarisation in a magnetic field and utilising the equilibrium spin polarisation. High-spin polarisation can thus only be achieved at very high fields and very low temperatures.

There are only a very few meson facilities in the world that can make high-precision μSR measurements: the Paul Scherrer Institute (PSI) in Villigen, Switzerland, which also currently has a very-low-energy muon beam; the Tri-University Meson Facility in Vancouver, Canada (TRIUMF), the Booster Muon facility at the high-energy research organisation KEK in Japan (soon to be replaced by J-PARC in Tokai, Japan) and ISIS at the Appleton Rutherford Laboratory in the United Kingdom. The experimental setup (continuous beam or pulsed; transverse, longitudinal, or zero field) and the nature of the beams differ at each of these facilities but, in general, they can all study the interaction of muons in their host material and get very detailed information about the magnetism in the host compounds.

μSR studies of superconductivity. The discovery of the high-critical-temperature cuprate superconductors led to an explosion of research aimed at discovering the underlying mechanism and understanding the properties of these novel materials. μSR has provided some key information about the magnetic properties of the cuprates and other novel superconducting systems. In particular, being able to map out the antiferromagnetic phase diagram for both LaSrCuO and YCaBaCuO has been extremely important in understanding the interplay of magnetism and superconductivity in the cuprates (Niedermayer *et al.*, 1998). More recently, μSR studies of very weak magnetism in YBaCuO (Sonier *et al.*, 2001) have highlighted the role of spatial inhomogeneities in the cuprates and strongly supports our understanding of the role of the inhomogeneous structure of the cuprates that is detailed in chapter 5. Finally, information about the symmetry of the order parameter, especially where it is suspected that there is time-reversal symmetry breaking, for example in the case of *p*-wave superconductivity, can be uniquely obtained by μSR measurements, because of their exquisite sensitivity to very weak magnetism (Luke *et al.*, 1998).

Resonant inelastic X-ray scattering. A spectroscopic technique that has developed over the last decade into a very powerful tool for studying excitations in condensed matter systems is resonant inelastic X-ray scattering (RIXS). This technique has become very important as the X-ray sources at the national facilities have been able to produce more and more photons over a wide spectrum of energies. At a typical light source, electrons are accelerated by a synchrotron and then injected into a storage ring, where they produce X-rays via Cherenkov radiation. The radiation, which is highly polarised, is then directed tangentially into various beamlines. These beamlines are designed for the experiments that are done at the ends. There are beamlines that originate in the straight sections of the ring and those that originate in the curved sections, and the characteristics of the beams are different, depending on where they originate from. The beamlines contain devices that can control the spectrum, the intensity, the beamline size and also the focus for focused beams. At the very end of the beamline is the experimental station, where the samples are placed and the scattering from the sample can be measured.

The advantage of using a synchrotron is the unprecedented range in energy and momenta it can explore and so it can look at almost all important excitations and especially those that are strongly correlated. Some of the unique features of this technique beyond its very large range of potential energies and momenta are that it can be made polarisation dependent, it is elemental and orbital specific, and it needs only very small samples.

RIXS uses incident photons that are resonant with one of the X-ray absorption edges of the material under exploration. This enhances the scattering rate, often by orders of magnitude. The RIXS process is a two-step process, as illustrated in Fig. 4.10.

The absorption of the source photon creates an excited intermediate state that has a core hole. From this excited state, a photon is emitted, and the final state is achieved. Simply speaking, the absorption process gives information about the unoccupied states, and the emission process provides information about the occupied states (see Fig. 4.10). In a typical experiment, information about the energy, momenta, polarisation, and intensity of the incident photon, combined with the same information from the emitted photon, can be used to fully characterise many of the key excitations in the material that is being investigated. Since both soft and hard X-ray sources can be configured to do RIXS, the range of excitations that can be studied is uniquely large. The only constraint is that the excitations should be overall charge neutral. However, despite this limitation, RIXS can probe the energy, momentum, and polarisation of any type of electron–hole excitation. This includes excitons in band metals, crystal-field excitations in strongly correlated materials, charge-density wave excitations, phonons, plasmons, orbital excitations, and magnetic excitations. Magnetic excitations are of some

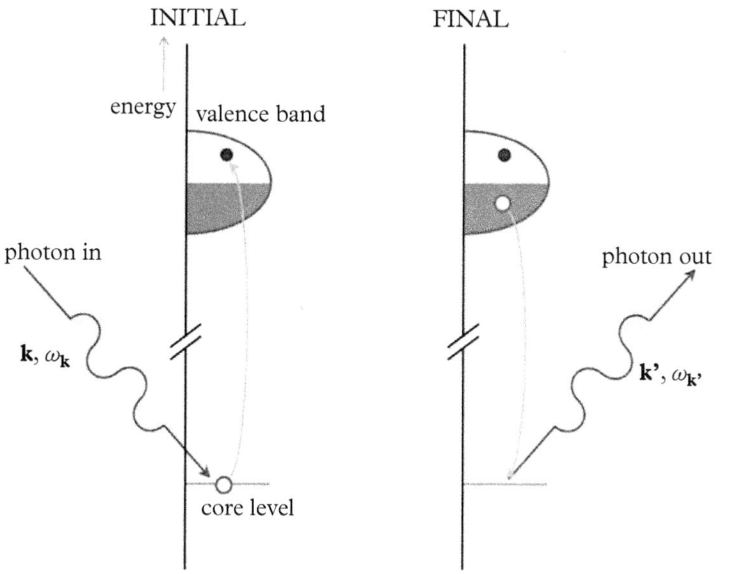

Figure 4.10 *X-rays from the source excite electrons from core levels into the valence band. An electron from an occupied site decays into the empty core level, and an X-ray is concomitantly emitted. This emitted photon has the energy and momentum of the valence excitation.*

particular interest and can be explored because photons can carry angular momentum (circularly polarised photons) that can be transferred to the electron's spin. It is important to note that, since spin is conserved, the polarisation of the beam can be used to distinguish magnetic excitation from other excitations. For a detailed review of this technique, see the article by Ament *et al.* (2011).

For this treatise, however, the excitations of most interest are superconducting excitations, and RIXS, again because of the control it enables over the frequency and the polarisation of the beam, can provide detailed information on the nature and symmetry of the pairing interaction in the superconducting state (Marra *et al.*, 2013). This is in addition to the fact that this technique is well suited to the study of the phonons that provide the pairing interaction in most superconductors, including the hydrides that are the highest T_c superconducting materials.

4.2 High-Pressure Techniques

The past several years have seen the emergence of high-pressure techniques for preparing superconducting materials that are not solids at ambient conditions. Although there are several techniques for putting materials under pressure, the one that has been the most successful in synthesising new superconducting compounds is the diamond anvil cell (DAC) technique (Eremets, 1996). This technique can be used to produce pressures

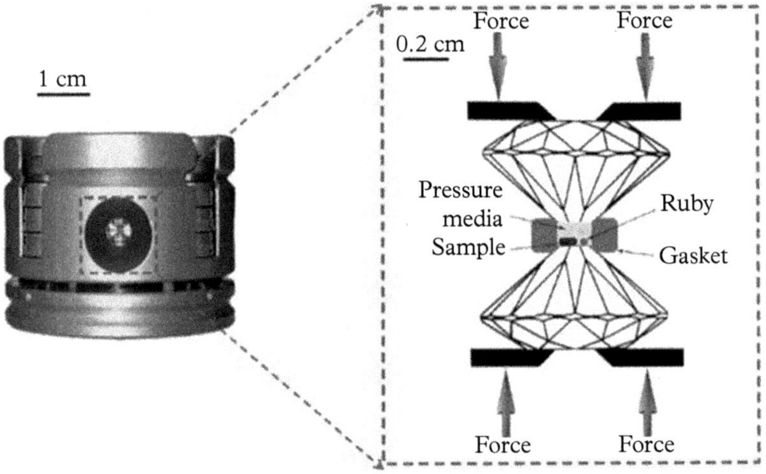

Figure 4.11 *A typical diamond anvil cell for very-high-pressure experiments to synthesise new superconducting phases, including the very-high-transition temperature hydrides.*

typically around 200–300 gigaPascals but, in principle, it can produce pressures greater than 700 gigaPascals or 7 million atmospheres! This technique can produce new phases of matter and has been used very successfully to produce new superconducting materials: H_3S at 203 K (Drozdov *et al.*, 2015) and LaH_{10} with a superconducting transition at 250 K (Somayazulu *et al.*, 2019).

In this technique, two high-quality, nearly identical diamonds are used to apply the pressure between their culet facets. The diamonds are mounted in a jig similar to that shown in Fig. 4.11.

The sample is placed between the polished culets of the diamond and confined by a gasket that prevents any material from being expelled out from between the diamonds. The gasket is usually a metal foil such as rhenium or tungsten. The gasket can also be a lighter material such as boron, boron nitride, or diamond itself. These lighter materials are useful when it is important to have X-rays pass through the gasket. The gasket is typically pre-indented by the diamonds and then a hole is drilled in the centre to form the sample chamber. Also inside the sample chamber is a small pressure-sensing material, often a ruby chip, whose fluorescence can be excited and observed through the diamond, as the diamond is transparent to the exciting as well as the fluorescent light. Although the pressure applied by the diamonds is uniaxial, this may be transformed to a uniform hydrostatic pressure by the use of a pressure-transforming medium that is also placed within the DAC. This medium can be a rare gas, a liquid-like methanol and ethanol, and, for the case of the hydrides, hydrogen gas. The diamond is transparent and so the material inside the gasket can be observed optically and also with X-rays. This allows a host of measurements to be made on the sample, including X-ray diffraction, optical absorption, Raman scattering, Brillouin scattering, and so on.

RF and microwave fields as well as magnetic fields can also be applied to the sample to allow measurements of nuclear magnetic resonance, electron paramagnetic resonance, and other magnetic measurements. Electrical leads can also be passed through the gasket, allowing measurements of resistance, the Hall effect, and other electromagnetic properties to be made. Lasers can illuminate the sample for measurements or for heating, whereby several thousand degrees kelvin can be obtained. This can be used to form new compounds or to study the materials in the DAC at high temperatures. The DAC can be mounted in a cryostat and cooled down to temperatures as low as a few millikelvins.

4.3 Preparation of Superconducting Thin Films and Tunnel Junctions

One of the most important techniques for both discovering and characterising superconductors is preparing the materials in the form of thin films, not only for characterising the fundamental properties of the superconductors but also for the preparation of tunnel junctions, particularly Josephson tunnel junctions, as the formation of tunnel junctions is one of the major applications of superconductivity. There are many ways to prepare thin films of superconducting materials; these fall into the commonly used categories of physical vapour deposition (PVD) and chemical vapour deposition (CVD). PVD methods that will be covered in this chapter include several sputtering techniques, ion beam deposition (IBD), pulsed laser deposition (PLD), and molecular beam evaporation (MBE). There are some other film techniques that are mostly used for thicker films; these include electrodeposition and some ceramic techniques like doctor blading and spin coating, in which a green slurry of the materials plus additives is used to lubricate and deflocculate the required constituents and then the materials are sintered into a continuous film. The sintering process evaporates all of the additives, leaving the compound of interest. This chapter will focus on the PVD and CVD techniques that are used for the majority of superconducting research and technology, especially Josephson junction technology. This chapter will not describe the electrodeposition or ceramic processing techniques, which are not in widespread use.

4.3.1 PVD Thin-Film Techniques

Sputtering. The most common PVD technique for preparing superconducting films is the sputtering process (see Fig. 4.12).

The sputtering process is typically performed in a stainless-steel vacuum vessel that can be vacuum pumped to a pressure below 10^{-6} Torr and sometimes (especially if the target materials are particularly sensitive to exposure to air or water vapor) to below 10^{-8} Torr. Inside the chamber is the target (or targets) of the materials to be sputtered. The target(s) are isolated from the chamber and connected to either a DC or RF voltage source that can provide voltages ranging from a few hundred to a few thousand volts.

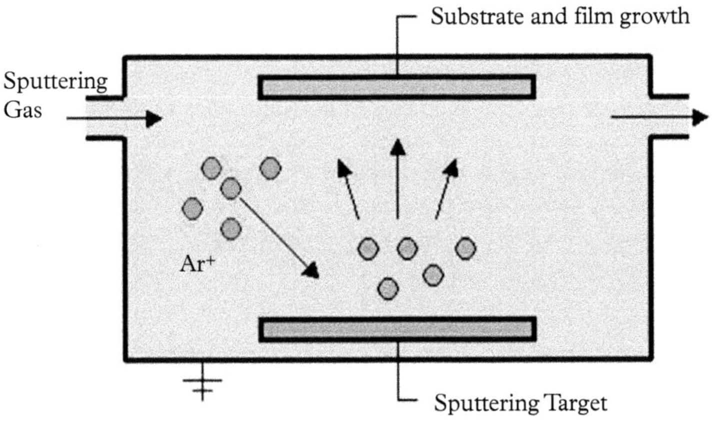

Figure 4.12 *A schematic of the sputtering process described in the text.*

In many of the sputtering systems that are used today, a strong magnet is placed behind the target. For example, if the target is a disc, the magnet is a ring. A flow of ultra-pure argon (or, in some cases, another noble gas) is bled into the chamber to a pressure of a few Torr. The high voltage ionises the argon and forms a plasma around (but not in direct contact with) the target and accelerates the argon towards the target; the collisions with the target cause material from the target to be expelled from the surface of the target and collected on the substrate, which is usually an insulating thin plate of silicon, or some other material on which the film of the superconductor can condense. In those cases in which there is a magnet behind the target, the plasma becomes more concentrated by the magnetic field, and the rate at which material is expelled is significantly enhanced. This type of sputtering is called magnetron sputtering. In general, the substrate is heated to temperatures as high as 1000 °C and rotated (for uniformity) so that the proper crystal structure can form as the films are being deposited. Sometimes the films are post-annealed to form the correct structure. This technique usually is used to prepare films that are not highly textured (grains not all aligned) but, with the right substrate and the right deposition conditions (temperature and growth rate), highly textured films have been fabricated using this method. Typically, this method is used to prepare high-quality films of elementary superconductors and alloys. RF sputtering in an ultra-high-vacuum environment produced the best quality films of niobium more than 40 years ago (see e.g. Wolf *et al.*, 1977).

Reactive ion sputtering. A variation on this straightforward sputtering process is called reactive ion sputtering. In this technique, a reactive gas like oxygen, nitrogen, or methane is slowly bled into the chamber along with the argon. These gases are also ionised and react strongly with the sputtered material from the target. Oxides, nitrides, and carbides can easily be formed using this reactive technique. Excellent films of NbN and NbCN have been fabricated this way (Francavilla *et al.*, 1987).

IBD. IBD is quite similar to sputtering but, instead of the argon being ionised by the voltage on the target, an ion gun placed opposite to the target provides the energetic ions or atoms. Ion guns are readily available, and the gun consists of an ionising filament and ion optics that are a series of plates at different voltages that accelerate the ions to the required energy (voltage), together with a neutralising filament that provides a source of electrons to neutralise the ions. The beam is directed towards the target. Both the energy of the beam and the beam intensity can be controlled independently (in conventional sputtering, they are closely coupled). Like sputtering, the collision of the argon atoms with the target causes target material to be expelled and collected on the substrate. The substrate can be heated and then rotated to ensure uniform films are formed with the right crystal structure. The benefit of this approach is that it yields better control of the sputtering process. It is difficult to model the dynamics of the plasma that is crucial for conventional sputtering, but an ion gun can be operated in a lower chamber vacuum, and the interaction of the atomic beam with the target can be modelled more accurately. A more advanced ion-beam technique uses a low-voltage ion gun (up to 100 V of ion acceleration) and biases the target to high voltage (300 V to 2000 V). The advantages of biasing the target are that only the target can be sputtered and the beam hits the target orthogonal to the target surface. This type of IBD offers even more control over the dynamics of the sputtering process than either conventional sputtering or conventional IBD. This specialised control is enabled by pulsing the target voltage. The pulses can be customised with respect to duration and, in fact, polarity, so that any charge formed on the target can be neutralised by the subsequent pulse or pulses. In fact, this biased-target ion-beam process maximises the sputter yield and provides a very uniform distribution of atoms or molecules hitting the substrate. This technique has recently produced some of the best quality Josephson junctions (see Cyberey *et al.*, 2019, and references therein).

PLD. An important technique that has emerged in the last few decades and has been very successful in preparing films of the high-transition-temperature cuprates is PLD. PLD has many similarities to IBD but some very significant differences. The substrate, as in sputtering, is inside a high-vacuum stainless chamber with a window that is transparent to the high-energy pulsed-laser frequency used for the deposition. There is a target (or targets) also in the chamber, in the path of the light from the laser (see Fig. 4.13). The laser typically is a high-power excimer pulsed laser that can produce nanosecond or shorter pulses at a pulse energy in the range of 0.5 to 10 J. A typical PLD laser is a KrF excimer laser with a wavelength of 235 nm. These lasers have a uniform energy profile that is very suitable for laser deposition. When the laser beam hits the target, it excites a plasma, since the energy of the ablated particles are energetic enough to be ionised. The ablated material consists of ions, atoms, molecules, and electrons and, if the geometry is correct, many of them strike the typically heated and rotating substrate where they are deposited. Depending on the detailed deposition parameters and substrate, the films that are grown can be amorphous, untextured polycrystalline, textured polycrystalline, or biaxially textured or single-crystal epitaxial. One of the first applications for PLD was to prepare films of the cuprate superconductors; this process is described in detail in the work by Venkatesan *et al.* (1988).

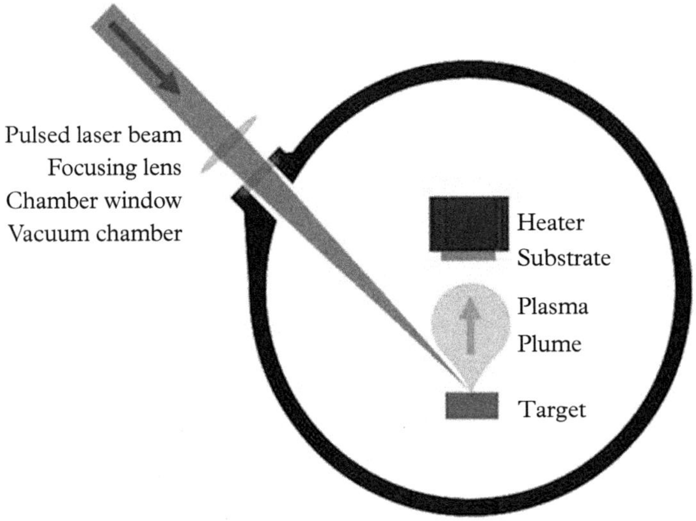

Figure 4.13 *Simple schematic of the pulsed laser deposition process.*

MBE. MBE is a deposition technique that is used to prepare very high-quality epitaxial films of many alloy and compound superconductors, as well as being a key deposition technique for high-quality semiconducting films. A schematic of such is deposition tool is shown in Fig. 4.14. This technique uses an ultra-high-vacuum environment that is created in a stainless-steel container that is pumped by a combination of pumps (diffusion pumps, turbo-pumps, ion pumps, cryo panels and cryo-pumps) that can achieve vacuums better than 10^{-9} Torr and usually one to two orders of magnitude below that (as low as 10^{-11} Torr). The residual pressure is measured by an ionisation gauge that is sensitive to 10^{-11} Torr. The chamber must be baked to get rid of residual water vapour and have a mass spectrometer to check for any residual contaminants. The materials that are to be deposited are placed in effusion cells that can be heated to temperatures sufficient to evaporate the material placed inside. In general, the effusion cells are customised for the particular material or class of material that is to be deposited. A beam flux monitor is used to measure the flux emanating from each cell. This is used to determine the rate from each cell and thus the correct proportions for each alloy or compound that is to be deposited. The substrate is placed on a holder that can be both rotated and heated. The layer-by-layer growth of the epitaxial film is monitored by reflection high-energy electron diffraction (RHEED), in which an electron gun is used to send a beam of electrons towards the substrate, which are then reflected onto a phosphorescent screen. The pattern that appears on the screen indicates the quality and type of growth that is occurring on the substrate. This growth technique has produced very high-quality films of most of the superconducting alloys and compounds that are of technological interest. Excellent examples of such film growth can be found in the publications by Bozovic *et al.* (1992) and Schlom (2015).

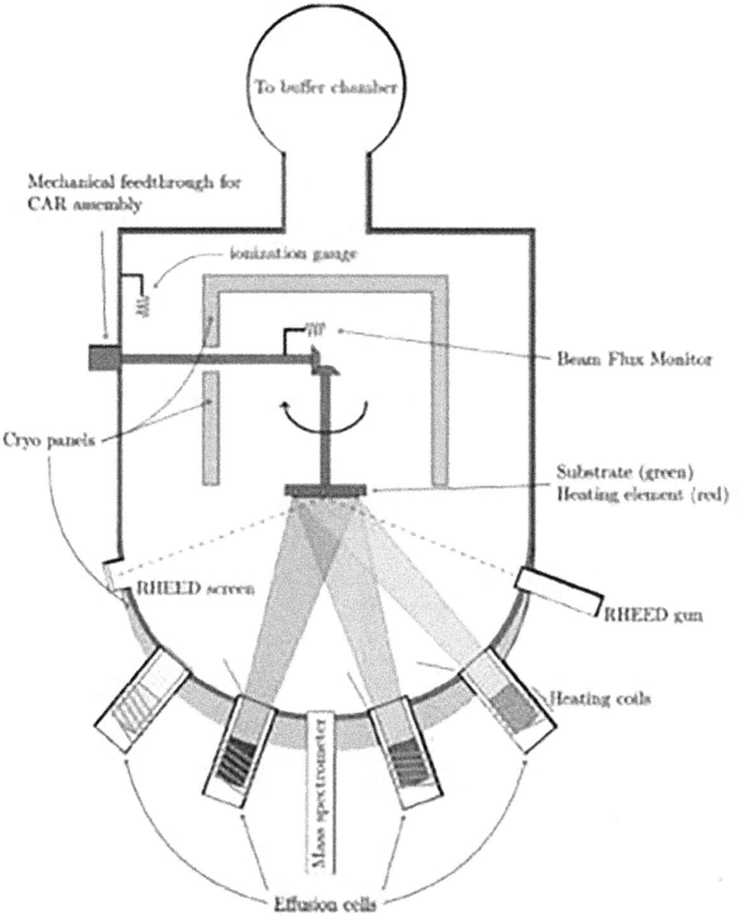

Figure 4.14 *A schematic of a typical molecular beam epitaxy system as described in the text.*

4.3.2 CVD Thin-Film Preparation

CVD is a technique used to prepare very high-quality films of many of the most significant technological materials, particularly for the semiconductor industry. It is the primary method for producing III–V compounds like GaAs, InAs, and GaN. It also can produce high-quality superconducting films, primarily the more complicated ones that require chemical reactions to form the superconducting compounds.

There are many formats for CVD growth that use different types of growth chambers to produce the films but two of the more common ones are shown in Figs 4.15 and 4.16.

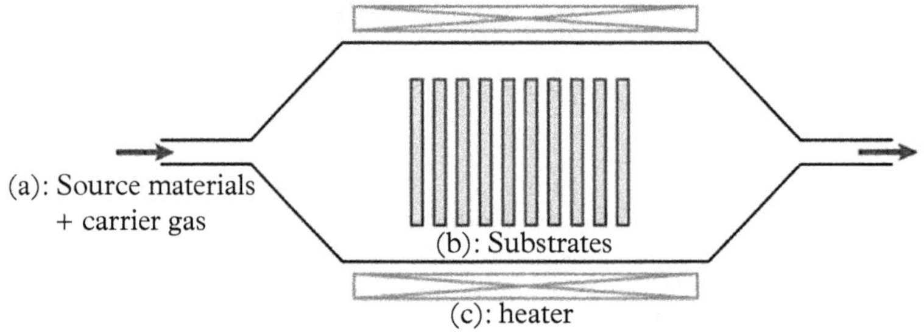

Figure 4.15 *Schematic of a simple hot-wall chemical vapour deposition process. The wall of the chamber is heated, and the substrates are heated by the wall.*

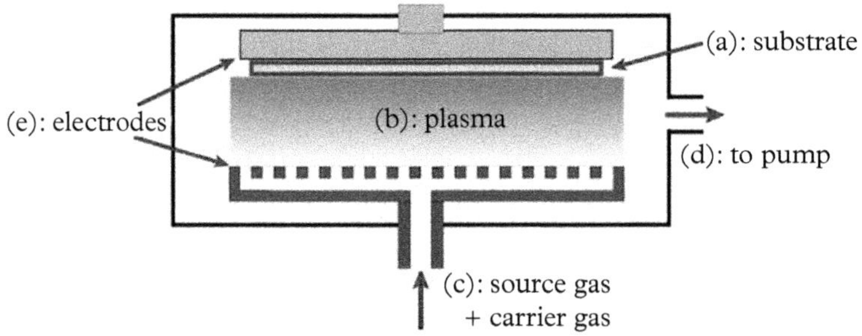

Figure 4.16 *A simple schematic of a plasma-enhanced chemical vapour deposition chamber that uses a plasma to enhance the deposition process so that the acceleration reaction rates allow for lower deposition temperatures.*

The CVD process involves the flow of one or more volatile organic precursors containing the element or elements of interest. The volatile precursors along with a carrier gas are vapourised (using a variety of techniques to form an aerosol) and then are passed through the CVD chamber in which they flow over and around the substrates. Because the substrates are heated, the gases either decompose or react on the substrate and form the films of interest. The volatile by-products of the decomposition or reaction are carried away by the constant flow through the chamber. There are many flavours of CVD, depending on, firstly, whether the substrates are radiatively heated by the walls of the chamber or directly heated by internal heaters; secondly, by how the vapour is formed, whether ultrasonically or by using an injector, like a fuel injector in an automobile; and, thirdly, by whether the process is enhanced by a plasma assist or not. There are other variations, including one method that uses a hot filament to help decompose the precursors and another that uses a laser to heat the substrates; the latter method called laser-enhanced CVD.

Examples of high-quality superconducting films produced by CVD can be found in the publications by Xu (2015) and Takano (2009).

4.4 Preparation of Josephson Tunnel Junctions

Josephson junctions are, in many respects, similar to the quasiparticle tunnel junctions described in chapter 1. The simplest manifestation of a Josephson junction consists of a cross pattern consisting of two superconducting films separated by a very thin oxide that forms a junction very similar to what is illustrated in Fig. 4.1. The main difference is that the current is a supercurrent of Cooper pairs rather than a quasiparticle current as for the junctions described in chapter 1. Because of the current and voltage phase relations due to the Josephson equations described in chapter 3, Josephson junctions have become extremely important technologically. Single Josephson junction devices or small numbers are typically prepared by sputtering and, by virtue of the Josephson current phase relations in a magnetic field, are extremely sensitive detectors of magnetic fields. These devices are called superconducting quantum interference devices, or SQUIDs (Jaklevic *et al.*, 1964). A DC SQUID consists of a superconducting ring with two Josephson junctions in the circumference. The flux within the ring is quantised in units of a fundamental constant called the flux quanta, Φ_0, which is exactly h/2e.

A schematic of a DC SQUID is shown in Fig. 4.17, along with the voltage versus flux that is enclosed in the SQUID ring. The right-hand side of this figure shows that the voltage across the SQUID, when the bias current just exceeds the critical current of the ring, is periodic in the flux and thus can be made extremely sensitive to small changes in the applied magnetic flux or, depending on the area of the ring, the applied magnetic field. State-of-the-art DC SQUIDS have sensitivities of femtotesla per square root hertz. That makes them among the most sensitive magnetometers. Josephson junctions have also become the voltage standard that defines the volt! The resonance between an applied ac current and the current produced by the ac Josephson effect causes very precise steps in the current–voltage characteristic of a Josephson junction, similar to that shown in Fig. 4.2. These steps are called Shapiro steps and are the basis for using Josephson junctions as a voltage standard. Because Josephson junctions can be used as switches analogous to transistors, they are being developed as an alternative to semiconductor technology. What has emerged is a superconductor

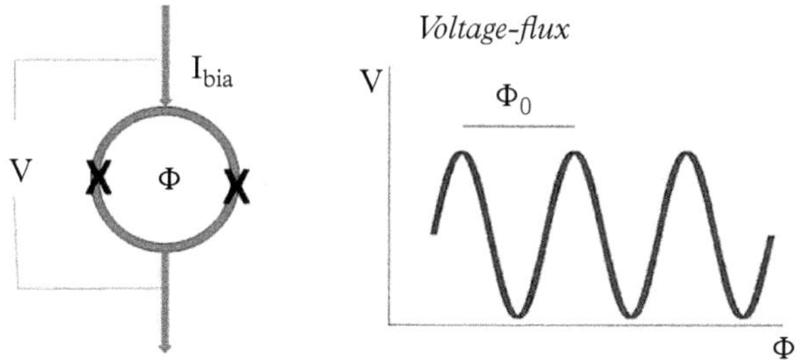

Figure 4.17 *The left side of this figure shows a superconducting ring with two Josephson junctions on the circumference. A dc bias current flows through the ring just above the critical current of the ring at zero flux. The right side of the figure shows the voltage across the ring as the applied flux is increased. The period of the voltage oscillations is Φ_0; the superconducting flux quantum is h/2e.*

fabrication technology very analogous to semiconductor technology that produced billions of transistors on a small chip. Already there are superconducting fabs that use similar processes to produce Josephson junction chips with hundreds of thousands of Josephson junctions on a chip on the scale of centimetres. Currently there are two major Josephson junction materials in use in these fabs. For 'conventional' superconducting circuits for high-performance classical applications and computing (see Kadin, 1999) the materials of choice are Nb as the superconductor and aluminium oxide as the insulating barrier. Superconductivity is a macroscopic quantum phenomenon, and there are unique aspects of a Josephson junction that make it very suitable for being a quantum bit, or qubit. Because magnetic flux is quantised in a magnetic ring, as described above, a superconducting ring with one or two Josephson junctions around the circumference, similar to the DC SQUID, becomes a nearly perfect qubit, and a quantum technology has grown around this fact (see Devoret, 2013).

4.5 Novel Experimental Techniques

4.5.1 The Nano-Assembly Technique and the Little Mechanism

Advances in nanotechnology over the last few decades enabled some remarkable discoveries to be made, using the ability to control the position of individual atoms, molecules, and nanostructures (see e.g. Zeppenfeld, Lutz, and Eigler, 1993). Particularly, the ability to assemble individual carbon nanotubes has made possible the demonstration of an attractive interaction between two electrons that is mediated by a third electron that is physically separated but coupled through an unusual electron–electron interaction that was first postulated by William Little and has consequently been labelled the Little mechanism (Little, 1964).

The experiment to demonstrate the attractive interaction between electrons was a tour-de-force of nano-assembly and nano-manipulation (Hamo *et al.*, 2016). Two single nanotubes were individually suspended between contacts and above an array of gates, each on their own microchip. One of the nanotubes , the 'system', had three contacts; the other nanotube, the 'polariser', had two contacts. The two microchips were mounted in a specially built scanning-probe microscope inside a dilution refrigerator. For each nanotube, the gates could be biased to produce a double quantum-well potential that could trap single electrons. The system could also trap electrons, or a single electron, in a well located between the second and third contacts; this region was used as a charge detector. The double-well potential was designed so that an electron could hop between the wells but avoid the contacts.

The two chips were mounted opposite to each other with the nanotubes facing but perpendicular to each other. This configuration is shown in Fig. 4.18.

This configuration within the probe allowed precise control of the distance and coupling between the polariser and the system; hence, coupling was obtained between the electrons trapped in the system and the polariser.

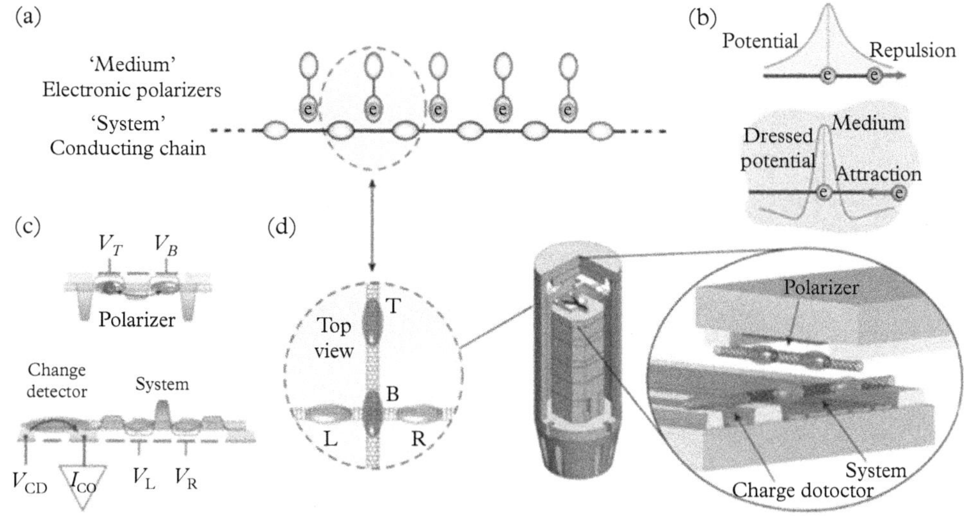

Figure 4.18 *(a) Schematic of the Little mechanism, in which a chain of electrons is coupled to physically isolated but electrically connected electrons that can act to polarise the electrons in the chain. (b) Illustration of a repulsive potential and an attractive potential where the there is a sign change in the potential. (c) Schematic of the two microchips: the upper one is the 'polariser', which has two contacts and is gated to provide a double quantum-well potential, and the lower one is the 'system', where there is a double well for the system, and a single well as the charge detector. (d) The left part of the figure is a schematic top view of the cross nanotubes, with the polariser on top and the system on the bottom. The middle is a schematic of how the experiment is mounted on the scanning nanoprobe, and the right side is an expanded view of the experimental setup, showing the two chips mounted facing each other with the nanotubes at right angles (Hamo et al., 2016).*

To perform the experiment, stability diagrams for electrons in each of the quantum wells were determined, as well as the condition required to populate and depopulate the quantum wells with single electrons. After a very detailed set of experiments and analysis, the authors were able to show very definitively that, depending on the position and coupling between an electron in the polariser and the electron(s) in the system, the interaction between electrons can be made attractive via the interaction of the electron in the system with an electron in the polariser whereby the electron in the polariser dresses the charge of the electron in the system. The conditions for flipping the sign of the interaction from repulsive to attractive are rather delicate, and there are certainly conditions under which the electron is overdressed so that the interaction is neither repulsive nor attractive.

This experiment demonstrated that the Little mechanism can be realised experimentally. However, it did not demonstrate a superconducting state in the nanotube. There is therefore still more work to be done to find a Little superconductor!

5

Materials I: High-T_c Copper Oxides

5.1 Introductory Remarks

The previous chapters of this book focused on the various mechanisms of superconductivity and spectroscopy, allowing the properties of the superconducting state to be studied. This chapter, as well as chapters 6–9 describe various materials. Of course, it is an unrealistic task to analyse the properties of all existing systems. The goal is to convey that the world of superconductivity is very rich. A number of materials with various remarkable manifestations of the phenomenon of superconductivity are described.

First, the properties of high-T_c oxides (cuprates) are described. The discovery of this new family was a real breakthrough, and this area of research still continues to pose real challenges. As is well known, it took almost fifty years to create the BCS theory, which explains the phenomenon of superconductivity. In some sense, history repeats itself. Cuprates were discovered approximately 35 years ago, in 1986, but the problem of the origin of the superconducting state in these materials is still controversial. The state of the art in this area, which is on frontiers in the field, is discussed in this chapter.

The milestone discovery of a new family of superconducting materials was made by Bednorz and Mueller in 1986 at IBM (Zurich, Switzerland). The search for new superconductors was guided by the experience of Alex Mueller, who is a leading expert in the area of ferroelectricity. The search focused on specific oxides because of their strong polarisability and polaronic effects. The synthesis of metallic ceramics (Michel and Raveau, 1984; see also Raveau *et al.*, 1991) and the theoretical study of the Jahn–Teller and polaronic effects (Hock *et al.*, 1983) played important roles in the search for new superconductors.

Prior to this discovery, superconductivity was a low-temperature phenomenon (see section 1.1 and Fig. 1.2). The search for new materials with higher T_c was mainly performed with Nb-based compounds, and the highest T_c, $T_c \simeq 22.3$ K was achieved for Nb–Ge. The discovery made by Bednorz and Mueller not only drastically increased the value of T_c but eventually resulted in an entirely different energy scale of the phenomenon of superconductivity. It also introduced a new class of materials: high-T_c copper oxides.

Superconducting State: Mechanisms and Materials. Vladimir Z. Kresin, Sergei G. Ovchinnikov, and Stuart A. Wolf, Oxford University Press (2021). © Kresin, Ovchinnikov, Wolf. DOI: 10.1093/oso/9780198845331.003.0005

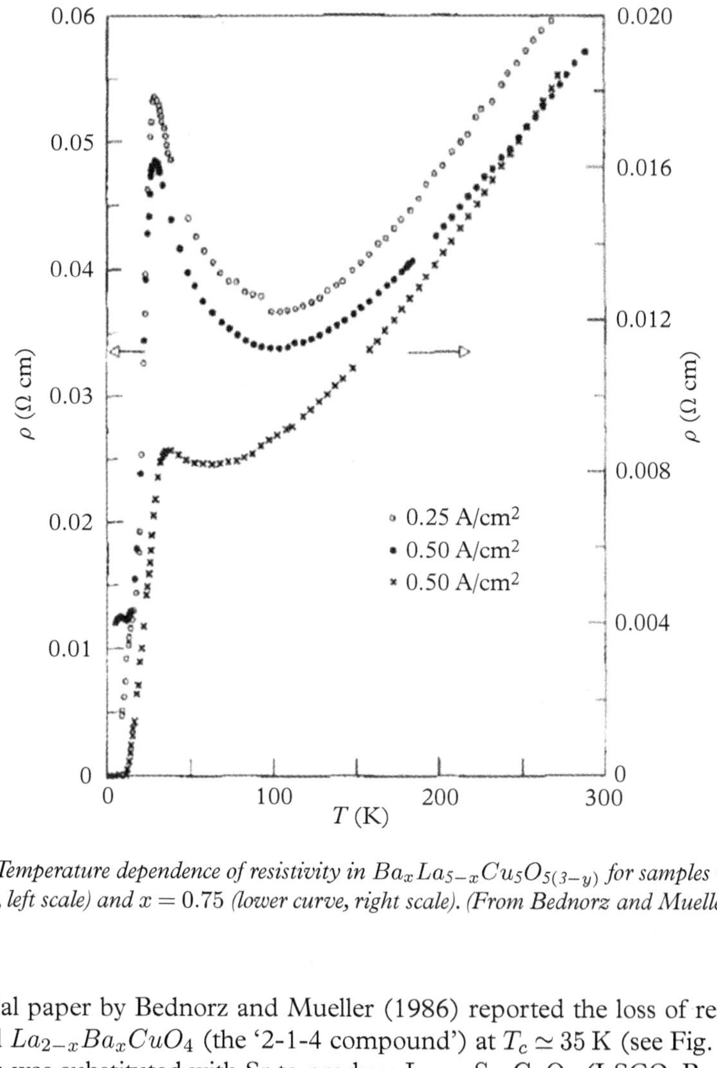

Figure 5.1 *Temperature dependence of resistivity in* $Ba_xLa_{5-x}Cu_5O_{5(3-y)}$ *for samples with* $x = 1$ *(upper curves, left scale) and* $x = 0.75$ *(lower curve, right scale). (From Bednorz and Mueller, 1986)*

The initial paper by Bednorz and Mueller (1986) reported the loss of resistance by the material $La_{2-x}Ba_xCuO_4$ (the '2-1-4 compound') at $T_c \simeq 35$ K (see Fig. 5.1).

Later, Ba was substituted with Sr to produce $\mathrm{La_{2-x}Sr_xCuO_4}$ (LSCO; Bednorz *et al.*, 1987), and the highest T_c for this class of cuprates was for $La_{1.85}Sr_{0.15}CuO_4$ ($\simeq 40$ K). The paper published in *Zeitschrift für Physik* (October, 1986) has the rather modest title 'Possible high T_c superconductivity in the Ba-La-Cu-O system'. Shortly after this publication, the Meissner effect was confirmed (Bednorz *et al.*, 1987a, b); Takagi *et al.*, 1987). Therefore, superconductivity in the family of copper oxides appears to be a real phenomenon, since the oxides display a loss of resistance along with the Meissner effect.

Bednorz and Mueller were awarded the Nobel Prize in 1987. There has never been such a short interval between a discovery and the award. The discovery resulted in a flurry of scientific activity: about 7,000 (!) articles regarding superconductivity were published during the next three years.

In this chapter, first, the main properties of cuprates will be described. Many of these properties are very different from those for usual superconductors, which were studied before 1986. Afterwards, the theoretical studies are discussed. It must be mentioned that the theory and, especially, the mechanism of high-T_c superconductivity is still controversial, but the main directions are presented below (see section 5.4).

5.2 Properties of the Normal State

5.2.1 Structure

All cuprates contain low-dimensional structural units (e.g. planes and chains). The key unit is the $Cu - O$ plain, where the pairing of carriers occurs. For example, the cuprate LSCO has the structure plotted in Fig. 5.2.

The structure contains $Cu - O$ planes and also the La layers, where some of the La ions are replaced by Sr ions. The layer is called a 'reservoir', since it provides the carriers which are transferred into the $Cu - O$ layer.

Early in 1987, Paul Chu and his collaborators (Wu *et al.*, 1987) discovered the new cuprate $YBa_2Cu_3O_{7-\delta}$ (YBCO; also known as abbreviated as '123'). Its value of $T_c \simeq 93$ K (!) was above the liquid nitrogen barrier of 77 K (Fig. 5.3). This breakthrough was very significant; it moved the phenomenon of superconductivity to a different scale. Moreover, this material appears to be most desirable among the cuprates for applications. YBCO has the structure plotted in Fig. 5.3. It also contains the CuO planes but, unlike LSCO, it has, as the reservoir layer, $Cu - O$ chains.

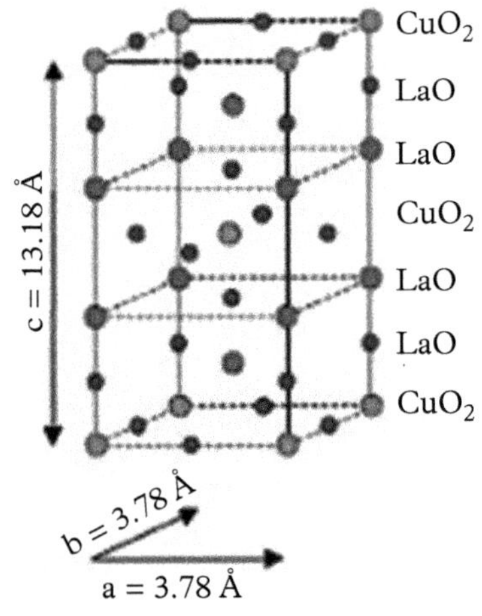

Figure 5.2 *Crystal structure of the $La_{2-x}Sr_xCuO_4$ compound (From Narlikar, 2014).*

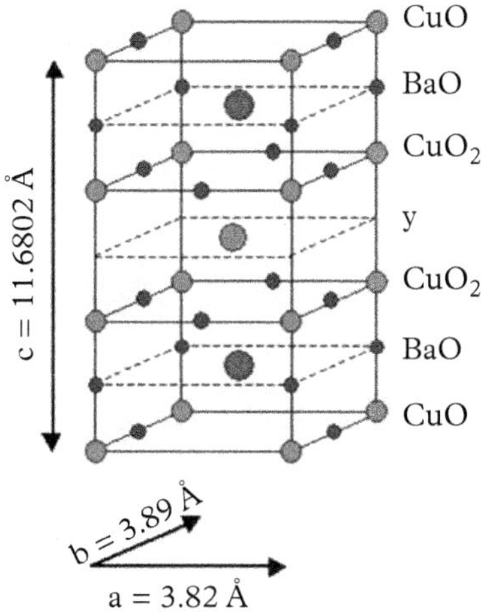

Figure 5.3 *The crystal structure of $YBa_2Cu_3O_7$, showing the copper–oxygen planes and the chain layer.*

The doping occurs chemically, by adding oxygen into $Cu - O$ chains.The subsequent charge transfer between the Cu ions on the plane and the added oxygen leads to an appearance of holes in the $Cu - O$ plane. The charge transfer between the plane and the chain occurs through an apical oxygen ion located between these structural units.

Afterwards, several other families of high-T_c cuprates were discovered. Among them $Bi_2Sr_2Ca_{n-1}Cu_nO_{2n+1}$ (Maeda *et al.*, 1988), $Tl_2Ba_2Ca_{n-1}Cu_nO_{2n+4}$ (Sheng and Hermann, 1988), and $HgBa_2Ca_{n-1}Cu_nO_{2n+3-\delta}$ (Schilling *et al.*, 1993). All of them can display the values of T_c above 100 K. It is interesting that an applied pressure can noticeably increase the value of T_c. One can see in Fig. 5.4 the pressure dependence of T_c for an Hg-based cuprate (Gao *et al.*, 1994). The value of $T_c = 164$ K at $P = 30$ GPa is the highest observed for the cuprates. At present, the record values of the critical temperature have been observed for hydrides under high pressure (see section 7.2). As for the ambient pressure, the highest T_c continue to be shown by the family of copper oxides.

5.2.2 Doping and the Phase Diagram

The cuprates are doped compounds. The parent material is an insulator and becomes metallic as a result of doping. The situation has some similarity with that in semiconductors. The level of doping could be different. As a consequence, the same is true for the carrier concentration. The most well-known method is so-called chemical doping. For example, the parent compound La_2CuO_4 is an insulator. However, the substitution $La \to Sr$ leads to an appearance of carriers (in this case, holes) and, therefore, to transition into metallic state, so that we are dealing with the doped compound LSCO. The change in x leads to a peculiar phase diagram (see below). The oxygen deviation from the

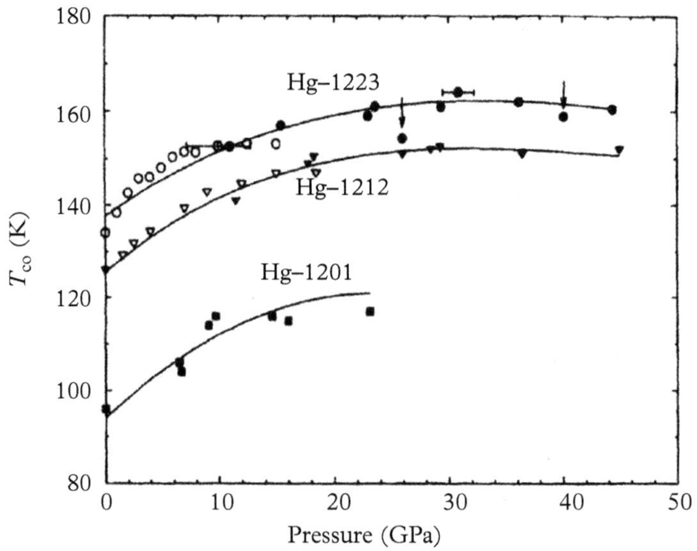

Figure 5.4 *Pressure dependence of the* T_c *of Hg1201,1212, and 1223. (From Gao et al., 1994)*

stoichiometric composition is another way of doping. For example, in $La_{2-x}Sr_xCuO_{4+y}$ the excess oxygen results in electron doping. This way of doping is especially used in the YBCO family of cuprates. The 'field effect' caused by an external electric field can lead to the formation of carriers near one surface and is the other method of doping (Mannhart *et al.*, 1991; Chu, 2002; Ahn *et al.*, 2003).

The left side of the phase diagram shows the behaviour of the electron-doped cuprates. Here the highest value of T_c is rather moderate ($T_c = 25\,\mathrm{K}$ for $NbCeCuO$). These materials play the role of a bridge between the usual superconductors and the high-T_c materials. As a whole, the temperature interval for the T_c of cuprates is broad, on the order of 100 K (from $T_c = 25$ K up to $T_c \simeq 130$ K; off-stoichiometric compounds have a T_c as low as several degrees).

Speaking of the spectroscopy of the cuprates in their normal state, one should, as usual, discuss one-particle and collective excitations (phonons, plasmons, magnons, etc.). The presence of a layered structure affects the spectra of all these excitations. Some of the peculiarities are described in section 1.2. Of course, the band structure and the dispersion laws for the excitations are essential for the optical, transport, magnetic, and so on, properties of the materials. They are of key importance for the pairing and its mechanism. Indeed, as we know, the collective excitations can form the 'glue' binding the Cooper pairs. Because of such an intimate connection between spectroscopy and the mechanisms of superconductivity, the main focus of this book, the discussion of the spectra of the cuprates will be presented below (section 5.4) in combination with the description of the state of the art of the problem of the mechanism of superconductivity in the cuprates. This topic is still controversial (see section 5.4.1).

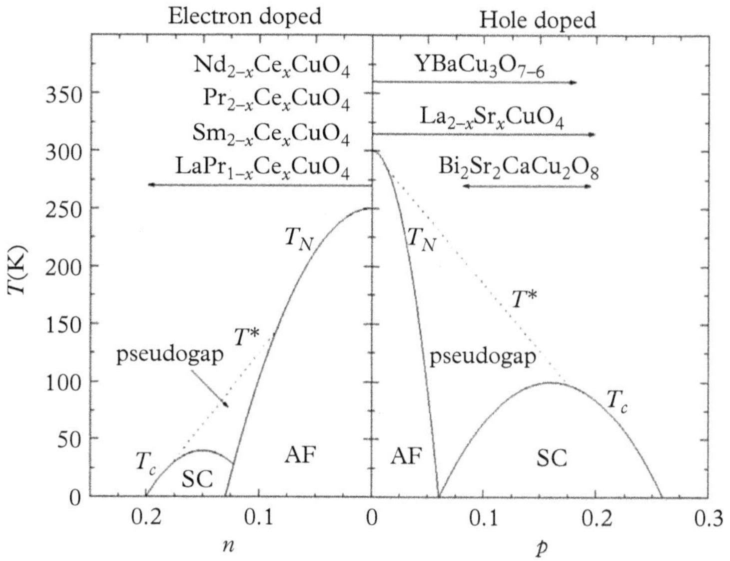

Figure 5.5 *The doping phase diagram for most of the cuprates, showing the antiferromagnetic region, the superconducting region, and the pseudogap region.*

All families of the cuprates are characterised by a similar phase diagram (Fig. 5.5). The parent compound is usually an antiferromagnetic insulator.

Upon doping, the Néel temperature decreases, and the sample makes a transition into the conducting state and then becomes superconducting. Its T_c increases with continuous doping until it reaches the maximum value. The normal state above T_c for the region $T_c < T_{c;max}$ (the so-called underdoped region) has a number of peculiar properties. This so-called pseudogap state will be described in more detail in sections 5.4.5 and 5.4.6 and in chapter 6. As for the region with decreasing T_c (the overdoped region), it represents the usual metallic state.

There are many two-dimensional (or quasi-two-dimensional) metals, but cuprates are very special materials (see the reviews by Mazin, 1989; Pickett, 1989; Dagotto, 1994; Kampf, 1994; Brenig, 1995; Scalapino, 1995; Shen and Dessau, 1995; Loktev, 1996; Ovchinnikov, 1997; Norman, 2014; Kordyuk, 2015). The most peculiar feature of the cuprates is the existence of the aforementioned ‘pseudogap’ state, which has been observed in the underdoped region. This state was observed in 1989, that is, shortly after the discovery of the high-T_c cuprates.

More specifically, the density of states appears to be drastically decreased (energy gap) near the Fermi level. This state has been dubbed the ‘pseudogap state’. Later, a number of other peculiar properties has been observed. The nature of the ‘pseudogap’ state is still a controversial issue; the same is true for the problem of the origin of high-T_c superconductivity (section 5.4). Probably, these two issues are interconnected.

We will discuss the state of art with the 'pseudogap' state in section 5.4.5, section 5.4.6, and chapter 6.

5.2.3 Nematicity

The term 'nematicity' comes from the physics of liquid crystals and means the violation of the rotational symmetry. An intriguing new electronic state, the electronic nematic (Kivelson *et al.*, 1998), has recently been revealed by a wide range of different techniques in various materials.

This spontaneous breaking of the rotational symmetry of the electron liquid concerning the crystal lattice has emerged as a seemingly ubiquitous characteristic of the normal state of many unconventional superconductors, having been observed in the cuprates (Ando *et al.*, 2002; Abdel-Jawad *et al.*, 2006; Daou *et al.*, 2010; Lawler *et al.*, 2010; Fujita *et al.*, 2014; Wu *et al.*, 2017, 2018), the ruthenates (Borzi *et al.*, 2007), Fe-based superconductors (Avci *et al.*, 2014; Fernandes *et al.*, 2014; Baek *et al.*, 2015; Watson *et al.*, 2015; Hosoi *et al.*, 2016; Licciardello *et al.*, 2019), heavy-fermion superconductors (Varma and Zhu, 2006; Okazaki *et al.*, 2011; Ronning *et al.*, 2017), and topological superconductors (Sun *et al.*, 2019).

The director vector (orientation of the nematicity direction) in some cuprates is not related to the crystal axes, as in tetragonal thin-film LSCO (Wu *et al.*, 2017, 2018), moreover, its orientation depends on temperature and doping; see Fig. 5.6, where the angle dependence of the transverse resistivity $\rho_T(\varphi)$ data is shown. These data are fitted by a simple expression $\rho_T(\varphi) = \rho_T^0 \sin[2(\varphi - \alpha)]$ with the two open parameters, the amplitude ρ_T^0, and the phase offset α. The longitudinal resistivity reveals very weak oscillations, just a few per cents, that appear to be random. For the orthorhombic substrate $NdGaO_3$, the thin-film LSCO has 1% orthorhombic distortion, and the angle α is pinned to one of the crystal axes at all temperatures. As in bulk LSCO single crystals with small (1.5%) orthorhombic distortion, the azimuth angle α between the director and the crystallographic [100] direction is equal to 45° (Ando *et al.*, 2002).

The anisotropy of magnetic and electronic properties in the CuO_2 planes (a violation of the fourfold C_4 rotational symmetry) has also been found also for $YBa_2Cu_3O_{7-y}$ (Cyr-Choiniere *et al.*, 2015; Sato *et al.*, 2017), $HgBa_2CuO_{4+y}$ (Murayama *et al.*, 2019), and $(Bi,Pb)_2Sr_2CaCu_2O_{8+y}$ (Ishida *et al.*, 2019). However, in YBCO, the fourfold rotational symmetry is already broken due to the orthorhombic crystal structure with one-dimensional CuO chains, and the violation of rotational symmetry is trivial. Thus, for tetragonal compounds like HgBaCuO and (Bi,Pb)2212, the formation of nematic order gives more information on specific electronic properties. For (Bi,Pb)2212 single crystals, the admixture of Pb is required to delete the superlattice modulation along the *b* axis, which is known for initial Bi2212 crystals.

As is the case for other hole-doped cuprates, the phase diagram of HgBaCuO contains the CDW, superconducting, and pseudogap regimes (Fig. 5.7). The short-range CDW order along [100]/[010], with a wavevector close to 0.28 reciprocal lattice units, forms a dome-shaped boundary inside the pseudogap regime (Tabis *et al.*, 2017). In zero magnetic field, CDW domains with a typical size of a few nanometres appear.

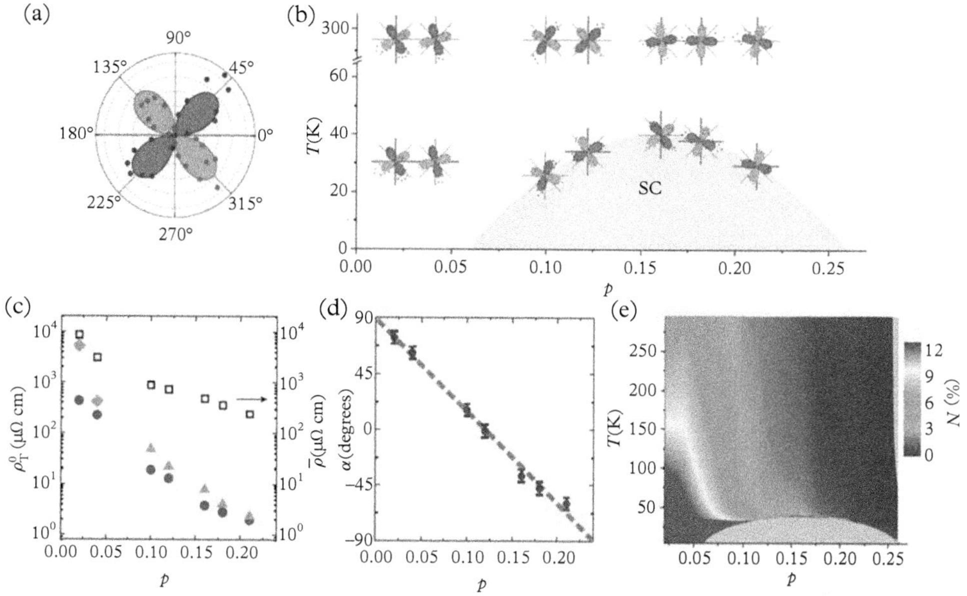

Figure 5.6 *Angle, temperature, and doping dependence of ρ_T. (a) An example of $\rho_T(\varphi)$ dependence in polar coordinates (for an LSCO film with $p = 0.12$ at $T = 295$ K). It resembles the 'cloverleaf'-shape characteristic of d-orbital symmetry. The radial distance measures the magnitude of ρ_T, with the positive values in blue and the negative in red. (b) The phase diagram, showing that the cloverleaf-shaped $\rho_T(\varphi)$ dependence is universal for the entire phase space investigated here; SC is the superconducting phase (for better visibility, the radial extents are all normalised to the same size). (c) The doping dependence of ρ (right scale) and ρ_T^0 (left scale): open black squares, longitudinal resistivity $\overline{\rho}(p)$ measured at $T = 295$ K; solid blue circles, $\rho_T^0(p)$ at $T = 295$ K; solid orange triangles, $\rho_T^0(p)$ at $T = T_c$ (midpoint); solid green diamonds, ρ_T^0 at $T = 30$ K is shown for samples with $p = 0.02$ and $p = 0.04$ that are non-superconducting. (d) The doping dependence of the offset angle α measured at $T = 295$ K. The angular resolution, and the upper limit on the error bar of α, is $\pm 5°$. € The doping and temperature dependence of the nematicity magnitude, $N = \rho_T^0/\overline{\rho}$. In the grey areas, the signal is below our noise floor. (From Wu et al., 2017)*

Measurements of the magnetic torque $\boldsymbol{\tau} = \mu_0 V \boldsymbol{M} \times \boldsymbol{H}$, where μ_0 is the permeability, V is the sample volume, and $\boldsymbol{M}$ is the magnetisation induced by external magnetic field $\boldsymbol{H}$, have revealed diagonal nematicity with B_{2g} symmetry, where the nematic director is oriented along the diagonal direction of the CuO_2 square lattice (Murayama *et al.*, 2019). For this nematicity, the components of the magnetic susceptibility tensors $\chi_{aa} = \chi_{bb}$ and $\chi_{ab} \neq 0$. The temperature dependence of nematicity is shown in Fig. 5.8.

In single crystals of $(Bi,Pb)_2Sr_2CaCu_2O_{8+y}$, where Pb was introduced to suppress the super-modulation along the *b* axes, the anisotropy of elastoresistance measurements has revealed the nematic order along the Cu-O-Cu bond (Ishida *et al.*, 2019). A similar bond order is known for the orthorhombic YBCO (Sato *et al.*, 2017), where the amplitude of the nematicity can be analysed in terms of excess nematicity defined

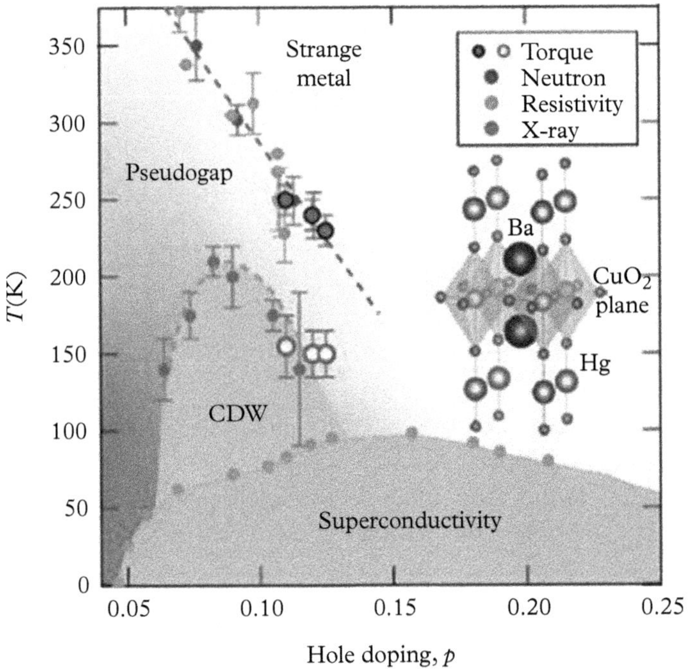

Figure 5.7 *Temperature-concentration phase diagram of Hg1201 cuprate. Orange filled circles represent T_c; the pseudogap temperature T^* is determined by neutron scattering (Li et al., 2011, blue filled circles,) and by resistivity measurements (Barisic et al., 2013, green filled circles). Purple filled circles show the charge density wave (CDW) onset temperature T_{CDW} determined by resonant X-ray diffraction (Tabis et al., 2017). Red filled circles represent the onset temperature of diagonal nematicity, which lies on T^* line (blue dashed line). Red open circles represent the temperature at which a suppression of the nematicity occurs, which is close to T_{CDW}. The inset shows the crystal structure of Hg1201. (From Murayama et al., 2019)*

as $\Delta\eta(T) = \eta(T) - \eta(T^*)$, where $\eta = (\chi_{aa} - \chi_{bb})/(\chi_{aa} + \chi_{bb})$. It has been shown that $\eta(T)$ is temperature independent above T^*. In a rather wide doping range, $\Delta\eta(T) \sim 5 \times 10^{-3}$ at $T \sim T^*/2$. A similar amplitude for the nematicity has been found for $(Bi,Pb)_2Sr_2CaCu_2O_{8+y}$ (Ishida *et al.*, 2019) and for Hg1201 (Murayama *et al.*, 2019), despite the different nematic directions.

Usually, nematicity is considered to be an electronic ordering below T^* that probably can explain the origin of the pseudogap state. It is interesting to analyse the overdoped region at the hole concentration p^* when $T^* \to 0$, where a quantum critical point is expected. At this point, the Lifshitz transition has been predicted (Ovchinnikov *et al.*, 2011b) that separates the normal Fermi-liquid regime for $p > p^*$ and the pseudogap state for $p < p^*$. Recently, the thermodynamic signatures of the quantum phase transition at p^* have been experimentally demonstrated by Michon *et al.* (2019). They measured the specific heat C of the cuprates $La_{1.8-x}Eu_{0.2}Sr_xCuO_4$ (Eu-LSCO) and

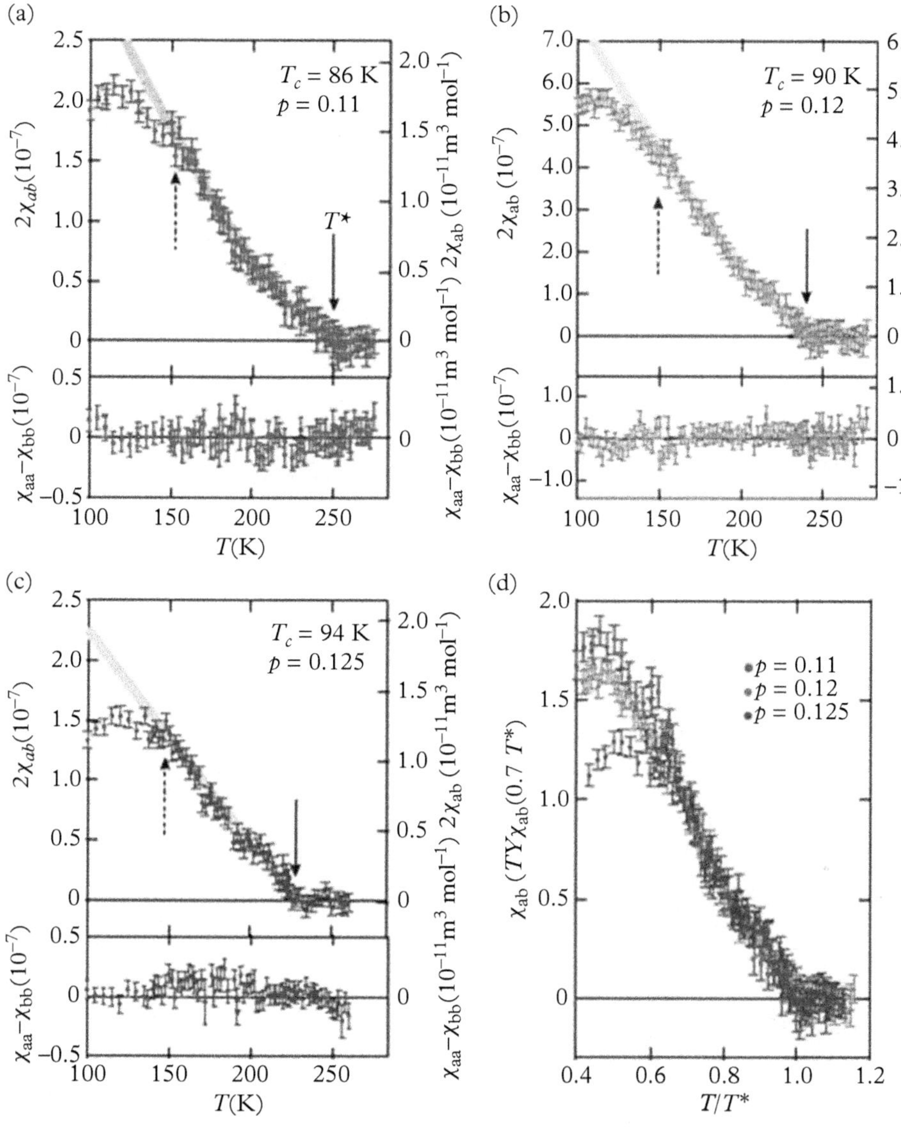

Figure 5.8 *Temperature dependences of $2\chi_{ab}$ (upper panel) and $\chi_{aa} - \chi_{bb}$ (lower panel) for hole concentrations (a) $p = 0.11$, (b) $p = 0.12$, and (c) $p = 0.125$. Solid arrows indicate the onset temperatures of χ_{ab}, which coincide well with T^*. Dashed arrows indicate the temperatures at which χ_{ab} deviates from the linear extrapolation from high temperature shown by the bold lines. These temperatures are close to T_{CDW}. (d) χ_{ab} normalised by the values at $T/T^* = 0.7$ for different doping levels plotted as a function of T/T^*. The data collapsed onto a universal curve, except for the low-temperature regime, where a suppression of χ_{ab} is observed, indicating a scaling behaviour. (From Murayama et al., 2019)*

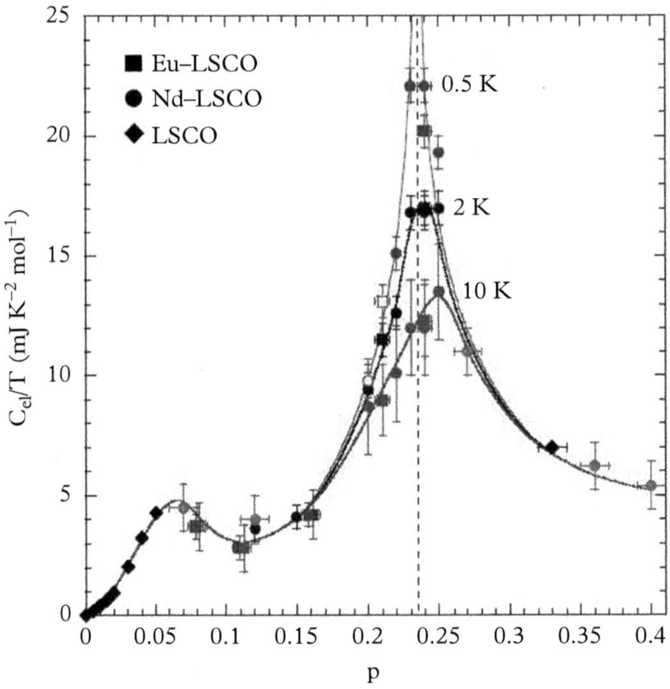

Figure 5.9 *Normal-state electronic specific heat of $La_{1.8-x}Eu_{0.2}Sr_xCuO_4$ (Eu-LSCO; squares) and $La_{1.6-x}Nd_{0.4}Sr_xCuO_4$ (Nd-LSCO; circles) at $T = 0.5$K (red), $T = 2$K (blue), and $T = 10$K (green). The vertical dashed line marks the pseudogap critical point p^* in Nd-LSCO. (From Michon et al. 2019)*

$La_{1.6-x}Nd_{0.4}Sr_xCuO_4$ (Nd-LSCO) at low temperature in a 15 Torr magnetic field large enough to suppress superconductivity, over a wide doping range across p^*. The two materials have the same crystal structure and phase diagram, with very similar boundaries for the pseudogap phase, $T^*(p)$, and the superconducting phase, $T_c(p)$. The electronic contribution to the specific heat C_{el}/T as a function of doping is strongly peaked at p^*, where it exhibits a log(1/T) dependence as $T \to 0$ (Fig. 5.9). This is strong evidence for the quantum critical point. Similar critical points were observed in the heavy-fermion superconductors (Löhneysen *et al.*, 1994) and the Fe-based superconductors (Walmsley *et al.*, 2013) at the points where their antiferromagnetic phase ends.

Summarising the discussions of nematicity, we can emphasise that it is a very interesting type of electronic order that has been found in many unconventional superconductors. Up to now, there is no clear understanding of this phenomena, and data are material dependent. Typically, it is considered that it is the order that may explain the pseudogap state in cuprates. Nevertheless, the epitaxial LSCO film data by Wu *et al.* (2017) provide the evidence that nematic order may exist beyond the pseudogap region; for example, the overdoped samples at 300 K are certainly above the T^*. Moreover, all the magnetic and transport measurements used to detect nematicity cannot be applied in the superconducting phase. So, the absence of their signal does not prove the absence of nematicity below T_c. Evidently, some other approaches should be used to check the possible electronic anisotropy in the superconducting phase. For example, in bulk crystals, the anisotropy of heat transport should be studied.

5.3 Properties of the Superconducting State

This section is concerned with the properties of the cuprates below T_c, in the superconducting state. Note that here we will be focusing on properties which are not directly related to the mechanism of pairing. The problem of the origin of high T_c in the cuprates will be discussed below, in section 5.4.6.

The cuprates are doped materials. Because of it, the value of the critical temperature, as well as values of other parameters (energy gap, critical field, etc.), depends on the level of doping (see Fig. 5.5). We will be focusing mainly on the region close to $T_{c;max}$, although some points related to other regions will be also stressed.

5.3.1 Coherence Length

A very important parameter is the coherence length ξ_0 defined by expression (2.13). For the usual superconductors, its value can be very large ($\sim 10^{-4}$ cm; see the discussion in section 2.1.6). As for the cuprates, the situation is entirely different: the coherence length is very short (on the order of 15–20 Å). This smallness is caused by a high value of T_c and a small value for the Fermi velocity.

Because of short coherence length, the cuprates belong to the 'clean' limit, since the mean free path for impurity scattering exceeds the coherence length. Its leads to a peculiar symmetry of the pairing order parameter and the appearance of a two-gap spectrum (see below).

5.3.2 The Critical Field H_{c2}

One of the most interesting and novel behaviours exhibited by the cuprates is the novel temperature dependence of the upper critical field H_{c2} for the cuprates and, in particular, the bismuth (2201) compound in the overdoped region. As we know, for conventional superconductors, the dependence $H_{c2}(T)$ has an almost linear increase near T_c and then saturates at $T > 0$ to some value $H_{c2}(0)$ (see Helfand and Werthamer, 1964, 1966). What is remarkable is that, for the cuprates, the upper critical field has a positive curvature at all temperatures that is exaggerated at low temperatures (Mackenzie *et al.*, 1993; Osofsky *et al.*, 1993); see Fig. 5.10.

The data show a sharp, almost linear increase towards $H_{c2}(0)$, whose value greatly exceeds that for usual systems. This unusual behaviour eluded explanation until it was realised that the dopants (in this case, oxygen ions, probably on apical sites) are also pair breakers with magnetic moments and, at temperatures close to T_c, suppress the critical

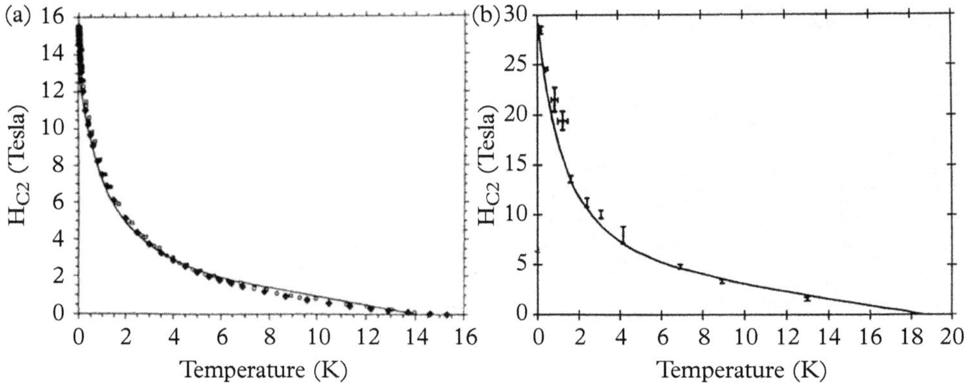

Figure 5.10 *Dependence* $H_{c2}(T)$ *for (a)* $Tl_2Ba_2CuO_6$ *and (b)* $Bi_2Sr_2CuO_6$*; solid points, experimental data (Mackenzie et al., 1993; Osofski et al., 1993); solid lines, theory. (Figure reproduced from Ovchinnikov and Kresin, 1996)*

temperature and the critical field so that the increase of H_{c2} with decreasing temperature is much smaller than in a conventional superconductor without pair breaking. The pair-breaking effect is caused by spin-flip scattering, because it is necessary to satisfy the spin conservation law. However, at low temperatures, the picture becomes different (Ovchinnikov and Kresin, 1996). The correlation between the magnetic moments becomes important, and the spin-flip scattering is frustrated. The spin-flip relaxation time becomes temperature dependent and increases as $T \to 0$. As a result, the superconducting state becomes less depressed (the 'recovery') and this leads to the observed increase in the value of the critical field.

5.3.3 The Two-Gap Spectrum

Section 3.2 contains a description of the two-gap scenario which is caused by the presence of two groups of electronic states (e.g. two overlapping energy bands). This phenomenon has never been observed in conventional superconductors. Indeed, because of the inequality $l < \xi$ (where l is the mean free path, and ξ is the coherence length), scattering by impurities mixes the electronic states, so we obtain the usual one-gap picture. However, the coherence length is short for the high-T_c cuprates, and this leads to possibility of observing a two-gap spectrum.

A remarkable system which displays the two-gap structure of the spectrum is the YBCO compound, which has two conducting subsystems: planes and chains. The most direct observation of two gaps was performed by Geerk *et al.* (1988) with the use of tunnelling spectroscopy (see also Santiago and Menezes, 1989). Indeed, two peaks in the density of states have been observed.

It is interesting that the second (smallest) gap can be induced in the chains in two ways (Kresin and Wolf, 1992): (1) via the usual two-band scenario (see section 3.2), where the second gap is induced by the phonon-mediated plane–chain transitions, and (2) via the proximity effect caused by the spatial separation of these units. These two ways have opposite impacts on T_c (see section 3.5); as a result, the value of T_c is determined mainly by intra-plane pairing. As was discussed above (section 3.3), the presence of magnetic impurities leads to the pair-breaking effect (for d-wave superconductivity, even non-magnetic impurities can act as pair breakers). As a result, one can observe the transition into a gapless superconducting state. In the present two-gap case, when the chains' spectrum has a small induced energy gap, the gapless state is created by a small concentration of the impurities, much smaller than for the one-gap superconductor. Note that a similar situation was observed for the high-T_c pnictides. More specifically, one can observe the formation of the gapless state for the $K_{1-x}Na_xFe_2As_2$ compound caused just by small disorder (Grinenko *et al.*, 2014).

Later, a two-gap spectrum was also observed for magnesium diboride, MgB_2 (see section 7.1.3) and for *Fe*-based pnictides (see section 7.3).

5.3.4 Symmetry of the Order Parameter

The concept of pairing with finite value of the orbital momentum l is not new. The case of d-wave symmetry ($l = 2$; we consider the singlet pairing) corresponds to the $d_{x^2-y^2}$ structure of the gap parameter. It is essential that we are dealing with the d-wave symmetry of the gap parameter in momentum, not real space; this is different from the usual quantum-mechanical picture of the electronic wave function with $l = 2$.

The symmetry of the order parameter in the cuprates has been the subject of intense study since its original discovery in 1986 (see e.g. reviews by Mahnhart and Chowdari, 2001, Van Hargingen, 2012, and Tsuei and Kirtley, 2000). Although this symmetry does not provide the key to the mechanism, it was still heavily studied, as it is an important feature of these materials. Especially, one should mention the phase shift experiments by Tsuei *et al.* (1994) and by Kouznetsov *et al.* (1997).

Experiments on tricrystal rings by Tsuei *et al.* (1994) provided very strong evidence that the order parameter has d-wave symmetry, especially in nearly optimally doped YBCO (Fig. 5.11). This experiment used flux quantisation in rings with zero, two, and three junctions that showed ½ flux quanta only in rings with three junctions. This result implied an order parameter that had a large d-wave component. A d-wave order parameter (in **k** space) has a zero in the k_x and k_y directions and a maximum at 45° to these axes.

One important consequence of pure d-wave symmetry is that there would be no Josephson tunnelling in the c direction between a purely d-wave superconductor and an s-wave superconductor because the net Josephson current from all of the lobes would sum to zero. A very important experiment was carried out by Kouznetsov *et al.* that measured a finite Josephson tunnelling in the c direction between Pb, which is a pure s-wave superconductor, and a twinned sample of YBCO. These results unambiguously proved that the symmetry of the order parameter was mixed with a clear s-wave component that changes sign across a twin boundary. The d-wave symmetry and

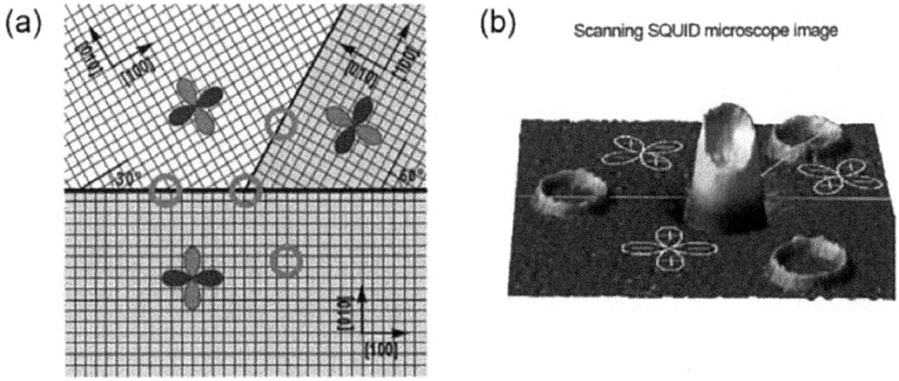

Figure 5.11 *Schematic of the d-wave testing tricrystal geometry. (b) Scanning superconducting quantum interference device (SQUID) microscope image of $YBa_2Cu_3O_{7-\delta}$ rings at 4.2 K and nominal zero magnetic field with the half-flux in the central three-junction ring. (From Tsuei et al., 1995)*

corresponding presence of the nodes of the gap parameter means that we are dealing with the absence of the gap in the density of states, although the gaplessness can be also observed as the impact of the pair breaking.

Experimentally, the *d*-wave symmetry is manifested in measurements of the temperature dependence of the penetration depth (Hardy *et al.*, 1993); the data show linear (non-exponential) behaviour at $T > 0$ K. The angular photoemission data (ARPES) also demonstrated the *d*-wave nature of the pairing (Shen *et al.*, 1993; Ding *et al.*, 1996).

Note that the *d*-wave structure of the gap parameter is very beneficial for many novel superconductors, including the cuprates. Indeed, as we know, the attraction between the carriers should overcome the Coulomb repulsion and, for conventional systems, this process is greatly helped by the logarithmic weakening of this repulsion (see eqn (2.27)). This factor is caused by the large value of the Fermi energy and, correspondingly, by the large scale for the coherence length. However, for the cuprates, the coherence length is rather short, and factor is not so effective. The *d*-wave order parameter is changing its sign. Because the equation for T_c has the structure $\Delta = \Sigma K \Delta$, it is clear that this feature makes it possible to obtain the solution, even if in some region the kernel K has a negative sign (the contribution of the attraction is smaller than the repulsion in this region).

An interesting experiment was performed by Razzoli *et al.* (2013). With the use of the photoemission (ARPES) technique, they observed the *s*–*d* evolution of the pairing gap as a function of doping. More specifically, the highly underdoped LSCO with $x \approx 0.08$ ($T_c \approx 20$ K) displays an *s*-wave gap, and the subsequent increase in doping leads to the *s*–*d* evolution, so that, at the optimum doping level ($x = 0.145$, $T_c = 33$ K), the sample displays the *d*-wave order parameter. A similar observation was reported for the underdoped $Bi_2Sr_2CaCu_2O_{8+\delta}$ compound (Vishik *et al.*, 2012). One can see that the observed *s*–*d* evolution correlates with the doping level and, correspondingly, with value of T_c and decreases in the value of the coherence length.

5.4 The Origin of High-T_c Superconductivity in Cuprates

5.4.1 General Comments

In accordance with the title of this book, the various mechanisms of superconductivity were initially described and then the focus shifted to the description of different materials. There are several important families of materials with well-established and generally accepted mechanisms of pairing (see chapters 7 and 8).

The situation with cuprates is entirely different. Despite the fact that it has been many years (35 years!) since their discovery, elucidation of the mechanism of high T_c in cuprates continues to be controversial. Many thousands of theoretical papers describing various potential mechanisms have been published; this number greatly exceeds the number of publications preceding the appearance of the BCS theory. The situation with the origin of high T_c in cuprates looks like an embarrassment for the scientific community, and it should be analysed. Copper oxides are compounds with a rather complex structure and many unusual properties. Their discovery was totally unexpected. Since almost nothing was known about these new materials, their initial theoretical study was a real 'golden' period for theoreticians. Indeed, it was possible to create various models and mechanisms. As a result, it was a period of intensive and fruitful development of the theory of superconductivity. It was shown that a high T_c value can be achieved through a number of various mechanisms. Of course, all of them cannot be applicable to copper oxides.

The key reason for the absence of a generally accepted mechanism of a high T_c in cuprates is not related to a shortage of theoretical treatments. The fundamental problem is the absence of key experiments, which can rule out some models and provide support for the relevant theory. In reality, it is difficult to devise an experiment that is sensitive to the pairing mechanism. Experimental studies usually provide information regarding such parameters as the energy gap, the critical field, and so on, but they are not affected directly by the pairing mechanism. Nevertheless, such experiments do exist. For example, the method developed by McMillan and Rowell, based on the special use of tunnelling spectroscopy (see sections 2.3.4 and 4.1.1), provided key support for the phonon mechanism of superconductivity in many so-called conventional materials. Without this milestone study, even today it would not be possible to discuss the mechanism of pairing in these materials. Another example is the experimental proof of the existence of the electronic mechanism, which was introduced by Little (see section 4.5.1).

Because of this peculiar situation, the following two sections discuss some existing and promising mechanisms. Such an approach is similar to that used by Fellini in his movie *8½*, which has more than one ending. According to one version (sections 5.4.2–5.4.4), the high-temperature superconducting state is provided by the electron–phonon mechanism, which is strengthened by the polaronic effect; the low-frequency plasmon modes also make a contribution. The second version is based on the consideration of strong correlations (sections 5.4.5–5.4.6). According to this approach, the superconducting state is provided by the joint contribution of the phonon and magnetic mechanisms.

5.4.2 The Electron–Phonon Interaction and the High-T_c State

As was shown in section 2.2.2, the electron–lattice interaction can, in principle, lead to high values of T_c. Of course, this statement alone does not provide the answer to the question about the nature of the superconducting state in cuprates, but it means that the electron–lattice interaction cannot be summarily ruled out as a potential mechanism. Let us describe the experimental data which provide a support for the phonon mechanism (see also the review by Kresin and Wolf, 2009).

As was described above (sections 2.3.4 and 4.1.1), tunnelling spectroscopy is a powerful tool which can provide key information about the pairing mechanism. It is very tempting to use tunnelling spectroscopy to study the nature of high-T_c superconductivity in the cuprates. However, there is a serious challenge. As we know, the coherence length, which is defined as $\xi = \hbar\nu_F/2\pi T_c$ (where ν_F is the Fermi velocity), is an important parameter; its value characterises the scale of pairing and can be visualised as the size of the pair. For usual superconductors, the value of ξ is rather large ($\sim 10^3 - 10^4$ Å) whereas, for the cuprates, it is quite small: $\xi = 15 - 20$ Å. The length scale for providing the tunnelling current at the interface between the superconductor and the insulator, that is, the depth over which the tunnelling current originates, is the pairing coherence length and, as noted above for conventional superconductors, this is a large quantity that greatly exceeds the thickness of the surface layer. In the cuprates, the coherence length is very short, and this makes the measurements difficult. Nevertheless, such experiments were performed. One of the first tunnelling experiments (Dynes *et al.*, 1992) was carried out to study YBCO. The inversion procedure carried out in this study resulted in the dependence $\alpha^2(\Omega)F(\Omega)$, shown in Fig. 5.12. The calculated value of the critical temperature was $T_c \approx 60$ K. This value lies below the experimental one, but is still quite high.

Break-junction tunnelling spectroscopy, which provides a high-quality contact, was employed by Aminov *et al.* (1994) and Ponomarev *et al.* (1999). They demonstrated (Fig. 5.13) that the current–voltage characteristic for the $Bi_2Sr_2CaCa_2O_8$ compound contains an additional substructure which strongly correlates with the phonon density of states; the phonon density of states was obtained by Renker *et al.* (1987, 1989) using inelastic neutron scattering. Such a correlation is a strong indication of the importance of the electron–phonon interaction.

The tunnelling conductance of $Bi_2Sr_2CaCu_2O_8$ was measured by Shiina *et al.* (1995), Shimada *et al.* (1998), and Tsuda *et al.* (2007). They also observed a correspondence of the peaks in d^2I/dV^2 and the phonon density of states. Moreover, the McMillan–Rowell inversion was performed, and the result was clearly supportive of the electron–phonon scenario. The spectral function $\alpha^2(\Omega)F(\Omega)$ contains two groups of peaks at $\Omega \approx 15$–20 meV and $\Omega \approx 30$–40 meV. The positions of the peaks correspond with a high degree of accuracy to the structure of the phonon density of states. The coupling constant λ appears to be equal to about 3.5; such strong coupling is sufficient (see eqn 2.34) to provide the observed value of T_c ($T_c \approx 90$ K).

Another tunnelling technique which appears to be a powerful tool in many studies is scanning tunnelling microscopy. This method is widely used in order to obtain

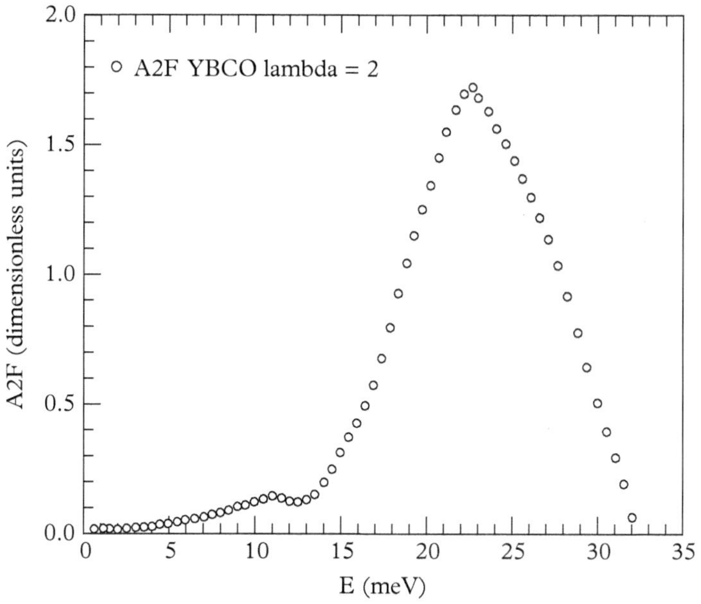

Figure 5.12 *Function* $\alpha^2(\Omega)F(\Omega)$ *for* $YBa_2Cu_3O_{7-\delta}$ *(YBCO). (From Dynes et al., 1992)*

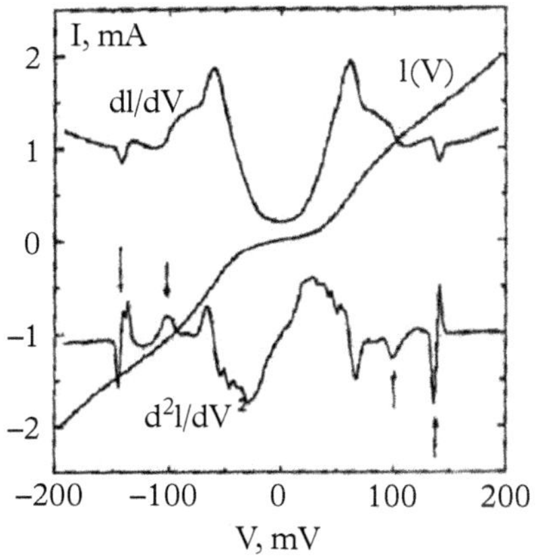

Figure 5.13 *Tunnelling* $I(V), dI/dV$*, and* d^2I/dV^2 *characteristics for a* $Bi_2Sr_2Ca_{n-1}Cu_nO_{2n+4+x}$ *break junction. (From Aminov et al., 1994)*

information about the local structure of the order parameter, its inhomogeneity, and so on. This type of tunnelling can also be used to perform a study that can probe the mechanism of high T_c. For example, Lee *et al.* (2006) used scanning tunnelling microscopy to carry out an analysis of the $Bi_2Sr_2CaCu_2O_{8+\delta}$ compound. As a part of the study, measurements of the tunnelling current and its second derivative, d^2I/dV^2

were performed. The locations of the peaks in the second derivative coincide with the position of specific phonon modes. This is a strong indication of the importance of the electron–phonon coupling.

In an interesting study performed by Shim *et al.* (2008), bicrystal tunnelling junctions were used to study $La_{1.84}Sr_{0.16}CuO_4$ films. The second derivative $\left(\partial^2 I/\partial V^2\right)$ data contained 12 peaks, and the positions of the peaks are in good agreement with Raman scattering data (Sigai *et al.*, 1990). The tunnelling data also agree with the picture obtained via neutron scattering. As a whole, tunnelling spectroscopy continues to be a powerful tool.

Another set of interesting data was obtained with the use of the photoemission technique (see above, section 4.1.3). After the discovery of high-T_c cuprates, the photoemission technique was developed as a powerful tool for obtaining information about the energy spectrum and the electronic structure of these novel materials. Photoemission experiments indicating the presence of substantial electron–phonon coupling were published by Lanzara *et al.* (2001). They studied different families of hole-doped cuprates, Bi2212, LSCO, and Pb-doped Bi2212, and they investigated the electronic quasiparticle dispersion relations. A kink in the dispersion around 50–80 meV was observed. This energy scale corresponds to the energy scale of some high-energy phonons; it is much higher than the energy scale for the pairing gap. Such a kink cannot be explained by the presence of a magnetic mode, because such a mode does not exist in LSCO, while the kink structure was also observed in this cuprate.

The structure observed by photoemission is consistent with the data on the phonon spectrum obtained by neutron spectroscopy. These measurements were also used in to obtain a crude estimate of the electron–phonon coupling, since the quasiparticle velocity in the low-temperature region is renormalised by the electron–phonon coupling constant $\lambda : v = v_b(1+\lambda)^{-1}$, where v_b is the bare (unrenormalised) velocity, which corresponds to the high-temperature region. This estimate indicates substantial electron–phonon coupling.

A different type of spectroscopy, so-called ultrafast electron crystallography, was employed by Gedik *et al.* (2007). The $La_2CuO_{4+\delta}$ compound was used, and doping by photoexcitation was performed. It is interesting to note that the number of photon-induced carriers per copper site was close to the density of chemically doped carriers in the superconducting compound. The study of time-resolved relaxation dynamics demonstrated the presence of transitions to transient states which are characterised by structural changes (i.e. a noticeable expansion of the c axis). The fact that such a large effect on the lattice is caused by electronic excitations is a strong signature of the electron–lattice interaction.

Let us also mention that a study of thermodynamic properties can also provide information about the pairing mechanism. This is due to the fact that the effective mass and the electronic heat capacity are renormalised by the electron–phonon interaction (see section 1.2.1 and eqn (1.28′). The presence of the second term in the expression for $\gamma(T)$ (eqn (1.28″)) reflects the fact that the moving electron becomes 'dressed' by the phonon cloud. As temperature increases, the 'cloud' becomes weaker, so that $\gamma(T)$ decreases. As a result, the measurements of electronic heat capacity at high temperatures and in the low-temperature region can be used to evaluate the value of the electron–phonon coupling constant λ, which determines T_c. Such measurements were performed by Reeves *et al.* (1993) for the YBCO compound. The main challenge was to evaluate the electronic contribution to the heat capacity at high temperatures where the heat capacity is dominated by the lattice. The lattice contribution was calculated by using the phonon density of states obtained by neutron scattering. As a result, the value $\lambda > 2.5$ was obtained, which means that there is strong electron–lattice coupling sufficient to provide the observed high T_c.

One should stress that the high-T_c superconducting state cannot be provided by weak electron–phonon coupling. Correspondingly, the analysis based on the BCS model, developed in a weak coupling approximation, is not applicable. The electron–phonon interaction should be strong, and strong coupling effects should be incorporated into the theory. In connection with this, it is important that the cuprates display the polaronic effects. As discussed in section 2.3.1, these effects lead to an effective increase of the electron–lattice interaction and are very beneficial for the pairing.

5.4.3 The Phonon Mechanism and the Polaronic States

After the discovery of high-T_c oxides, many scientists excluded the electron–phonon interaction from the list of potential mechanisms. Partly it was due to the natural temptation to introduce something new and exciting. In addition, there was the conviction that, despite the phonon mechanism being successful as an explanation for the superconducting state in conventional materials, it is not sufficient to explain the observed high values of T_c. The 'legend' that the value of T_c caused by the exchange by phonons cannot exceed 30 K was born and it still alive among some scientists. This 'legend' was shaken by the discovery of MgB_2 (see section 7.1.3) with a higher value for the critical temperature. And, of course, the discovery of the superconductivity in hydrides (see section 7.2), with a record value of T_c which is close to room temperature and caused by the phonon mechanism, speaks for itself.

We discussed above (see section 2.2.2 and, especially, section 2.3.5) the question about the upper limit of T_c. In connection with it, one should note that the high value of T_c cannot be provided by the weak electron–phonon coupling. It requires the presence of the strong interaction. It turns out that the value of the electron–phonon coupling constant is effectively increased by the polaronic effect (see section 2.3.1). As a result, we are dealing with the electron–phonon mechanism, which looks as the 'old' scenario, but with a key new aspect: its effect is increased, thanks to the peculiar polaronic effect. Its application to the copper oxides will be discussed in the next section. Note that, because of the polaronic effect, the total wave function cannot be written as a product of the electronic and ionic functions, and this means that the electronic and the nuclear motions cannot be separated.

The polaronic state: the isotope effect for T_c. As we know, the value of the critical temperature strongly depends on the carrier concentration n, that is, on the doping level: $T_c \equiv T_c(n)$. The value of the carrier concentration is affected by the isotope substitution and, more

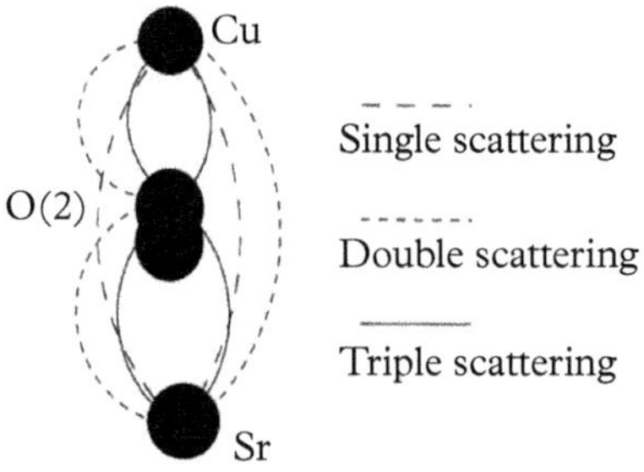

Figure 5.14 *'Double-well' structure for the apical oxygen. (From Haskel et al., 1997)*

specifically, by the dynamic polaronic effect which occurs if an ion is characterised by not one but two close potential minima. Such a polaronic effect, indeed, leads to the observed isotopic dependence (sections 3.5.7 and 3.5.8).

Let us start with an apical oxygen ion, placed between the planes and chains in YBCO (Fig. 5.14). The importance of the apical oxygen was stressed by Mueller (1990). The cuprates are doped materials, and the charge transfer occurs through this ion. The key observation was reported by Haskel *et al.* (1997). Namely, the 'double-well' structure of the apical oxygen ion (Fig. 5.14) has been observed with the use of an X-ray-absorption technique.

The in-plane oxygen ions can also display the polaronic effect (the site-selective isotope substitution was described by Zech *et al.* (1994)). It is important for an analysis of the electron–lattice intercalation (see section 2.3.1), but it also affects the dynamics of the doping. This is observed for the $Y_{1-x}Pr_xBa_2Cu_3O_7$ samples studied by Keller (2003) and Khasanov *et al.* (2003) and is caused by the mixed valence state of the Pr ions.

The polaronic isotope effect is described by eqn (3.96). One can see directly from this equation that $\alpha_{na} \propto (\partial T_c/\partial n)$ and, therefore, at the optimum doping ($T_c \equiv T_{c;max}$), the isotope coefficient should have a minimum value; this, indeed, corresponds to the observed picture (Crawford *et al.*, 1990; Babushkina, 1991; Franck, 1991). A decrease in the doping level leads to the finite value of the derivative and to a rapid increase in the value of the isotope coefficient; its value can, in principle, exceed that for the BCS theory.

Equation (3.96) contains experimentally measured quantities and can be compared with the experimental data. Such an analysis for a whole temperature range has been performed by Weyeneth and Mueller (2011); see Fig. 5.15 and also the work by Weyeneth (2017). One can see remarkable agreement between the theory and experimental data for several cuprates. Such an agreement provides key support for the polaronic concept.

More data on the isotope effect are given in Fig. 5.16, where the dependence of the isotope effect on the position of the oxygen reveals the dominant participation of the in-plane oxygen ions in the formation of superconductivity, although the apical oxygen does not contribute to the isotope effect.

The dynamic polaronic state is characterised by the presence of two minima for the ionic motion (in our case, for the motion of the oxygen ion). In such a case, the total wave function is a sum of two terms (see section 1.2.2, eqn (1.30)). It is essential that the total

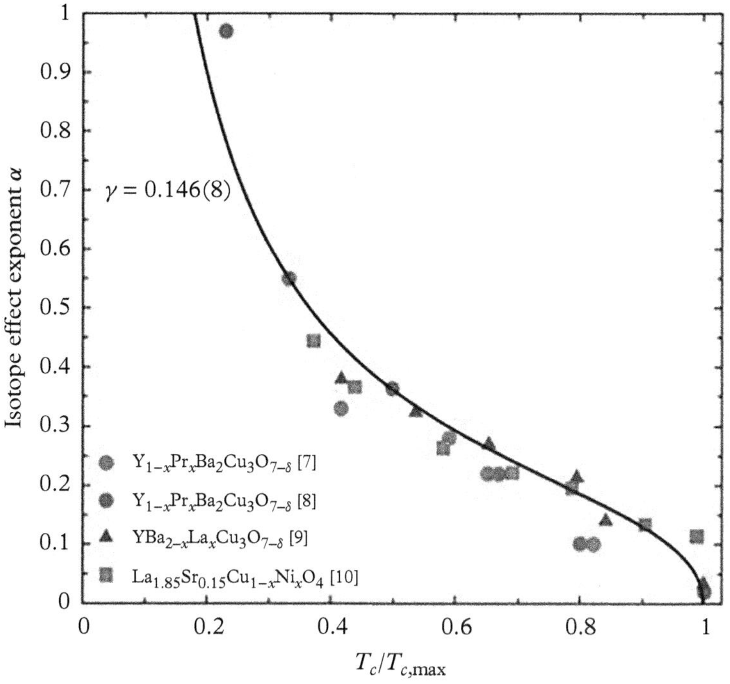

Figure 5.15 *The oxygen isotope-effect exponent* α *as a function of* $T_c/T_{c,max}$ *for several cuprates; experimental points from Weyeneth and Mueller (2011); solid line represents theory, eqn (3.96).*

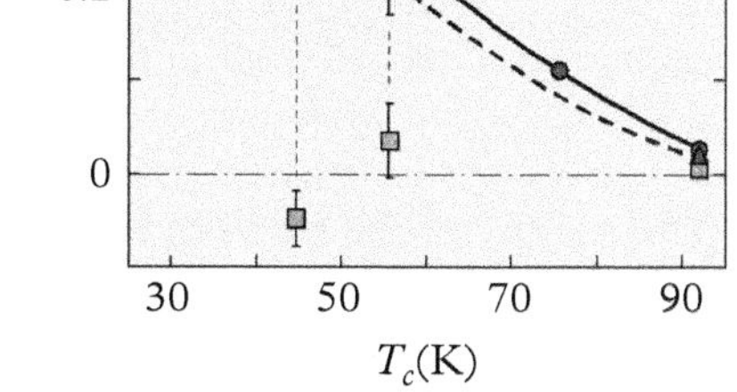

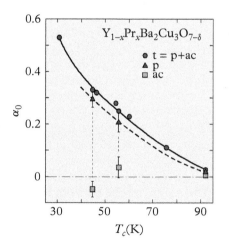

Figure 5.16 *The oxygen site-selective isotope-effect exponent as function of hole concentration. (From Khasanov et al., 2003)*

wave function is not the product of the electronic and ionic wave function. Therefore, as mentioned above, because of the polaronic effect, the electronic and the ionic motion cannot be separated.

The isotopic effect for the penetration depth. The carrier concentration affects not only T_c but also the value of the penetration depth (section 3.5.8). That is why one can expect that the penetration depth should also display the isotopic dependence. This is an unusual effect, which has never been observed for conventional superconductors. This effect is specific for the doped materials only and is due to the dynamic polaronic effect. The cuprates represent the most important family of such superconductors.

The isotope effect for the penetration depth was predicted by Kresin and Wolf (1995) and later observed (Zhao *et al.*, 1997; Khasanov *et al.*, 2004); see the reviews by Kresin and Wolf (2009) and Keller (2003, 2005). The isotopic dependence of the penetration depth δ at $T = 0\,K$ is described by eqns (3.101) and (3.102). Let us first note that, unlike the isotopic dependence of T_c (eqn (3.96)), the isotope coefficient β is not small at $T = T_{max}$.

Equation (3.102) establishes the relation between the isotope coefficients describing the impacts of the isotope substitution on the T_c and the penetration depth. Since, in the underdoped region, the dependence $T_c(n)$ is close to linear, one can expect, based on eqn (3.102), that, in this region, $|\alpha| \simeq 2|\beta|$. Such a dependence is in good agreement with the empirical Uemura plot (Uemura *et al.*, 1989; also see the discussion in Khasanov *et al.*, 2006). This agreement also provides a support for the polaronic scenario.

5.4.4 Dynamic Screening, the Plasmon Mechanism, and the Coexistence of Phonon and Plasmon Mechanisms

The impact of the electron–lattice interaction does not exclude the additional contribution to the pairing which comes from other electronic or magnetic excitations. The electronic mechanisms (exchange by excitons or acoustic plasmons) were discussed in sections 2.5 and 2.7. As we know, many novel superconducting systems (e.g. cuprates, MgB_2, and princites) belong to the family of layered (quasi-two-dimensional) conductors and are characterised by strongly anisotropic transport properties. One can raise the following interesting question: why is layering a favourable factor for superconductivity? One can show (see section 1.2.3) that layered conductors have a plasmon spectrum that differs fundamentally from that of three-dimensional metals. In addition to a high-energy 'optical' collective mode, the spectrum also contains an important low-frequency part ('electronic' sound; see Figs 1.8 and 1.9). The screening of the Coulomb interaction is incomplete, and the *dynamic* nature of the Coulomb interaction becomes important. The contribution of the plasmons in conjunction with the phonon mechanism may lead to high values of T_c.

Because of peculiar nature of the plasmon spectrum and, more specifically, because of the presence of low-frequency acoustic modes, the experimental observation of these collective excitations has been a real challenge. Indeed, as was noted in section 1.2.2, the usual method, namely electron energy loss spectroscopy (EELS), does not provide

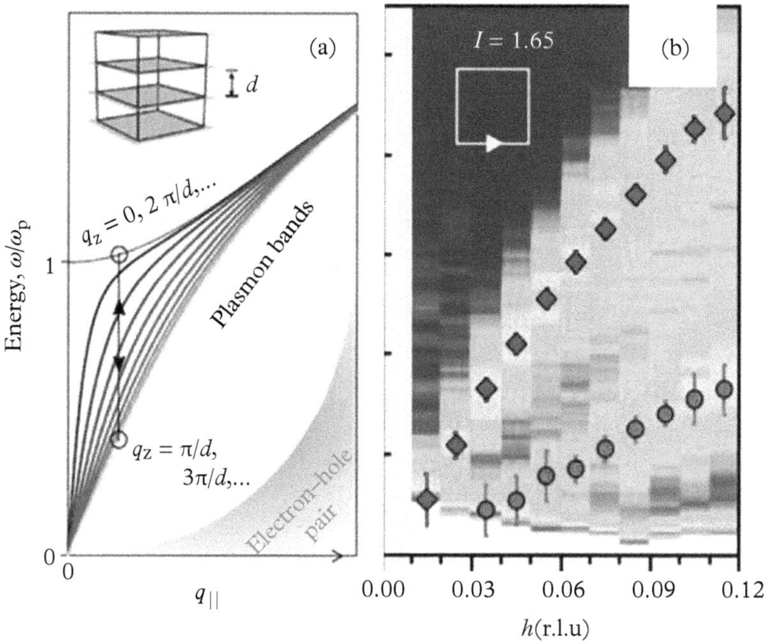

Figure 5.17 *(a) Plasmon dispersion as a function of in- and out-of-plane momentum transfer. (b) Resonant inelastic X-ray scattering intensity map for $La_{1.89}Ce_{0.11}CuO_{4-x}$. Square and circles correspond to charge density excitations (plasmons) and paramagnons, respectively. (From Hepting et al., 2018)*

sufficient information, since it is unrealistic to analyse the propagation of low-energy electronic beams inside a metallic system. This challenge is directly related to the study of plasmons and their impact on the pairing in copper oxides.

Recently, this problem was solved, and acoustic plasmons were observed experimentally via RIXS. The first such observation was described in the paper by Hepting *et al.* (2018), where these modes were observed in the electron-doped copper oxides $Nd_{2-x}Ce_xCuO_4$, where $x = 0.15$; one can see directly (Fig. 5.17) the acoustic plasmon and magnetic modes.

As was noted in sections 1.2.3 and 4.1.3, the RIXS technique allows one to measure the dispersion of the acoustic mode, that is, its dependence on the in-plane and out-of-plane momenta. Although the electronic motion in the layered system of interest is almost two dimensional (the small interlayer hopping leads to the existence of a small gap at $q_{II} = 0$, where q_{II} is the in-plane momentum), the Coulomb interaction is three dimensional, and it leads to the three-dimensional nature of the observed spectrum and its corresponding dependence on q_z. An additional advantage of the method is that it makes it possible to separate the charge (plasmons) excitations and magnons. It is due to difference dependence of the scattering on the polarisation of the incoming radiation.

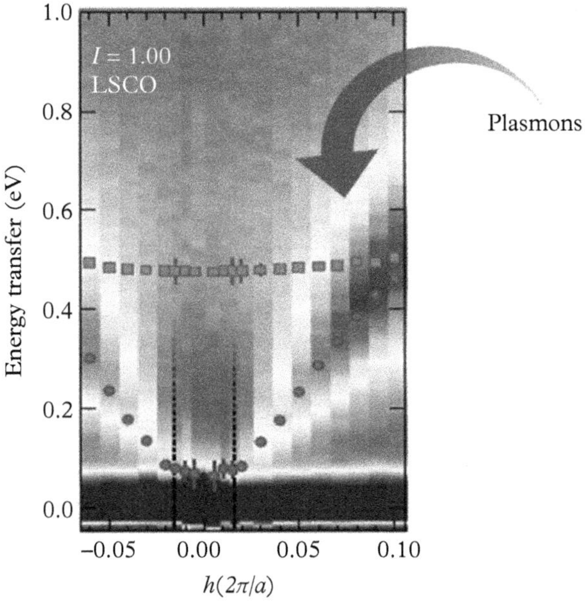

Figure 5.18 *Plasmon excitation in $La_{2-x}Sr_xCuO_4$. (From Zhou, 2020)*

The next important key contribution was made by Zhou (2020). The acoustic plasmon modes for the hole-doped cuprates, more specifically, for LSCO and BI2201 copper oxides, were observed. The data are displayed in Fig. 5.18; one can see the optical and acoustic plasmon modes and magnetic excitations. The evaluation of T_c can be performed with the use of eqns (2.53) and (2.54) and the corresponding calculation of the polarisation operator (Bill *et al.*, 2003). The use of the RPA approximation is justified by the large value of the dielectric constant ε_{M}. One can use the following values of the parameters for $La_{1.85}Sr_{0.15}CuO_4$: the interlayer distance is $L = 6.5$ A, the Fermi vector $k_{\mathrm{F}} = 3.5.10^7 \mathrm{cm}^{-1}$, and $\varepsilon_{\mathrm{M}} = 5 - 10$.

A detailed analysis based on eqn (2.54) shows that the impact of dynamic screening is different for various layered systems. For example, for the metal-intercalated nitrides (see section 7.4.3), the plasmon contribution dominates. The contribution is not so large, but is still essential (about 40%) for some of the organic superconductors (see section 8.1.2). As for the cuprates, the plasmon contribution is not so crucial but is noticeable: about 20% of the observed value of T_c is due to acoustic plasmons. The main role is played by phonons, and their impact leads to a high value of T_c.

The method based on precise infrared measurements (see section 4.1.2), developed by Little *et al.* (1999, 2007), allows one to reconstruct the junction $\alpha^2(\Omega)F(\Omega)$ for an energy interval above the region of phonon energies. According to the analysis, the pairing in cuprates is provided by both phonon and electronic channels.

5.4.5 Strong Correlations, Electronic Structure, and ARPES

The electronic structure of cuprates in the normal phase. The ab initio electronic structure calculations are dominated by the density functional theory (DFT). DFT has revolutionised this field (Fulde, 2011; Jones, 2011). However, as Fulde pointed out:

> DFT is a ground state theory and therefore it is of little surprise that it fails, in particular for strongly correlated systems, when low-energy excitations are calculated from it. The failure is inherent and independent of any approximations which are made for the potential in the Kohn-Sham equation.
>
> (Fulde, 2011, p. xviii)

Indeed, the conventional DFT approach to the electronic structure of the tetragonal undoped La_2CuO_4 (Mattheiss, 1987) reveals an incorrect metallic state with a half-filled $Cu(3d)$–$O(2p)$ band with two-dimensional character and a nearly square Fermi surface that is unstable for the formation of the insulator CDW of the SDW phase (Pickett, 1989). At first glimpse, the formation of the band insulator in the CDW phase due to the Peierls transition below T_p or the SDW antiferromagnetic phase below T_N solves the problem of the La_2CuO_4 insulator ground state, but it is not so. In such approaches, above the critical temperatures T_p or T_N, one will find again the metallic state, while La_2CuO_4 is the insulator above the Néel temperature. Recently developed, the most elaborated SCAN meta-GGA scheme (Sun *et al.*, 2015) makes it possible to calculate the antiferromagnetic insulator for undoped La_2CuO_4 and the normal metal with the large Fermi surface for the overdoped $La_{1.75}Sr_{0.25}CuO_4$ (Furness *et al.*, 2018). Again, the finite-temperature-transition antiferromagnetic insulator–paramagnetic insulator for the undoped La_2CuO_4 cannot be obtained. Thus, the conventional ab initio band theory approach based on DFT can be reliable only in the overdoped region of the phase diagram at $T = 0$.

Up to now, there is no ab initio scheme that makes it possible to calculate the electronic properties of cuprates at small and optimal doping or/and temperature. That is why the microscopic model approach is inevitable to treat strong electron correlations. The best options are obtained using the hybrid schemes like local density approximation plus DMFT (LDA+DMFT; Anisimov *et al.*, 1997; Lichtenstein and Katsnelson, 1998; Held *et al.*, 2001; Kotliar *et al.*, 2006) or local density approximation plus generalised tight binding (LDA+GTB; Gavrichkov *et al.*, 2000; Korshunov *et al.*, 2004, 2005; 2012). Due to the same reason, the interaction of electrons with phonons and acoustic plasmons also has to be reconsidered to take into account the effects of strong electron correlations. Some steps in this direction have been made within the polaronic version of the generalised tight binding (GTB) approach, p-GTB (Makarov *et al.*, 2015; Shneyder *et al.*, 2018).

The most well-known electronic structure model for CuO_2 layers is the two-dimensional Hubbard model on a square lattice. Why is it this model? When it can be used in cuprates? Here we will discuss the general multiband strongly correlated approach and consider how the Hubbard and the $t - \mathcal{J}$ models appear as the low-energy effective models (Ovchinnikov and Shneyder, 2007). Soon after the beginning of the cuprates study, the three-band *p–d* model has been proposed to describe the hybridised Cu(3*d*)–O(2*p*) band with two-dimensional character (Emery, 1987; Varma *et al.*, 1987; Zhang and Rice, 1988). This model includes the copper and oxygen orbitals and captures the main features of the electronic structure near the Fermi level. Later, polarised X-ray absorption spectroscopy (XAS; Pompa *et al.*, 1991) and EELS (Romberg *et al.*, 1990) showed a sizeable (10–15%) occupancy of the Cu $d_{3z^2-r^2}$ orbitals in several investigated cuprates. To account for observed states, the three-band model has to be further enlarged by including the Cu $d_{3z^2-r^2}$ orbitals (Gaididei and Loktev, 1988; Weber, 1988). These five-band models include, in the general case, several strong intra-atomic Coulomb interaction parameters inside the Cu^{+2} ion and the O^{-2} ion, as well as the nearest-neighbour copper–oxygen Coulomb interaction with the Hamiltonian

$$
\begin{aligned}
H &= H_d + H_p + H_{pd} + H_{pp}. \\
H_d &= \sum_r H_d(r), H_d(r) = \sum_{\lambda\sigma}\left[(\varepsilon_{d\lambda}-\mu)\,n_{r\lambda}^{\sigma} + \frac{U_d}{2} n_{r\lambda}^{\sigma} n_{r\lambda}^{-\sigma}\right] \\
&\quad + \sum_{\sigma\sigma'}\left(V_d n_{r1}^{\sigma} n_{r2}^{\sigma'} - J_d d_{r1\sigma}^{+} d_{r1\sigma'} d_{r2\sigma'}^{+} d_{r2\sigma}\right) \\
H_p &= \sum_i H_p(i), H_p(i) = \sum_{\alpha\sigma}\left[(\varepsilon_{p\alpha}-\mu)\,n_{i\alpha}^{\sigma} + \frac{U_p}{2} n_{i\alpha}^{\sigma} n_{i\alpha}^{-\sigma}\right] \\
&\quad + \sum_{\sigma\sigma'}\left(V_p n_{i1}^{\sigma} n_{i2}^{\sigma'} - J_p p_{i1\sigma}^{+} p_{i1\sigma'} p_{i2\sigma'}^{+} p_{i2\sigma}\right) \\
H_{pd} &= \sum_{<i,r>} H_{pd}(i,r), H_{pd}(i,r) \\
&= \sum_{\alpha\lambda\sigma\sigma'}\left(T_{\lambda\alpha} p_{i\alpha\sigma}^{+} d_{r\lambda\sigma} + h.c. + V_{\lambda\alpha} n_{r\lambda}^{\sigma} n_{i\alpha}^{\sigma'} - J_{\alpha\lambda} d_{r\lambda\sigma}^{+} d_{r\lambda\sigma'} p_{i\alpha\sigma'}^{+} p_{i\alpha\sigma}\right) \\
H_{pp} &= \sum_{<i,j>}\sum_{\alpha\beta\sigma}\left(t_{\alpha\beta} p_{i\alpha\sigma}^{+}\, p_{j\beta\sigma} + h.c.\right)
\end{aligned}
\tag{5.1}
$$

Here H_d and H_p are the local energies of *d*- and *p*-electrons, and H_{pd} describes the interatomic *p–d* hopping $T_{\lambda\alpha}$, Coulomb $V_{\lambda\alpha}$, and exchange $J_{\lambda\alpha}$ interactions. The last term, H_{pp}, contains the interatomic oxygen–oxygen hopping t_{pp}. We consider different *d*-orbitals λ and *p*-orbitals α, concerning the crystal-field splitting. The five-bands *p–d* model includes the $d_{x^2-y^2}$, $d_{3z^2-r^2}$ orbitals of copper, the $p_{x/y}$ orbitals of in-plane oxygen, and the p_z orbital of apical oxygen. For simplicity, the intra-atomic Coulomb $U_d\,(U_p)$, $V_d\,(V_p)$, and exchange $J_d\,(J_p)$ matrix elements are assumed to be the same for all orbitals, this simplification is not principal, and realistic matrix elements with orbital dependence can be trivially incorporated. All parameters of the Hamiltonian are assumed to be known either from ab initio calculations or from experimental data. For the three-band model, a constrained-density-functional approach has been used to calculate the

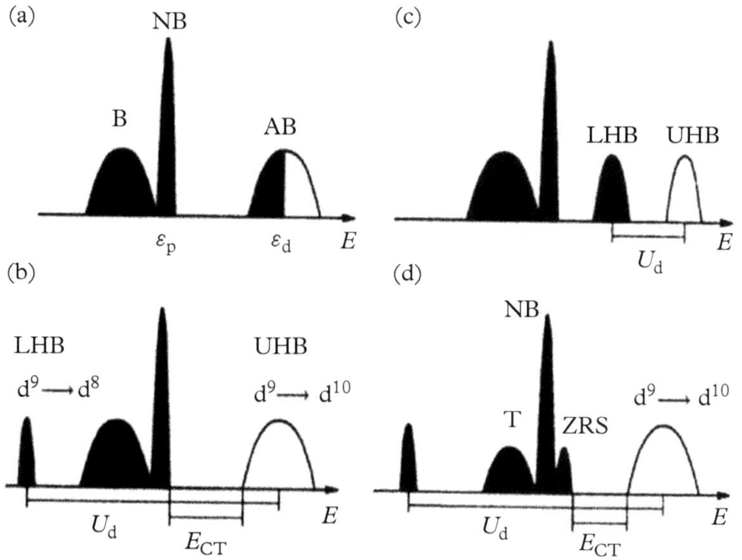

Figure 5.19 *Possible variants of the electronic structure in the three-band p–d model: (a) Metal. (b) Mott–Hubbard insulator with $U_d < \Delta$. (c) Charge-transfer insulator with $U_d > \Delta$. (d) Charge-transfer insulator with split Zhang–Rice singlet and triplet states; AB, antibonding; B, bonding; E_{CT}, charge-transfer gap; LHB, lower Hubbard band; NB, non-bonding; T, triplet; UHB, upper Hubbard band; ZRS, Zhang–Rice singlet. (From Horsch and Stephan, 1993)*

model parameters by Schluter *et al.* (1988) and McMahan *et al.* (1989). An important parameter of the *p–d* multiband model is the charge transfer energy $\Delta = \varepsilon_d - \varepsilon_p$. A scheme of possible variants of the electronic structure of cuprates within the three-band model is shown in Fig. 5.19.

For the five-band model, the Hamiltonian parameters have been found with a similar approach by Korshunov *et al.* (2004) and Korshunov *et al.* (2005). For both models, the local Coulomb parameter $U_d \sim 10\ eV$ is the largest energy indicating strong electron correlations.

As in section 2.7, we may consider two opposite limits of the parameters for the five-band model. In the weak correlation limit, it is possible to treat the Coulomb interaction within the conventional perturbation theory and obtain a renormalised Fermi liquid that disagrees with the cuprate physics in the pseudogap area of the phase diagram. In the regime of strong electron correlations, several perturbation approaches on $t/U \ll 1$ start from the Hubbard (1965) paper using the projection operators or the Hubbard X-operators (see section 2.7.4 and appendix D). These approaches treat the strong local Coulomb interaction exactly, and the interatomic hopping by a perturbation method. The main conclusion of these perturbation methods is that an electron in a strongly correlated system is considered to be the sum of fermionic quasiparticles with the charge e, spin ½, renormalised energy, and spectral weight distributed over all quasiparticles

(Fulde, 1991). It is not so trivial keeping in mind some indications for the spinon–holon view on the electron that are based on the exact solution of the one-dimensional Hubbard model in the infinite U limit (see section 2.7.4). Application of the Hubbard ideas to realistic multiorbital models results in the GTB method (Ovchinnikov and Sandalov, 1989; Ovchinnikov, 1994; Gavrichkov *et al.*, 2000). It is related to the exact Lehman representation for the single-electron Green function (Lehmann, 1954), which, at zero temperature, can be written as

$$G_\sigma(k,\omega) = \sum_m \left(\frac{A_m(k,\omega)}{\omega - \Omega_m^+} + \frac{B_m(k,\omega)}{\omega - \Omega_m^-} \right) \tag{5.2}$$

where the quasiparticle energies are given by

$$\Omega_m^+ = E_m(N+1) - E_0(N) - \mu, \quad \Omega_m^- = E_0(N) - E_m(N-1) - \mu$$

and the quasiparticle spectral weight is equal to

$$A_m(k,\omega) = |\langle 0, N | a_{k\sigma} | m, N+1 \rangle|^2, \quad B_m(k,\omega) = |\langle m, N-1 | a_{k\sigma} | 0, N \rangle|^2$$

Here $|m, N\rangle$ is the m-th eigenstate of the system with N electrons,

$$H|m,N\rangle = E_m|m,N\rangle$$

The index m enumerates the quasiparticle with spin $S = 1/2$, electrical charge e, energy Ω_m^+ (Ω_m^-), and spectral weight A_m (B_m). Equation (5.2) may be considered to be a sum over different quasiparticle bands, with m being the quasiparticle band index. At finite temperature, the Lehmann representation for the Green function can be written as

$$G_\sigma^R(k,\omega) = \sum_{mn} W_n \frac{A_{mn}(k,\omega)}{\omega - \Omega_{mn}^+ + i0} \left(1 + e^{-\Omega_{mn}^+/T}\right) \tag{5.3}$$

Here, the quasiparticle as the excitation between the initial state n and the final state m has the energy

$$\Omega_{mn}^+ = E_m(N+1) - E_n(N) - \mu$$

and the statistical weight of a state $|n\rangle$ is determined by the Gibbs distribution function with the thermodynamic potential Ω:

$$W_n = \exp\{(\Omega - E_n + \mu N)/T\}$$

At a non-zero temperature, both the ground state $|0, N\rangle$ and the excited states $|n, N\rangle$ are partially occupied. Thus, a quasiparticle is enumerated by a pair of indexes (m, n). It is a single-electron excitation in a multielectron system due to addition of one electron to the initial N-electron state $|n, N\rangle$ with the formation of the final $(N+1)$-electron state $|m, N+1\rangle$. The electron-addition energy is $\Omega^{+}_{\text{mn}} = E_m(N+1) - E_n(N) - \mu$, while the electron removal energy is $\Omega^{-}_{\text{mn}} = E_m(N) - E_n(N-1) - \mu$. The Landau Fermi-liquid quasiparticle is a specific case of eqn (5.3) with only one quasiparticle close to the Fermi level. For a free electron with energy ε_0, the quasiparticle energies are equal to $\Omega^{+} = \Omega^{-} = \varepsilon_0 - \mu$. The perturbation theory that realised these ideas for the lattice will be discussed in the next section.

The electronic structure within the LDA+GTB multielectron approach. In practical calculations, the Lehmann representation is useless because the multielectron eigenstates $|m, N\rangle$ for the crystal are not known. The GTB method is the cluster perturbation realisation of the multielectron ideas of the Lehmann representation (Ovchinnikov, 2003b); it has been suggested to treat all Coulomb interactions within the CuO_4 (CuO_6) unit cells exactly with the perturbation treatment of the intercluster interactions and hopping. Later, the hybrid LDA+GTB scheme was suggested to treat the compound with its model parameters found from DFT calculations (Korshunov *et al.*, 2004; Korshunov *et al.*, 2005; see appendix D. In this approach, the total Hamiltonian is rewritten as the sum of all intracluster terms H_f, $H_0 = \sum H_f$, and the sum of all intercluster terms H_{fg}, $H_{int} = \sum H_{fg}$. The numerical exact diagonalisation of the local part provides the exact multielectron eigenenergies and eigenstates, $H_f|m\rangle = E_m|m\rangle$. As in section 2.7.4, we introduce the Hubbard operators $X_f^{m,m} = |m\rangle\langle n|$, which are determined by a set of exact intracluster eigenstates. Then the local part of the total Hamiltonian becomes the diagonal $H_f = \sum_m E_m X_f^{m,m}$ and describes the energy of the multielectron terms $|m\rangle$ in different configurations. For cuprates, the hole language is more appropriate, and the relevant configurations are (1) the hole vacuum $|0\rangle$ with the energy E_0 corresponding to the filled shell $d^{10}p^6$, (2) the single-hole molecular orbital $|p\sigma\rangle$, with the orbital p being bound and unbound combinations of the mixed copper and oxygen orbitals with b_{1g} or a_{1g} symmetry with spin $s = 1/2$ and spin projection $\sigma = -1/2, +1/2$, and (3) the two-hole eigenstates, which may be either singlets with $s = 0$ denoted as $|ns\rangle$, or spin triplets $|mM\rangle$ with spin projection $M = -1, 0+1$. Here n and m enumerate different singlets and triplets with two holes. The most important two-hole ground singlet $|1s\rangle$ is formed not only by the well-known Zhang–Rice singlet d^9p^6 (near 50%) but also by the two holes on oxygen ($d^{10}p^5p^5$) and the two holes on copper (d^8p^6), which also have rather noticeable contributions.

The Hilbert space of the exact eigenstates for the CuO_6 cluster is shown schematically in Fig. 5.20. At zero temperature in the undoped case, only the ground state $|b_{1g,\sigma}\rangle$ is occupied and all other terms are empty. Different single-particle excitations from the occupied ground state $|b_{1g,\sigma}\rangle$ may be fermions with charge e and spin $S = 1/2$. The only one-hole annihilation (electron-addition) quasiparticle with the energy Ω_c forms the conductivity band in La_2CuO_4, while many hole-addition (electron-annihilation) quasiparticles participate in the formation of a valence band with a complicated structure. The quasiparticle with the energy Ω_{V1} involves the two-hole singlet, while the quasiparticle with the energy Ω_{V2} involves the two-hole triplet states. Sometimes such excitations are called singlet or triplet bands; we stress here that all these excitations are fermions

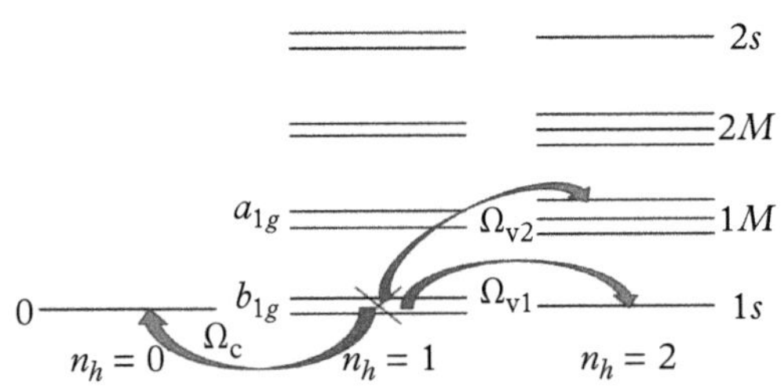

Figure 5.20 *The energy-level scheme of the Hilbert space of the CuO_6 cluster eigenstates with the hole numbers $n_h = 0, 1, 2$. The cross indicates the ground state $|b_{1g}\rangle$ of the $d^9p^6 + d^{10}p^5$ configuration in the undoped cuprate. Arrows show single-electron excitations. (From Gavrichkov and Ovchinnikov, 2008)*

with spin ½. The matrix elements in the spectral weights A and B in the Lehmann representation (5.2) provide the conservation of charge and spin for each quasiparticle.

In principle, we may also consider excitations from empty initial to empty final states. For example, the hole addition to empty $|a_{1g,\sigma}\rangle$ with the final two-hole state $|1s\rangle$ is the fermion quasiparticle, but the filling factor F, which appears due to the X-operator algebra (see appendix D and section 2.7.4) and determines the quasiparticle spectral weight, is zero for excitation between two empty states. We call such a quasiparticle a virtual excitation. It is physically clear that the probability of the excitation from empty state to empty state should be zero. When the occupation changes due to finite temperature, doping, or external irradiation, the virtual excitations may acquire non-zero spectral weight and form the in-gap states. Thus, the quasiparticle band structure in the GTB approach is not rigid; it depends on external conditions.

The next step in GTB is to include effects of the intercluster hopping and interactions. Analysis of the diagram expansion within the Hubbard operator diagram technique (Zaitsev, 1975) results in the exact generalised Dyson equation (Ovchinnikov and Val'kov, 2004) for fermionic quasiparticles with electrical charge e and spin ½ (see appendix D, eqn (D8)). Thus, the electron is the linear combination of excitations between different multielectron terms, similar to the Lehmann representation. They were called Hubbard electrons in the GTB approach, to emphasises their difference from free electrons. The largest difference with the conventional Dyson equation is the presence of the strength operator $P(k,E)$ both in the numerator of the quasiparticle matrix Green function, where it determines the spectral weight or the oscillation strength, and in the denominator in the kinetic energy term $P(k,E)t(k)$ to renormalise the quasiparticle band structure together with the self-energy $\Sigma(k,E)$. The simplest approximation is the Hubbard I mean field one where $\Sigma(k,E) = 0$ and $P^{m,n} = \delta_{m,n} F_n$, where the filling factor $F_n = \langle X^{p,p}\rangle + \langle X^{q,q}\rangle$ for the quasiparticle band index $n = (p,q)$.

In the undoped paramagnetic case, all three quasiparticles shown by arrows in Fig. 5.20 have the filling factor $F_{n\sigma} = 1/2$. Nevertheless, in the antiferromagnetic phase, the filling factor is spin and temperature dependent: $F_{n\sigma} = 1/2 + 2\sigma\langle S^z\rangle$. The temperature-dependent LDA+GTB band structure of the undoped insulator La_2CuO_4 is shown in Fig. 5.21 (Makarov and Ovchinnikov, 2015). In the antiferromagnetic state

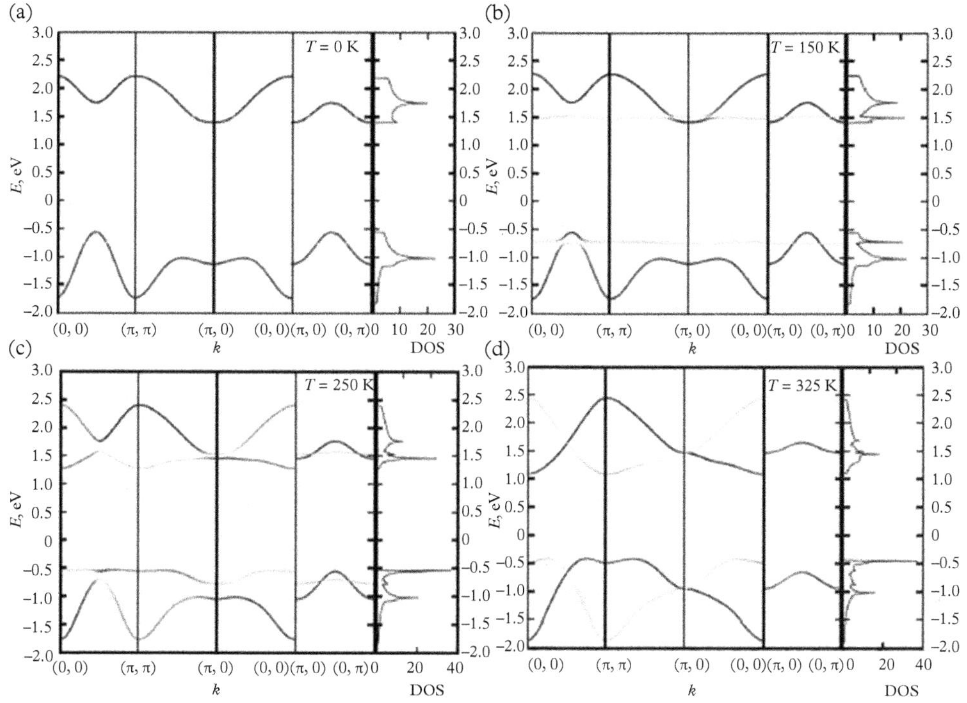

Figure 5.21 *Temperature-dependent band structure and density of states in undoped La_2CuO_4. (From Makarov and Ovchinnikov, 2015)*

with the two sublattices A and B, the spin-up and spin-down $|b_{1g,\sigma}\rangle$ levels are split proportional to the sublattice magnetisation $\langle S^z \rangle$, which is calculated self-consistently with the Néel temperature $T_N = 325$ K . At finite temperature, some shadow bands with a small spectral weight appear that have zero spectral weight at $T = 0$. At $T = 150$–200 K, they form the almost dispersionless flat bands near the top of the valence band and the bottom of the conductivity band. These flat bands result in very narrow peaks in the density of states at the band edges. The top of the valence band at $T = 0$ is in the $(\pi/2, \pi/2)$ point and smoothly transforms to (π, π) in the paramagnetic phase. The temperature dependence of the insulator gap in the undoped La_2CuO_4 is shown in Fig. 5.22.

One important positive feature of the LDA+GTB band structure of the undoped La_2CuO_4 is the insulator in the paramagnetic phase that corresponds to experimental data and has not been reproduced by any conventional DFT approach.

Now we can discuss how the multiband *p–d* model is related to the simpler single-band Hubbard model or the *t–J* model. If we neglect all the excited states in Fig. 5.20, the remain four states ($|0\rangle$, $|b_{1g,\sigma}\rangle$, and $|1s\rangle$) form the basis of the effective single-band Hubbard model, so the energy scale that determines the applicability of the Hubbard model is given by min $(\varepsilon_{a1g} - \varepsilon_{b1g}, E_{1M} - E_{1s})$. In this effective Hubbard model, the quasiparticle Ω_c will be the low Hubbard band of holes, and Ω_{v1} will be the upper

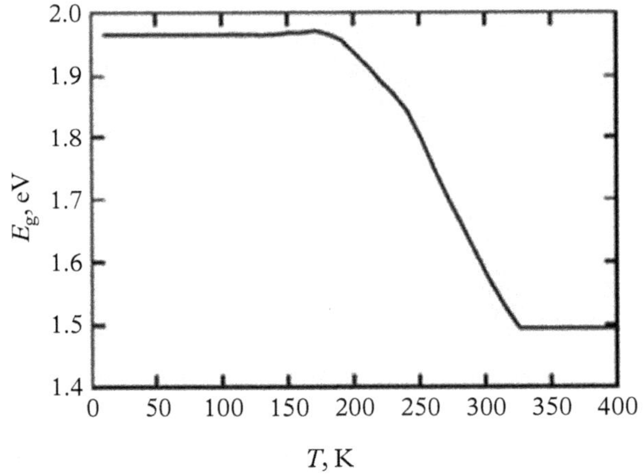

Figure 5.22 *Temperature dependence of the insulator gap in undoped* La_2CuO_4*. (From Makarov and Ovchinnikov, 2015)*

Hubbard band of holes, and the effective $U_{eff} = \Omega_c + \Omega_{v1}$. Further reduction of the Hubbard model to the low-energy $t - J$ model was given in section 2.7.4.

Below we will discuss the doping and temperature-dependent electronic structure of LSCO, assuming that the model parameters are fixed and only the hole concentration $n_h = 1 + x$ varies. All papers on the Hubbard and related models use this assumption. For a qualitative view, it is adequate but, for a quantitative comparison to experimental data, the doping dependence of the interatomic hopping parameter t_{ij} should be considered, due to the lattice parameter changes with doping (Gavrichkov *et al.*, 2007).

For the electronic structure and properties of the hole-doped cuprates, the effect of short-range charge and spin correlations is known to be important, as was discussed in section 2.7.4. The simplest approximation for the self-energy, eqn (2.95), where only the static spin correlations are taken into account, results in the doping-dependent Fermi surface with two Lifshitz transitions with changes in Fermi surface topology (see chapter 2, Fig. 2.20; Korshunov and Ovchinnikov, 2007). The ARPES experiments have revealed another picture, the doping- and temperature-dependent pseudogap. It will be considered in chapter 6.

The spin-fermion model. There is one more approach to the electronic structure and properties of cuprates that also takes into account the strong electronic correlations but considers the copper *d*-holes and the oxygen *p*-holes as two different bands; it is the spin-fermion model (Barabanov *et al.*, 1988; Emery and Reiter, 1988; Prelovsek, 1988; Stechel and Jennison, 1988; Zaanen and Oles, 1988; Matsukawa and Fukuyama, 1989). In the spin-fermion model, the relevant collective excitations are spin fluctuations, peaked at or near the antiferromagnetic momentum (π,π). The corresponding spin-fluctuation approach (Abanov *et al.*, 2003; Monthoux *et al.*, 2007; Scalapino, 2012; Val'kov *et al.*, 2017) naturally explains the *d*-wave symmetry of the superconducting state induced by the magnetic pairing and yields a non-Fermi-liquid behaviour of fermionic self-energy and optical conductivity (Hartnoll *et al.*, 2011; Chubukov *et al.*, 2014) in a rather wide frequency range, even when the magnetic correlation length is only a few lattice spacings. This approach does describe the temperature and doping dependence of the London penetration depth (Dzebisashvilli and Komarov, 2018) and precursors to magnetism (Schmalian *et al.*, 1998; Kyung *et al.*, 2006; Sedrakyan and Chubukov, 2010).

Polarons in GTB. Now we will discuss the polaronic effect of the electron–phonon interaction and how it may be incorporated in the GTB approach. It is well known that, in the adiabatic theory of the electron–ion interaction in solids, the bare ionic vibrations are high-energy plasmons and, due to the electron–phonon interaction, these plasma oscillations are renormalised in acoustic waves (Migdal, 1960; see also Kresin and Wolf, 2009, for a review). Due to several different ions in the high-T_c cuprates, there are also many optical phonon modes. As was shown by Giustino *et al.* (2008) by the ab initio calculations of the electron–phonon self-energy in cuprates, the largest contribution is provided by the following three modes: buckling, breathing, and apical breathing. All are related to the oxygen vibrations around copper ions. To incorporate this type of electron–phonon interaction in the GTB calculations, Makarov, Shneyder *et al.* (2015) developed the polaronic version of GTB, p-GTB. The total Hamiltonian includes electron, optical-breathing-phonon and electron–phonon interaction terms:

$$H = H_e + H_{ph} + H_{e-ph} \tag{5.4}$$

Here the electronic part H_e is represented by the three-band *p–d* model Hamiltonian, and the phonon part H_{ph} includes the optical mode of the in-plane breathing oxygen vibrations, neglecting its small dispersion. We write the phonon Hamiltonian in the local representation (where g denotes copper sites with vibrating in-plane oxygen neighbour ions; see Piekarz *et al.*, 1999, and Shneyder *et al.*, 2018):

$$H_{ph} = \sum_{g} \hbar\omega \left(f_g^+ f_g + \frac{1}{2} \right)$$

The electron–phonon interaction results from the modulation of the local electron energy ε_d by a changing crystal field (the diagonal contribution) and modulation of the amplitude of the *p–d* hopping t_{pd} (the off-diagonal electron–phonon interaction). The electron–phonon interaction Hamiltonian may be written as

$$H_{el-ph} = \sum_{g\sigma} M_d \left(f_g^+ + f_g \right) d_{g\sigma}^+ d_{g\sigma} + \sum_{g\delta\sigma} M_{pd\sigma} \left(f_g^+ + f_g \right) \left(d_{g\sigma}^+ p_{g+\delta,\sigma} + H.C. \right)$$

Following the GTB approach, we start with the exact diagonalisation of the local part of the total Hamiltonian (5.4), including a sum of the independent CuO_4 clusters. In the hole vacuum sector with $n_h = 0$, there is no electron–phonon interaction and the eigenstates are just the product of multielectron and multiphonon wavefunctions

$|0,v\rangle = |d^{10}p^6\rangle|v\rangle$, where $|v\rangle$ is a multiphonon state with $v = 0, 1, \ldots, N_{max}$ phonons. The parameter $N_{max} \gg 1$, it was found from the condition that the addition of one more phonon does not change the energy by more than 1%. Depending on parameters, it was equal to 100–200. For the single-hole $n_h = 1$ and the two-hole $n_h = 2$ sectors of the Hilbert space, we will take into account only the lower polaron states formed by the spin doublet $|b_{1g}\rangle$ and the singlet $|1s\rangle$ shown in Fig. 5.20. The reason for such an approximation is the different energy scales for the electronic terms shown in Fig. 5.20 (about 1 eV) and the polaronic terms (of the phonon frequency order). Then the single-hole and the two-hole polaronic eigenstates look like

$$|\sigma, i\rangle = \sum_{v=0}^{N_{max}} \left(c^d_{i,v} d^+_\sigma + c^b_{i,v} b^+_\sigma\right) |0,v\rangle \tag{5.5}$$

$$|2,j\rangle = \sum_{v=0}^{N_{max}} \left(c^{ZR}_{j,v} \frac{1}{\sqrt{2}} \left(d^+_{-\sigma} b^+_\sigma - d^+_\sigma b^+_{-\sigma}\right) + c^{dd}_{j,v} d^+_{-\sigma} d^+_\sigma + c^{bb}_{j,v} b^+_{-\sigma} b^+_\sigma \right) |0,v\rangle$$

The eigenenergies and the coefficients in the eigenstates are obtained numerically during the exact diagonalisation. The multielectron energy-level scheme is similar to one shown in Fig. 5.20, with the difference being that each term in this figure is split in $N_{max} + 1$ subterms of the Franck–Condon excitations. One example is shown in Fig. 5.23. The

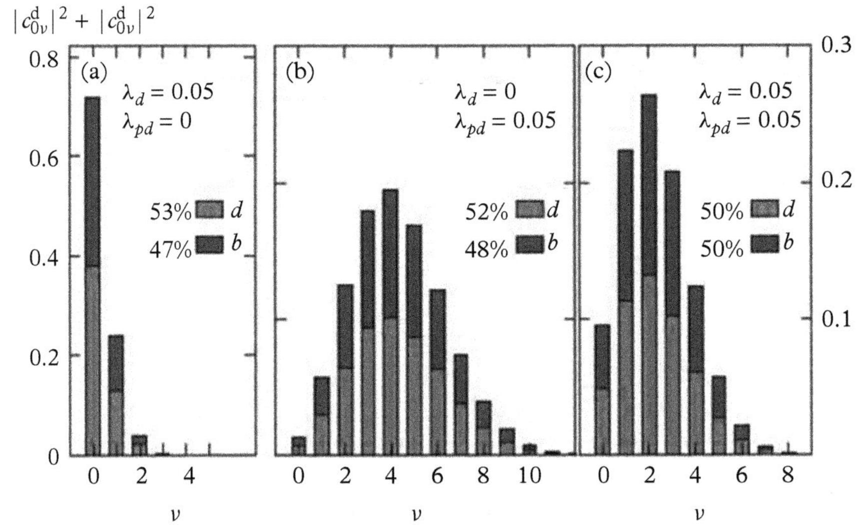

Figure 5.23 *The spectral weight distribution of the ground single-hole term $|\sigma, 0\rangle$ over the multiphonon states for different sets of the diagonal and off-diagonal electron–phonon interaction parameters λ_d and λ_{pd}. Dark and light grey indicate the oxygen p and copper d contributions, respectively. (From Shneyder et al., 2018)*

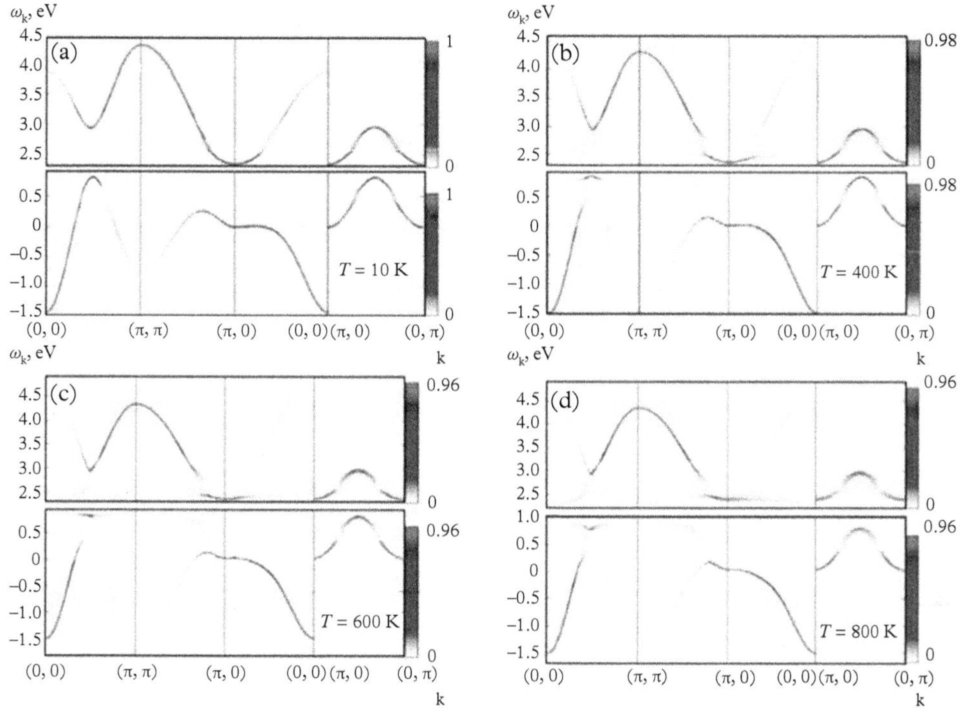

Figure 5.24 *Evolution of the band structure of Hubbard polarons in undoped La_2CuO_4 with increasing temperature in the regime of weak electron–phonon interaction with equal diagonal and off-diagonal dimensionless coupling $\lambda_d = \lambda_{pd} = 0.025$. At $T = 400$ K and above, the formation of a flat band at the top of the valence band is seen near the points $k = (0, 0)$ and $k = (\pi, \pi)$. (From Shneyder et al., 2018)*

weak diagonal electron–phonon interaction with the zero off-diagonal one in Fig. 5.23a results in a weak polaronic effect, so the main contribution to the spectral weight has no phonons. In contrast, the similar weak off-diagonal electron–phonon interaction with the zero diagonal one in Fig. 5.23b results in a strong polaronic effect with maximal weight for the four-phonon configuration. When both of the electron–phonon interaction contributions are treated as equal, as in Fig. 5.23c, they partially compensate each other, with the maximum shifted to two phonons.

After the exact diagonalisation of the intracell part of the Hamiltonian, the Hubbard operators are constructed in a similar way as described in appendix D and in chapter 2 for the conventional Hubbard model. The p-GTB version of GTB is the procedure in the basis of local multielectron and multiphonon states (Makarov, Shneyder *et al.*, 2015; Shneyder *et al.*, 2018). An electron is the linear combination of the quasiparticles that are called Hubbard polarons. Their band structure is temperature dependent due to thermal occupation of the multiphonon sublevels (see Fig. 5.24).

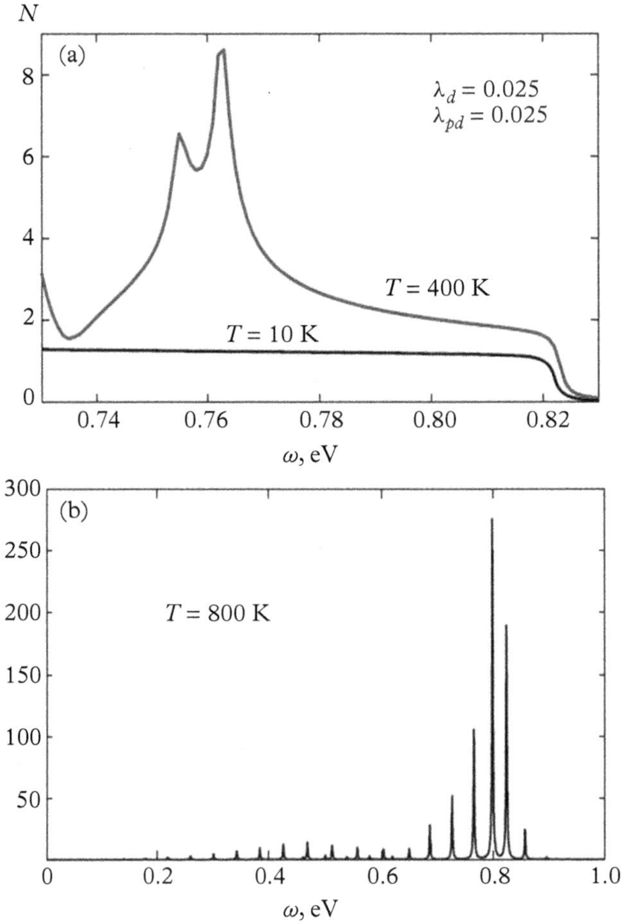

Figure 5.25 *Temperature-dependent density of states in undoped cuprate near the top of the valence band (a) and spectral function linewidth for* $k = \left(\frac{\pi}{2}, \frac{\pi}{2}\right)$ *enlarged by the Franck–Condon resonances (b) for* $\lambda_d = \lambda_{pd} = 0.025$. *(From Shneyder et al., 2018)*

The strongest temperature-induced transformation of the band structure occurs near the top of the valence band, the area which is important for hole-doped cuprates. Strong renormalisation of the electronic density of states with increasing temperature is shown in Fig. 5.25a. The other temperature-induced effect, shown in Fig. 5.25b, is the increase in the spectral function line width for the wave number $k = \left(\frac{\pi}{2}, \frac{\pi}{2}\right)$. This effect can be compared with the ARPES data from Shen *et al.* (2004) for undoped $Ca_2CuO_2Cl_2$. As is clear from Fig. 5.25b, the mechanism of the line width broadening is related to the Franck–Condon resonances.

Numerically, the increasing line width for the weak electron–phonon interaction in Fig. 5.25b is smaller than the experimental value; it agrees with the ARPES data for a stronger electron–phonon interaction when $\lambda_d = \lambda_{pd} = 0.1$. The Franck–Condon mechanism of broadening has been discussed previously by Shen *et al.* (2007). A similar effect of the electron–phonon interaction on the spectral function linewidth was studied early on within the $t - \mathcal{J}$ model by Bonca and Trugman (2001), Ishihara and Nagaosa (2004), and Mishchenko and Nagaosa (2004); see also the review by Mishchenko (2009).

Pseudogap and short-range correlations. Initially, the pseudogap was found by nuclear magnetic resonance experiments in $YBCO_{6+x}$ at small doping, where the maximum of the temperature dependence in the inverse relaxation time $1/T_1T$ was found at $T = 130$–150 K (Horvatić *et al.*, 1989; Warren *et al.*, 1989). Later, the gap was found for the spin excitations with the wavenumber $Q = (\pi/a, \pi/a)$ via inelastic neutron scattering (Rossat-Mignod *et al.*, 1991). The pseudogap has been found also in the infrared optical spectra, in the Raman spectra, and in the transport properties; see the review by Puchkov *et al.* (1996). Finally, the pseudogap is directly seen in the ARPES data; for example, in Bi-2212 cuprate, a decrease in the hole concentration results in the opening of a gap on the Fermi surface along the line $(\pi/a, 0)$–$(\pi/a, \pi/a)$ (Marshall *et al.*, 1996). The discussion of the pseudogap state based on the ARPES data can be found in the review by Kordyuk (2015).

The microscopic origin of the pseudogap is still unclear, with a lot of theoretical models invented, especially models of a spin/charge order (Millis and Norman, 2007; Yao *et al.*, 2011). The formation of low-temperature orthorhombic (LTO) and low-temperature tetragonal (LTT) stripes has been found by the extended X-ray absorption fine structure (EXAFS) method (Biankoni and Missori, 1994). The CDW long order at the peculiar hole concentration $x = 1/8$ or short-ordered CDW/SDW stripes around this concentration has been discussed by Tranquada (2014); see also the work by Fradkin *et al.* (2015). We will discuss here the short-range spin correlations effect on the electronic structure in the regime of strong electron correlations to demonstrate one possible mechanism for pseudogap formation. Certainly, this mechanism is not the only one; the total picture is still far from being clear. The recent Raman scattering data demonstrate the formation of the CDW at some temperature $T_{\mathrm{CDW}} < T^*$ for three different cuprates, Hg-1223, Y-123, and Hg-1201 (Loret *et al.*, 2019).

Close to the Fermi surface at small excitation energy the multiband Hubbard model may be reduced to the single-band Hubbard model that, in the limit of strong correlations, allows further reduction to the $t - \mathcal{J}^*$ model with three-site correlated hopping or to the simpler $t - \mathcal{J}$ model that neglects three-site correlated hopping. A comparison of these three models can be found in chapter 2 (see Fig. 2.15).

The LDA+GTB approach as well as the standard DFT approach results in the electronic band structure of quasiparticles with an infinite lifetime. Finite lifetime effects result from quasiparticle scattering that is beyond the Hubbard I mean field approximation in the dispersion equation (eqn (D9) in appendix D), which neglects the self-energy effects. As was discussed in section 2.6.4, the most important effects beyond the mean field approximation are related to those of quasiparticle scattering on spin and charge fluctuations. The static contribution of the short-range antiferromagnetic fluctuations to the self-energy given by eqn (2.95) with a self-consistent calculation of the spin correlation functions up to the ninth neighbours provides the doping-dependent quasiparticle band structure and the Fermi surface with two Lifshitz quantum phase transitions shown in chapter 2 in Fig. 2.20. Nevertheless, this picture contradicts the experimental ARPES data that have revealed doping-dependent pseudogap behaviour; see chapter 6. This means that short-order fluctuation dynamics should be also taken into account.

CPT with increasing cluster size makes it possible to incorporate all of the correlations inside the cluster exactly with the mean field Hubbard I approximation for the intercluster interactions. In chapter 2, Fig. 2.24, we showed the pseudogap formation obtained with 12-site CPT by Senechal and Tremblay (2004). The largest cluster in the CPT treatment of the Hubbard model up to now is a 16-atom 4×4 cluster. The doping- and temperature-dependent electronic structure within CPT will be discussed in chapter 6, in comparison with ARPES.

To conclude the discussion of when the Hubbard and the more general multiband Hubbard-like models can be useful, we want to mention that it is possible to study some properties of the cuprates' normal-state phase diagram and some specific mechanisms of superconducting pairing. Nevertheless, it is evident that pure electronic models alone cannot describe both normal and superconducting properties. Indeed, for superconductivity, phonon pairing (and maybe plasmon pairing) should be considered as being on equal footing with the electronic mechanism; the isotope effect is the one with the strongest argument. Describing the normal-state complicated phase diagram requires at least considering the electron–phonon interaction to understand the CDW effects. The CPT calculations of the charge correlation functions with 12- and 16-atom clusters within the Hubbard model (without the electron–phonon interaction) demonstrate only trivial and strongly decreasing charge fluctuations with doping (Kuzmin *et al.*, 2019). Moreover, the temperature dependence of the normal-state electronic properties like the pseudogap (which will be discussed in chapter 6) calculated by CPT (Kuzmin *et al.*, 2020) qualitatively corresponds to the trend found from ARPES measurements but, due to the large excitation energy between electronic terms, quantitative agreement is impossible without phonons and the electron–phonon interaction.

Concluding our discussion of the electronic structure of cuprates in the normal phase, we want to emphasise that electrons in the regime of strong electron correlations are Hubbard fermions, which are more complicated quasiparticles than those known from a conventional Fermi-liquid approach. Their properties are formed under the strong effect of the short-range antiferromagnetic order and short-range charge order. Otherwise, electrons are polarons, renormalised by the strong spin and charge fluctuations simultaneously.

ARPES: Experimental data and theoretical analysis. ARPES directly measures the occupied part of the single-particle spectral function (Damascelli *et al.*, 2003) with ever-increasing energy and momentum resolution with the electron wave vector in-plane projection. The cuprates are well suited for the ARPES technique because of their quasi-two-dimensional electronic structure. The $Bi_2Sr_2CaCu_2O_{8+y}$ (Bi2212) and $Bi_2Sr_{2-x}La_xCuO_{6+y}$ (Bi2201) families provide the most reliable data, due to having

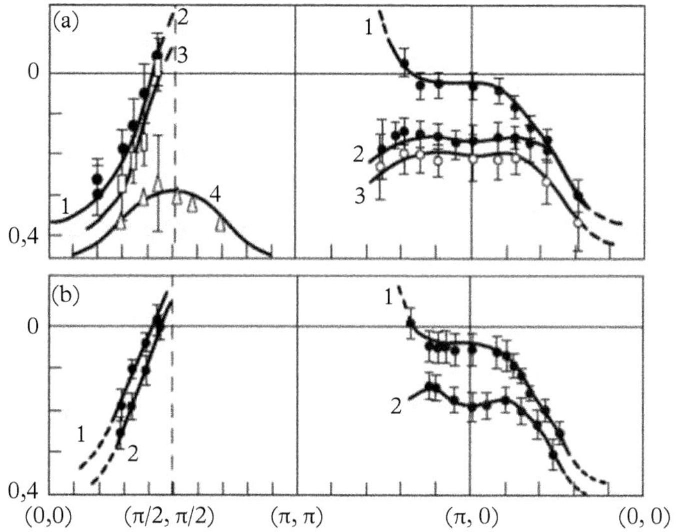

Figure 5.26 *(a) Electron dispersion for $Bi_2Sr_2Ca_{1-x}Dy_xCu_2O_{8+y}$ films with different hole concentrations: (1) 1% Dy, $T_c = 85$ K; (2) 10% Dy, $T_c = 65$ K; (3) 17.5% Dy, $T_c = 25$ K; and (4) 50% Dy, insulator. (b) Bi2212 crystals annealed in (1) air, $T_c = 85$ K, and (2) argon, $T_c = 67$ K. (From Marshall et al. 1996)*

pristine cleaved surfaces that protect the low-energy bulk electronic structure, owing to the weak van der Waals forces between the two Bi–O planes. In the absence of reliable ab initio electronic structure calculations, the information obtained from ARPES data on the electron dispersion is of primary importance.

First of all, the ARPES data prove the existence of the Fermi surface in the normal state, while its shape and size appear to be doping and temperature dependent, unlike the normal Fermi liquid (Fig. 5.26). ARPES has revealed that, instead of a conventional Fermi surface, there are arcs in the pseudogap state for the underdoped cuprates.

Second, the ARPES measurements directly confirm the $d_{x^2-y^2}$ symmetry of the superconducting gap (Fig. 5.27) and found the doping and temperature dependence of the gap amplitude. Over the past two decades, experiments have improved tremendously, allowing more precise information about the electronic structure, including the gap functions, to be obtained; see recent reviews by Hashimoto *et al.* (2014), Kordyuk (2015), and Vishik (2018). Both the superconducting gap and the pseudogap with a different angle and temperature dependences are distinguished. In the underdoped Bi-2212 single crystal with $T_c = 92$ K, the antinodal spectral function $A(k_{AN}, E)$ shows that there is a gap at the Fermi level that does not close at $T = T_c$ and survives as a pseudogap up to $T = T^* = 190$ K (Fig. 5.28).

Recently, three different sets of ARPES data beyond the simple energy distribution curve and the momentum distribution curve spectra have given strong support for the complicated temperature dependence of the pairing gap and the pseudogap (Reber *et al.*, 2012; Kondo *et al.*, 2013; Kondo *et al.*, 2015; see also Vishik, 2018). For the temperature

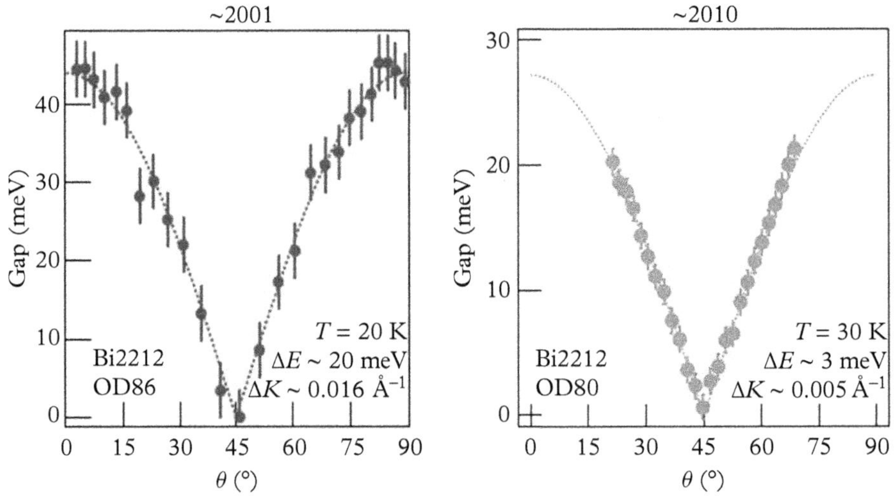

Figure 5.27 *The angle dependence of the* $d_{x^2-y^2}$ *symmetry superconducting gap: (left) measurements taken in 2001; (right) measurements taken in 2010. A strong improvement in the experiment resolution is evident. (From Hashimoto et al., 2014)*

below some pairing temperature $T < T_{pair}$, the nodal-point Fermi surface with a pairing gap and a pseudogap can be distinguished. Point nodes are typical for *d*-wave pairing, so the point nodes are related to some pairing at $T < T_{pair}$ (in chapter 6, the notation $T_c{}^*$ will be introduced for the pairing above T_c, but here we follow the authors' notation; see Fig. 5.29). In the region $T_{pair} < T < T^*$, there is a Fermi arc near the nodal direction, and a pseudogap near the antinodal direction. Above T^*, the Fermi surface restores as in a normal metal. The question of whether it is a normal Fermi liquid or some other strange metal has not been solved yet; see the recent discussion by Reber *et al.* (2019).

Recent progress in increasing energy and momentum resolution provides the possibility of ARPES developing as a true self-energy spectroscopy. Such an example is given by Li *et al.* (2018). Figures 5.30 and 5.31 present the raw data sets and the matching two-dimensional fits from a $(Bi, Pb)_2Sr_2CaCu_2O_8$, lightly underdoped, $T_c = 85$ K superconductor. The Pb doping removes the 5×1 superstructure that can add contamination to the antinodal portions of the spectra, and the experimental photon energies 9 eV have been judiciously chosen to select only the antibonding band from the bilayer split-band structure. The data were taken over a rather large energy range to capture the large-energy-scale features as well as the low-energy scale features such as superconducting gaps.

The conventional Gor'kov formalism is used to analyse the ARPES data in both the normal and the superconducting phases with the matrix electron Green's function

$$G(k,\omega) = \frac{1}{(\omega - \sum)^2}\begin{pmatrix} \omega - \sum + \xi_k & -\phi \\ -\phi & \omega - \sum - \xi_k \end{pmatrix} \tag{5.6}$$

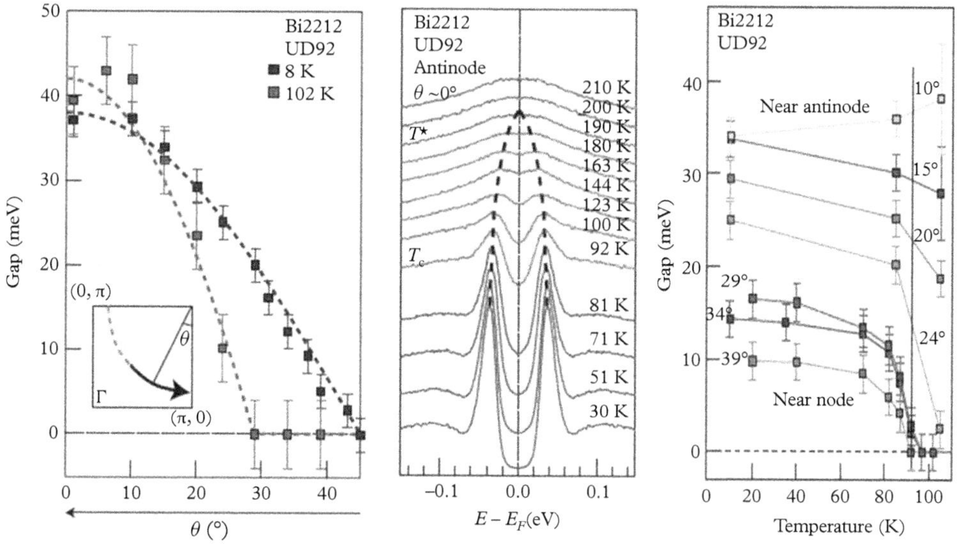

Figure 5.28 *Gaps for Bi2212 underdoped crystal at* $T_c = 92$ K *and below and above* T_c*. The superconducting gap at* $T = 8$ K *closes in the nodal point, while the pseudogap at* $T = 102$ K *exists for some finite range of angles; zero pseudogap means the finite-size Fermi arc near the nodal direction (left panel). The temperature dependence of the symmetrised spectra at the antinode (middle panel) shows the pseudogap closing at* T^**. The temperature dependence of the gap size at various momenta is shown in the right panel. The near nodal gap closes at* T_c *following BCS-like behaviour, whereas the gap around the antinode does not close at* T_c*. (From Hashimoto et al., 2014)*

and the ARPES spectral function $A(k,\omega) = -ImG_{11}(k,\omega)/\pi$. Here ξ_k is the bare-band dispersion, the complex self-energy $\Sigma = \Sigma' + i\Sigma''$, and $\phi = Z(\omega)\Delta$ is the pairing order parameter with the superconducting gap Δ and $Z(\omega) = 1 - \Sigma'/\omega$. Assuming some phenomenological function for the imaginary self-energy of the marginal Fermi-liquid form by Varma *et al.* (1989),

$$\Sigma''(\omega) = \lambda\sqrt{\omega^2 + (\pi k_B T)^2} + \Gamma_0$$

(more specifically, the power-law liquid form by Reber *et al.* (2019)), Li *et al.* (2018) have found the real part of the self-energy using the Kramer–Kronig relation. Thus, the real and imaginary parts of the self-energy (or self-energy spectroscopy) have been reproduced with 14 fitting parameters for different temperatures. As the authors explained:

> In essence, we were using a conventional quasiparticle-like approach for both the normal and superconducting states, with the exception that there is a complicated and large self-energy term Σ that captures the interactions (and the departure from true Landau quasiparticle physics). That both the superconducting state and normal-state spectra can

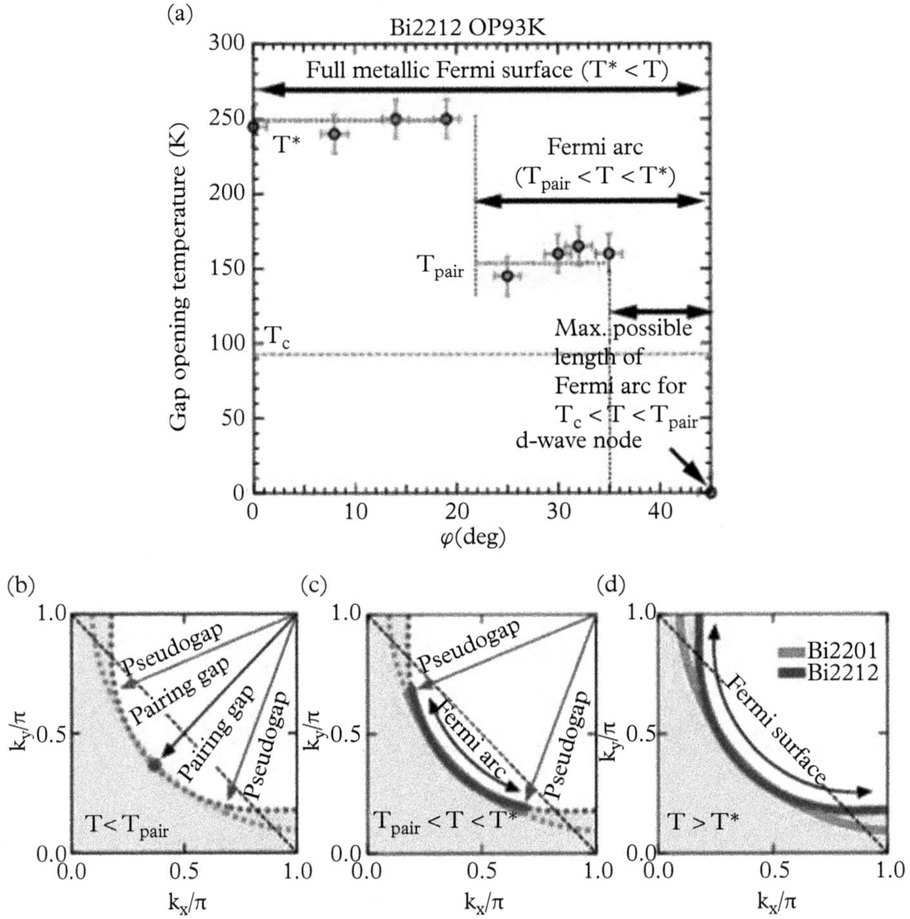

Figure 5.29 *(a) Temperatures where gaps appear at the node and the antinode. (b–d) Fermiology in three temperature regimes: (b) the Fermi surface in point nodes, the pairing gap, and the pseudogap at* $T < T_{pair}$*; (c) the Fermi arcs and the pseudogap at a higher temperature* $T_{pair} < T < T^*$*; and (d) a full Fermi surface above* T^**. (From Kondo et al., 2013)*

> be so well described within this semi-conventional approach was in itself a surprising finding: the generally broad structures and large backgrounds observed in ARPES spectra of cuprates have often been described as being so far outside the realm of conventional physics that such semi-conventional quasiparticle-like approaches were considered untenable. (Li *et al.*, 2018, p. 2).

The data from Fig. 5.30 were taken near the middle of the zone (partway between the node and antinode; cut shown in Fig. 5.30s) and reveal the evolution of the various features as a function of temperature from deep in the superconducting state into the normal state. The key new information obtained in this approach is the self-energy

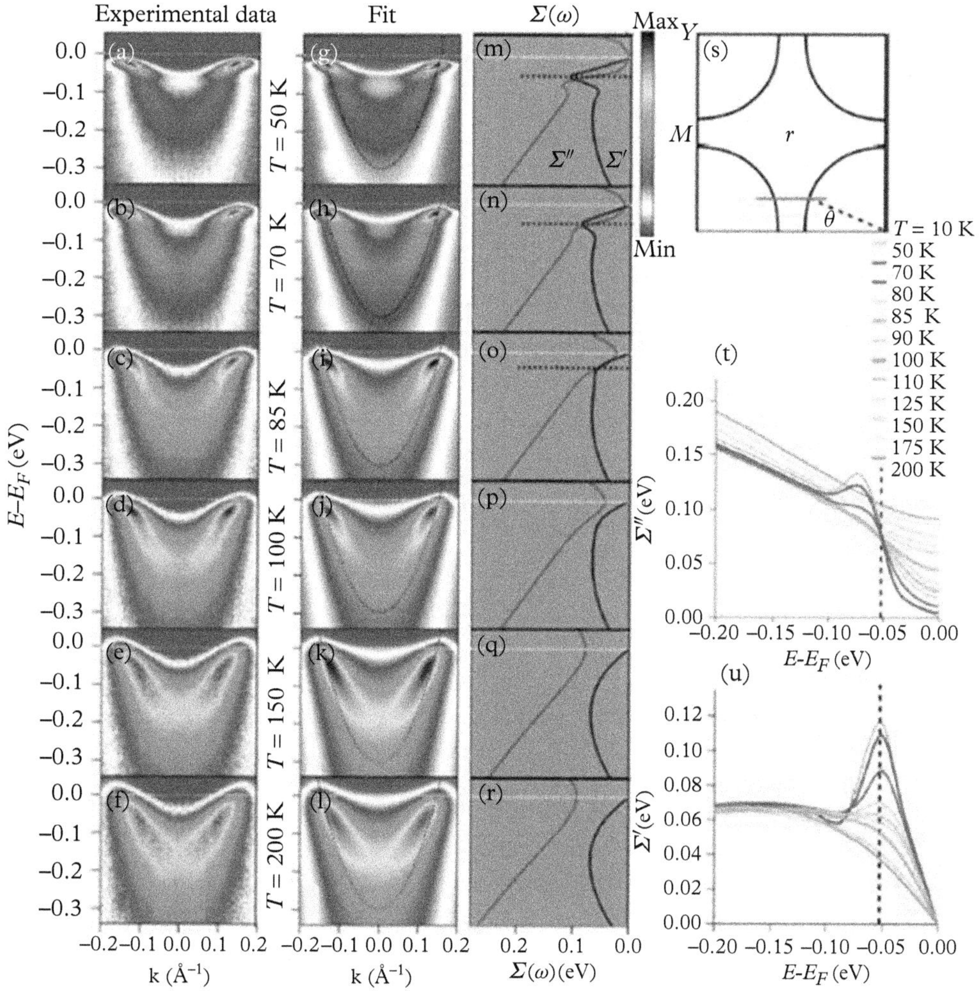

Figure 5.30 *Temperature-dependent ARPES spectra and self-energies from a lightly underdoped ($T_c = 85$ K) $(Bi, Pb)_2Sr_2CaCu_2O_8$ superconductor. (a–f) Measured spectra for the wavevector shown in panel s for different temperatures. (g–l) Two-dimensional fits, with the main unknown parameters being the superconducting gaps, the electronic self-energies, and the bare electron band, where the dotted curve is the bare-band dispersion. (m–r) Corresponding self-energies extracted from the fit spectra on the left. (t) The imaginary part of the self-energy (or scattering rate) that is strongly reduced near the Fermi energy as temperature is lowered. Along with this is a strong kink or mass enhancement in the real part of the self-energy (u). (From Li et al., 2018)*

$\Sigma = \Sigma' + i\Sigma''$, which was found to vary dramatically as a function of energy and temperature and from node to antinode, as shown in Fig. 5.30t and u, and Fig. 5.31. Upon cooling from high temperature, the main evolution of both parts of the self-energy and the pairing gap begins at T_{pair} above T_c, as can be seen in Fig. 5.31a. The pairing

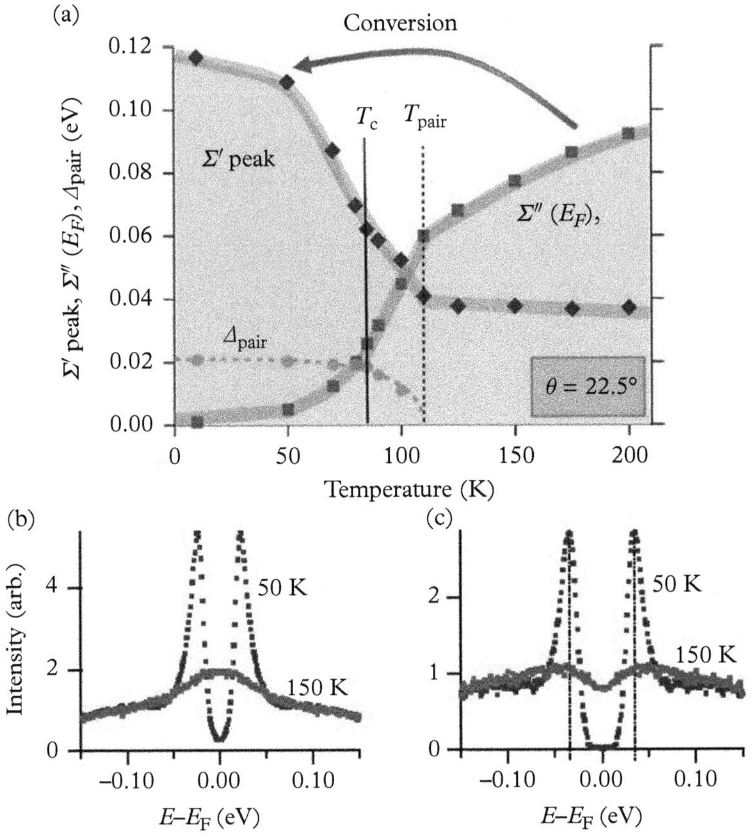

Figure 5.31 *(a) Temperature dependence of the gaps and self-energies of the midzone cut (shown in Fig.5.30s); black diamonds represent the real part Σ' along the dashed line in Fig. 5.30u, $\Sigma''(E_F)$ is shown by gray squares, and the circles show superconducting gaps extracted from two-dimensional fits. Upon cooling from high temperature, the main evolution of the parameters begins at T_{pair}, not at T_c. Symmetrised energy distribution curves at k_F of the (b) midzone (Fermi surface angle at 22.5°) and (c) near antinodal (32°) spectra at 50 K and 150 K. The energy distribution curve of the near antinodal spectrum at 150 K shows a strong depletion of the spectral weight near the Fermi level (the pseudogap), whereas the midzone one does not show any gap feature at 150 K. (From Li et al., 2018)*

above T_c will be discussed in the chapter 6. The symmetrised energy distribution curves are shown in Fig. 5.31 at k_F of the midzone (Fermi surface angle at 22.5°; panel b) and near the antinodal (32°) spectra at 50 K and 150 K (panel c). The energy distribution curve of the near antinodal spectrum at 150 K shows a strong depletion of the spectral weight near the Fermi level (pseudogap), whereas the midzone one does not show any gap feature at 150 K.

The results by Li *et al.* (2018) also bring into question the concept of adiabaticity, that is, the assumption that the electrons are much faster or have a higher energy scale than the bosons. Their data show that the electron energy at the antinode is ~40 meV, which is nominally the same energy as the bosonic degrees of freedom most discussed as coupling to the electrons (neutron resonance mode at ~40 meV (Fong *et al.*, 1999), the B_{1g} phonon at ~40 meV (Cuk *et al.*, 2004), and the LO phonon at ~55 meV (Lanzara *et al.*, 2001)).

Summarising the discussion of the ARPES data concerning the pseudogap state, one may note that the electrons in the nodal parts of the Brillouin zone have metallic properties, while, in the antinodal parts, they are similar to the electrons in disordered semiconductors.

It should be noted that a conventional type of ARPES data analysis usually uses the symmetrical Lorentzian line shape for the momentum distribution curve that is standard in the Fermi-liquid approach. There are some indications in the literature for an asymmetric momentum distribution curve (Matsuyama and Gweon, 2013), which has been found in both underdoped and optimally doped LSCO (Yoshida *et al.*, 2007) and $Ca_{2-x}Na_xCuO_2Cl_2$ (Shen *et al.*, 2005).

In the limit of strong electron correlations, the standard perturbation theory that results in the Green's function (5.6) cannot be justified. The generalised Dyson equation in this limit (see eqn (D8)) contains a new function, the so-called strength operator, in addition to the self-energy (Ovchinnikov and Val'kov, 2004). The analysis of the ARPES data in the strong correlation limit has revealed both symmetric and asymmetric contributions (Ovchinnikov *et al.*, 2014). In the underdoped cuprates, a strongly correlated regime seems to be more appropriate.

The concentration- and temperature-dependent electronic spectral function has been calculated recently for the Hubbard model within CPT by Kuzmin *et al.* (2020). At zero temperature for different hole concentrations, it is possible to perform the CPT calculation with exact diagonalisation of the 16-atom 4×4 clusters as the unit cell of the supercell two-dimensional Hubbard model, while, for finite temperature at fixed doping, the maximal cluster size is 12 atoms. The overview of the doping- and temperature-dependent momentum distribution curve is shown in Fig. 5.32.

For comparison, the generalised mean field with the static spin correlation function contribution to the self-energy (Korshunov and Ovchinnikov, 2007) and the doping-dependent Fermi surface with two Lifshitz transitions is shown in Fig. 5.32a. As can be seen in Fig. 5.32b and c, the exact treatment of the intracluster short-range correlations up to the ninth coordination sphere transforms continuous Fermi contours in the arcs. Similar transformations for the doping-dependent Fermi surface at $T = 0$ have also been obtained through cluster-DMFT (Civelli *et al.*, 2005; Stanescu and Kotliar, 2006; Haule and Kotliar, 2007; Ferrero *et al.*, 2009; Sakai *et al.*, 2009), via the composite-operators approach (Avella, 2014). Three different regimes of electronic structure, the doped Mott insulator (sector selective phase), the momentum-space differentiated Fermi liquid, and the nearly isotropic Fermi liquid, have been found with increasing doping concentration via DCA (Gull *et al.*, 2008, 2009, 2010). A similar transformation with temperature has not yet been found.

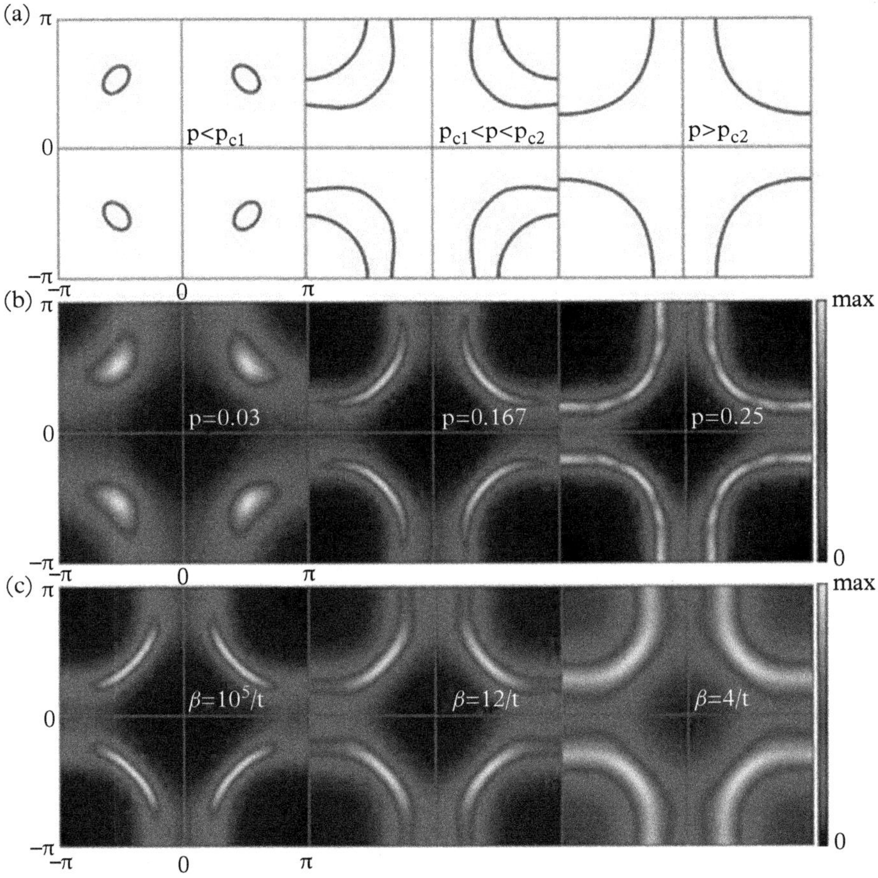

Figure 5.32 *(a) The schematic representation of the mean field Fermi surface for three values of doping divided by two Lifshitz transitions. (b) The cluster perturbation theory (CPT) result using a* 4×4 *cluster at* $T = 0$ *for similar doping values. (c) The CPT result using a 12-site cluster at* p $= 0.167$ *for different inverse temperatures* $\beta = 1/kT$. *The doping and temperature evolutions of the electronic spectra and the pseudogap are similar. (From Kuzmin et al., 2020)*

The left column in Fig. 5.32 demonstrates that, at low doping/temperature, the spectral weight in the antinodal point at zero Fermi angle (see Fig. 5.28) is practically zero. More precisely, when $A(AN,0) \ll A(N,0)$, where AN indicates the antinodal spectral weight, and N indicates the nodal spectral weight, this area of low doping and temperatures may be called a strong pseudogap regime. The middle column in Fig. 5.32 demonstrates that, at higher doping /temperatures, the pseudogap persists but the antinodal spectral weight is increasing, so that $A(AN,0) \leq A(N,0)$; this area may be called a weak pseudogap regime. Finally, with further increases in doping/temperature, the Fermi surface is restored in the whole Brillouin zone at the line $T^*(x)$, which separates the normal Fermi-liquid behaviour at $T > T^*(x)$ from the pseudogap regime at $T < T^*(x)$. Thus, instead of producing two sharp Lifshitz quantum phase transitions at $T = 0$, as would be obtained with the mean field approach, CPT results in two smooth crossovers, both with increasing doping and temperature. The precision of modern CPT is not enough to distinguish whether the transition from the pseudogap state into the normal Fermi-liquid state at the $T^*(x)$ line is a smooth crossover or a phase transition. At least, at zero temperature and for doping x^*, where $T^*(x^*) \to 0$, the measured logarithmic peculiarity of the electronic specific heat (Fig. 5.9) indicates the phase transition.

These transformations of the electronic structure are related to evolution of the spin correlation function with temperature and doping (Fig. 5.33). While, for zero doping and temperature, the spin correlation function changes its sign in a manner similar to that for the antiferromagnetic order $C_1 < 0$, $C_2 > 0$, $C_3 > 0$, $C_4 < 0$, $C_5 > 0$, and so on, with increasing doping and temperature, only $C_1 < 0$ and $C_2 > 0$ demonstrate

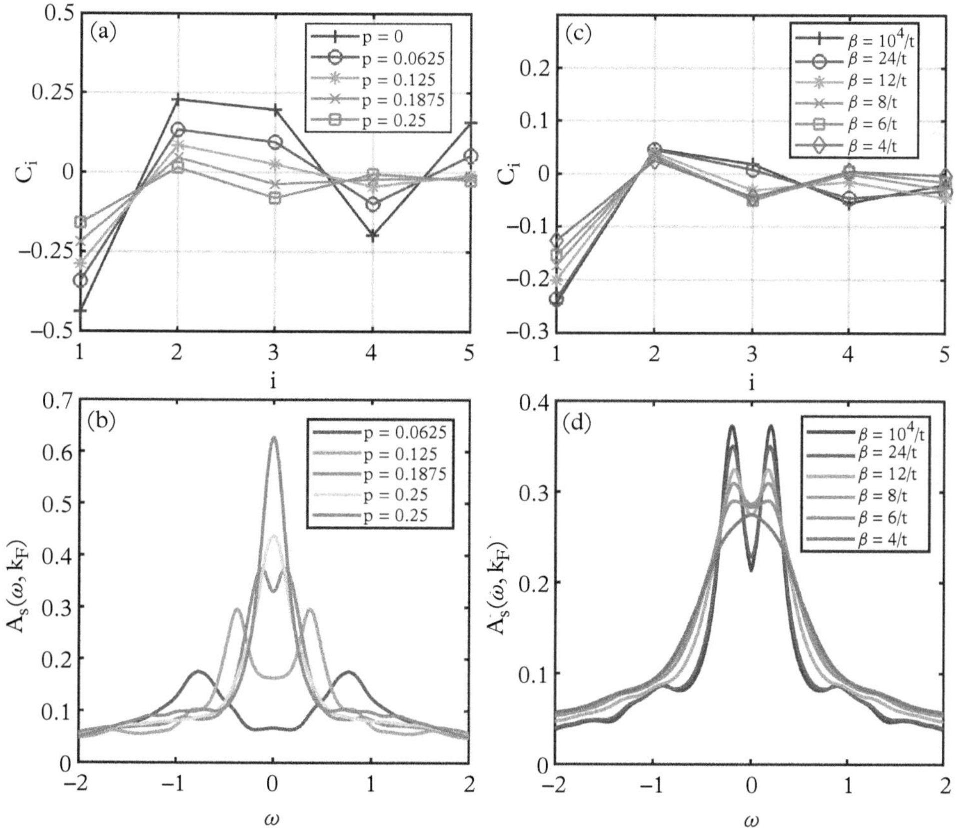

Figure 5.33 *Spin correlation functions* $C_i = \langle S_0^z S_i^z \rangle$ *up to the fifth coordination sphere for different (a) doping levels, (c) temperatures, and (b, d) symmetrised antinodal spectral functions. (From Kuzmin et al., 2020)*

the antiferromagnetic order; more distant correlations violate the antiferromagnetic short -range order. Formation of the pseudogap in the symmetrised spectral function (Fig. 5.33b and d) is characterised by a flat energy dependence, separating the local minima from the local maxima. From Fig. 5.33, it is clear that smooth crossover from a weak pseudogap to a strong pseudogap with doping at $T = 0$ occurs for $p_{c1} \cong 0.15$, and from a weak pseudogap to a normal Fermi liquid for $p_{c2} \cong 0.20$. Considering the change of the symmetrised spectral function for doping, $p = 0.167$, with temperature, for $\beta = 4/t$ (where t is the nearest-neighbour hopping parameter), a single peak typical for the normal Fermi liquid is seen. The dip at the Fermi level almost disappears at $\beta = 6/t$, which is a sign of the pseudogap formation temperature T^* (Norman *et al.*, 1998; Kanigel *et al.*, 2006). A weak pseudogap appears at $\beta = 12/t$, while, at $\beta = 24/t$, a strong pseudogap appears. In summary, with an increase in temperature, a strong pseudogap smoothly transforms into a weak pseudogap and then, at T^*, transforms into a Fermi liquid.

As concerns the charge correlation functions, the same CPT calculations within the Hubbard model have revealed only a trivial fast decrease in correlations with distance, for all doping and temperature values. It is clear that, to study possible charge density correlations, the electron–phonon interaction should be involved simultaneously with electron correlations.

5.5 Theoretical Models of the Mechanism of High T_c in Cuprates: Phonon versus Magnetic Pairing or Phonon and Magnetic Pairing Together?

So far in this book, we have discussed three possible mechanisms for the high T_c in cuprates, and there are more in the literature. The phonon mechanism, together with polaronic effects and low-energy plasmons, was discussed in detail in chapter 2. In this chapter, we will discuss data on contributions to magnetic pairing besides that from static antiferromagnetic coupling. Plakida and Oudovenko (2014) have considered superconductivity in the two-dimensional extended Hubbard model with a large intersite Coulomb repulsion V in the limit of strong correlations $U \gg t$ to elucidate the spin-fluctuation mechanism of high-temperature superconductivity.

The theory is based on the Mori projection technique in the equation of motion method for the Green's functions in terms of the Hubbard operators. In the two-subband regime for the Hubbard model for $U > 6t$, the spin–electron kinematic interaction evolves from complicated commutation relations for the Hubbard operators (Hubbard, 1965). This interaction brings about the weak exchange interaction $J = 4t^2/U$, due to the interband hopping and, at the same time, the intraband hopping results in a much stronger kinematic interaction $g_{sf} \sim W \gg J$ of electrons with spin excitations. Here $W = 4t$ is the half-bandwidth of the free electrons. Since the intersite Coulomb repulsion $V > J$ destroys the superconductivity induced by the antiferromagnetic exchange interaction, the spin-fluctuation pairing in the second order of the kinematic interaction should be taken into account. Plakida and Oudovenko (2014) have calculated the doping dependence of superconducting T_c for various values of U and V and have shown that, as long as V does not exceed the kinematic interaction, that is, $V < W$, the d-wave pairing is preserved. A similar conclusion for the $t - \mathcal{J}$ model had been obtained earlier by Plakida and Oudovenko (1999) and by Prelovsek and Ramsak (2005).

Let us start with ruling out the mechanisms that violate some of the principal aspects of cuprate physics. First of all, now we have solid arguments that the Fermi surface exists (being more complicated than normal metals); these arguments result mainly from data from ARPES and quantum oscillations in a strong magnetic field. This fact excludes the bipolaronic models and any other models with boson charge carriers.

At the second stage of our analysis, let us get together the key facts from experimental data that must be explained by the theory. We consider the following main facts:

1 the isotope effect
2 the quasi-two-dimensional crystal structure
3 the strong electron correlations

Like any crystal, cuprates have phonons and electron–phonon interactions, so the BCS-type electron–phonon mechanism of pairing is inevitable. The isotope effect just emphasises the necessity of the phonon mechanism. The electron–phonon interaction theory of pairing without restriction of weak coupling was discussed in section 2.2 and, for cuprates, it is given in the review by Kresin and Wolf (2009). Polaronic effects appear to be important in the regime of strong electron–phonon coupling. Nevertheless, a hard question to the BCS-type electron–phonon mechanism concerns the d-wave symmetry of pairing. Why is it not s-wave? Usually, phonon pairing in an s-channel is stronger than in a d-channel.

The quasi-two-dimensional crystal structure of cuprates results in a drastic change in the plasmon structure and transforms the high-frequency three-dimensional plasma oscillations into acoustic ones with the linear dispersion law; thus, its contribution to the superconducting pairing may be comparable to that from phonon coupling (see section 2.6). The other effect of the layer crystal structure of cuprates is their magnetic quasi-two-dimensional behaviour. The exchange interaction of coppers spins $S = 1/2$ within the Heisenberg model results in the three-dimensional long-range antiferromagnetic order for undoped cuprates like La_2CuO_4, which is quickly destroyed by low doping. Nevertheless, the strong in-plane exchange interaction provides short-range antiferromagnetic correlations that have been well documented with many experimental methods, including neutron scattering. The antiferromagnetic correlations may themselves provide the superconducting pairing (see section 2.7).

Strong electron correlations in the CuO_2 layers result from the specific crystal chemistry of $3d$ metal oxides: due to the correlation effect, all undoped cuprates are Mott–Hubbard insulators rather than half-filled band metals. It is the strong electron correlations that hinder the development of the theory of superconductivity in cuprates. That is why, in discussing the magnetic mechanism of pairing, we will restrict ourselves to the strongly correlated version given in section 2.7.4.

Now let us ask ourselves whether any of the main three mechanisms, phonon, plasmon, and magnetic, alone can explain at least qualitatively the main features of the high-T_c superconductivity in cuprates. We do not know how to explain the d-wave symmetry of the superconducting phase in the phonon and plasmon approaches; it has to be postulated that the coupling in the d-channel in eqn (2.14′) with $l = 2$ is larger than the s-wave coupling with $l = 0$. For conventional weakly correlated electrons, usually the s-wave coupling is dominant. We also do not know how to explain the shape of the $T_c(x)$ superconducting dome dependence within the phonon and the plasmon mechanisms.

Both of these questions, the gap symmetry and the dome shape, can be answered by postulating a magnetic mechanism of pairing in the regime of strong correlations. Nevertheless, the isotope effect requires the electron–phonon interaction to be considered with magnetic pairing on equal footing as well as the general argument of the presence of the electron–phonon interaction in every material. The main problem of strong correlation physics is that there is no convenient perturbation theory, so the non-perturbative complicated approaches have to be used. In general, this area of physics is far from providing a reliable and complete solution.

Below, we present some of the results from the simultaneous treatment of strong electron correlations and the electron–phonon interaction. As was discussed in section 5.4.5, a polaron version of the multielectron GTB approach can be constructed (Makarov *et al.*, 2015; Shneyder *et al.*, 2018). It results in the temperature-dependent band structure of the Hubbard polarons shown in Fig. 5.24. Up to now, the theory of superconductivity in such an approach has not been developed. Magnetic pairing and renormalisation of the normal-state electron dispersion have been described by Shneyder and Ovchinnikov (2006) within the $t - \mathcal{J}$ model with additional phonon coupling; the BCS-like equation for the d-wave gap has the total coupling

$$\lambda_{tot} = \frac{1-x}{2} J + \lambda_{ph} \theta \left(|\varepsilon_q - \mu| - \omega_D \right) \tag{5.7}$$

Here $\mathcal{J}$ is the antiferromagnetic exchange interaction, x is the concentration of doped holes, $\lambda_{ph} = \left(\frac{(1+x)(3+x)}{8} - \frac{3C_{01}}{4} \right) G$, C_{01} is the nearest-neighbour spin correlation function (which is negative, due to the short-order antiferromagnetic correlations and increases the phonon coupling), and G is the electron–phonon coupling. The θ function restricts the pairing to the electron energy region near the Fermi level of the Debye frequency width, while the exchange interaction is effective for all electrons in the band. Three phonon modes with the strongest electron–phonon interaction have been considered: Cu-O octahedral breathing, buckling modes, and the Cu-apical O breathing mode according to Giustino *et al.* (2008). Due to the symmetry of the apical O oscillations normal to the CuO2 plane, this mode did not contribute to the in-plane electron–phonon interaction effects, while the buckling-mode support and the breathing mode give a negative contribution to the d-wave pairing:

$$G = g^2_{buckl}/\omega_{buckl} - g^2_{breath}/\omega_{breath} \tag{5.8}$$

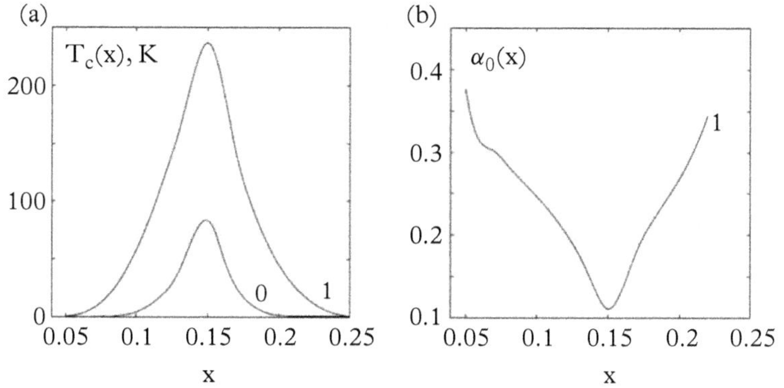

Figure 5.34 *(a) The critical-temperature dependence and (b) doping dependence of the oxygen isotope exponent for the effective electron–phonon interaction parameter G/J, which is indicated next to the respective curves. (From Ovchinnikov and Shneyder, 2010)*

The absence of the apical mode contribution in the effective electron–phonon interaction parameter G agrees with the site-selective isotope substitution data shown in Fig. 5.16. The similar conclusions on the opposite contributions of the breathing and buckling modes to the d-wave pairing have also been obtained by Bulut and Scalapino (1996), Piekarz *et al.* (1999), and Honerkamp *et al.* (2007).

In the BCS-like mean field gap equation, all of the model parameters were calculated within the LDA+GTB approach but the electron–phonon coupling G. The oxygen isotope effect for the different ratio of G/J may be positive (negative) when the buckling-mode pairing is more (less) than the pair-breaking breathing phonon contribution to the total electron–phonon coupling G. The ratio G/J was fitted to obtain the $\alpha(x)$ dependence (Fig. 5.34) in accordance with the experimental data. The positive $\alpha(x)$ values with a minimum are obtained only when $G > 0$ (Fig. 5.35a). The value $G = J$ makes it possible to obtain at least a qualitative agreement to the experimental data in Fig. 5.15. Without a magnetic contribution to pairing (the upper curve in Fig. 5.35b), the pure electron–phonon mechanism cannot reproduce the doping dependence of the isotope-effect exponent while, for $G/J = 1.1$ (the lower curve), the $\alpha(x)$ dependence looks similar to the experimental data shown in Fig. 5.34 (Shneyder and Ovchinnikov, 2009).

Two contributions to the critical temperature T_c have been calculated, from magnetic pairing alone for $G/J = 0$, and both phonon and magnetic pairing simultaneously for $G/J = 1$. We want to mention that both contributions are not a sum of two independent effects; due to the presence of the nearest-neighbour spin correlation function in λ_{ph} (eqn (5.7)), there is some more deep interrelation of magnetic and phonon pairing for strongly correlated electrons. Finally, it is clear from the left side of Fig. 5.34 that magnetic and phonon contributions to the T_c value are of the same order of magnitude.

Despite the clear physical picture obtained by Ovchinnikov and Shneyder (2009, 2010), their theory is the mean field theory that cannot be rigorously justified. That is why further theoretical studies are required that will take into account phonon, plasmon, and magnetic pairing with adequate treatment of the polaronic effects and the effects

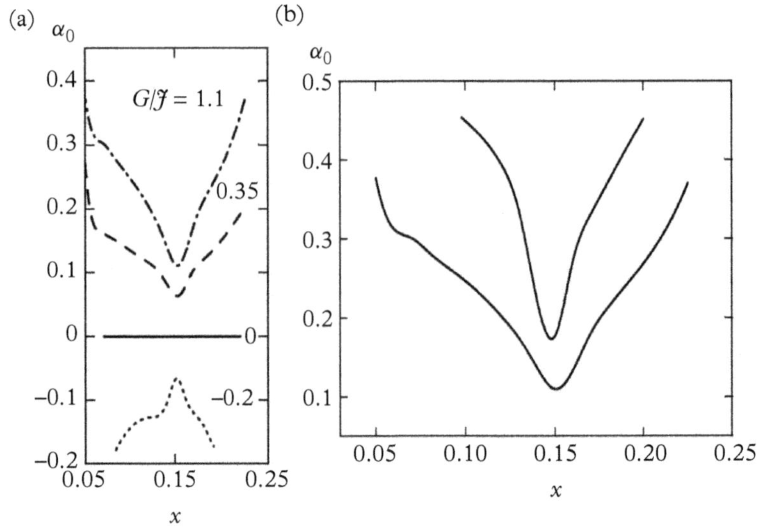

Figure 5.35 *The hole concentration dependence of the isotope effect on* T_c *for different values of the phonon G to magnetic* $\mathcal{J}$ *coupling ratio* ($G/\mathcal{J}$)*. (a). Phonon coupling may be positive as well as negative (see eqn 5.8); only positive G provides the shape with a minimum at optimal doping, and* $G/\mathcal{J} = 1.1$ *is found from fitting the experimental data. (b). For* $\mathcal{J} = 0$ *(upper curve), the shape of the isotope exponent versus doping did not correspond to experimental data, while, for* $G/\mathcal{J} = 1.1$*, there is a qualitative agreement with experimental data. (From Ovchinnikov and Shneyder, 2010)*

of strong electron correlations. The combination of magnetic and phonon pairing has also been considered by Plakida and Oudovenko (2013). They studied the extended Hubbard model in the limit of strong correlations by also taking into account the nearest-neighbour Coulomb repulsion V and electron–phonon coupling. It was found that the d-wave pairing with high T_c is mediated by the strong kinematic interaction of electrons with spin fluctuations. Contributions coming from the weak Coulomb repulsion V and phonons turned out to be small, since only the $l = 2$ harmonics of the interactions give a contribution to the d-wave pairing.

In summary, we still have no definite answer for what the main mechanism of pairing in cuprates is. So, we will use Fellini's idea from the movie *8.5* and discuss several possible endings for the story of superconductivity in cuprates:

A Phonons with polaronic effects together with acoustical plasmons in the strong coupling regime provide the main mechanism of pairing.

B Magnetic pairing in the regime of strong electron correlations alone explains all of the main features of the d-wave pairing in cuprates.

C Phonons, plasmons, and magnetic pairing of polarons work together.

The advantages and disadvantages of each scenario have been discussed above.

6

Inhomogeneous Superconductivity and the 'Pseudogap' State of Novel Superconductors

The high-T_c oxides, as well as some other superconducting systems, display many properties in the normal state above T_c which are drastically different from those for conventional materials. This unusual normal (it is called 'normal' because there is finite resistance) state was dubbed the 'pseudogap' state. The first observation of this state was reported in 1989, that is, shortly after the discovery of high-T_c superconductivity. Nuclear magnetic resonance measurements demonstrated the presence of an energy gap for spin excitations (Alloul *et al.*, 1989; Warren *et al.*, 1989).

Subsequent studies reveal a number of other features. It is clear that, above T_c, the sample is in a peculiar state which is intermediate between fully superconducting and normal (see Table 6.1). Indeed, like any normal metal, the sample displays a finite resistance. In addition, such a key feature as macroscopic phase coherence does not persist above T_c. However, one can observe some features typical for the superconducting state, such as an energy gap, anomalous diamagnetism, and the isotope effect. The observation of the anomalous diamagnetism (the Meissner effect; see section 6.1.1) is especially important. Note also that the manifestation of the 'pseudogap' state depends on the doping level, being the strongest for the underdoped region.

6.1 'Pseudogap' State: Main Properties

Let us first compare the behaviour of novel superconductors above T_c with that for usual systems; see Table 6.1, which directly contrasts the two classes.

The most fundamental feature is the anomalous diamagnetism (the Meissner effect), usually observed below the critical temperature. As for the region above T_c, the magnetic response of conventional metals is relatively small and almost temperature independent.

As we know, superconductors have a finite resistance above T_c, whereas, below the critical temperature, they are in a dissipationless state ($R = 0$). In addition, ac transport behaves differently above and below T_c, namely, above T_c, with high accuracy (see e.g.

Superconducting State: Mechanisms and Materials. Vladimir Z. Kresin, Sergei G. Ovchinnikov, and Stuart A. Wolf, Oxford University Press (2021). © Kresin, Ovchinnikov, Wolf. DOI: 10.1093/oso/9780198845331.003.0006

Table 6.1 *'Pseudogap' state versus conventional superconducting and normal states.*

	Superconducting state ($T < T_c$)	**Normal state ($T > T_c$)**	**"Pseudogap" state ($T_c^* > T > T_c$)**
Resistance	R = 0	R ≠ 0	R ≠ 0
Energy gap	$\Delta \neq 0$	$\Delta = 0$	$\Delta \neq 0$
Anomalous diamagnetism	Yes	No	Yes
Macroscopic phase coherence	Yes	No	No
Josephson effect	Yes	No	"Giant" effect
Isotope effect	Yes	No	Yes
Impedance Z	ReZ ≠ ImZ	ReZ = ImZ	ReZ ≠ ImZ

Landau *et al.*, 2004): ReZ = ImZ (where Z is the surface impedance). In contrast, in usual superconductors, below T_c, one can observe the strong inequality Re$Z \neq$ ImZ.

The superconducting state is also characterised by macroscopic phase coherence. For example, such a remarkable phenomenon as the Josephson effect is directly related to this feature.

As one can see from Table 6.1, the 'pseudogap' state appears to be rather peculiar. Let us describe its properties in more detail.

6.1.1 Anomalous Diamagnetism above T_c

Usual normal metals display relatively weak response to a small external magnetic field. Indeed, the electronic gas is characterised by low Pauli paramagnetism. The magnetic susceptibility of real metals consists of several contributions (see e.g. Ashcroft and Mermin, 1976), and the resulting response might be diamagnetic, but the total susceptibility is almost temperature independent. The situation above T_c in the cuprates appears to be drastically different. Unusual magnetic properties of the 'pseudogap' state have been observed with use of various techniques.

Scanning SQUID microscopy was used by Iguchi *et al.* (2001) to study the underdoped LSCO compound. This technique allows one to create a local magnetic image of the surface, its 'magnetic' map. The critical temperature of the underdoped LSCO films was $T_c \approx 18$ K. A peculiar inhomogeneous picture has been observed: the film contains diamagnetic domains, and their presence persists up to 80 K(!). The total size of the diamagnetic regions grows as the temperature decreases (Fig. 6.1). As a result, the diamagnetic response appears to be strongly temperature dependent; this is a very unusual feature of the materials.

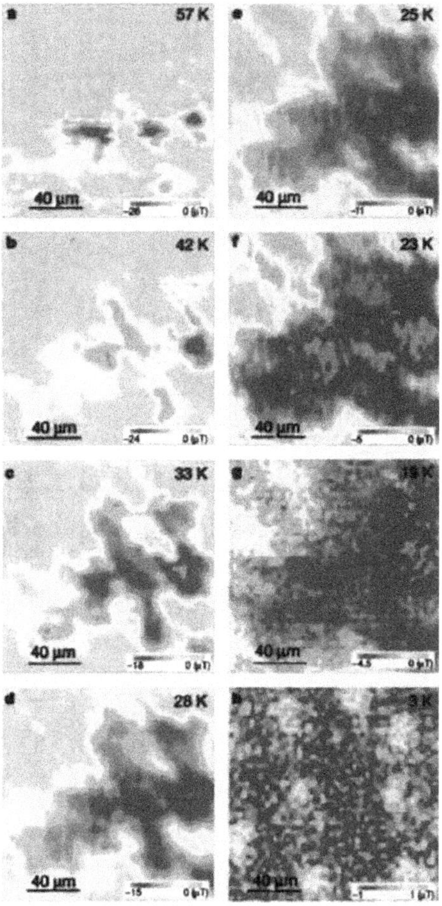

Figure 6.1 *Development of magnetic 'islands' with temperature (magnetic imaging of* $La_{2-x}Sr_xCuO_4$ *films).*

A very interesting experiment was described by Sonier *et al.* (2008); they measured the μSR relaxation. This method allows one to measure directly the local magnetic field. As we know, μsR spectroscopy (μsR stands for muon spin relaxation; see chapter 4) is the experimental technique based on the implantation of spin-polarised muons into the sample (see e.g. the reviews by Keller (1989) and Schenck (1985)). In many respects, this technique is similar to the electronic paramagnetic resonance method and, even more, to the nuclear magnetic resonance method. The method allows one to obtain information about the local magnetic field, since it affects the spin precession. The study by Sonier *et al.* (2008) is of special interest, because the μsR method, unlike scanning tunnelling microscopy, is a *bulk* measurement. According to the data, one observes a strong magnetic response above T_c. More specifically, the relaxation process

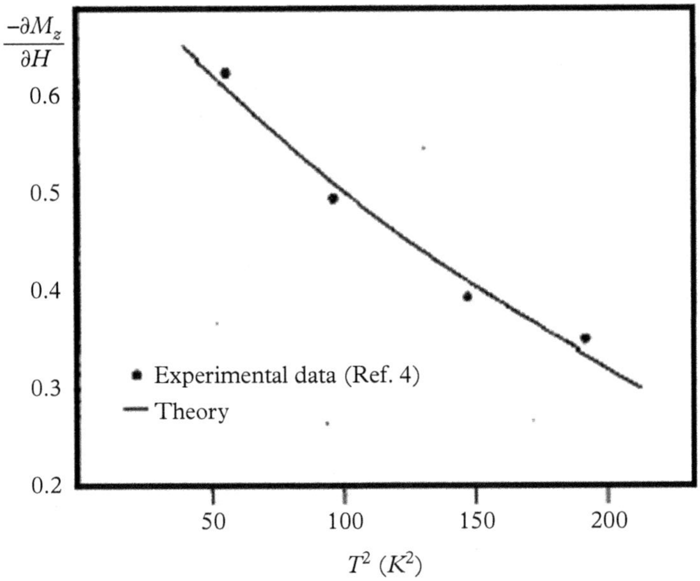

Figure 6.2 *Diamagnetic susceptibility for the $Tl_2Ba_2CuO_{6+\delta}$ (T_c = 15 K): experimental data; solid line indicates theory.*

is described by the function $G(t) = \exp[-(\Lambda t)^{\beta}]$, where β = const. The YBCO and LSCO samples have been studied. The relaxation was compared with that for Ag, which is not a superconducting metal. If the state at $T > T_c$ is a usual normal metal, one should expect behaviour similar to that for Ag. However, according to Sonier *et al.* (2008), the behaviour of Λ for LSCO and YBCO at $T > T_c$ is different: the quantity $\Lambda - \Lambda_{ag}$ strongly depends on temperature, and this corresponds to the presence of magnetic regions above T_c.

A strong temperature-dependent diamagnetic response has been also observed by Bergemann *et al.* (1998) by using the torque magnetometry technique for the overdoped $Tl_2Ba_2CaO_{+\delta}$ compound above $T_c \approx 15$ K. As in the case for LSCO, the diamagnetic moment was also strongly temperature dependent (Fig. 6.2). Torque magnetometry was also employed recently by Wang *et al.* (2005) to study the Bi2212 compound. Similarly, a diamagnetic response was observed. It is essential that the analysis ruled out fluctuations as a key source of the observed diamagnetism.

6.1.2 Energy Gap

The presence of an energy gap above T_c has been observed using various experimental methods. Even the title 'pseudogap state' reflects the existence of the gap structure. In connection with this, it is worth noting that this title is confusing, since we are dealing not with a 'pseudogap' but with a real gap, that is, with real dip of the density of states in the low-energy region.

The energy gap ε is a fundamental microscopic parameter. As we know, $\varepsilon = 0$ in the normal phase and opens up at T_c. One should note, however, that the energy gap is, indeed, an important parameter, but its presence, unlike the order parameters, is not a crucial factor for superconductivity. For example, one can observe 'gapless' superconductivity (see section 3.3.2), caused by the pair-breaking effect, for example, by the presence of localised magnetic moments.

Let us start with tunnelling spectroscopy, which allows one to perform the most detailed and reliable study of the gap spectrum. The data obtained by Renner *et al.* (1998) for an underdoped Bi2212 crystal are shown in Fig. 6.3. The scanning tunnelling microscopy of the crystals cleaved in vacuum was employed. One can directly see the dip in the density of states (energy gap) which persists above $T_c \approx 83$ K and stays up to ~200 K (!). One should stress that the gap structure changes *continuously* from the superconducting region $T < T_c$ to the 'pseudogap' state $T > T_c$; there is no noticeable change at T_c. This can be considered an indication that the gap structure above T_c is related to superconducting pairing.

Tunnelling spectroscopy was also employed by Tao *et al.* (1997). They concluded that the gap observed at $T > T_c$ reflects the presence of the paired electrons above T_c.

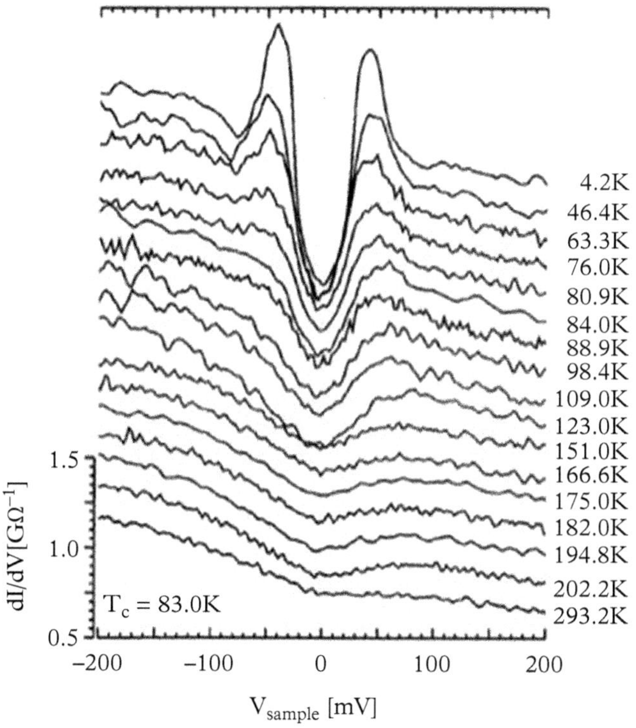

Figure 6.3 *Tunnelling conductance of states for the underdoped Bi2212 crystal.*

Interlayer tunnelling spectroscopy was employed by Suzuki *et al.* (1999); the presence of an energy gap at $T > T_C$ has also been confirmed.

The presence of an energy gap for spin excitations has been established using nuclear magnetic resonance. Actually, as was mentioned above, the 'pseudogap' state was initially observed using this method. The gap has been observed in YBCO for different nuclei in both Knight-shift and spin-relaxation rate experiments for ^{89}Y (Alloul *et al.*, 1989), ^{63}Cu (Warren *et al.*, 1989; Walstedt and Warren, 1990), and ^{17}O (Takigawa *et al.*, 1989). It follows also from optical data (Fig. 6.4).

Photoemission spectroscopy has also revealed the presence of an energy gap at $T > T_c$ (see the reviews by Shen and Dessau, 1995 and Randeria and Campuzano, 1997). For example, it has been demonstrated that the energy gap persists in an underdoped sample of Bi2212 at $T > T_c$ (Loeser *et al.*, 1996 and Ding *et al.*, 1996 come to a similar conclusion). Photoemission spectroscopy (Kanigel *et al.*, 2008) has revealed the presence of disconnected Fermi arcs, some of which correspond to those in a normal metal. One can also observe gapped regions which correspond to the superconducting state. The picture is consistent with pairing above T_C.

We described above the data which presents direct spectroscopic observations of the gap structure above T_c. A gap in the spectrum can also be inferred from heat capacity data. One should note that the measurements carried out by Loram *et al.* (1994) and Wade *et al.* (1994) were some of the first observations of the 'pseudogap' state. The

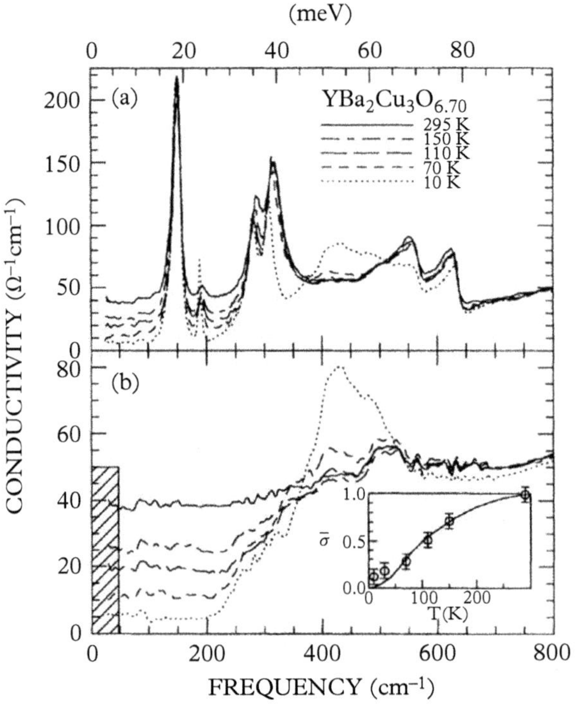

Figure 6.4 *The optical conductivity of $Yba_2Cu_3O_{6.7}$ along the c-axis.*

measurements of the Sommerfeld constant $\gamma(T)$ display a loss of entropy caused by the gap structure. The data for the energy gap $\Delta(T)$ are derived from values of the electronic entropy $S(T, x)$ for $0.73 < x < 0.97$. Again, it has been observed that the energy gap persists for $T > T_c$, and the effect is especially strong for underdoped samples.

6.1.3 The Isotope Effect

Another interesting property of the 'pseudogap' state is the strong isotope effect. This effect has been observed by Lanzara *et al.* (1999) for LSCO using X-ray absorption near-edge spectroscopy (XANES). The effect has been also observed by Temprano *et al.* (2000) for the $HoBa_2Cu_4O_8$ compound. The slightly underdoped $HoBa_2Cu_4O_8$ sample was studied by neutron spectroscopy. As we know (see e.g. the review by Mesot and Furrer, 1997), the opening of the gap, which could be associated with the 'pseudogap', affects the relaxation rate of the crystal-field excitations. The isotopic substitution $^{16}O \rightarrow {}^{18}O$ leads to a drastic change in the value of the pseudogap temperature T_c^* ($T_c^* \approx 170$ K $\rightarrow T_c^* \approx 220$ K). Such a large isotope shift corresponds to a large value of the isotope coefficient. Note that, unlike the case for typical superconductor, its value is negative.

6.1.4 The 'Giant' Josephson Effect

Tunnelling measurements demonstrating the presence of pairing states above T_c were performed by Zhou *et al.* (2019). They fabricated high-quality LSCO/La_2CuO_4/LSCO tunnelling junctions, where the La_2CuO_4 forms the tunnelling barrier. Various doping levels ($x = 0.1, x = 0.12, x = 0.14$, and $x = 0.15$), up to the optimal value, were studied. According to the study, the energy gap is not lost at T_c but continues smoothly into the pseudogap region. The analysis of the observed noise shows that that the picture of the pairing above T_c cannot be explained by fluctuations. This demonstrates that intrinsic pairing exists above the resistive transition.

The 'giant' Josephson proximity effect is another interesting phenomenon observed in the 'pseudogap' region above T_c (Bozovic *et al.*, 2004). In this study, $La_{0.85}Sr_{0.15}CuO_4$ films ($T_c \approx 45$ K) were used as electrodes, and the underdoped LaCuO compound ($T_c' \approx 25$ K) formed the barrier, which was prepared in the c-geometry (coherence length $\xi_c \approx 4$A). The measurements were performed at $T_c' < T < 35$ K, so the barrier was in the 'pseudogap' state. Since $T > T_c'$, we are dealing with a *SNS* junction (where S stands for superconductor, and N stands for normal conductor). As is known, for such a junction, the thickness of the barrier should not exceed the coherence length, which is of order of ξ_c. However, the Josephson current was observed for thicknesses of the barrier up to 200 A (!). Such a 'giant' effect cannot be explained using conventional theory. We will discuss this effect in detail in section 6.4.3.

6.1.5 Transport Properties

The microwave properties and ac transport high-T_c oxides have been studied by Kusco *et al.* (2002). It was shown that, above T_c, that is, in the normal state, $\mathrm{Re}Z \neq \mathrm{Im}Z$, where Z

is the surface impedance. This is an unusual property, since, in ordinary normal metals, with a high degree of precision, the real and imaginary parts of the impedance are equal (see e.g. Landau *et al.*, 2004), that is, $\mathrm{Re}Z_n \approxeq \mathrm{Im}Z_n$. The observed inequality $\mathrm{Re}Z \neq \mathrm{Im}Z$ is typically observed in superconducting materials.

Interesting data on thermal conductivity for various high-T_c compounds were described by Sun *et al.* (2006). The measurements performed for different doping levels reveal the absence of universal dependence for the thermal flow and indicate the importance of strong inhomogeneity in the cuprates.

6.2 The Inhomogeneous State

As was described above, intensive experimental studies reveal a number of unusual features of the cuprates above the resistive transition T_c. We described anomalous diamagnetism (which strongly depends on temperature), an energy gap structure, a strong inequality $\mathrm{Re}Z \neq \mathrm{Im}Z$ (where Z is the surface impedance), a 'giant' Josephson proximity effect, and an isotope effect on T_c^*.

The key issue, which is still controversial, is related to the nature of the 'pseudogap' state. The main question is whether this state is a normal metal or a superconductor. At first sight, the observation of finite resistance provides the answer to this question. Nevertheless, as was described above (see Table 6.1), in reality, the situation is more complicated.

The most fundamental fact is the observation of the Meissner effect above $T_c = T_c^{res.}$ ($T_c^{res.}$ corresponds to the transition into a dissipationless macroscopic state with zero resistance). It is essential that the contribution of fluctuations is not sufficient to explain the data for the underdoped region (Caretta *et al.*, 2000; Lasciatari *et al.*, 2002); indeed, the temperature scale for T_c^* is very large. Therefore, we are dealing with a serious challenge. Namely, one needs to explain the coexistence of finite resistance and anomalous diamagnetism.

The properties of the 'pseudogap' state are caused by the intrinsic inhomogeneity of the metallic phase (Ovchinnikov *et al.*, 1999, 2001, 2002; Kresin *et al.*, 2006; Kresin and Wolf, 2012).

6.2.1 The Qualitative Picture

Consider an inhomogeneous superconductor, so that $T_c = T_c(r)$. The system contains a set of superconducting regions 'islands' embedded in a normal metallic matrix (Fig. 6.5). The properties of such system correspond to the 'pseudogap' state. Indeed, the normal metallic matrix provides finite resistance whereas the existence of the superconducting 'islands' leads to the diamagnetic moment and energy gap structure.

As was mentioned above (section 6.1.1), the presence of diamagnetic 'islands' has been observed directly by Igushi *et al.* (2002); see Fig. 6.1. The superconducting 'islands' are embedded in the normal metallic matrix. As a result, we are dealing with the interface of the superconductor with the normal metal, and the proximity effect (see section 8.4)

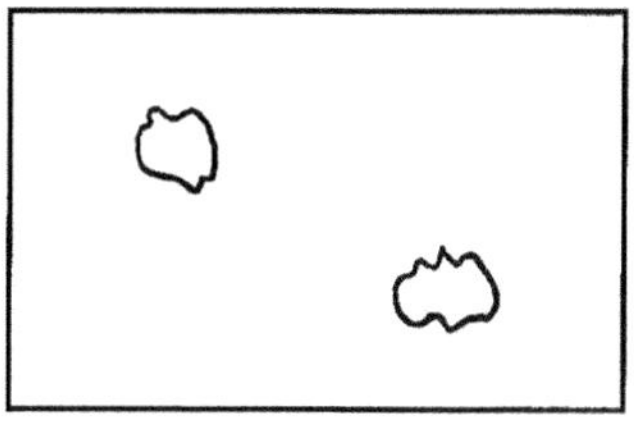

Figure 6.5 *Inhomogeneous structure: 'islands' are characterised by values of T_c's higher than the matrix.*

plays a crucial role. The proximity effect determines a minimum length scale of the superconducting regions which is of the order of the coherence length ξ_0. Indeed, if a superconducting 'island' is smaller than ξ_0, its superconducting state is totally depressed by the proximity effect between the superconducting region and the normal metallic phase.

As temperature decreases towards T_c, the size of the superconducting regions increases, as does the number of 'islands'. The critical temperature T_c corresponds to the percolation transition, that is, to the formation of a macroscopic superconducting region (an 'infinite cluster', in terms of the percolation theory; see e.g. Shklovskii and Efros, 1984, and Stuffer and Aharony, 1992), and to phase coherence and dissipationless superconducting phenomena.

In conventional superconductors, the resistive and Meissner transitions occur at the same temperature, T_c. The picture in the 'pseudogap' state is different. The resistive and Meissner transition are split. The Meissner transition (the appearance of the diamagnetism) occurs at T_c^*, whereas the resistive transition, that is, the transition to the macroscopic dissipationless state, takes place at T_c, and $T_c < T_c^*$. Note that in the papers by Ovchinnikov *et al.* (1999, 2001), the notations T_c^{res} and T_c^{Meis} were used. Here we use only the notations T_c and T_c^*, so that $T_c \equiv T_c^{\text{res}}$, and $T_c^* \equiv T_c^{\text{Meis}}$.

It is important to note also that the 'pseudogap' state in the region $T_c < T < T_c^*$ is not a phase-coherent one; each superconducting 'island' has its own phase. At $T = T_c$, a macroscopic superconducting region is formed and, below T_c, we are dealing with a macroscopically phase-coherent phenomenon.

6.2.2 The Origin of Inhomogeneity

As was noted above, the presence of superconducting 'islands' embedded in a normal metallic matrix implies an inhomogeneity of the compound. There are two possible scenarios for such an inhomogeneous structure:

1. an inhomogeneous distribution of pair breakers
2. an inhomogeneous distribution of carriers leading to spatial dependence for the coupling constant

Both scenarios lead to an inhomogeneous superconductivity.

Pair breaking can be caused by localised magnetic moments (Abrikosov and Gor'kov, 1961;see section 3.3.1). Qualitatively, pair breaking can be visualised in the following way. A Cooper pair consists of two carriers with opposite spins (for singlet pairing; this is the case for both *s*- and *d*-wave scenarios). A localised magnetic moment acts to align both spins in the same direction and this leads to pair breaking. It is also known, that for *d*-wave pairing, non-magnetic impurities are also pair breakers.

As was discussed in section 3.3.1, a pair-breaking effect leads to a depression in T_c. Therefore, a non-uniform distribution of pair breakers makes the critical temperature spatially dependent: $T_c \equiv T_c(\boldsymbol{r})$. Such a distribution is caused by the statistical nature of doping. The region which contains a larger number of pair breakers is characterised by a smaller value for the local T_c.

Note that the pseudogap phenomenon is strongly manifested in the underdoped region. At optimum doping, the distribution of pair breakers becomes more uniform and, as a result, $T_c^* \simeq T_c$. The 'pseudogap' state at $T > T_{c;opt}$ is characterised by the gap caused not by pairing but by some other competing order, for example, short-range SDW or CDW (see section 5.4.5). In the underdoped state, the spectrum is affected by both contributions.

As was mentioned above, an inhomogeneity can also be caused by an inhomogeneous distribution of carriers (see Ovchinnikov *et al.*, 2001). Note that the superconducting state of inhomogeneous systems has been discussed previously (see e.g. Ghosal *et al.*, 1998, 2001; Cho *et al.*, 2019).

The picture described is directly related to the concept of phase separation. This concept was introduced by Gor'kov and Sokol (1987) shortly after the discovery of the high T_c oxides and then was studied in many papers (e.g. Sigmund and Mueller, 1994). The concept implies the coexistence of metallic and insulating phases. The picture of the 'pseudogap' state described above is the next step in the inhomogeneous scenario. Namely, as well as the mixture of metallic and insulating phases, the metallic phase itself is inhomogeneous; we are dealing with the coexistence of normal and metallic regions within the metallic phase.

6.2.3 The Percolative Transition

As was mentioned above, the transition at $T \equiv T^{res}$ corresponds to the formation of a macroscopic superconducting phase. This transition is of a percolative nature (Ovchinnikov *et al.*, 1999; Mihailovich *et al.*, 2002; Alvarez *et al.*, 2005). In terms of percolation theory, the transition to the dissipationless macroscopic state corresponds to the formation of an 'infinite' cluster. The percolative nature of the transition at $T = T_c$ is due to the statistical nature of doping. The picture is similar to that introduced in manganites, which represent another family of doped oxides (see chapter 9). Manganites (e.g. $La_{0.7}Sr_{0.3}MnO_3$) are characterised by the presence of ferromagnetic metallic regions embedded in the low-conducting paramagnetic matrix above $T_c \equiv T_{Curie}$, where T_{Curie} is the Curie temperature. At $T_c = T_{Curie}$, one can observe a percolative transition to the macroscopic ferromagnetic metallic state. The transition from the 'pseudogap' to the macroscopic dissipationless state is also of a percolative nature.

6.2.4 Inhomogeneity: Experimental Data

We describe above (section 6.1.1) an interesting study of the La-based compound (Igushi *et al.*, 2001) performed using the scanning tunnelling microscopy technique with magnetic imaging (Fig. 6.1), which has directly demonstrated the presence of diamagnetic 'islands' embedded in a normal matrix and shows the percolative picture as $T \to T_c$. The same can be said about μSR spectroscopy (Sonier *et al.*, 2006), which is also described in section 6.1.1.

Nanoscale inhomogeneity was described in an interesting paper by Zeljkovic *et al.* (2012). In addition to observing the inhomogeneous picture in $Bi_{2-y}Sr_{2-y}CaCu_2O_{8+x}$, the authors were able to correlate the structure of the 'pseudogap' state with different types of oxygen dopants. It turns out that the apical oxygen vacancies are especially effective. This is an interesting observation, because we know that the apical oxygen is very important for the charge transfer, and such a defect, indeed, can act as a strong pair breaker.

One key study using scanning tunnelling spectroscopy was performed by Gomes *et al.* (2007) and Parker *et al.* (2010). The measurements were performed on $Bi_2Sr_2CaCu_2O_{8+\delta}$ samples at finite temperature, so that the energy spectrum was measured above T_c, in the 'pseudogap' region. These data have provided crucial information about local values of the gap and its evolution with temperature. The study of this evolution has led to the conclusion that the observed gap spectrum does indeed correspond to superconducting pairing. It is essential that the distribution of gaps turns out to be strongly inhomogeneous. As a whole, the presence of pairing persists for temperatures which greatly exceed those of T_c, especially for the underdoped samples. For example, for the sample with the doping level $x \approx 0.1$ ($T_c \approx 75$ K), the pairing persists up to $T_c^* \approx 180$ K. It is also interesting that, from these measurements, one can determine the local values of the gap $\Delta_{\text{loc.}}$ and T_p^{loc}. One can conclude that the ratio $2\Delta_{\text{loc}}/T_p^{loc}$ appears to be equal to $2\Delta_{\text{loc}}/T_p^{loc} \approx 7.5$. Such a large value of this ratio greatly exceeds that for the usual superconductors (according to the BCS theory, $2\Delta/ T_c \approx 3.52$) and corresponds to very strong electron–phonon coupling (see section 3.5). An inhomogeneous structure with superconducting regions ~3 nm in size was observed by Lang *et al.* (2002).

Measurements of the temperature dependence of the critical current and the vortex structure (Darhmaoui and Jung, 1998; Yan *et al.*, 2000; Jung *et al.*, 2000) have revealed a strong inhomogeneity in YBCO and TBCCO samples. The presence of two different phases and an inhomogeneous structure of the order parameter has also been demonstrated.

As was stressed above, one should distinguish the intrinsic critical temperature T_c^* and the resistive T_c so that $T_c \equiv T_c^{res}$. According to the approach described, the pairing gap is related to T_c^*, since the Cooper pairing occurs at first at this temperature. That is why the changes in the value of T_c^* lead to corresponding changes in the gap value. As for

$T_c \equiv T_c^{res}$, this temperature describes the percolation transition to the macroscopic superconducting state and is not related directly to the pairing interaction. Correspondingly, changes in T_c should not affect the gap value. It is interesting that precisely this picture has been observed in Zn-substituted $Bi_2Sr_2CaCu_2O_{8+y}$ by Lubashevsky *et al.* (2011) The values of T_c^* and T_c were changed independently (e.g. the value of T_c was modified by Zn substitution). Indeed, it has been observed that the value of the gap measured using ARPES was sensitive to changes in T_c^* but was not affected by changes in the resistive T_c.

6.3 Energy Scales

The real picture in the cuprates is complicated as we are dealing with three different energy scales (Kresin *et al.*, 2004, 2006) and, correspondingly, three characteristic temperatures (we denote them T_c, T_c^*, and T^*).

6.3.1 The Highest Energy Scale (T^*)

The highest energy scale, which we have labelled T^* (~5.10^2 K) corresponds to the appearance of the phase separation. We are dealing with a peculiar inhomogeneous structure of the compounds. For example, for YBCO, the formation of the chains occurs at T^*.

An energy gap could open in the region below T^*. This gap is not related to the pairing but, as was mentioned above, there are many other reasons for the appearance of a gap. For example, the presence of a chain structure in YBCO is consistent with a CDW and, correspondingly, with a gap on part of the Fermi surface. Nesting of states might lead to a CDW instability in other compounds as well.

An intrinsic inhomogeneity is the important property of the compound; this is due to the statistical nature of doping and is manifested in phase separation (Gor'kov and Sokol, 1987; see also the book edited by Sigmund and Mueller, 1994). This property implies the coexistence of metallic and insulating phases. The periodic stripe structure (Bianconi, 1994a, b; Tranquada *et al.*, 1995, 1997; Zaanen, 2000) also appears below T^*.

6.3.2 The Diamagnetic Transition (T_c^*)

If the compound is cooled down below T^*, then, at some characteristic temperature we have labelled T_c^* ($T_c^* \approx 2.10^2$ K), one can observe a transition into the diamagnetic state.

The characteristic temperature T_c^* corresponds to the appearance of superconducting regions embedded in a normal metallic matrix (Fig. 6.5). The presence of such superconducting clusters('islands') leads to a diamagnetic moment, whereas the resistance remains finite, because of the normal matrix. As for the energy gap, the coexistence of pairing and a CDW determine its value below T_c^*. It is remarkable that the superconducting state appears at a temperature T_c^* which is much higher than the resistive T_c. This value of T_c^* corresponds to the real transition to the superconducting state (one can call it an 'intrinsic critical temperature'; see Kresin *et al.*, 1996).

Strictly speaking, the experimentally measured value of T_c^* lies below the intrinsic critical temperature, because of the impact by the proximity effect. Nevertheless, T_c^* is an important experimentally measured parameter. It corresponds to the appearance of diamagnetic 'islands' and reflects the impact of pairing. The superconducting phase appears, at first, as a set of isolated 'islands'.

The picture of different energy scales, T^* and T_c^*, as just described is in total agreement with interesting experimental data obtained by Kudo *et al.* (2005). In their study, the impact of an external magnetic field was studied by out-of-plane resistive measurements. According to the study, there are, indeed, two characteristic temperatures (Kudo *et al.* dubbed them T^* and T^{**}; $T^* > T^{**}$). The behaviour of the resistivity appears to be independent of the magnetic field in the region $T^* > T > T^{**}$ but strongly affected by the field at $T < T^{**}$. According to Kudo *et al.*, the state formed below T^{**} is related to superconductivity. The characteristic temperatures T^* and T^{**} directly correspond, respectively, to the energy scales T^* and T_c^* (in our notation $T^{**} \equiv T_c^*$) introduced above.

The picture for several energy scales with T^* and T_{coh} (in our notation, $T_{coh} \equiv T_c^*$) was discussed also by Chatterjee *et al.* (2011). According to them, the pair correlation is manifested below T_{coh}. In many aspects, this picture totally corresponds to that described in this section. A similar picture was described by Kaminski *et al.* (2014). They performed an angular photoemission study and introduced the characteristic temperature T_{pair}; the pairing occurs below this temperature, which lies below T^* and noticeably above T_c (the resistive transition; see the next section).

6.3.3 The Resistive Transition (T_c)

As the temperature is lowered below T_c^*, new superconducting clusters appear (Fig. 6.1) and existing clusters form larger 'islands'. This is a typical percolation scenario. At some characteristic temperature (T_c), a macroscopic superconducting phase is formed (the 'infinite' cluster, in terms of the percolation theory; see e.g. Shklovskii and Efros, 1984). The formation of a macroscopic phase at T_c leads to the appearance of a dissipationless state ($R = 0$).

It is important also to stress that, in the region $T_c^* > T > T_c$, each 'island' has its own phase, so that there is no phase coherence for the whole sample. Macroscopic phase coherence appears only below T_c.

Therefore, there are three different energy scales and, correspondingly, three characteristic temperatures, T^*, T_c, and T_c^* (Fig. 6.6).

The value of T_c is lower than T_c^* because of local depressions caused by the pair-breaking effect and an inhomogeneous distribution of pair breakers (dopants). It is interesting to note that the value of T_c^* is close to an intrinsic value of the critical temperature. This value is noticeably higher than the resistive T_c.

To conclude this section, let us stress again that the inhomogeneous distribution of pair breakers (dopants) along with local depressions in the value of critical temperature leads to a spatial dependence of T_c, that is, $T_c(\mathbf{r})$. The value of T_c^* is close to an 'intrinsic' critical temperature.

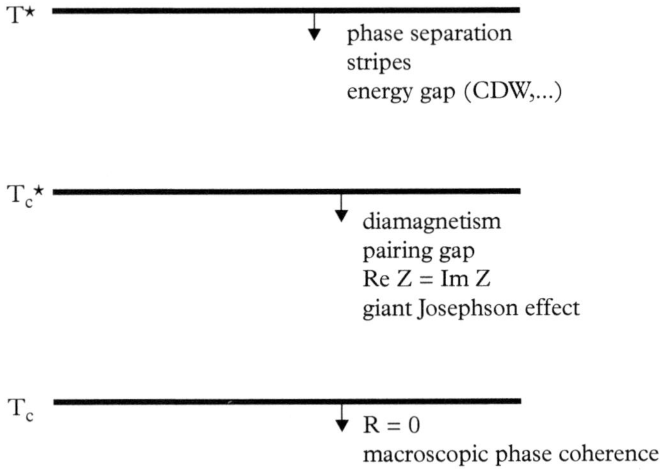

Figure 6.6 *Energy scales.*

What are manifestations of the high-temperature superconducting state? Of course, the presence of the normal matrix at $T > T_c$ excludes the possibility of observing a state with zero dc resistance ($R = 0$). But one can observe a number of various phenomena (see section 6.1) which will be analysed in the next section.

6.4 Theory

In this section, we are going to present the theoretical analysis of the main features of the pseudogap state: diamagnetism, ac properties, and the 'giant' Josephson proximity effect. The analysis should take into account the inhomogeneity of the structure. There are two essential factors:

- the proximity effect, since the superconducting regions are embedded into normal metallic matrix, and
- the pair-breaking effect.

6.4.1 General Equations

The inhomogeneity of the system is a key ingredient of the theory. Because of it, it is convenient to use a formalism describing the compound in *real* space. That is why we have employed the method of integrated Green's functions, which was developed by Eilenberger (1968) and independently by Larkin and Ovchinnikov (1969); see also the review by Larkin and Ovchinnikov (1986).

The main equations have the form

$$\alpha\Delta - \beta\omega + \frac{D}{2}\left(\alpha\partial_{\boldsymbol{r}}^2\beta - \beta\partial_{\boldsymbol{r}}^2\alpha\right) = \alpha\beta\Gamma \qquad (6.1)$$

$$|\alpha|^2 + |\beta|^2 = 1 \qquad (6.1')$$

$$\Delta = 2\pi \mathrm{T}\,|\lambda| \sum_{\omega>0} \beta \qquad (6.1'')$$

Here α and β are the usual and pairing Green's functions, respectively, averaged over energy, Δ is the order parameter, $\Gamma \equiv \tau_s^{-1}$ is the spin-flip relaxation time, and λ is the pairing coupling constant. Because of the inhomogeneity, all of these quantities are spatially dependent. In addition, $\partial_\pm = \partial_{\boldsymbol{r}} \pm 2ie\boldsymbol{A}$, where $\boldsymbol{A}$ is the vector potential, and $\partial_{\boldsymbol{r}} = (\partial/\partial\boldsymbol{r})$. We consider the 'dirty' case, so that D is the diffusion coefficient. These equations contain the spatially dependent functions α, β, and Δ. The method is very effective for the treatment of spatially dependent properties.

6.4.2 Diamagnetism

The Cu-O layers contain superconducting 'islands', and their presence leads to an observed diamagnetic moment. Because of the dependence $T_c(\boldsymbol{r})$, the size of the superconducting region occupied by the 'islands' decreases as temperature is increased. As a result, one can observe strongly dependent diamagnetism.

Let us describe the evaluation of the diamagnetic moment (Ovchinnikov *et al.*, 1999). Based on eqns (6.1)–(6.1″), one can calculate the order parameter $\Delta(\boldsymbol{r})$ and then the current $j(\boldsymbol{r})$. Then, one can calculate the magnetic moment, since the magnetic moment for an isolated cluster is

$$M_Z = L \int d\boldsymbol{\rho}\,[\boldsymbol{\rho}\boldsymbol{j}]_z \qquad (6.2)$$

Here L is the effective thickness of the superconducting layer; the axis z has chosen to be perpendicular to the layers, and $\boldsymbol{\rho}$ is perpendicular to OZ.

Assume that the sample contains a sufficient amount of magnetic impurities so that $\tau_s T_c{}^\circ \ll 1$; as a result, $T_c \ll T_c{}^\circ$, where T_c is the average value of the critical temperature, and $T_c{}^\circ$ corresponds to the transition temperature with no magnetic impurities. In this case, with the use of eqns (6.1)–(6.1″), we obtain

$$\Delta = 2\pi T\,|\lambda| \sum_{\omega>0} \left(\Gamma + \omega - \frac{D}{2}\,\partial_-^2\right)^{-1} \left\{\Delta - \frac{\omega}{2}\beta_0|\beta_0|^2 + \frac{D}{4}\beta_0\partial_-^2|\beta_0|^2\right\} \qquad (6.3)$$

The order parameter can be found in the form $\Delta = C\Delta_0$, where $C \equiv C(T)$, and Δ_0 is the solution of the equation

$$\left[\Gamma - (D/2)\,\partial_-^2\right]\Delta_0 = (\Gamma_\infty + \gamma)\,\Delta_0 \tag{6.4}$$

Here Γ_∞ is the value of Γ outside of the 'island', and γ is the minimum eigenvalue. As a result, one arrives at the following equation:

$$\ln\left(T_c^0/T\right) = \psi\left[0.5 + \frac{\Gamma_\infty + \gamma}{2\pi T}\right] - \psi(0.5) + \frac{C^2}{12\Gamma_\infty^2}\frac{\left(\Delta_0^{*2}, \Delta_0^2\right)}{\left(\Delta_0^*, \Delta_0\right)} \tag{6.5}$$

Here ψ is the Euler function, and the notation (f, g) corresponds to the scalar product of the functions. The transition temperature T_c is determined by the equation which can be obtained from eqn. (6.5) if we insert $C = 0$ and $\gamma = 0$:

$$\ln\left(T_c^0/T_c\right) = \psi\left[1/2 + (\Gamma_\infty/2\pi T_c)\right] - \psi(1/2) \tag{6.6}$$

which is the well-known pair-breaking equation (Abrikosov and Gor'kov, 1961; see section 3.3.1). Equation (6.5) is the generalisation of eqn (6.6) for the inhomogeneous case.

The current density is described by the following expression (Larkin and Ovchinnikov, 1969):

$$j = -ievD\pi T\sum_\omega\left(\beta^*\partial_-\beta - \beta\partial_+\beta^*\right) \tag{6.7}$$

As a result, we can obtain the following expression for the magnetic moment of an isolated cluster:

$$j = -\frac{ievDC^2}{2\pi T}\psi'\left(\frac{1}{2} + \frac{\Gamma_\infty + \lambda_1}{2\pi T}\right)\left(\Delta_0^*\partial_-\Delta_0 - \Delta_0\partial_+\Delta_0^*\right) \tag{6.8}$$

Here, $\kappa = \int d\boldsymbol{p}p^2\Delta_0^2$, where Δ_0 is the solution of eqn (6.4), and the vector potential has been chosen as $\boldsymbol{A} = \frac{1}{2}\left[\boldsymbol{H}\,\boldsymbol{r}\right]$. Note also that, because the cluster size is smaller than the penetration depth, one can neglect the spatial variation of the magnetic field.

Consider the most interesting case when the variation of the amplitude $\delta\Gamma = \Gamma_\infty - \Gamma$ has the form

$$\delta\Gamma(\boldsymbol{r}) = \begin{cases}\delta\Gamma(\rho) & \rho < \rho_0\\ 0 & \rho > \rho_0\end{cases} \tag{6.9}$$

where ρ_0 is the 'island' radius.

With the use of eqns (6.4)–(6.6), one can obtain the following expression for the magnetic moment:

$$M_z = -A\left(\tilde{B} - \tau^2\right) H \tag{6.10}$$

Here

$$A = \left(8\pi^2 e^2 vDT_c^2/\Gamma_\infty\right) \rho_0^2 z_c^{-4} n\left(\tilde{x}_{3;2}\tilde{x}_{2;1}/\tilde{x}_{1;4}\right); \tilde{B} = B + 1 \tag{6.11}$$

$$\alpha = T_c^2\left(\Delta_0^*, \Delta_0\right)\left(\Delta_0^{*2}, \Delta_0^2\right)^{-1}; \; \tau = T/T_c; \; B = -6\lambda_1\Gamma_\infty/(\pi T_c)^2$$

and

$$\lambda_1 = -\delta\Gamma + 0.5\mathrm{D}(z_0/\rho_0)^2 \tag{6.12}$$

where n_s is the concentration of the superconducting clusters, $\tilde{x}_{n;i} = \int_0^{z_0} dx \cdot x^n J_0^i(x)$, and z_0 is the lowest zero of the Bessel function; $z_0 \approxeq 2.4$. If $\delta\Gamma \ll \Gamma_\infty$, then $T_c^* \gg T_c$.

The value of λ_1 (and, therefore, the value of $\tilde{\beta}$) depends on an interplay of two terms. The first term reflects the impact of pair breaking, and the second term describes the proximity effect. It is natural that the impact of the proximity effect increases with a decrease in the size of the inhomogeneity ρ_0.

One can see directly from eqn (6.10), that it is possible to observe a noticeable diamagnetic moment. Indeed, if we assume the realistic values $p_F = 10^{-20}$ gcm sec^{-1}, $l = 40$ Å (l is a mean free path: $D = v_F l/3$), T_c=10 K, Γ_∞=10^2 K, $\delta\Gamma = 50$ K (we put $k_B = 1$), $\rho_0 = 80$, and $n_s \approxeq 0.1$, we obtain the following values for the parameters: $A \approxeq 10^{-5}$, $B = 3$, and $\gamma = 5$ K. Then, for example, at $T = 11$ K, one can observe $\chi_D = M_Z/H = -3\times10^{-5}$; this contribution greatly exceeds the usual paramagnetic response of a normal metal, $\chi_P \approxeq 10^{-6}$.

A diamagnetic response can be observed in the region $(T/T_c) < \tilde{B}$. This is natural, since the influence of the proximity effect (see e.g. Gilabert, 1977) to depress the superconductivity grows with a decrease in the size ρ_0 of the superconducting grain.

6.4.3 Transport Properties: The 'Giant' Josephson Effect

The dc transport properties of the inhomogeneous system above T_c are determined by the normal phase, since only this phase can provide a continuous path. The situation with ac transport is entirely different, and the superconducting 'islands' make a direct contribution to the ac conductivity and to surface impedance.

As we know, the real and imaginary parts of the surface impedance of a normal metal are almost equal (see e.g. Landau *et al.*, 2004). Indeed, the surface impedance Z is determined by the relation:

$$Z = \left(\frac{\omega}{4\pi\sigma}\right)^{1/2} \exp\left(-i\pi/4\right) \tag{6.13}$$

For normal metals, the difference between ReZ and ImZ is negligibly small and is connected with the dependence: $\sigma(\omega) = \sigma_0(1 - i\omega\tau_{tr})^{-1}$; in our case, $\omega\tau_{tr} \ll 1$. The situation in superconductors is entirely different (see e.g. Tinkham, 1996). The same is true for the 'pseudogap' state. Indeed, a metallic compound which contains superconducting 'islands' is characterised by a strong inequality: $\mathrm{Re}(Z) \neq |\mathrm{Im}(Z)|$. This can be shown theoretically (Ovchinnikov and Kresin, 2002) and measured experimentally (Kusco *et al.*, 2002) for an $HgBa_2Ca_2Cu_3O_{8-\delta}$ compound at $T > T_c$.

Let us discuss the situation with the Josephson current. We mentioned above an interesting experimental study of *SNS* Josephson junctions (Bozovic *et al.*, 2004). This phenomenon cannot be explained by the usual theory of *SNS* proximity junctions.

We focus on the especially interesting case of $SN'S$ junctions where the electrodes are high-T_c superconducting films (e.g. $La_{0.85}Sr_{0.15}CuO_4$ or $YBa_2Cu_3O_7$) and the barrier N' is made from an underdoped cuprate; T_c' is the critical temperature of the underdoped barrier, and T_c is the critical temperature of the electrode. The generally accepted notation N' emphasises the difference between an $SN'S$ and a typical *SNS* junction (then $T_c^{\mathrm{N}} = 0$ K), so that $T_c' > T_c$. Here we consider temperatures when the barrier is in the normal resistive state because $T > T_c'$. The use of the underdoped cuprate as a barrier is beneficial for various device applications because the structural similarities between the electrodes S and the barrier N' eliminate many interface problems.

The 'giant' phenomenon is manifested in a finite superconducting current through the $SN'S$ Josephson junction with a thick barrier, so that $L \gg \xi_{N'}$ (where L is the thickness of the barrier, and $\xi_{N'}$ is the proximity coherence length). The configuration is such that the layers forming the barrier N' are parallel to the electrodes so that the Josephson current flows in the c-direction. Then the coherence length is a very short $\xi_c \approx 4$ A, so that we are dealing with the 'clean' limit.

This type of junction in LSCO material was studied by Bozovic *et al.* (2004). In this study, $La_{0.85}Sr_{0.15}CuO_4$ films ($T_c \approx 45$ K) were used as electrodes, and the underdoped LaCuO compound ($T_c' \approx 25$K) formed the barrier. The atomic-layer-by-layer molecular beam epitaxy technique was used for these junctions and it provides atomically smooth interfaces. The barrier was prepared in the c-axis geometry. As was noted above, the coherence length $\xi_C \approx 4$ Å. The measurements were performed at $T_c' < T < 35$ K. The Josephson current was observed for thickness of L up to 200 Å(!). Such a 'giant' effect cannot be explained with use of the conventional theory. Indeed, as we know (see e.g. Barone and Paterno, 1982), the amplitude of the Josephson current for the *'clean' limit* is

$$j_m = j_0 \exp\left(-L/\xi_N\right) \tag{6.14}$$

The thickness of the barrier L should be comparable with the barrier coherence length ξ_N, and this condition is satisfied for conventional Josephson junctions. The picture described above for the junctions with the cuprates is entirely different, since $L \gg \xi_{N'}$.

The superconducting current in the c-direction occurs via an intrinsic Josephson effect between the neighbouring layers (see Kleiner *et al.*, 1992; Kleiner and Muller, 1994). If the barrier contains several homogeneous normal layers, then the Josephson current through such a barrier is practically absent.

To understand the nature of the 'giant' Josephson proximity effect, it is very important to stress that the barriers we are considering are formed by underdoped cuprates. As a result, the barriers are not in the usual normal state but the 'pseudogap' state; indeed, $T_c' < T < T_c^*$. For example, according to the study by Iguchi *et al.* (2001; see section 6.1.1) of LSCO (where $x \approxeq 0.1$; $T_c \approxeq 18$ K) in which the stoichiometry is close to that for the sample used by Bozovic *et al.* (2004) as the barrier, this compound has a value of $T_c^* \cong 80$ K whereas $T_c' = T_c^{res} \cong 20$ K describes the resistive transition to the dissipationless state. Since the diamagnetic moment measured by Iguchi *et al.* persists up to T_c^*, the question of the origin of the 'giant' proximity effect is directly related to the general problem of the nature of the 'pseudogap' state.

This effect can be explained (Kresin *et al.*, 2003) by the approach described in this chapter, based on the intrinsically inhomogeneous structure of the compound. According to the model, the CuO layers forming the N' barrier contain superconducting 'islands', and these 'islands' form the path for the Josephson tunnelling current.

For typical *SNS* junctions, the propagation of a Josephson current requires the overlap of the pairing functions F_L and F_R (see e.g. Kresin, 1986); F_R and F_L are pairing Gor'kov functions for left- and right-side electrodes. This overlap is caused by the penetration of F_L and F_R ('proximity') to the N barrier. For the system of interest here, the situation is quite different. Each 'island' has its own pairing function with its own phase. As a result, the Josephson current is caused by the overlap of F_L and F_1, F_1 and F_2, and so on, where F_1 corresponds to the 'island' located at the layer nearest to the left electrode, and so on. The superconducting 'islands' form the network with the path for the superconducting current.

The propagation of a Josephson current through the $SN'S$ junction requires the formation of a channel between the electrodes. The transport of the charge in superconducting cuprates in the c-direction is provided by the interlayer Josephson tunnelling (the intrinsic Josephson effect). Therefore, the Josephson current through the barrier is measurable because of the superconducting state present in the layers.

The transfer of the Josephson current in this model implies that the electrons tunnel inside of the layers between the superconducting 'islands' until one of them appears to be close to some 'island' in the neighbouring layer. Then the next step, namely the interlayer charge transfer via the intrinsic Josephson effect occurs, and so on. As a result, the chain formed by the superconducting 'islands' provides the Josephson tunnelling between the electrodes, and the path represents a sequence of superconducting links. It is important to note that the amplitude of the total current is determined by the 'weakest' link in the chain.

The density of the critical current is determined by the equation

$$j = A \int \exp(-r/\xi)\, dP \tag{6.15}$$

or

$$j = (A/\xi) \int_0^\infty dRP \exp(-R/\xi) \tag{6.15'}$$

where $\xi \equiv \xi_{11}$ is the in-plane coherence length, R is the distance between the 'islands' on the same layer, $A \propto n_s j_{c\perp}$, $n_s \equiv n_s(T)$ is the concentration of the superconducting region, so that $S_{sup} = n_s S$ is the area occupied by the superconducting phase, S is the total area of the layer, $j_{c\perp}$ is the amplitude of the Josephson interlayer transition, and P is the probability of the formation of a chain with length R for the links, so that $P = p^{m-1}$ (see e.g. Stuffer and Aharony, 1982), m is the number of layers forming the barrier, and p is the probability for two neighbouring in-plane 'islands' to be separated by a distance $r < R$.

Assume that p is described by a Gaussian distribution, that is,

$$p = (c/\pi\delta)^{1/2} \xi^{-2} \int_0^R dr\, r \exp\left\{-\delta^{-1}\left[(r/\xi) - n_s^{-1/2}\right]^2\right\} \tag{6.16}$$

where δ is the width of the distribution, and $c =$ const. Then the integral (6.16) can be calculated by the method of steepest descent, and we obtain

$$j_m = \tilde{A} f(m) \exp\left(-1/n_s^{1/2}\right) \tag{6.17}$$

where $\tilde{A} =$ const. Equation (6.17) can be written in the form

$$j_{\max} = j_0 \left(-1/\eta(T)\right) \tag{6.18}$$

Here

$$j_0 = j_{\max}(T_C), \quad \eta(T) = n_s^{1/2}\left(1 - n_s^{1/2}\right)^{-1}$$

$$f(m) \simeq \exp\left[-\delta \ln^{1/2}\left(m\tilde{l}\right)\right], \quad \tilde{l} = (\pi\delta)^{-1/2}$$

We assumed that $\delta \ll 1$, and $m \gg 1$.

One can see directly from eqns (6.17) and (6.18) that the current amplitude depends strongly on temperature and is determined mainly by the dependence of the area occupied by the superconducting 'islands' on T: $n_s \equiv n_s(T)$. In addition, there is a weak dependence of j_{max} on the barrier thickness.

The dependence $n_s(T)$ is different for various systems and is determined by the function $T_c(r)$, that is, by the nature of the doping. Note that, near T_c^*, the value of $n_s(T)$ is very small and the current amplitude is negligibly small. However, the situation is different in the intermediate temperature region and in the region $T \ll T_c^*$, which is not far from T_c. This is true for the data by Bozovic *et al.* (2004): $T_C' = 25$ K and

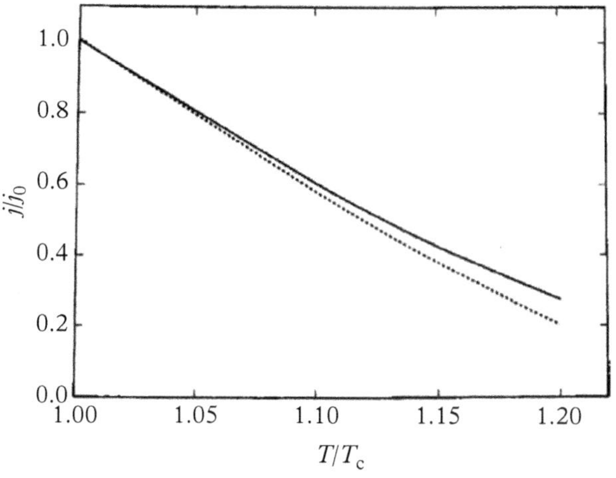

Figure 6.7 *Dependence of the Josephson current on temperature: dotted line indicates experimental data; solid line indicates theory.*

$T'_C < T < 35$ K. For example, the value $T = 30$ K is relatively close to T'_c but much below $T^* = 80$–100 K. At $T = 30$ K, there are many superconducting 'islands', so the value of $n_s(T)$ is relatively large. At temperatures close to T_c, one can use eqn (6.18) with $\rho = \alpha(t-1)^{\nu}$, where $t = T/T_c$ and $a =$ const (we have chosen $a = 10$). One can see (Fig. 6.7) that such a dependence with $\upsilon = 1.3$ is in a good agreement with the experimental data. Note that this value for υ is close to the value of the critical index for the correlation radius in the percolation theory (see e.g. Shklovskii and Efros, 1984).

In principle, one can use a junction with a barrier grown in the ab direction, so that the c-axis is parallel to the S electrodes. Then the path contains SNS junctions formed by the 'islands' with metallic N' barriers. Since $\xi_{ab} \gg \xi_c$ ($\xi_{ab} \approxeq 20 - 30$Å), one should expect an even larger scale for the 'giant' Josephson proximity effect with thickness L up to 10^3 Å. The calculation performed by Covaci and Marsiglio (2006) also supports the conclusion that the 'giant' proximity effect benefits from the presence of superconducting regions.

Therefore, the 'giant' Josephson proximity effect is also caused by the intrinsic inhomogeneity of the cuprates. The 'giant' scale of the phenomenon is provided by the presence of 'superconducting' islands embedded in the metallic matrix and forming the chain transferring the current. The use of superconductors in the 'pseudogap' state as barriers represents an interesting opportunity for 'tuning' the Josephson junction on a 'giant' scale.

6.4.4 The Isotope Effect

As mentioned above, according to an interesting experiment (Temprano *et al.*, 2000), the 'pseudogap' state is characterised by a strong isotope effect, that is, by a large shift in T_c^* caused by isotope substitution. In that experiment, the authors studied the $HoBa_2Cu_4O_8$

compound with a value of $T_c^* \approx 170$ K. The isotopic substitution $^{16}O \to {}^{18}O$ leads to a drastic change in the value of T_c^*. One can observe the following change: $T_c^* \approx 170\text{ K} \to T_c^* \approx 220$ K (!). We are dealing with a giant isotope effect; in addition, its value is negative.

Such an isotopic dependence of T_c^* is also a consequence of the presence of the superconducting regions and reflects the fact that the superconducting pairing persists above the resistive transition. It is interesting that the isotope coefficient has a negative sign. This unusual feature is consistent with the dynamic polaron model of the isotope effect (see section 9.7 and eqn (9.12)). Indeed, a strong non-adiabaticity (axial oxygen in YBCO is in such state) results in a peculiar polaronic isotope effect:

$$\alpha = \gamma \frac{n}{T_c^*} \frac{\partial T_c^*}{\partial n} \tag{6.19}$$

Based on eqn (6.19), one can explain why the isotope coefficient has a negative value. Indeed, $\alpha \propto \partial T_c^* / \partial n$ and $(\partial T_c^* / \partial n) < 0$: an increase in doping in the underdoped region leads to decrease in the value of T_c^* (at optimum doping, $T_c^* \approxeq T_c$, and this is due to an increase in the number of dopants, that is, pair breakers, so that the distribution of pair breakers becomes more uniform).

6.5 Other Systems

Intrinsic inhomogeneity is an essential feature of the high-T_c oxides, and this feature is manifested in a peculiar 'pseudogap' behaviour. However, the scenario is a more general one and the 'pseudogap' state can be observed in other inhomogeneous systems as well. Let us describe some of these systems.

6.5.1 Borocarbides

Borocarbides represent an interesting family of novel superconductors, because they allow us to study an interesting interplay between superconductivity and magnetism (see e.g. Canfield *et al.*, 1998, Schmiedeshoff *et al.*, 2001). According to data by Lascialfari *et al.* (2003), the borocarbide YNi_2B_2C displays precursor diamagnetism above T_c ($T_c = 15.25$ K). An analysis of magnetisation data taken with a high-resolution SQUID magnetometer led to the conclusion that the unusual results were caused by an inhomogeneity of the compound and by the presence of superconducting droplets with a local value of the critical temperature higher than the usual T_c corresponding to a macroscopic dissipationless current. The presence of the superconducting isolated droplets is due to an inhomogeneity which is similar to that observed in the cuprates and described in this chapter. Indeed, an unconventional temperature dependence of the critical fields H_{c1} and H_{c2} was observed (Suh *et al.*, 1996; Schmiedeshoff *et al.*, 2001). This dependence can be explained by the presence of pair breakers (Ovchinnikov and Kresin, 2000). A statistical distribution of pair breakers leads to a spatial dependence of the critical temperature, $T_c \equiv T_c(r)$, and to the inhomogeneous 'pseudogap' picture (Fig. 6.5).

6.5.2 Granular Superconductors: The Pb + Ag System

Granular superconducting films were already being studied intensively before the discovery of high-T_c superconductivity (see e.g. Simon *et al.*, 1987, and Dynes and Garno, 1981). These films also represent inhomogeneous superconducting systems. Such inhomogeneous films could display a diamagnetic moment above T_c. It would be interesting to carry out a study of their magnetic properties.

An interesting study of the Pb + Ag system was described by Merchant *et al.* (2001). An electrically discontinuous (insulating) Pb film was covered with an increasing thickness of Ag (Fig. 6.8). The Ag acts to couple the superconducting Pb grains via the proximity effect. The resistive transition, as well as tunnelling spectra, has been taken on a series of these films. The most insulating film has no resistive transition but a full Pb gap, as revealed by the tunnelling spectra. This gap is reduced as Ag is added, reflecting the decrease in the mean field T_c of the Pb grains. At some point, the composite film becomes continuous and superconducting, with a low resistive transition temperature. The evolution of the mean field transition temperature and the resistive transition temperature with increasing thickness of Ag mimics the phase diagram of the cuprates with doping. The mean field transition temperature resembles the pseudogap onset temperature, and the resistive T_c resembles the superconducting transition temperature, with the mean field transition temperature lying above the resistive transition.

The results of the Pb/Ag artificial inhomogeneous superconductor model the behaviour of the cuprates. The cuprates are doped substitutionally and inhomogeneously. At some concentration of doping, there are regions with a high-enough concentration of carriers to locally superconduct and therefore reduce the low-energy density of states. The evolution of these islands into a percolating dissipationless state would resemble the percolating proximity coupling described above. It is then not surprising that the phase diagrams would be nearly identical.

Granular superconductors are by their very nature inhomogeneous materials that in many respects behave in unexpected ways. Typically, granular superconductors consist of small grains with sizes comparable to or smaller than the coherence length. These grains are often in a matrix that is normal or insulating but the grains themselves are close enough so that they couple by either *SNS* or *SIS* (where *I* stands for insulator) coupling. Depending on the coupling strength and the temperature, these superconducting islands can behave like zero-dimensional superconducting particles above T_c with strong fluctuations (Deutscher *et al.*, 1973; Wolf and Lowrey, 1977) or as

Figure 6.8 *Pb/Ag proximity system.*

arrays of junctions whose superconducting properties mimic a single very-high-quality junction, despite there being many junctions in the granular array (Wolf *et al.*, 1979).

The inhomogeneous nature of granular 'conventional' superconductors allows them to serve as a model system for the behaviour of superconductors that become superconducting as a consequence of doping. This includes the cuprates as well as the Fe-based superconductors and the superconducting semiconductors (e.g. Nb-doped $SrTiO_3$). For a random granular superconductor that contains grains of different sizes and varying coupling strengths, the transition from the normal state to the superconducting state is not sharp but, rather, broad, reflecting the inhomogeneities inherent in the system. As a granular sample (particularly one that contains *SIS* coupling) is cooled from a high temperature, the highest-T_c grains (which may not be the largest; see below) begin to superconduct, and a superconducting gap opens up on these grains but only locally. At this point, the resistance is still dominated by the intergranular resistance, and the overall resistivity does not show any sharp, significant feature. As the temperature is lowered further, more grains become superconducting, and neighbouring superconducting grains can couple and form small networks. At this point, a small Meissner effect can be observed, since there can be a shielding current over finite regions of the sample, but long-range superconductivity is still not present. As the temperature is lowered even further, superconductivity percolates and the resistance drops to zero, well below the superconducting transition of the highest-T_c grains. This scenario has been clearly observed in NbN grains embedded in a BN matrix (Simon *et al.*, 1987). Similar results were obtained by Merchant *et al.* (Merchant *et al.*, 2001) using grains of Pb in an Ag matrix where, for a range of grain-to-grain coupling strengths, the authors clearly observed a pseudogap region and then a transition to a fully superconducting state. They pointed out the very distinct analogy with the pseudogap region above the resistive transition in the cuprates.

Another very important aspect of granular materials is the enhancement of T_c that occurs for some films. A good example of this effect is for Al films, for which it was observed that the critical temperature $T_c \approx 2.1$ K (Shukhareva, 1963; Strongin *et al.*, 1965). An increase was also observed for granular systems. Although bulk Al has a transition temperature of 1.2 K, granular Al that consists of Al grains embedded in an Al oxide matrix can have a transition temperature as high as 3.1 K (Deutscher *et al.*, 1973). The origin of this enhancement is not fully understood but the leading candidate for this enhancement is due to size quantisation effects that lead to a discreet energy spectrum rather than a continuous spectrum (Kresin and Tavger, 1966; Parmenter, 1968; Shanenko *et al.*, 2006). This explanation is supported by spectroscopic and T_c measurements (Alekseevskii and Vedeneev, 1967; Guo *et al.*, 2004). Another possible explanation has been proposed (Deutscher, 2005, 2012) which relates the enhancement to the interplay of correlation effects and Kondo physics which occurs in the vicinity of the metal–insulator transition, that is, as the films become more resistive. Note that, in principle, such an increase could be explained by the change in the phonon spectrum, but this explanation has been ruled out by the heat capacity measurements on Al nanoparticles (Hock *et al.*, 2011).

It has been demonstrated that the value of the Debay frequency for these nanoparticles is similar to that for bulk Al.

6.6 Ordering of Dopants, and the Potential for Room Temperature Superconductivity

The 'pseudogap' scenario and its manifestation, which is especially strong in the underdoped region of the cuprates, implies that the transition into a superconducting state is a two-step process.

The superconducting regions appear below some characteristic temperature T_c^* (we call it an 'intrinsic' critical temperature); its value greatly exceeds the usual resistive $T_c \equiv T_{c;\text{res}}$. For example, for an underdoped LSCO compound ($T_c \approx 25$ K), $T_c^* \approx 80$ K. For a YBCO compound, $T_c^* \approx 250$ K, that is, it is close to room temperature. Decreasing the temperature towards $T_c \equiv T_{c;\text{res}}$ leads to an increase in the number of 'islands' and to an increase in their size. The second step is the transition at T_c, which is of a percolative nature and corresponds to the formation of a macroscopic superconducting phase (an 'infinite cluster', in the terminology of the percolation theory) capable of carrying a macroscopic supercurrent. It is obvious that the set of superconducting clusters embedded in a normal matrix structure is unable to carry a superconducting current. Such a structure reflects the intrinsic inhomogeneity of the sample, which is caused by the statistical nature of doping.

It would be attractive to 'order' the superconductive regions and, instead of statistically distributed superconducting 'islands', have a continuous superconducting phase; then, one could expect the supercurrent to flow at high temperatures. In other words, it would mean an effective increase in the value of resistive T_c. We describe below the concept of 'ordering of dopants' which looks promising in achieving this goal.

The delocalised carriers whose presence is responsible for the metallic state and, correspondingly, the superconducting state, are created by doping. At the same time, the superconducting pairing could be depressed by the pair-breaking effect (see section 3.3.1). The dopants play a double role: they provide delocalised carriers and are also responsible for the pair breaking. The statistical nature of doping leads to a random distribution of dopants and, at relatively low doping, the spatial distribution is rather broad. As a result, regions with a smaller number of dopants have a larger value for the critical temperature; these regions form the superconducting 'islands' inside of the normal matrix.

An increase in T_c can be achieved by a procedure that allows for a special ordering of defects (Wolf and Kresin, 2012). To clarify the concept, let us consider a specific example, namely, the YBCO compound. The doping of the parent $YBa_2Cu_3\,O_6$ sample is provided by adding oxygen into the chain layer. The mixed-valence state of the in-plane Cu leads to the plane-chain charge transfer and the appearance of a hole, initially on the Cu site. Because of diffusion, the hole enters the system of delocalised carriers responsible for the metallic and, correspondingly, superconducting behaviour: superconductivity is caused by the pairing of such holes. The added oxygen ion and the corresponding in-plane Cu form the defect mentioned above, with its pair-breaking impact.

This situation could be strongly affected by specific ordering of the oxygen ions. More specifically, let us consider the thin film formed by a layer-by-layer deposition technique, where the film is growing in the c-direction. The film should be built as a set of columns. The top layer could be imagined as a set of strips. An additional oxygen should be placed inside of specific columns. As a result, these columns would contain chains with a composition close to $YBa_2Cu_3O_7$. The chains would be parallel to the interstrip boundaries. Such columns could be called the reservoir columns, since holes are created in these areas. The holes would be delocalised and diffuse into the neighbouring columns, which would be free of defects and represent the high-T_c regions. Such a structure could be built, for example, using a nano-implantation technique (see e.g. Schenkel *et al.*, 2010; Toyli *et al.*, 2010) with appropriate annealing or other methods for providing an ordered defect structure similar to what has been described here.

As a whole, the film would then contain alternating reservoir and high-T_c columns. Because of the presence of the oxygen defects, the reservoir columns would have a value for T_c lower than that in the neighbouring columns, which would be free of defects. It is expected that the high-T_c columns would have values of T_c close to the intrinsic critical temperature, T_c^*, that, in fact, could be close to room temperature. Moreover, such a structure would provide a continuous path for the supercurrent, and this would mean an effective increase in the resistive T_c.

Note that the value of T_c in these high-T_c columns could be depressed by the proximity effect with the reservoir columns, and this would lead to some limitation on the width of the high-T_c columns, W_H. Indeed, we have the S_L–S_H proximity system where L and H correspond to the reservoir and the high-T_c columns ($T_{ch} > T_{cl}$). We know that the scale of the proximity effect is of order of the coherence length, ξ_H, of the high-T_c region. To minimise the impact of the proximity effect, the width of the high-T_c column, W_H, should be larger than ξ_H.

However, the value of ξ_H is rather small. Indeed, $\xi_H \approx \hbar v_F/2\pi T_{c,H}$. If we take the values $v_F \approx 10^7$sm/sec, and $T_c \approx$ (2.0–2.5) 10^2 K, we obtain $\xi_H \approx$ 5–10A. Because there is such a small value for the coherence length, the condition $W_H \gg \xi_H$ is perfectly realistic. Note that similar ordering could be effective for an increase in T_c for other cuprates, for example, LSCO or Bi2212 compounds.

The proposed method is conceptually analogous to the observed increase in T_c for the cuprates that is caused by pressure (see e.g. the work by Schilling and Klotz (1992) and Hochheimer and Etters (1991) and the analysis by Kresin *et al.* (1996)), by an applied field (Mannhart *et al.*, 1991; Ann *et al.*, 2006), or by the photoinduced effect (Yu *et al.*, 1992; Pena *et al.*, 2006). Indeed, external pressure, radiation, or an applied field affects the doping without creating defects, that is, pair breakers. The specifics of the structure proposed above is that the carriers appear in a region (the high-T_c columns) which is free of defects; they are produced in a different spatially separated column. As was stressed above, we observe the impact of such separation in the existing cuprates (in the 'pseudogap state'), especially in the underdoped region, where the superconducting clusters (high-T_c regions) are embedded into a normal matrix. Thanks to such a separation, one can observe the anomalous diamagnetism (Meissner effect) at temperatures higher than the resistive T_c. The proposed ordering leads to a noticeable increase of the critical temperature for the resistive transition and thus a much higher effective transition temperature.

The ordering leads to an increase in the size of superconducting clusters; as a result, the percolative transition occurs at higher temperature. This effective increase could be enhanced in a material with a larger dielectric constant (Mueller and Shengelaya, 2013). More specifically, Mueller and Shengelaya proposed creating a multilayer structure, so that the underdoped high-T_c oxide thin film (in the range of ~1–10 nm) is sandwiched between the high-dielectric-constant insulator layers; their presence reduced the Coulomb repulsion between the superconducting 'islands'. The examples of such materials are $SrTiO_3$ (Mueller and Burkard, 1979, with $\varepsilon \simeq 10^4$), or other ferroelectrics, like $Sr_{1-x}Ba_xTiO_3$ or $Pb(Zr_xTi_{1-x})O_3$ ($\varepsilon \simeq 3 \times 10^2$).

An additional increase in T_c can be achieved with the use of isotope substitution (Mueller, 2012). Indeed, as is noted below (section 9.2.3), one can observe a large negative isotope effect and as a result, the $O_{16} \to O_{18}$ substitution leads to an increase on the order of 30 K (!).

The ordered doping manifested itself in experiments by Liu *et al.* (2006). The ordering of apical oxygen has been observed for an $Sr_2CuO_{3+\delta}$ superconductor ($T_c \simeq 75$ K); this superconductor is characterised by a peculiar feature, namely the apical oxide sides are not fully occupied, and the hole doping occurs through the charge transfer between the apical oxygen and the Cu–O plane. The additional oxygen can be deposited on these sides in an ordered way and, as a result, an increase in T_c up to 95 K was observed.

A correlation between structure and its ordering and superconducting properties, that is, a noticeable rise in T_c, has been described by Fratini *et al.* (2010) and Pokkia *et al.* (2011). X-ray radiation was used to create ordered superconducting regions in the La_2CuO_{4+y} samples, and it was accompanied by an increase in T_c ($T_c \simeq 33.5$ K $\to$ $T_c \simeq 41.5$ K). The onset of high-quality superconductivity was associated with two-dimensional defect ordering.

6.7 Remarks

The approach to the 'pseudogap' state described in this chapter is rather general in the sense that it is valid for any pairing force (phonons, magnons, plasmons, excitons, etc.), but, nevertheless, we are dealing (in the region $T_c < T < T_c^*$) with real Cooper pairs, so that T_c^* is the 'intrinsic' critical temperature. Moreover, each superconducting 'island' has its own phase. It is an essential feature of this picture that the superconducting regions are embedded in a normal metallic matrix which provides normal dc transport. As a result, the proximity effect between the superconducting regions and the normal metal plays an important role.

Until recently, the presence of inhomogeneities was considered the signature of a poor-quality sample (except for the 'pinning' problem). However, we think that the situation is similar to that in the history of semiconductors. Indeed, initially the presence of impurities in these materials was considered a negative factor (they were called 'dirty' semiconductors). But, later, when scientists developed tools allowing the precise control of the impact of various impurities (donors and acceptors), it became clear that the presence of impurity atoms is a critical ingredient; even the language had changed and sounds more 'respectful' ('doped' semiconductors). The analogy between inhomogeneous novel superconductors and semiconductors is even stronger, because we are dealing with doping for both classes of materials.

7

Materials (II)

7.1 Conventional Superconductors

7.1.1 Ordinary Bulk Materials

First, we will discuss the properties of ordinary superconductors; sometimes they are called conventional materials. All of them are described by BCS physics with *s*-wave pairing, which is mediated by phonons (with weak up to strong coupling). The coherence length is large and noticeably exceeds the lattice period.

Initially, the appearance of unconventional superconductivity was connected to the discovery of high-T_c cuprates. But, later, the field was split into two areas of focus and became uncertain in many aspects. For example, hydrides with a record value of T_c (see section 7.2) are phonon-mediated high-T_c materials. On the other hand, manganese diborides, in which the T_c is also provided by the phonon mechanism, are characterised by a two-gap electronic spectrum.

The majority of ordinary superconductors have relatively low values of T_c and a large coherence length. The highest value of T_c among single elements is $T_c = 9.2\,\mathrm{K}$ for niobium. A special place is occupied by the so-called A-15 materials. Their properties will be discussed in the next section.

7.1.2 A-15 Superconductors

The A-15 materials form an important family of superconductors. Many scientists have focused on these materials for a long period of time. More specifically, we will focus on intermetallic compounds with the formula A_3B, where A is the transition metal and B is some other element. The corresponding structure was discovered by Boren (1933); chromium silicide (Cr_3Si) was the first A_3B compound. The structure of this compound, which is typical for A-15 materials, is shown in Fig. 7.1. The main feature is the presence of three chains formed by the transition metals. These low-dimensional structures are the key components of the materials. The reduced dimensionality is provided by the relatively short distance between the A transition metals.

Superconducting State: Mechanisms and Materials. Vladimir Z. Kresin, Sergei G. Ovchinnikov, and Stuart A. Wolf, Oxford University Press (2021). © Kresin, Ovchinnikov, Wolf. DOI: 10.1093/oso/9780198845331.003.0007

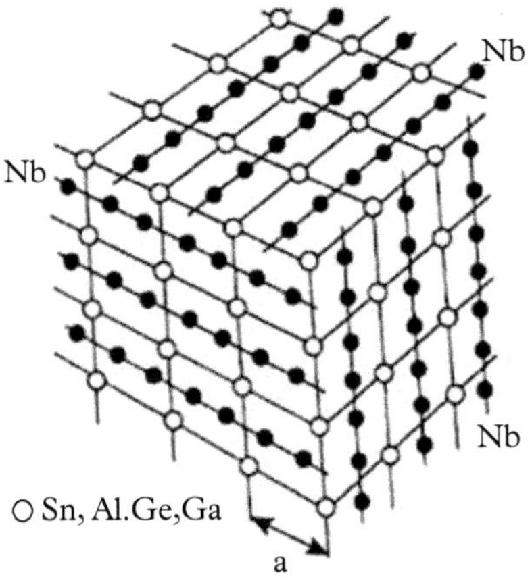

Figure 7.1 *Quasi-one-dimensional metal chains in A-15 superconductors.*

The breakthrough discovery of V_3Si, with its record high value of $T_c \simeq 17\,\text{K}$, was reported by Hardy and Hulm (1953). During the next 25 years, the highest T_c values observed had been for members of the A-15 family.

The structure presented in Fig. 7.1 is an idealised one. There are many factors affecting the properties of various A-15 compounds, including the T_c value (see the reviews by Muller, 1980; Testardi, 1975; Narlikar, 2014; and Steward, 2015). At present, more than 50 different A-15 compounds are known. Among them, the MO_3Al compound has a low $T_c \simeq 0.6$ K. At the same time, more than 15 of them have T_c values exceeding 10 K. Among these compounds, Nb_3Sn ($T_c \simeq 18$ K), which was discovered by Mattias *et al.* (1954), deserves special mention. This material is still important for applications such as the creation of a high magnetic field.

The highest T_c value among A-15 materials has been observed for Nb_3Ge films (Gavaler *et al.*, 1974; Testardi *et al.*, 1974; $T_c \simeq 23$ K). This value was the highest until the discovery of the high-T_c state for some other A-15 compounds were studied by Labbe and Friedel (1966). They stress that the large electronic density of states, which is due to the existence of narrow energy bands formed by the *d*-states of transition metals, is a key factor (Fig. 7.2). This large density of states was also manifested in the heat capacity data (Junod, 1982; also see Steward, 2015).

According to eqns (2.24) and (2.24′), the value of the density of states affects the strength of the electron–phonon coupling. For example, the Nb_3Sn compound is characterised by a large value for the coupling constant $\lambda \simeq 1.8; 2\varepsilon(0)/T_c \simeq 4.4$ (Wolf *et al.*, 1980; Wolf, 2012). Note also that the presence of the one-dimensional chains formed by transition metals leads to the $(E)^{-1/2}$ singularity in the density of states. The presence of the narrow band results in the peculiar situation in which the characteristic electronic energy scale is less than the width of the phonon spectrum (see section 2.5). Then the pre-exponential factor is of electronic origin, and it noticeably affects the value of the isotope coefficient.

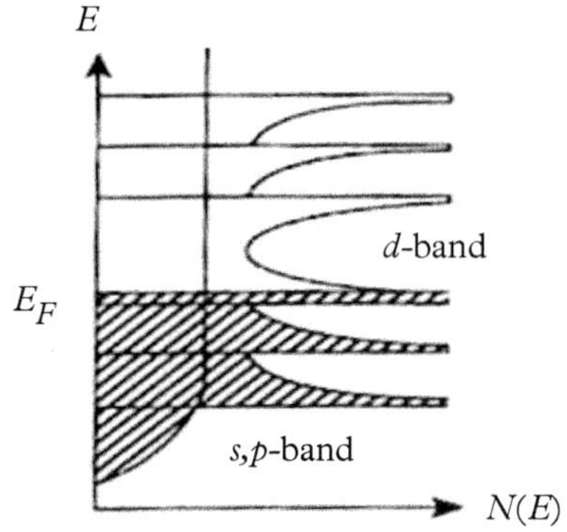

Figure 7.2 *A scheme of the A-15 electron energy structure with narrow d-bands.*

7.1.3 Magnesium Diboride

The discovery of high-T_c cuprates (see chapter 5) has been one of the key events in the history of superconductivity. This breakthrough not only transferred the field into a different energy scale (up to $T_c \simeq 170$ K) but also led to the appearance of novel superconducting materials that are drastically different from the usual A-15 compounds, including fullerides (discovered in 1992; see section 8.1.4) and magnesium diboride (MgB_2), whose properties will be described in this section.

The paper by Nagamatsu *et al.* describing the superconducting state of MgB_2 with $T_c = 39$ K was published in 2001 and has attracted a lot of interest (see the review by Muranaka and Akimutsi, 2012). If this paper had been published 16 years earlier, in 1985, it would have been a sensation. Indeed, in 1985, the highest observed T_c value was $T_c \simeq 23$ K (for Nb_3Ge).

The crystal structure of MgB_2 is depicted in Fig. 7.3. It can be seen that the crystal has a layered structure in which the boron layers are sandwiched by metallic Mg layers. The lattice is a hexagonal one $\left(a = 3.08\ \text{Å}, c = 3.51\ \text{Å}\right)$, with the space group *P6/mmm* (see Jones and Marsh, 1954).

It should be noted that many novel superconducting systems are characterised by the presence of low-dimensional structural units (e.g. planes, chains, and clusters), such as high-T_c cuprates (section 5.2.1), pnictides (section 7.3.4), fullerides (section 8.1.4), graphene (section 8.1.5), and so on.

The band structure of MgB_2 is rather peculiar (Fig. 7.4). It contains two different groups of bands: one of them corresponds to the in-plane bands (σ-bands), and the second group describes the antisymmetric P_2 bands (π-bands). The classification of the σ- and π-electronic states will be discussed later on (see section 8.2.2 and appendix C).

MgB_2 is an intermetallic compound. Indeed, unlike simple metals, it contains two kinds of atoms but is characterised by an ordered crystal structure. The superconducting transition is plotted in Fig. 7.5. The T_c value ($T_c = 39$ K) is the highest among undoped

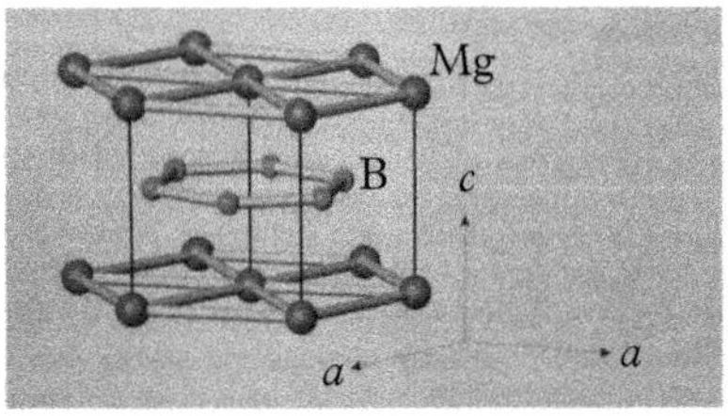

Figure 7.3 *The MgB_2 layered crystal structure.*

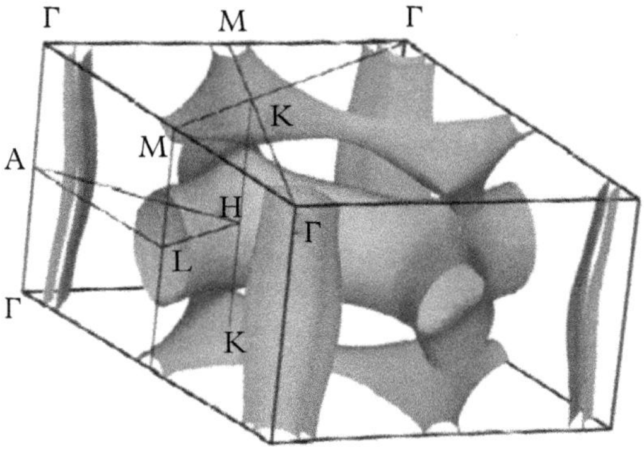

Figure 7.4 *The MgB_2 Fermi surface formed by σ-bands and π-bands.*

solids at ambient pressure. Speaking of superconductivity, its mechanism is conventional for MgB_2.

More specifically, it is caused by the interaction of electrons with E_{2g} phonons (in-plane vibrations). Historically, this development was shocking for many scientists, who believed the 'legend' that phonon-mediated superconductivity has an upper limit for T_c close to 30 K (see the discussion in section 2.4). Later, this 'legend' was completely destroyed after the discovery of superconducting hydrides with record values of T_c close to room temperature (see section 7.2).

In addition to the phonon mechanism as the origin of the pairing in MgB_2, it should be noted that, according to the photoemission data, this is an *s*-wave superconductor. Therefore, in many aspects, we are dealing with a rather conventional (BCS) picture.

However, there is one essential feature of the superconducting state that makes it different from many usual materials. Namely, it displays a two-gap electronic spectrum. Such a structure was observed experimentally with the use of photoelectron spectroscopy (Tsuda *et al.*, 2001; Souma *et al.*, 2003). The gaps were assigned to the σ- and π-bands. They are closed at the same T_c, as it should be for two-gap superconductivity (see section 3.2), and their values near $T = 0$ K are as follows: $\Delta_\sigma = 5.5 \pm 0.5$ meV and $\Delta_\pi = 1.5 \pm 0.5$ meV (Souma *et al.*, 2003), and $\Delta_\sigma \simeq 5.5 \pm 0.4$ meV and $\Delta_\pi = 2.2 \pm 0.4$ meV (Tsuda *et al.*, 2001). The values of these data, because of their accuracy, can

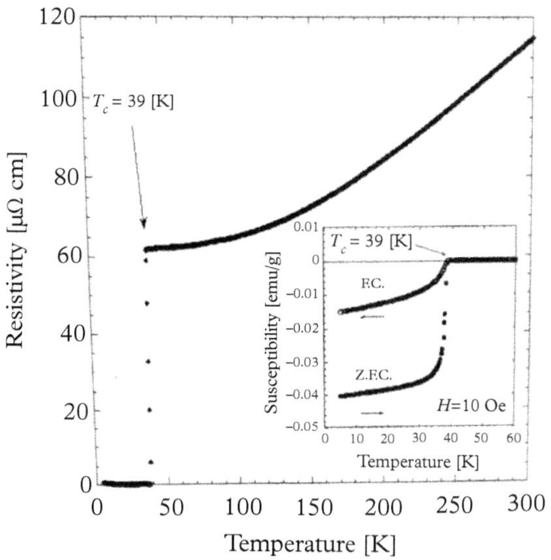

Figure 7.5 *The electrical and magnetic properties of MgB_2 (From Nagamatsu et al., 2001).*

be considered to be close to each other. The experiments performed with the use of scanning tunnelling microscopy (Iavarone *et al.*, 2002) also confirm the presence of two energy gaps ($\Delta_1 \simeq 2.2$ meV, $\Delta_2 \simeq 7$ at $T = 4.2$ K). The temperature dependence of the gap was also measured, and the authors observed that both gaps vanished at the same T_c value.

The presence of two gaps was also confirmed by heat capacity measurements (Bouquet *et al.*, 2001; Wang *et al.*, 2001). As is well known (see section 3.2), the temperature dependence of the electronic heat capacity at $T \to 0$ is determined by the smaller energy gap (Δ_π for MgB_2), whereas, at higher temperatures, the total electronic heat capacity is the sum of both contributions.

The experimental data are in good agreement with the results of theoretical studies which were performed shortly after the initial discovery of superconductivity in MgB_2 (Liu *et al.*, 2001; Choi *et al.*, 2002). According to theoretical analysis, the two-gap structure is caused by strong anisotropy of the electron–phonon matrix element. A large difference in the values of the gaps, together with this factor, leads to a decrease in the average value typical for many conventional superconductors and to the appearance of the two-gap structure.

Very interesting studies have been performed with the use of scanning tunnelling microscopy (Eskildsen *et al.*, 2002). First, the scanning tunnelling microscopy measurements demonstrated the presence of two energy gaps. The values of the gaps appear to be very different: the value of the gap corresponding to the σ-band is $\Delta_\sigma \simeq 75.5$ K, whereas that for the π-band is $\Delta_\pi \simeq 17.5$ K. The gaps are closed at the same T_c; the temperature dependence of the gaps is plotted in Fig. 7.6.

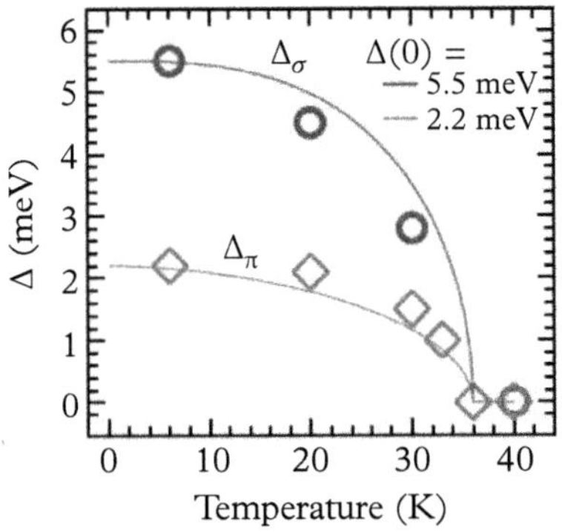

Figure 7.6 *Two superconducting gaps in MgB_2 formed by σ-bands and π-bands. (From Tsuda et al., 2003)*

Of note, in accordance with the general theory of two-gap superconductivity (see section 3.2), $2\Delta_\sigma/T_c \simeq 4.2$, whereas $2\Delta_\pi/T_c \simeq 0.45$. In other words, for one gap, this ratio is larger than $(2\Delta/T_c)_{BCS} \simeq 3.5$, while, for the second gap, it is smaller than the BCS value.

The paper by Eskildsen *et al.* (2002) also describes the behaviour of MgB_2 in an external magnetic field. The corresponding measurements allowed for the determination of the value of the critical field $H_{C2} \simeq 3.1$ T. The important coherence length (ξ) parameters were as follows: $\xi^\sigma \simeq 13$ nm for the σ-band, and $\xi^\pi \simeq 51$ nm for the π-band. The mean free path appears to be of the same order of magnitude. Therefore, the material is in the moderate clean limit. This factor, along with a large difference in the gap amplitudes, allows us to understand why the two-gap scenario persists for MgB_2 without the mixing typical for all usual superconductors (see section 3.2). More specifically, the band structure of the material is characterised by the presence of two σ-bands and two π-bands. However, four gaps are not observed, because the two σ-gaps are mixed and form a single σ-gap Δ_σ; similarly, a single gap Δ_π is observed. In the absence of further mixing, a two-gap picture is seen.

The study of the vortex structure along with a large difference between the values of the gaps Δ_σ and Δ_π leads to the conclusion that the phenomenon of induced two-gap superconductivity is present (see section 3.2). In general, two-gap superconductivity is described by three coupling constants; in this case, two diagonal coupling constants, $\lambda_{\sigma\sigma}$ and $\lambda_{\pi\pi}$, and one non-diagonal constant, $\lambda_{\sigma\pi}$, which describes the interband transitions. Therefore, for MgB_2, $\lambda_{\pi\pi} = 0$; the π-band is intrinsically normal, and the pairing described by the gap parameter Δ_π occurs thanks to the interband transitions. The ratio $\lambda_{\sigma\pi}/\lambda_{\sigma\sigma}$ is rather small, and this leads to the observation of a small value for the induced gap Δ_π. Overall, as stressed by Eskildsen *et al.* (2002), all of the main properties of MgB_2, such as a high T_c value, a critical magnetic field, and so on, are determined by the σ-band. The presence of the smaller gap Δ_π is not so crucial, although it determines the behaviour of the electronic heat capacity of $T \to 0$ K.

7.2 Hydrides: High Pressure

7.2.1 Introduction

The idea of room temperature superconductivity was envisioned shortly after superconductivity was discovered. However, achieving that end has proved to be rather difficult (see section 1.1).

Nevertheless, the possibility of observing room temperature superconductivity has recently become perfectly realistic. And, indeed, after more than a hundred years of intensive development, scientists have finally been able to observe the phenomenon of superconductivity at room temperature (Snider *et al.*, 2020). The material (carbonaceous sulfur hydride) is in the superconducting state up to the temperature $T_c = 288$ K (near 15 °C). The effect has been observed at high pressure, $P = 267 \pm 10$ GPa. We discuss in this chapter the properties of this remarkable family of high T_c materials, superconducting hydrides. As for the different paths taken towards room temperature superconductivity, they are summarised later on (see section 8.5).

7.2.2 History

The situation of the new high-T_c family, the hydrides, is entirely different from that of the other well-known class of high-T_c materials, the cuprates (see chapter 5). The discovery of high-T_c superconductivity in cuprates (Bednorz and Mueller, 1986) was totally unexpected, and the discussion about the mechanism responsible for it still continues. The experimental observation of a high T_c in hydrides (Drozdov *et al.*, 2015) was preceded by theoretical investigations and key predictions (see e.g. the review by Duan *et al.*, 2019). All of the predictions were based on the phonon mechanism, which was generally accepted shortly after its experimental discovery.

The original idea was expressed by Ashcroft in 1968. The idea is based on the well-known BCS expression (eqn (2.21)) for T_c. Since the phonon frequency $\tilde{\Omega} \propto (M)^{-1/2}$, where M is the ionic mass, the material formed by hydrogen, that is, the lightest ion, has a large phonon frequency; this factor is beneficial for reaching a high T_c value. However, solid hydrogen can only be formed under high pressure (Wigner and Huntington, 1935); as a result, high-pressure physics has become a major tool for studying the superconducting state in hydrogen and hydrogen-containing compounds (hydrides).

It is a very serious challenge to obtain metallic hydrogen as it requires a very high pressure to produce. However, there has been very noticeable progress in this direction (see the next section). Nevertheless, the transition of metallic hydrogen into the superconducting state requires even higher pressures and has not been observed yet.

The initial experiments on hydrides (e.g. the study of SiH_4 by Eremets *et al.*, 2008) demonstrated that hydrides can indeed display a superconducting state, although the value of T_c they obtained was relatively modest ($T_c \simeq 17$ K). The key theoretical analysis was carried out by Li *et al.* (2014) and Duan *et al.* (2014). The theoretical prediction stimulated an experimental investigation, which found that sulphur hydride had a superconducting state at $T_c = 203$ K (!) (Drozdov *et al.*, 2015). This was the most significant breakthrough since the discovery of high-T_c oxides (Bednorz and Müller, 1986; see chapter 5).

If it is assumed that a sample of some material is in the superconducting state, what kinds of rigorous evidence are needed to show that this is, indeed, the case? It is well known that the superconducting state is characterised by a number of peculiar properties different from those in a normal metal. The most remarkable characteristic among them is the absence of electrical resistivity. However, the drastic decrease in resistivity below some temperature is a strong indication but not necessarily strong evidence of the transition, because it is impossible to measure experimentally the total absence of resistance. In reality, there are number of factors and various transitions that are accompanied by a noticeable change in resistance. To rigorously determine whether or not a superconducting transition is present, the behaviour of the material in an external magnetic field should be studied. The observation of the Meissner effect is the most rigorous manifestation of the transition but at least a decrease in the T_c value with an increase in the magnetic field should be observed. It is important to note that Drozdov *et al.* (2015) not only observed a drastic decrease in the resistance but also studied the magnetic response (see also the review by Eremets and Drozdov, 2016).

The observation of a high-T_c superconducting state in sulphur hydrides generated a series of intensive theoretical studies (Akashi, 2015; Errea *et al.*, 2015; Bianconi and Jarlborg, 2015; Durajski, 2015; Bernstein *et al.*, 2015; Papaconstantopoulos, 2015; Akashi, 2016; Errea *et al.*, 2016).

Sulphur hydride was the first hydride with a high T_c value. Afterwards, it was natural to anticipate even higher T_c values for other hydrides. And, indeed, in 2018 was reported that the La-H compound displays a $T_c \approx 250 - 260$ K (Somayazulv *et al.*, 2018; Drozdov *et al.*, 2019; see section 2.4). Such values are really close to room temperature. The recent discovery of superconductivity in the S-C-H compound (Snider *et al.*, 2020), with a $T_c = 288$ K (near 15 °C) at high pressure ($P = 267 \pm 10$ GPa) was the first observation of the phenomenon of superconductivity at room temperature.

7.2.3 Metallic Hydrogen

As noted above, the initial focus was on the study of metallic hydrogen and its potential for having a high T_c value. This expectation was caused by a high frequency of vibration of the hydrogen ion. However, hydrogen first must become metallic, and the transition from the insulating state into the metallic state requires a high pressure (Wigner and Huntington, 1935).

The phase diagram for hydrogen is rather complicated (Fig. 7.7). It was a very non-trivial task even to make this diagram. The phase diagram of any material is based on the knowledge of its structure for different phases. However, X-ray diffraction, which is the usual technique for structure determination, is inapplicable here because scattering by hydrogen is very weak. The diagram shown in Fig. 7.7 is based on Raman and infrared measurements along with resistivity data. The low-temperature phases I–III were observed at relatively low pressures ($\leq$150 GPa; see the review by Mao *et al.*, 1994).

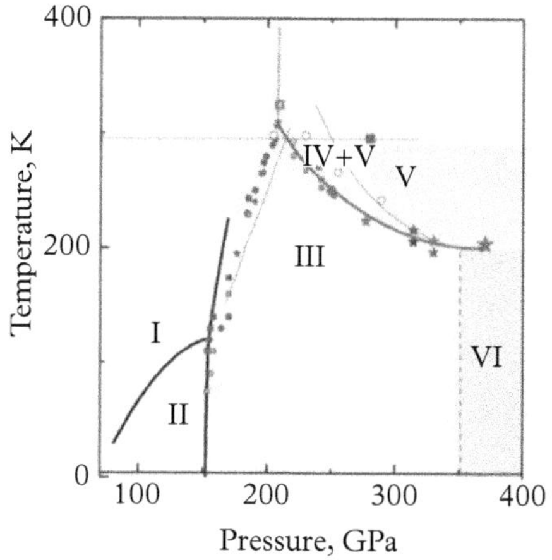

Figure 7.7 *The pressure-temperature phase diagram for hydrogen. (From Mao et al., 1994)*

One must distinguish between the molecular and the atomic structures of hydrogen. Correspondingly, there have been two reports on the metallisation of hydrogen, each of them describing different paths. One of the papers describes the observation of such a transition through the atomic structure (Dias and Silveira, 2017). In contrast, according to Eremets *et al.* (2016), the transition from the insulating state into the metallic one for molecular hydrogen occurs through a semimetal intermediate stage, that is, through creating an increasing overlap of valence and conducting energy bands. This overlap grows with an increase in pressure and leads to the formation of the metallic state. The authors of the two studies have reported their observations of various manifestations of the metallic state. One can note that the metallic state, by its definition, is characterised by the presence of the Fermi surface. One can expect that one of the usual probes (e.g. the de Hass–van Alphen oscillation) will demonstrate the presence of the Fermi service; such an observation will prove that we are dealing with the metallic state. However, a detailed description of the phases of hydrogen and the transition is beyond the scoop of this book.

The described properties were observed at pressures ~360 GPa. As for the superconducting state of metallic hydrogen, theoretical calculations (McManon *et al.*, 2012) show that a high-T_c state of pure metallic hydrogen requires pressures on the order of ~500 GPa. Hopefully, this phenomenon will be observed in the near future.

7.2.4 The Impact of Pressure: New Structures

The main properties of high-T_c hydrides are described below. But, initially, the following question must be asked: why does high pressure appear to be such a key tool? The factors that make the use of high pressure and its impact so unique will be discussed first.

Increasing pressure is not a new experimental technique (see section 4.2). The first well-known manifestation of this method is that it affects the value of the lattice parameter. This factor leads to some modifications of the phonon spectrum and, therefore, to some changes in the T_c, but this effect is relatively small.

The second factor that leads to a larger impact, specifically in layered materials such as cuprates (see chapter 5), is related to the change in distance between the layers, which is caused by an external pressure. If the carriers in these layered materials are provided by doping (the cuprates represent the most important case of such materials), then the doping and, correspondingly, the carrier concentration could be strongly affected by the pressure. If the carrier's 'reservoir' is located in a different layer, then such a case is present. The most important example of such a system is the YBCO compound, in which the carriers in the Cu–O plane are provided by the charge transfer from the layer containing the Cu–O chains. A decrease in the distance between these layers leads to an increase in the carrier concentration n in the basic unit, namely in the C–O plane. Because of the dependence $T_c(n)$, a noticeable change in T_c (on the order of ~20 K) can be observed (see the reviews by Chu, 2017 and Kresin *et al.*, 1996).

Finally, the third and most important factor that plays a key role in superconductivity in hydrides is the structural transition induced by pressure, which is a thermodynamic parameter. Such a phase transition causes a new structure to appear; this structure is stable only within some interval of pressures and is unstable at ambient pressure. The stability can be theoretically determined by calculating the enthalpy-difference curves.

These new structures created by applying a high pressure can display superconducting properties, and some of them can have a high T_c value. This is exactly the case for the high-T_c hydrides, which are described below.

7.2.5 Hydrides: Main Properties

Unlike solid hydrogen, the unit cell of hydride compounds contains heavy element ions in addition to hydrogen. The properties of various hydrides will be described in the following sections. Here, the main features common to all such superconducting compounds are introduced.

The Mechanism of superconductivity. Starting from the fundamental question about the origin of superconductivity in hydrides, it is generally accepted that the superconducting state in these compounds is provided by the electron–phonon interaction. This conclusion follows from the theoretical development and the prediction preceding the experimental discovery. The analysis was based on the theory of pairing caused by phonon exchange (see section 2.2). Another key factor is the large value of the isotope coefficient for the hydrogen $\rightarrow$ deuterium substitution (H $\rightarrow$ D), which appears to be close to its optimum value ($\alpha_{max} = 0.5$).

This statement about the mechanism is fundamentally important. The existence of a real experimental setting displaying the superconducting state with $T_c = 203$ K (!) for sulphur hydride and $T_c \simeq 250$–260 K (!) for lanthanum hydride, due to the phonon mechanism, is bringing to an end the 'legend' (see section 2.3.5) that the phonon mechanism is able to create the pairing only below some limit corresponding to a relatively low temperature range.

According to the expressions for T_c (see eqns (2.10), (2.22), (2.30), and (2.34)), the high T_c value can be caused by a high phonon frequency and a large value for the coupling constant. These factors determine the record high-T_c values for hydrides.

As noted above, metallic hydrogen has promise for reaching a high value of T_c because of the small value of the hydrogen ion mass. Nevertheless, the search among various compounds containing hydrogen and heavy ions (hydrides) is even more promising. Indeed, metallic hydrogen only has acoustic phonon modes. It is well known that the electron–phonon coupling constant λ is defined by the expression (1.27). This expression contains the phonon density of states $F(\Omega)$, and $F(\Omega) \propto q^2 dq/d\Omega = q^2(d\Omega/dq)^{-1}$. Speaking of acoustic phonons, the main contribution to the pairing comes from the short-wave region, where the derivative $d\Omega/dq$ (q is the phonon momentum) becomes small and leads to an increase in the phonon density of states. The presence of two (or more) ions in the unit cell leads to the appearance of additional optical phonons. As for the optical modes, their frequency weakly depends on q, almost in the whole region of q, and this leads to a large value of $F(\Omega)$. It can be seen directly from eqn (1.27) that an increase in the phonon density of states is a favourable factor that increases the strength of the electron–lattice coupling. Moreover, the presence of heavy ions provides the existence of additional acoustic modes, and this also increases the coupling constant.

In general, the phonon mechanism is described by eqns (2.17) and (2.17′), which, at $T = T_c$, have the form of eqns (2.18) and (2.18′), respectively. For the majority of hydrides studied, it is essential that the unit cell contain two ions with very different mass values. As a result, the phonon spectrum can be separated into two groups (see section 3.5.3). The high-frequency optical modes describe the vibrational motion of light hydrogen ions, whereas the acoustic modes correspond to the motion of heavy ions. The high value of T_c is provided mainly by the optical modes and the motion of light hydrogen ions. But just the presence of these ions is not sufficient. There should be strong coupling to the optical modes, and this is provided by a large number of hydrogen ions (ligands) in the unit cell. Indeed, as shown in Fig. 7.10 below, the hydrides with high T_c values contain a rather dense 'forest' of optical phonons.

The phonon spectrum of hydrides is rather broad (up to ~200 meV (!)). Therefore, evaluation of T_c should be performed with considerable care. This is especially true if a coupling constant is to be introduced and an analytical expression for T_c is to be obtained; it is useful to have such an expression to study various dependences of T_c. However, the usual approach described in section 2.2, based on the use of a *single* coupling constant and a *single* characteristic phonon frequency, is too crude to obtain accurate results. It is more suitable to use the concept with two coupling constants, λ_{opt} and λ_{ac}, which are defined by eqn (2.37), and two characteristic frequencies, $\tilde{\Omega}_{opt}$ and $\tilde{\Omega}_{ac}$, which are defined by eqn (2.37′).

Equation (2.36) allows the evaluation of the T_c value. The most interesting case is when $\lambda_{opt} \gg \lambda_{ac}$ and $\lambda_{opt} \gtrsim 1$. Then T_c is determined by eqn (2.39), where T_c^0 is the T_c value for the case when electrons interact only with the optical modes.

As a whole, the T_c value strongly depends on the interplay of the optical and the acoustic modes, that is, on the relative intensity of the 'forest' of optical modes. If

$\lambda_{ac} \gg \lambda_{opt}$ and $\lambda_{ac} \lesssim 1$, then the T_c value is rather low. Overall, depending on the strength of the coupling constants and their relative interplay, any T_c value can be obtained so that the regions of high and low T_c values can be bridged within the framework of the same phonon mechanism.

The two-gap spectrum. Each hydride has a rather complex phase diagram. As mentioned above, an increase in pressure leads to a set of structural transitions. Pressure is the thermodynamic parameter, and the structure is stable within some pressure interval; it corresponds to the minimum enthalpy. The question of the nature of the phase transition is very interesting and will be discussed in the next section.

There are specific structures for various hydrides (H_3S, LaH_{10}, HfH_{10}, etc.; see below), corresponding to the maximum T_c. Because of the high T_c value, the coherence length appears to be short. For such a 'clean' case, the presence of a two-gap structure for the electronic spectrum is expected (see section 3.2). In addition, an increase in pressure within this phase will demonstrate that, at some pressure, T_c has a maximum. Therefore, a continuous increase in pressure leads to a relatively slow decrease in T_c. The reason for such behaviour should be explained.

The isotope effect. Because of the peculiar structure of the phonon spectrum with its separation into two groups, the isotope effect for an H $\rightarrow$ D substitution plays a special role. Its value directly reflects the relative contribution of the optical modes. If the T_c value is determined mainly by the optical modes, the value of the isotope coefficient is close to $\alpha_{max} \simeq 0.5$ (its value is described by eqn (3.90)). In the opposite case, when the acoustic modes make the main contribution to the pairing, the value of α is small because the H $\rightarrow$ D substitution does not noticeably affect the value of T_c. The method using two coupling constants appears to be very useful (see below).

High T_c: Main criteria. As was noted above, the high-T_c state of the hydrides is provided by the electron–phonon mechanism. The family of superconducting hydrides is rather large. Based on the studies of several hydride families, several rules can be formulated (Hie *et al.*, 2020) to enable a successful search for high-T_c hydrides. These rules are as follows:

1. The crystal structure must have high symmetry, as this leads to an increase in the degeneracy of electronic states and, correspondingly, in an increase in the density of states.
2. No H_2-like molecular unit should be present.
3. There should be a large H contribution to the total electronic density of states at the Fermi level.
4. There should be strong coupling of the electrons on the Fermi surface with high-frequency optical phonons.

H_3S-clathrate, which has a six-fold cubic structure, LaH_{10}, which also has a six-fold cubic structure, and HfH_{10} hydride, which has a layered hexagonal structure, are examples of compounds that satisfy these four criteria (see below and section 2.7).

To end this section, it should be stressed that the recent progress and efforts of scientists performing ab initio calculations have played a key role in the study of high-T_c hydrides. It is impossible to overestimate the importance of evaluating the stable structures at various pressures as well as determining the band structures, phonon spectra, and the spectral function $\alpha^2(\Omega)\,F(\Omega)$. Several groups of theoreticians have been working in several countries, and their analyses and collaborations with experimentalists deserve the highest credit.

Below, several families of superconducting hydrides are described. Initially, materials in which the phenomenon of a high T_c has been observed are the main focus. Afterwards, several interesting and promising predictions are discussed.

7.2.6 Sulphur Hydrides

High-T_c phases. The first focus will be on sulphur hydrides. This is not a random choice. The preceding theoretical development (see section 2.2) and the experimental breakthrough have attracted a lot of attention. This resulted in a flourish of activity and a detailed study of S-H compounds.

Experimentally, Eremets and his collaborators (Drozdov *et al.*, 2015; see the review by Eremets and Drozdov, 2016) observed a drastic drop in resistance near a particular temperature; its value can be assumed to be the critical temperature. More specifically, the measured resistance was at least two orders of magnitude below that of pure copper. As is well known, it is unrealistic to observe experimentally a total absence of resistance. Therefore, it is crucial to study the behaviour of a sample in an external magnetic field as well. And, indeed, the value of T_c shifts downward in the presence of such a field. Direct observation of the Meissner effect is a very serious challenge because of the very small size d of the sample ($\sim$100 μm). Nevertheless, Troyan *et al.* (2016) have observed magnetic field expulsion for a system consisting of a thin Sn film placed inside of a bulk system. The Meissner effect was also observed by Huang *et al.* (2016); see also the review by Duan *et al.* (2019) describing ac magnetic susceptibility measurements.

An increase in pressure leads to structural transitions and a corresponding change in the value of T_c. The structural transformations are the key component of the observed path to higher values of T_c.

Next, the two phases with the highest T_c values will be the focus. One of them (R3m; Fig. 7.8a, b) has a value for T_c of approximately 120 K, and the other ($Im\overline{3}m$; Fig. 7.8c) has the highest value T_c of approximately 200 K. Below, the former structure is referred to as the 'low-T_c phase' (although this name sounds ironic for $T_c \approx 120$ K), and the second structure is referred to as the 'high-T_c phase'.

The phase diagram shows that the pressure dependence of T_c is rather peculiar. Indeed, Fig. 7.9 demonstrates that T_c is strongly dependent on the applied pressure (Einaga *et al.*, 2016). It increases from ~100–120 K up to the record ~200 K over the relatively narrow pressure interval of 125–150 GPa.

Note that, once the T_c reaches its maximum value, it decreases upon a further rise in the applied pressure. The decline is rather slow; therefore, the T_c dependence is strongly asymmetric relative to $T_{c;max}$. This peculiar dependence, $T_c(P)$ (sharp increase up to $T_c \approx 200$ K, followed by a slow decrease), is explained below.

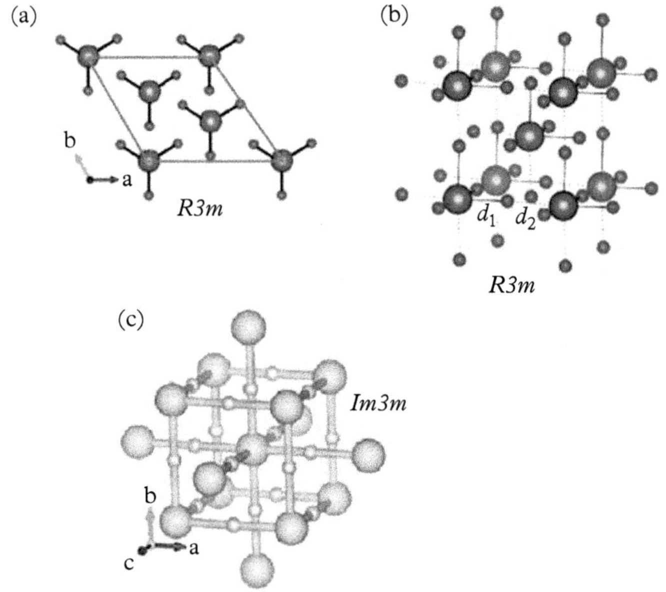

Figure 7.8 *Crystal structure of sulfur hydride at different pressures.*

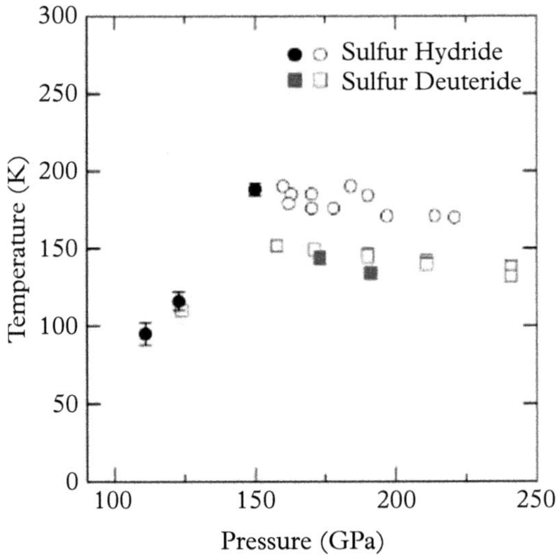

Figure 7.9 *Pressure dependence of T_c (From Einaga et al., 2016).*

An increase in pressure leads to the following important transformation (Duan *et al.*, 2014; see Fig. 7.8):

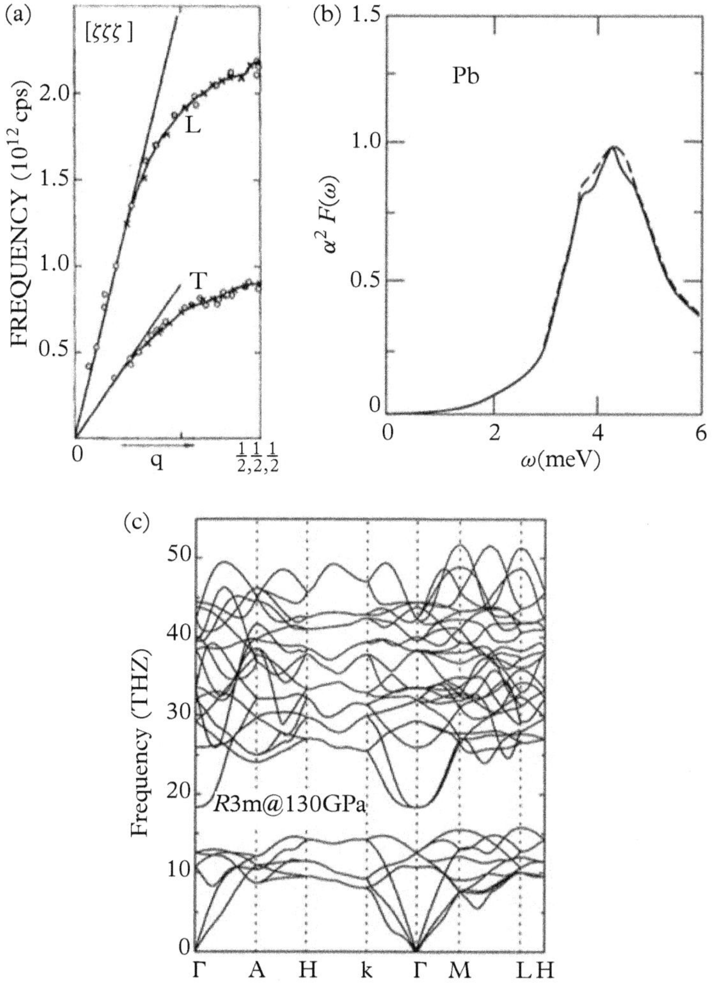

Figure 7.10 *A complex structure of hydrides phonon spectrum (From Einaga et al., 2016).*

$$3\ H_2S \rightarrow 2H_3S + S \tag{7.1}$$

According to the data from advanced spectroscopy (Goncharov *et al.*, 2017), which makes it possible to monitor the evolution of a structure with pressure, the transformation (7.1) occurs at $P \gtrsim 40$ GPa. The structures $R3m\,(T_c \approx 120\ \text{K})$ and $Im\overline{3}m(T_c = 203\ \text{K})$ contain H_3S units.

Critical temperature. As discussed above, the evidence for the phonon mechanism of superconductivity was in accordance with the theoretical predictions and was validated experimentally. The observation of a strong isotope effect also supported this conclusion.

Indeed, the value of the isotope coefficient in the high-T_c phase of the H_3S compound ($T_c = 203$ K) is rather large: $\alpha = 0.35$ (Drozdov *et al.*, 2015).

Nevertheless, evaluation of the T_c value should be performed with considerable care. In fact, the phonon spectrum of hydrides, including H_3S, is rather broad (up to ~200 meV (!)) and has a complex structure (see Fig. 7.10c). In addition to the usual acoustic low-frequency branches, many optical ones can be seen. The high-frequency optical modes describe the motion of light hydrogen ions, and their presence is the key component of the scenario. However, it should be noted that the acoustic branches corresponding to the vibrations of sulphur ions also make a noticeable contribution (see below). Their relative contribution can be seen from Fig. 7.11, since the curve $\lambda(\omega)$ directly reflects the contribution of different frequency regions to the electron–phonon coupling.

In order to evaluate the values of these parameters, the calculations performed by Duan *et al.* (2014) can be used. Their values are different for various phases (for $R3m, T_c \approx 120$ K; for $Im\,\overline{3}m$, $T_c = 203$ K; see Fig. 7.7). For the 'low-T_c' phase $R3m$, the following can be obtained: $\lambda_{opt} \approx 1$, $\lambda_{ac} \approx 1$, $\lambda_{opt} \simeq \lambda_{ac}$, $\tilde{\Omega}_{opt} = 105\,\text{meV}$, and $\tilde{\Omega}_{ac} = 26\,\text{meV}$. Since, in this case, $T_c < \tilde{\Omega}_{ac} \ll \tilde{\Omega}_{opt}$, the usual BCS approximation with the addition of the renormalisation function can be used: $Z = 1 + \lambda_T$, $\lambda_T = \lambda_{opt} + \lambda_{ac}$ (see eqn (2.38)). Then $T_c \approx 120$ K is obtained.

Turning to the more interesting case of the high-T_c $Im\,\overline{3}m$ structure and the use of Fig. 7.11, the following is obtained for this structure:

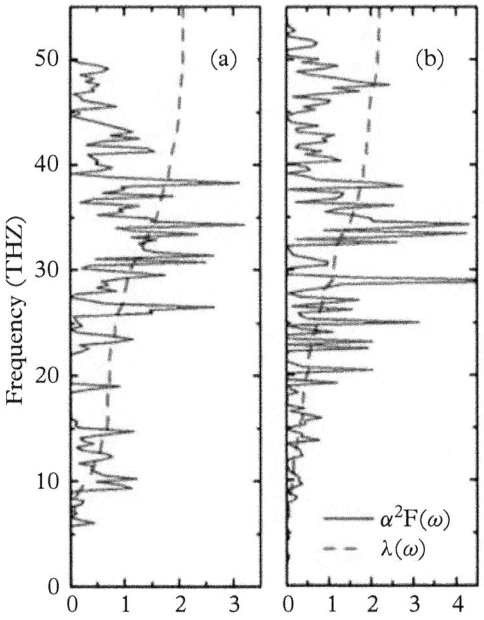

Figure 7.11 *The electron–phonon coupling spectra of hydrides (From Duan et al., 2014).*

$$\lambda_{opt} = 1.5, \lambda_{ac} = 0.5; \tilde{\Omega}_{opt} \approx 1700 \text{ K}, \tilde{\Omega}_{ac} \approx 450 \text{ K}$$

In this case, the BCS approximation is not applicable because the electron–phonon interaction with optical phonons is strong. It is interesting that the total coupling constant is approximately similar to that for the 'low-temperature' phase *Rm3m* ($\lambda_T \approx 2$). However, the redistribution of the interaction towards an increase in the coupling to high-frequency optical modes plays a key role. Indeed, in this case, expressions (2.39) and (2.40) should be used. After a simple calculation, $T_c \approx 215$ K is obtained. This value is close to that measured by Drozdov *et al.* (2015).

The fact that the coupling constant λ_{opt} in the cubic phase is so large is the key factor underlying the observed high value of T_c ($T_c \approx 203$ K). Qualitatively, this comes about due to the ability of sulphur to retain several hydrogen atoms in its proximity, that is, the presence of many light ligands near the sulphur atoms. There are six (!) ligands in the high-T_c phase (see Fig. 7.9). This can be understood as a manifestation of a mixed valence of sulphur.

Note that eqn (2.39) is valid if $\lambda_{opt} \gg \lambda_{ac}$. The values $\lambda_{opt} = 1.5$ and $\lambda_{ac} = 0.5$ satisfy this condition; therefore, the approach is self-consistent. Note also that the value of T_c^o (see eqn (2.23); $T_c \equiv T_c^0 = T_c^{opt}$) is equal to $T_c^{opt} \approx 170$ K, whereas ΔT_c^{ac}, corresponding to the second term in eqn (2.39), appears to be equal to $\Delta T_c^{ac} \approx 45$ K.

Dividing the phonon spectrum and, correspondingly, the electron–phonon interaction into two parts turns out to be rather fruitful. First of all, the value of λ_{opt} is within the range of applicability of eqn (2.39). Moreover, the relative contributions of the optical and acoustic branches of the phonon spectrum to T_c can be evaluated. For the high-T_c phase, the contribution from the optical phonons comprises ~80%, and only ~20% is due to the acoustic part. The impact of the acoustic phonons is noticeably smaller than that of the optical branches (45 K vs. 170 K), but it is still essential.

A similar method for evaluation of the T_c value can be used to analyse other hydrides (see below). Next, another fundamental property, namely the isotope effect, will be analysed. Its presence is the main manifestation of the phonon mechanism of high-T_c superconductivity in hydrides.

The isotope effect. The substitution of deuterium for hydrogen ($D \rightarrow H$) noticeably affects the T_c value. Observation of this isotope effect is of fundamental importance, since it proves that (i) the high-T_c state is caused by an electron–phonon interaction, and (ii) the high-frequency hydrogen modes determine the value of T_c. Indeed, the optical modes are mainly due to the motion of hydrogen, whereas, for the acoustic modes, the participation of sulphur ions prevails. Therefore, the magnitude of the isotope shift for the deuterium-for-hydrogen substitution indirectly reflects the relative contributions of each group (optical vs. acoustic modes) on the observed T_c.

The isotope coefficient α is defined by the relation $T_c \propto M^{-\alpha}$ (see chapter 3 and section 5.1). As mentioned previously (section 2.6), it is convenient to use the concept of two coupling constants (see chapter 3 and section 5.4). The value of the isotope coefficient in the high-T_c phase can be calculated with the use of eqn (3.90). With $\lambda_{opt} \approx 1.5$, $\lambda_{ac} \approx 0.5$, and $\tilde{\Omega}_{opt} \approx 450$ K, $\alpha \approx 0.35$ is obtained, which is in rather good agreement with the experimental data shown in Fig. 7.9. For T_c in the 'low-T_c' phase, the expression shown in eqn (2.38) should be used. Then, with the use of eqn (3.89), it is determined that $\alpha \approx 0.25$, which is noticeably smaller than that for the high-T_c phase.

The value of the isotope coefficient in the high-T_c phase is relatively large, reflecting the fact that the pairing in this phase is dominated by the optical H-modes, whereas in the 'low-T_c' phase, the contributions of the optical and acoustic modes are comparable. The impact of the isotopic substitution in the region of lower T_c values is weaker than that in the high-T_c phase. A smaller value of α corresponds to a larger role played by the optical phonons in the cubic high-T_c phase.

The value of the isotope coefficient can be affected by anharmonicity (Errea *et al.*, 2015), polaronic effects, and the dependence of the chemical potential μ^* on the frequency $\tilde{\Omega}_{opt}$.; the last contribution is on the order of $(\mu^*/\lambda_{opt})^2$ and is small. Nevertheless, the main conclusion that the value of the isotope coefficient depends on the pressure and is different in distinct phases remains valid and reflects the relative contributions of the optical and acoustic modes.

The phase diagram: Structural transitions. As discussed above and shown in Fig. 7.9, the T_c value drastically increases in the relatively narrow pressure interval of 125–150 GPa. Indeed, the increase is rather sharp (from $T_c \approx 120$ K up to $T_c \approx 200$ K). This increase is directly related to the phase transition from the *R3m* structure (Fig. 7.8a, b) to the high-T_c $Im\overline{3}m$ cubic structure (Fig. 7.8c). It is important to clarify the nature of this phase transition. In general, this might be a first-order or a second-order phase transition. Another possibility is that it is a 2.5-order phase transition. A first-order transition corresponds to a jump in the value of major thermodynamic functions, for example the thermodynamic potential, whereas a second-order transition is accompanied by a jump in the derivatives of such functions, for example by a jump in the heat capacity (the thermodynamic potential changes continuously). As for a 2.5-order transition (sometimes called a Lifshitz transition), it corresponds to a topological change in the structure of the Fermi surface.

It turns out that the question regarding the nature of the *R3m* ($T_c = 120$ K) $\to$ $Im\overline{3}m$ ($T_c = 203$ K) transition is related to another puzzle, namely the behaviour of the T_c in the high-T_c cubic phase. As shown in Fig. 7.9, an increase in pressure following the transition into this phase is not accompanied by a further increase in T_c. Instead, a slow decrease in the value of T_c is observed.

Next, the nature of structural transitions will be discussed. As mentioned above, an increase in pressure leads to a sequence of such transitions. An important example of such a transition is a sharp increase in the value of T_c ($T_c \approx 120\,\mathrm{K} \to T_c \approx 200\,\mathrm{K}$) for sulphur hydride in the narrow pressure interval between $P \simeq 125$ GPa and $P \simeq 150$ GPa. The crystal lattices of the high-T_c ($T_c \simeq 200$ K) and low-T_c ($T_c \simeq 120$ K) phases belong to different symmetries: cubic $Im\overline{3}m$ symmetry for the high-T_c phase, and trigonal *R3m* symmetry for the low-T_c phase (Fig. 7.8). According to studies carried out by Gor'kov and Kresin (2016, 2018), the phase transition into the high-T_c state is of the first order, which can explain the non-monotonic dependence $T_c(P)$, that is, the slow decrease in T_c at $P > P_{max}$ (P_{max} corresponds to the maximum T_c value).

The nature of the phase transition can be analysed initially with the use of the Landau theory of phase transitions (see Landau and Lifshitz, 1980) and group theory. According to the Landau theory, the thermodynamic potential of the high-symmetry phase $\Phi(T, P)$ can be expanded in the series

$$\Phi = \Phi_0 + A\eta^2 + B\eta^3 + C\eta^4 + \cdots \tag{7.2}$$

Analysis of the coefficients leads to the conclusion that the $R3m \rightarrow Im\overline{3}m$ transition is of the second order (since the coefficient $B = 0$). However, there is one more factor that modifies such a conclusion. The thing is that the thermodynamic potential contains one more factor. It turns out that the Fermi surface of the cubic phase $Im\overline{3}m$ contains a flat region at a finite value of the electronic momentum. Then the electron–phonon interaction leads to phonon softening (Afanasjev and Kagan, 1963; Varma and Simons,1983; Kresin, 1984a) and competition between the pairing and CDW instability. The existence of flat regions at finite momenta can be described as the appearance of pockets on the Fermi surface, which contains regions with an increased density of states. This leads to the two-gap scenario (see below). The most interesting case is when $T_c > T_{CDW}$ (see also below, and section 2.7). The coupling of electronic states to the lattice, especially, intra-pocket states, leads to structural changes and can transform a second-order transition into a first-order transition. The addition to the thermodynamic potential in the $Im\overline{3}m$ phase describing this interaction (which is called quadratic striction) can be written in the form

$$H_{str} = -q \int \hat{U}(r) \boldsymbol{S}_H^2(r) d\boldsymbol{r} \tag{7.3}$$

where $\hat{U}(r)$ is the strain tensor and $\boldsymbol{S}_H$ is the displacement of a hydrogen ion from its equilibrium position in the middle between the sulphurs in the $Im\overline{3}m$ phase (Fig. 7.9) this is the so-called stretching mode. By analogy with the analysis by Larkin and Pikin (1969) and Barzykin and Gor'kov (2009), the transition is described by the equation

$$\lambda u = \lambda^2 B \frac{3x}{4Kx} |x|^{1/2} \tag{7.4}$$

where $x = \tau + \lambda u$, $\tau = T - T_0$, $\lambda = -2q/a$, a is the S–S closest distance in the $Im\overline{3}m$ phase, and K is the elastic moduli. Equation (7.4) has three solutions corresponding to a jump in the state at $T = T_0$. The decrease in pressure in the high-T_c $Im\overline{3}m$ phase leads to the disappearance of the flat regions on the Fermi surface at $T = T_0$ and, correspondingly, to the absence of the term (eqn (7.3)) in the $R3m$ phase. This jump is a signature of the first-order transition.

The slow decrease in T_c is caused by the presence of small pockets on the Fermi surface. These pockets exist as an addition to the usual (large) Fermi surface; their energy scale is below the energy of optical phonons. It is essential that pockets are absent in the 'low-T_c' phase (the $R3m$ structure). Because of their appearance in the high-T_c phase, it is tempting to relate the high value of T_c to the presence of these pockets. However, such an explanation should be ruled out because it contradicts the experimental data, namely the observation of the isotope effect (see section 2.3.4). Indeed, if these pockets play a leading role, then there should be a pre-factor of an electronic origin in the expression for T_c. This is similar to the van Hove scenario (section 2.7.4). If the pre-factor does not contain phonon energy, the isotope effect is absent. Therefore, the pockets should play a secondary role and serve as a small addition to the strong electron–phonon interaction caused by a large Fermi surface.

Note also that if one is trying to assign the leading role to pockets, then it is clear that the on-pocket interactions should be rather strong in order to provide high T_c values. However, in this case, the rigorous treatment is not known because of violation of the 'Migdal theorem' (see Sec.1.2.1). Fortunately, the above-described calculation, assuming the prevailing role of usual, large energy bands with strong electron–phonon coupling, provides a rigorous description. Moreover, there is good agreement with the experimental data.

Nevertheless, the small pockets with their weak interactions with phonons are manifested in the spectroscopy of the materials. Namely, their presence leads to the appearance of a second energy gap in the electronic spectrum. Therefore, the two-gap model should be applied to study their impact (see section 3.2). In this case, three coupling constants can be introduced: λ_L, which is responsible for strong electron–phonon interactions on the large band; $\lambda_P \ll 1$, describing weak coupling on the pockets; and $\lambda_{LP} \ll 1$, describing the transitions from large-band electrons to the pairing states on the pockets. The coupling constants λ_L, λ_P, and λ_{LP} are described by eqns (2.24) and (2.24′), and the parameter λ_{LP} contains the matrix element describing the corresponding transitions caused by the electron–phonon interaction. Because of the interband transitions, the system has the common temperature of the superconducting transition T_c. In addition, their presence is beneficial for superconductivity.

Assuming that the sharp increase in T_c (from $T_c \approx 120$ K to $T_{c;max} \approx 200$ K) is the result of the first-order structural transition into the high-T_c cubic phase, then this phase is characterised by the coexistence of a wide band (responsible for a large part of the Fermi surface) and small pockets. As mentioned above, the interaction between the large band and the pockets leads to a shift in the transition temperature: $\Delta T_c = T_c - T_{c0}$.

Based on the two-gap model, it is easy to show that the shift ΔT_c is proportional to the density of states on the pocket: $v_p(E_F) \propto m_{poc} p_{F;poc}$ where m_{poc} and $p_{F;poc}$ are the effective mass and momentum for the pockets' states, respectively, and T_{c0} is the T_c value in the absence of the pockets. It is essential to determine whether the pockets appear instantly, as a result of a discontinuous first-order transition, or whether the transition into the high-T_c phase is either of the second order or of a topological nature, in which case the pockets' size would grow continuously with a further increase in pressure.

The first-order transition is accompanied by the emerging singularity in the density of states in the form of pockets. The further increase in pressure leads to shrinking of the pockets, with an effective decrease in the Fermi momentum $p_{F;poc}$ and the corresponding decline of the two-gap picture. This explains the observed slow decrease in T_c after the transition; the small scale of the decrease in T_c at $P > P_{cr}$ is related to the small values of λ_P and λ_{LP}.

The two-gap spectrum and its evolution with pressure, including the decrease in the amplitude of the second gap at $P > P_{cr}$, should be confirmed by future tunnelling experiments. The presence of the second energy gap will manifest as the second peak in the density of states.

The development of tunnelling spectroscopy is an important forthcoming project. In fact, tunnelling measurements under pressure were carried out by Zavaritskii *et al.* (1971) to study the properties of a conventional superconductor with strong coupling, Pb. This method has not been used so far for hydrides. Because of the large T_c values and the energy gap, the tunnelling $I(V)$ needs to be measured for a wider energy interval than that used for conventional superconductors. Similar measurements have been performed for high-T_c cuprates by Aminov *et al.* (1994) and Ponomarev *et al.* (1999), who used the break junction technique, and by Lee *et al.* (2006), who used scanning tunnelling spectroscopy. Therefore, tunnelling spectroscopy should be applicable for sulphur hydrides as well. This technique will make it possible to reconstruct the $\alpha^2(\Omega)F(\Omega)$ function as well as determine the Coulomb pseudopotential μ^*. Tunnelling spectroscopy also can be employed to measure other important parameters of the system, including the energy gap, and to observe the multigap structure and its evolution with pressure.

7.2.7 Lanthanum Hydrides

The La-H_{10} compound displays high T_c value ($T_c \simeq 250$–260 K). The high-T_c superconducting state in La-H_{10} has been observed independently in two laboratories (Somayazulu *et al.*, 2018; Drozdov *et al.*, 2019). Both groups observed a drastic decrease in resistivity. A shift towards lower values of T_c in an external magnetic field also has been observed (Drozdov *et al.*, 2019).

It is interesting that the structure of this compound is different from that for the H_3S material. As described above (see Fig. 7.8), the structure of the high-T_c phase of sulphur hydride $(Im\overline{3}m)$ is cubic. As for LaH_{10}, its structure in the high-T_c phase (the so-called $Fm\overline{3}m$ phase) is shown in Fig. 7.12. The $Fm\overline{3}m$ phase has the so-called clathrate structure, which, indeed, is very different from the cubic one.

The clathrate structure was introduced by Wang *et al.* (2012) as a result of the study of the hydride compound CaH_6 (see below, and section 2.4). The phase diagram of LaH_{10} describing the phase transformation as a function of the applied pressure is presented in Fig. 7.13 (Sono *et al.*, 2020). The most interesting part of this diagram is the phase $Fm\overline{3}m$ (at $P > 200$ GPa), which displays the highest observed T_c. Song *et al.* (2020) studied this phase in detail. Their method of study was similar to that used to study the $Im\overline{3}m$ phase of the H_3S compound (see section 2.7).

Figure 7.14 shows the phonon dispersion and phonon density of states in the $Fm\overline{3}m$ phase of LaH_{10}. As in the approach used in the previous section, the two coupling constants λ_{opt} and λ_{ac} as well as the characteristic frequencies $\tilde{\Omega}_{\text{ac}}$ and $\tilde{\Omega}_{\text{opt}}$ can be calculated (see eqns (2.37) and (2.37′)). The quantities can be evaluated with use of the spectral function $\alpha^2(\Omega)F(\Omega)$ and the electron–phonon integral $\lambda(\Omega)$, shown

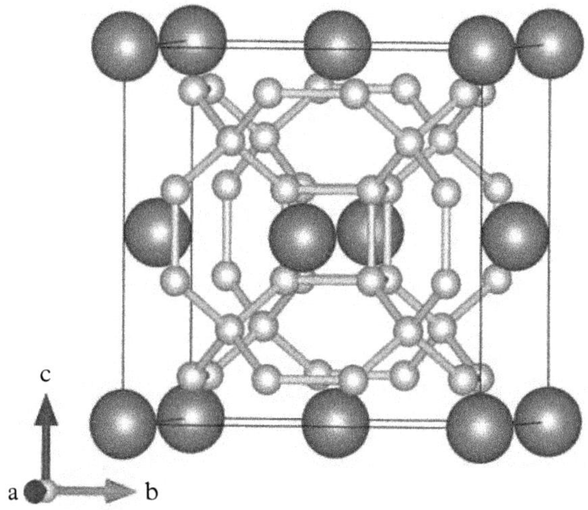

Figure 7.12 *The high-T_c phase structure of LaH_{10} (the $Fm\overline{3}m$ phase). (From Sono et al., 2020)*

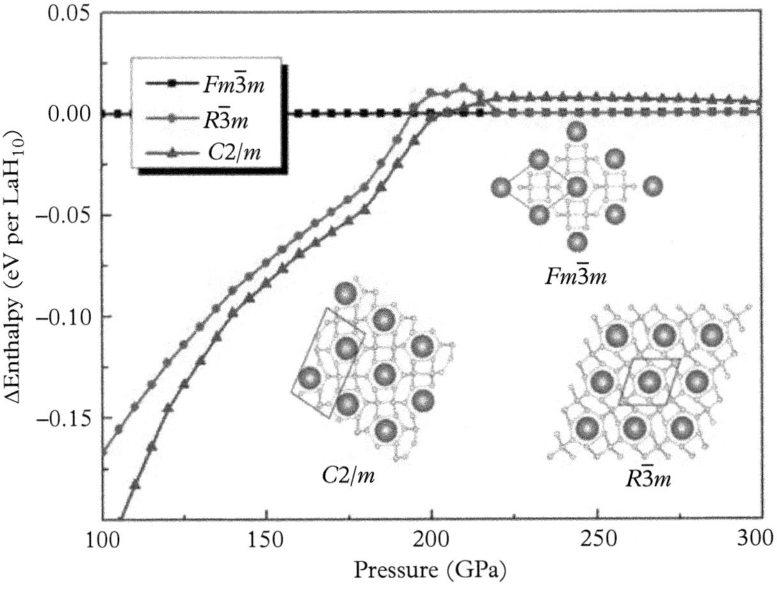

Figure 7.13 *The phase diagram of LaH_{10} under applied pressure. (From Sono et al., 2020)*

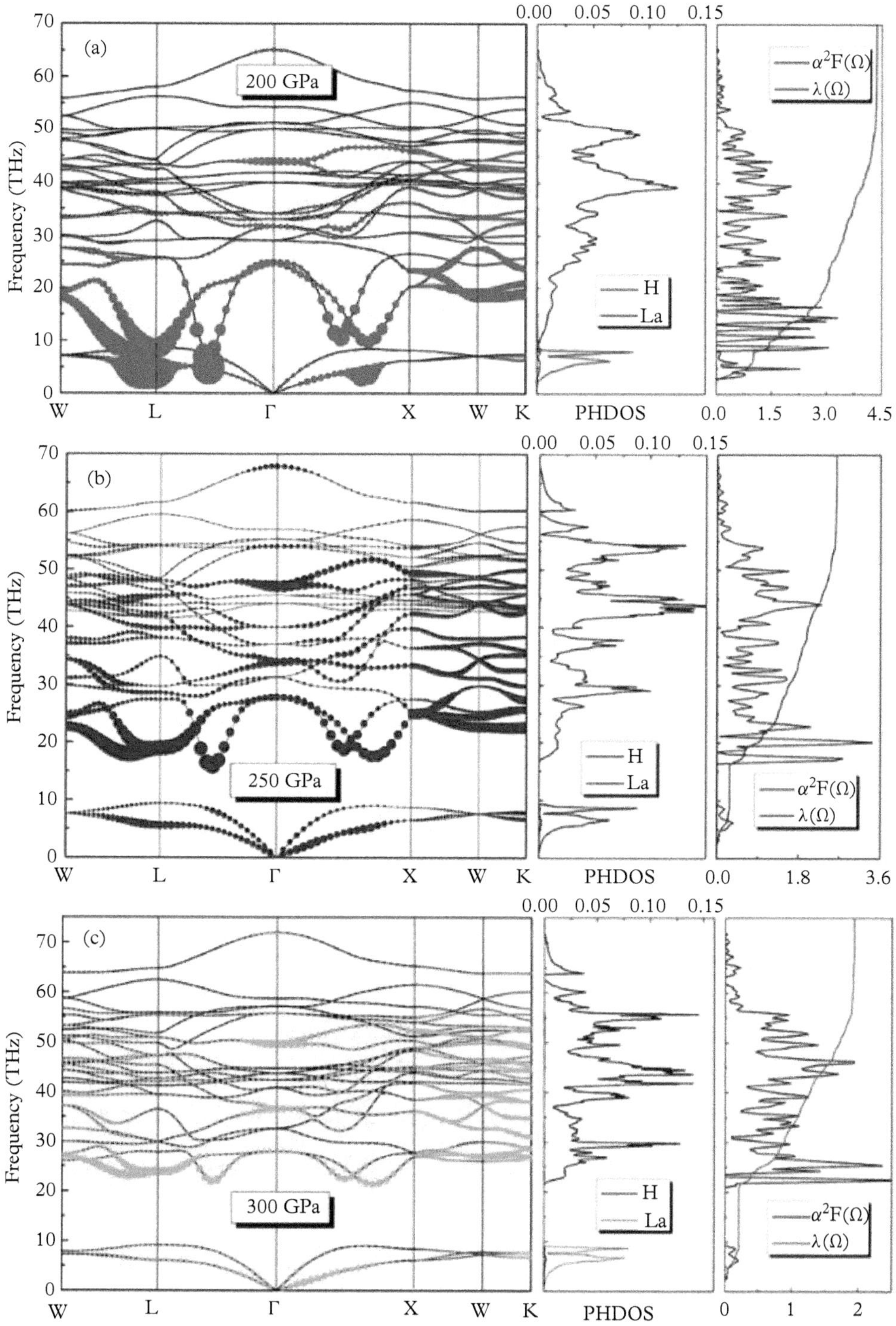

Figure 7.14 *The phonon dispersion and phonon density of states at different pressure in the $Fm\overline{3}m$ phase of LaH_{10}. (From Sono et al., 2020)*

in Fig. 7.14. Their values depend on the pressure. For example, for $P = 200$ GPa, one obtains $\tilde{\Omega}_{ac} = 2.2 \times 10^2$ K and $\tilde{\Omega}_{opt} = 1.25 \times 10^3$ K; $\lambda_{ac} = 1.1$ and $\lambda_{opt} = 3.25$. The phonon density of states is shown in Fig. 7.14; accordingly, one can select $\Omega_1 = 8.3$ THz.

With the use of these parameters, the T_c value can be evaluated. Since $\lambda_{opt} \gg \lambda_{ac}$, eqn (2.39) can be used. Here, T_c^0 is the value of T_c caused by the interaction with optical modes only. The proper equation can be used to calculate the value of T_c^0. It can be seen directly from Fig. 7.14 that the value of λ_{opt} is rather large: $\lambda_{opt} = 3.25$; it is much larger than that for the $Im\overline{3}m$ phase of the H_3S compound. For such a large value of λ_{opt}, the McMillan–Dynes expression for T_c^0 is not applicable (it is valid for $\lambda \lesssim 1.5$ only, but it is applicable for the $Im\overline{3}m$ phase of the H_3S compound, see section 2.2.2). A different expression, valid at large values of λ_{opt}, should be used (eqn (2.41)).

With the use of eqns (2.39) and (2.41), the values of T_c at various pressures can be calculated. Specific T_c values are presented in Table 7.1. As shown in this table, the maximum value of T_c turns out to be equal to $T_c \simeq 254$ K at $P = 200$ GPa.

The values of T_c^0 that are entered into the expression for T_c can be determined from eqn (2.41): $T_c \simeq 240$ K for $P = 200$ GPa, $T_c \simeq 236$ K for $P = 250$ GPa, and $T_c \simeq 200$ K for $P = 300$ GPa (assuming $\mu^* = 0.13$). The calculated values of T_c depend on the parameters introduced above as well as on μ^*. For example, for $\mu^* = 0.1$, one obtains $T_c \simeq 270$ K (at $P \simeq 200$ GPa). On the whole, the values of T_c obtained by Song *et al.* (2020) are in good agreement with the experimental data.

As shown in Table 7.1, the value of T_c has a maximum value at $P = 200$ GPa, and then it decreases with a further increase in pressure. This picture is similar to that observed for the $Im\overline{3}m$ phase (see above), and the explanation (Song *et al.*, 2020a) is also similar. The first-order phase transition into the $Fm\overline{3}m$ phase is accompanied by the appearance of small pockets on the Fermi surface. An increase in pressure leads to a slow decrease in T_c (c.f. section 2.6).

Equation (3.90) can be used to evaluate the isotope coefficient for the $H \to D$ substitution in the $Fm\overline{3}m$ phase. Using the parameters shown in Table 7.2, one obtains $\alpha \approx 0.45$. This value is larger than that for the $Im\overline{3}m$ phase of the H_3S compound (see section 2.7), meaning that the relative contribution of the optical modes into the pairing for the $Fm\overline{3}m$ phase is larger than that for the $Im\overline{3}m$ phase.

The focus here is on the high-T_c $Fm\overline{3}m$ phase. From the phase diagram (Fig. 7.13), it can be seen that the preceding phase (at $P < 200$ GPa) has a rather complex structure.

Table 7.1 *The coupling constants λ_{opt} and λ_{ac} in $Fm\overline{3}m$ LaH_{10} at different pressures, and the values of the critical temperature*

P(GPa)	Ω_1(THZ)	λ_{opt}	λ_{ac}	$\tilde{\Omega}_{opt}$ (10^3K)	$\tilde{\Omega}_{ac}$ (10^2K)	T_c (K, $\mu^* = 0.13$)
200	8.3	3.25	1.1	1.25	2.2	254
250	14.7	2.35	0.3	1.6	2.9	244
300	20	1.75	0.2	1.8	3.0	210

Table 7.2 *The isotope coefficient of the $Fm\bar{3}m$ phase in LaH_{10} at different pressures*

P(GPa)	a
200	0.45
250	0.47
300	0.46

It contains a mixture of several phases (the 'Magneli' phases; see Akashi *et al.*, 2016) and is described by the percolation scenario. The picture observed by Drozdov *et al.* (2019) at $P < 200$ GPa reflects the presence of several phases. It also contains some clusters containing the $Fm\bar{3}m$ phase. The picture is similar to that for the 'low-T_c' state of sulphur hydrides (see above) in the region, close to the $R3m$–$Im\bar{3}m$ transition.

7.2.8 Calcium Hydrides

In the previous sections, hydrides with a high-T_c superconductivity predicted theoretically and observed experimentally are described. Below, several interesting predictions are the main focus. Based on the history of this field, one can expect that the introduced materials will definitely be produced and that their high-T_c superconducting state will be observed. Each observation will require serious obstacles to be overcome and will be an important achievement, but eventually the growth of a remarkable family of high-T_c superconducting hydrides will be witnessed.

The structure depicted in Fig. 7.15 is different from that for sulphur hydrides. This predicted structure was recently established for the LaH_{10} compound, which at present displays the high T_c value ($T_c \approx 250$ K): the high-T_c value for LaH_{10} as well as for YH_{10} hydrides was predicted.

Starting with the spectral function $\alpha^2(\omega)F(\omega)$ evaluated for calcium hydride by Wang *et al.* (2012) (see Fig. 7.15b), the T_c value for CaH_6 can be evaluated with eqn (2.39). Indeed, in accordance with the approach described in section 2.2.6, the electron–phonon interaction can be separated into two parts. From Fig. 7.15, it can be determined that $\lambda_{opt} \approx 2.1$, $\lambda_{ac} \approx 0.6$, $\tilde{\Omega}_{opt} = 820\ \text{sm}^{-1}$, and $\tilde{\Omega}_{ac} = 350\ \text{sm}^{-1}$. Correspondingly, one can write $T_c = T_c^0 + \Delta T_c^{ac}$; $T_c^0 \approx T_c^{opt}$ is determined by the contribution of the optical modes. However, because of the large value of λ_{opt}, the MacMillan–Dynes expression for T_c^0 (eqn (2.23)) is not applicable (it is valid for $\lambda_{opt}, \preccurlyeq 1.5$). The situation is similar to that for the high-T_c phase of lanthanum hydride (section 7.2.7). The analytical expression that is valid for any value of the coupling constant (eqn (2.41)) can be used. With the use of this equation and the parameters for CaH_6, $T_c \approx 230$ K is obtained. This value contains the contributions of the optical (T_c^0) and acoustic (ΔT_c) modes: $T_c^0 \approx 180$ K and $\Delta T_c \equiv \Delta T_c^{ac} \approx 50$ K. Therefore, the optical and acoustic modes contribute 78% and 22%, respectively, to the total value of T_c. As for the isotope coefficient, the value

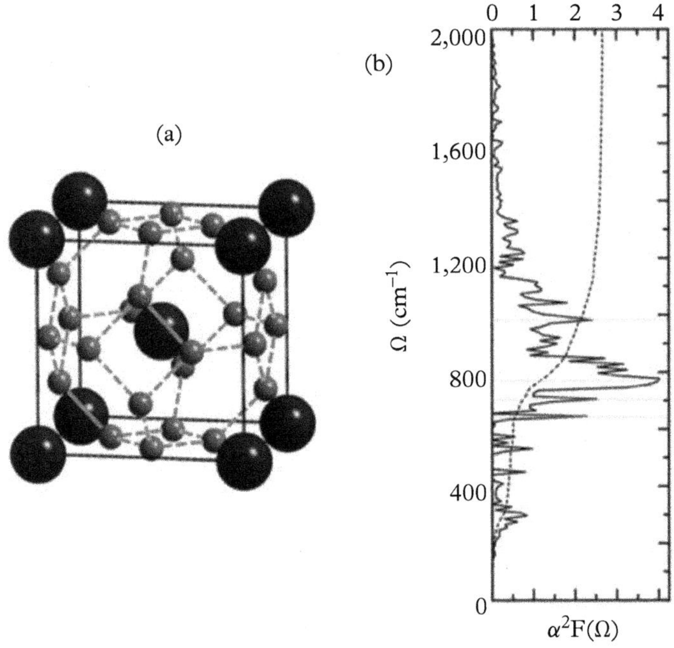

Figure 7.15 *(a) Structure and (b) spectral function of CaH_6. (After Wang et al., 2012)*

$\alpha \approx 0.36$ can be obtained from eqn (2.90). This value is similar to that for the high-T_c phase of sulphur hydrides.

The compound MgH_6, which is similar to CaH_6, was also studied (Feng *et al.*, 2015), and the value of $T_c \approx 260$ K at $P \gtrsim 300$ GPa was predicted (see also the review by Bi *et al.*, 2019).

7.2.9 Hydrogen 'Penta-Graphene-Like' Structure

A new class of high-T_c hydrides, with structures entirely different from those of H_3S and LaH_{10} compounds (see above), was introduced and studied in detail by Xie *et al.* (2020). This class contains many new hydrides. Among them, the hexagonal HfH_{10} hydride appears to be the most interesting compound. It has $T_c \approx 235$ K at $P \simeq 250$ GPa. This compound (*P6₃/mmc* phase) has a layered structure (Fig. 7.16). In addition, the H ions form H pentagons with an additional sub-lattice formed by Hf ions.

These ions act as electron donors, which increase the electronic density of states at the Fermi level. The density of states is comparable with that for LaH_{10}, and it is not a coincidence that both compounds have large coupling constants and high T_c values.

The phonon spectra and the spectral function $\alpha^2(\Omega)F(\Omega)$ for the $P6_3/mmc$ phase of the HfH_{10} compound are presented in Fig. 7.17. It can be seen that a large number of H ions leads to the dense 'forest' formed by the dispersion curves for the optical modes,

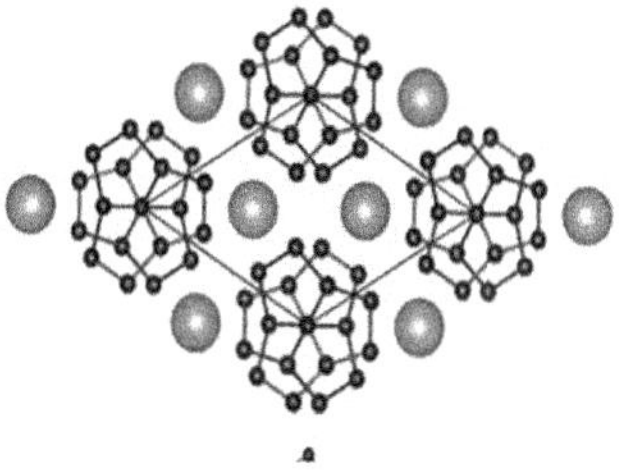

Figure 7.16 *The 'penta-graphene-like' structure of hydrogen. (After Xie et al., 2020)*

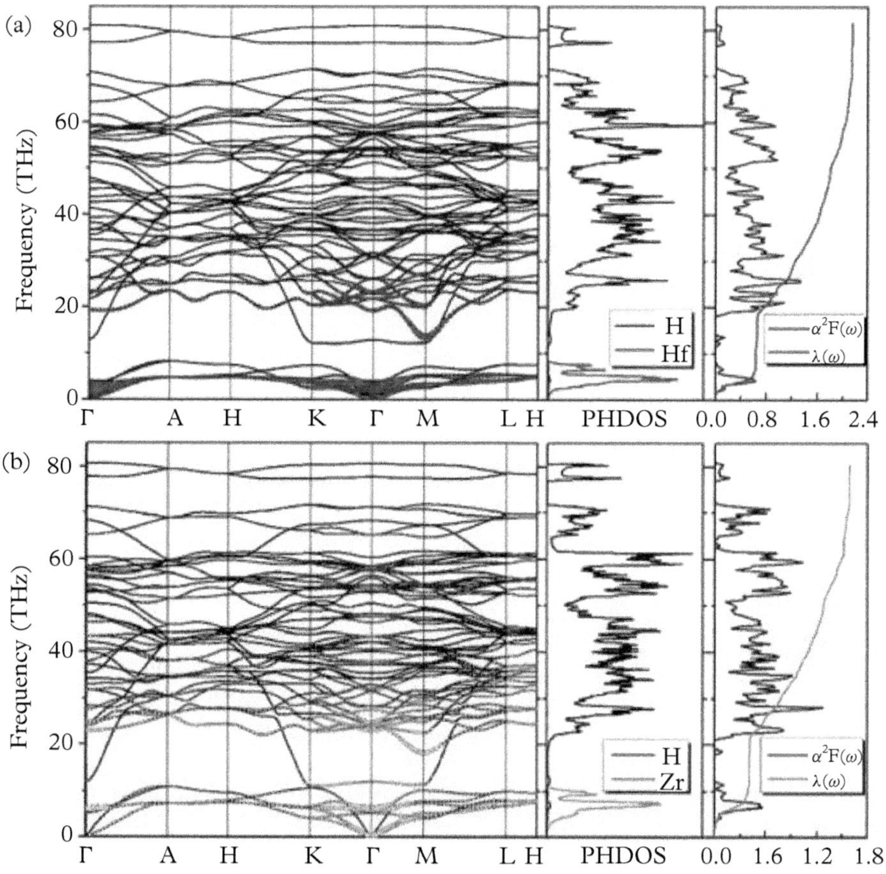

Figure 7.17 *The phonon spectra and the spectral function $\alpha^2(\Omega)F(\Omega)$ for the $P6_3/mmc$ phase of (a) HfH_{10} and (b) ZrH_{10} compounds at $P = 300$ GPa. (After Xie et al., 2020)*

and to strong coupling to these modes. With the use of the spectral function $\alpha^2(\Omega) F(\Omega)$ and the Eliashberg equation (eqn (2.18)), the T_c value can be evaluated.

The evaluation of T_c can be performed in two ways. One of them is based directly on the linear equation at $T = T_c$. The value of the phonon frequency in the phonon propagator $D(\omega_n - \omega_{n'}; \Omega)$ is replaced by some average value (see the discussion in section 2.2.1). Afterwards, the method of two coupling constants (section 2.2.2.) can be used, allowing calculation of T_c.

The 'penta-graphene-like' layered structure forms a new addition to the cubic H_3S and clathrate hydride (e.g., LaH_{10}, CaH_6; see above) families. All three families can display T_c values higher than $T_c = 200$ K. In addition, the new family is characterised by the presence of a transition metal, layered structure, and interesting physics with additional doping, as well as a corresponding increase in the density of states, which are provided by this transition metal.

7.2.10 Tantalum Hydrides

In this section, the family of tantalum hydrides (e.g. TaH_2, TaH_4, and TaH_6) is described. This family does not display record-high values of T_c, but its theoretical analysis makes it possible to study the evaluation of the main parameters. The Ta–H system was theoretically analysed by Zhuang *et al.* (2017). As mentioned above, the spectral function $\alpha^2(\omega) \mathrm{F}(\omega)$ was evaluated for three different compounds: TaH_2, TaH_4, and TaH_6. Using the method of two-coupling constants, the values of T_c and the isotope coefficient for this family of hydrides can be calculated, thus allowing the interplay of the optical and acoustic phonon modes to be seen (Kresin, 2018).

Starting with the hydride TaH_6, the spectral function is presented in Fig. 7.18. The phonon spectrum can be split into two regions: region I ($\Omega < 15$ THz) and region II ($\Omega > 15$ THz). With the use of eqn (2.36a), the characteristic frequencies can be calculated: $\tilde{\Omega}_{opt} \approx 1440$ K and $\tilde{\Omega}_{\mathrm{ac}} \approx 220$ K. Assuming here and below that $\mu^\star = 0.1$, the curve shown in chapter 5 in Fig. 5.6c allows the values of the coupling constants to be determined. It can be seen that $\lambda_{\mathrm{opt.}} \approx 1.1$ and $\lambda_{\mathrm{ac.}} \approx 0.4$. With the use of eqn (2.39), the value of T_c can be calculated: $T_c \approx 140$ K for the TaH_6 compound. Note that $T_c^{\circ} \approx 115$ K and $\Delta T_{c;ac} \approx 25$ K.

As the next step, the value of the isotope coefficient can be evaluated. Since the optical modes mainly correspond to the motion of hydrogen ions whereas the acoustic modes describe the motion of heavy Ta ions, the value of the isotope coefficient for the $\mathrm{H} \to \mathrm{D}$ substitution reflects the relative contribution of the high-frequency optical modes. The expression shown in eqn (2.40) can be used, obtaining $\alpha \approx 0.4$ for TaH_6. It is close to the limiting value of 0.5, meaning that the optical modes are dominant for the pairing in the TaH_6 system. If the optical modes are dominant, the values of T_c and α are consistent with the assumption that eqns (2.29) and (2.40) are valid.

Turning our attention to the TaH_4 system, using an approach similar to that performed above for the TaH_6 compound, one can evaluate $\tilde{\Omega}_{opt} \approx 2200$ K and $\tilde{\Omega}_{ac} \approx 265$ K, and also $\lambda_{\mathrm{opt.}} \approx 0.35$ and $\lambda_{\mathrm{ac.}} \approx 0.35$. It can be seen that the values of the coupling constants are almost equal; for such a case, eqns (2.29) and (2.40) are not applicable. However,

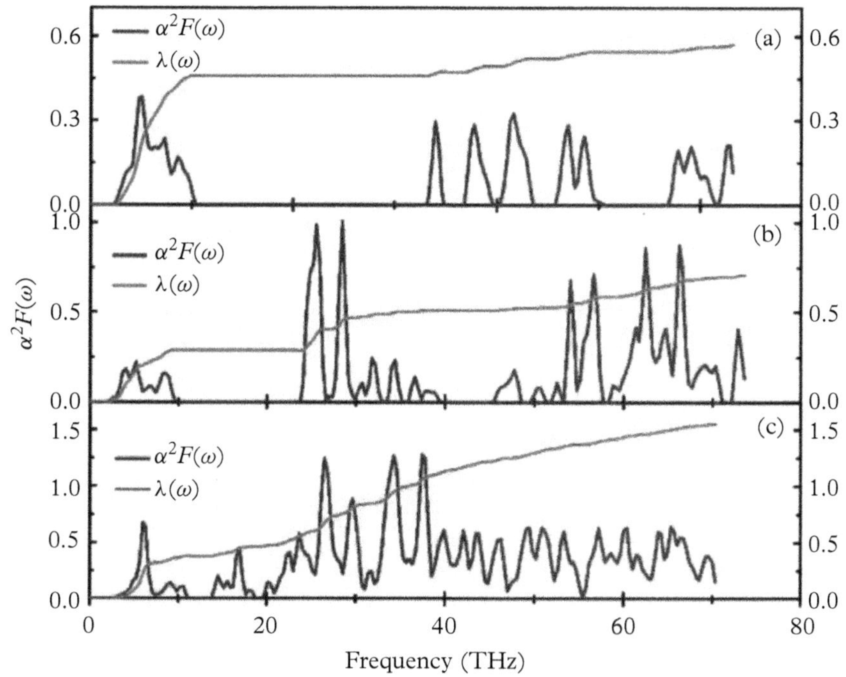

Figure 7.18 *The spectral functions for (a)TaH_2, (b) TaH_4, and (c) TaH_6 (From Kresin, 2018).*

in this case, the strength of the coupling is relatively small, and the usual logarithmic approximation can be used. Taking into account the presence of two peaks at $\tilde{\Omega}_{opt}$ and $\tilde{\Omega}_{ac}$ (c.f., the analysis of superconductivity for the *R3m* structure ($T_c \approx 100$ K) in sulphur hydrides in section 7.2.6), then $T_c \approx 36$ K is obtained.

The isotope coefficient can be evaluated from eqn (2.37) and the relation (see eqn (3.42a))

$$\alpha = 0.5\left(\tilde{\Omega}_{opt}/T\right)\left(\partial T_c/\partial\tilde{\Omega}_{opt}\right)$$

Then

$$\alpha = 0.5\left(\lambda_{opt}/\lambda_T\right)$$

And, for TaH_4, its value is $\alpha \approx 0.25$. Finally, for TaH_2, it can be determined from Fig. 7.18a that $\tilde{\Omega}_{opt} \approx 2600$ K and $\tilde{\Omega}_{ac} \approx 70$ K. As for the coupling constants, $\lambda_{opt.} \approx 0.1$ and $\lambda_{ac.} \approx 0.45$. Then T_c can be determined from eqn (2.37) and is equal to $T_c \approx 3.5$ K. The isotope coefficient for the H $\rightarrow$ D substitution appears to be small: $\alpha \approx 0.1$.

It can be seen from Table 7.3 that an increase in the T_c value for the sequence $TaH_2 \rightarrow TaH_4 \rightarrow TaH_6$ is caused by redistribution of the electron–phonon interaction between

Table 7.3 *Coupling and isotope-effect parameters for several Ta hydrides*

Hydride	λ_{total}	$\lambda_{opt.}$	$\lambda_{ac.}$	λ_o / λ_a	T_c	α
TaH_2	0.55	0.1	0.45	0.22	3.5K	0.1
TaH_4	0.7	0.35	0.35	1	36K	0.25
TaH_6	1.5	1.1	0.4	2.75	140K	0.4

the optical and acoustic modes. This redistribution is also reflected in the values of the isotope coefficient for the H $\rightarrow$ D substitution. This value increases with an increase in T_c. The strength of λ_{opt} is directly related to the increase in the number of hydrogen ion modes, which increases with an increase in the number of H ions in the unit cell. Thus, the interplay of the introduced coupling constants is directly related to the behaviour of T_c.

7.2.11 H_2O

Finally, the properties of the most well-known hydride, H_2O will be discussed. Its potential superconducting state was examined in a theoretical paper (Flores-Livas *et al.*, 2016). The authors considered the combination of various channels of metallisation (pressure and doping). H_2O can be transferred into the solid phase (ice) by applying pressure (up to $P \approx 150$ GPa). The crystal continues to be an insulator at this pressure. The sample can also be doped by nitrogen. As a result of such doping under pressure, the material becomes metallic and even superconducting, with a rather high value of $T_c \approx 60$ K.

Note that, at ambient pressure and also at low pressure (up to $P \approx 110$ GPa), the oxygen ion has four hydrogen neighbours (so called phase I). Two of these neighbours are covalently bonded with oxygen and form the H_2O molecule, and the other two ions form additional hydrogen bonds. The lengths of the bonds are different, and the structure is asymmetric. However, at higher pressures (of the order of $P \approx 300$ GPa), the so-called ice-X phase is formed, and it is characterised by symmetric O–H bonds (Goncharov *et al.*, 1999).

At $P \approx 150$ GPa, the ice crystal is still in the insulating state. To prompt a transition into the metallic state, the use of doping is needed. It has been determined that nitrogen is the best dopant. The substitution of nitrogen for oxygen leads to hole conductivity. In addition, the transition into the metallic state is accompanied by changes in the phonon spectrum. All of these changes provide the transition into the superconducting state. The calculations show that the best values of the superconducting parameters correspond to relatively low doping (4–6%). This leads to the value of $T_c \approx 60$ K. Of course, this value is below $T_c \approx 203$ K, which was observed for the H_3S phase of sulphur hydride, but it is still very high.

The idea of combining high pressure and doping is elegant and looks promising. However, future experiments are needed to confirm this interesting prediction.

7.2.12 High T_c at Lower Pressure

Room temperature superconductivity is a perfectly realistic phenomenon. It follows from many theoretical studies, and the first such an observation was described recently (see the next section). But its presence requires high pressure. The next challenge is to find a hydride which displays room temperature superconductivity at lower pressure (the final goal: at ambient pressure). The first step in this direction was proposed by Song *et al.* (2020b). According to this study, there exists a strong correlation between the presence of the f-states on the Fermi surface and an appearance of the nesting states. The existence of the nesting states leads to possibility of a charge wave structural transition and corresponding 'softening' of the phonon spectrum (Afanas'ev and Kagan, 1963; Kresin, 1984). If the value of the critical temperature T_c for the transition into the superconducting state is higher than $T_{c;\mathrm{CDW}}$, then this 'softening' mechanism leads to an increase in T_c (see eqn (2.24) and section 2.2.2). As a result, hydrides containing Yb and Lu (they are lanthanides with sodalite cage, which is beneficial for superconductivity) display interesting properties. More specifically, for YbH_6, T_c is about 145 K at a pressure of only $P = 70$ GPa. The most remarkable case is the LuH_6 compound, which, at pressure $P = 100$ GPa, is a superconductor with $T_c = 273$ K (!).

7.2.13 Ternary Hydrides

Li_2MgH_{16} as a potential room temperature superconductor. Hydrides, described above, contain two ions in the unit cell. As a next step, hydrides with a larger number of such ions should be considered. A prominent example of such a study has been described by Sun *et al.* (2019). They introduced Li_2MgH_{16} as a potential room temperature superconductor. This compound should display the superconducting state with $T_c \simeq 473$ K (!) at $P \simeq 250$ GPa. This material, like LaH_{10} and CaH_6 (see above), has a clathrate structure. Interestingly, the scenario leading this ternary hydride to such a high-T_c state appears to be similar to that for doped ice, which is described in the previous section. The compound MgH_{16} contains many hydrogen ions, making it favourable for a high T_c value. However, this is not the case, since many of these ions form H_2 molecules and this factor strongly diminishes their contribution to the pairing; as a result, $T_c \simeq 73$ K. The addition of Li ions breaks up this molecular unit; this is equivalent to electronic doping and drastically increases the value of T_c.

The phonon spectrum of $LiMgH_{16}$ ($Fd\bar{3}m$ structure) contains a very dense 'forest' of optical phonons (Fig. 7.19). Figure 7.19 shows the spectral function $\alpha^2(\Omega)F(\Omega)$ and the dependence $\lambda(\Omega)$. The value of the total coupling constant $\lambda_T = \lambda_{opt} + \lambda_{ac}$ is large: $\lambda_T = 3.35$. The main contribution to the pairing is provided by optical phonons: $\lambda_{opt} \simeq 2.1$. It leads to the value of $T_c \simeq 473$ K. Such a large T_c value is in total agreement with the conditions formulated above (section 2.7).

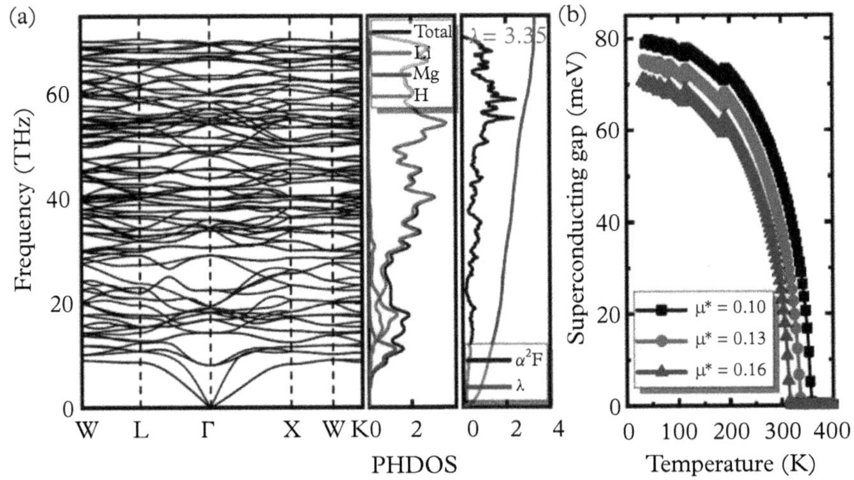

Figure 7.19 *The phonon spectrum, spectral function, and calculated energy gap for $LiMgH_{16}$. The phonon spectrum of $LiMgH_{16}$ ($Fd\overline{3}m$ structure) contains a very dense 'forest' of optical phonons. (After Sun et al., 2019)*

The first room temperature superconductor. Recently, even after the completion of a whole manuscript, we have learned about the observation of the first room temperature superconductor. This observation was recently described in the paper 'Room-temperature superconductivity in a carbonaceous sulfur hydride' by Snider *et al.* (2020). More specifically, the material is in the superconducting state up to the temperature $T_c = 288$ K (near 15 °C). The effect has been observed at high pressure $P = 267 \pm 10$ GPa. The structural unit contains $H_2S + H_2 + CH_4$ subunits. Interestingly, the two first components form, under pressure, the hydride SH_3, the first high-T_c hydride (see above). The ideal stoichiometry, probably, is $(H_2S)(CH_4)H_2$.

The subunits are bound by the van der Waals forces. Note that the unit's structure is unknown, and its evaluation is a non-trivial problem, since the usual method, X-ray spectroscopy, is not efficient in this case because of the light masses of hydrogen and carbon. The superconducting state is manifested in a drastic drop of resistivity, a decrease in T_c with an increase in an external magnetic field, and the Meissner effect. The last factor is the crucial one, since the Meissner effect provides direct evidence for the superconductivity.

The search for materials which display room temperature superconductivity at a lower pressure (the goal: ambient pressure) is the next important challenge (see section 7.2.12).

7.3 Pnictides

Fe-based superconductors are iron-pnictide compounds whose superconducting properties were discovered in 2008 (Kamihara *et al.*, 2008; see the review by Hosono, 2012).

This new type of superconductor is based on conducting layers of iron and pnictide (typically arsenic (As) and phosphorus (P)) and seems to show promise as the next generation of the high temperature superconductors. Suddenly, in addition to the famous copper-based superconductors, researchers had a new class of materials exhibiting the macroscopic quantum phenomenon of superconductivity at high temperatures, and it looked as though the road to room temperature superconductivity might be smoother because of the chance to compare and contrast these two systems. The discovery of Fe-based superconductors signalled, in the minds of many, the transition from the 'copper age' to the new 'iron age'.

Ten years over the maximal T_c in the bulk Fe-based superconductors is 56 K in $Sr_{0.5}Sm_{0.5}FeAsF$ (Wu *et al.*, 2009). Nowadays, the road to room temperature superconductivity is mostly related to the high pressure H_3S and metal hydrides; nevertheless, searching for new superconductors with high T_c at ambient pressure is still an active field of investigation. All these problems became more acute after the experimental observation of superconductivity with $T_c \sim 80$–100 K in monolayers of FeSe (epitaxial films) grown on the $SrTiO_3$ substrate (and several similar compounds). At present, we can speak of a 'new frontier' in studies of high-temperature superconductivity (Wang *et al.*, 2012). The interface superconductivity will be discussed in chapter 8, here we will discuss the bulk Fe-based superconductors.

The family of Fe-based superconductors is quite large. It includes various Fe pnictides and Fe-chalcogenides (pnictogens are elements of group V, which includes N, P, As, Sb, and Bi; chalcogens are elements from group VI, which includes O, S, Se, and Te). Examples of Fe-pnictides are the 1111 systems RFeAsO (R is the rare earth element), the 122 systems XFe_2As_2 (X is the alkaline earth metal), and 111 systems compounds like LiFeAs. Examples of Fe-chalcogenides are FeSe, FeTe, and $A_xFe_{2-y}Se_2$ (where A = K, Rb, Cs). The crystallographic structures of various families of Fe-based superconductors are shown in Fig. 7.20. The square lattice of Fe is the basic element. Fe is surrounded by As or P situated in the tetrahedral positions within the first subclass and by Se, Te, or S within the second subclass.

The Fe *d*-orbitals are significantly overlapped and, apart from that, the out-of-plane pnictogens or chalcogens are well hybridised with the t_{2g} subset of the Fe *d*-orbitals, and all of them contribute to the Fermi surface. A minimal model is then the significantly multiband model. In this regard, the Fe-based materials have more similarity to ruthenates and magnesium diboride than to cuprates. Both cuprates and Fe-based materials in most cases become superconducting when doped, that is, some atoms are replaced by others and, consequently, the potential is changed at sites where the replacement was made. In this regard, disorder is an inherent part of the observed picture of superconductivity, and one has to have a clear-eyed understanding of its role and impact on the features of studied systems.

At first glance, the phase diagrams of cuprates and many Fe-based superconductors are similar. In both cases, the undoped materials exhibit the antiferromagnetism, which vanishes with doping; the superconductivity occurs at some non-zero doping and then disappears, so that T_c forms a 'dome'. While in cuprates the long-range ordered Néel phase vanishes before the superconductivity occurs, in Fe-based materials the

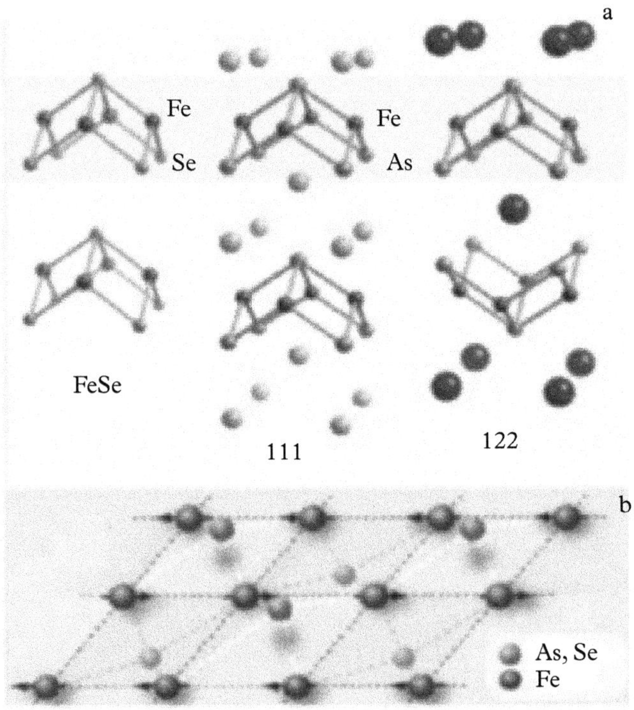

Figure 7.20 *(a) The crystal structure of the simplest Fe-based superconductor. (b) The structure of the conducting plane of Fe ions and pnictogens. Arrows show the directions of spins for typical ordering in the antiferromagnetic phase. (From Sadovskii, 2016)*

competition between these orders can take several forms. In LaFeAsO, for example, there appears to be a transition between the magnetic and the superconducting states at a critical doping value, whereas, in 122 systems ($BaFe_2As_2$ and alike), the superconducting phase coexists with the magnetism over a finite range and then persists to higher doping (see Fig. 7.21). It is tempting to conclude that the two classes of superconducting materials show generally very similar behaviours, but there are profound differences as well.

The first striking difference is that the undoped cuprates are the Mott insulators, while the Fe-based materials are metals. This suggests that the Mott–Hubbard physics of the half-filled Hubbard model is not a good starting point for pnictides, although some authors have pursued strong coupling approaches. It does not, of course, exclude the effects of correlations in Fe-based materials, but they may be moderate or small. In any case, DFT-based approaches describe the observed Fermi surface and the band structure reasonably well for the whole phase diagram, unlike the situation in cuprates, especially in undoped and underdoped regimes.

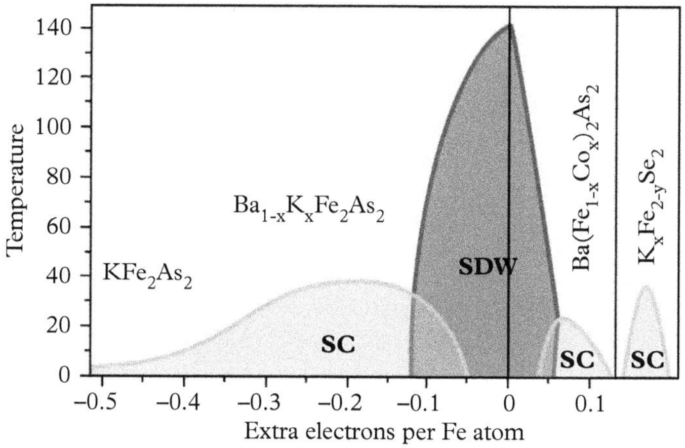

Figure 7.21 *Phase diagram of the 122 family of ferro-pnictides, complemented by the 122(Se) family as a generalised phase diagram for the Fe-based superconductors. (From Kordyuk, 2012)*

The second important difference pertains to the normal state properties. The underdoped cuprates reveal pseudogap behaviour in both one-particle and two-particle charge and/or spin excitations, while similar robust behaviour is absent in Fe-based materials. There is a wide variety of unusual features of the pseudogap state in cuprates. For example, a strange metal phase near optimal doping in hole-doped cuprates is characterised by linear-T resistivity over a wide range of temperatures. In Fe-based materials, different temperature–power laws for resistivity, including the linear-T dependence of the resistivity for some materials, have been observed near the optimal doping and have been interpreted as being due to the multiband physics and interband scattering (Golubov *et al.*, 2011). There are, however, indications of a pseudogap formation in densities of states of some pnictides' bands (Kuchinskii and Sadovskii, 2008; Kordyuk, 2015).

The mechanism of doping deserves additional discussion. Doping in cuprates is accomplished by replacing one of the spacer ions with another one with different valences, as in $La_{2-x}Sr_xCuO_4$ and $Nd_{2-x}Ce_xCuO_4$ or adding extra out-of-plane oxygen, as in $YBa_2Cu_3O_{6+y}$. The additional electron or hole is then assumed to dope the plane in an itinerant state. In Fe-based materials, the nature of doping is not completely understood (similar phase diagrams are obtained by replacing the spacer ion or by the in-plane substitution of Fe with Co or Ni). For example, $LaFeAsO_{1-x}F_x$, $Ba_{1-x}K_xFe_2As_2$ and $Sr_{1-x}K_xFe_2As_2$ belong to the first case, while $Ba(Fe_{1-x}Co_x)_2As_2$ and $Ba(Fe_{1-x}Ni_x)_2As_2$ belong to the second one. Whether these heterovalent substitutions dope the FeAs or FeP plane as in the cuprates was not initially clear (Sawatzky *et al.*, 2009) but now it is well established that they affect the Fermi surface, consistent with the formal electron-count doping (Nakamura *et al.*, 2011). Another mechanism to vary the electronic and magnetic properties is via isovalent doping with phosphorous in $BaFe_2(As_{1-x}P_x)_2$ or ruthenium in $BaFe_2(As_{1-x}Ru_x)_2$.

'Dopants' can act as potential scatterers and change the electronic structure because of the difference in ionic sizes or simply by diluting the magnetic ions with non-magnetic ones. In Fe-based materials, therefore, some of the doping mechanisms connected with the changes in the transition metal layer. But, crudely, the phase diagrams of all Fe-based materials are quite similar, challenging workers in the field discuss a systematic structural observable which correlates with the variation of T_c. Among several proposals, the height of the pnictogen or chalcogen above the Fe plane has frequently been noted as playing some role in the overall doping dependence (Kuroki *et al.*, 2009; Kuchinskii *et al.*, 2010).

Fe-based superconductors are quasi-two-dimensional materials with the conducting square lattice of Fe ions. The Fermi level is occupied by the $3d^6$ states of Fe^{2+}. It was found in the early DFT calculations (Lebegue, 2007; Mazin *et al.*, 2008; Singh and Du, 2008), which are in a quite good agreement with the results of quantum oscillations and ARPES. All five orbitals, $d_{x^2-y^2}$, $d_{3z^2-r^2}$, d_{xy}, d_{xz}, and d_{yz}, are near or at the Fermi level. Within the five-orbital model (Graser *et al.*, 2009), which correctly reproduces DFT band structure (Cao *et al.*, 2008), the Fermi surface is composed of four sheets: the two hole pockets around the (0; 0) point, and the two electron pockets around (π; 0) and (0; π) points. Such k-space geometry results in the possibility of the SDW instability due to the nesting between the hole and electron Fermi surface sheets at the wave vector $Q = (\pi; 0)$ or $(0; \pi)$. Upon doping x, the long-range SDW order is destroyed.

If electrons are doped, then, for the large x, the hole pockets disappear, leaving only the electron Fermi surface sheets that are observed $K_xFe_{2-x}Se_2$ and FeSe monolayers (Liu *et al.*, 2012). Upon an increase in hole doping, a new hole pocket appears around the (π; π) point and then the electron sheets vanish. KFe_2As_2 corresponds to the latter case. The ARPES experiment confirms that the maximal contribution to the bands at the Fermi level comes from the d_{xz}, d_{yz}, and d_{xy} orbitals (Kordyuk, 2012). At the same time, the presence of a few pockets and the multiorbital band character significantly affect the superconducting pairing.

Unlike the case for cuprates, for which there is d_{x2-y2} symmetry of the gap for all of the different families, no consensus on any universal gap structure has been reached for Fe-based superconductors, even after several years of intensive research on high-quality single crystals. There is strong evidence that small differences in electronic structure can lead to the strong diversity in superconducting gap structures, including nodal states and states with a full gap at the Fermi surface. Knight shift measurements have been performed on several Fe-based materials, including $Ba(Fe_{1-x}Co_x)_2As_2$ (Ning *et al.*, 2009), $LaFeAsO_{1-x}F_x$ (Grafe *et al.*, 2008), $PrFeAsO_{0.89}F_{0.11}$ (Matano *et al.*, 2008), $Ba_{1-x}K_xFe_2As_2$ (Matano *et al.*, 2009), LiFeAs (Li *et al.*, 2010), and $BaFe_2(As_{0.67}P_{0.33})_2$ (Nakai *et al.*, 2010). It was found that the Knight shift decreases in all crystallographic directions. This effectively excluded the triplet symmetries such as the p-wave or the f-wave. As concerns singlet pairing, in the three-dimensional tetragonal system, the group theory allows only five one-dimensional irreducible representations according to how the order parameter transforms under rotations by $\pi/2$ and other operations of the tetragonal group: A_{1g} ('s-wave'), B_{1g} ('d-wave', $x^2 - y^2$), B_{2g} ('d-wave', xy), A_{2g} ('g-wave', $xy(x^2 - y^2)$), and E_g ('d-wave', xz; yz).

Several direct experiments provide shreds of evidence against the *d*-wave in Fe-based superconductors. The Josephson current in the c-direction when the studied sample is coupled to a known *s*-wave superconductor would confirm the *s*-type of the former. Exactly such current was observed in the 122 single crystals (Zhang *et al.*, 2009). Another piece of evidence comes from the absence of the so-called anomalous Meissner effect (or Wohlleben effect; Hicks *et al.*, 2009). This effect appears in polycrystalline samples with a random orientation of grains. It was predicted at the beginning of the cuprates era by Geshkenbein and Larkin (1986) and since then it has been routinely observed only in the *d*-wave superconductors. The Wohlleben effect appears because the response to a weak external magnetic field is paramagnetic, that is, opposite to the standard diamagnetic response of an *s*-wave superconductor. This happens because half of the weak links have a zero phase shift, and the other half have the π phase shift.

Note that, due to the multivalley Fermi surface, several possibilities for the *s*-wave gaps are possible. Gaps with the same symmetry may have different structures. Figure 3.20 from section 3.3.3 illustrates this for *s*-wave symmetry. The generalised A_{1g} (extended *s*-wave symmetry) probably involving a sign change of the order parameter between Fermi surface sheets or its parts (Hirschfeld *et al.*, 2011). The fully gapped *s*-states without nodes on the Fermi surface differ only by a relative gap sign between the hole and electron pockets, which is positive in the s_{++} state and negative in the $s\pm$ state. In addition to a global change of sign, which is equivalent to a gauge transformation, we can have individual rotations on single pockets and still preserve symmetry. These states are called the 'nodal $s_{\pm}$' ('nodal s_{++}') and are characterised by the opposite (same) averaged signs of the order parameter on the hole and electron pockets. Nodes of this type are sometimes described as 'accidental', since their existence is not dictated by symmetry, unlike the symmetry nodes of the *d*-wave gap. In general, the extended *s*-wave gap may be written as (here $\alpha = a, b$ denotes the Fermi surfaces centred at (0, 0) and $(\pi, 0)/(0, \pi)$ points, respectively; see eqn (3.67))

$$\Delta^{\alpha}(\mathbf{k}) = \Delta_0^{\alpha} + \frac{1}{2}\Delta_1^{\alpha}(\cos\, k_x a + \cos\, k_y a)$$

where the first term is the spherical *s*-gap and the second one is the anisotropic first harmonic. For equal signs of Δ_0^a and Δ_0^b, we have the s_{++} state if $\Delta_1^{\alpha} < \Delta_0^{\alpha}$, and the nodal s_{++} if $\Delta_1^{\alpha} > \Delta_0^{\alpha}$. Similarly, for opposite signs of Δ_0^a and Δ_0^b, we have the $s_{\pm}$ state if $\Delta_1^{\alpha} < \Delta_0^{\alpha}$, and the nodal $s_{\pm}$ if $\Delta_1^{\alpha} > \Delta_0^{\alpha}$. The $s_{++}$and $s_{\pm}$ states all have the same symmetry, that is, the sign of neither changes if the crystal axes are rotated by $\pi/2$. In contrast, the *d*-wave state changes sign under such rotation.

Let us discuss the mechanisms of superconductivity in Fe-based superconductors. The electron–phonon interaction by itself (without Coulomb repulsion) leads to the s_{++} gap (Boeri *et al.*, 2008). Up to now, there is still no reliable experimental data on the isotope effect in the Fe-based superconductors to verify the phonon mechanism of pairing. The electron–phonon mechanism of Cooper pairing is considered to be insufficient to explain the high values of T_c in these systems (Sadovskii, 2008). The preferable mechanism is assumed to be the pairing due to the exchange of antiferromagnetic fluctuations. The repulsive nature of this interaction leads to a picture of $s_{\pm}$ pairing with different signs of the superconducting order parameter (gap) on the hole-like (around the Γ point in the center of the Brillouin zone) and the electron-like (around the M points in the corners of the zone) sheets of the Fermi surface (Hirschfeld *et al.*, 2011). Due to the multiband electronic structure of Fe-based superconductors, impurity scattering may result in some non-trivial effects that are absent in the single band Abrikosov–Gor'kov theory. These were discussed in section 3.3.

However, when we consider systems based on the FeSe monolayers, this picture becomes inconsistent, as the observed topology of the Fermi surfaces with a complete absence of any 'nesting' electron-like and hole-like sheets, or even with a total absence of hole-like Fermi surfaces, contradicts this picture. There is simply no obvious way to produce well-developed spin (antiferromagnetic) fluctuations. According to Sadovskii (2016), the values of $T_c \sim 40$ K in the intercalated FeSe layers can be achieved, in principle, by increasing the density of states at the Fermi level. At the same time, it is also clear that an increase in T_c to values more than 65 K, as observed in the FeSe monolayers on $SrTiO_3$ and $BaTiO_3$ substrates, cannot be explained along these lines. It is natural to assume that such increase is somehow related to the nature of the $SrTiO_3$ ($BaTiO_3$) substrate, for example, with the additional pairing interaction of carriers in the FeSe layer appearing due to their interaction with some kind of elementary excitations in the substrate, in the spirit of the 'excitonic' mechanism, as was initially proposed by Ginzburg, 1970. This aspect of the interface superconductivity will be discussed above in chapter 8.

At the end of this chapter, we will discuss the material science aspects. Although around a good decade has passed after the discovery of Fe pnictides/chalcogenides superconductors, the improvement in the quality of the material is still in progress. In particular, good-quality, single crystals that are large enough are very much in demand. Generally, the travelling floating zone/Bridgman technique has been used for growing Fe chalcogenide single crystals (Sales *et al.*, 2009; Ma *et al.*, 2014). Maheshwari *et al.* (2018) succeeded in growing flux-free large single crystals of FeTe and reported the heat capacity and a Mossbauer study analysis of self-flux grown FeTe.

Interestingly, the ground state of both undoped cuprates and Fe pnictides/chalcogenides is magnetic in nature. There is a broad consensus that the magnetic order disappears before the onset of superconductivity in these systems. FeTe is not a superconductor, but it has antiferromagnetic ordering below 70 K (Li *et al.*, 2009), which becomes superconducting by Te site doping of Se/S with a T_c of up to 15 K (Fang *et al.*, 2008; Subedi *et al.*, 2008; Yeh *et al.*, 2008; Hu *et al.*, 2009). The superconductivity in Fe chalcogenides has been increased up to 50 K with the intercalation of favourable alkali metals (Ying *et al.*, 2012).

FeTe, the parent compound of Fe chalcogenides, shows drastic changes in electrical resistivity as well as in magnetic susceptibility at 70 K, which belong to antiferromagnetic ordering along with a first-order structural phase transition (Li *et al.*, 2009). This first-order phase transition is clearly seen in heat capacity measurements at 70 K (Maheshwari *et al.*, 2018). The T_c of an optimally doped superconducting Fe chalcogenide, that is, $FeTe_{1/2}Se_{1/2}$, rises up to 37 K with the application of external pressure (Masaki *et al.*,

2009), although no superconducting phase occurs for undoped $FeTe_{0.92}$ under high pressures of even up to 19 GPa (Okada *et al.*, 2009).

7.4 Other Systems

7.4.1 Ruthenates and the Magnetic Superconductors, Ruthenocuprates

The interest in ruthenates increased after the discovery of high-temperature superconductivity in cuprates. In addition to the fabrication of new Y-, Bi-, Tl-, and Hg-cuprates with a $T_c > 100$ K in 1987, a search for other oxide superconductors began worldwide. The only known $Ba_{1-x}K_xBiO_3$ has a cubic structure different from that of cuprates; see the comparison with cuprates by Anshukova *et al.* (1997). Among oxides isostructural to cuprates, Sr_2RuO_4, with a $T_c \sim 1$ K, proved to be the only known superconductor of that type at that time (Maeno *et al.*, 1994). Being isostructural with the antiferromagnetic insulator La_2CuO_4, the ruthenate Sr_2RuO_4 has no long-range magnetic order and possesses metallic and superconducting properties without doping, unlike La_2CuO_4; see the review by Ovchinnikov (2003).

It was soon found that ruthenates form an extremely interesting class of materials. Firstly, the superconductivity in Sr_2RuO_4 was found to be exotic, with triplet Cooper pairs with $S = 1$ and a finite orbital angular momentum in a direct analogy with the superfluid phases of ^{3}He. Secondly, there is a strong competition of ferromagnetic, antiferromagnetic, orbital, and superconducting ordering, as well as metal–insulator transitions, which manifest themselves most vividly in the complex phase diagram for Ca_{2-x} Sr_xRuO_4. Thirdly, the double-layered $Sr_3Ru_2O_7$ ruthenates provide an example of a new class of quantum critical points. Fourthly, the rutheno-cuprates exhibit coexistence of magnetism and superconductivity, which will be discussed below. A new family of magnetic superconductors based on the high-T_c cuprates was discovered in 1997 (Felner *et al.*, 1997). It is remarkable that the magnetic ordering temperature, T_N, is noticeably higher than superconducting T_c, more specifically, for $R_{1.4}Ce_{0.6}RuSr_2Cu_2O_{10-\delta}$ (RCeRuSCuO, where R $\equiv$ Gd or Eu). For example, for GdCeRuSCuO, one can observe that $T_c \simeq 42$ K, whereas $T_N \simeq 122$ K. EuCeRuSCuO is another example of such a compound ($T_c \simeq 32$ K; $T_N \simeq 180$ K). One can see that $T_N/T_c \gg 1$ (~4). The transition into the superconducting state is rather sharp.

One might think that the observed coexistence of magnetism and superconductivity is a manifestation of the presence of two different phases with corresponding phase separation. However, a detailed study performed by Felner *et al.* (1998), which used scanning tunnelling microscopy and the Mossbauer technique, has shown that we are dealing with a single phase (see Fig. 7.22). The Mossbauer measurements proved that the magnetism occurs on the Ru site. As for the scanning tunnelling microscopy, it shows that the pairing is taking place on the Cu–O plane. Note also that the resistance measurements at $T > T_c$ (e.g. at $T \simeq 45$ K for GdCeRuSCuO) show an absence of any spectroscopic gaps for any small bias, regardless of tip position. This observation clearly indicates that

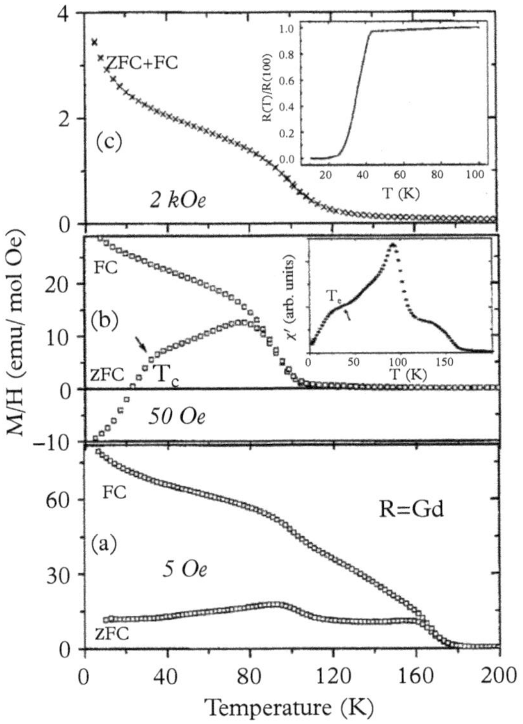

Figure 7.22 *Coexistence of superconductivity and magnetism for the* $Gd_{1.4}Ce_{0.6}RuSr_2Cu_2O_{10-x}$ *compound; FC, field cooling; ZFC, zero-field cooling. (From Felner et al., 1997)*

the gaps observed below T_c are caused by the pairing and, correspondingly, they show that we are dealing with the transition into a superconducting state.

The properties of another magnetic superconductor, $RuSr_2GdCu_2O_8$ (Ru-1212), have been studied by Chu *et al.* (2000). For this material, the temperature of the magnetic transition (T_{cm} = 133 K) is also noticeably higher than $T_c \simeq 50$ K. One can see (Fig. 7.23) that the external magnetic field shifts T_c towards its lower values; such an impact of the magnetic field is natural for the superconducting transition. One can also observe an interesting interplay of the Meissner states and the spontaneous vortex states.

7.4.2 Heavy Fermions

From the Kondo effect to heavy fermions. Long ago, Meissner and Voigt (1930) communicated the first observation of an increase of the electrical resistivity of nominally pure Cu upon cooling to about 2 K. Later, the resistivity minimum was found in a number of simple metals like Cu and Au. It was realised that the low-temperature increase in the temperature dependence of the resistivity $\rho(T)$ was caused by a tiny concentration of magnetic (mainly Fe) impurities (Daybell and Steyert, 1968).

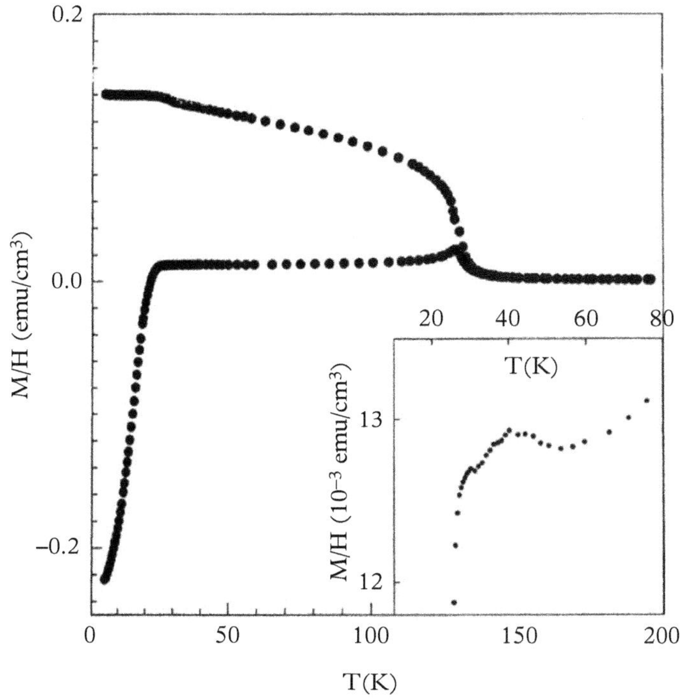

Figure 7.23 *Coexistence of superconductivity and magnetism for the* $RuSr_2GdCu_2O_8$ *compound; expanded scale near the transition. (From Chu et al., 2000)*

This phenomenon was explained by Kondo (1964) by taking into account the dynamical degrees of freedom of the $3d$ spin of the magnetic impurity and the sharpness of the Fermi surface. He derived a dominating contribution to the scattering rate of the conduction electrons with a logarithmic temperature dependence, in line with the experimental observations. As shown by renormalisation group analysis (Wilson, 1983), the effective antiferromagnetic coupling between the localised $3d$-electron spin and the spins of the conduction electrons becomes stronger as the temperature is lowered. This results in the formation of a so-called Kondo singlet, in which conduction electron states are entangled with the localised $3d$-impurity states.

A direct manifestation of the Kondo singlet is seen in the formation of a Kondo resonance at the Fermi energy E_F, well below the characteristic Kondo temperature T_K. For $T \ll T_K$, a local Fermi liquid with a non-magnetic ground state develops (Nozieres, 1974). This Fermi-liquid formation is illustrated nicely by the linear temperature dependence of the incremental specific heat of both CuCr ($T_K \sim 1$ K) and CuFe ($T_K \sim 20$ K); see the review by Steglich and Wirth (2016). It was found that $C_{imp} = \gamma T$, where the coefficient γ exceeds the Sommerfeld coefficient of the electronic specific heat of pure Cu by three to four orders of magnitude.

Intensive studies on the canonical Kondo alloy (La, Ce)Al_2 (Maple, 1973) revealed a plethora of phenomena, such as re-entrant superconductivity (Riblet and Winzer, 1971; Steglich and Armbrüster, 1974), and a giant Seebeck effect (Moeser, 1974). An increase of the Ce concentration in this quasi-binary alloy system from less than 1% to 100% allows for a detailed study of the interplay between antiferromagnetic correlations and the Kondo effect, all the way up to the antiferromagnetic Kondo lattice system $CeAl_2$. The latter exhibits the enhanced Sommerfeld coefficient $\gamma = 0.135\,\mathrm{J\,K^{-2}\,mol^{-1}}$, which rises to $1.7\,\mathrm{J\,K^{-2}\,mol^{-1}}$ in the absence of antiferromagnetic order (Bredl *et al.*, 1978). This suggests that, in the Kondo lattice system $CeAl_2$, the Kondo resonances are as pronounced as those in a dilute Kondo alloy such as CuFe. Interestingly, in the paramagnetic compound $CeAl_3$, the Sommerfeld coefficient $\gamma = 1.6\,\mathrm{J\,K^{-2}\,mol^{-1}}$ hints at an effective mass m^* of the charge carriers at E_F, which exceeds the mass of the bare electron by a factor of the order of 1000 (Andres *et al.*, 1975). The strongly enhanced value of m^* illustrates a correspondingly reduced Fermi velocity v_{F^*}, which is about three orders of magnitude smaller than the Fermi velocity in a simple metal. As a result of Bloch's theorem, the charge carriers in a Kondo lattice system have the same quantum numbers (for charge and spin) as the bare electrons and are termed 'heavy electrons' or 'heavy fermions'. Such Kondo lattices are prototypes of strongly correlated electron systems. Thus, an unconventional normal state is the first peculiarity of the heavy fermion materials.

Unconventional superconductivity. The first heavy fermion superconductor with a T_c below 0.6 K, $CeCu_2Si_2$, was discovered by Steglich *et al.* (1979). Since its discovery, over 30 heavy fermion superconductors have been found in materials containing rare earth or actinide elements (see the review by Steglich and Wirth, 2016) with maximal $T_c = 18.5$ K in $PuCoGa_5$ (Sarrao *et al.*, 2002). Given the antagonistic relationship of superconductivity and magnetism, the discovery of bulk superconductivity in $CeCu_2Si_2$ was more than surprising. In this tetragonal compound, a periodic arrangement of 100% of the magnetic Ce^{3+} ions appears to be a prerequisite for the occurrence of superconductivity; in contrast, the non-magnetic reference system $LaCu_2Si_2$ is not a superconductor (Steglich *et al.*, 1979). Also, superconductivity in $CeCu_2Si_2$ becomes fully suppressed upon doping with quite small amounts of non-magnetic impurities (Spille *et al.*, 1983). Due to the huge electron effective mass, the renormalised Fermi velocity is of the same order as the sound velocity. Therefore, the electron–phonon interaction is not retarded and, consequently, the predominant direct Coulomb repulsion between charge carriers is not screened in the low-temperature phase of $CeCu_2Si_2$. For this reason, an alternative Cooper-pairing mechanism must be operational.

As the heavy-fermion metals $CeAl_3$ and $CeCu_2Si_2$ appeared to be phenomenologically related to the heavy Fermi liquid ^{3}He (Leggett, 1975), it was natural to assume that the superconductivity in the latter compound was of the triplet (odd parity) variety (Anderson, 1984). However, the temperature dependence of the upper critical field of $CeCu_2Si_2$ indicated strong Pauli limiting below $\simeq 0.5\ T_c$, that is, singlet (even-parity) superconductivity (Assmus *et al.*, 1984). The huge initial slope of the upper critical field curve (Assmus *et al.*, 1984), which again demonstrates that massive Cooper pairs are formed by the heavy charge carriers, has become a characteristic of heavy fermion superconductivity.

While most of the superconductors of this kind exhibit a spin-singlet, typically *d*-wave, pairing state, there exist a few prime candidates for heavy-fermion triplet superconductivity, for example, UPt_3 (Tou *et al.*, 1998) and UNi_2Al_3 (Ishida *et al.*, 2002) as well as the ferromagnets (pressurised) UGe_2 (Saxena *et al.*, 2000), URhGe (Levy *et al.*, 2005), and UCoGe (Huy *et al.*, 2007).

Of particular interest is the discovery of superconductivity in materials that lack a crystallographic center of inversion symmetry, like $CePt_3Si$ (Bauer *et al.*, 2004). Such a lack of inversion symmetry allows for mixing between even- and odd-parity pair states, where the degree of mixing apparently depends on the strength of the antisymmetric spin–orbit coupling (Gor'kov and Rashba, 2001; Agterberg *et al.*, 2006).

One of the most interesting questions in this field concerns the microscopic origin of superconductivity in heavy-fermion metals. As early as 1986, antiferromagnetic fluctuations had been considered to act as the glue in heavy-fermion superconductors (Miyake *et al.*, 1986; Scalapino *et al.*, 1986). The first experimental observation along this line was a significant decrease in the magnetic neutron-scattering intensity below $T_c = 0.5$ K in UPt_3 (Aeppli *et al.*, 1989b). For further discussion of the interplay between heavy-fermion superconductivity and incipient antiferromagnetic order, we recommend the reviews by White *et al.* (2015) and Steglich and Wirth (2016).

The coexistence of magnetism and superconductivity in heavy fermions. More than 60% of the presently known heavy-fermion superconductors exhibit a non-Fermi-liquid low-temperature normal state, which, in most cases, derives from a nearby antiferromagnetic quantum critical point. The latter defines a continuous phase transition at absolute zero temperature between the two ground states, an antiferromagnetically ordered and a paramagnetic ground state, tuned by a suitable non-thermal control parameter δ, such as an external or chemical pressure or a magnetic field (Gegenwart *et al.*, 2008). In fact, in heavy-fermion metals, unconventional superconductivity often develops close to antiferromagnetic quantum critical points. This is exemplified in the temperature–pressure phase diagram for $CePd_2Si_2$, which displays a narrow superconducting dome centred around the critical pressure $P_c = 2.8$ GPa, at which antiferromagnetic order apparently disappears in a smooth way (Mathur *et al.*, 1998). As seen in the inset in Fig. 7.24, the low-T resistivity obeys a non-Fermi-liquid power law with an exponent of $\simeq 1.2$ over almost two decades in temperature.

For some heavy fermion superconductors, superconductivity and magnetism can coexist on the phase diagram; see, for example, the phase diagram for $CeRhIn_5$, which was found from studies of the temperature dependence of the heat capacity under high pressure (Fig. 7.25).

The coexistence of superconductivity and magnetism has been discussed theoretically within microscopic approaches by Sacramento (2003) and Alvarez and Yundurain (2007). Within the periodic Anderson model, the same f-electrons of Ce are shown to be responsible for the antiferromagnetic and d-wave superconductivity(Val'kov and Zlotnikov, 2012, 2013).

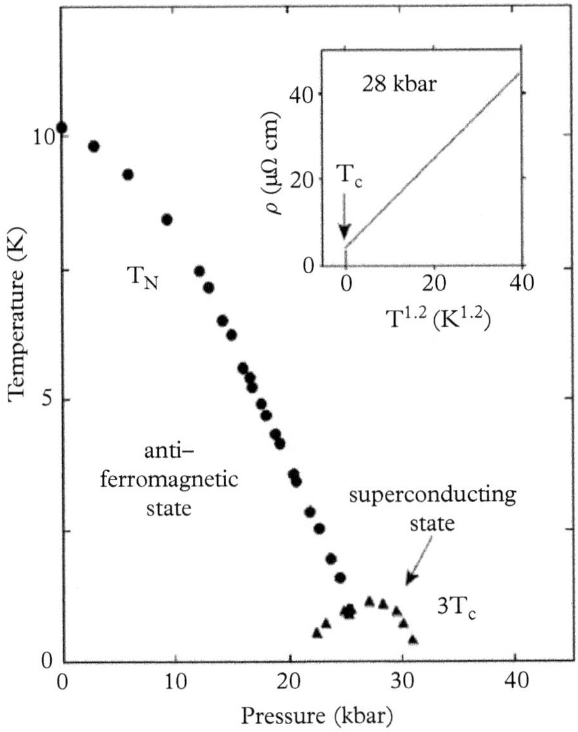

Figure 7.24 *Temperature–pressure phase diagram for $CePd_2Si_2$. Around the critical pressure $P_c \sim 28$ kbar, where the antiferromagnetic order is expected to be suppressed, a superconducting dome appears. Inset: At P_c, the resistivity varies with temperature as $T^{1.2}$ over almost two decades. (From Mathur et al., 1998)*

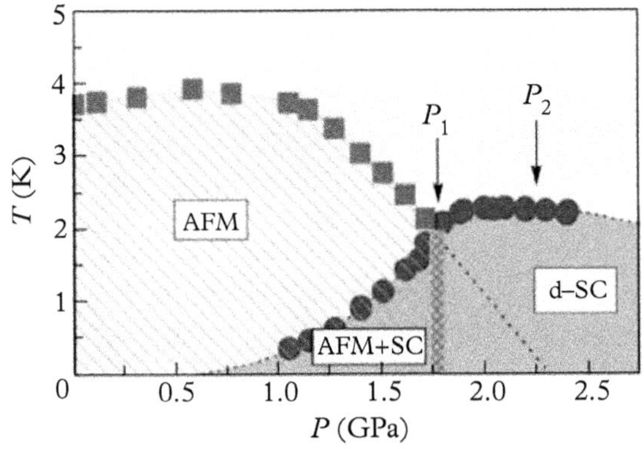

Figure 7.25 *Phase diagram for $CeRhIn_5$. (From Park et al., 2008)*

7.4.3 Intercalated Nitrides: Self-Supported Superconductivity

A new and interesting family of layered superconductors was discovered in 1998 by Yamanaka *et al.* The composition of this material is rather complex and has the formula $Li_x(THF)_yHfNCl$, where THF is tetrahydrofuran (C_4H_8O). The Li ions and THF are intercalated between the Cl layers. This intercalation leads to a large interlayer distance $d(d = 18.7$ Å; see Fig. 7.26). The stability is provided by weak van der Waals forces between the Cl layers.

The most remarkable feature of this member of the so-called intercalated-nitride family is its superconductivity with a relatively high critical temperature value ($T_c \simeq 26$ K). This value does not seem to be high by present standards, but it is still higher than that for the A-15 compounds ($T_c \simeq 22.3$ K for Nb–Ge; see section 7.1.2). Before the discovery of cuprates, the A-15 compounds used to have the highest T_c values.

The most interesting study focused on the mechanism of superconductivity in intercalated nitrides. In this case, the absence of magnetic ions excludes magnetic interactions as the mechanism of pairing. In addition, there is no evidence for the presence of strong correlations, and the material can be treated by the Fermi-liquid theory. At the same time, as was demonstrated by the experimental measurements, the electron–phonon interaction is not strong enough to provide the observed values of T_c (Tou *et al.*, 2001a, b, 2003). The value of the electron–phonon coupling constant λ can be determined from measurements of the electronic effective mass m^* or the Sommerfeld constant (see section 1.2.1), since $m^* = m_0(1+\lambda)$, where m_0 is the band value of the electronic mass. Based on the measurements of magnetic susceptibility and thermodynamic data, it can be concluded that $\lambda \ll 1$; such a small value can provide $T_c \lesssim 1$ K, that is, one much below the observed value of $T_c \simeq 26$ K.

Note also that, if the superconducting state is provided in the phonon-mediated mechanism, then it would be natural to assume, by analogy with MgB_2 (section 7.1.3)

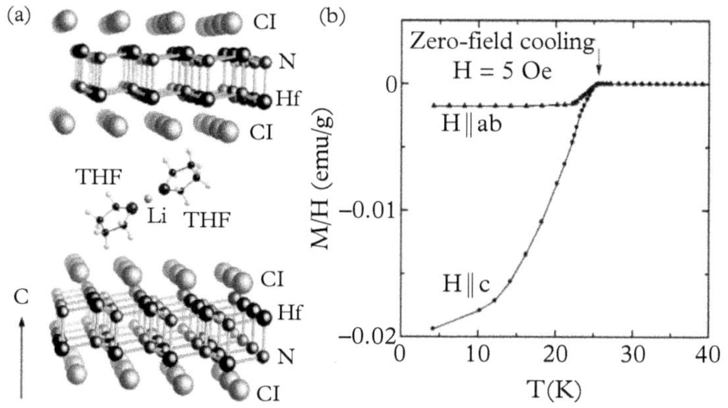

Figure 7.26 *(a) Structure and (b) magnetic properties of the intercalated nitrides $Li_x(THF)_yHfNCl$. (From Yamanaka et al., 1998).*

and hydrides (section 7.2), that the key role is played by the vibrations of N ions. However, measurements of the isotope coefficient for N substitution reveal a very small value of $\alpha \simeq 0.07$ (Tou *et al.*, 2003).

These studies lead to the conclusion that the superconducting state for the compound Li_x(THFNCl) cannot be provided by the electron–phonon interaction. This conclusion is also in agreement with the theoretical calculations (Weht *et al.*, 1999; Heid and Bohnen, 2005).

Because the compound of interest has a layered structure, it is natural to consider the pairing caused by the exchange of acoustic plasmons (see section 2.6). These low-frequency collective excitations can make a noticeable contribution. Indeed, the superconducting state in the described nitrides with $T_c \simeq 26$ K is provided by such a plasmon mechanism (Bill *et al.*, 2003). As mentioned above, the phonons can lead to a low value of T_c ($\precsim 1$ K). The phonons play an important role: their contribution allows the static Coulomb repulsion (μ^*) to be overcome. The dynamic part of the Coulomb interaction (plasmons) leads to an increase in the value of the critical temperature up to $T_c \simeq 26$ K. For the compound $Li_{0.45}(THF)_{0.3}$HfNCl, the following set of parameters obtained from the experimental data and band-structure calculations can be used: the interlayer distance $L = 18.7$ Å, $\left(\frac{m^*}{m_e}\right) \simeq 0.6$, $\mu^* \simeq 0.1$, and $\varepsilon_M = 1.8$ (the large value of ε_M leads to the applicability of RPA; see eqn (4.6)). As a result, one can obtain the value $T_c \simeq 24.5$, which is close to the value observed experimentally.

It is remarkable that the superconducting state in these layered materials is provided by collective modes ('acoustic' plasmons) of the same electrons that are undergoing the superconducting transitions. This is the first system observed with self-supported superconductivity.

7.4.4 Tungsten Oxides

The oxides occupy a special place in condensed matter physics. Indeed, oxides form the most interesting family of ferroelectric materials. Manganites, with their unique magnetic properties (see chapter 9), are also oxides. And, finally, the copper oxides display the highest T_c values for the transition into the superconducting state (at ambient pressure). That is why, shortly after the milestone discovery of copper oxides, scientists started to search for other oxides that are different from cuprates but also display the high T_c state. Reich and Tsabba (1999) have reported regions with $T_c \simeq 90$ K for Na-doped WO_3 (see also Levi *et al.*, 2000). Recently, Schengelaya and Mueller (2019) have proposed the use of the oxygen-deficient tungsten oxide WO_{3-x}. A detailed description of the study was presented by Schengelaya *et al.* (2020). It was established that the compound $WO_{5.9}$ contains the most promising composition. In this case, the superconducting regions (clusters) are embedded in a normal matrix. In other words, the system is described by a percolative picture similar to that described in section 6.2. In such a case, the resistivity data are not very informative, and the transition is determined by magnetic measurements. The value of T_c appears to be equal to 80 K. It is remarkable that the intercalation of lithium raised this value up to 94 K (!). This value is above the temperature of liquid nitrogen. In addition, electron paramagnetic resonance

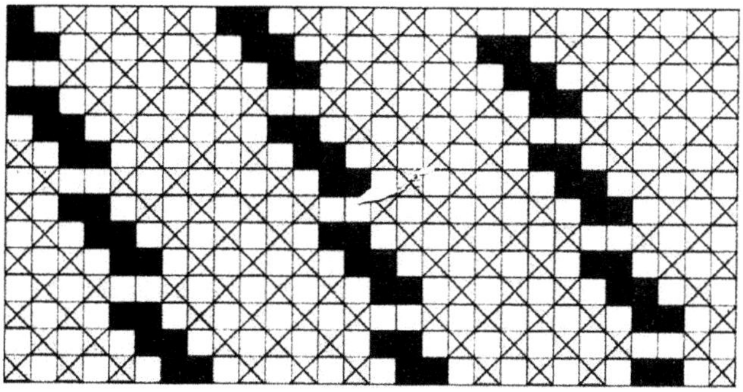

Figure 7.27 *Idealised structure of the Magneli phase $WO_{2.9}$. (From Shengelaya et al., 2020)*

measurements were carried out. In reality, we are dealing with a Magneli phase (1949) picture of filamentary superconductivity when only a small portion of the material is in the superconducting state (Fig. 7.27). According to the author's analysis, manifestations of bipolaronic superconductivity are present (see section 2.3.3). It is interesting that, in this case, as for the cuprates, an electron-doped high-T_c superconductivity state exists.

7.5 Topological Superconductors

In the gapped superconducting state, all negative energy states are fully occupied.[1] Thus, as for insulators, one can define topological numbers for the occupied states. Depending on the dimension and symmetries of the system, various topological numbers can be introduced. In the broadest sense, a superconductor is regarded as being topological if any such topological numbers are non-zero (Sato and Ando, 2017).

Unconventional superconductors often have nodal superconducting gaps, where the nodes themselves have topological numbers. From the above definition, such unconventional superconductors are topological, and they may be called weak topological superconductors. In a narrower sense, a fully opened gap is required, in addition to a non-zero topological number, for topological superconductivity. In this case, no gapless bulk excitation exists and the system may be called a strong topological superconductor. Similar to the case of a quantum Hall state, the transport properties of such a system at low temperature are purely determined by topologically protected gapless excitations localised at a boundary or at a topological defect.

Topological superconductors are characterised by a non-trivial topological invariant, a term that has become well known from the theory of the quantum Hall effect (Niu *et al.*, 1985; Volovik, 2009). The topological invariant reveals the existence or absence of topologically protected edge states, including Majorana modes, with energy lying inside a bulk superconducting gap. Such states are localised on the boundaries of a

[1] This section was written in collaboration with Anton Zlotnikov.

superconducting sample. These effects are similar to the ones described in topological insulators (Hasan and Kane, 2010), in which edge subgap states appear inside an insulating gap.

There are broadly two categories of topological superconductors: intrinsic ones and artificially engineered ones (Sato and Ando, 2017). Intrinsic topological superconductors are those in which a topologically non-trivial gap function naturally shows up. As already discussed, superconductors having odd-parity pairing are, in most cases, topological. Also, non-centrosymmetric superconductors, in which parity is not well defined, can be topological when the spin-triplet component is strong. It is possible to get a topological superconductor by doping a topological insulator. We will discuss concrete examples of such intrinsic topological insulators in this chapter. One can also artificially engineer topological superconductivity in a hybrid system consisting of a metal or semiconductor in proximity with a conventional *s*-wave superconductor. This category of topological superconductors is currently attracting significant attention due to its potential as a building block of Majorana-based qubits for topological quantum computation.

It is thought that topological superconductors include the following:

1. *p*-Wave and chiral *p*-wave superconductors (Read and Green, 2000; Kitaev, 2001). Possible candidates in this category include Sr_2RuO_4 (Mackenzie and Maeno, 2003; Das Sarma *et al.*, 2006) and the uranium heavy-fermion superconductors UGe_2, UCoGe, and URhGe (Sau and Tewari, 2012; Mineev, 2017).
2. Chiral *d*-wave superconductors (Volovik, 1997), such as certain *d*-wave superconductors with broken time-reversal symmetry, and superconductors with triangular layers, for example, Na_xCoO_2 (Takada *et al.*, 2002; Baskaran, 2003).
3. Proximity-effect systems topological insulator–superconductors (Fu and Kane, 2008).
4. Non-centrosymmetric superconductors (Sato and Fujimoto, 2009).
5. Doped topological insulators. The first material of this kind was $Cu_xBi_2Se_3$, with superconductivity up to $T_c \sim 4$ K, which was discovered by Hor *et al.*, 2010. Such material yields promising ground for the search for topological superconductivity, for the following reason: since the topological surface states are present in topological insulators even when carriers are doped (as long the doping is not too heavy), one would expect that the superconductivity in the bulk material would lead to proximity-induced superconductivity of the surface, where two-dimensional topological superconductivity should be found. Another example is Bi_2Se_3, for which a high-quality single crystal has been grown via the self-flux method (Awana *et al.*, 2017). The as-grown crystals are large (a few centimeters) in size and have a metallic conductivity down to 2 K and a high, non-saturating, positive magnetoresistance to the tune of >240% at 2 K in an applied field of 14 Torr. The high magnetoresistance of this compound (250%, 2 K, 14 Torr) also exhibited quantum oscillations. The same group has also grown single crystals of doped $Nb_{0.25}Bi_2Se_3$, with magnetic measurements indicating that

superconductivity exists at below 2.5 K, with lower and upper critical fields at 2 K that are of around 50 Oe and 900 Oe, respectively (Sharma *et al.*, 2020). This is in conformity with earlier results by Hoshi *et al.* (2017).

6. Semiconductor–superconductor heterostructures with strong spin–orbit interaction in a magnetic field (Lutchyn *et al.*, 2010; Oreg *et al.*, 2010; Sau *et al.*, 2010). The experimental realisation is semiconducting InAs or InSb nanowires contacted with superconducting electrodes or covered by superconducting Al (Mourik *et al.*, 2012; Deng *et al.*, 2016; Zhang *et al.*, 2018).
7. Superconducting systems with non-collinear magnetic ordering, including magnetic nanoparticles on the superconducting substrate (Choy, 2011; Klinovaya *et al.*, 2013) and magnetic superconductors (Martin and Morpurgo, 2012; Lu and Wang, 2013; Val'kov *et al.*, 2018).

Aside from chiral d-wave superconductors, all of the other systems mentioned support Majorana bound states, which are special topologically protected edge states with zero excitation energy. Existing quasiparticle excitations with zero energy in such systems are often called Majorana fermions. Majorana fermions in solid-state systems have attracted considerable interest not only due to new surface effects in superconductors but also because of their non-Abelian statistics, which has led to the idea of decoherence-free quantum computations (Nayak *et al.*, 2008).

Interest in topological superconductivity has grown considerably since the proposals of Read and Green (2000) and Kitaev (2001) considering p-wave superconductivity. The simplest model is a one-dimensional tight-binding chain of spinless fermions with hopping and p-wave superconducting pairings between fermions on the nearest sites (Kitaev, 2001). Such a model supports Majorana modes and has a solution at special parameters when Majorana modes are localised exactly on the first and last sites of the chain. Nevertheless, clear experimental evidence for p-wave superconducting systems has not been obtained yet. Therefore, different heterostructures composed of a conventional s-wave spin-singlet superconductor and a semiconductor with spin–orbit coupling in an external magnetic field have been proposed (Lutchyn *et al.*, 2010; Oreg *et al.*, 2010; Sau *et al.*, 2010). It is assumed that effective p-wave superconductivity can be realised in such systems, due to the spin–orbit interaction and the magnetic field.

The phase diagram in a variable-magnetic-field h-chemical-potential μ for a one-dimensional nanowire with Rashba spin–orbit interaction (parameter α) and proximity-induced s-wave superconducting pairings (parameter Δ) is shown in Fig. 7.28. Parameter t describes the electron hopping. The red lines show the conditions for which the gap in the energy spectrum for periodic boundary conditions is closed and the topological invariant (Majorana number) is changed. The topological phases are formed inside these lines, and the trivial phases are formed outside these lines. Blue lines show the conditions of existence of Majorana fermions with exactly zero energy in a nanowire with 30 sites. It can be seen that almost all Majorana modes appear in topologically non-trivial phases, in agreement with their bulk-boundary correspondence. It should be noted that, when

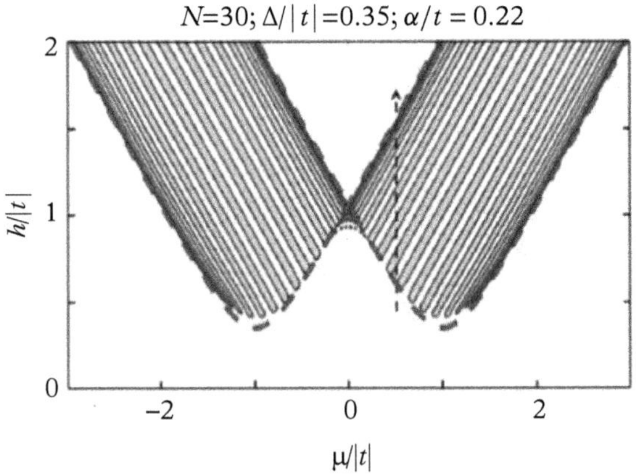

Figure 7.28 *The topological phase diagram of a closed nanowire in the plane (chemical potential–external magnetic field). The area of the non-trivial topological phase is shown by blue stripes. (From Val'kov et al., 2019b)*

the zero energy is realised, the fermion parity of the ground state is changed. Regions with negative fermion parity are shaded in Fig. 7.28.

Experimental progress has been achieved for InAs and InSb semiconducting nanowires epitaxially coated by superconducting aluminium (Mourik *et al.*, 2012; Deng *et al.*, 2016; Zhang *et al.*, 2018). Such a device is shown in Fig. 7.29. Side gates and contacts are Cr/Au (10 nm/100 nm). The Al shell thickness is ~10 nm. The substrate is *p*-doped Si, acting as a global back-gate, covered by 285 nm SiO_2. The two tunnel-gates are shorted externally as well as the two super-gates. The magnetic field direction is aligned with the nanowire axis for all measurements. It is believed that the quantised zero-bias peak at zero-bias voltage relates to Majorana fermions. It can also be seen in Fig. 7.29 that such peak is observed in a certain interval of magnetic fields, as the theory predicted.

It should be noted that such experiments can give only indirect evidence of the formation of Majorana modes, and the zero-bias peak can be caused by other mechanisms (Moore *et al.*, 2018; Pan *et al.*, 2020). Even so, semiconducting nanowires with induced superconductivity are often used to propose different devices supporting Majorana modes (Val'kov *et al.*, 2019).

It has been shown that the spiral magnetic field or helical magnetic ordering can play the same role in the mechanism of the Majorana modes formation as spin–orbit interaction together with a uniform magnetic field (Braunecker *et al.*, 2010; Martin and Morpugo, 2012). Therefore, other examples of topological superconducting systems have been proposed: (i) a chain of magnetic nanoparticles with arbitrary directions of magnetisation on a superconducting substrate (Choy *et al.*, 2011), (ii) nanowires with induced superconductivity in a spatially non-uniform magnetic field (Kjaergaard *et al.*, 2012), (iii) quasi-one-dimensional systems with RKKY exchange interaction and helical magnetic order in contact with a superconductor (Braunecker and Simon, 2013;

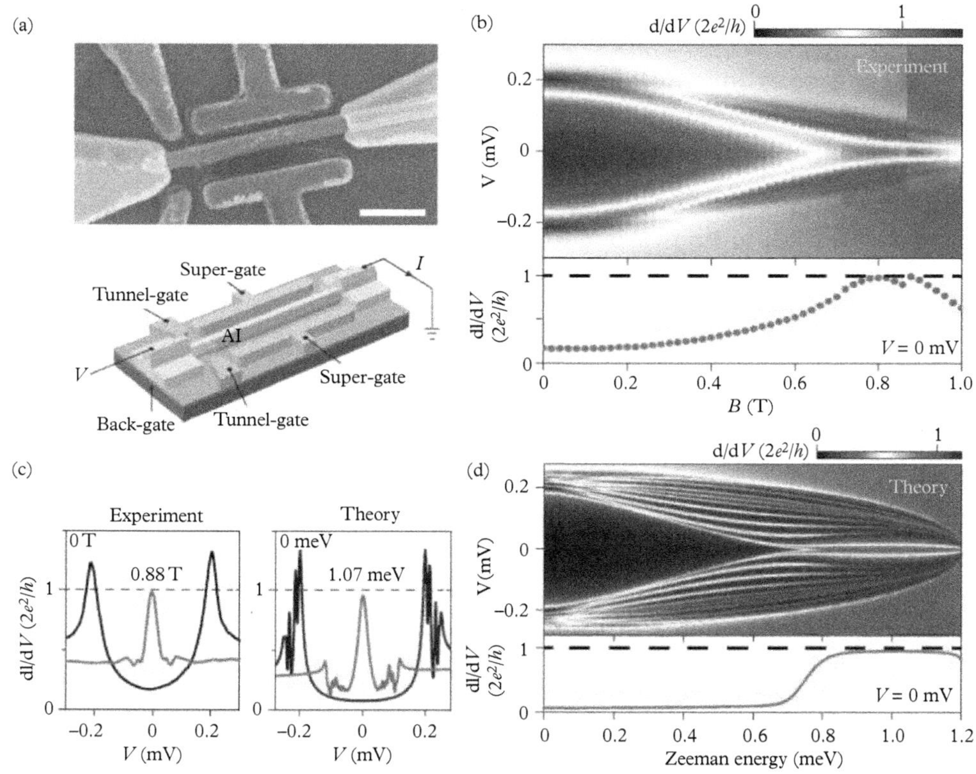

Figure 7.29 *Quantized Majorana zero-bias peak. (a) False-colour scanning electron micrograph of device A (upper panel) and its schematics (lower panel). The scale bar is 500 nm. (b) Magnetic-field dependence of the quantised zero-bias peak in device A with the zero-bias line-cut in the lower panel. The super-gate (tunnel-gate voltage is fixed at − 6.5 V (− 7.7 V), while the back-gate is kept grounded. The temperature is 20 mK unless specified. (c) Comparison between experiment and theory. Left (right) panel shows the vertical line-cuts from b (d) at 0 T and 0.88 T (1.07 meV). (d) Majorana simulation of device A, assuming the chemical potential* $\mu = 0.3$ *meV, the tunnel barrier length* = 10 *nm, the tunnel barrier height* = 8 *meV, and the superconductor–semiconductor coupling* = 0.6 *meV. A small dissipation broadening term (∼30 mK) is introduced into all of the simulations to account for the averaging effect from finite temperature and a small lock-in excitation voltage (8* μ*V). (From Zhang et al., 2018)*

Klinovaja *et al.*, 2013), and (iv) magnetic superconductors with helical or non-collinear spin ordering (Martin and Morpugo, 2012; Lu and Wang, 2013; Val'kov and Zlotnikov, 2019).

A chain of Fe atoms on the superconductor Pb has been experimentally studied by scanning tunnelling microscopy (Nadj-Perge *et al.*, 2013; Ruby *et al.*, 2015; Pawlak *et al.*, 2016; Feldman *et al.*, 2017). Firstly, it was thought that the magnetic order of the chain was ferromagnetic (Nadj-Perge *et al.*, 2013). Then it was observed that the Majorana modes were caused by the spin–orbit interaction in Pb. Nevertheless, it should be noted that even an almost-ferromagnetic ordering in a chain with a small helix angle can induce the second mechanism of formation of Majorana modes (Ruby *et al.*, 2015). It should be noted that, in an attempt to find the Majorana state, measurements of vortices in different superconducting heterostructures have been carried out via scanning tunnelling microscopy (Sun *et al.*, 2016).

The high-resolution conductance maps and topographic measurements on the same Fe chain, obtained by Feldman *et al.*, (2017) are shown in Fig. 7.30. The Majorana modes are expected at the ends of the chain where the zero-bias peak is implemented. The samples were cooled to about 200 mK. Nevertheless, the zero-bias peak reaches a maximum value of about 0.16 e^2/h. This value is still smaller than the predicted one for Majorana fermions, suggesting that the temperature of the measurements is still large compared to the tunnelling energy scale.

Majorana modes can be also found in two-dimensional systems (Sato and Fujimoto, 2009; Sato *et al.*, 2010; Martin and Morpugo, 2012). For simplicity, the cylindrical (quasi-one-dimensional) geometry of superconducting systems is often considered when the periodic boundary conditions are applicable in one direction of the two-dimensional

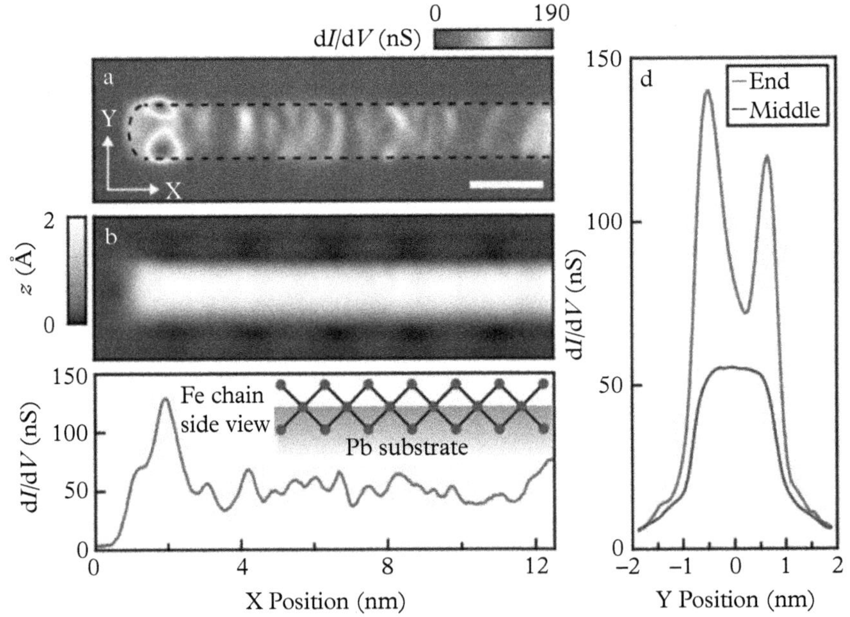

Figure 7.30 *Double eye pattern and spatial dependence of the zero-bias peak. (a, b) Map of zero-bias conductance (a) along a chain and (b) its corresponding topography. Scale bar: 2 nm. (c, d) dI/dV as a function of position (c) along and (d) transverse to the Fe chain. Data in panel c are averaged over the 1.6 nm chain width. Inset: Schematic of a zigzag Fe chain partially embedded in the substrate. In panel d, the end spectrum (red) is averaged over the 1 nm length of the double-eye feature, and the mid-chain spectrum (purple) is averaged over the remainder of the chain. (From Feldman et al., 2017)*

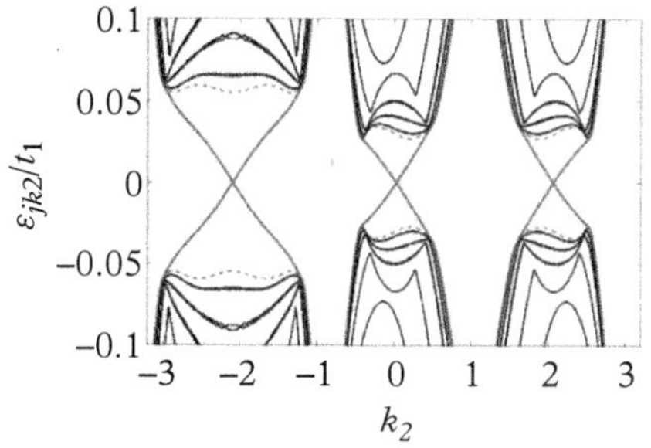

Figure 7.31 *The Fermi-type excitation spectrum of the stripe with a triangular lattice in the coexistence phase of chiral superconductivity and 120° spin ordering. (From Val'kov and Zlotnikov, 2019)*

lattice while there are open boundaries in the other direction. The dependences of several branches of the excitation spectrum of the stripe with a triangular lattice in the coexistence phase of chiral superconductivity and 120° spin ordering on the quasi-momentum are shown in Fig. 7.31 (Val'kov and Zlotnikov, 2019).

The dashed line denotes the boundary of the bulk excitation spectrum at periodic boundary conditions, along with both directions of the triangular lattice. It can be seen that the excitation spectrum in this case has an energy gap, which is denoted by the dashed line. Edge states appear inside the spectrum gap (shown by the thick red line) in the cylindrical geometry. Thin black solid lines selectively show the branches lying in the region of the bulk spectrum. At $k_2 = -2\pi/3$, the zero-energy Majorana modes are realised.

In the conclusion, we would like to mention (following the review by Sato and Ando, 2017) that, in intrinsic topological superconductors, investigations of novel phenomena associated with surface/edge Majorana fermions will be an important subject. The first step would be to study heat and spin transport properties, for which Majorana fermions will be responsible. Also, the quantisation of thermal Hall conductivity would be intriguing as a novel manifestation of topology. In the future, conceiving a Majorana qubit based on intrinsic topological superconductors for quantum computation would be an interesting and potentially important direction. Needless to say, discoveries of new types of intrinsic topological superconductors to widen our knowledge of topological materials remain important.

In hybrid systems, since the possibility of engineering topological superconductivity has already been confirmed, the interest is shifting towards controlling and reading the occupation of a Majorana zero-mode, which is the basis of qubit operation. The most exciting prospect is the demonstration of non-Abelian statistics associated with a Majorana zero-mode (Read and Green, 2000). Such an experiment requires a braiding operation involving at least four Majorana fermions (Ivanov, 2001) and confirmation of the flipping of a qubit. This is technically very demanding, but already concrete setups and protocols have been proposed (Aasen *et al.*, 2016). Another exciting prospect of the hybrid systems is the potential to implement the 'surface code' (Bravyi and Kitaev, 2001), a protocol to realise scalable quantum computation with built-in error corrections. The surface code can, in principle, be realised by using non-topological quantum states, but it has been argued that using Majorana fermions will make it much simpler to realise the surface code (Vijay *et al.*, 2015; Landau *et al.*, 2016).

Regarding the theory of topological superconductors, one of the important remaining problems is the complete understanding of topological crystalline superconductivity. Real materials have a variety of crystal symmetries, which could enrich their topological structures. For instance, it was revealed recently that non-symmorphic space group symmetries such as glide reflection and screw rotation make it possible to have novel Z_2 and Z_4 topological phases with Mobius-twisted surface states (Teo *et al.*, 2008). To clarify whether other exotic topological surfaces may exist, more systematic studies are needed. Finally, revealing hitherto-unknown topological phenomena is of particular interest. In this regard, a systematic search based on general frameworks such as topological quantum field theories might give universal and tractable predictions for novel topological phenomena.

8

Materials (III)

8.1 Organic Superconductivity

8.1.1 History

Organic superconductors form a relatively young family of superconducting materials: the phenomenon of organic superconductivity was observed for the first time in 1980 (Jerome *et al.*, 1980), in the complex material bis-tetramethyl-tetraselenafulvalene-hexfluorophosphate ($(TMTSF)_2PF_6$; $T_c \simeq 0.9$ K). Since this first observation, the field of organic superconductivity has grown very noticeably (see Fig. 8.1; also see Jerome and Greuzet, 1987; Kresin and Little, 1990; Ishiguro and Ito, 1996; Ishiguro, 2001; Jerome, 2012).

The concept of organic superconductivity was introduced by Little (1964), who developed the model of the organic polymer (see section 2.5); this prediction was confirmed by the latest discoveries. The organic superconductors, like oxides, including high-T_c oxides, and princides, are complex systems. Synthesis plays a key role; as a result, this field overlaps solid state physics, materials science, and organic chemistry.

Progress in the area of organic superconductivity was based on previous developments leading to the synthesis of organic conductors. Indeed, the usual bulk organic materials are insulators. The possibility of organic metallicity was studied by London (1937), who analysed aromatic organic molecules. These molecules (see section 8.2.2) contain delocalised electrons (so-called π-electrons), and this feature (delocalisation) make these electrons similar to conduction electrons in metals. It is worth noting that the analysis of these molecules resulted in the formulation of the famous London equations, which are a key ingredient of the physics of superconductivity.

The observation of high conductivity in the perylene bromide complex (Akamutsu *et al.*, 1954) led to the development of the field of organic metals. The next important step was the creation of a stable organic metal formed by TTF-tetracyanoquinodimethane (TCNQ) units (Coleman *et al.*, 1973; Ferraris *et al.*, 1973). The unit is a combination of two flat molecules (Fig. 8.2).

All this development was the precursor for the discovery of organic superconductivity. As mentioned above, the first such superconductor was created in 1980 (Jerome *et al.*,

Superconducting State: Mechanisms and Materials. Vladimir Z. Kresin, Sergei G. Ovchinnikov, and Stuart A. Wolf, Oxford University Press (2021). © Kresin, Ovchinnikov, Wolf. DOI: 10.1093/oso/9780198845331.003.0008

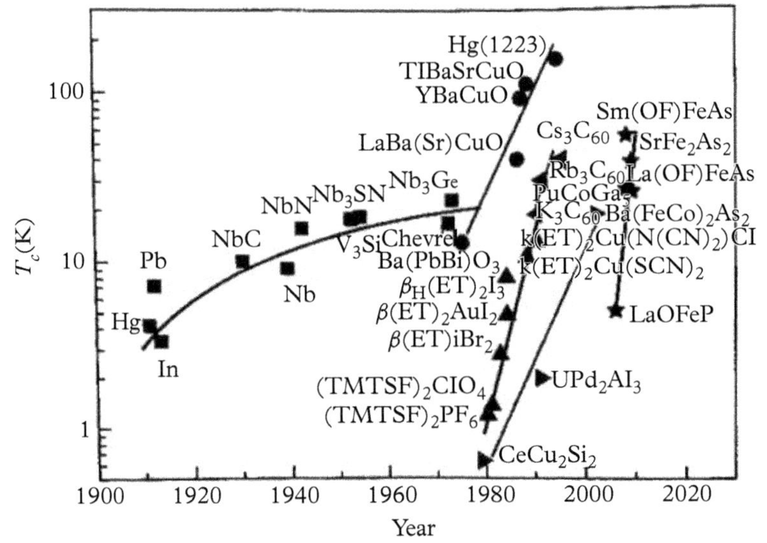

Figure 8.1 *Evolution of T_c in various series of superconductors over the years. (Figure reproduced from Jerome, 2012)*

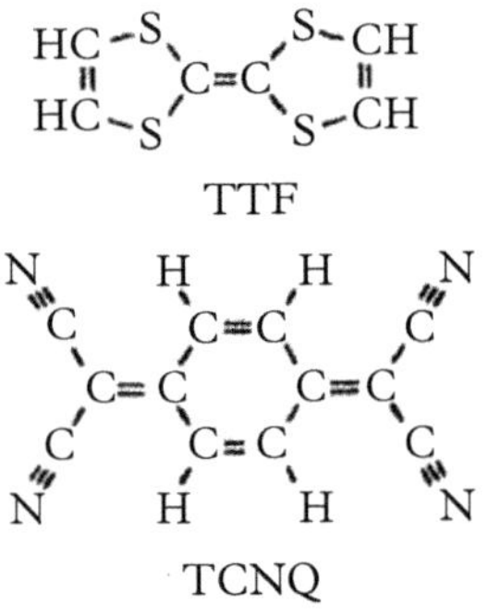

Figure 8.2 *Tetrathiafulvalene and tetracyanoquinodimethane molecules. (Figure reproduced from Ishiguro et al., 2001)*

1980) with $T_c \simeq 0.9$ K under a pressure of around 9 kbar. Later, many new compounds with higher values of T_c have been synthesised.

An important addition to the family of organic superconductors has been formed by doped fullerines (fullerides, see below and section 1.3; Haddon *et al.*, 1991; Hebard *et al.*, 1991; Holczer *et al.*, 1991; Rosseinsky *et al.*, 1991). The fulleredes belong to this family, because the conductivity is provided by the presence of carbon atoms and, correspondingly, by π-electrons. The studies of recently obtained low-dimensional (two-dimensional) organic material, graphene, have also attracted a lot of attention (section 1.4).

8.1.2 The TMTSF and ET Families: Structure and Properties

Let us discuss various types of organic superconductors. Firstly, we will describe the $(TMTSF)_2X$ systems (the so-called Bechgaard salts; Bechgaard *et al.*, 1979), where X stands for an electron-acceptor molecule such as PF_6, AsF_6, ClO_4, ReO_4, and so on.

As mentioned above, superconductivity of the $(TMTSF)_2X$ system with $X \equiv PF_6$ was the first observation of organic superconductivity (Jerome *et al.*, 1980). The systems of the TMTSF molecule can be seen in Fig. 8.3.

The crystal formed by $(TMTSF)_2X$ units is highly anisotropic, so that we are dealing with a quasi-one-dimensional crystal. For example, if $X \equiv ClO_4$, the ratio of the conductivities at room temperature in the ab-plane is $\sigma_a/\sigma_b \simeq 25$, and $\sigma_a/\sigma_b \simeq 200$ for $X \equiv PbO_6$. Correspondingly, the electron-transfer energies are $t_a = 0.25$ eV, $t_b = 0.0025$ eV, $t_c = 0.0015$ eV. The molecules form columns, and the carriers move in the easiest way along these columns (along the a-axis).

The quasi-one-dimensional structure leads to nesting at the Fermi surface. The typical phase diagram ($X \equiv AsF_6$) is plotted in Fig. 8.4. In the low-temperature region at ambient pressure, the compound is an SDW insulator. The SDW instability is consistent with the nesting feature and with the quasi-one-dimensional nature of the structure.

One should stress that an external pressure greatly affects the properties of organic crystals. It is not surprising, because we are dealing with molecular crystals, and the intermolecular coupling is rather weak, especially in the c-direction. One can see from Fig. 8.4 that the pressure strongly decreases the temperature of the metal–insulator transition. As a result, the metallic phase persists to a lower temperature, thus making it possible to observe the transition into a superconducting state. Only the compound with $X \equiv ClO_4$ becomes superconducting at ambient pressure.

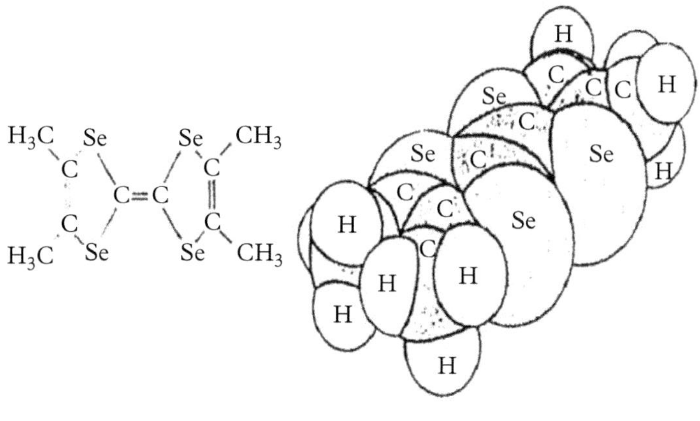

Figure 8.3 *Structure of a tetramethyl-tetraselenafulvalene molecule. (Figure reproduced from Ishiguro et al., 2001)*

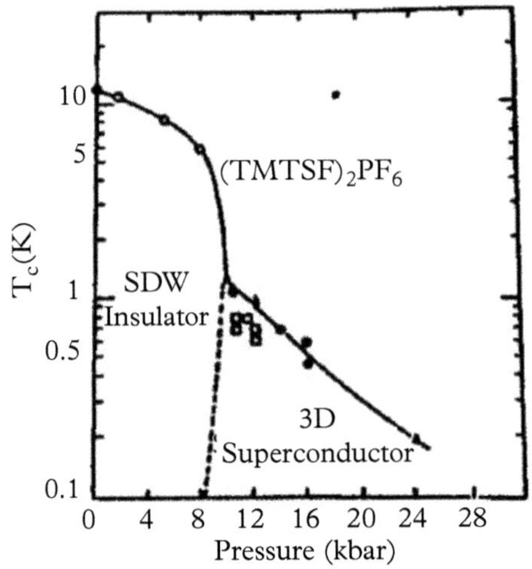

Figure 8.4 *Phase diagram of bis-tetramethyl-tetraselenafulvalene-hexfluorophosphate. (Figure reproduced from Ishiguro et al., 2001)*

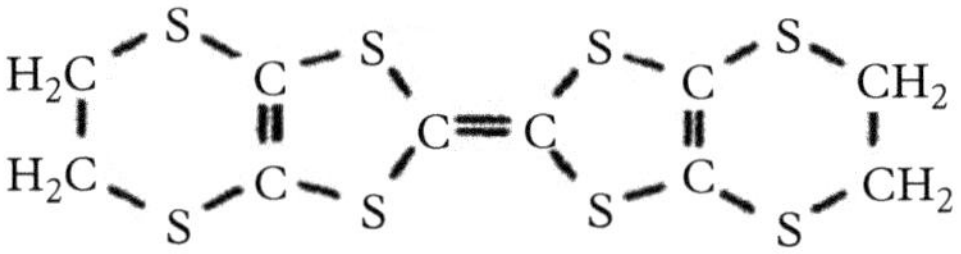

Figure 8.5 *An ethylenedithiotetrathiafulvalene molecule. (Figure reproduced from Ishiguro et al., 2001)*

As for some major superconducting parameters, they have the values $2\Delta(0)/T_c \simeq 3.7$, $c_s/\gamma T_c \simeq 1.67$. These values are close to those for the conventional superconductors (see section 2.1.4).

Another group of organic superconductors is also characterised by low-dimensional structures. We are talking about ethylenedithiotetrathiafulvalene (ET) salts. The basic molecule, also sometimes denoted as BEDT-TTF, is depicted in Fig. 8.5.

Unlike the $(TMTSF)_2X$ family, ET salts have a quasi-two-dimensional structure (see Fig. 8.6). Correspondingly, the Fermi surface contains a cylindrical part. Note also that the in-plane coherence length is of the order of $\xi_{\parallel} \simeq 6.10^2$Å, whereas $\xi_{\perp} \simeq 30$ Å. This anisotropy reflects the fact that the pairing mainly occurs in the layers, and the interlayer coupling is rather weak. This is confirmed by the observation of the internal Josephson coupling, that is, the transfer in the c-direction below T_c is dominated by Josephson interlayer tunnelling (Ito *et al.*, 1994; Muller, 1994).

One should also mention the unusual dependence of the critical temperature on non-magnetic scattering. For example, superconductivity was suppressed in $(TMTSF)_2X$, $X \equiv ClO_4$ with its allowing (Tomic *et al.*, 1983). This observation excludes the usual *s*-wave pairing state (see section 2.1.7) and it means that we are dealing with

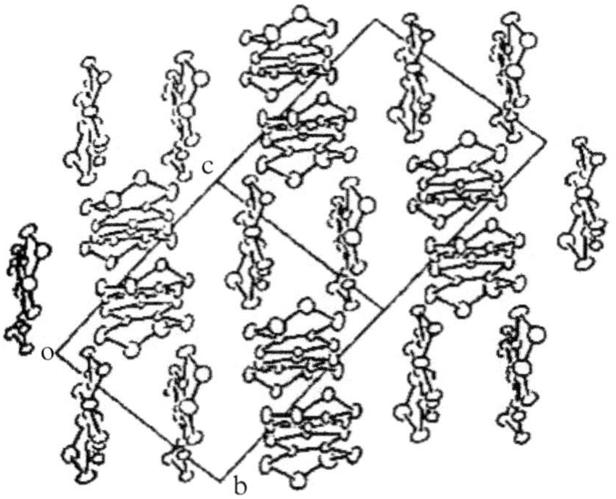

Figure 8.6 *Arrangement of the ethylenedithiotetrathiafulvalene (ET) molecules of* κ-$(ET)_2I_3$. *(Figure reproduced from Ishiguro et al., 2001)*

an unconventional pairing (e.g. with d-wave symmetry; see Ichimura *et al.*, 2008). A similar conclusion follows from the measurements of the penetration depth (Carrington *et al.*, 1999). Another possibility is the appearance of the inhomogeneous Fulde–Ferrel–Larkin–Ovchinnikov (FFLO) state (Ferrell and Fulde, 1964; Larkin and Ovchinnikov, 1964). The manifestation of such a state is discussed by Varelogiannis (2002), Singleton and Mielne (2002), Cho *et al.* (2009), and Croutory and Buzdin (2013).

There has been great progress in understanding the mechanism of superconductivity in organic materials, but some controversy still remains. There are strong arguments that, for the $(TMTSF)_2X$ compound, the magnetic mechanism (spin fluctuations) is the dominant one (see Jerome, 2012). Indeed, there is a strong proximity between SDW and superconducting phases (see Fig. 8.4). The superconducting state arises below some critical pressure near the SDW ground state. It is natural to expect the magnetic fluctuations to play a dominant role in the pairing. A detailed study performed with the use of various methods (see the review by Jerome, 2012) along with theoretical analysis, for example, by Duprat and Bourbonnais (2001), support this conclusion.

On the other hand, low-frequency phonon modes can also contribute to the pairing. Experimental studies supporting such a conclusion were performed (Hawley *et al.*, 1986; Nowack *et al.*, 1986; Gray *et al.*, 1987) on some $(ET)_2X$ superconductors. The point-contact tunnelling spectroscopy proposed by Yanson (1974) was employed (see the review by Ishiguro *et al.*, 2001). This point of view has been also supported by heat capacity measurements (Elsinger *et al.*, 2000); the picture is well described by strong coupling theory (see section 2.3) with $\lambda \simeq 2.5$, so that $2\epsilon(0)/T_c \simeq 5.4$, and s-wave pairing is present. The electronic heat capacity decreases exponentially, in accordance with the picture of a fully gapped order parameter.

As a whole, the study of this fundamental problem is not complete. It might be that the pairing in some superconducting materials (e.g. in $(TMTSF)_2X$) is dominated by the contribution of spin fluctuations, whereas, for example, in $(ET)_2X$, it is caused by the electron–phonon interaction. Perhaps we are dealing with some combination of mechanisms. One can expect that this question will be resolved in the near future.

8.1.3 Intercalated Materials

There are a number of materials which are intrinsically insulators or semiconductors but which become superconducting upon the intercalation of specific metallic ions. The most remarkable example of such systems is the family of fullerenes; we will discuss their properties in the next section.

As a first example of such organic systems, let us mention intercalated graphite. The first observation of the superconducting transition in the alkali-metal-doped-graphite K_8C ($T_c \simeq 140$ mK) was described in 1965 (Hannay *et al.*, 1065). The highest value of T_c for such systems ($T_c \simeq 11.5$ K) was observed recently by Emery *et al.* (2005) for Ca-doped graphite.

A drastic increase in T_c, up to $T_c \simeq 18$ K, has been observed for a K-doped picene compound (Mitsuyashi *et al.*, 2010; see also the review by Kubozono, 2011). Picene is an aromatic hydrocarbon.

The crystal structure of K_x picene can be seen in Fig. 8.7. The herringbone structure contains low-dimensional units. It is interesting that the superconducting phase is formed for $X \simeq 3$, and the value of T_c is very sensitive to the doping level X. Indeed, there is no superconductivity for $X \lesssim 2$. The phase with $T_c \simeq 7$ K corresponds to $X \simeq 2.9$, whereas the much higher value, $T_c \simeq 18$ K (!) has been observed for the compound K_x picene with $X = 3.3$. The intercalated potassium atoms not only donate electrons but also, according to calculations (Kosugi *et al.*, 2009) provide strong hybridisation of electronic orbitals and modify the electronic structure of the undoped crystal.

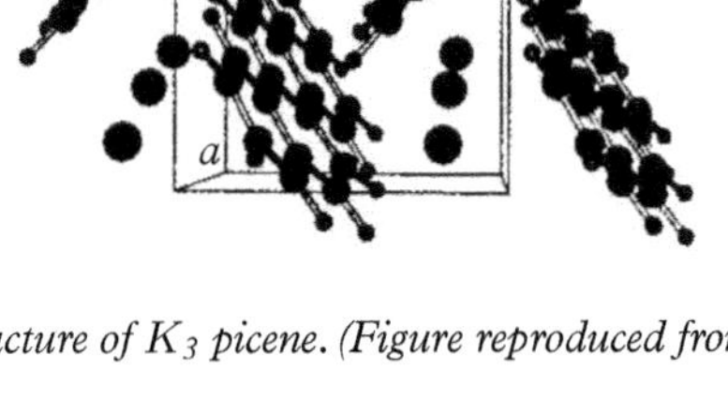

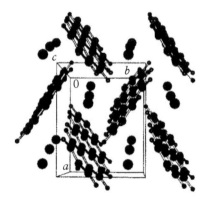

Figure 8.7 *Crystal structure of K_3 picene. (Figure reproduced from Kubozono et al., 2011)*

Aromatic molecules such as picene have a strongly developed π-electron system of delocalised electrons (see below and sections 1.4 and 3.6). The search for intercalated compounds containing other aromatic molecules looks very promising.

8.1.4 Fullerides

The building block of the fullerides is the C_{60} cluster (Fig. 8.8). Therefore, this organic material is an example of a cluster-based compound (see below and chapter 3). The C_{60} cluster (the so-called Buckminster fullerene) has an almost spherical shape.

The C_{60} clusters aggregate and form a cubic crystal with the lattice constant $a \simeq 14$ Å. Such a crystal is a semiconductor with a direct band gap which is of the order of 1.9 eV.

The situation changes drastically if alkali metals are intercalated into the space between the clusters. These metallic ions act as donors, so that the clusters' lowest unoccupied orbital (the HUMO orbital) becomes filled. As a result, the crystal becomes a metal (Haddon *et al.*, 1991) and, below some critical temperature T_c, makes the transition into a superconducting state (see the review by Hebard, 1992).

It is interesting that the normal resistance above T_c increases with temperature, but there is no usual saturation (the Iotte–Regel limit) corresponding to the situation when the mean free path is of the order of the lattice period. This means that the electron–lattice scattering is greatly affected by the intracluster vibrational modes.

The superconducting state was observed for the K_3C_{60} compound (Hebard *et al.*, 1991) with $T_c \simeq 18$ K. This was followed by similar observations (Holczer *et al.*, 1991; Rosseinsky *et al.*, 1991) for $Rb_3C_{60}(T_c \simeq 28$ K) and for Cs_2RbC_{60} with $T_c \simeq 33$ K. The highest value of the critical temperature for the fullerides was observed for Cs_3C_{60} (Tanigaki *et al.*, 1991). Under pressure ($\simeq 15$ kbar), the superconducting state persists up to $T_c \simeq 40$ K (Palstra *et al.*, 1995). At present, this is the highest observed value of T_c for organic superconductors.

The pairing in C_{60}-based materials is provided by the electron–phonon interaction. It follows from the isotope effect measurements. Indeed, the isotope substitution (Ramirez *et al.*, 1992) leads to a value for the isotope coefficient which is close to $\alpha \simeq 0.4$, which is close to an ideal BCS value. Probably, the main contribution comes from the intracluster

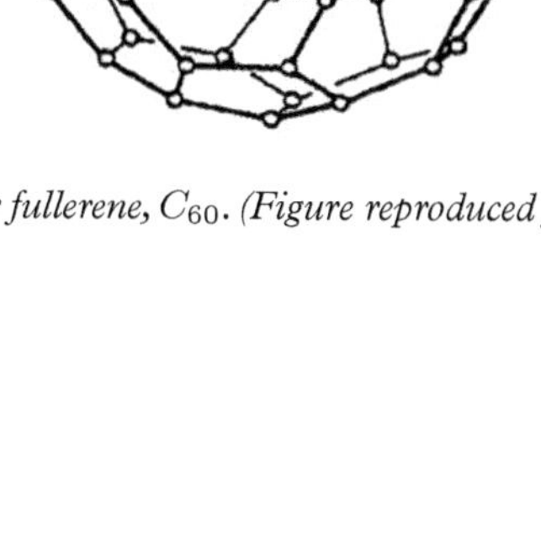

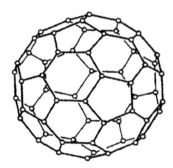

Figure 8.8 *Buckminster fullerene, C_{60}. (Figure reproduced from Ishiguro et al., 2001)*

vibrations. The spectrum of the intracluster vibrations extends up to $\Omega_{max} \simeq 0.2$ eV. If the main contribution comes from the intermediate frequency range (Varma *et al.*, 1991; Schluter *et al.*, 1992), it would correspond to the intermediate strength of the interaction.

It is also interesting to note that the fullerides are characterised by a small value for the Fermi energy. For example, $E_F \simeq 0.2$ eV for Rb_3C_{60}, and $E_F \simeq 0.3$ eV for K_3C_{60}. As a result, the condition $\widetilde{\Omega} \ll E_F$ (Migdal theorem; see section 2.4) does not hold in this case. Such a strongly non-adiabatic situation deserves special study.

In principle, there could be a peculiar scenario of superconductivity for cluster-based materials, namely, the π-electrons can form the Cooper pairs on an isolated cluster. For example, for the C_{60} cluster, such a pairing was described by Bergomi and Jollicoeur (1994) and Jollicoeur (1998). Then the transport between such clusters is provided by Josephson tunnelling. Such a mechanism, which is entirely different from the usual band picture and realistic for cluster-based materials (see below section 8.3), was introduced by Friedel (1992) after the discovery of superconductivity in fulleredes. However, the impact of external pressure on T_c, studied by Sparn *et al.* (1992), allows us to conclude that the superconducting state in C_{60} materials is provided by pairing of the delocalised band states formed by the overlap of the electronic wave functions corresponding to the neighbouring clusters. Indeed, the value of the critical temperature decreases under the pressure. This decrease is caused by decreasing the intercluster distance. In turn, it results in a larger overlap of the electronic wave functions and a corresponding broadening of the energy bands. As a result, the density of states v_F becomes smaller, which leads to a decrease in the value of the coupling constant λ (see eqn (3.23)) and, correspondingly, to the observed smaller value of T_c.

8.1.5 Graphene

The study of the two-dimensional organic material graphene has recently attracted a lot of attention. In reality, graphene is not a new substance; it has been studied theoretically for a long time. Indeed, we know that graphite, which is a well-known material used for writing by all of us, contains a set of parallel graphene layers bounded by van der Waals forces. However, a recent breakthrough occurred when this ideal two-dimensional system was experimentally isolated (Novoselov and Geim, 2004). This achievement was followed by very intensive study; about 5000 (!) scientific papers were published during next five years. Indeed, graphene displays unusual optical properties, transport properties, magnetic properties, and so on, which very different from these for conventional three-dimensional solids. A description of these properties is beyond the scope of this book (see e.g. the reviews by Castro Neto *et al.*, 2009 and Das Sarma, 2011). Here we restrict ourselves to a brief description of the most fundamental properties of graphene and discuss some questions related to its potential manifestation of the superconducting state.

Graphene, as a real organic material, contains carbon ions forming a relatively simple two-dimensional structure (Fig. 8.9). This so-called honeycomb structure can be visualised as being built from benzene rings (see appendix C) and then stripped of its hydrogen ions. The benzene ring is the basic unit for many organic materials

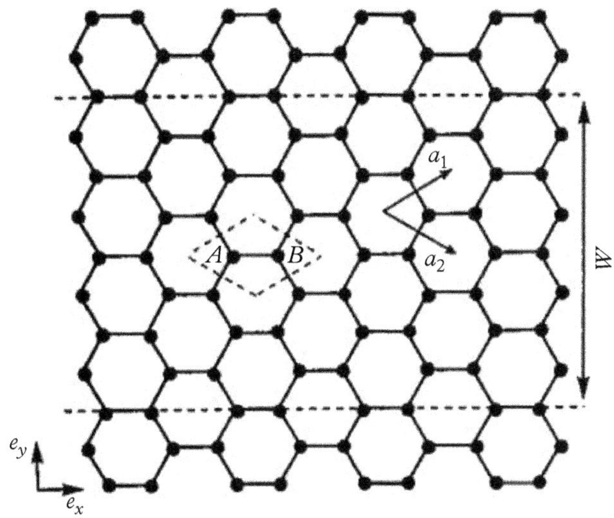

Figure 8.9 *Schematic picture of a graphene lattice. The dashed lines display the boundaries of the Bravais lattice unit cell. (Figure reproduced from Journal of Low Temp. Phys., 2018)*

(see appendix C). Carbon nanotubes can also be viewed as a two-dimensional system which is obtained from graphene by rolling it along some direction (rolled-up glinders). As mentioned above, graphite is a set of graphene layers interacting through van der Waals forces. As for fullerene (Fig. 8.9), it is also a carbon allotrope, that is, it belongs to the family of systems containing carbon ions, all of which have different structures. Fullerene (C_{60}) is a cluster (the properties of such clusters are described below: also see chapter 2) which can be viewed as a zero-dimensional system obtained from graphene by rolling it up and then introducing some pentagons into the hexagonal lattice.

The electronic spectrum of graphene is its most remarkable feature. It is drastically different from that for usual solids and is described by the following linear law:

$$E(K) = \pm v_F K \tag{8.1}$$

This expression is valid at small values of the momentum K. The sign '$\pm$' corresponds to the electron and hole parts of the spectrum (Fig. 8.10a). The higher-order corrections can be neglected at small K; more specifically, at $K \ll K_B$, K_B are corresponding reciprocal lattice vectors. Such a cut-off vector K_C for graphene is $K_C \simeq 0.25\ \text{nm}^{-1}$. This unusual dispersion law (8.1) was obtained theoretically by Wallace (1947), McClure (1957), and Slonczewski and Weiss (1957), that is, before its experimental production.

The parameter v_F describes the slope of the electronic dispersion line and is called 'Fermi velocity'; its value is $v_F \simeq 10^9$ cm/sec.

The structure (Fig. 8.10) represents the combination of two triangular lattices with two ions in the unit cell. In the reciprocal lattice, it corresponds to two points in the

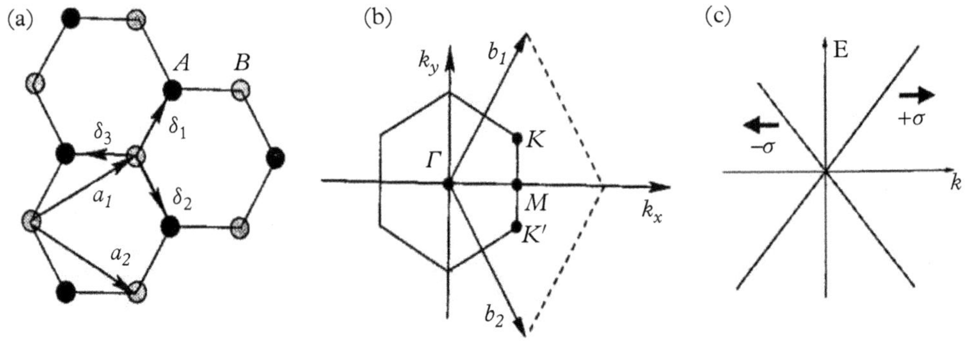

Figure 8.10 *(a) The lattice and Brillouin zone; K, K′ are the Dirac points. (b) Energy dispersion:* $E = v_F|\boldsymbol{k}|$.

Brillouin zone, so-called Dirac points. The presence of these points can be treated as a double degeneracy, since it reflects the presence of two valleys in the momentum space. This double degeneracy can be described as a pseudospin quantum number $\vec{\sigma}$ and, as a result, we now deal with the Dirac massless equation, which is well known in quantum electrodynamics (see e.g. Berestetskii *et al.*, 1971):

$$iv_F\vec{\sigma}\nabla\Psi = E\Psi \tag{8.2}$$

The electronic wave function can be constructed, as usual for carbon-based compounds, as a sum of atomic orbitals. As a result, one can introduce the σ-states, which are located in the planes and provide strong σ-bonds (this leads to the formation of the rather strong structure of graphene), and π-states, which are analogous to the states in the conduction energy band.

It is remarkable that the valence and the conduction bands of graphene do not overlap and there is no energy gap between them. They cross at a single point, and this leads to a number of peculiar properties.

Let us now discuss the question about the possibility of an appearance of the superconducting state in graphene or in the related structures. Let us first note that the situation appears to be similar to that for fullerides (see section 8.1.4). Indeed, fullerides are not superconducting but this state can be induced by intercalation. Similarly, graphene is not intrinsically a superconductor except in the case of bilayer graphene (see below), but it can be transferred into the superconducting state by some external factors. Let us discuss some possibilities.

As noted above, graphene is built from the benzene rings with the hydrogen ions stripped off. The removal of these ions is not beneficial for the pairing. Indeed, as we know very well from the recent progress with the hydrides (see section 7.2), the hydrogen ions provide the presence of high-frequency phonon modes and this presence leads to the formation of the superconducting state, even with a high value of T_c. In connection with this, one can propose the following for the graphene-like systems.

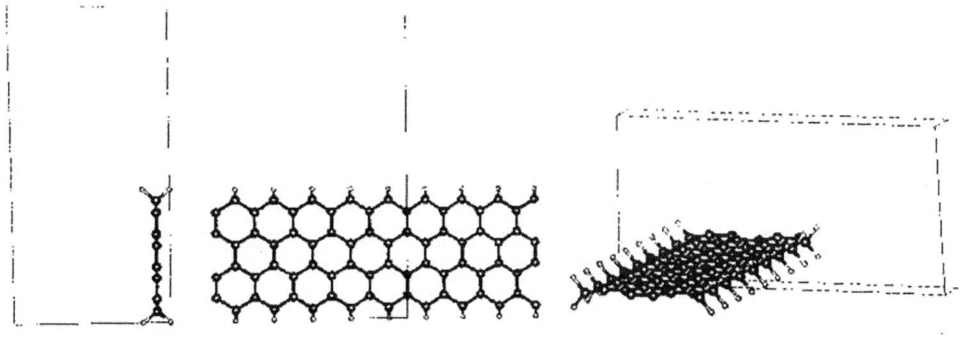

Figure 8.11 *Graphene nanoribbon of thickness $M = 3$.*

Firstly, we can search for the superconducting state in graphene nanoribbons (GNRs; Fig. 8.11). The ribbons represent two-dimensional stripes with the variable width M ($M = 1, 2, 3, \ldots$) depending on the number of benzene rings in the transverse direction. The edges contain hydrogen ions, which play a role similar to that in hydrides. One may think that it is better to use nanoribbons with the smallest width ($M = 1$). Indeed, in such a case, the contribution of the hydrogen ions looks optimum. However, in this case, one must take into account the quantisation in the transverse direction. As a result, the energy gap for this direction becomes rather large and effectively we are dealing with a one-dimensional system. Then the fluctuations can destroy the pairing (see section 3.6). The problem could be overcome by increasing the width of the GNR and studying GNR2 and GNR3 systems (the number in the end corresponds to the width, which is represented by the M; see above), or by placing the hydrogen ions outside of the plane. Another possibility is similar to that studied for organic polymers (section 2.5). Namely, one can build double- (or triple-)layer ribbons; the small charge transfer between the layers diminishes the impact of the fluctuations.

When the hydrogen ions are located above (or below) the graphene plane, the system is called graphene. In this case, the valence state of the in-plane carbon ion becomes different: instead of a π-electron orbital, a hydrogen bond is created. Another possible way to create the superconducting state in graphene is to intercalate the material with some light atoms.

This case could be analogous to the high-T_c superconducting state in MgB_2 (see section 7.1.3). The presence of light boron ions leads to the appearance of high-frequency phonon modes and, in combination with strong electron–phonon coupling, leads to a high value of T_c. A similar situation occurs with doped diamond, which is another carbon allotrope. Doping by boron makes it possible to transform diamond into a metal (Ekimov *et al.*, 2004). The synthesis was performed at high pressure and high temperatures. As a result, this discovery has resulted in the appearance of the area of superhard superconducting materials (see e.g. Dubitsky *et al.*, 2005; Blasé *et al.*, 2009). The boron in this case served as the source of the hole doping (similarly to the MgB_2

case). One should note that the problem of the superconducting state in semiconductors has a long history (Gurevich *et al.*, 1962; Cohen, 1964, 1969; Gor'kov, 2017; Gastiasoro *et al.*, 2019). The intercalation of boron transformed diamond into a metal which displays a relatively high value for the critical temperature: $T_c \simeq 4$ K. Theoretical studies (Boeri *et al.*, 2004; Xiang *et al.*, 2004) have demonstrated that the boron ions, in addition to providing hole doping, enable the presence of high-frequency optical modes with noticeable electron–phonon coupling ($\lambda \simeq 0.4$ for $\sim 2.8\%$ constant of B); the observed value of T_c is caused by these factors. An interesting theoretical study was also carried out by Savini *et al.* (2010). They considered graphene, which is a fully hydrogenated graphene with an sp^3-bonded system, as a two-dimensional analogue of diamond. The reduced dimension leads to larger electronic density of states, compared to the three-dimensional case; this is typical for size-quantising systems. As a result, boron doping would lead to even higher values of T_c (~ 80 K) than for the doped diamond.

Another possible way to induce the superconducting state in graphene is to use the proximity effect. As is known, the proximity effect (see section 3.4), that is, an appearance of superconductivity in the normal metal that is caused by the tunnelling constant (S–N) between the metal and the superconductor, occurs between systems which are weakly connected (tunnelling) without any modification of these materials. According to an analysis by Chevkez *et al.* (2018), Pb particles placed on graphene layers satisfy this condition, so that these Pb islands continue to be freestanding and their electronic structure is unperturbed by the conduct. This should make it possible to induce the superconducting state in graphene with the use of the proximity effect with the superconducting Pb particles. The effect can be detected by the scanning tunnelling microscopy technique. Instead of Pb, one can use metallic nanoclusters. Some of these (see section 8.3) are characterised by high values for the critical temperature (e.g. $T_c \simeq 120$ K for Al_{66} nanoclusters; they contain $N = 198$ delocalised electrons). If such clusters can be placed on the graphene surface without perturbance of their electronic structure, then the relatively high-T_c superconducting state can be induced in the graphene. The proximity contact can also be effective in the case when a metallic particle (or nanocluster) is placed between two graphene layers.

Speaking of bilayer systems, one should also mention the interesting studies on twisted bilayer graphene (see e.g. the reviews by Rozhkov *et al.*, 2016 and Das Sarma *et al.*, 2011). Such a system is made by stacking one graphene layer on top of another with some twist angle. This is an example of the so-called Moire pattern. It is remarkable that, at the specific twist angle $\theta = 1.05^\circ$ (it can be called the 'magic' angle), the sample displays the superconducting state with $T_c \simeq 1$K. This value of the critical temperature is relatively high. Indeed, according to theoretical studies (e.g. Morell *et al.*, 2010; Bistritzer and MacDonald, 2011; Heikkila and Volovik, 2016; Volovik, 2018), the electronic energy band near this twist angle is flat and this leads to a low value for the Fermi energy ($E_F \sim 10$ meV), so that the value of T_c is high relative to E_F.

At present, graphene and graphene-like systems like twisted bilayer graphene are undergoing intensive studies, and one can expect a number of interesting developments in the near future.

8.2 Small-Scale Organic Superconductivity

8.2.1 BETS

Let us start with the organic superconductor $(BETS)_2GaCl_4$, where BETS stands for bis(ethylenedithio)tetraselenafulvalene (Kobayashi *et al.*, 1995). The bulk sample of this superconductor has a $T_c \simeq 8$ K and a layered structure. The building blocks are the BETS molecules (Fig. 8.12) and, in addition, the $GaCl_4$ molecules, which accept electrons. The crystal is formed by double-stacked BETS molecules, and the structure contains the chains formed by these units. The electron delocalisation and, correspondingly, the bulk conductivity occurs because the two BETS molecules act as donors, transferring electric charge to the acceptor $GaCl_4$ molecule; as a result, a half-filled orbital is formed.

A study performed with the use of scanning tunnelling microscopy shows that the sample and the chains contain molecular islands of different sizes (Clark *et al.*, 2010). The scanning tunnelling microscopy technique allows to determine the presence of the superconducting energy gap and measure its value. It turns out that the gap can be observed even for two pairs of molecules. It means that the pairing, according to Clark *et al.* (2010), can be observed for the system which contains only four (!) molecules. Such a small scale is interesting in itself, but is also promising for future applications in nano-electronics.

8.2.2 Aromatic Molecules

Aromatic molecules comprise another group of small-scale organic systems that display Cooper pairing. Their properties and the manifestations of their pair correlations

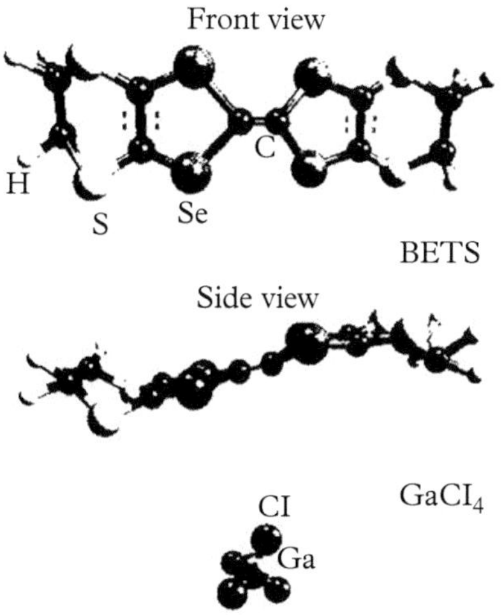

Figure 8.12 *Chemical structure of bis(ethylenedithio)tetraselenafulvalene and $GaCI_4$. (Reproduced from Clark et al., 2010)*

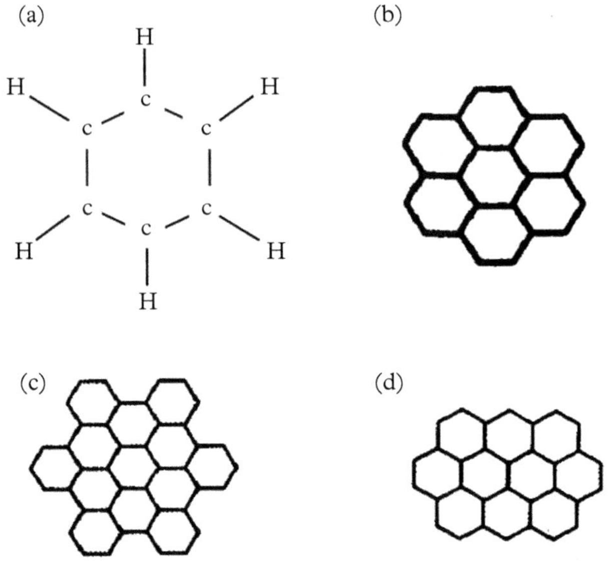

Figure 8.13 *Molecules with a delocalised electron: (a) benzene; (b) coronene; (c) hexabenzocoronene; (d) ovalene.*

exemplify those of low-dimensional nanosystems. In reality, aromatic molecules could also be discussed in the next section, which is devoted to the pairing in the nanosystems, so this topic is on the overlap of two fields: organic superconductivity and nano-physics.

Figure 8.13 displays examples of the so-called conjugated hydrocarbons. As one can see from this figure, the benzene molecule (see appendix C) is the major building block of these structures. We will focus below on the π-electron states; these states can be separated (see appendix C) and form a group of levels that are similar to the conduction band in solids.

The π-electrons move in a field created by the so-called σ-core; the σ-core is formed by an ionic system and other electrons. Each carbon atom supplies one such delocalised π-electron (e.g. the hexbenzcoronene molecule contains 32 π-electrons). As noted above, the π-electrons are similar to the conduction electrons in metals, only, in this case, we are dealing with a finite Fermi system. As π-electrons can form Cooper pairs, the molecules are in the superconducting state (Kresin, 1967; Kresin *et al.*, 1975; Kresin and Ovchinnikov, 2020). Of course, an isolated aromatic molecule is a small system where features such as conductivity and zero resistance do not exist. But pair correlation similar to that in usual superconductors exists and manifests itself in various phenomena.

As an example, we consider the coronene molecule, $C_{24}H_{12}$ (Fig. 8.13c); it contains 24 π-electrons. Firstly, let us evaluate the electronic energy spectrum of the molecule (in the absence of pair correlation). We use the method described in appendix C for the benzene molecule. This method is based on the symmetry of the molecule (D_{6h}) and

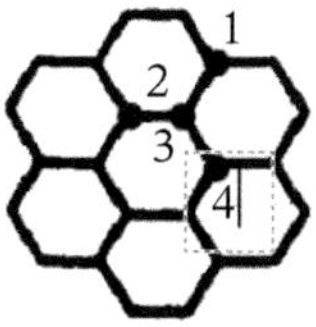

Figure 8.14 *A coronene molecule; the set of four carbon ions is indicated (from Kresin and Ovchinnikov, 2020).*

the analogy of its periodicity with respect to the rotation with the period $\varphi_s = \pi/3$ and that of the Bloch function in solids.

For the coronene molecule, let us select the set of four carbon ions (Fig. 8.14). One can see that, by consecutive rotation of this set relative to the axis with the period $\pi/3$, we cover the whole structure of the molecule. This allows us to seek the total electronic wave function as the sum

$$\Psi(\boldsymbol{r}) = \sum_s e^{im\varphi^s} \left\{ \alpha\Psi^{(1)} + \beta\Psi^{(2)} e^{i\varphi_2} + \gamma\Psi^{(3)} e^{i\varphi_3} + \mu\Psi^{(4)} e^{i\varphi_4} \right\}_s \tag{8.3}$$

Here $\varphi^s = s\widetilde{\varphi}$, $\widetilde{\varphi} = \pi/3$; $s = 1,2,\ldots,\psi^{(i)}$ corresponds to the i-th ion in the set, and $\varphi_2, \varphi_3, \varphi_4$, are the relative phase factors. Expression (8.3) is similar to that for the benzene molecule, but it is also different, because of the more complicated structure of the molecule C_6H_6.

Substituting expression (8.3) into the Schrödinger dinger equation, we obtain the secular equation

$$\|S_{ik}\| = 0, \;\; i,k = 1,2,\ldots,4 \tag{8.4}$$

Here $S_{11} = \mathrm{E} - 2\mathrm{T_z}\cos m\widetilde{\varphi}$, $S_{ii} = \mathrm{E}$, (we have selected $\varepsilon_0 = 0$), $S_{12} = S^*_{21} = -\exp(i\varphi_2)$ $\mathrm{T_z} = S_{2k} = S^*_{k2} = -\exp[i(\varphi_k - \varphi_2)]$; $S_{34} = S_{43} = -\exp[im\widetilde{\varphi} + \varphi_4 - \varphi_3]$, $S_{13} = S_{31} = S_{14} = S_{41} = 0$.

Equation (8.4) can be written in the form

$$Z^4 - 2Z^3\cos m\widetilde{\varphi} - 4Z^2 - 4Z\cos m\widetilde{\varphi} + \left(1 + 4\cos^2 m\widetilde{\varphi}\right) = 0$$

Here $Z = E/T_z$.

The energy spectrum (in units T_z) is shown in Fig. 8.15. It is interesting that the spectrum is symmetrical relative to the chemical potential ($\varepsilon_0 = 0$). One can see also the distribution of all 24 π-electrons.

The value of the parameter T_z can be taken from the study of the benzene molecule ($\simeq$350 meV). The excitation energy refers to the chemical potential placed in the middle of the HOMO–LUMO interval (where LUMO stands for singly occupied molecular orbital) and appears to be of the order of ~175 meV. This value is noticeably below the vibration energy of the hydrogen ion (~200–250 meV). The presence of high-frequency

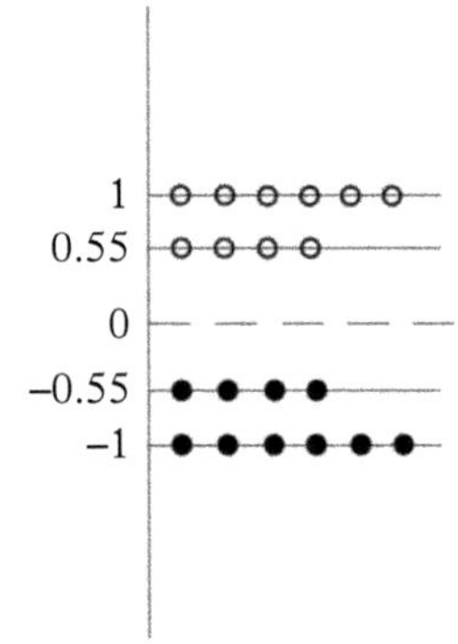

Figure 8.15 *Electronic energy spectrum for a coronene molecule.*

hydrogen ions is the key factor which leads to the pairing of π-electrons being perfectly realistic. The situation is similar to that in the hydrides (see section 7.2), where the pairing is caused by the interaction of the conduction electrons with high-frequency hydrogen phonon modes.

Let us discuss the manifestations of the pairing, that is, describe the experimental observations that, when analysed, lead to the conclusion that the π-electrons do, indeed, form Cooper pairs.

The energy spectrum plotted in Fig. 8.15 is evaluated directly from the Schrödinger equation, without taking into account the pairing correlation. One can see that the quantities $\omega_{o-o'}$ and $\omega_{o'-o''}$, which correspond to the excitations of the first and second levels, respectively, are of the same order of magnitude. However, according to the experimental data, the picture is different. For a coronene molecule, $\omega_{o-o'} \simeq 22.5 \times 10^3\ \text{cm}^{-1}$ and $\omega_{o'-o''} \simeq 5.5 \times 10^3\ \text{cm}^{-1}$ (Clar, 1964). This large difference can be explained by the pair correlation and by the presence of the corresponding energy gap, which affects the energy spectrum and leads to a drastic increase in the value of the lowest level.

It is interesting that, for the ion $C_{24}H_{13}{}^{-}$, which contains an odd number of electrons, one can observe large reduction of the $0-0'$ separation (Hotjtiak and Zandstra, 1960). This is natural, since this ion contains an unpaired electron (the odd–even effect).

Another manifestation is related to the behaviour of the molecule in an external magnetic field. We know that the orbital part of the total Hamiltonian describing the impact of such a field,

$$\widehat{H} = \widehat{H}_p + \widehat{H}_d$$

$$\widehat{H}_p = \frac{ie}{2m} \int \psi^+ (\nabla \boldsymbol{A} + \boldsymbol{A}\nabla) \psi d\mathbf{r} \tag{8.5}$$

$$\widehat{H}_d = -\frac{e^2}{2m} \int \psi^+ \psi \mathrm{A}^2 d\mathbf{r} \tag{8.5'}$$

contains two terms. The first term corresponds to the paramagnetic part of the magnetic moment, whereas the second term describes the diamagnetic part.

The paramagnetic term can be written in the form

$$\langle \widehat{H}_p \rangle = \frac{ie}{2m} H_i \sum_{\lambda,\lambda'} \rho'_{\lambda,\lambda'} \widehat{M}_{i;\lambda,\lambda} \tag{8.6}$$

where $\widehat{M}_i = -i[\mathbf{r}\nabla]_{\mathrm{i}}$ (see eqn (8.5)), and ρ' is the change in the density matrix (Migdal, 1960, 1967) caused by the perturbation $\widehat{V}$ (in our case, $\widehat{V} = (ie/4mc)\,\widehat{\mathbf{H}}\widehat{\mathbf{M}}$):

$$\rho'_{\lambda,\lambda'} = \frac{\left(\xi_\lambda, \xi_{\lambda'} + \Delta^2 - \varepsilon_\lambda \varepsilon_{\lambda'}\right) V_{\lambda\lambda'}}{2\varepsilon_\lambda \varepsilon_{\lambda'} \left(\varepsilon_\lambda + \varepsilon_{\lambda'}\right)} \tag{8.6$'$}$$

The total orbital magnetic susceptibility, which can be evaluated with use of the relation $\chi = -\partial^2 (\Delta E)/\partial H^2$, is the sum of two terms, so that $= \chi_P + \chi_D$, where

$$\chi_P = \frac{e^2}{8m^2} \sum_{\lambda,\lambda'} \frac{\xi_\lambda, \xi_{\lambda'} + \Delta^2 - \varepsilon_\lambda \varepsilon_{\lambda'}}{2\varepsilon_\lambda \varepsilon_{\lambda'} \left(\varepsilon_\lambda + \varepsilon_{\lambda'}\right)} \left| \widehat{M}_{z;\lambda\lambda'} \right|^2 \tag{8.7}$$

If the order parameter $\Delta = 0$, we obtain the expression for the so-called Van Vleck paramagnetism (1932):

$$\chi_P = \frac{e^2}{8m^2} \sum_{\lambda,\lambda'} \frac{1}{|\xi_\lambda| + |\xi_{\lambda'}|} \left| \hat{M}_{z;\lambda\lambda'} \right|^2; \quad \xi_\lambda \xi_{\lambda'} < 0 \tag{8.7$'$}$$

We consider the case when the field is perpendicular to the plane of the molecule (the z-axis). Expression (8.7) follows directly from eqns (8.6) and (8.6′). The expression for the diamagnetic term can be easily obtained from eqn (8.5′). As is known, in usual metals, the terms χ_P and χ_P almost cancel each other (one can observe only a small, residual Landau diamagnetism). As for the superconductors, the presence of the gap parameter leads to the suppression of the paramagnetic term and to a well-known anomalous diamagnetism (Meissner effect). As for molecular systems, the study of their properties requires evaluation of both terms, χ_P and χ_D, since they are of the same order of magnitude. However, the aromatic molecules of interest display strong diamagnetic response (e.g. for coronene, $\chi \equiv \chi_D \simeq -2.4 \times 10^{-4}$ cm^3/mol), which means that the paramagnetic term is small, because of the impact of the pairing order parameter Δ.

The manifestation of pairing in aromatic molecules is similar to that for nucleons (protons and neutrons) in atomic nuclei; this is a well-established concept in nuclear physics (see e.g. Ring and Schuck, 1980, and Broglia and Zelevinskii, 2013). This is not surprising because, in each case, we are dealing with a finite Fermi system. Both systems (nuclei and aromatic molecules) interact similarly with radiation. As we known, spectroscopy of usual bulk superconductors is greatly affected by the opening of the energy gap (at $T = T_c$). As for the systems we discuss in this section, their finite size leads to the discrete nature of the spectrum even in the absence of pair correlation. However, as discussed above, because of the pairing, the energy spacing between the ground state and the first excited state (the so-called 0–0′ transition) greatly exceeds the energy of the 0′–0″ transition between the first and the second excited states. In addition, one can observe the odd–even effect (see e.g. Matveev and Larkin, 1997), that is, the lowest value of the frequency of the absorbed light is much smaller for the system with an odd number of electrons, because of the presence of one unpaired particle.

Note that the pair correlation of nucleons leads to a decrease in the value of the momenta of inertia relative to the solid configuration. This phenomenon is similar to anomalous diamagnetism because, as we know, the transition to the rotating coordinate system is equivalent to the appearance of an external magnetic field (see e.g. Landau and Lifshitz, 2000).

The key ingredient leading to the pairing of π-electrons in the molecules of interest is the presence of hydrogen ions with high values of the phonon frequency for the corresponding modes. The scenario is similar to that for the high-T_c hydrides (section 7.2) or some graphene-like systems (e.g. nanoribbons; see section 8.1.5). Since the pairing is caused by the electron–phonon coupling to these hydrogen modes, the value of the gap parameter and, therefore, the structure of the electronic spectrum are sensitive to $H \to D$ isotope substitution. Such an isotope effect should be observed, and this observation would reflect the presence of the pairing phenomenon for the π-electron system.The pairing of π-electrons impacts the properties of biologically active systems. This aspect will be discussed later on (see section 10.3).

The π-electrons in the aromatic molecules form a two-dimensional finite Fermi system. They represent the superconducting nanosystem. The pairing can be observed also in three-dimensional nanosystems, more specifically, in some metallic nanoclusters. This phenomenon will be discussed in the next chapter.

8.3 Pairing in Nanoclusters, and Nano-Based Superconducting Tunnelling Networks

This section contains a description of the superconducting state in metallic clusters. It turns out that the pairing can be strong, if the cluster parameter satisfies certain, but perfectly realistic, conditions. Clusters constitute a new family of high-temperature superconductors. In principle, T_c can be increased to room temperature.

An advanced study of nanoparticles was carried out by Tinkham and his collaborators (see e.g. Tinkham *et al.*, 1995; see also the review by von Delft and Ralph, 2001); the tunnelling spectroscopy was employed. However, these nanoparticles were relatively large, ~50–100 Å in diameter, that is, they contain ~10^4–10^5 electrons. Here we focus on much smaller nanoparticles, clusters containing ~10^2 delocalised electrons. The interest in such systems was prompted by the milestone discovery (Knight *et al.*, 1984; also see Friedel, 1999) of the shell structure of the spectrum in metallic nanoclusters.

8.3.1 Clusters, and Shell Structures

Main parameters. Clusters are the aggregates of atoms (or molecules) that have the structure A_k, where k is the number of atoms A (e.g. Na_k, Ga_k); see Fig. 8.16. The number k of atoms can vary from 2 to any large value, and therefore the study of clusters makes it possible to track the evolution from isolated atoms to bulk solids.

Metallic clusters contain delocalised electrons, which are formed from the valence electrons of the atoms; the number N of delocalised electrons is the main parameter of the cluster. For simple monovalent metallic clusters, such as Na_k and Li_k, we have $N = k$. For more complicated clusters, $N = vk$, where v is the number of valence electrons per atom; for example, $N = 135$ for Al_{45} clusters, because each Al atom contains three valence electrons. With the use of mass spectroscopy, specific clusters can be selected.

Energy shells. The shell structure is a key feature of nanoclusters (see e.g. reviews by de Heer, 1993; Kresin and Knight, 1998; Frauendorf and Guet, 2001; and Issendorff, 2011). As was noted above, Knight and his collaborators (1984) discovered that the electronic states in many clusters form energy shells similar to those in atoms. The electronic states in atoms are classified by values of orbital momenta (s, p, ... shells). A similar picture is valid for clusters, although, unlike atoms, clusters do not have point-like positive centres; the ions form a positive core (see Fig. 8.16).

The shell structure in metallic clusters is also similar to those in atomic nuclei: proton and neutron states form energy shells. This is a well-known concept in nuclear physics (see e.g. Ring and Schuck, 1980). Shell structure was discovered initially for simple monovalent alkali clusters, such as Na and K (see e.g. the work by Knight *et al.*, 1984 and Wriggle *et al.*, 2002). Later, this phenomenon was also observed for Al, Ga, In, Zn, and Cd clusters (Schriver *et al.*, 1990; Pellarin *et al.*, 1993; Baguenard, 1993;

Figure 8.16 *Cluster.*

Katakuse *et al.*, 1986; Ruppel and Rademann, 1992; Lermé *et al.*, 1999). It is interesting that the presence of energy shells is not a universal feature. For example, for Nb clusters, the spectrum does not display a shell structure.

The presence of shell structure is manifested in the appearance of so-called magic numbers with $N = N_m$ (e.g. $N_m = 20, 40, 58, \ldots, 138, 168, \ldots$). Such 'magic' clusters possess completely filled electronic shells and, like 'inert' atoms, are the most stable (Fig. 8.17).

It is essential that 'magic' clusters have spherical symmetry with good accuracy. Correspondingly, their electronic states are classified by the orbital momentum l and the radial quantum number n. Note also that, because of their finite size, the electronic energy spectrum is discrete.

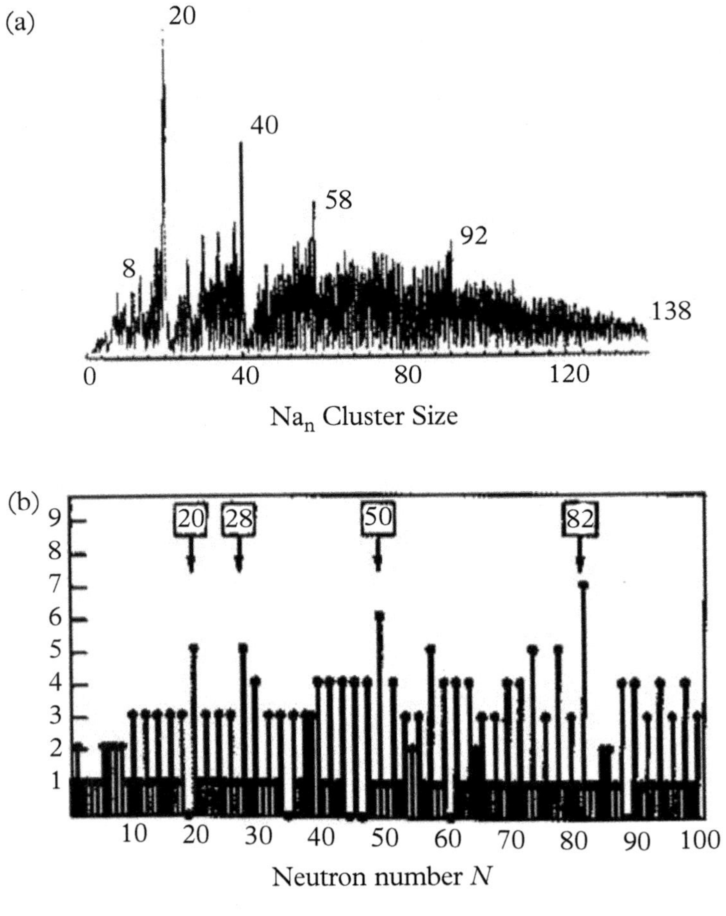

Figure 8.17 *The peaks in stability correspond to 'magic' numbers: (a) in clusters; (b) in nuclei.*

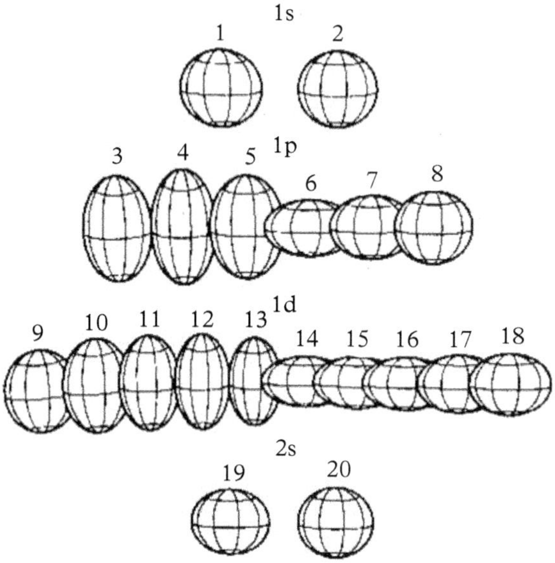

Figure 8.18 *Cluster shapes for various values of N.*

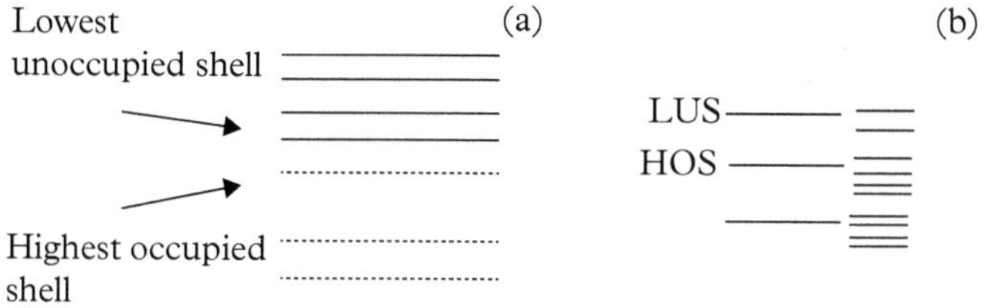

Figure 8.19 *Energy spectra for (a) a 'magic' cluster and (b) in the case of a nearly complete shell.*

If the shell is not complete, the cluster undergoes a Jahn–Teller distortion, and its shape becomes ellipsoidal. More specifically, as electrons are added to the lowest unoccupied shell (LUS), the shape deforms from spherical and becomes prolate. The energy levels are the four degenerate levels $(m, -m)$ and $(s_z, -s_z)$. The deformation splits the degenerate levels into sublevels (Fig. 8.18) labelled by the projection of the orbital momentum (m). An interesting and unique correlation between the number of electrons and the energy spectrum can therefore be observed; this is a remarkable feature of metallic clusters.

Half-filling and its crossing. If we continue to add electrons, then the shape initially continues to be prolate but, after crossing the half-shell threshold, one can observe the transfer to the oblate configurations (Fig. 8.19). This conclusion follows from ab initio calculations showing that, after crossing the half-filling, the energy minimum corresponds to the oblate configuration. However, in reality the picture near half-filling is more complicated and represents a superposition of the prolate and oblate structure (see appendix B). First of all, indeed, it is hard qualitatively to envision such an abrupt transformation (from prelate to oblate configuration) of a rather large system driven by adding just one electron (e.g. $Nb_{27} \rightarrow Nb_{28}$). Moreover, the conclusion about the sharp transition contradicts the experimental data on spectroscopy. Indeed, it is known that the adsorption spectrum peaks at the plasma frequency (see the review by Kresin, 1992). That is, for the 'magic' clusters, one can observe just one peak in the spectrum. As for the specific ellipsoidal structure of the prolate configuration and its symmetry (see Fig. 8.19), the intensities of the observed two peaks should be different and satisfy the ratio 2:1. For the oblate structure, the ratio should be the inverse (1:2). However, near the half-filling, the intensities of these peaks are almost equal (Borggreen, 1993).

In reality, the transition in the half-filling region is of a continuous nature. More specifically, the prolate and oblate structures have close values of energy and form the quasi-resonance states (see appendix B). This means that the cluster's state is the superposition of the prolate and the oblate structures. In the time-dependent picture, which, qualitatively, is more transparent, the cluster oscillates between these two structures (Kresin, 2008). This scenario can manifest itself in the possible picture of the pairing for some clusters (see section 8.3.2).

8.3.2 Pair Correlation

The presence of a shell structure leads to the possibility of the electrons displaying the phenomenon of pair correlation; in other words, the electrons form Cooper pairs. This possibility was discussed by Knight (1987) and especially by Friedel (1992). Below, we follow the approach developed by Kresin and Ovchinnikov (2006). Afterwards, various aspects of pairing in clusters were discussed by Croitoru *et al.* (2011), Garcia-Garcia *et al.* (2011), and Lindenfeld *et al.* (2011).

The qualitative picture. As we know, Cooper pairs in bulk superconductors are formed by electrons with opposite momenta and spins. For the clusters of interest, the momentum is not a quantum number, and the Cooper pairs are formed by electrons with opposite projections of angular momenta (m, $-m$). The pairing strongly affects the cluster's energy spectrum. We focus on 'magic' clusters; the impact is also strong for clusters with slightly incomplete shells, where the excitation energy in the absence of pairing can be rather small. As a whole, one should stress that the strength of pair correlation varies for different clusters. Correspondingly, the critical temperature T_c and the energy gap are strongly dependent upon the cluster's parameters, its shape, the strength of the coupling, and so on. Some specific cases will be described below. It is important to stress that the superconducting state is not a universal feature of all metallic clusters (this is not surprising; the same is true for bulk superconductors). However, and this is main conclusion, in some special, but perfectly realistic situations, one can obtain very large values of T_c.

Qualitatively, high values of T_c can be understood in the following way: consider the highest occupied shell which is highly degenerate (large l; the degeneracy $g = 2(2\,l + 1)$); this means that it contains many electrons (Fig. 8.19); this feature can be viewed as a sharp peak in the density of states at the Fermi level. The situation is similar to that introduced and described by Labbe *et al.* (1967) for A-15 materials; the presence of a peak in the density of states (the van Hove singularity) results in a noticeable increase in T_c; see section 7.1.2.

Let us discuss one more point. As was noticed by Anderson (1959a), the superconducting state in small particles can manifest itself if the pairing energy gap ε exceeds or is comparable to the energy spacing caused by the finite size of the particle, that is, if $\varepsilon \gtrsim \delta E$, where $\delta E \approx E_F/N$ is the average energy level spacing, and ε is the energy gap. Based on the numbers $E_F \approx 10^5$K and $\varepsilon \approx 2T_c \approx 10$ K, one can conclude that the number of electrons, *N*, should be relatively large: $N \approx 10^4$.

As a result, if we are concerned with small nanoclusters, $N \approx 10^2$–10^3, it might seem that they do not display superconducting properties, because the average level spacing greatly exceeds the pairing energy gap. But such an estimation can be challenged. Indeed, for larger T_c (e.g. $T_c \approx 10^2$ K for cuprates; for nanoclusters, T_c can be even higher: see below), the energy gap is much larger than 10 K, and this leads to a noticeable decrease in the limiting value of *N*. More importantly, the estimation by Anderson (1959a) is based on the assumption that the energy levels are approximately equidistant. This assumption can be justified for nanoparticles which do not display shell structure, for example, for Pb nanoparticles (see e.g. Zolotavin and Guyot-Slonnest, 2010). However, the discovery of the shell structure in many clusters totally invalidates this assumption: the pattern of electronic states is very different from an equally spaced level distribution (see e.g. Fig. 8.19). Instead, they contain highly degenerate energy levels, or groups of very close levels.

In many aspects, the picture of pairing is similar to that in atomic nuclei (see section 10.1). In both cases (nuclei and clusters), we are dealing with finite Fermi systems and shell structures. The pairing states are labelled by similar quantum numbers $(m, -m)$. The manifestation of pairing also has similarities (see below). However, the origin of the phenomenon is different: the pairing in clusters is caused by the electron–vibration coupling, that is, the mechanism is similar to that in usual bulk superconductors. Therefore, the pairing scenario in nanoclusters represents an interesting intermediate case between the nuclei and solids; it has some overlap with the phenomena observed for these entirely different states of matter.

Main equations and critical temperature. Let us write down a general equation describing the pairing in a metallic cluster. Such an equation for the pairing order parameter $\Delta(\omega\text{n})$ has the following form:

$$\Delta(\omega_n) Z = \eta \frac{T}{2V} \sum_{\omega_{n'}} \sum_{s} D(\omega_n - \omega_{n'}) F_s^+(\omega_{n'}) \tag{8.8}$$

Here $\omega_n = (2n+1)\pi T$; $n = 0, \pm 1, \pm 2, \ldots$;

$$D\left(\omega_n - \omega_{n'}, \widetilde{\Omega}\right) = \widetilde{\Omega}^2 \left[(\omega_n - \omega_{n'})^2 + \widetilde{\Omega}^2\right]^{-1}$$

$$F_s^+(\omega_{n'}) = \Delta(\omega_{n'}) \left[\omega_{n'}^2 + \xi_s^2 + \Delta^2(\omega_{n'})\right]^{-1} \tag{8.8$'$}$$

are the vibrational propagator and the Gor'kov pairing function, respectively; $\xi_s = E_s - \mu$ is the energy of the *s*-th electronic state referred to the chemical potential μ, where $\widetilde{\Omega}$ is the characteristic vibrational frequency; V is the cluster volume; $\eta = \langle I \rangle^2 / M\widetilde{\Omega}^2$ is the so-called Hopfield parameter (see e.g. Grimvall, 1981); $\langle I \rangle$ is the electron–ion matrix element averaged over electronic states involved in the pairing; M is the ionic mass (we are using the harmonic approximation); and Z is the renormalisation function (see section 2.6). We employ the method of the thermodynamic Green's function, which similar to that for the phonon mechanism in bulk solids (see section 2.2 and the appendix).

The electron–vibration interaction is considered the major mechanism of pairing. The general equation (8.8) explicitly contains the vibrational propagator. We use this equation because we want to go beyond the restriction $T_c \ll \widetilde{\Omega}$ (the weak coupling approximation). Equation (8.8) looks similar to that in the theory of strong coupling superconductivity (see section 2.1, eqns (2.16) and (2.17)), but is different in two key aspects. Firstly, it contains a summation over discrete energy levels E_S, whereas, for a bulk superconductor, one integrates over a continuous energy spectrum (over ξ). Another important difference is that, unlike the case for a bulk superconductor, here we are dealing with a finite Fermi system, so that a number of electrons N is fixed. As a result, the position of the chemical potential differs from the Fermi level E_F and is determined by the values of N and T. Specifically, one can write

$$N = \sum_{\omega_n;s}\sum_{s} G_s(\omega_n)\, e^{i\omega_n\tau}\ _{|\tau\to 0} = \sum_{s}\left(u_s^2\varphi_s^- + \nu_s^2\varphi_s^+\right) \tag{8.9}$$

Here $G_S(\omega_n)$ is the thermodynamic Green's function

$$u_s^2, v_s^2 = 0.5\left(1 \mp \frac{\xi_s}{\varepsilon_s}\right); \varphi_s^{\mp} = \left[1 + \exp\left(\mp\frac{\varepsilon_s}{T}\right)\right]^{-1};$$

$$\varepsilon_s = \sqrt{\xi_s^2 + \varepsilon_{0;s}^2} \tag{8.9$'$}$$

where $\varepsilon_{0;s}$ is the pairing gap parameter for the *s*-th level, and $\varepsilon_{0;s}$ is the root of the equation $\varepsilon_{0;s} = \Delta(-i\varepsilon_s)$. Since $\xi_s = E_s - \mu$, eqn (8.9) determines the position of the chemical potential for the given number of electrons N as well as the dependence $\mu(T)$. These dependences are important for the study of pairing in clusters. Equations (8.8) and (8.9) are the main equations of the theory making it possible to evaluate the critical temperature and the energy spectrum.

In a weak coupling case ($\eta/V \ll 1$ and, correspondingly, $\pi T_c \ll \widetilde{\Omega}$), one should put in eqn (8.8), $D = 1$, recovering the BCS scheme. Note also that the Coulomb term μ^* can be included in the usual way. As was noted above, eqn (8.8) contains a summation over all discrete electronic states. For 'magic' clusters which have a spherical shape, one can replace summation over states by summation over the shells $\sum_s = \sum_j g_j$, where g_j is the shell degeneracy $g_j = 2(2l_j + 1)$, and l_j is the orbital momentum. If the shell is incomplete, the cluster undergoes a Jahn–Teller deformation, so that its shape becomes ellipsoidal, and the states '*s*' are classified by their projection of the orbital momentum $|m| \leq l$, so that each level contain up to four electrons (for $|m| \geq 1$).

Based on eqn (8.8), one can evaluate the critical temperature: at $T = T_c$, one should put $\Delta = 0$ in the denominator of the expression (8.8′). Note also that, since $\lambda_b = \eta\nu_b$, λ_b and ν_b are the bulk values of the coupling constant and the density of states, respectively, and η is the Hopfield parameter (see eqn (8.8)), one can write

$$\Delta(\omega_n) Z = \tilde{\lambda}_b \pi T \sum_j \sum_{\omega'_n} g_j \frac{\tilde{\Omega}^2}{\tilde{\Omega}^2 + (\omega_n - \omega'_n)^2} \Delta(\omega_{n'}) \left[\omega_{n'}^2 + \Delta^2(\omega_{n'}) + \xi_j^2\right]^{-1} \Big|_{T=T_c}$$

$$\tilde{\lambda}_b = \lambda_b (2\pi\nu_b V)^{-1} = (2/3\pi) \lambda_b (E_F/N) \tag{8.10}$$

Indeed, $v_b = m^* p_F / 2\pi^2 = (3/4) n/E_F$, since $n \propto p_F{}^3$, where n is the carrier concentration; for clusters, the value of n is close to its bulk value (Ekardt, 1984). Note also that the characteristic vibrational frequency $\tilde{\Omega}$ is close to its value for bulk materials. Indeed, the pairing is mediated mainly by the short-wavelength region of the vibrational spectrum, which is similar for clusters and bulk systems. In addition, the equality $\tilde{\Omega} = \Omega_D$ follows directly from the heat capacity measurements performed by Hock *et al.* (2011).

As noted above, eqn (8.10), the equation for Z, and eqn (8.9) are the main equations that allow us to evaluate the value of the critical temperature. It is essential that these equations contain the experimentally measured parameters N (the number of delocalised electrons), E_F, $\tilde{\Omega}$, the degeneracy g_i, λ_b, and the energy spectrum (ξ_i). As a result, one can calculate the values of the critical temperature for various clusters. It is convenient to write eqn (8.10) in the dimensionless form:

$$\phi_n = \sum_{n'} K_{nn'} \phi_{n'} \tag{8.10′}$$

$$K_{nn'} = g\tau_c \sum_s \left\{ \left[1 + (\tilde{\omega}_n - \tilde{\omega}'_n)^2\right]^{-1} - \delta_{nn'} Z \right\} \left(\tilde{\omega}_{n'}^2 + \tilde{\xi}^2\right)^{-1} \Big|_{T_c}$$

Here

$$\phi_n = \Delta(\omega_n) \tilde{\Omega}^{-1}, \tilde{\omega}_n = \omega_n \tilde{\Omega}^{-1}, \tilde{\xi}_j = \xi_j \tilde{\Omega}^{-1}, g = \lambda_b \left(4\pi\tilde{\Omega} v_b V\right)^{-1}$$

The calculation can be performed with use of the matrix method (see section 3.4.5). One can demonstrate (Kresin and Ovchinnikov, 2006) that, for perfectly realistic values of the parameters, a high value of T_c can be obtained. For example, if we consider the 'magic' cluster with the typical parameter values $\Delta E = E_H - E_L = 65$ meV, $\tilde{\Omega} = 25$ meV, $m^* = m_e$, $k_F = 1.5 \times 10^8$ cm^{-1}, $\lambda_b = 0.4$, the radius $R = 7.5$ Å, and $g_H + g_L = 48$ (e.g. $l_H = 7$, $l_L = 4$), we will arrive at a value of $T_C \approxeq 10^2$ K(!). Note that a large degeneracy of the LUS and the highest occupied shell (HOS) plays a key role.

Critical temperature: various clusters. The value of T_c is very sensitive to the cluster parameters. The most favourable case corresponds to:

1 a cluster with large values of the orbital momentum l for the HOS and the LUS and, hence, with the large degeneracies g_{HUS} and g_{LOS} (qualitatively, this corresponds to a large effective density of states), and

2 a relatively small energy spacing between these shells.

An increase in T_c can be achieved by changing the parameters (ΔE, $\widetilde{\Omega}$, etc.) in the desired direction. For example, for $\Delta E \approxeq 0.2$ eV, $\lambda_b = 0.5$, $m \approxeq 0.5m_e$, $R \approxeq 5.5$ Å, $\widetilde{\Omega} = 50$ meV, and $g_H + g_L = 60$, we obtain $T_c \approxeq 240$ K (!). In principle, T_c can be increased to room temperature.

Based on eqns (8.8) and (8.9), once can calculate the values of T_c for specific clusters. The results vary noticeably depending on their parameters. For example, consider the cluster Al_{56}. It contains $N = 168$ electrons, since each Al atom contains three valence electrons. This is a 'magic' cluster, with the HOS corresponding to $l = 7$. Indeed, according to Schriver *et al.* (1990), for $N = 168$, one can observe a sharp drop in the ionisation potential. The next 'magic' cluster contains $N = 198$ electrons and corresponds to the hybridisation of three shells with (i) $l = 4$, $n = 2$, (ii) $l = 2$, $n = 2$, and (iii) $l = 0$, $n = 4$; as a result, $g_H + g_L = 60$. According to Schriver *et al.* (1990), the spacing between the HOS and the LUS is $\Delta E \approx 0.1$ eV. Here, as usual, $\Delta E = E_L - E_H$, $E_L \equiv E_{LUS}$, where E_{LUS} corresponds to the LUS, that is, to the unoccupied energy level of the next 'magic' cluster (in this case, to Al_{66}, with $N = 198$). With use of the parameters for Al_{56} clusters ($R \approx 6.5$ Å, $\tilde{\Omega} = 350$ K, $\lambda_b = 0.4$, $m^* = 1.4m_e$, $k_F = 1.7 \cdot 10^8$ cm^{-1}), we obtain the following value of T_c for these clusters: $T_c \approxeq 90$ K (!). One can expect high values of the critical temperature ($T_c \simeq 100 - 120$ K) for the 'magic' cluster Al_{66}.

As for Ga_{56} clusters, we need to use a different set of parameters ($\widetilde{\Omega} = 270$ K, $\lambda_b = 0.4$, $m^* = 0.6m_e$, $k_F = 1.710^8$ cm^{-1}), and we obtain $T_c \approxeq 145$ K (!). It greatly exceeds the value of T_c for the bulk sample ($T_{c;B} \approxeq 1.1$ K).

In a similar way, one can calculate T_c for clusters with slightly incomplete shells. Such clusters undergo a Jahn–Teller distortion, which leads to the splitting of the degenerate levels. The splitting is described by the following expression (see e.g. Migdal, 2000; Landau and Lifshitz, 1977):

$$\delta E_l^m = 2E_l^{(0)} U r_l^m \tag{8.11}$$

$$r_l^m = 2\left[l(l+1) - 3m^2\right](2l+3)^{-1}(2l-1)^{-1}$$

Here U is the deformation potential, $E_l^{(0)} = E_H = E_{HOS}$. For example, for Zn_{83} clusters ($N = 166$), one can obtain a value of $T_c \approxeq 105$ K.

A special situation might occur for the clusters with near half-filled shells. The cluster configuration oscillates between the prolate and oblate structures (see appendix B and the discussion above, section 8.2.1). The electronic states are characterised by the quantum number 'm' (the projection of the orbital momentum) and, for the region close to half-filling, the shell (HOS) is far from being completely filled. Therefore, we are dealing with large orbital degeneracy. The aforementioned oscillations (the dynamic John–Teller effect) serve to remove this degeneracy. But the degeneracy can be removed by different channels. Indeed, the dynamic John–Teller effect competes with the electron–phonon interaction and, as a result, with the Cooper pairing phenomenon.

The scenario where the pairing prevails is perfectly realistic. It turns out that, for the cluster, it is energetically more favourable to make a transition into a spherical configuration with pairing correlation between the electrons with opposite values of m (and spins). The values of T_c (and the energy gap) can be calculated with the use of eqns (8.10×) and (8.9). It is very important that, unlike the case for 'magic' clusters, the cluster with a near-complete shell, in the present case, a near half-full HOS, the location of the chemical potential μ is entirely different. For 'magic' clusters, μ is in the middle of the HOS–LUS interval. For the present case, μ is lead near the top of the HOS, and this is beneficial for the pairing. For example, for the cluster Zn_{76} ($N = 152$), one can obtain $T_c \simeq 115$ K (!). As for the energy gap, which determines the scale of the pairing, $\varepsilon_0 \simeq 345$ K. This value exceeds that for the vibrational motion, which provides the scale for the dynamic John–Teller effect. As a result, the gain in energy ($\sim 2\varepsilon_0$) leads to this channel being dominant for some clusters.

The energy spectrum: fluctuations. Based on eqns (8.8) and (8.9), one can also study the impact of pairing on the energy spectrum. Here one should remember that, because of the finite size, the electronic spectrum is discrete even in the absence of pair correlation. That is why, unlike the bulk case, where the pairing leads to an appearance of the discrete feature of the spectrum (energy gap), for nanoclusters, pairing is manifested in the change of the energy spectrum. Namely, one can observe an increase in the spacing between the ground state and the excited state, especially at low temperatures near $T = 0$ K. For 'magic' clusters, the pairing is essential if it leads to a noticeable increase in the LUS–HOS spacing. The spectrum is determined by the equation

$$\varepsilon_{os} = \Delta\left[i(\xi_s^2 + \varepsilon_{0;s}^2)\right)^{1/2} \tag{8.12}$$

where ε_{os} is the gap parameter, and Δ and ξ_s are defined as in eqn (8.8). For example, for a Cd_{83} cluster, one can obtain the following value for the excitation energy: $\Delta\varepsilon_{min} =$ 34 meV. This value greatly exceeds that for the same cluster in the absence of pairing: $\Delta\varepsilon_{min} = 6.5$ meV.

A serious question can be raised about the role of fluctuations for the investigated transition of nanoclusters into the pairing state. Note that, in this context, it would be incorrect to consider the clusters as zero-dimensional systems. Indeed, the large values of T_c result in a relatively small coherence length, in fact, one comparable to the cluster size. The broadening of the transition $\delta T_c/T_c$, due to fluctuations, can be rigorously evaluated (Kresin and Ovchinnikov, 2006). One can show that $(\delta T_C/T_C) \approx 5\%$. This width noticeably exceeds that for bulk superconductors, but is still relatively small.

8.3.3 The Cluster-Based Tunnelling Network: Macroscopic Superconductivity

Above we discussed various manifestations of pairing in isolated clusters. Let us now address the interesting question about the possibility of creating macroscopic supercurrents based on Josephson tunnelling between nanoclusters.

Josephson tunnelling between clusters needs to be analysed with considerable care (Ovchinnikov and Kresin, 2010). Indeed, unlike usual bulk superconductors, the clusters are characterised by discrete energy spectra. Moreover, one should take into account the fact that the tunnelling itself splits the energy levels and leads to the formation of symmetric and asymmetric terms. The tunnelling can be treated with use of the tunnelling Hamiltonian and we obtain the expression (cf. chapter 3, eqn (3.8′))

$$j_m = \frac{e\hbar^3}{4m^2} \sum_{\nu,\nu'} |T_{\nu\nu'}|^2 |\Delta|^2 W^{-3} \tag{8.13}$$

where

$$W = \left[|\Delta|^2 + (\Delta E)^2_{LH} \right]^{1/2}$$

$T_{\nu\nu'}$ is the tunnelling matrix element, and $(\Delta E)_{LH}$ is energy space between the HOS and the LUS. Unlike the bulk case, for the cluster, with its discrete energy levels, it would be incorrect to replace the summation in eqn (8.13) by integration. The summation should be carried out for each selected cluster. As an important factor, one should take into account the degeneracy of the shells. This degeneracy leads to a drastic increase in the amplitude of the current. For example, for the cluster with the set of parameters

$$\Delta E_{LH} = 65 \text{ meV}, \quad \tilde{\Omega} = 25 \text{ meV}, \quad m^* = 0.75 m_e$$

$$K_F = 1.5 \times 10^5 \text{ cm}^{(-1)}; \text{the radius } R = 6 \text{ Å},$$

$$G_H = 30, G_L = 18 \ (e.g., L_H = 7, \ l_L = 4)$$

one can obtain the following value for the ratio of the current amplitudes of the nanocluster and bulk superconductor, $\eta = \left(j^{\text{cl}}_{\text{max}} / j^{\text{bulk}}_{\text{max}}\right)$: $\eta \simeq 10^2$.

Therefore, indeed, one can observe the drastic increase in the current amplitude relative to the usual bulk case. Even larger increase can be observed for the clusters Al_{46} ($N = 138$), or some Al clusters with close number of the valence electrons. The same is true for the Al_{66} clusters ($N = 198$). These clusters are of special interest, because the manifestations of the pair correlation have been observed in them (see below, section 2.7). In addition to there being a regular tunnelling channel, the discrete nature of the electronic spectra leads to possibility of charge transfer through an additional channel, namely through resonant tunnelling. Via Hamiltonian formalism, this can be evaluated with the use of the total current operator. It is essential that this additional channel leads to an additional increase in the current amplitude.

We discussed above the Josephson tunnelling between two nanoclusters. Based on such tunnelling, one can build tunnelling networks (Fig. 8.20). At first, one can consider the network transferring the dc Josephson current. Such a network was described in chapter 3 (see section 3.1.2).

One can also consider the cluster-based network transferring ac current. Such a network was described in chapter 3 (see section 3.1.2), where it was shown that such a network can be synchronised.

The presence of a Coulomb blockade competes with the Josephson energy. But one can show that, for relatively short intercluster distances (~10 Å), Josephson energy greatly exceeds the impact of the Coulomb repulsion.

Let us also stress one point. As we know, the Josephson current results from the phase difference between superconducting electrodes (in our case, between clusters). However, the phase of a cluster does not have a specific value; this follows from the uncertainty relation between the phase and the number of particles. As a result, the phase is not defined for each of the clusters. Nevertheless, the phase difference that enters the expression for the Josephson current is defined. This fact was stressed by Cobert *et al.* (2004). As usual, the phase difference is determined by the value of the transmitted current.

The network would be made of clusters deposited on a surface. This implies that the cluster structure and the energy spectrum are not destroyed by the surface–cluster interaction. The search for such a substrate is a serious challenge in the area of nano science. The progress with the so-called soft landing technique based on the use of special organic substrates (see e.g. Meirwes-Broer, 2000; Duffee *et al.*, 2007) makes building such a network realistic. One can expect that this problem will be solved in the near future.

A new type of three-dimensional crystal (cluster metals) can be created by using metallic nanoclusters as building blocks. The lattice of such a crystal is formed by clusters instead of the usual point-like ions. Such solids are analogous to molecular crystals. The presence of pair correlation in an isolated cluster leads to Josephson charge transfer between neighbouring clusters; as a result, such a cluster-based three-dimensional crystal would display macroscopic superconductivity. The concept of such a superconducting crystal was introduced by Friedel (1992).

An interesting study of Ga-based cluster metals was described by Hagel *et al.* (2002) and by Bono *et al.* (2007). A crystal was formed by Ga_{84} clusters ($N = 252$; each Ga atom contains three valence electrons). The superconducting transition at $T_c \approx 7.2$ K has been observed; note that HOS–LUS spacing for such a cluster is rather large, and there

Figure 8.20 *A cluster-based tunnelling network.*

is no reason to expect a very high value of T_c. The observed value of T_c is noticeably higher than $T_{c;b} \approx 1$ K for the bulk Ga.

In principle, one can foresee a crystal synthesis out of different clusters (e.g. Ga_{56}) with rather strong internal pair correlation, and this leads to macroscopic superconductivity with higher values of the critical temperature.

Thus, charge transfer between clusters via Josephson coupling can give rise to strong, macroscopic, high-temperature superconducting currents.

8.3.4 Experimental Observation

A cluster-based tunnelling network transferring Josephson current at high temperature would serve as a direct experimental manifestation of Cooper pairing in nanoclusters. However, such a network has not been built, because of the above-mentioned problem with the 'soft' landing of clusters. That is why we should search for other experimental observations, more specifically, for manifestations of pairing occurring in isolated clusters. Of course, one should not look for an absence of resistance. Moreover, since the cluster size is much smaller than the penetration depth, the observation of the Meissner effect is rather complicated. In reality, the critical evidence can be obtained from the data on spectroscopy.

Pairing leads to a strong temperature dependence of the excitation spectrum. Below T_c and especially at low temperatures close to $T = 0$ K, the excitation energy is strongly modified by pairing and noticeably exceeds that in the region $T > T_c$.

A change in the excitation energy should be experimentally observable and would represent a strong manifestation of pair correlation. Nevertheless, we are talking about a rather difficult experiment. It requires generating beams of isolated metallic clusters at different temperatures in combination with mass selection.

Such an experimental study was performed by Halder *et al.* (2015) and by Halder and Kresin (2015). With the use of photo-ionisation measurements, the spectral manifestation of pair correlation in Al_{66} clusters was observed. The choice of these clusters was not occasional. Indeed, the cluster ($N = 198$) is large, and the values of its parameters is very favourable for high values of T_c: the overlay of the three highest occupied energy shells ($L = 4, L = 2, L = 0$) leads to total degeneracy $G_H = 30$. The next LUS corresponds to $N = 132$ ($L = 8$) with the degeneracy $G_L = 34$. With the use of the values $\tilde{\Omega} = 325$ K, $m^*/m_e = 0.6$, and $\lambda_b = 0.4$ (this is the value of the coupling constant for amorphous Al with $T_c \simeq 6$ K (Grimvall, 1981)), one can theoretically obtain $T_c \simeq 125$ K. Note, in addition, that the ionic core of the cluster Al_{66} is rather spherical, so that the impact of the ionic structure (field effects) on the cluster shape is small.

Photo-ionisation is described by the yield function $Y(\omega)$ (number of photoelectrons for the photon frequency ω):

$$Y(\omega) = \int_{-\infty}^{\infty} dE\, M(E) f(E) D(E)$$

Here M is the matrix element, f is the Fermi distribution function, and D is the density of states, so that $D(\omega) \propto \partial Y(\omega)/\partial\omega$. The photo-ionisation yield was measured experimentally at different temperatures. Bumps at $T \sim 120$ K were observed for Al_{66} clusters (Fig. 8.21). It is interesting that such bumps have never been observed before. Indeed, they can be seen for specific clusters and in a limited temperature range. For example, for comparable Cu clusters, these bumps do not appear. Moreover, even for Al clusters, one can see these bumps only for selected clusters, like Al_{66}. It is clear that these bumps correspond to an electronic transition.

The derivation $\partial Y(\omega)/\partial\omega$ is proportional to the electronic density of states, and one can see directly (Fig. 8.21) that the bumps corresponds to an appearance of the peaks in the density of states of some temperatures, and the amplitude of these peaks grows with a decrease in T. Note that a similar picture was observed for $Bi_2Sr_2CaCu_2O_{8+\delta}$ (Nepjuko *et al.*, 1997; Park *et al.*, 1994), and it was caused by the transition into the superconducting state and the opening of the energy gap at T_c. The value of the gap grows with decreasing temperature and this leads to an increase in the peak amplitude. The same scenario was observed experimentally for Al_{66} clusters. The formation of the bump-like structure is caused by an increase in the density of states near the gap. Indeed, the formation of the gap leads to a 'compression' of the energy levels, since the total number of electronic states is conserved.

One can see directly from the data plotted in Fig. 8.21 that the transition into the superconducting state occurs in the region close to $T_c \simeq 120 - 140$ K. This value noticeably exceeds that for bulk AI ($T_c^6 \simeq 1.1$ K for the usual structure, and $T_c^{am} \simeq 6$ K for an amorphous solid) and this is caused by presence of the shell structure of the electronic spectrum and the corresponding large degeneracy for Al_{66}.

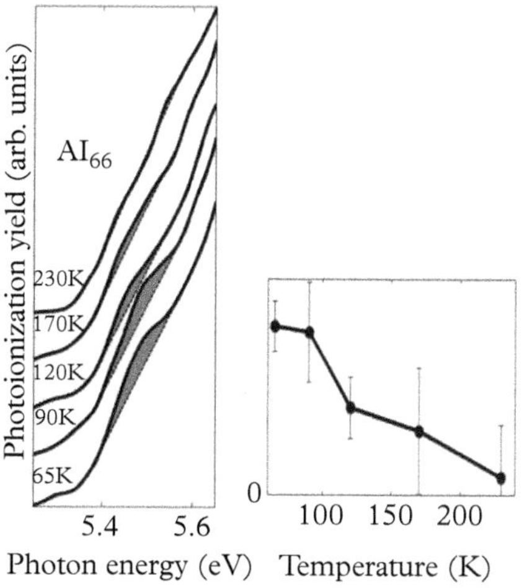

Figure 8.21 *Photo-ionisation yield plot for Al66 clusters: (a) a special bump-like structure; (b) area under the bump as a measure of the intensity variation of the peak as a function of temperature (From Halder et al., 2015).*

The pairing can be observed in large clusters. It is not by chance that Cao *et al.* (2008) observed a jump in heat capacity for Al_{45}^- ions. They developed a special so-called multicollisional induced dissociative method that made it possible to measure the heat capacity of an isolated cluster. Their study shows that, for selected clusters, for example, for Al_{45}^- ions, a peak in heat capacity can be observed. It is essential that the peak is not a universal feature and can be observed only for selected clusters. The peak was observed at $T \approx 200$ K (!); the result is highly reproducible. As we know, a jump in heat capacity is the signature of a phase transition. The Al clusters are not magnetic, and therefore a magnetic transition can be excluded. Special measurements of mobility also exclude any structural transition. It is natural to assume that we are dealing with a transition into a superconducting state.

The theoretical and, especially, experimental studies of the pairing in nanoclusters, as a consequence of the presence of the shell structure of their electronic spectra, should be continued. As mentioned above, the spectroscopy of such clusters is rather complicated. It is required the creation of cluster beams with their controlled temperature, along with mass spectroscopy (selection of specific clusters) and measurements of their spectra. Nevertheless, one can expect future progress in this interesting field.

8.4 Interface Superconductivity

The first detailed theoretical analysis of interface superconductivity was carried out by Allender *et al.* (1973). They considered the interface of metallic and semiconductor materials (see section 2.5.2) and analysed the 'excitonic' mechanism, that is, the pairing at the interface caused by the exchange of excitonic excitations in the semiconductor. The analysis was proceeded by the Little's treatment of the pairing through the electronic excitations and its expansion to the two-dimensional systems by Ginzburg (see section 2.3.1). It is important that the pairing can be provided by phonon modes of both materials.

Initially, the interface region and, more specifically, the confined system of delocalised electrons between two insulators, $LaAlO_3$ and $SrTiO_3$, was generated by Ohtomo *et al.* (2002) and Ohtomo and Hwang (2004). A detail study (Thiel *et al.*, 2006; Reyren *et al.*, 2007; Cen *et al.*, 2009) with use of transport, magnetic, and scanning tunnelling microscopy measurements demonstrated the presence of the superconducting state in this layer with $T_c \sim 200$ mK. The transition in such a two-dimensional state should be analysed with considerable care because of competing with the Berezinskii (1972)–Kosterlitz–Thoulers (1972; BKT) transition. The parameters, the superconducting layer, and, especially, the carrier concentration, can be controlled by an external electric field ('field effect').

A detailed study of the alkaline–earth iron arsenide $CaFe_2As_2$ (Ca112) has resulted in interesting confirmation of the presence of the interface superconductivity with $T_c \simeq 25$ K (Zhao *et al.*, 2016; Deng *et al.*, 2018; Chu *et al.*, 2019). Depending on the annealing and consequent cooling procedures, one can prepare two phases (p_1 and p_2) that are different in their structures. Through special low-temperature annealing (at $T \sim 350°$ C), one can prepare a sample with a mixture of these two phases. None of these phases is superconducting down to 2 K. However, the interface region which appears in random-stacking boundaries, displays the superconducting state with $T_c \simeq 25$ K (!).

The most impressive case of interface superconductivity corresponds to the FeSe/$SrTiO_3$ system (Fig. 8.22). The composition itself deserves a special description. Indeed, FeSe is made as the single–unit–cell thin film, is the simplest, purest, and two-dimensional member of the Fe-based family of superconducting materials. Its study is beneficial also for the analysis of pnictides (section 7.3) and can help in complete determination of the origin of their superconductivity. As for $SrTiO_3$, this semiconductor has been and continues to be a subject for detailed analysis (Cohen, 1969; Gor'kov, 2017; Gastiasoro *et al.*, 2019); its value of T_c depends on the carrier concentration and is reaching $T_c \simeq 0.3$ K. As for the bulk FeSe, its intrinsic value of $T_c \simeq 8.5$ K (Hsu *et al.*, 2008).

The FeSe/$SrTiO_3$ system, which consists of a thin FeSe film placed on an $SrTiO_3$ substrate, displays the superconducting state with $T_c \simeq 65$ K (!). It is natural that this remarkable observation (Wang *et al.*, 2012) generated a lot of experimental and theoretical studies. First of all, it has been established (He *et al.*, 2013) that this superconducting state is not just feature of thin FeSe film; the participation of $SrTiO_3$ is also a key ingredient. As we know, the cuprates and similar systems are doped materials, and doping is realised by chemical substitution. In the present case, the doping is provided by the annealing process. In addition, one can demonstrate that the carriers are electrons. A detailed study of magnetic properties, including the observation of the Meissner effect, has been carried out by Deng *et al.* (2014); this observation is the most rigorous evidence that we are dealing with two-dimensional interface superconductivity.

As for the mechanism of superconductivity for this system, which contains two materials, it is very attractive to involve the excitonic mechanism, that is, to explain the phenomenon by the presence of electron–hole excitations across the interface. However, detailed theoretical calculations (Sadovskii, 2016) show that, although the channel for

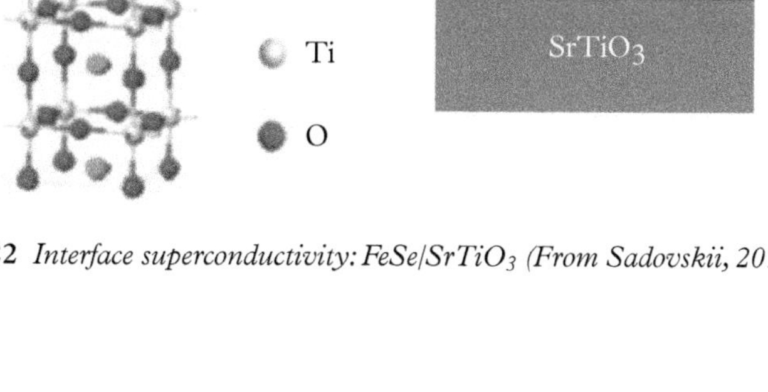

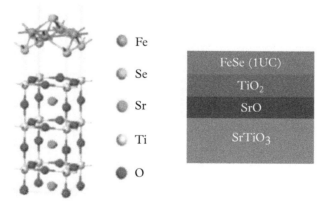

Figure 8.22 *Interface superconductivity: FeSe/$SrTiO_3$ (From Sadovskii, 2016).*

the pairing can make some contribution, it is not sufficient to reach such a high value of T_c. Rather, electron–phonon interaction plays a key role. Xiang *et al.* (2012) stressed the point that $SrTiO_3$ has a remarkable property: this material is a 'quantum paraelectric' insulator (Mü eller and Burkard, 1979). The presence of the responding low-frequency phonon mode (ferroelectric phonons: the relative motion of Ti and O ions) leads to strong electron–phonon coupling. Indeed, according to the measurements by He *et al.* (2013), $2\varepsilon_0/T_c \simeq 6-7$ (where ε_0 is the energy gap). This value greatly exceeds that for the usual BCS superconductors ($2\varepsilon_0/T_c \simeq 3.5$, see eqn (2.9)) and strongly indicates that the electron–phonon interaction strength is, indeed, large. Speaking of the superconducting state of the $FeSe/SrTiO_3$ system, the high value of T_c is caused by the strong interaction of the electrons in the FeSe film with the soft phonon modes in $SrTiO_3$, that is, by the 'softening' mechanism (Xiang *et al.*, 2012).

Interface superconductivity is relatively young area of research, and one can expect a number of new developments. The high-T_c superconducting state can be provided not only by the 'softening' mechanism, but also by the presence of high-frequency phonon modes in hydrides (section 7.2). The high-frequency optical modes in combination with strong coupling, provided by the presence of many hydrogen ligands, leads to a high value of T_c. One can envision the creation of new substrate (instead of $SrTiO_3$) containing hydrogen ions. Note that high pressure in the case of hydrides plays a dual role. The thing is that hydrides at ambient pressure are not in a metallic state. Their transition into the metallic state occurs at some pressure. In addition, the specific structure (such as I_3M for hydrides) with a large number of hydrogen ligands is also created with the use of high pressure. For the case of interface superconductivity, one of the materials (e.g. FeSe) is already in a metallic state, and one can hope that the appearance of high-T_c superconductivity would not require such high pressure.

8.5 Room Temperature Superconductivity: Paths, Systems, and Challenges

8.5.1 General Comments

The dream of attaining room temperature superconductivity was envisioned shortly after the discovery of the phenomenon in 1911. This goal is perfectly realistic and there have been several paths taken to reach room temperature superconductivity. Recently, scientists observed the phenomenon of room temperature superconductivity in a real experiment (see below and section 7.2).

Let us stress initially that, according to the present state of the microscopic theory, the room temperature superconductivity can exist, and any factors limiting the value of the critical temperature below room temperature are absent. The general problem of creating the system which is in the superconducting state at room temperature is connected with the nature of the phenomenon. Indeed, this phenomenon is of a quantum-mechanical nature, and we are dealing with a macroscopic quantum effect. We described in section 3.1 such phenomena as flux quantisation, Josephson tunnelling, and so on, and they directly manifest the quantum-mechanical laws. And the task of observing such a phenomenon at high temperature for the many-electron system is far from being trivial. The thermal motion washes out the quantisation of single electronic states and, usually, we observe a classical picture. That is why superconductivity has been always observed at the so-called extremal conditions. The usual condition is low temperature, so that the transition into the superconducting state for conventional materials occurs at the values of the critical temperature, which are much below the room temperature.

Another extreme condition is created by the high external pressure P, with values of P much higher, relative to ambient pressure (section 7.2). It is known that the impact of pressure, which is a thermodynamic parameter, is similar to lowering the temperature: the pressure depresses a thermal disorder and makes it possible to stabilise the structures, which are thermodynamically unstable at usual conditions.

Nanoclusters form another 'extreme' system. Here the quantisation is due to a relatively small number of electrons (the 'finite Fermi system'), that is, to the small size of the clusters. It is not by chance that they are called 'artificial atoms'. Such a fundamental feature as the presence of the electronic shell structure makes clusters similar to atoms with corresponding quantisation values. Of course, the pairing in an isolated nanocluster is not the macroscopic phenomenon. This effect should be combined with another quantum phenomenon, Josephson effect. The network of such tunnelling junctions will make possible the transfer of macroscopic superconducting current at high temperatures.

Speaking of high-T_c cuprates, the discovery transfers the phenomenon of superconductivity into different energy scale. Their key property is the presence of layered structure, that is, the low-dimensional structural units. The reduced dimensionality is a factor beneficial for superconductivity, since the size quantisation leads to an effective increase in the electronic density of states. The same is true about the interface superconductivity and the quasi-one-dimensional system proposed by W. Little for realisation of the electronic mechanism of pairing.

8.5.2 Promising Directions

One can name several promising materials. These include the new high-T_c oxides, obtained with modulated doping; hydrides under high pressure; nanoclusters; low-dimensional systems containing different electronic groups and affected by the Coulomb interactions between them; systems with lossless currents in the ground state; and systems displaying interface superconductivity.

We have previously discussed, in different chapters, the properties of these materials. Here we want just to present a brief summary, reflecting the state of art in the search for room temperature superconductivity. The ordering of the topics below is rather chaotic, since the breakthrough may occur in any of these directions, or others not mentioned here.

Below we focus on scientific paths for finding room temperature superconductivity. In other words, we focus on the possibility of building the real experimental setup, allowing us to observe the phenomenon of superconductivity at room temperature. The problem of practical applications will be the next challenge. Indeed, if one can only see room temperature superconductivity under high pressure, then it is hard to imagine immediate application. Nevertheless, this would be a fundamental breakthrough. Moreover, there have been many times when some discovery initially looked like a purely academic achievement but later became an important deal in our everyday life. Definitely, the same will happen with room temperature superconductivity and its applied aspect. But, as mentioned above, here we focus on existing scientific challenges.

High T_c oxides and tungsten oxides. As we know, the oxides comprise a very special class of materials. Indeed, the discovery of the high-T_c superconducting copper oxides (cuprates: see sections 1.1 and 5.3) transferred the field of superconductivity to a higher energy scale, one much close to room temperature superconductivity. The magnetic oxides (manganites; see section 9.2) form the most remarkable magnetic system. The ferroelectrics are also oxides. It is interesting that oxygen is the key ingredient of such superconducting, magnetic, and ferroelectric systems; this is even a challenge for solid state chemistry. It is not by chance that the (highly successful) search for high-T_c superconductivity in oxides was started by K. A. Mueller, who was a leading expert in ferroelectricity and understood the dynamics of lattices containing oxygen ions.

We previously described the main properties of tungsten oxides (WO_{3-x}), (see section 7.4.4). Recent key developments have resulted in the appearance of a new family of high-T_c oxides (Schengelaya and Mueller, 2019; see also Schengelaya *et al.*, 2019), namely, the WO_{3-x} compounds. The electron doping is provided by the oxygen depletion. The compound $WO_{2.9}$ has a $T_c \simeq 80$ K. Lithium intercalation increased the T_c up to the value $T_c \simeq 94$ K. This is the highest observed value of the critical temperature for non-cuprates at ambient pressure. Such a high value of T_c for the beginning of a new family of oxides is very promising, and one can expect further progress.

Modulated doping and an increase in T_c in the cuprates. As is known, the state above T_c for underdoped cuprates turns out to be rather unusual. It is termed the 'pseudogap state' and is characterised by finite resistance along with the presence of an energy gap and the Meissner effect. Therefore, it combines the properties of both superconducting and normal states. This picture reflects the inhomogeneity of the system (see section 6.1). More specifically, above T_c (we call this resistive T_c), the sample contains superconducting regions ('islands') embedded in a normal metallic matrix. This matrix provides normal conductivity, whereas the presence of the 'islands' leads to the gapped state and anomalous diamagnetism.

The high-T_c cuprates are doped systems. The dopants create carriers (usually 'holes'), which become delocalised around the sample. At the same time, pair breaking occurs in the proximity of the pair breaker, and this is the factor depressing T_c. The modulated doping is a special method (see section 6.2) that makes it possible to dope along specific regions ('channels'). The holes created by doping move and, as a result, the pairing is not affected by pair breaking; effectively, it creates macroscopic regions which are not affected by pair breaking and, correspondingly, have a higher T_c.

Hydrides: The first room temperature superconductor. The family of hydrides (see section 7.2), that is, materials which contain hydrogen and heavy ions, is the most promising for reaching the room temperature superconductivity state.

The observation of $T_c = 203$ K (!) in the sulphur hydrides is the most important breakthrough since the milestone discovery of the high-T_c cuprates. As a result, the goal of reaching room temperature superconductivity has become perfectly realistic. Indeed, it was clear from previous theoretical and experimental studies (see section 7.2) that sulphur hydrides represent a large family of compounds, and it was realistic to expect further progress in reaching room temperature superconductivity. Later, $T_c \simeq 250$ K (!) was observed for the LaH_{10} compound (Liu *et al.*, 2018; Drozdov *et al.*, 2019). The high value of $T_c \simeq 243$ K was reported recently by Kong *et al.* (2019), so that, indeed, we are dealing with a family of high-T_c hydrides; the observed values are close to room temperature. To our delight, the goal of reaching room temperature superconductivity was achieved during the preparation of this book, and described in the paper by Snider *et al.* (2020);its title 'Room-temperature superconductivity in a carbonaceous sulfur hydride' speaks for itself. This is a three-component (S-C-H) hydride which continues to be in the superconducting state up to $T_c = 288$ K (near 15 °C).

The original idea was proposed by N. Ashcroft (1968): material containing hydrogen, which is the lightest ion, has high values for the phonon frequency, and this factor is beneficial for reaching high values of T_c. However, solid hydrogen can be formed only under high pressure; as a result, high-pressure physics became a major tool in the search for the superconducting state in hydrogen and hydrogen compounds. However, metallic hydrogen can transition into the superconducting state under pressures on the order of 600 GPa, and this goal has not been achieved yet.

The various compounds (hydrides) containing hydrogen and other heavy ions are even more promising relative to pure hydrogen. Indeed, metallic hydrogen has only acoustic phonon modes, whereas the presence of at least one more heavy ion in the unit cell leads to the appearance of optical phonon modes. These modes are dominated by the light hydrogen motion and are characterised by high frequency. Moreover, these optical modes have a large phonon density of states $F(\Omega)$ in a rather broad range of the phonon momenta, and this increases the electron–phonon coupling. In addition, the superconducting state in hydrides is benefited from the presence of acoustic phonon modes caused, mainly, by the motion of heavy ions.

The high-T_c state of the hydrides is caused by the electron–phonon mechanism, which is manifested in the noticeable isotope shift for H $\rightarrow$ D substitution, with values of the isotope coefficient close to $\alpha_{\max} = 0.5(\alpha \approx 0.35)$. This is the fundamental result, which brings to the end a 'myth' born many years ago. According to this 'myth', the electron–phonon coupling has an upper limit for $T_c \approx 30$ K (?). This unfounded statement (see section 2.2) was totally ruled out after the observation of almost room temperature superconductivity with the use of a real experimental setup and is caused by strong interaction of electrons with high-frequency optical phonons.

One can expect that other hydrides can display similar or even higher values of the critical temperature. For example, according to theoretical analysis, this can be expected for the YH_6 compound (Liu *et al.*, 2017), CaH_6 (Wang *et al.*, 2012), Li_2MgH_{16} (Sun *et al.*, 2019), and LuH_6 (Song *et al.*, 2020b). At present, it is clear that the hydrides form the first family of room temperature superconducting materials.

Nanosystems. Nanocluster-based materials (section 6.2) represent an example of another system with values of T_c, which potentially can reach at room temperature. The metallic nanocluster A_n is a set of atoms (e.g. Al_{56}, Zn_{23}) with its valence electrons becoming delocalised. These delocalised electrons form a finite Fermi system and, because of the finite size, the clusters are characterised by discrete electronic energy spectra.

A very important feature of nanoclusters is the presence of the so-called shell structure of their electronic spectra. Energy shells are similar to those for atoms ($s, d, \ldots$ shells) and for atomic nuclei. If the energy shell is filled ('magic' clusters; they are similar to 'inert' atoms), then the shape of the cluster is almost spherical. Then the states are classified by the values of the orbital momenta 'L' and their projection 'm', so that there is a well-known degeneracy, $G = 2(2L + 1)$. The Cooper pairs in nanoclusters are formed by electrons with opposite projections of the angular momentum $(m, -m)$. In many aspects, the picture of pairing is similar to that in atomic nuclei (see section 10.1).

Some specific clusters have very large values for the critical temperature T_c. Qualitatively, this can be explained as follows. If the HOS is highly degenerate, it contains many electrons; this can be seen as a sharp peak in the density of states at the Fermi level. The value of T_c is very sensitive to the cluster parameters. The most favourable case corresponds to:

1 a cluster with large values of the orbital momentum L for the HOS and LUS shells (the main contribution comes from these shells), and
2 a relatively small energy spacing between the HOS and the LUS.

The cluster containing $N = 198$ electrons (e.g. Al_{66}, Ga_{66}, or Zn_{99}) represents such a special case. Indeed, it is a large 'magic' cluster, and the values of its parameters are very favourable for high values of T_c. In addition, its ionic core is rather spherical, and the total degeneracy is large. Speaking of nanoclusters, one should distinguish two aspects. The first is the manifestation of the pair correlation for an isolated cluster, and the second is the observation for the network formed by many superconducting nanoclusters. The second possibility is especially attractive, since it leads to cluster-based macroscopic superconductivity.

A change in the energy spectrum at $T \leq T_c$ should be experimentally observable and would represent a strong manifestation of the pair correlation. Such a change was observed for Al_{96} clusters ($T_c \simeq 120$ K) with use of the photo-ionisation method (see section 8.2.2). Even larger effects can be observed in some selected other clusters. If a cluster is placed on a surface, then its shell structure is damaged by the cluster–surface interaction. It is very important to find a substrate which will keep the shell structure intact. Then it would be possible to build a cluster-based tunnelling network that would make it possible to transmit a high-temperature macroscopic superconducting current. The aromatic organic molecules (section 8.2.2) and quantum dots represent other nanosystems: their study is also of definite interest.

The ground state with lossless current. As we know, electrons in atoms move without any losses, even in the ground state, so that we are dealing with 'lossless current' on an atomic scale, similar to the superconducting current. However, the picture is entirely different on the macroscopic scale. All known bulk materials do not transfer any current in their ground state. The current state can be created by an applied voltage. This feature is caused by the $t \to -t$ symmetry; as a result, the distribution function depends only on the absolute value of electron velocity, not on its direction.

However, one can try to build a system which is remarkable by its violation of the $t \to -t$ symmetry. The ground state of such a system is characterised by the presence of lossless current. The search for such a remarkable system was described in chapter 3 (see section 3.1.5), and the model of one such system is presented in chapter 3 in Fig. 3.12. This search is a serious challenge, but the direction is realistic and very promising.

The 'electronic' mechanism. There are a number of directions which are promising and might lead to the observation of room temperature superconductivity. The phenomenon of room temperature superconductivity can be provided by an electronic mechanism proposed in a pioneering paper by Little (1964). The method is based on the presence of two electronic subsystems (see section 2.5) and they are characterised by energy scales that are larger than those in the phonon mechanism. One should note that the usual experimental studies aimed at an analysis of specific properties of superconductors could be complicated but are still rather straightforward. A real and serious challenge is to perform an experimental study allowing us to determine the mechanism of pairing for some specific sample.

The remarkable, state-of-the-art experiment (Hamo *et al.*, 2016; see section 4.5.1) demonstrates the presence of the electron–electron attraction mediated by the second electronic subgroup. Two nanotubes were used; one of them is the main conducting system, whereas the second nanotube (polariser) provides an additional interaction between the electrons in the first nanotube. The distance between these nanotubes can be controlled. It has been demonstrated that the intensity of the interaction depends on the size of the second group; for a smaller size, one can expect the main system to be capable of displaying a high-T_c superconducting state.

Interface superconductivity. The interface of two materials can display peculiar properties, including the phenomenon of interface superconductivity (see section 8.4). As a result, we are dealing with two-dimensional superconductivity; such a state is rather peculiar. The carrier concentration, which is a main parameter, can be controlled, for example, by an external electric field (the 'Field' effect). It is interesting that the behaviour of the gap as a function of the charge-carrier depletion turns out to be similar to that in the high-T_c cuprates. A well-known example is the $LaAlO_3$–$SrTiO_3$ interface. The superconducting state can be observed for the very low value of the carrier concentration, which is much smaller than that for bulk materials.

The system, which contains a single FeSe layer placed on $SrTiO_3$, displays a transition into the superconducting state at $T_c \simeq 65$ K (!). The impact of the substrate is very large, since the value of T_c for an isolated FeSe is $T_c \simeq 8$ K. A study of the interface superconductivity has started relatively recently, and this direction is very promising.

8.5.3 Conclusion

The search for room temperature superconductivity has become one on the frontier of modern condensed matter physics. We discussed above several different directions aimed at one goal: the observation of room temperature superconductivity. As one can see, the spectrum of the search is rather broad. The study of hydrides with high pressure looks like the most promising direction. But, taking into account the history of physics and its lessons, one can also expect some interesting, unexpected developments.

One can state only that the phenomenon of room temperature superconductivity is perfectly realistic, and it will be observed in a near future. At present, the main focus is on the search for the fundamental phenomenon, room temperature superconductivity, without thoughts of its future application. And again, based on history, one can say that, if the phenomenon of room temperature superconductivity is observed, then ways to use it for various applications will be found. It is hard to imagine the changes in future technology when London's law will be more important than Ohm's law.

At the same time, it is interesting to release the imagination and try to paint some pictures from the future. That is why we conclude this section by citing the following prediction made by W. Little (1965):

> The magnet floats freely above the sheet supported entirely by its magnetic field. The field is unable to penetrate the superconductor and so provides a cushion on which the magnet rests. It is easy to imagine hovercraft of the future utilising this principle to carry passengers and cargo above roadways of superconducting sheet, moving like flying carpet without friction and without material wear or tear. We can even imagine riding on magnetic skis down superconducting slopes and ski jumps – many fantastic things would become possible.

9 Manganites

9.1 Introduction

This chapter is concerned with the properties of the so-called manganites. Manganites are not superconductors. They are magnetic materials, so one might well ask why a book on superconductivity would contain a chapter on manganites. However, the inclusion of such a chapter is justified because both cuprates and manganites are mixed-valence compounds, and there is a lot of similarity between them.

Manganites are named after the manganese ion, which is a key ingredient of the compounds. Their chemical composition is $A_{1-x}R_xMnO_3$; usually, $A \equiv La$, Pr, or Nd, and $R \equiv Sr$, Ca, or Ba. These materials were first introduced by Jonner and van Santen in 1950. Unlike the case for the usual ferromagnetics, the transition of manganites to the ferromagnetic state (at $T = T_C$, where T_C is the Curie temperature) takes place at finite 'doping', $x \neq 0$, and is accompanied by a drastic increase in conductivity. This simultaneous transition from an insulating to a metallic and magnetic state is one of the most remarkable fundamental features of these materials.

One year later, Zener (1951) explained this unusual correlation between magnetism and transport properties by introducing a novel concept, the so-called double-exchange mechanism. Zener's pioneering work was followed by more detailed theoretical studies by Anderson and Hasegawa (1955) and de Gennes (1960).

A revival of interest in the manganites and their properties came about after the remarkable discovery of the colossal magnetoresistance effect by Jin *et al.* (1994). The very name of the phenomenon originates from the observation of a thousand-fold (!) change in the resistivity of La–Ca–Mn–O films near $T = 77$ K in the presence of an applied magnetic field, $H \approx 5$ Torr.

It is important to mention that the discovery of colossal magnetoresistance in the magnetic oxides (manganites) was preceded by the discovery of high-temperature superconductivity in copper oxides (cuprates) by Bednorz and Mueller (1986). As was noted above, despite the obvious difference in the two phenomena (superconductivity vs. ferromagnetism), there is a deep analogy between the two classes of materials. Both classes are doped oxides. The parent (undoped) compounds (e.g. the cuprate $LaCuO_4$ or the manganite $LaMnO_3$) are antiferromagnetic insulators. It is 'doping' that leads to

Superconducting State: Mechanisms and Materials. Vladimir Z. Kresin, Sergei G. Ovchinnikov, and Stuart A. Wolf, Oxford University Press (2021). © Kresin, Ovchinnikov, Wolf. DOI: 10.1093/oso/9780198845331.003.0009

the insulator–metal transition for both systems. Of course, there are profound differences between these compounds, but the discovery and subsequent intensive study of the high-T_c cuprates was very beneficial for the progress in understanding manganites. It is worth noting that the discovery of the colossal magnetoresistance effect was made possible through the use of high-quality thin films. The preparation of such films (Chahara *et al.*, 1993) was based on a method developed for high-temperature superconducting oxides.

The study of manganites remains very active (see e.g. the reviews by Coey *et al.*, 1999; Tokura, 2003; and Gor'kov and Kresin, 2004). We focus here on the fundamentals of these materials and on the similarities between their properties and those of the high-T_c superconducting oxides. Below we follow the analysis by Gor'kov and Kresin (2004).

9.2 Electronic Structure and Doping

9.2.1 Structure

Let us start from an undoped (parent) compound, $LaMnO_3$. Consider its cubic perovskite structure (Fig. 9.1). The Mn^{3+} ions are located at the corners, and the La ion at the centre of the unit cell. Each Mn^{3+} ion is caged by the O^{2-} octahedron (Fig. 9.2); locally, this forms a MnO_6 complex, with the Mn ion in the symmetric central position surrounded by six light oxygen ions.

As was noted above, the $LaMnO_3$ crystal contains Mn^{3+} ions. The valence state of Mn is determined by the simple neutrality count, since the La ion has a '+3' valence state, and each oxygen ion is in the O^{2-} valence state.

The electrons in a free Mn atom form the incomplete d-shell $(\ldots)3d^5\,4s^2$, where $(\ldots) \equiv 1s^2 2s^2 2p^6 3s^2 3p^6$ (see e.g. Landau and Lifshitz, 1977). Therefore, the Mn^{3+} ion, with the d-shell $(\ldots)\,3d^4$, contains four d-electrons.

The fivefold orbital degeneracy is split by the cubic environment into two terms: t_{2g} and e_g. The t_{2g} level contains three electrons that form the 't-core'. The last d-electron (the e_g electron) is well separated in energy and forms a loosely bound state (see Fig. 9.3). This electron plays a key role in conducting and other properties of manganites (see

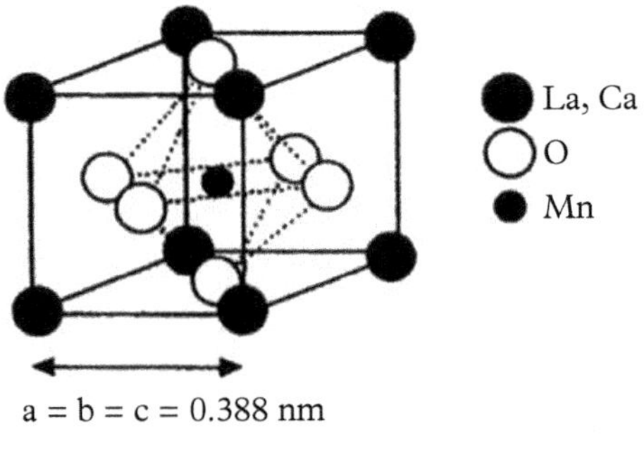

Figure 9.1 *The parent compound* $LaMnO_3$, *shown as its unit cell.*

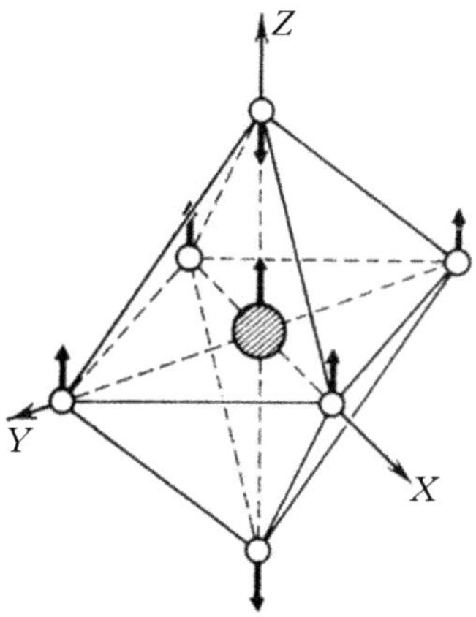

Figure 9.2 *The* MnO_6 *octahedron.*

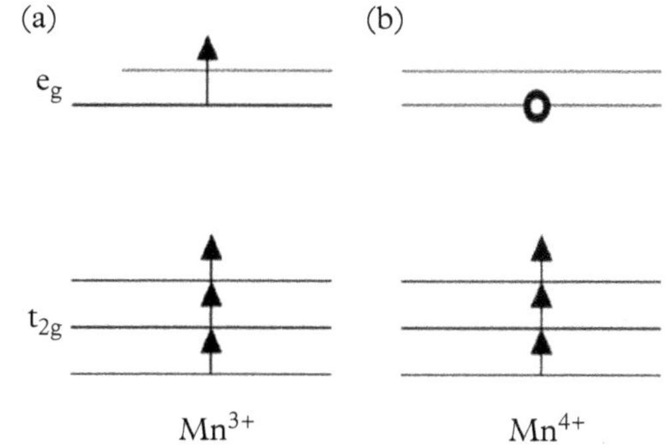

Figure 9.3 *(a) The d-shell of the* Mn^{3+} *ion; the total spin of the t-core is* $S = 3/2$*; the* e_g *degenerate level is split for clarity. (b) The d-shell of the* Mn^{4+} *ion, showing the o-hole; the* $La^{3+} \to Sr^{2+}$ *substitution leads to the* $Mn^{3+} \to Mn^{4+}$ *transition.*

below) as well as in determining its magnetic order. The analysis of its behaviour in the lattice is a major topic in microscopic theory.

Hund's rule demands that the three *d*-electrons forming the '*t*-core' have the same spin orientation; as a result, the localised '*t*-core' has the total spin $S = 3/2$. The e_g electron is also affected by the same strong Hund's interaction. Therefore, its spin must be polarised along the same direction as for the *t*-core.

It is also essential that the e_g term be a double-degenerate one. As a result, we meet with the situation in which the Jahn–Teller effect becomes an important factor that can lead to lattice instability.

The parent compound $LaMnO_3$ is an insulator, and its transition to the conducting state is provided by doping; the scenario is similar to that in the cuprates. The doping is realised through a chemical substitution, for example $La^{3+} \to Sr^{2+}$, that is, by placing a divalent ion into the local La^{3+} position. The $La^{3+} \rightarrow Sr^{2+}$ substitution leads to a change in the valence of the manganese ion: the four-valent Mn ion loses its e_g electron (Fig. 9.3). The absence of the electron can be described as the creation of a hole. The Sr^{2+} ion is located at the centre of the cubic cell. As for the hole itself, it is spread out over the unit cell, being shared by eight Mn ions.

As a result, one obtains the crystal $La_{1-x}Sr_xMnO_3$, where a number of La ions have been randomly substituted by Sr ions. Even though it has holes, the crystal, at first, continues to behave as an insulator. In other words, each hole remains localised on the scale of one unit cell. In this concentration range, localisation corresponds to the formation of local polarons. Such an insulating state is preserved with an increase in doping up to some critical value $x = x_c \approx 0.16 - 0.17$.

At $x = x_c$ the material makes a transition into the conducting (metallic) state. This can be directly seen from the data obtained by Urushibara *et al.* (1995; see Fig. 9.4). Note that the conductivity of the best samples of the Sr-doped films at low T is where the order $= 10^4 - 10^5\ \Omega^{-1}\ \text{cm}^{-1}$, that is, we are now dealing with a typical metallic regime.

It is remarkable that the transition at $x = x_C$ is also accompanied by the appearance of the ferromagnetic state. The correlation between conductivity and magnetism is the fundamental feature of manganites, and we address this subject below.

So far, we have concentrated on manganites in the low-temperature region and the evolution of their properties with doping. Take now a sample that is in the metallic

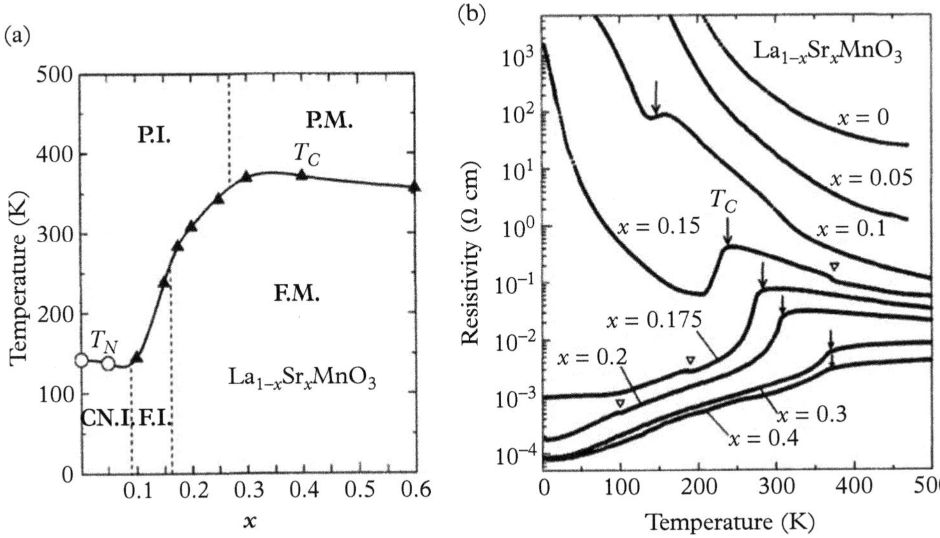

Figure 9.4 *Phases of manganites* ($La_{1-x}Sr_xMnO_3$)*: (a) phase diagram; (b) evolution of the dependence* R(T) *with doping; AI, antiferromagnetic insulator (A-phase); FM, ferromagnetic metal; FI, ferromagnetic insulator; PI, paramagnetic insulator; PM, disordered paramagnetic with large resistance;* T_N, Néel's *temperature. (Urishibara et al., 1995)*

ferromagnetic state with a fixed carrier concentration, for example $x = 0.3$, and then increase the temperature. The ferromagnetic state persists up to the Curie temperature $T_C \approx 170$ K. Above this temperature, the compound makes a transition into the paramagnetic state with much higher resistivity. Once again, one sees that there is a correlation in the electronic property that manifests itself in an almost simultaneous change (at $T = T_C$) in both conductivity and magnetisation.

9.2.2 Magnetic Order

The type of magnetic structure is determined by the doping level. The parent compound, $LaMnO_3$, belongs to the so-called antiferromagnetic insulating phase. The ferromagnetic ordering in layers is combined with an antiferromagnetic order in a direction perpendicular to the layers (the A-structure; see Fig. 9.5). The value of the Néel temperature (see Fig. 9.4a) for this antiferromagnetic phase is relatively low: $T_N \approx 150$ K.

The doping eventually leads to the appearance of the three-dimensional ferromagnetic state (for $x \gtrsim 0.16$). This state persists up to $x \approx 0.5$ for $(LaSr)MnO_3$, and then the crystal, while continuing to be in the metallic state, can change its magnetic structure, which turns into a *metallic* A–state. The magnetic order in this state is similar to that of the parent (underdoped) compound (Fig. 9.5), but it has metallic conductivity. Such a compound is a natural spin valve system. Indeed, this 'giant' magnetoresistance, as is known, was initially observed by using a special artificial multilayer structure. In contrast, metallic manganites with a magnetic A-structure are natural three-dimensional systems which display giant magnetoresistance. In addition, such material can be used to make a new type of Josephson junction: the superconductor–antiferromagnet–superconductor Josephson junction (see section 9.7.2).

9.2.3 The Double-Exchange Mechanism

Here we qualitatively discuss the nature of the observed ferromagnetic spin alignment in manganites. As noted above, the concept of the double-exchange mechanism was introduced by Zener (1951) almost immediately after the discovery of manganites.

Here the situation is opposite to that of the cuprates. The mechanism of high-T_c superconductivity is still a controversial issue, although the phenomenon was discovered in 1986. As for manganites, the nature of their unusual magnetism and simultaneous transition into metallic and ferromagnetic states was explained rather shortly thereafter.

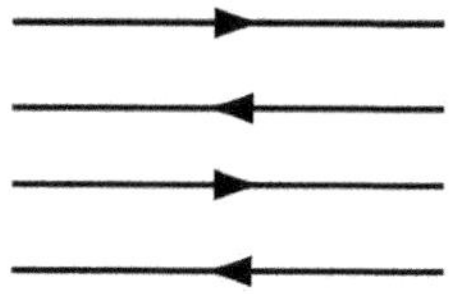

Figure 9.5 *The A-structure.*

Let us here discuss the double-exchange mechanism. If one of the Mn^{3+} ions becomes four-valent ($Mn^{3+} \rightarrow Mn^{4+}$; this occurs as a result of doping, e.g., $La^{3+} \rightarrow Sr^{2+}$ substitution), a hole appears on this site. This allows for another e_g electron localised initially at the neighbouring Mn^{3+} ions to jump into the new vacant place (such hopping corresponds to the hole moving in the opposite direction). But, as was noted above, the e_g electron is spin polarised (Fig. 9.3) because of the Hund's interaction with its t-core. The total spin of each t-core is equal to $S = 3/2$, but their mutual orientations on different sites are independent. The relative orientation of the spins of the e_g electron and the 'vacant' t-core is the crucial factor for the hopping, because of the strong Hund's interaction. Indeed, imagine that the direction of the spin of the core for the Mn^{4+} ion is opposite to that for the e_g electron of the neighbouring Mg^{3+} ion. Then the hopping is forbidden. At the same time, such hopping, like any increase in the degree of delocalisation, is energetically favourable. In other words, it provides an increase in kinetic energy, and the ground state of the ferromagnetically ordered system (in which all spins are polarised along one direction) lies below the paramagnetic state. As a result, the t-cores become ferromagnetically coupled, and this, in turn, favours hopping by the e_g electrons.

This simple picture qualitatively describes the origin of the ferromagnetism in manganites and demonstrates the direct interdependence between hopping and ferromagnetic ordering. Namely, this correlation leads to the observed interplay between conductivity and magnetism. However, a quantitative analysis requires a rigorous treatment of hoping along with the Hund's interaction and will be described below.

It is important to emphasise that the charge transfer in the conducting ferromagnetic manganites is provided by spin-polarised electrons (called spin majority electrons). Such a conductor is different from a usual metal where the spins of the conduction electrons occur in both directions. Because of such spin specifics, the conducting state in manganites is called 'half-metallic'. It should be stressed that, in a half-metal, there are no spin minority electrons at the Fermi level.

9.2.4 Colossal Magnetoresistance

A 'huge' magnetoresistance effect has been observed in the ferromagnetic metallic film $La_{0.67}Ca_{0.33}MnO_3$ (Jin *et al.*, 1994). The magnetoresistance is defined as

$$\Delta R/R_H = [R(T,H) - R(T,0)]/R(T,H) \tag{9.1}$$

The magnetoresistance has a sharp peak near $T \approx 190$ K, that is, near the Curie temperature for the Ca-doped manganites, and the change in resistivity caused by the application of a magnetic field ($H \approx 5$ Torr) is very large: $\Delta R/R_H \approx -1.3 \times 10^3$ (!).

Such a drastic change in resistivity is caused by the aforementioned correlation between magnetic ordering and conductivity and, therefore, is directly related to the double-exchange mechanism realised in manganites. Indeed, below the Curie temperature T_C, conductivity has a metallic band mechanism, whereas, above T_C, the mechanism of conductivity bears a polaronic hopping character. The presence of an external magnetic field is a favourable factor that makes it possible to establish ferromagnetic ordering at temperatures higher than T_C ($H = 0$). The magnetic order triggers the conductivity increase (through the double-exchange mechanism), and this leads to a large resistivity change: $R_H = R(T, H) - R(T, 0)$. The shift is negative, that is, it corresponds to the transition into ferromagnetic metallic state.

The 'colossal' magnetoresistance effect greatly exceeds in its value the 'giant' magnetoresistance effect (see e.g. the reviews by Fert, 2007; von Helmholt *et al.*, 2008; and Parkin, 1995). The giant magnetoresistance effect is observed mainly in artificial multilayers systems with alternating magnetic structure and is used in many applications. It reaches the value $R_H/R_0 \approx 50\%$. The scale of the colossal magnetoresistance phenomenon is much larger.

The natures of colossal magnetoresistance and giant magnetoresistance are entirely different. It is interesting, nevertheless, that, as mentioned above, a metallic manganite with an A-structure (say, in $La_{2-x}Sr_xMnO_3$, with $x > 0.5$) forms a natural giant magnetoresistance system, as this might lead to the giant magnetoresistance effect in manganites.

9.3 Percolation Phenomena

9.3.1 Low Doping and the Transition to the Ferromagnetic State at Low Temperatures

Substitution creates a hole located inside one unit cell. Coulomb attraction prevents the hole from delocalising, namely, from spreading throughout the whole lattice. An increase in doping leads to an increase in the number of unit cells containing holes. In addition, if Sr substitution occurs for two or more neighbouring units, a larger cluster forms, and the holes become delocalised over the clusters. From Zener's double-exchange mechanism, we expect that the spins of the Mn ions have ferromagnetic alignment inside each cluster.

The random character of the Sr substitution leads to a statistical (chaotic) distribution of these clusters. As a result, the growth of the clusters can be treated by means of the percolation theory (see e.g. Shklovskii and Efros, 1984; Deutscher, 1987; Stuffer and Aharony, 1992).

As noted above, an increase in the doping level of a $La_{1-x}R_xMnO_3$ crystal ($R \equiv Sr$, Ba, ...) leads to an increase in the clusters' size and their overlap. Finally, at some critical value of $x = x_c$ (the percolation threshold), the system forms an 'infinite' cluster piercing the whole sample. In other words, at $x = x_c$, one first sees the appearance of connected islands of a *macroscopic* metallic ferromagnetic phase. Like other phase transitions, the percolative insulator–metal transition in manganites can be characterised by some critical indexes. This theoretical approach to the transition in manganites was first introduced and developed by Gor'kov and Kresin (1998, 1999; see the review by Gor'kov and Kresin, 2004; see also Dzero *et al.*, 1999, 2000). Subsequently, this approach has been employed in many papers (e.g. Jaime *et al.*, 1999; Dagotto *et al.*, 2001).

The approach based on the percolation concept implies that the system is intrinsically inhomogeneous. The inhomogeneity may manifest itself through 'phase separation'.

The phenomenon of 'phase separation' corresponds to the simultaneous existence of mutually penetrating sub-phases. As was noted above, the progress in the theory of manganites was strongly influenced by the preceding discovery and by studies of high-temperature superconductivity in copper oxides. For instance, the concept of phase separation (the coexistence of insulating and metallic sub-phases) was first suggested by Gor'kov and Sokol (1987) in relation to the high-T_c superconducting oxides. The idea received further development and experimental attention in many papers (for a review, see e.g. Sigmund and Mueller, 1994). The phase separation in manganites was analysed by Nagaev (1996). An ab initio study by Pineiro *et al.* (2012) showed that the nanoscale doping inhomogeneity appears to be stable.

The percolation can be interpreted as a more subtle case of phase separation where 'phases' coexist on a mesoscopic scale or even as interweaving clusters. Macroscopic properties are determined, mainly, by one sub-phase, where there are also inclusions of another phase. At $x > x_c$, metallic paths are formed and gradually become more and more dominant, but the insulating regions are also present.

9.3.2 The Percolation Threshold

Experimentally, the metallic behaviour of a compound at low T sets in at the critical value $x = x_c \approx 0.16$ (Urushibara *et al.*, 1995). At cooling, the temperature at which conductivity sharply increases almost coincides with the onset of the ferromagnetism, especially in good-quality samples. For example, for the $La_{0.67}Ca_{0.33}MnO_3$ sample ($T_C = 274$ K) studied by Heffner *et al.* (1996), these temperatures coincide to within 1 K (see also the work by Schiffer *et al.*, 1995 and the review by Coey *et al.*, 1999). Such closeness of the onset temperatures is a strong indicator in favour of the double-exchange mechanism.

Let us discuss the threshold value $x_c \approx 0.16$. Recall that material may be prepared by various methods but always at high temperatures. As a result, positions of the atom R, which substitutes for a parent atom, are completely random. A divalent atom R locally creates a 'hole' localised over adjacent Mn sites. The Coulomb forces in the dielectric phase keep the 'hole' close to the negative charge at R^-. When the concentration is small, average distances between R ions are large, so the holes remain isolated, forming trapped states (polarons).

Consider the concentration at which the nearest-neighbour R atoms start, in accordance with the percolation picture, to form infinite clusters piercing the whole crystal. Usually, one attacks percolation via one of two discrete mathematical models of the cubic lattice: the 'site' model and the 'bonds' model. In the site model, a hole is localised at a single centre. The critical concentration for the site model depends on the type of lattice; for a simple cubic structure, $x_C = 0.31$ (see e.g. Shklovskii and Efros, 1984). However, this is not the case for manganites. Although ionic substitution does take place at the centre of the cubic unit, the hole formed is spread out over several Mn sites around the R^-ion. At the same time, the charge transfer due to the formation of the larger cluster occurs only along Mn–O–Mn bonds. Hence, the picture of a critical cluster constructed from the R^- ions is not correct; such an initial (nucleating) cluster is not a point-like formation and already has a finite size ('thickness'). The size is even bigger at a dielectric constant large enough to weaken the Coulomb attraction to the R^- ion. According to numerical calculations (Scher and Zallen, 1970), this circumstance (i.e. the involvement of a scale of a few lattice constants into the percolation problem) strongly decreases the value of the critical concentration for percolation. It is interesting that now the critical value x_C depends only on the spatial dimension and appears to be invariant for all lattices. For the three-dimensional case (see e.g. Shklovskii and Efros, 1984)

$$x^{3D} \approx 0.16 \tag{9.2}$$

The site problem, when corrected by a finite hole radius, becomes similar to 'continuum' percolation. As to the continuous limit, it is only natural that independent numerical studies (Shklovskii and Efros, 1984) lead to close values for x_C. Therefore, the variant of the percolation theory at $x_C \approx 0.16$ describes the threshold above which the formation of an infinite cluster first takes place. In its manifestation, it corresponds to the appearance of a new macroscopic state, that is, in our case, to the transition into a metallic ferromagnetic state.

It is remarkable that the experimentally observed value of the critical concentration $x_C \approx 0.16$ corresponds, indeed, with a good accuracy, to the value obtained in the framework of percolation theory.

Measurements for $La_{1-x}Sr_xMnO_3$ have been performed by Urushibara *et al.* (1995). The analysis of the phase diagram for $La_{1-x}Ca_xMnO_3$ (Schiffer *et al.*, 1995; see also Coey *et al.*, 1999) shows that the values of the critical concentration for different manganites are close to the value (9.2).

The percolative description means the situation when one phase manifests itself as tiny inclusions ('islands') embedded into another, the dominant macroscopic phase. Here lies the difference between the percolative picture and that of the macroscopic electronic phase separation.

9.3.3 Increase in Temperature and the Percolative Transition

Above we described the evolution of the phase diagram, including the metal–insulator transitions as driven by doping at low temperatures. Let us consider a compound with a fixed level of doping. For example, manganite with $x = 0.3$ has a well-established metallic ferromagnetic state in the low-temperature region. An increase in temperature leads to the transition (at $T = T_C$) to the paramagnetic and low-conducting state. This transition can be also treated by means of the percolation theory, and one can apply ideas similar to those described above. Indeed, the high-temperature resistivity ($T > T_C$) is very large and one may approximately take the conductivity here as $\sigma(T > T_C) = \rho^{-1}(T > T_C) \approx 0$. The fast increase in $\sigma(T)$ at $T < T_C$ is then expected to correspond to

$$\sigma(T < T_c) \propto (T_c - T)^{\gamma} \tag{9.3}$$

where γ is a critical index.

The statement that the transition at T_C is also of a percolative nature implies the intrinsic inhomogeneity of the sample, that is, the phase separation, which is quite similar to what was discussed earlier.

It is essential to emphasise the additional similarities between the manganites and the cuprates. The statistical nature of doping leads to the percolative transitions observed for both systems. As was described above (chapter 6) the transition of the cuprates at $T_c = T_c^{res}$ into the macroscopic dissipationless state is of a percolative nature. It is similar to that in manganites for their transition at T_c into a ferromagnetic metallic state.

9.3.4 Experimental Data

The theoretical approach based on the percolation theory has strong experimental support. To start with, the experimentally measured critical concentration $x_C \cong 0.16$ for the $La_{1-x}Sr_xMnO_3$ compound, and a close value for other manganites, is in excellent agreement with the value predicted by the three-dimensional percolation theory (see eqn (9.2)). This agreement is a direct quantitative indication in favour of the percolative nature of the metal–insulator transition.

Percolation always implies some inhomogeneity of the system. Let us first consider the low-temperature region and trace in more detail the doping dependence. The neutron-pulsed experiments by Louca and Egami (1997, 1999) directly indicate the presence of such local inhomogeneities. These experiments probed the local arrangements of the oxygen octahedra. The presence of such insulating inclusions in $La_{1-x}Sr_xMnO_3$ is seen up to $x \approx 0.35$, that is, well above x_C, in the metallic region. The inhomogeneous structure has also been observed with the use of tunnelling spectroscopy (Becker *et al.*, 2002).

Note also that the value of magnetisation M depends on the doping level: $M \equiv M(x)$. The observed dependence is close to the simple relation $M(x) = (4 - x)\mu_B$, but it is *less* than this value (see Coey *et al.*, 1999); this indicates that an admixture partly made up of the 'non-metallic' (non-ferromagnetic) phase still persists at these concentrations. This fact and the concentration range agree well with the value $x \approx 0.35$ estimated from the neutron experiments (Louca and Egami, 1999).

At the same time, there are data which indicate the existence of metallic ferromagnetic regions below x_C, inside the insulating phase. The presence of such metallic islands should manifest itself in the linear (electronic) term in the heat capacity. Indeed, according to Okuda *et al.* (1998), the finite value of Sommerfeld's constant at $x < 0.16$ was observed in $La_{1-x}Sr_xMnO_3$.

Interesting results for the $La_{1-x}Zn_xMnO_3$ compound were reported by Felner *et al.* (2000). The authors measured the magnetisation and magnetic susceptibility and observed an additional ferromagnetic signal at $T = 38$ K for $x = 0.05$ and $x = 0.1$. The samples with $x = 0.05$ and $x = 0.1$ are in the insulating state, but the interesting results described above were explained by assuming that, in accordance with the percolation picture, the samples contained ferromagnetic metallic clusters. The presence of such clusters leads to the additional signal.

As discussed above, the transition at $T = T_C$ from the low-conducting high-temperature state to the metallic ferromagnetic state, in turn, can be treated in terms of percolation. Therefore, one should expect to observe insulating paramagnetic inclusions below T_C. The phenomenon, indeed, has been directly demonstrated via the scanning tunnelling microscopy experiments conducted at the surface of $La_{1-x}Ca_xMnO_3$ ($x \approx 0.3$) by Fath *et al.* (1999). The scanning tunnelling microscopy images were taken in magnetic fields between 0 and 9 Torr. The insulating regions were observed at temperatures below bulk T_C(!). With an increase in the magnetic field, insulating regions shrink and convert into metallic ones (as they should for the double-exchange mechanism), although some insulating regions survive even at fields as high as 9 Torr.

A detailed analysis of the local structure can be performed with use of X-ray adsorption fine-structure spectroscopy (see Li *et al.*, 2015). According to Booth *et al.* (1996, 1998), above T_c one can observe not only localised states, but also free carriers. These results strongly support the percolation picture.

Additional experimental support comes from Mössbauer spectroscopy measurements (Chechersky *et al.*, 1999). A strong paramagnetic signal has been observed at $T > T_C$. A decrease in temperature leads, at $T < T_C$, to the appearance of a strong signature of the ferromagnetic state, the 'six peaks' structure in the Mössbauer signal. However, even at $T < T_C$, the paramagnetic signal persists down to $T \cong 20$ K, which is much below T_C.

For some manganites, for example $Nd_{1-x}Sr_xMnO_3$, one can observe (at $x \gtrsim 0.5 - 0.6$) a transition to the charge-ordered state (see e.g. Mahendiran *et al.*, 1999). It is interesting that the transition between this and the ferromagnetic metallic states is also described by percolation theory (Kukumoto *et al.*, 1999). We focus here on the lower doping regions because of their strong similarities with the cuprates.

According to the data, obtained by Billinge *et al.* (2000) using X-ray diffraction, a $La_{1-x}Ca_xMnO_3$ crystal above T_C, in the paramagnetic phase, is characterised by three different bond lengths: b_1 for Mn–O, b_2 for Mn^{4+}–O, and b_3 for Mn^{3+}–O, where $b_2 < b < b_3$. Below T_C, close to $T = 0$ K, there is only one bond length, b_1. Interestingly, at temperatures that are higher than 0 K but are still below T_C, one can observe the appearance of b_2 and b_3, in accordance with the percolation scenario.

All these data (see also the review by Gor'kov and Kresin, 2004) strongly support the approach based on the percolative picture.

9.3.5 Large Doping

The three-dimensional metallic ferromagnetic phase of $La_{1-x}Sr_xMnO_3$ persists up to $x \approx 0.5$. A further increase in doping leads to a rather sharp transition (at low-enough temperature) to the so-called metallic A-phase. This phase is also metallic, as far as the low-temperature conductivity is concerned (similar to the ferromagnetic phase at smaller x) but it has a different magnetic structure. Namely, it consists of metallic ferromagnetic layers with the magnetisation orientation alternating in the direction perpendicular to the layers.

This metallic A-phase persists up to $x \approx 0.55$. For a compound with such a high concentration, it might be more convenient to consider first the opposite end of the phase diagram, $x = 1$. The limit $x = 1$ described the compound $RMnO_3$ (e.g. $SrMnO_3$).

This material is an insulator and contains only Mn^{4+} ions. Such compound does not contain e_g electrons at all. Starting from this end, one can describe the phase diagram as the result of substitutions A → R (e.g. SrLa), that is, as the electron-doped compound. The composition of the sample can be written as $La_ySr_{1-y}MnO_3$. At $y = 0.45$ (which corresponds to $x = 0.55$ in the 'hole-doping' picture), there is a transition to the metallic state. Because the antiferromagnetic ordering is along the c-direction (where the c-axis has chosen to be perpendicular to the layers), we start dealing with almost two-dimensional transport (hopping in the c-direction is spin forbidden, thanks to the Zener double-exchange mechanism).

It is rather temping to interpret the metal–insulator transition at $y = 0.45$ as a percolative transition in the two-dimensional model. Indeed, as was noted in the previous section, the percolation threshold depends on the dimensionality of the system; for the hole-doping (three-dimensional) case, $x_C = 0.16$. According to percolation theory, for the two-dimensional case (see Shklovskii and Efros, 1984), $y_C = 0.45$. This value for the variant is in remarkable agreement with the experimentally observed value $y_C \approx 0.45$ ($x_C{}^{\text{large}} \approx 0.55$) for the $La_{1-x}Sr_xMnO_3$ compound.

9.4 Main Interactions and the Hamiltonian

As was noted above, the e_g electron is the key player in the physics of manganites. Its hopping provides both the mechanism of conductivity and the ferromagnetism (the double-exchange mechanism).

One should also add that, because two neighbouring octahedra share a common oxygen ion (along the Mn–O–Mn bond), the deformations are not independent and, when we speak of the Jahn–Teller effect below, we always mean a so-called *cooperative* Jahn–Teller effect (see e.g. Kaplan and Vechter, 1995), that is, the corresponding structural change of a whole lattice.

The total Hamiltonian for the e_g electron has the form

$$\widehat{H} = \sum_i \left(\sum_{i,\delta} \hat{t}_{i,i+\delta} - J_H \sigma \cdot S_i + g\hat{\tau}_i Q_i + J_{el} Q_i^2 \right) \tag{9.4}$$

The first term describes the hopping process from a site i to its nearest neighbour $i + \delta$. Note that, because of the double degeneracy of the e_g-level, $\hat{t}_{i,i+\delta}$ becomes a 2×2 matrix.

The second term represents the Hund's interaction. The Hund's coupling is rather strong ($J_H S \simeq 1.0 - 1.5$ eV); this is the largest energy scale in the theory.

The third term is the Jahn–Teller interaction. It manifests in the spontaneous lattice distortion (Fig. 9.6); Q_i are the vibrational modes (Kanamori, 1961), and $\hat{\tau}$ is the pseudospin matrix (Kugel and Khomskii, 1982). The importance of the Jahn–Teller term

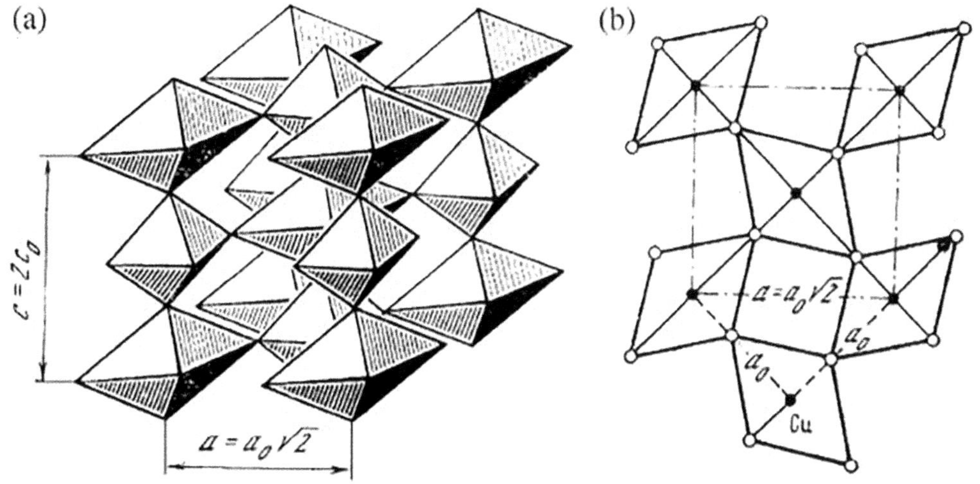

Figure 9.6 *Jahn–Teller deformation of the octahedron: (a) antiferrodistortion (the three-dimensional picture); (b) the top view.*

was stressed by Millis *et al.* (1995). The last term accounts for the 'elastic' energy part of the local Jahn–Teller mode.

As was mentioned above, in this approach, the largest energy scale in the Hamiltonian (9.4) corresponds to the Hund's interaction (~1.0–1.5 eV). As for the hopping and the Jahn–Teller terms, they have a competitive strength, that is, the same order of magnitude, so that $t \sim g \sim 0.1$ eV, with $t, g << \mathcal{J}_H$.

The hopping term in (9.4) leads to delocalisation of the e_g electron and, correspondingly, results in the band picture. Having a large value for the Hund's coupling constant $\mathcal{J}_H$ is an important feature of manganites. The double degeneracy of the e_g orbitals plays an important role in manganites' properties. We call this the two-band scenario (see below).

The fact that the band approach can capture the main physics of manganites is far from being obvious. Another approach often uses explicitly the generalised local Hubbard model to account for the strong on-site electron–electron interactions. Actually, instead of having a direct Hubbard on-site interaction $U > 0$, which hinders the double occupancy of a single Mn site, one may consider the local Jahn–Teller effect as an alternative way to describe the same physics. Indeed, a single electron positioned on the degenerate e_g orbital on the Mn site will cause a local lattice distortion, reducing the energy of the system. On the other hand, having *two* electrons on the same site does *not* lead to Jahn–Teller instability, and the Jahn–Teller energy gain is not realised. Therefore, locally, it is always more favourable energetically to have a single electron on a given site. We should also stress the large values of the Hund's coupling ($\mathcal{J}_H \sim 1.0$–1.5 eV) that is responsible for the spin alignment of all electrons on a given Mn site.

Based on the Hamiltonian (9.4), one can find the electron energy spectrum and, afterwards, study numerous electronic properties of the system.

9.5 The Ferromagnetic Metallic State

9.5.1 The Two-Band Spectrum

Let us find the single-electron spectrum of the three-dimensional metallic ferromagnetic manganite as it follows from the Hamiltonian (9.4); the results apply at $0.16 < x < 0.5$ (Dzero *et al.*, 2000). One can use the tight binding approximation, assuming that the overlap of the electronic wave functions for neighbouring Mn ions is rather small.

The Bloch electronic wave function in the tight binding approximation, as usual, may be chosen in the form

$$\Psi = \sum_n e^{i\boldsymbol{p}\boldsymbol{a}_n} \psi(\boldsymbol{r} - \boldsymbol{a}_n) \tag{9.5}$$

The summation is over all lattice centres $\boldsymbol{a}_n$,$\boldsymbol{p}$ is the quasi-momentum, and Ψ denotes the column formed by a localised (atomic) wave function (see below). This is the only difference that comes from the fact that the state of the e_g electron is the double-degenerate one.

In order to calculate the one-electron spectrum, one should choose the local basis set. It is convenient to use the normalised basic set of the form (Gor'kov and Kresin, 1998, 2004)

$$\psi_1 \propto z^2 + \varepsilon x^2 + \varepsilon^2 y^2$$
$$\psi_2 \equiv \psi_1^* \tag{9.6}$$

where $\varepsilon = \exp(2\pi i/3)$. This choice allows us to account for the cubic symmetry of the initial lattice.

In principle, one can pick up another basis which is often used in literature, namely the set of real functions

$$\varphi_1 \propto d_{z^2} \equiv 3z^2 - \left(x^2 + y^2\right)$$
$$\varphi_2 \propto d_{z^2 - y^2} \equiv x^2 - y^2 \tag{9.7}$$

There is a simple connection between these two sets:

$$\psi_1 = (\varphi_1 + i\varphi_2)/\sqrt{2}$$
$$\psi_2 \equiv \psi_1^* \tag{9.8}$$

Of course, the expression for the energy spectrum does not depend on the choice of the local basic set.

The equation of motion for an electron determining its energy spectrum is as follows (note that, for the metallic phase, the spectrum is determined by the hopping term only;

cf. Anselm, 1982):

$$E\begin{Bmatrix}\psi_1(\mathbf{r}_i)\\ \psi_2(\mathbf{r}_i)\end{Bmatrix}=\sum_{\mathbf{a}_n}\hat{T}_{i,i+n}\begin{Bmatrix}\psi_1(\mathbf{r}_i+\mathbf{a}_n)\\ \psi_2(\mathbf{r}_i+\mathbf{a}_n)\end{Bmatrix} \tag{9.9}$$

$$\hat{T}_{ij}=\begin{Bmatrix}T^{11} & T^{12}\\ T^{21} & T^{22}\end{Bmatrix} \tag{9.9'}$$

As a result, one can obtain the following expression for the energy spectrum:

$$E_{1,2}(\boldsymbol{k})=-\,|A|\,(c_x+c_y+c_z\pm R) \tag{9.10}$$

where

$$c_i=\cos(k_i a), i=x,y,z\;|\mathrm{A}|\approxeq 0.16\mathrm{eV} \tag{9.11}$$

The constant $|A|$ is determined from the experimentally measured parameters (this is the single constant which enters the theory), a is the lattice period, and

$$R=\left(c_x^2+c_y^2+c_z^2-c_xc_y-c_yc_z-c_zc_x\right)^{1/2} \tag{9.12}$$

One can see that the spectrum consists of the two branches. The two-band structure of the electron spectrum is an important feature of metallic manganites. It naturally explains the observed asymmetry of the phase diagram (hole vs. electron doping). Moreover, it is essential for a detailed description of various features of the compounds and, especially, their optical properties (see below and section 5.4).

As we assumed above, the Hund term is the largest one, so that, in what follows, we use the strong inequality $\mathcal{J}_H >> |A|$. Therefore, all branches of the electronic spectrum are shifted up or down by the energy $\pm J_H S$, depending on the e_g- and t_{2g}-mutual spin orientation. Itinerant spins for each of the branches would merely split into two by adding the $\pm J_H\langle S\rangle$ energy term. Therefore, the lowest two bands correspond to the parallel orientation of the e_g- and local t_{2g}-spins:

$$E^L_{1,2}=-J_H\langle S\rangle-|A|\,(c_x+c_y+c_z\pm R) \tag{9.13}$$

where c_i and R are defined by eqns (9.11) and (9.12). There are also two upper bands:

$$E^v_{1,2}=J_H\langle S\rangle-|A|\,(c_x+c_y+c_z\pm R) \tag{9.14}$$

However, the ground state is always formed by making use of the two lowest bands, $\varepsilon^L_{1,2}$.

With the use of eqns (9.12) and (9.14), one can estimate the total width of the spectrum ΔE:

$$W=6\,|A| \tag{9.15}$$

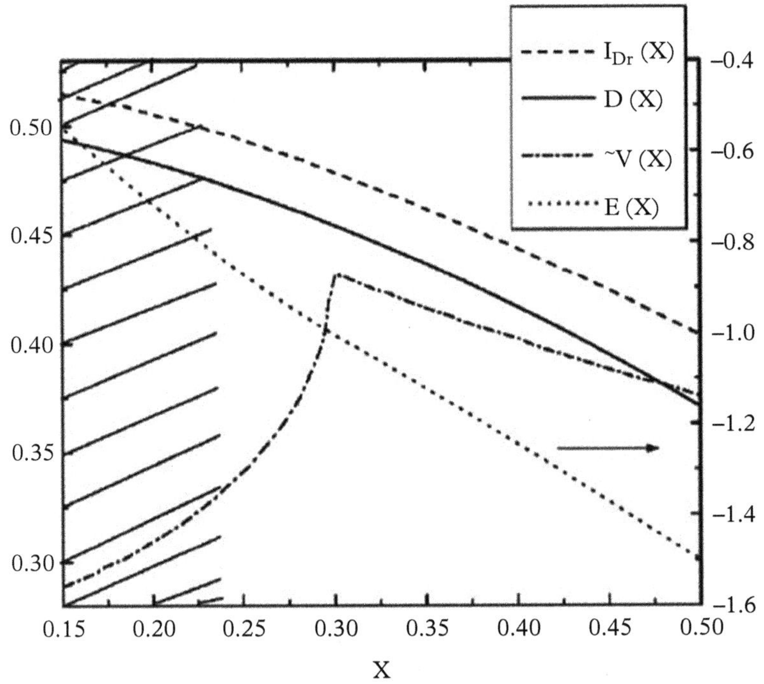

Figure 9.7 *The dimensionless Fermi level (in units of* $|A|$*)* $E(x)$*, the density of states* $\overline{v}(x)$*, the spin coefficient* $D(x)$*, and the Drude conductivity* $I_{Dr}(x)$*, plotted as a function of concentration,* x*. The meaning of the different lines is indicated in the inset. The shaded area shows the percolative regime.*

In accordance with eqn (9.11), ΔE is of the order of 1 eV. In practice, at all concentrations, the Fermi level lies at lower energies. Therefore, in metallic manganites, we are dealing with relatively narrow energy bands. Using eqns (9.10)–(9.12) as the spectrum, one can calculate the concentration dependence of the main parameters of manganites (Fig. 9.7).

9.5.2 Heat Capacity

As is known, there are several contributions to the low-temperature heat capacity. In ferromagnetic manganites, in addition to the common electronic ($\propto T$) and phonon ($\propto T^3$) terms, there is also a contribution from spin waves (magnons). Their dispersion law ($\omega \propto k^2$) leads to the contribution $C_{mag}.\propto T^{3/2}$. In addition, there is a kink in the density of states at $x \approx 0.3$ (Fig. 9.7). That kink takes its origin from the fact that this concentration is the point of the 2.5 'Lifshitz' singularity (Lifshitz, 1960; also see Blanter *et al.*, 1994).

As was just mentioned, the total magnon contribution to specific heat is proportional to $T^{3/2}$. The proximity to the Lifshitz '2.5' transition results in appearance of a term in the electronic specific heat term with the same T dependence. This observation makes the procedure of extracting the 'pure' magnon $\propto T^{3/2}$ terms less transparent.

The spectrum (9.13) can be used to calculate the density of states per single spin $N(x)$ and then the usual linear electronic term $C = \gamma T$ with the Sommerfeld constant $\gamma = \hbar^2 N/3$. With use of the value at $x = 0.3$ (see Fig. 9.6) and the value $|A| \approx 0.16$, we obtain $\gamma \approx 6$ mJ/mole K^2. This value is in a reasonable agreement with the experimental data (see e.g. Viret *et al.*, 1997).

9.5.3 Isotope Substitution

As we know, molecules as well as various complex systems can display the dynamic John–Teller effect. We discussed this phenomenon above as the origin of a peculiar isotope effect in superconductors. It turns out that manganites also display an isotopic dependence of the critical temperature for the transition into the three-dimensional ferromagnetic metallic state.

It turns out that the oxygen isotope substitution ($O^{16} \to O^{18}$) leads to a large shift in the Curie temperature (Zhao *et al.*, 1996; Franck *et al.*, 1998; Gordon *et al.*, 2001; see also the review by Belova, 2000), $T_c \propto M^{-\alpha}$; for instance, for $La_{0.8}Ca_{0.2}MnO_3$, the isotope coefficient is quite large and is equal to $\alpha \approx 0.85$. Like high-T_c oxides, this shift is also caused by the dynamic polaronic effect (Kresin and Wolf, 1998).

The oxygen ion is located between the Mn ions, and its dynamics is directly involved in the process of charge transfer. Seated just between two Mn ions, the oxygen ion, in addition, is characterised by *two* minima (Fig. 9.8). This picture is similar to that in the high-T_c superconducting oxides. The presence of two minima leads to the appearance of symmetric and asymmetric energy terms and corresponding energy-level splitting.

The energy splitting corresponds to inverse time for the charge hopping between the two Mn ions $\left(\Delta E \propto \tau^{-1}\right)$. Since the ferromagnetic ordering is caused by the electron hopping, the splitting makes it possible to estimate the value of the critical temperature

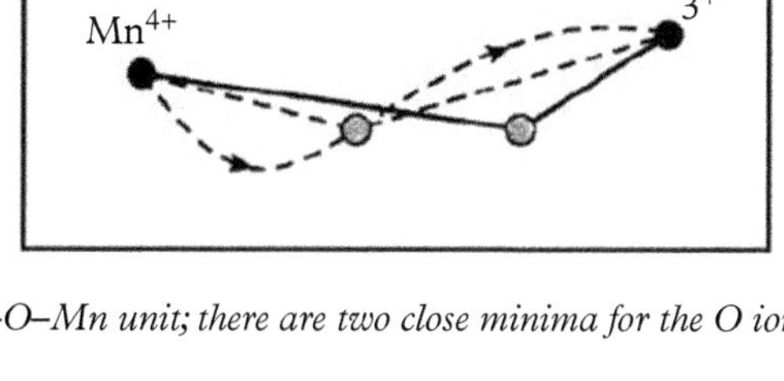

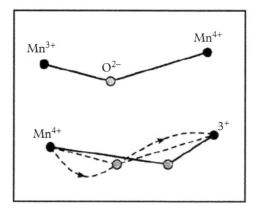

Figure 9.8 *The Mn–O–Mn unit; there are two close minima for the O ion. (Sharma et al., 1996)*

for the ferromagnetic ordering; in other words, it is natural to assume that $T_c \propto \Delta E$ (Millis *et al.*, 1996).

The expression for the splitting can be written in the form

$$\Delta E \propto L_0 F \tag{9.16}$$

where L_0 is the electronic factor (at $\overline{R} = \vec{R}_0$), and F is the Franck–Condon factor (see section 1.2.2).

The electronic factor L_0 determines the hopping parameter, A (and, therefore, T_c in terms of the usual double-exchange model) with the *frozen* lattice. One can see from eqn (9.16) that the account of the O ion dynamics leads to the appearance of the additional Franck–Condon factor. The obvious inequality, $F < 1$, leads to a decrease in the energy splitting. Moreover, the presence of the Franck–Condon factor leads to the oxygen isotope effect (see below).

Let us add some qualitative remarks. The charge transfer for the extra electron can be visualised as a multistep process: firstly, the electron makes a transition from the Mn^{4+} site to the oxygen; then the oxygen goes over to another minimum, from where the electron jumps into to the other Mn ion. Therefore, actually, the charge transfer includes the important additional step of the oxygen moving between the two minima. Namely, this additional step leads to a drastic increase, described by the Franck–Condon factor, in the characteristic time for the jump between the two Mn ions. Naturally, this factor also decreases the strength of the ferromagnetic coupling, and, consequently, T_c. In the traditional language, the charge transfer is accompanied by a transition to another electronic term; this process is similar to the Landau–Zener effect (see e.g. Landau and Lifshitz, 1977).

As noted above, the dependence on the mass of the oxygen isotope is brought into the problem through the appearance of the Franck–Condon factor. Indeed, this factor contains the overlap of the nuclear wave functions and, therefore, depends on the nuclear mass M (in our case, the mass of the O ion). The Franck–Condon factor can be written in the form $F \sim \exp(-\kappa^2)$, $\kappa = (\Delta\rho/2\rho)$, where $\rho = (\hbar/M\Omega)^{1/2}$ is the vibration amplitude. If we take the value $\kappa \approx 1.25$–1.50 (which is consistent with the data from Sharma *et al.*, 1996), we obtain $F \approx 0.18$–0.20. The value of the isotope coefficient appears to be in good agreement with the experimental data. Indeed, defining the isotope coefficient by the relation $T_c \propto \Delta_E M^{-\alpha}$ one can rewrite it as $\alpha = -(M/\Delta M)(\Delta T_c/T_c)$; with $T_c \propto \Delta E$, we obtain

$$\alpha = (\mathrm{M}/\Delta\mathrm{M})\,(\Delta T_c/T_c) = -\,(\mathrm{M}/\Delta\mathrm{M})\,(\Delta\mathrm{F}/\mathrm{F}) \tag{9.17}$$

where $M = M - M^*$; we used the relation $T_c \propto \Delta E$ and eqn (9.16). Based on the expression for the Franck–Condon factor, we obtain $\alpha = 0.5\kappa^2$. Using the value $\kappa \approx 1.25$–1.50, we obtain values in the range of 0.8–1.0, in good agreement with the value obtained by Zhao *et al.* (1996).

The charge transfer in the metallic manganites is accompanied by the motion of the oxygen ion between two minima: the electron becomes 'dressed' by this motion. This dynamic polaronic effect affects the value of the critical temperature.

9.5.4 Optical Properties

The study of the optical properties of metallic manganites is of special interest, because it is necessary to take into consideration the two-band structure of the spectrum. For manganites, the interband transitions appear to be the key factor in the infrared region.

The ac conductivity is described by the well-known Kubo–Grinwood expression

$$\begin{aligned}\sigma_{ij} = -\frac{e^2\hbar^2}{V\omega}\sum_{k,k'} f_0(\boldsymbol{k})\left[1 - f_o(\boldsymbol{k}')\right] \\ \times \langle\Psi(\boldsymbol{k})|\hat{\boldsymbol{v}}|\Psi(\boldsymbol{k})\rangle\langle\Psi(\boldsymbol{k}')|\hat{v}|\Psi(\boldsymbol{k}')\rangle \\ \times \left[\delta\left(\varepsilon(\boldsymbol{k}') - \varepsilon(\boldsymbol{k}) - \omega\right) - \delta\left(\varepsilon(\boldsymbol{k}') - \varepsilon(\boldsymbol{k}) + \omega\right)\right]\end{aligned} \tag{9.18}$$

where $f_o(\boldsymbol{k})$ is the Fermi distribution function, $\boldsymbol{k}$ is a quasi-momentum, $\hat{\boldsymbol{v}}$ is a velocity operator, and $\Psi(\boldsymbol{k})$ is the two-column electron wave function. For a cubic crystal, $\sigma_{ij} = \sigma_{ji}$.

To start with, we determine the velocity operator, $\hat{\boldsymbol{v}} = \dot{\boldsymbol{r}}$ (see Lifshitz and Pitaevsky, 1999):

$$\hat{\boldsymbol{v}}(\boldsymbol{k}) = \frac{1}{\hbar}\frac{\partial\varepsilon_l(\boldsymbol{k})}{\partial\boldsymbol{k}} + \frac{i}{\hbar}\left[\varepsilon_l(\boldsymbol{k}) - \varepsilon_l'(\boldsymbol{k})\right]\langle l\boldsymbol{k}|\hat{\boldsymbol{\Omega}}|l'\boldsymbol{k}\rangle \tag{9.19}$$

The off-diagonal operator $\hat{\Omega}$ is defined by the relation

$$\langle l\boldsymbol{k}|\hat{\boldsymbol{\Omega}}|l'\boldsymbol{k}\rangle = i\int u_{\boldsymbol{k}}^{*l'}(\boldsymbol{r})\frac{\partial u_{\boldsymbol{k}}^l}{\partial\boldsymbol{k}}\,d^3\boldsymbol{r} \tag{9.20}$$

and $\nu_k^l(r)$, the periodic Bloch functions.

Since l, l' are the band indexes, the second term in (9.19) explicitly includes the interband transitions. The light absorption comes about due to these contributions. These are the direct transitions, and the importance of these transitions for light absorption is simply due to the negligibly small value of the photon's momentum. As a result, conservation of energy and momentum can be simultaneously satisfied (in the clean limit) only for the interband transitions.

The matrix element for the interband transition is

$$\langle l\boldsymbol{k}|\widetilde{\boldsymbol{\Omega}}|l'\boldsymbol{k}\rangle = i\frac{a}{\hbar}\frac{\sqrt{3}}{4}\frac{(-\sin k_x)(c_y - c_z)}{|t_{12}|^2} \tag{9.21}$$

As a result, one arrives at the following expression for the 'optical' (interband) contribution:

$$\sigma_{opt}(\omega, x) = \frac{3\pi e^2}{a\hbar} \frac{1}{\tilde{\omega}^3} \int \frac{d^3\mathbf{p}}{(2\pi)^3} \sin^2 p_x (c_y - c_x)^2 \times n(\varepsilon_+(\mathbf{p}))[1 - n(\varepsilon_-(\mathbf{p}))] \delta(\tilde{\omega} - 2R(\mathbf{p})) \quad (9.22)$$

The interband transitions affect the optical conductivity (see e.g. Okimoto and Tokura, 2000), which is also very sensitive to the sample quality.

9.6 The Insulating Phase

9.6.1 The Parent Compound

The parent compound $AMnO_3$ (e.g. $LaMnO_3$) is an insulator and an antiferromagnet. In addition, its magnetic ordering is in the so-called A-structure, with a relatively low Néel temperature ($T_N \approx 140$ K). Then the ferromagnetic ordering in layers is combined with an antiferromagnetic order in the direction perpendicular to the layers. It turns out that even the properties of the parent manganite can be understood in the framework of the two-band picture (Gor'kov and Kresin, 1998; Dzero *et al.*, 2000). Therefore, the band approach represents a unified description of the low-temperature properties of manganites, applicable to all phases (metallic and insulating).

Let us first make several preliminary remarks. The 'right' stoichiometric end ($x = 1$) of the phase diagram for the $A_{1-x}R_xMnO_3$ compound, that is, for example, for $SrMnO_3$, is also an insulator, and this is not surprising. Indeed, in this case, all of the manganese ions are Mn^{4+}, the itinerant e_g electrons are absent, and the ions contain only the strongly bound t_{2g} groups.

The situation with the parent compound, $AMnO_3$ (the 'left' side of the phase diagram, $x = 0$) is entirely different, because of the presence of loosely bound e_g electrons.. These electrons are responsible for the metallic conductivity in doped manganites. Nevertheless, despite the presence of such electrons, the parent material behaves as an insulator. One can show that the two-band picture previously introduced (section 7.5) but extended to incorporate the complete Hamiltonian, that is, including the Jahn–Teller effect, allows us fully understand the insulating ground state.

One more remark is as follows: one might think that this limiting case ($x = 0$) should also be treated, like $x = 1$, in the framework of the localised picture; this would mean that the Mn^{3+} e_g shells should behave as localised e_g orbitals. And, indeed, the pertinent properties of manganites are often interpreted in terms of a generalised microscopic Hubbard model (see e.g. Kugel and Khomskii, 1973). The key feature of the Hubbard model is the assertion that, for two electrons to be placed on the same site, the energy cost is very high (the famous Hubbard $U > 0$ duc to the on-site Coulomb repulsion!). However, the Hubbard Hamiltonian approach has been challenged by Gor'kov and Kresin (1998); see also the work by Dzero *et al.* (2000). Firstly, there are experimental motivations for such a challenge. For instance, it was shown experimentally that, in the doped manganites $La_{1-x}Sr_xMnO_3$, at rather low concentrations, say, $x = 0.2$, the system may display excellent metallic behaviour at low temperatures (Lofland *et al.*, 1997). Meanwhile, the nominal number of e_g electrons per Mn site, $N = 1 - x < 1$, was changed by only one-fifth in this study. In addition, from the theoretical point of view, it is, of course, clear that dealing with the strictly atomic d-orbitals would be a strong oversimplification. If a Mn ion is placed into the oxygen octahedron environment, the e_g terms are formed by the whole ligand, so that the 'pure' d-functions become considerably hybridised with the surrounding oxygen states (e.g. see the discussion by Anderson, 1959b and by Pickett and Singh, 1996). Hence, the electronic polarisation would undoubtedly reduce the magnitude of the 'Hubbard'-like (on-site) interactions. Recall also, as was stated previously (section 7.4), that the Jahn–Teller instability also makes the situation when two electrons occupy the same site unfavourable. A study of the optical properties of $La_{1-x}Ca_xMnO_3$ at various values of x (Luke *et al.*, 1998) demonstrated the presence of the Jahn–Teller splitting. Finally, for the Jahn–Teller effect (which itself is nothing but another form of the Coulomb interaction) to arise, there is no need to use a scenario with Mn^{3+} localised states.

Let us describe the band insulator qualitatively; for a more detailed analysis, see the work by Gor'kov and Kresin (2004). The parent compound $AMnO_3$ has, magnetically, the A-structure, that is, the neighbouring layers have opposite magnetisation. As a result, the transport in this compound would have a two-dimensional nature, since hopping in the c-direction is spin forbidden. Each two-dimensional unit cell contains one delocalised electron. In the usual one-band picture, such a system would stay metallic since, at Pauli band filling, the electrons will occupy half of the band. However, the double-exchange mechanism, together with the large value of J_H, causes all of the electrons have only one spin direction. Therefore, a single band in a 'half-metallic' picture would be fully occupied. However, in manganites, e_g electrons occupy not one but two bands. If we neglect the Jahn–Teller term, the spectrum will be described by eqn (9.13) and this should lead again to a metallic, not insulating, state. Therefore, an analysis based on the Hund's (double-exchange) and hopping terms only is not sufficient, since it would produce a metallic state. Things are different if one takes into account static deformations caused by the Jahn–Teller interaction. Indeed, as the result, the 'superstructure' imposed by the Jahn–Teller deformations, as shown in Fig. 9.6, causes the two-dimensional Brillouin zone now to be double-folded, that is, reduced by a factor of 2. After a superlattice is imposed at the Jahn–Teller collective transition, the same number of electrons may completely fill up the reduced Brillouin zone, producing an insulator.

9.6.2 Low Doping and Polarons

The previous section is concerned with the parent compound, that is, with the undoped manganites. The chemical substitution (e.g. La $\rightarrow$ Sr) of La^{+3} by the divalent ion leads to the formation of a 'hole' in the unit cell (see section 2.1). Experimentally, at a low doping level, the material remains an insulator. This fact that, at small concentrations, doped manganites preserve the insulating state raises an interesting problem.

The insulating behaviour at light doping means that the introduced 'holes' remain localised. One obvious reason for this is certainly the Coulomb attraction to the doping ion (Sr^{2+}) that prevents the hole from immediately joining the top of one of the conduction bands. However, even if the Coulomb forces were screened on large distances in the presence of a finite hole concentration, the strong electron–phonon interaction originating from the local Jahn–Teller term may significantly change the band characteristic. Local distortions may result in forming self-consistent trapping centres for holes/electrons, that is, creating a new type of 'carrier', the so-called 'polaron'. In the three-dimensional case, for a band carrier to become trapped in the polaronic state (usually with a much heavier effective mass), energy is needed to overcome the energy barrier separating the band and the trapped polaronic states. With an increase in polaron density, the latter should merge gradually into the band spectrum. The situation is more interesting and less trivial in case of a two-dimensional electronic spectrum. It is worth saying a few words about this scenario, since, as we have just discussed, for the parent $LaMnO_3$ with its A-type magnetic structure, the conducting network of the MnO planes bears two-dimensional features.

In the two-dimensional case, the energy barrier for polaron formation may be equal to zero. Correspondingly, carriers may be either itinerant or localised (having heavy masses), depending on the numerical value of some parameter, C. This parameter characterises the competition for the energy gain between the gain in elastic energy (see the linear and quadratic terms in eqn (9.4)) due to the Jahn–Teller distortion and the kinetic energy gain due to the finite band width, which is proportional to t. If the value

$$C \approx g^2/J_{el}t \tag{9.23}$$

exceeds a threshold, usually of the order of unity, the doped hole would inevitably go into a trapped state (Rashba, 1982; Toyozawa and Shinozuka,1983).

In simple terms, eqn (9.23) tells us whether, due to lattice deformation, the hole energy goes below the bottom of the band and thus remains 'localised'. Recall again that, for the A-structure, the transport has mainly a two-dimensional nature. Since, experimentally, at low doping, manganites initially remain in the insulating state, we conclude that criterion (9.23) favours the absence of the potential barrier for the localisation of introduced holes; in other words, holes are trapped in 'heavy' polarons.

The polaronic picture and criterion (9.23) only make sense in the limit of low-enough carrier concentrations. An increase in doping leads, eventually, to percolation and the phase separation picture described in section 6.2. We focus here on the low-temperature region. It is worth nothing, however, that a study performed with the use of a combination of X-ray spectroscopy and photoemission spectroscopy (Mannella *et al.*, 2004) demonstrated that the state of disorder above T_c is consistent with the presence of Jahn–Teller distortion and the polaronic picture.

At the percolation threshold $x_c \approxeq 0.16$, an itinerant conduction network develops, leading to a transition into the metallic (and ferromagnetic) ground state. As we have shown before, somewhat above the percolative concentration threshold, the macroscopic metallic phase formed can again be described in terms of band theory (see section 7.5).

9.7 The Metallic A-Phase and the Superconductor–Antiferromagnet–Superconductor Josephson Effect

9.7.1 Magnetic Structure

We discussed above the properties of the ferromagnetic metallic phase of the manganites $A_{1-x}R_xMnO_3$. This phase occupies the doping region $0.16 < x < 0.4$–0.5.

The study of $La_{1-x}SrMnO_3$ (Akimoto *et al.*, 1998; Tokura and Tomioko, 1999; Izumi *et al.*, 2000; Moritomo *et al.*, 2001) has led to a remarkable observation that increased doping ($0.5 < x < 0.55$) gives rise to the appearance of a metallic antiferromagnetic phase, the so-called A-phase. For the A-phase, the core (t_{2g}) spins are aligned ferromagnetically in each MnO plane (e.g. the ab-plane) and antiferromagnetically along the axis perpendicular to the planes (the c-axis; Fig. 9.5). In the ferromagnetic phase, all of the electrons, including the e_g ones, are fully polarised (the 'half-metallic' state). Now the 'half-metallic' state is only realised inside each of the ab-planes of the A-phase.

Therefore, the magnetic A-structure combines the ferromagnetic order in the layers with the antiferromagnetism in the c-direction. Such a magnetic order leads to highly anisotropic transport, because charge transfer in the c-direction is spin forbidden.

9.7.2 The Josephson Contact with an A-Phase Barrier

As is known, the superconducting state is characterised by macroscopic phase coherence. The Josephson effect (1962; see e.g. Barone and Paterno, 1982, and Tinkham, 1996) represents the most remarkable manifestation of this coherence. A Josephson contact is comprised of superconducting electrodes separated by a tunnelling barrier. The dc current flows through the contact without any external voltage and is described by the equation

$$J = j_0 \cos\varphi \tag{9.24}$$

where $\varphi = \varphi_1 - \varphi_2$ is the phase difference of the superconducting order parameters on each side of the barriers. The most common case is the S–I–S Josephson junction, where S denotes the superconductor and I denotes a thin insulating barrier. Another well-known case is the S–N–S junction; here, N denotes a normal metal. The amplitude of the superconducting current for such a junction (see e.g. Barone and Paterno, 1982, and Kresin, 1986) is proportional to

$$\mathrm{j}_0 \propto \exp\left(-\mathrm{L_N}/\xi_\mathrm{N}\right) \tag{9.25}$$

if $L_N > \xi_N$. Here ξ_N is the normal coherence length, defined as $\xi_N = v_F/2T$ (the 'clean' case; Clarke, 1969; see also section 3.4). One can see from eqn (9.25) that, for the effect to be observable, the thickness of the barrier L_N should not noticeably exceed the coherence length ξ_N.

Let us consider the barrier formed by a magnetic metal. It is well understood that, for two superconductors with singlet (*s*-wave or *d*-wave) pairing, a ferromagnetic barrier (S–F–S junction, where F denotes the ferromagnetic barrier) would present a strong obstacle. Indeed, the Josephson current is the transfer of a Cooper pair with the spins of its two electrons in opposite directions. The exchange field in the ferromagnetic barrier is trying to align the spins in the same direction, and this leads to the pair-breaking effect. The situation is entirely different for the case of tunnelling through an antiferromagnetic metallic system (S–AFM–S, where AFM denotes the antiferromagnetic metallic barrier; Gor'kov and Kresin, 2001, 2002). Unlike the case with a ferromagnetic barrier, where pair breaking takes place, superconducting currents might penetrate through an antiferromagnetic metallic barrier much more easily.

Let us consider a junction in a geometry that would allow the Josephson current to flow along the ferromagnetic layers. The layers should be weakly coupled electronically. This is important, because the Josephson current represents the transfer of *correlated* electrons. A manganite in the metallic A-phase would form such a barrier, as an A-phase manganite is a natural spin valve system (see e.g. Kawano *et al.*, 1997).

To demonstrate this effect, we assume that the barrier is thick enough so that we can neglect phenomena taking place in immediate proximity to the boundary. Then one can use the interface Hamiltonian in the form

$$H = V\Delta^{(i)}\psi^{+}(i)\psi(i) \tag{9.26}$$

Here $\Delta^{(i)}$ are superconducting order parameters on each side; i, ψ^{+}, and ψ are the field operators for the carriers inside the barrier; and V is a tunnelling matrix element at the boundary of the barrier. Integration along the contact surface is assumed.

As a next step, it is practical to evaluate not the current itself directly but instead the 'surface' contribution to the thermodynamic potential, $\delta\Omega$, which is caused by the presence of the barrier separating two bulk superconductors. The current is then determined as the derivative $\delta\Omega/\delta\varphi$, where $\varphi = \varphi_1 - \varphi_2$ is the phase difference between the two superconductors, so that $\Delta^{(1)} = |\Delta^{(1)}|\exp(i\varphi_1), \Delta^{(2)} = |\Delta^{(2)}|\exp(i\varphi_2)$. The amplitude of the Josephson current, j_m, turns out to be proportional to the matrix element of the Cooper diagram:

$$K(1,2) = |V|^2 \sum_{\omega_n} \Delta^{(1)}_{\sigma\sigma'} G_{\sigma\sigma''}(1,2;\omega_n) G_{\sigma'\sigma'''}(1,2;-\omega_n) \Delta^{*(2)}_{\sigma''\sigma'''} \tag{9.27}$$

For a singlet superconductor, $\hat{\Delta}$ is a matrix of the form $\Delta^{(i)}_{\sigma\sigma'} = \Delta^{(i)}(i\sigma_y)_{\sigma\sigma'}$. The thermodynamic Green's function is employed, and summation over repeating spin indexes is assumed. To properly evaluate the current through the magnetic barrier, one should pay special attention to the spin structure of eqn (9.27), because the Green's functions $G_{\sigma'\sigma''}$ and $G_{\sigma''\sigma'''}$ inside the barrier are not diagonal in spin indices. The calculation leads to the following general expression for the amplitude of the Josephson current:

$$j_m = r\gamma\pi T \sum_{\omega_n > 0} \int dp_z \exp(-\omega_n L/v_F) \exp\left(iLt_{\|} M/Sv_{\perp}\right)$$

$$\gamma = (1 - M/S)^2;\ r \propto |V|^2 |\Delta_1|^2 |\Delta_2|^2 \tag{9.28}$$

Assume that the width of the barrier $L >> \xi_N$, where ξ_N is the coherence length for the normal layer, eqn (9.25). Then one can keep only the first term of the sum in eqn (9.28). In the tight binding approximation, $t_{\|}\ (p_z) = t_0 \cos(p_z d)$, where d is the interlayer distance. Integrating over p_z, we arrive at the following expression:

$$j_m = j_m^0 e^{-\frac{L}{\xi_N(T)}} J_0\left(\beta M/S\right) \tag{9.29}$$

Here $\xi_N(T)$ is the coherence length inside the barrier, $j_m^0 = [1 - (M/S)]^2$, $\beta = (t_0/T_c)(L/\xi_0)$; $\xi_0 = hv_F/2\pi T_c$; and $v_F = v_F^0$ is the maximum value of the component of the Fermi velocity along L. It is essential that $\beta >> 1$; indeed, for manganites, $t_0 >> T_C$ and $L >> \xi_0$. If the canting M/S is not negligibly small, one can use an asymptotic form of the Bessel function, $J_0(x)$, and we obtain

$$j_m \approx (\pi\beta M/2S)^{-\frac{1}{2}} j_m^0 e^{-\frac{L}{\xi_N(T)}} \cos\left(\beta M/S - \pi/4\right) \tag{9.30}$$

Equation (9.30) is valid if $L >> \xi_N(T)$. Therefore, the antiferromagnetic barrier does transfer the Josephson current; in this case, the exchange field does not break the Cooper pairs.

It is also very interesting that the mutual magnetic orientations of layers can be controlled by an external magnetic field. Thus, the antiferromagnetic structure can be transformed into the ferromagnetic configuration. A complete antiferromagnetic-to-ferromagnetic reorientation would have a drastic impact on the system, causing the Josephson current for the ferromagnetic case to be forbidden ($j_m = 0$). This follows from eqn (9.29); indeed, then $M = S$ and $j_m = 0$. Therefore, one can change the current with an external magnetic field.

It is very interesting that, according to eqn (9.30), one observes the oscillatory dependence of the amplitude of the Josephson current on an external magnetic field: $j_0 \propto \cos(M/S - \pi/4)$. Canted moments may be induced by even a weak field and would result at $\beta >> 1$ in rapid and large ('giant') changes in the Josephson current ('giant' magneto-oscillations for the Josephson contact with the presence of the A-structure as a weak link). Note that the oscillating dependence was theoretically obtained by Buzdin *et al.* (1982) for the ferromagnetic barrier; they considered the oscillations caused by a *change in the barrierthickness*. As for the antiferromagnetic barrier, one can observe this effect as being due to the slight variation in the external magnetic field.

9.8 Discussion: Manganites versus Cuprates

Manganites, like cuprates, are complex oxides, mixed-valence compounds. Both systems are characterised by rich phase diagrams and the fact that the state is greatly affected by doping.

It has been demonstrated that, in manganites, an insulator-to-metal transition to the ferromagnetic and metallic state at finite doping represents a peculiar type of the transition that must be described in terms of percolation (Gor'kov and Kresin, 1998; see also the review by Gor'kov and Kresin, 2004). The percolative approach is also applicable for the transition at the Curie temperature. A similar picture can be observed for the cuprates (see section 6.4). The systems are intrinsically inhomogeneous, and we are dealing with phase separation.

As was demonstrated above, many results could be obtained analytically. It is clear that such an approach has a serious advantage not only because it allows one to gain an additional insight but also because the calculations are tractable, whereas the numerical results are sometimes contradictory.

A sensible quantitative theory of manganites that self-consistently explains the main peculiarities of manganites can be built by using the relatively simple Hamiltonian (9.4), which contains, as the key ingredients, a hopping term and Hund's and Jahn–Teller interactions. The band approach utilising all these interactions (the generalised two-band model) provides an adequate and unifying approach. Note that the corresponding two-gap scenario is essential for a cuprate like YBCO (see section 5.3).

Both systems (cuprates and manganites) are greatly affected by the Jahn–Teller instability and polaronic effects. The oxygen ions display dynamic Jahn–Teller effects and this leads to peculiar isotopic dependences. Manganites and cuprates share many similarities. Thus, for example, it was not by accident that de Gennes, who previously had made important contributions to both fields, that is, superconductivity and manganites (de Gennes, 1960, 1966), made an effort to explain the origin of high-T_c superconductivity in the cuprates by using their similarity with manganites (Plevert, 2011).

However, as was noted above (section 9.2), the situation in cuprates is opposite to that in the manganite case. For manganites, the unusual transition into the ferromagnetic and metallic state was explained shortly after its discovery (Jonner and van Santen, 1950; Zener, 1951). With the cuprates, on the other hand, the mechanism of high T_c is still a controversial issue, and a general consensus has not yet been achieved.

10

Superconducting States in Nature

The absence of resistivity is the most well known but not the only feature of superconductors. They are characterised by peculiar magnetic properties, optical properties, transport properties, and so on. Pair correlation is the key ingredient underlying all these properties of the superconducting state. It turns out that the phenomenon of pair correlation affects the behaviour of entirely different systems in nature, so one can see it as a special state of matter.

10.1 Pair Correlation in Atomic Nuclei

Let us start from the pair correlation of nucleons in atomic nuclei. This phenomenon was introduced shortly after the appearance of the BCS theory (see e.g. Broglia and Zelevinskii, 2013). Initially, we will describe some of the relevant properties of nucleons. Afterwards, the pair correlation of nucleons will be discussed.

10.1.1 Nucleons

Protons and neutrons form two groups of nucleons. Each proton, as well as each neutron, is characterised by spin $s_z = 1/2$, so that, for example, the protons in atomic nuclei form the Fermi system. It is similar to the electronic system in metal. But, unlike the electrons in bulk metal, both Fermi systems in nuclei contain a finite number of fermions and form the so-called finite Fermi systems. Electrons in metallic nanoclusters (section 8.2) also form finite Fermi systems, like those in atomic nuclei. A weak attraction between electrons leads to the formation of Cooper pairs. The interaction between the nucleons is much stronger, so that one can expect the existence of the pair correlation in nuclear matter.

The finite size leads to strong impact of the surface on various properties and, first of all, on the energy spectrum of nucleons. The first manifestation is that the energy spectrum is discrete. Moreover, the one-particle spectrum is strongly affected by surface deformations.

The so-called shell model is a very important part of the nuclear physics (see e.g. Goeppert-Mayer, 1949; Ring and Schuck, 1980). According to this model, each nucleon

Superconducting State: Mechanisms and Materials. Vladimir Z. Kresin, Sergei G. Ovchinnikov, and Stuart A. Wolf, Oxford University Press (2021). DOI: 10.1093/oso/9780198845331.003.0010

moves in the self-consistent field formed by all other nucleons; this field is the same for all nucleons and depends on the distance from the centre of the nuclei. Of course, this picture is, in large degree, based on the Pauli principle, which strongly diminishes the interaction between fermions, and it makes the picture of isolated particles moving in some average field formed by other particles appear realistic.

The states of nucleons (e.g. neutrons) form energy shells similar to those in atoms. The electronic states are classified by the orbital momenta l, their projection m, and spins $(\pm 1/2)$. As for the radial part of the wave function, it is described by models (the oscillator or the potential box). The so-called magic numbers (2, 8, 20, 28, 40, etc.) correspond to complete shells. An important role, especially for large values of orbital momenta, is played by spin–orbit interaction. The 'magic' nuclei are characterised by relatively large stability, so that its dependence on the number of nucleons is not monotonic.

Note that the shell structure of a metallic nanocluster (section 8.3), which also an example of finite Fermi systems, is similar to that for nuclei. One should stress only that, for the description of nanoclusters, one can use the adiabatic approximation, which allows us to classify rigorously the energy spectra (electronic and vibrational). Moreover, the forces (Coulomb interaction) are studied much better than interactions between nucleons. As a result, the theory of nanoclusters is more advanced and can be developed at the microscopic level.

As was noted above, 'magic' nuclei have a spherical symmetry. Additional nucleons fill the next, unoccupied shell, and this leads to surface deformation. This can be described by the following expression:

$$R(\theta,\varphi) = R_0 \left\{1 + \beta\,(5/16\pi)^{1/2} \left[\cos\gamma\ (3\cos^2\theta - 1) + 3^{1/2}\sin\gamma\,\sin^2\theta\ \cos 2\varphi\right]\right\} \tag{10.1}$$

Here $R(\theta,\varphi)$ is the length of the vector pointing from the centre to the surface, R_0 is the radius of the spherical nucleus, and θ and φ are the spherical coordinates for the fixed coordinate system. The quantities γ and β are the so-called Hill–Wheeler coordinates (1953). These coordinates describe the scale of deformation (β) and the degree of mixture of the prolate and oblate configurations (γ). The value $\gamma = 0$ corresponds to the prolate configuration (where the axis z has been chosen to be the symmetrical axis), see eqn (10.1), and $\gamma = \pi/3$ describes the oblate configuration (with x as the axis of symmetry; see Fig. 10.1. Intermediate values of γ correspond to triaxial deformation.

Because of these shapes (Fig. 10.1), the shape of the nuclei oscillates between two quasi-resonant states (see appendix B). Indeed, the coexistence of different shapes was studied for ${}^{43}_{16}S_{27}$ (Sarazin *et al.*, 2000; Gaudefroy *et al.*, 2009; Fortenato, 2009). This phenomenon is also similar to that in metallic nanoclusters.

10.1.2 Pair Correlation

As mentioned above, the concept of pair correlation in atomic nuclei was introduced soon after the creation of the BCS theory (Bohr *et al.*, 1958; Belyaev, 1959; Migdal,

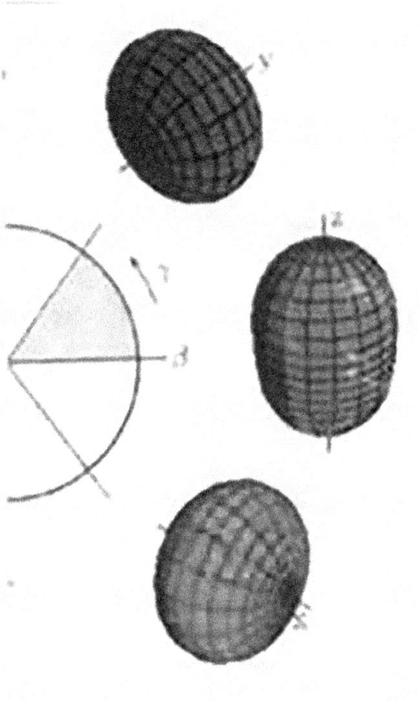

Figure 10.1 *Nuclear configurations.*

1959). The first question related to this phenomenon is, what is the manifestation of the pairing of nucleons? Of course, the discussion about an absence of resistance does not make any sense in this context. In reality, the pairing is manifested in the spectroscopy of nuclei. First of all, one should mention the odd–even effect. The presence of the energy gap means that the energy required for the transition of the nuclei into an excited state should, at least, exceed the value of the gap. This is the case, when the nuclei (e.g. a system of protons) contains even number of nucleons.

The situation when nuclei contain an odd number of nucleons is different. In this case, the system has one unpaired nucleon and, as a result, it can absorb much less energy. The even–odd effect can be observed independently for the neutron and proton systems. For example, for neutrons, it is described by the following so-called pairing-gap parameter:

$$\delta_n = \pm \mathrm{F}; F = 0.5\left[E_B(Z,N+1) - 2E_B(Z,N) + E_B(Z,N-1)\right]$$

Here the sign '+' corresponds to an odd number of neutrons, and the sign '−' to an even number, N is the number of neutrons in the nuclei, and Z is the number of protons. The dependences of the pairing-gap parameters for odd and even numbers of neutrons are discussed by Bertsch (2013), who has shown that odd-N nuclei are less strongly bound than their even-N neighbours. On average, the relative decrease of energy for odd-N nuclei is about 1 MeV. The effect is especially strong near the 'magic' numbers.

Similarly, one can introduce the parameter δ_p, which describes the pairing for a proton system. The dependences for odd-P and even-P parameters look similar to those for the neutron system. However, the difference near the 'magic' number is not large as for the neutrons. At present, an explanation for the difference between the pairing of neutrons versus that of protons is absent. It probably requires a better understanding of nuclear forces, so this problem is a challenge for a future study. As a whole, the odd–even effect is a fundamental manifestation of the pairing phenomenon in atomic nuclei.

The next manifestation of the pair correlation is its impact on the values of the momenta of inertia of deformed nuclei. These values can be determined from the measurements of rotational spectra of nuclei. It turns out that these measurements provide values which are noticeably smaller than those obtained from the simple calculation performed for rigid structures. This looks strange, because the rigid model should be applicable for a description of a system with strong interactions and such a high density.

The pair correlation of nucleons, and the corresponding pairing energy gap parameter 2Δ, are the key ingredients allowing to explain the data on momenta of inertia (Migdal, 1967; Bohr and Mottelson, 1974). The best qualitative explanation is based on the two-liquid model. Because of the pairing, the system of nucleons can be visualised as containing two components, a superconducting (or superfluid, by analogy with He II) one and a normal one. The superfluid component does not contribute to the moment of inertia, and its value is determined by the presence of the normal 'part'. We placed the word 'part' in quotation marks because the situation is similar to that in liquid He II below the critical point. The presence of two components does not mean that we are dealing with real mixtures of two liquids. In reality, one can talk about two types of motion describing the behaviour of superfluid helium. As a result, the observed moment of inertia for the rotating liquid He II is less than for the usual liquid. The calculated value of the moment of inertia for atomic nuclei depends on the value of the energy gap $2\mathit{\Delta}$. For infinite Δ, the value of the moment of inertia is equal to zero, whereas its value is equal to that for 'solid' nuclei if $\Delta = 0$.

Speaking of the pairing interactions, one should make two points. There exists the contribution of low-frequency surface deformations. This contribution is similar to that of phonons for usual superconductors. Note also that, as a whole, we are dealing with strong nuclear forces. Their impact is diminished by the Pauli principle but, nevertheless, there is reason to restrict the interaction by two-body interactions. The absence of any small parameter leads to comparable contribution of three-body interactions, and so on. In connection with it, it is interesting that the existence of α-particles, which are observed from radioactive decay and contain two protons and two neutrons, allows us to assume the importance of specific four-particle correlations in nuclear matter. But a more detailed picture of such correlations requires a deeper understanding of nuclear forces, and this is an interesting challenge for future studies.

10.2 Pair Correlation and Astrophysics

After describing the impact of pair correlations on the properties of atomic nuclei, let us turn our attention to the entirely opposite scale, which is related to the topics studied in astrophysics.

In 1932 Landau predicted the existence of stars containing nuclear matter (see also Landau, 1965). The explicit prediction of a neutron star was made by Baade and Zwicky (1934); it is remarkable that this prediction was made just two years after the discovery of neutrons by Chadwik (1932). According to Baade and Zwicky (1934), there exists 'the transition from ordinary stars into neutron stars, which in their final stages consist of closely packed neutrons'. The first experimental observation was reported in 1968 (Hewish *et al.*, 1968; see also Bell Burnell, 2007). Initially, the observed object was named LGM ('little green men'; this name was given to creatures travelling in flying saucers). Later, this object was given the name 'pulsar', after their signature pulsating signals. They are unusual stars, because they contain dense nuclear matter, mainly neutrons. On the scale of astronomy, their size is relatively small (~10 miles), but, because of their high density, their mass is comparable with that of the sun (~1.4 solar mass). As a whole, a neutron star can be visualised as a giant nucleus with a size similar to that for a relatively large city.

Despite the giant scale, the strong interaction between neutrons leads to their pair correlation, similar to that in atomic nuclei. Let us discuss the manifestations of this phenomenon. It is interesting that the same fundamental phenomenon, pair correlation, can manifest itself differently, depending on the subject of interest. For usual superconductors, one can observe the absence of resistance, the Meissner effect, and so on. For atomic nuclei, it is manifested in the odd–even effect and the peculiar behaviour of the momenta of inertia (see section 10.1). As for the neutron star, the main manifestations are connected with the heat capacity of the superconducting state and also with its vortex structure.

As was previously described (see chapter 2), the presence of an energy gap leads the electronic heat capacity to decrease exponentially with a decrease in temperature, that is, more rapidly relative to the usual linear law. As a result, the value of the heat capacity is rather low. With respect to neutron stars, this feature is important for the proper study of the star's evolution (see the review by Yakovlev and Pethick, 2004). The simple equation that makes it possible to analyse the coating of the pulsar has the form

$$\frac{dE_{\text{th}}}{dt} = -c_{\text{V}}\frac{dT}{dt} \tag{10.2}$$

Here E_{th} is the star's total thermal energy, and c_{v} is its specific heat. The value of $(\partial E_{\text{th}}/\partial t)$ can be determined from measurements of the photon and neutrino luminosities. A detailed analysis demonstrates that the pairing does indeed lead to the observed sharp cooling of the neutron star in the Cassiopeia A supernova (Page, 2013).

The radio signals coming from pulsars are extremely regular, but sometimes one can observe jumps in the periods; these are called 'glitches.' The appearance of these 'glitches' and the relaxation after them is affected by the presence of vortices in the superconducting nuclear matter: there exists friction between the normal component and the superfluid component.

In conclusion, let us quote Freeman Dyson (1970):

> I myself find that the most exciting part of physics at the present moment lies on the astronomical frontier, where we have just had an unparalleled piece of luck in discovering the pulsars. Pulsars turn out to be laboratories in which the properties of matter and radiation can be studied under conditions millions of times more extreme than we previously had available to us . . . the pulsars will, during the next 30 years, provide crucial tests of theory in many parts of physics ranging from superfluidity to general relativity.

10.3 Biologically Active Systems

We previously discussed (section 8.2) the superconducting state in organic system, including aromatic molecules. All biologically active systems contain carbon atoms. As a result, these systems, or some sets of carbon atoms, contain delocalised π-electrons. The picture is similar to that in hydrocarbon compounds and, more specifically, in aromatic molecules (see section 8.2.2). The π-electrons are similar to conduction electrons in metals; as π-electrons can form Cooper pairs, we can see pair correlation in biologically active systems.

There are a number of phenomena which can benefit from the pair correlation. For example, an important class of phenomena, the so-called redox reactions, represent usually processes of the transfer of *pairs* of electrons. The name 'redox' is a shortened form of 'reduction–oxidation'. The redox reaction describes the process of charge transfer between two atoms, which is accompanied by the loss of two electrons from one of them (the reducing agent) and the gain of electrons (the oxidising agent) by another one. The transfer of electronic *pairs* is due to the double valence of the oxygen (see also the discussion in the book by Marouchkine, 2004).

In such a case, the charge transfer occurs as a manifestation of the tunnelling effect. Since we are dealing with the transfer (tunnelling) of two electrons, one can consider an important channel for such a transfer, namely Josephson tunnelling (section 3.1.2). It is essential that the probability of the Josephson tunnelling be the same as for the one-particle propagation through the barrier. As a result, the tunnelling through the Josephson contact is favourable for the transfer of a pair of electrons, as is required for the redox reaction. This Josephson tunnelling is similar to that for a nanocluster-based network or a network formed by aromatic molecules, for example, coronene molecules (see section 8.2.3).Then we dealing with a charge transfer between two systems with discrete energy levels. Such tunnelling can proceed through a resonance channel, or with the use of additional degrees of freedom (e.g. vibrations); then the energy conservation requirement is satisfied.

An interesting experiment by De Vault and Change (1966) was concerned with protein reaction rates and their temperature dependence $K(T)$; the reaction was triggered by photon absorption. At $T > 100K$, they observed the usual Arrhenius behaviour with $K(T) \propto \exp(-E_{el}/T)$. However, at $T < 100K$, the reaction rate appeared to be temperature independent. The absence of the temperature dependence can be explained by the presence of a different channel for the reaction, namely by the tunnelling effect (Grigorov and Chernavskii, 1972; Goldstein and Bialek, 1983).

The problem of charge transfer and the pairing state in the DNA molecule has attracted a lot of attention. This is not by chance, because of fundamental importance of this molecule. The conduction through the DNA molecule was reported by several authors (e.g. Fink and Schonenberger, 1999; Chepellianskii *et al.*, 2011; Simchi *et al.*, 2013). There is the question of the proper choice of electrodes to provide the required contact. For example, Ga nanoparticles can serve as such electrodes.

A superconducting state, induced via the proximity effect (see section 3.4), was observed by Kasumov *et al.* (2001) and discussed by Simchi *et al.* (2014).

Note also that a large scale of pair correlation could be important to the information transfer in biological systems. A review by Mikheenko (2019) contains an interesting analysis of possible superconductivity in the brain. The presence of such a correlated state could be important for the transfer of information. As for the dissipationless superconducting current, it can serve as the key ingredient for units providing memory. As is known, it is difficult to perform any measurements in living organisms, which are active at the time of experiment. For the experiments described by Mikheenko, the animals' fast-delivered brain slices were used. The transport measurements, made with the use of specially prepared graphene electrodes, demonstrated that the I–V characteristics are consistent with the picture of the superconducting charge transfer. Of course, these studies, as well as those described in other papers and the aforementioned review, represent a first step in this interesting and important direction.

Let us conclude this section with a quote from the book *Quantum Biochemistry* by Pullman and Pullman (1963):

> All the essential biochemical substances which we are related to or perform the fundamental functions of the living matter are constituted of completely or, at least, partially conjugated systems... The essential fluidity of life agrees with the fluidity of the electronic cloud in conjugated molecules. Such systems may thus be considered as both the cradle and the main backbone of life.

Appendix A
The Dynamic Jahn–Teller Effect

As we know, a system containing degenerate electronic states is unstable. The degeneracy is removed via the Jahn–Teller effect, and, as a result, one can observe the splitting of initially degenerate energy levels. One should distinguish between static and dynamic Jahn–Teller effects.

The static Jahn–Teller effect (Fig. A.1) is caused by the electron–lattice interaction. This interaction leads to the so-called Jahn–Teller static distortion, so that the ions move to different equilibrium positions. As a result, we are dealing with an ionic configuration which has a lower symmetry, and this leads to the removal of the initial degeneracy. The scale of the distortion is determined by the interplay of the electron–lattice interaction and the static energy terms. The shift is energetically favourable (static distortion energy) and, as a result, the distorted configuration corresponds to a new stable state. It is essential that the distortion energy exceed the zero-point energy of the corresponding vibrational mode. Indeed, only in this case can the system be permanently distorted (the static Jahn–Teller effect).

However, if this is not the case, then we are dealing with the dynamic picture (the dynamic Jahn–Teller effect; see e.g. Salem, 1966). Namely, the system displays a dynamic interconversion between different distorted forms, that is, it oscillates between energy equivalent distortions. There could be an intermediate case which can be treated as an interplay of the static and dynamic Jahn–Teller effects (see Fig. A.1).

The so-called tunnelling splitting is a typical example of the dynamic Jahn–Teller effect (see Bersuker, 2006). Indeed, as we know, the tunnelling between two potential boxes leads to the splitting of initially degenerate (or quasi-degenerate) states, that is, to the removal of the degeneracy. The apical oxygen in the cuprates also displays such an effect, and the same is true for the oxygen ion in manganites. For both these cases, the dynamic Jahn–Teller effect is manifested in the peculiar isotope dependence. We call it the 'polaronic isotope effect'. In this case, we are talking about the oxygen ions tunnelling between two equilibrium minima. The total wave function is written as the sum of the products of the electronic and vibrational wave functions, describing the isolated boxes. As a result, unlike the usual adiabatic picture, the electronic and lattice degrees of freedom appear to be not separable. This is a key ingredient of the dynamic polaron picture.

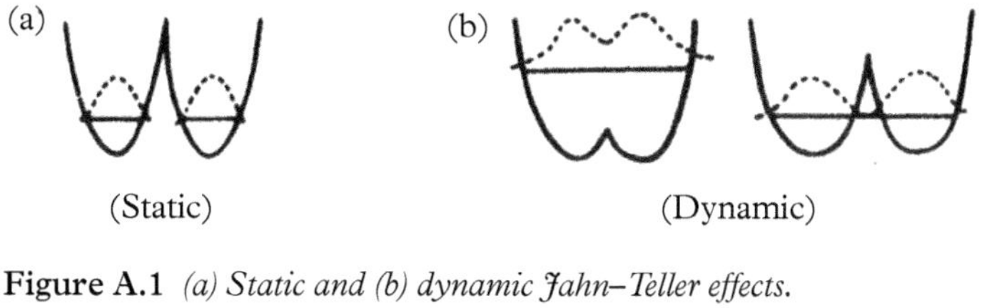

Figure A.1 *(a) Static and (b) dynamic Jahn–Teller effects.*

It is interesting that the formation of the well-known energy bands in solids is the manifestation of the dynamic Jahn–Teller effect. Indeed, if we employ the tight binding approximation, then the total electronic wave function is a superposition of atomic wave functions (orbitals) describing the electronic state of an isolated ion. Assume, for simplicity, that we are dealing with the monoatomic lattice. Then, in the zero-order approximation, when all ions do not interact and are totally isolated, we have the degeneracy of electronic states (many ions with the same energy levels). This degeneracy is removed by the electronic tunnelling between neighbouring ions, which is described by the tunnelling operator $\hat{T} \sim \int d\mathbf{r}\ \psi_i \left[V - v\left(\mathbf{r}\right)\right] \psi_{i+1}$ (see section 1.2.1). As a result, the levels of isolated ions are split and form the electronic energy bands, so that the degeneracy is removed. This splitting is the manifestation of the dynamic Jahn–Teller effect. A similar splitting occurs in the hydrocarbon molecules, for example, the benzene molecule (see appendix C). Note that the diabatic representation (section 1.2.2) is a very convenient tool for describing the dynamic Jahn–Teller effect.

Appendix B
Finite Fermi Systems: Quasi-Resonant States

Various finite Fermi systems, such as atomic nuclei, metallic nanoclusters, and hydrocarbon molecules (e.g. benzene ions) display a remarkable correlation between the number of fermions and the structure of the system. For example, metallic nanoclusters for small shell filling are characterised by degeneracy of the electronic states and this leads to the Jahn–Teller distortion towards the ellipsoidal shape (see appendix A). If the number of electrons is such that almost half of the shell is filled, then the energies of two possible ellipsoidal configurations, prolate and oblate, are almost equal. Then the degeneracy can be removed by the oscillations between these two configurations (the dynamic Jahn–Teller effect). The prolate and oblate structures form the quasi-resonant states and, in the time-dependent picture, we are dealing with oscillations between them.

The coexistence of various shapes is a well-known phenomenon in nuclear physics (see section 10.1). The nucleons form energy shells similar to those in nanoclusters, and one can observe the oscillations between the prolate and the oblate configurations of the nuclei. The manifestations of the presence of quasi-resonant states in nuclei and clusters are similar (Kresin and Friedel, 2011).

Consider now the benzene molecule, as an example of the hydrocarbon system. The molecule has D_{6h} symmetry (see appendix C), and its π electronic states are described by the quantum number m (quasi-projection of orbital momentum). Nevertheless, the benzene molecule does not undergo the Jahn–Teller distortion, because the upper π-level contains four electrons and is completely filled. The situation is different for the benzene ion formed by the removal of one π-electron, $C_6H_6^+$. In this case, the π-electron vacancy leads to electronic degeneracy. Then we are dealing with two different configurations of the structure of the ion. The distance between two neighbouring carbon ions alternates (in quantum chemistry, it corresponds to the alternation of single and double bonds). The structure of $C_6H_6^+$ corresponds to two possible quasi-resonant configurations.

The scenario with quasi-resonant states is different from that with isomers. Isomers correspond to different nanosystems, while quasi-resonant states describe, for example, the same cluster, or the same nuclei. Isomers represent a special case of quasi-resonant states (tunnelling is absent; $\hat{T} = 0$).

Appendix C
The π and σ Electronic States: The Benzene Molecule

Hydrocarbon molecules, which contain carbon and hydrogen atoms, form a very important class of organic systems. We focus here on the properties of the benzene molecule, C_6H_6, which is the basic unit of these molecules. Each carbon atom contains four valence electrons; each of these electrons is delocalised and moves in the field formed by the ionic system and other electrons.

Using a method similar to that in the quantum theory of solids, one can demonstrate that the electronic energy spectrum for such an electron contains groups of levels (branches) that are similar to the valence and conducting energy bands in solids. It allows us to separate the electronic states into π and σ groups; which not only differ in their spatial distribution but also correspond to different parts of the energy spectrum. The symmetry of the molecule and an analogy with the Bloch states in solids allows us to evaluate the spectrum in a rather simple way.

As was noted above, the benzene molecule C_6H_6 belong to the family of hydrocarbons, which are of key importance among organic systems. This molecule plays a role similar to that for hydrogen in atomic physics. Many systems, like aromatic compounds, graphene, fullerenes, and so on (see sections 8.1 and 8.2) contain the benzene molecule as the major building block. The parameters of the molecule are indicated in Fig. C.1. The molecule is formed by the carbon and hydrogen atoms. The valence electrons are not bound to the ions. It is known that these electrons are separated into two groups:

1) π-electrons, which are delocalised and move along the so-called σ core, and
2) σ-electrons; the ionic system and these electrons form the so-called σ-core, which creates the potential for the delocalised motion of the π-electrons.

This separation into π and σ groups works rather well in molecular spectroscopy, quantum chemistry, and so on. The energy spectrum of the π-electron system is usually calculated with the use of the Hückel method (1931); see, for example, the review by Higasi *et al.* (1965). Below we describe a simpler method based on the symmetry of the benzene molecule and the analogy with the description of electronic states in solids.

Note also that the separation of the electronic states into the π and σ groups should be proven rigorously; this point was stressed, for example, by Higasi *et al.* (1965). According to them, the evidence should be left to physicists; as for chemists, it is sufficient to know that this separation works as a useful concept.

Initially let us focus on the π electronic states and their energy spectrum. This means that we assume that these states can be separated from the σ states (this statement will be proven rigorously below). Our goal is to evaluate the π electron energy spectrum. In addition, the method of evaluation will be used for description of a whole system of delocalised valence electrons.

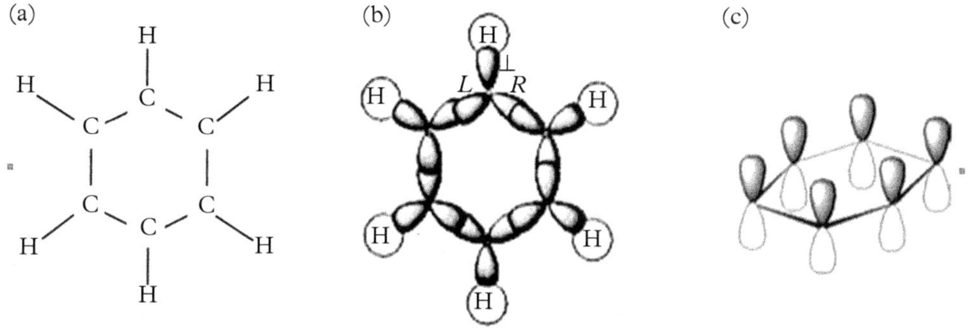

Figure C.1 *The benzene molecule: (a) the C_6H_6 structure and (b, c) the electronic orbitals.*

C.1 The π-Electron System and Its Energy Spectrum

The benzene molecule C_6H_6 contains six π-electrons corresponding to p_z electronic wave functions (orbitals) perpendicular to the plane of the molecule. The number of π-electrons is equal to that of carbon ions. The electronic tunnelling between the orbitals leads to π-electrons being delocalised; each of them is moving in the field formed by the σ-core and other π-electrons. This tunnelling leads to the overlap of the π electronic wave functions (it corresponds, in quantum chemistry terminology, to the formation of the so-called π-electron bond).

As mentioned above, the π-electron energy spectrum is usually evaluated with the use of the so-called Huckel method. More specifically, the total π-electron wave function is presented as a sum (the linear superposition) of six atomic wave functions ψ_z (ψ_z orbitals). Then one can write down the non-linear secular equation; its solution represents the spectrum of interest.

According to group theory, the molecule has D_{6h} symmetry. Therefore, it is invariant with respect to the rotation on $\pi/3$ relative to the z-axis (where the z-axis has chosen to be perpendicular to the plane and goes through the centre of the molecule).

As we know, because of the translational symmetry in solids, the electronic wave function depends on the quasi-momentum $\boldsymbol{p}$. Correspondingly, the so-called Bloch function in the tight binding approximation can be written in the following form (see e.g. Aschroft and Mermin, 1976):

$$\Psi(r, \boldsymbol{r}) = \sum_s e^{i\boldsymbol{p}\boldsymbol{r}_s} \varphi(\boldsymbol{r}_s) \tag{C.1}$$

where $\boldsymbol{p}$ is the quasi-momentum (e.g. $|p_x|_{\max} = \pi/a$, where a is the corresponding lattice period).

Let us go back to the benzene molecule and the analogy with the description of solids; one can then write the expression for the one-particle π electronic wave function. If this molecule had rotational symmetry (relative to the z-axis), then it would be appropriate to introduce the projection of angular momentum m on the z-axis ($m = 0, 1, 2, \ldots$; as usual, we put $\hbar = 1$). For a real molecule, which does not have rotational symmetry but is periodic with a $\pi/3$ rotation, it is reasonable to introduce the quasi-projection of the orbital momentum. This quantity is similar to quasi-momentum in crystals and reflects the periodicity with respect to the $\pi/3$ rotation. Correspondingly, the electronic wave function can be sought in the form

$$\Psi(\boldsymbol{r}) = \sum_s e^{im\varphi_s}\psi_s \tag{C.2}$$

Equation (C.2) is analogous to that used in solid state physics (cf. eqn (C.1)). Here ψ_s is the electronic wave function for the isolated carbon atom, $\varphi_s = \tilde{\varphi}s$ (Fig. C1b,c), and

$$\tilde{\varphi} = \pi/3;\ \ m = 0, \pm 1, \pm 2, 3 \tag{C.3}$$

can be called the quasi-projection of the angular momentum (it is analogous to the quasi-momentum), $s = 1, \ldots, 6$. Note that the states $m = 1, 2$ are double orbitally degenerate. The lowest group of levels, similar to the first Brillouin zone in solids, is restricted by the values $3 \geq |\mathrm{m}| \geq 0$. As a result, the lowest π group contains 12 electronic states (including those with opposite spin).

Substituting expression (C.2) into the Schrödinger equation $\hat{H}\Psi = E\Psi$ with the Hamiltonian $\hat{H} = \hat{T}_{\boldsymbol{r}} + V$ (where $\hat{T}_{\boldsymbol{r}}$ is the electronic kinetic energy operator, and V is the total potential energy), one can evaluate the energy spectrum. Note that the Hamiltonian can be written in the form

$$\hat{H} = \hat{h} + \tilde{V} \tag{C.4}$$

where $\hat{h} = \hat{T}_{\boldsymbol{r}} + v$ describes an isolated ion, and

$$\tilde{V} = V - v(r_s) \tag{C.4$'$}$$

is the effective potential leading to the inter-ionic charge transfer. After substituting (C.4) into the Schrödinger equation, let us multiply both sides by ψ_1^*. It leads to the following expression for the energy spectrum:

$$E_\pi = \varepsilon_0 + \mathrm{T_z}\cos m\tilde{\varphi} \tag{C.5}$$

Here ε_0 is the electronic energy of an isolated C atom (one can select that $\varepsilon_0 = 0$), T_z is the matrix element

$$T_z = \int \psi_z^{(1)*}\tilde{V}\psi_z^{(2)}d\tau \tag{C.6}$$

and $\tilde{V}$ is defined by eqn (C.4′). We neglected by all direct overlaps of the electronic functions from different ions; as for the integrals containing $\tilde{V}$, one keeps only those from the neighbouring ions. The matrix elements T_z describe the tunnelling between such ions. This description is analogous to that used in solid state theory (see e.g. Sommerfeld and Bethe, 1933; Ziman, 1960; Aschroft and Mermin, 1976) for a description of the energy bands in the tight binding approximation.

The energy spectrum (C.5) contains four energy levels classified by the values of the quasi-projection m = 0, 1, 2, 3. Two of them (m = 0, 1) are characterised by positive values of the energy (recall that $\varepsilon_0 = 0$). In addition, because of the above-mentioned orbital degeneracy, two of them ($m = 1, 2$) contain four electrons each (including the spin degeneracy). As a whole, there are 12 π electronic states. The six π-electrons can be distributed, in accordance with the Pauli principle and occupy the lowest two levels, whereas the two upper levels are vacant; this allows the intragroup transitions.

C.2 The Benzene Molecule: π and σ States

Let us consider all the valence electrons of a whole benzene molecule. Such a molecule C_6H_6 contains six carbon atoms. An isolated C atom has the electronic structure $1s^2 2s^2 2p^2$ (see e.g. Landau and Lifshitz, 1977). The transition of one $2s$ electron into close $2p$ state leads to the configuration $1s^2 2s 2p_x 2p_y 2p_z$ with four valence electrons. As is known, it is convenient to introduce linear combinations of the atomic wave function and, as a result, we are dealing with four orthogonal wave functions (so-called atomic orbitals; see Fig. C.1b, c).

Since each carbon ion has four valence electrons, we are dealing with a total of 30 valence electrons (24 electrons from six carbon atoms, and six electrons from the H atoms). Each of these electrons is delocalised in the field formed by the ionic core containing six C^{+4} ions and six protons (H^+) and by other electrons.

As above, one can employ the tight binding approximation; then the electronic wave function can be written in the following form (cf. eqn (C.2)):

$$\Psi(\boldsymbol{r}) = \sum_s e^{i\tilde{m}\varphi_s}\left[\alpha_{\tilde{m}}\psi_L^s + \beta_{\tilde{m}}\psi_R^s + \mu_{\tilde{m}}\psi_\perp^s + \eta_{\tilde{m}}\psi_H^s + \gamma_{\tilde{m}}\psi_Z^s\right] \tag{C.7}$$

Equation (C.7) is analogous to that used in the solid state theory, for example, for the case of degenerate energy levels (see e.g. Anselm, 1982; see also Dewar, 1969). A similar equation is used in the theory for manganites (see section 9.4, where the band structure is described in the tight binding approximation) and reflects the degeneracy of the Mn ions.

Substituting expression (C.7) into the Schrödinger equation $\hat{H}\Psi = E\Psi$ and multiplying it by $\psi_i^{(1)}$, we obtain the general secular equation. The roots of this equation determine the electronic spectrum of the molecule or, more specifically (see above), various groups (branches) of energy levels. As for the calculations in section C.1, we neglect the integrals containing direct products of electronic functions from neighbouring ions. We keep only the terms containing the integrals $\int \psi_i \hat{H} \psi_k d\boldsymbol{r}$; here $i, k = \{R, L, \perp, H, Z\}$ are the orbitals for the neighbouring ions. As for the Hamiltonian, it is described by eqns (C.4) and (C.4′). Since the Schrödinger equation is multiplied by the functions $\psi_i^{(1)}$, the secular equation will contain the matrix elements T_{ik} of the tunnelling operator $\hat{T} \equiv \tilde{V} = V - \nu(\boldsymbol{r}_s)$ with the wave functions $\psi_i^{(2)}, \psi_i^{(5)}$ of the neighbouring ions. Let us also remember that we have selected $\varepsilon_0 = 0$.

It is interesting that the secular equation has a special line (and column) corresponding to multiplying the Schrödinger equation by the function $\psi_z^{(1)}$. Because the atomic orbital $\psi_z^{(1)}$ is asymmetric with respect to reflection in the plane of the molecule, whereas other orbitals and the operator $T \equiv \tilde{V}$ are symmetrical, all such tunnelling matrix elements are equal to zero, except $T_z = \int \psi_z^{(1)} \tilde{V} \psi_z^{(2)} d\boldsymbol{r}$ and $\int \psi_z^{(1)} \tilde{V} \psi_z^{(5)} d\boldsymbol{r}$, with corresponding phase factors. As a result, this line of the secular equation can be easily separated and it will give us one of the roots of the equation, which corresponds to the group of levels with the dispersion, as described by eqn (C.5). In this way, we obtain the group of 12 levels, which is the π-electron branch. Therefore, the separation of the π-electron states is caused by the asymmetric nature of the ψ_z atomic orbitals, unlike the symmetry of the three other in-plane orbitals and the tunnelling operator.

The residual (reduced) secular equation has the form

$$|S_{ik}| = 0, i, k = 1,2,3,4 \tag{C.8}$$

or, more specifically,

$$S_{ik} = \begin{vmatrix} E & -T_p e^{-im\tilde{\varphi}} & T_{p;\perp} e^{im\tilde{\varphi}} & 0 \\ -T_p e^{im\tilde{\varphi}} & E & T_{p;\perp} e^{im\tilde{\varphi}} & 0 \\ -T_{p;\perp} e^{-im\tilde{\varphi}} & -T_{p;\perp} e^{-im\tilde{\varphi}} & E - T_\perp \cos m\tilde{\varphi} & -T_{\perp;H} \\ 0 & 0 & -T_{\perp;H} & E - \varepsilon_H \end{vmatrix} = 0$$

Here $\tilde{\varphi} = \pi/3; m = 0, \pm 1, \pm 2, 3$ (see eqn (C.3)), ε_H is the electronic term of the hydrogen ion, and the quantities $T_p, T_{p;\perp}, T_\perp$, and $T_{\perp;H}$ are defined by the relations

$$T_p \equiv T_{RL}^{1;2} \equiv T_{LR}^{1;6} = \int \psi_R^{(1)} \tilde{V} \psi_L^{(2)} \, d\mathbf{r} \tag{C.9}$$

$$T_\perp = \int \psi_\perp^{(1)} \tilde{V} \psi_\perp^{(2)} \, d\mathbf{r} \tag{C.9'}$$

$$T_{\perp;H} = \int \psi_H^{*(1)} \tilde{V} \psi_\perp^{(1)} \, d\mathbf{r} \tag{C.9''}$$

$$T_{p;H} = \int \psi_R^{*(1)} \tilde{V} \psi_\perp^{(2)} \, d\mathbf{r} \tag{C.9'''}$$

where $\tilde{V}$ is defined by eqn (C.4). The value of the energy E is referred to the upper occupied electronic term of an isolated carbon atom (we put $\varepsilon_0 = 0$).

The matrix elements T_z, T_P, $T_\perp$, $T_{\perp;H}$, and $T_{p;\perp}$ describe various channels for electronic tunnelling between the neighbouring ions. This tunnelling leads to the splitting of the electronic levels of an isolated C ion. It is natural to assume that $T_p, T_z \gg T_\perp; T_{\perp;H}; T_{p;\perp}$.

As mentioned above, the secular equation allows us to determine the electronic energy spectrum for the benzene molecule. The energy levels are the roots of this equation. The secular equation is non-linear and, as a result, it has several roots. Because of additional dependence on m (this dependence on m is similar to the dependence of the energy on the quasi-momentum in solids), we are dealing with several energy branches; the energy levels in each branch are classified by the quantum number m, so these branches are analogous to energy bands in solids.

More specifically, the general secular equation has five roots, and this means that we are dealing with five groups (branches) of energy levels (see below). Each group contains four levels. Two of them are orbitally non-degenerate (m = 0, 3) and each of them can accommodate two electrons with opposite spins. In addition, there are two double-degenerate levels (m = 1, −1; 2, −2). Therefore, each branch can accommodate 12 electrons.

It is essential that, because of antisymmetric nature of the π electronic wave function (which is antisymmetric with respect to reflection in the plane of the molecule and all quantities $S_{i5} = S_{5i} = 0$), one branch can be determined immediately which contains the tunnelling between the electronic ψ_z states belonging to neighbouring C ions. This makes it possible to distinguish the group E_π for levels, which is described by eqn (C.5).

Other branches can be determined from the reduced secular equation. This equation has the form (C.8), but with I, k = 1, ..., 4. Since the terms $T_{p;\perp}$ and $T_{\perp;H}$ are relatively small (see above), one can, in a first approximation, put the terms $S_{13} = S_{31} = 0$, and $S_{14} = S_{41} \approx 0$. Then the determination of the roots is reduced to the solution of two simple equations. Then it is easy to find

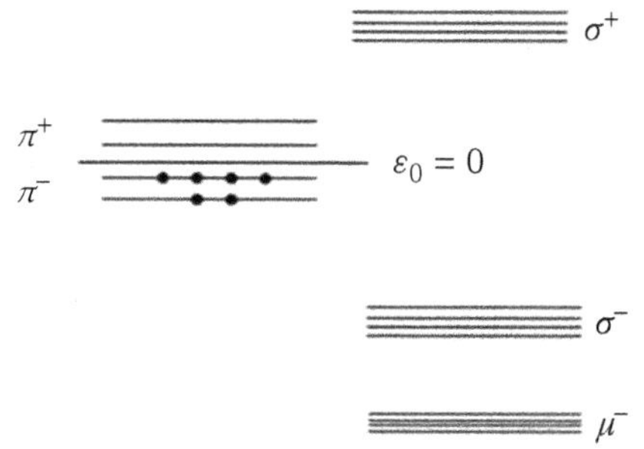

Figure C.2 *The spectrum of the benzene molecule.*

$$E_\sigma{}^{\pm} = \pm T_p \tag{C.10}$$

These groups form the σ^- $(= -T_p)$ and $\sigma^+(=T_p)$ energy levels. If we take into account small additional terms containing $T_{p;\perp}$, $T_{\perp;H}$, and $T_\perp$, then this energy level will be slightly split (e.g. $\sim (T_{p;\perp}/T_p)^2$ for the σ levels), and the sublevels will be classified by m.

Finally, the energy levels E_4 and E_5 are described by the following dispersion relation:

$$E_\mu^- = \varepsilon_H - 2T_\perp \cos m\tilde{\varphi}; \;\; E_\mu^+ \simeq 0 \tag{C.11}$$

These levels form the groups μ^+ and μ^-. As a result, there are five branches of energy levels. As was noted above, each branch contains four energy levels and can accommodate 12 electrons, whose states differ in the value of the quasi-projection of angular momentum m, and spin. One can see in Fig. C.2 the structure of the electronic energy spectrum for the benzene molecule. The values of the parameters differ in various sources, but the general structure of the spectrum is reflected in Fig. C.2.

The total number of valence electrons is 30. Based on the Pauli principle, these electrons can be distributed among the energy levels plotted in Fig. C.2. As a result, all of the lowest levels (all E_μ^- levels, all E_σ^- levels and two E_π levels) are occupied. The upper two E_π energy levels and all of the E_μ^+ and E_σ^- levels are vacant.

The difference between π- and σ-electrons is not in delocalisation versus localisation. Indeed, the wave functions for all valence electrons, including the σ states are delocalised. The difference is that the π states and σ states correspond to different parts of the energy spectrum. The σ branch is totally occupied, while the π branch is partially occupied. For the π branch, intra-branch transitions are permitted. That is why studies of aromatic molecules focus mainly on the properties of π-electrons (see e.g. Higasi *et al.*, 1965; Dalem, 1966; Atkins and Friedman, 2010). The σ branch is similar to valence (totally occupied) energy bands in solids, whereas the π electronic branch is similar to conductance bands.

Appendix D
The Multi-Electron Tight Binding Methods GTB and LDA+GTB

D.1 Introduction

The multi-electron GTB method has been developed to calculate the band structure of strongly correlated materials like cuprates, manganites, and cobaltites (Ovchinnikov and Sandalov, 1989; Gavrichkov *et al.*, 2000). It is a version of the CPT approach. The initial GTB version was based on the multiband Hubbard model Hamiltonian with parameters fitted from experimental data (Ovchinnikov and Petrakovsky, 1991). Later the hybrid version LDA+GTB was developed with parameters calculated within the DFT approach (Korshunov *et al.*, 2005). As any other CPT approach, the GTB method starts with the exact diagonalisation of the intracell part (H_c) of the multi-electron Hamiltonian and treats the intercell part with perturbation theory. Thus, it includes a realisation of the Lehmann representation inside one unit cell with all local quasiparticle energies and spectral weights calculated via exact diagonalisation. The total LDA+GTB procedure consists of the following steps:

- Step I: LDA. Calculation of the LDA band structure, and construction of the Wannier functions with the given symmetry and computation of the one- and two-electron matrix elements of the tight binding Hamiltonian with the local and nearest-neighbour Coulomb interactions.
- Step II: Exact diagonalisation. Separation of the total Hamiltonian H into the intra- and intercell parts, $H = Hc + Hcc$, where Hc represents the sum of the orthogonal unit cells, $Hc = \Sigma_f H_f$. The exact diagonalisation of a single unit cell term H_f and construction of the Hubbard X-operators $X_f^{pq} = |p\rangle\langle q|$ using the complete orthogonal set of eigenstates $\{|p\rangle\}$ of H_f.
- Step III: Perturbation theory. Within the X representation, local interactions are diagonal and all intercell hoppings and long-range Coulomb interaction terms have the bilinear form in the X operators. Various perturbation approaches known for the Hubbard model in the X representation can be used. The most general one includes treatment within the generalised Dyson equation obtained by the diagram technique for the Hubbard operators (Ovchinnikov and Val'kov, 2004).

Below we discuss each step in detail.

D.2 Step I: LDA

LDA provides us with a set of Bloch functions $|\psi_{\lambda \boldsymbol{K}}\rangle$ (where λ is the band index) and band energies $\varepsilon_\lambda(\boldsymbol{k})$. For example, the LDA band structure calculation for La_2CuO_4 and Nd_2CuO_4 was done within the TB-LMTO-ASA (for tight binding, linear muffin-tin orbital, atomic sphere approximation) method (Andersen *et al.*, 1986). Using the Wannier function formalism (Anisimov *et al.*, 2005) or the NMTO method (Andersen and Saha-Dasgupta, 2000), we obtain the single-electron energies ε_λ and hopping integrals $T_{fg}^{\gamma\gamma'}$ of the multiband tight binding model (Korshunov *et al.*, 2004, 2005):

$$H = \sum_{f,\lambda,\sigma} (\varepsilon_\lambda - \mu) n_{f\lambda\sigma} + \sum_{f \neq g} \sum_{\lambda\lambda',\sigma} T_{fg}^{\lambda\lambda'} c_{f\lambda\sigma}^\dagger c_{g\lambda'\sigma}$$
$$+ \frac{1}{2} \sum_{f,g,\lambda\lambda'} \sum_{\sigma_1,\sigma_2,\sigma_3,\sigma_4} V_{fg}^{\lambda\lambda'} c_{f\lambda\sigma_1}^\dagger c_{f\lambda\sigma_3} c_{g\lambda'\sigma_2}^\dagger c_{g\lambda'\sigma_4} \tag{D1}$$

where $c_{f\lambda\sigma}$ is the annihilation operator in the Wannier representation of a hole at the site f with the orbital λ and the spin σ, and $n_{f\lambda\sigma} = c_{f\lambda\sigma}^\dagger c_{f\lambda\sigma}$. Note that the number and symmetry of the Wannier functions are determined by the energy window that we are interested in. For cuprates, the two e_g orbitals of copper, and the oxygen p orbitals covalently bonding with them, are relevant. For the cobaltite $LaCoO_3$, with competition between the high- and low-spin configurations, all five of the d orbitals and three of the p oxygen orbitals are important (Ovchinnikov *et al.*, 2011c). The values of the Coulomb parameters $V_{fg}^{\lambda\lambda'}$ are obtained by LDA supercell calculations following Gunnarsson *et al.* (1989). For Cu in La_2CuO_4, the Hubbard parameter U and the Hund's exchange J_H are equal to 10 eV and 1 eV, respectively (Anisimov *et al.*, 2002).

D.3 Step II: Exact Diagonalisation

In transition metal (Me) oxides, the unit cell may be chosen as the MeO_n (where $n = 6, 5, 4$) cluster and usually there is a common oxygen shared by two adjacent cells. All other ions provide the electroneutrality and contribute to the high-energy electronic structure. In the low-energy sector, they are inactive. Before the exact diagonalisation calculations, we solve the problem of the non-orthogonality of the oxygen molecular orbitals in adjacent cells. For the $\boldsymbol{\sigma}$ bonding of the $3d$ metal e_g electrons and a_{1g}, b_{1g} oxygen orbitals, this problem is solved explicitly using the diagonalisation in k-space (Raimondi *et al.*, 1996).We have used the same procedure for the t_{2g} orbitals. Such orthogonalisation results in the renormalisation of the hopping and Coulomb matrix elements in eqn (D1). Later we will work with the renormalised parameters. After the orthogonalisation, the Hamiltonian (D1) can be written as the sum of intracell and intercell contributions:

$$H = H_c + H_{cc}, \quad H_c = \sum_f H_f, \quad H_{cc} = \sum_{f,g} H_{fg} \tag{D2}$$

with orthogonal states in different cells described by H_f. The exact diagonalisation of H_f gives us the set of eigenstates $|p\rangle = |n_p, i_p\rangle$ with the energy E_p. Here, n_p is the number of electrons per unit cell, and i_p denotes all other quantum numbers like spin, orbital moment, and so on. We perform the exact diagonalisation with all possible excited eigenstates, not the Lancoz procedure.

How can we determine which configurations are relevant? They are found from the local electroneutrality. Let us consider $LaMeO_3$, with Me being a $3d$ element as an example. The ionic valence is $La^{3+}Me^{3+}O_3{}^{2-}$; thus, the $3d$ cation is Me^{3+} (d^4 for Mn^{3+} and d^6 for Co^{3+}). Due to the covalence, the ground state of the unit cell is given by the hybridisation of $d^np^6 + d^{n+1}p^5$ configurations (in the spectroscopic notation $d^n + d^{n+1}\underline{L}$, where $\underline{L}$ means a ligand hole (Zaanen and Sawatzky, 1990). The electron addition process results in a d^{n+1} subspace of the Hilbert space with a mixture of the $d^{n+1} + d^{n+2}\underline{L}$ configuration. Similarly, electron removal results in a d^{n-1} subspace with a $d^{n-1} + d^n\underline{L} + d^{n+1}\underline{L}^2$ mixture. Thus, for a stoichiometric compound, there are three relevant subspaces of the Hilbert space: d^{n-1}, d^n, and d^{n+1}. For each subspace, exact diagonalisation provides a set of multi-electron states $|n,i\rangle$ with energy $E_i(n)$, $i = 0, 1, 2, \ldots, N_n$.

Within this set of multi-electron terms, the charge-neutral Bose-type excitations with the energy $\omega_i = E_i(n) - E_0(n)$ are shown by the vertical wavy lines in Fig. D.1. Electron addition excitations (local Fermi-type quasiparticles) have energies $\Omega_{c,i} = E_i\,(n+1) - E_i(n)$. Here, the index c means that quasiparticles form the empty conductivity band. Similarly, the valence band is formed by the electron-removal Fermi-type quasiparticles with energies $\Omega_{v,i} = E_0(n) - E_i\,(n-1)$ (a hole creation with energy $E_i\,(n-1) - E_0(n)$ is shown in Fig. D.1). This multi-electron language has been used in spectroscopy (see e.g. Zaanen and Sawatzky, 1990). The proper mathematical tool for studying both the local quasiparticle and their intercell hoppings is given by the Hubbard X operators

$$X_f^{pq} = |p\rangle\langle q| \tag{D3}$$

with the algebra given by the multiplication rule $X_f^{pq}X_f^{rs} = \delta_{qr}X_f^{ps}$ and by the completeness condition $\sum_p X_f^{pp} = 1$. The last two equations reflect the fact that X operators are the projective operators. The operator X_f^{pq} describes the transition from the initial state $|q\rangle$ to the final state $|p\rangle, X^{pq}|q\rangle = |p\rangle$. The important property of X operators is that any local operator is given by a linear combination of X operators. Indeed, $\hat{O}_f = \hat{I}\cdot\hat{O}\cdot\hat{I} = \sum_{p,q}|p\rangle\langle p|\hat{O}_f|q\rangle\langle q| = \sum_{p,q}\langle p|\hat{O}_f|q\rangle X_f^{pq}$. The commutation rule for Hubbard operators follows from the X-operator

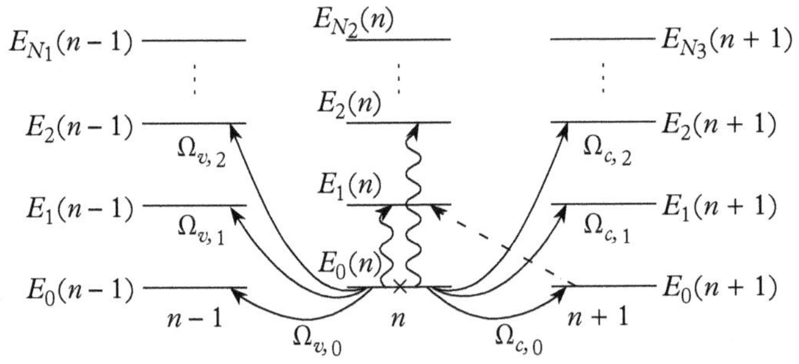

Figure D.1 *The occupied ground term $E_0(n)$, marked by the cross and excited terms $E_i(n)$ of the d^n configuration as well as the electron-removal (electron-addition) d^{n-1} (d^{n+1}) sectors of the Hilbert space with energies $E_i\,(n-1)\,(E_i\,(n+1))$. The vertical wavy lines show local Bose-type excitons. Solid lines with arrows show hole (or electron) removal (index v), and electron addition (index c) Fermi-type excitations. The dashed line shows the virtual Fermi-type excitation from $E_0\,(n+1)$ to $E_1(n)$. (From Korshunov et al., 2012)*

algebra and is rather awkward. Nevertheless, if $n_p - n_q$ is odd ($\pm1, \pm3$, etc.) then X^{pq} is called a quasi-fermionic operator; if $n_p - n_q$ is even ($0, \pm2$, etc.), X^{pq} is called a quasi-bosonic operator.

We will use the simplified notation

$$X_f^{pq} \rightarrow X_f^{\alpha_m} \rightarrow X_f^m, \tag{D4}$$

where the number m enumerates the excitations and plays the role of the quasiparticle band index; each pair (p, q) corresponds to some vector $\boldsymbol{\alpha}(p, q) = \boldsymbol{\alpha}_m$, which is called the 'root vector' in the diagram technique (Zaitsev, 1976). In this notation, a single-electron (single-hole) creation operator is given by a linear combination of the Hubbard fermions, $c_{f\lambda\sigma} = \sum_m \gamma_{\lambda\sigma}(m) X_f^m$, $\gamma_{\lambda\sigma}(p,q) = \langle p | c_{f\lambda\sigma} | q \rangle$. In the X representation, the intracell part of the Hamiltonian is diagonal:

$$H_c = \sum_{f,p} (E_p - n_p\mu) X_f^{pp} \tag{D5}$$

$$H_{cc} = \sum_{f \neq g} \sum_{m,m'} t_{fg}^{mm'} X_f^{m\dagger} X_g^{m'} \tag{D6}$$

where the matrix elements are equal to

$$t_{fg}^{mm'} = \sum_{\sigma\lambda\lambda} T_{fg}^{\lambda\lambda'} \gamma_{\lambda\sigma}^*(m) \gamma_{\lambda',\sigma}(m')$$

All intra-atomic *d–d* Coulomb interactions are included in H_c and are treated exactly. The dominant nearest-neighbour part of the *p–d* and *p–p* Coulomb interactions is also included in H_c and gives a contribution to the energies E_p, while a small part of it (~10%) provides the intercell Coulomb interaction that is also bilinear in the X operators:

$$H_{cc}^{Coul.} = \sum_{f \neq g} \sum_{p,q,p'q'} V_{fg}^{p,q,p'q'} X_f^{pq} X_g^{p'q'}$$

D.4 Step III: Perturbation Theory

The characteristic local energy scale is given by the effective Hubbard parameter $U_{eff} = E_0(n+1) + E_0(n-1) - 2E_0(n)$ that is given by the difference of the initial $d^n + d^n$ and the excited $d^{n-1} + d^{n+1}$ configurations (Zaanen and Sawatzky, 1990). The same energy can be obtained as the local gap between the conductivity and valence bands, $U_{eff} = \Omega_{c,0} - \Omega_{v,0}$. Depending on the ratio of the bare Hubbard U and the charge excitation energy $\Delta_{pd} = \varepsilon_p - \varepsilon_d$, U_{eff} may represent the Mott–Hubbard gap for $U < \Delta_{pd}$ or the charge transfer gap E_{CT} for $U > \Delta_{pd}$.(Zaanen *et al.*, 1985). The intercell hopping and interaction result in the dispersion and decrease the energy gap, $E_g < U_{eff}$. The intercell hoppings, eqn (D6), and the non-local Coulomb interactions can be treated by perturbation theory. We would like to emphasise that, in the X representation, the perturbation H_{cc} has exactly the same structure as the hopping Hamiltonian in the conventional Hubbard model. That is why the accumulated experience of the Hubbard model study in the X representation can be used here.

The single-electron Green's function for a particle with momentum $\boldsymbol{k}$, energy E, spin σ, and orbital indices λ and λ', $G_{\boldsymbol{k}\sigma}^{\lambda\lambda'}(E) \equiv \langle\langle a_{\boldsymbol{k}\lambda\sigma} | a_{\boldsymbol{k}\lambda'\sigma}^+ \rangle\rangle_E$, is given by a linear combination of the Hubbard operator's Green functions $D_{\boldsymbol{k}\sigma}^{mn}(E) = \langle\langle X_{\boldsymbol{k}\sigma}^m | X_{\boldsymbol{k}\sigma}^{n*} \rangle\rangle_E$:

$$G_\sigma(\boldsymbol{k},E)=\sum_{mn}\gamma_{\lambda\sigma}(m)\gamma^*_{\lambda'\sigma}(n)D^{mn}_{\boldsymbol{k}\sigma}(E) \tag{D7}$$

The question is how to determine the X-operator Green's function. One of the usual ways to treat perturbations is the diagram technique. For X operators, with their awkward algebra, there is no standard Wick's theorem. Nevertheless, a generalised Wick's theorem was proven long ago by Westwanski and Pawlikovski (1973). The first convenient version of the diagram technique was formulated by Zaitsev (1976). The general rules of the diagram technique for X operators are described in detailed by Ovchinnikov and Val'kov (2004), who obtained the generalised Dyson equation

$$\hat{D}(\boldsymbol{k},E)=\left[(E-\Omega_m)\,\delta_{mm'}-\hat{P}(\boldsymbol{k},E)\,\hat{t}(\boldsymbol{k})+\hat{\Sigma}(\boldsymbol{k},E)\right]^{-1}\hat{P}(\boldsymbol{k},E) \tag{D8}$$

Here $\Omega_m=E_p(n+1)-E_q(n)$ is the m-th quasiparticle local energy. Besides the self-energy matrix $\hat{\Sigma}_{\mathbf{k}\sigma}(E)$, the unconventional term $\hat{P}_{\mathbf{k}\sigma}(E)$, called 'the strength operator', appears due to the X-operator algebra. A similar term had been known for the spin-operator diagram technique (Bar'yakhtar *et al.*, 1984). This term determines the quasiparticle oscillation strength (spectral weight) as well as the renormalised bandwidth. The Hubbard I solution is obtained by setting $\Sigma=0$ and $P^{m,n}=\delta_{m,n}F_n$, where $F_n=\langle X^{p,p}+X^{q,q}\rangle$. We call F_n 'the occupation factor' because it is determined by a sum of the occupation numbers of the initial and final states for the excitation from $|q>$ to $|p>$; it provides a non-zero spectral weight for the quasiparticle excitation between at least partially filling eigenstates and gives zero spectral weight for excitations between empty states. The dispersion equation for the quasiparticle band structure of the Hubbard fermions in this case is given by

$$\det\left\|\delta_{mn}\left(E-\Omega_m\right)/F_n-t^{mn}\left(\boldsymbol{k}\right)\right\|=0 \tag{D9}$$

The dispersion equation (D9) is similar to the conventional tight binding dispersion equation but, instead of the single-electron local energy ε, we have the local quasiparticle energy Ω_m. This is why we call this approach the GTB (for 'generalised tight binding') method, or the multi-electron tight binding method.

The occupation numbers $\langle X^{pp}\rangle$ are calculated self-consistently via the chemical potential equation. The change of the concentration n_e redistributes the occupation numbers and, due to the occupation factors, changes the quasiparticle band structure.

In the GTB method, the intracell Green function can be found exactly:

$$G^{\lambda\lambda'}_{o\sigma}(E)=\sum_m|\,\gamma_{\lambda\sigma}(m)\,|^2\delta_{\lambda\lambda'}D^m_{0\sigma}(E) \tag{D10}$$

where $D^m_{0\sigma}(E)=F_m/\left(E-\Omega_m+i\delta\right)$. Comparing this exact local Green's function with the Lehmann representation (Ovchinnikov, 2003b), we can say that the electron here is a linear combination of local (Hubbard) fermions with the quasiparticle energy Ω_m and the spectral weight $|\,\gamma_{\lambda\sigma}(m)|^2 F_m$. It is exactly the same language as in the Lehmann representation. The difference is that it is realised locally, and both the quasiparticle energy and the spectral weight are calculated explicitly. Thus, the LDA+GTB method is a perturbative realisation of the Lehmann representation.

It should be stressed that the LDA+GTB bands are not the single-electron conventional bands. There is no single-particle Schrödinger equation with the effective potential that gives the LDA+GTB band structure. These quasiparticles are excitations between different multi-electron terms. The LDA+GTB bands depend on the multi-electron term occupation numbers through the strength operator and self-energy. There is no rigid band behaviour from the very beginning; the band structure depends on doping, temperature, pressure, and external fields.

Several examples of LDA+GTB calculations for doping- and temperature-dependent band structure in cuprates, manganites, and cobaltites can be found in the review by Korshunov *et al.* (2012).

Appendix E
Methods of Quantum Field Theory

Study of the superconducting state and related phenomena usually cannot be performed with the use of just the perturbation theory. Even the starting point, the Cooper instability leading to the formation of bound pairs, cannot be treated as a perturbation (see the discussion in section 2.1). Usually, we are dealing with rather strong interactions, which requires performing a summation of an infinite number of terms. In order to meet such a challenge, special methods that originated in quantum field theory have been employed. For example, in many chapters of this book, we use the diagrammatic technique, which was developed initially by R. Feynman in quantum electrodynamics. Of course, a detailed description of these methods is beyond the scope of this book. We only introduce in this appendix the main quantities used in the present many-particle theory and some basic relations. For a more detailed study, we recommend the excellent books by Abrikosov *et al.* (1975), Lifshitz and Pitaevskii (1980), and Mahan (1993). The pioneering work that allows us to use the methods of quantum field theory in solid state theory, and in statistical physics in general, was performed by Galitsky and Migdal (1958).

Let us focus initially on the ground state ($T=0$). The first key quantity describing the many-electron system is the one-particle electronic Green's function defined as

$$G(x_1,x_2) = -i\langle T\tilde{\psi}_\alpha^+(x_1)\tilde{\psi}_\beta(x_2)\rangle \tag{E.1}$$

Here $\tilde{\psi}_\alpha(x)$ are the ψ operators in the second quantisation representation (see e.g. Landau and Lifshitz, 1977); $x=\{\boldsymbol{r},t\}$; the time dependence corresponds to the so-called Heisenberg representation of the ψ operator; α and β are the spin indices; and T is the ordering operation, so the operators should be arranged from left to right in order of decreasing time. The brackets mean the average over the ground state.

By knowing the Green's function, one can learn about a number of important properties of the system of interest. First of all, the density matrix is directly related to the Green's function:

$$\rho_{\alpha\beta}(\boldsymbol{r}_1,\boldsymbol{r}_2) = -\frac{i}{N}G_{\alpha\beta}(t_1,\boldsymbol{r}_1;t_1+0,\boldsymbol{r}_2) \tag{E.2}$$

As is known, the average value of any quantity describing a single particle can be determined with use of the density matrix. Note also that the particle density for fermions is equal to

$$n(x) = -\lim_{\substack{\mathbf{r}'\to r\\ t'\to t+0}} G_{\alpha\alpha}(x,x') \tag{E.3}$$

As for the current density, it can be also determined from the Green's function:

$$j(x) = -m^{-1} \lim_{\substack{\mathbf{r}' \to \boldsymbol{r} \\ t' \to t+0}} (\nabla_x - \nabla_{x'}) G(x,x') \tag{E.3$'$}$$

Note also that, in the absence of external fields, the Green's function of a homogeneous system depends on the differences $\boldsymbol{r} - \boldsymbol{r}'$ and $t - t'$. The Fourier component of the Green's function of non-interacting fermions has the form

$$G^{(0)}(\mathbf{p},\omega) = \frac{1}{\omega - \varepsilon_0(\mathbf{p}) + i\delta \operatorname{sign}(|\mathbf{p}| - \mathbf{p}_\mathrm{F})} \tag{E.4}$$

It is essential that the use of the Green's function method allow us to determine the energy spectrum of the system. As is known (Landau, 1956; see also Lifshitz and Pitaevskii, 2002), the excited state of many-body system, which is not far from its ground state, can be described as a gas of quasiparticles. The poles of the Green's function determine the energy levels of the quasiparticles, that is, the energy spectrum of the system.

In addition to the one-particle Green's function, eqn (E.1), one can introduce the Green's function $F(x,x')$ describing the Cooper pairing (Gor'kov, 1958):

$$F(x,x') = -i\langle N|T\psi_\alpha(x_1)\psi_\beta(x_2)|N+2\rangle \tag{E.5}$$

This function describes the transition of two electrons into the pairing condensate. One can also define the function $F^+(x,x')$ containing two ψ^+ operators.

Another approach that makes it possible to study the many-body system at finite temperatures is based on the thermodynamic Green's function formalism. This method was introduced by Matsubara (1955) and then developed by Abrikosov *et al.* (1959). One can introduce the replacement $t \to \mathrm{i}\tau$; $-T^{-1} \le \tau \le T^{-1}$. Expansion in a Fourier series leads to the analysis of the component $G(\boldsymbol{p},\omega_n)$, where $\omega_n = (2n+1)\pi T$ (where $n = 0, \pm1, \pm2, \ldots$). Such a Green's function for non-interacting electrons has the form (cf. eqn (E.4))

$$G(\mathbf{p},\omega_n) = \frac{1}{i\omega_n - \varepsilon_0(\mathbf{p}) + \mu} \tag{E.6}$$

One can show that the values of the thermodynamic Green's function coincide with the values of the so-called retarded Green's function in the discrete points of the imaginary axis ω. The thermodynamic Green's functions can be used to evaluate such important parameters as the critical temperature. That is why it is used in many sections of this book.

Evaluation of the Green's functions for the system of interacting particles is usually performed with the use of a special diagrammatic method that was initially introduced in quantum field theory by R. Feynman and later developed by F. Dyson. The diagrammatic method can be used to evaluate the usual time-dependent Green's function at $T = 0$ K, while its modified version makes it possible to obtain an expression for the thermodynamic Green's functions. The Green's functions are denoted as in Fig. E.1.

The equations for the usual and the pairing Green's functions are presented in chapter 2, Fig. 2.3 (for the phonon mechanism of superconductivity). They contain the so-called self-energy parts displayed in chapter 1, Fig. 1.3, and chapter 2, Fig. 2.2. The broken lines correspond to the phonon

Figure E.1 *(a) The usual Green's function G (the thickness of the line is different for non-interacting and interacting electrons). (b) The pairing Green's function F^+.*

Figure E.2 *Random phase approximation for the vertex.*

Green's function (or phonon propagator); the analytical expression for this is written in chapter 2 as eqn (2.15) (in the thermodynamic Green's function formalism).

The Green's functions in the superconducting state are described by the diagrammatic equations in Fig. 2.3 and, correspondingly, by the analytical expression

$$G(\mathbf{p},\omega_n) = \frac{i\omega_n + \xi_{\mathbf{p}}}{\omega_n^2 + \xi_{\mathbf{p}}^2 + \Delta^2(\omega_n)} \tag{E.7}$$

$$F(\mathbf{p},\omega_n) = \frac{\Delta(\omega_n)}{\omega_n^2 + \xi_{\mathbf{p}}^2 + \Delta^2(\omega_n,\mathbf{p})} \tag{E.7$'$}$$

Here $\xi_{\mathbf{p}}$ is the electron energy referred to the chemical potential, and $\Delta(\omega_n,\mathbf{p})$ is the pairing self-energy part. The equation for this self-energy part is presented in Fig. 2.2, or, analytically, is written in accordance with eqn (2.17).

One can also introduce the two-particle Green's function (see chapter 1, Fig. 1.7). Its poles also determine the energy spectrum of the system. Among them, of course, there are the poles of the one-particle Green's function but, in addition, there are the poles corresponding to the so-called collective excitations caused by the interactions between the electrons. The most important type of such excitations are plasmons (see section 1.2.3 and Fig. 1.7). The special case when one can perform the summation of all the essential diagrams is called RPA. Then the total vertex Γ is described by the diagrammatic equation shown in Fig. E.2.

Alternatively, it can be described by

$$\Gamma = V + V \sqcap \Gamma$$

Here

$$\sqcap = \sum_n \int G(\mathbf{p},\omega_n)\; G(\mathbf{p}+\mathbf{k},\omega_n+\omega_m)\, d\mathbf{p}$$

is the so-called polarisation operator. Its calculation makes it possible to evaluate the spectrum of plasmons and its dispersion, as was described in section 1.2.3.

As was noted above, this appendix is by no means a replacement for a study of the methods of quantum field theory in solid state physics. Above, we have indicated several excellent books describing these methods in all detail. Here we have only presented several key points of the modern theory of many-body systems.

Bibliography

Aasen, D., Hell, M., Mishmash, R., Higginbotham, A., Danon, J., Leijnse, M. (2016), *Phys. Rev. X* 6, 031016.

Abanov, A., Chubukov, A., Schmalian, J. (2003), *Adv. Phys.* 52 (3), 119.

Abdel-Jawad, M., Kennett, M., Balicas, L., Carrington, A., Mackenzie, A., McKenzie, R., Hussey, N. (2006), *Nat. Phys.* 21, 821.

Abrikosov, A. (1957), *JETP* 5, 114.

Abrikosov, A., Gor'kov, L. (1959), *JETP* 8, 1090.

Abrikosov, A., Gor'kov, L. (1961), *JETP* 12, 1243.

Abrikosov, A., Khalatnikov, I. (1959), *Adv. Phys.* 8, 45.

Abrikosov, A., Gor'kov, L., Dzyaloshinski, I. (1959), *JETP* 8, 182.

Abrikosov, A., Gor'kov, L., Dzyaloshinski, I. (1959), *JETP* 9, 636.

Abrikosov, A., Gor'kov, L., Dzyaloshinski, I. (1975), *Methods of Quantum Field Theory in Statistical Physics*, Dover, NY.

Acosta, V., Bouchard, L., Budker, D., Folman, R., Lenz, T., Maletinsky, P., Rohner, D., Schlussel, Y., Thiel, L. (2019), *J. Supercond. Nov. Magn.* 32, 85.

Aeppli, G., Bishop, D., Broholm, C., Bucher, E., Siemensmeyer, K., Steiner, M., Stusser, N. (1989b), *Phys. Rev. Lett.* 63, 676.

Aeppli, G., Hayden, S., Mook, H., Fisk, Z., Cheong, S., Rytz, D., Remeika, J. P., Espinosa, G., Cooper, A. (1989), *Phys. Rev. Lett.* 62, 2052.

Afanas'ev, A., Kagan, Y. (1963), *JETP* 16, 1030.

Agterberg, D., Frigeri, P. A., Kaur, R. P., Koga, A., Sigrist, M. (2006), *Physica B* 378, 351.

Ahn, C., Triscone, J., Mannhart, J. (2003), *Nature* 424, 1015.

Akamutsu, H., Inokuchi, H., Matsunaga, Y. (1954), *Nature* 173, 168.

Akashi, R., Kawamura, M., Tsuneyuki, S. Nomura, Y., Arita. R. (2015), *Phys. Rev. B* 91, 224513.

Akashi, R., Sano, W., Arita, R., Tsuneyuki., R. (2016), *Phys. Rev. Lett.* 117, 075503.

Akhiezer, A., Pomeranchuk, I. (1959), *JETP* 9, 605.

Akimoto, T., Maruyama, Y., Maritomo, Y., Nakamura, A. (1998), *Phys. Rev. B* 55, 5594.

Alekseevskii, N., Vedeneev, S. (1967), *JETP Lett.* 6, 302.

Alexandrov, S., Ranninger, J. (1981), *Phys. Rev. B* 23, 1796.

Allen, P., Cohen, M. (1970), *Phys. Rev. B* 1, 1329.

Allen, P., Dynes, R. (1975), *Phys. Rev. B* 12, 995.

Allender, D., Bray, J., Bardeen, J. (1973), *Phys. Rev. B* 7, 1020.

Alloul, H., Bobroff, J., Gabay, M., Hirschfeld, P. (2009), *Rev. Mod. Phys.*, 81, 45.

Alloul, H., Mahajan, A., Mendels, P., Riseman, T., Yoshinari, Y., Collin, G., Marucco, J. (1995), in *Proceedings of the 7th Workshop on High-Temperature Superconductivity*, Miami.

Alloul, H., Ohno, T., Mendels., P. (1989), *Phys. Rev. Lett.* 67, 3140.

Altshuler, B., Reizer, M., Varlamov, A. (1983), *JETP* 57, 1329.

Alvarez, G., Mayr, M., Moreo, A., Dagotte, E. (2005), *Phys. Rev. B* 71, 014514.

Alvarez, J., Yundurain, F. (2007), *Phys. Rev. Lett.* 98, 126406.

Ambegaokar, V., Baratoff, A. (1963), *Phys. Rev. Lett.* 11, 104.

Ament, L., van Veenendaal, M., Devereaux, T., Hill, J., van den Brink, J. (2011), *Rev. Mod. Phys.* 83, 705.

Aminov, B., Hein, M., Miller, G., Piel, H., Ponomarev, Y., Wehler, D., Boockholt, M., Buschmann, L., Gunherodt, G. (1994), *Physica C* 235, 1863.
Andersen, O., Pawlowska, Z., Jepsen, O. (1986), *Phys. Rev. B* 34, 5253.
Andersen, O., Saha-Dasgupta, T. (2000), *Phys. Rev. B* 62, 16219(R).
Anderson, P. (1959a), *J. Phys. Chem. Sol.* 11, 59.
Anderson, P. (1959b), *Phys. Rev.* 115, 2.
Anderson, P. (1963), *Sol. State Phys.* 14, 99.
Anderson, P. (1984), *Phys. Rev.* B 30, 1549.
Anderson, P. (1987), *Science* 235, 1196.
Anderson, P. (1990), *Phys. Rev. Lett.* 64, 1839.
Anderson, P. (1997), *Adv. Phys.* 46, 3.
Anderson, P., Hasegawa, H. (1955), *Phys. Rev.* 100, 675.
Anderson, P., Rowell, J. (1963), *Phys. Rev. Lett.* 10, 230.
Anderson, P., Brinkman, W. (1973), *Phys. Rev. Lett.* 30, 1108.
Ando, T., Flower, A., Stern, F. (1982), *Rev. Mod. Phys.* 54, 437.
Ando, Y., Segawa, K., Komiya, S., Lavrov, A. (2002), *Phys. Rev. Lett.* 88, 137005.
Andreev, A. (1964), *JETP* 19, 1998.
Andres, K., Graebner, J., Ott, H. R. (1975), *Phys. Rev. Lett.* 35, 1779.
Anisimov, V., Kondakov, D., Kozhevnikov, A., Nekrasov, I., Pchelkina, Z., Allen, J., Mo, S.-K., Kim, H.-D., Metcalf, P., Suga, S., Sekiyama, A., Keller, G., Leonov, I., Ren, X., Vollhardt, D. (2005), *Phys. Rev. B* 71, 125119.
Anisimov, V., Korotin, M., Nekrasov, I., Pchelkina, Z., Sorella, S. (2002), *Phys. Rev. B* 66, 100502(R).
Ann, C., Bhattacharya, A., Di Ventra, M., Eckstein, J., Frisbie, C., Gershenson, M., Goldman, A., Inoue, I., Mannhart, J., Millis, A., Morpurgo, A., Natelson, D., Triscone, J.-M. (2006), *Rev. Mod. Phys.* 78, 1185.
Anselm, A. (1982), *Introduction to Semiconductor Theory*, Prentice-Hall, NY.
Anshukova, N., Golovashkin, A., Ivanova, L., Rusakov, A. (1997), *Phys. Usp.* 40, 843.
Aoki, D., Huxley, A., Ressouche, E., Braithwaite, D., Flouquet, J., Brison, J., L'hotel, E., Paulsen, C. (2001), *Nature*, 413, 613.
Aoki, D., Nakamura, A., Honda, F., Li, D., Homma, Y., Shimizu, Y., Sato, Y. J., Knebel, G., Brison, J., Pourret, A., Braithwaite, D., Lapertot, G., Niu, Q., Vališa, M., Harima, H., Flouquet, J. (2019), *J. Phys. Soc. Jpn* 88, 043702.
Ashcroft, N. (1968), *Phys. Rev. Lett.* 21, 1748.
Ashcroft, N. (2004), *Phys. Rev. Lett.* 92, 187002.
Ashcroft, N., Mermin, N. (1976), *Solid State Physics*, Holt, NY.
Aslamazov, L., Larkin, A. I. (1968), *Sov. Sol. State Phys.* 10, 875.
Aslamazov, L., Larkin, A. I. (1974), *JETP* 40, 321.
Assmus. W., Herrmann, M., Rauchschwalbe, U., Riegel, S., Lieke, W., Spille, H., Horn, S., Weber, G., Steglich, F., Cordier, G. (1984), *Phys. Rev. Lett.* 52, 469.
Atkins, P., Friedman, R. (2010), *Molecular Quantum Mechanics*, Oxford University Press, Oxford.
Avci, S., Chmaissen, O., Allred, J. M., Rosenkranz, S., Eremin, I., Chubukov, A. V., Bugaris, D. E., Chung, D. Y., Kanatzidis, M. G., Castellan, J.-P., Schlueter, J. A., Claus, H., Khalyavin, D. D., Manuel, P., Daoud-Aladine, A., Osborn, R. (2014), *Nat. Comm.* 5, 3845.
Avella, A. (2014), *Adv. Cond. Mat. Phys.* 2014, 515698.
Awana, G., Sultana, R., Maheshwari, P. K., Goyal,· R., Gahtori, B., Gupta, A., Awana, V. P. S. (2017), *J. Supercond. Nov. Magn.* 30, 853.

Baade, W., Zwicky, F. (1934), *Phys. Rev.* 45, 138.
Babushkina, N., Inyushkin, A., Ozhogin, V., Taldenkov, A., Kobrin, I., Vorobeva, T., Molchanova, L., Damyanets, L., Uvarova, T., Kuzakov, A. (1991), *Physica C* 185, 901.
Baek, S., Efremov, D., Ok, J., Kim, J., van den Brink, J., Büchner, B. (2015), *Nat. Mat.* 14, 210.
Baguenard, B., Pellarin, M., Bordas, C., Lerme, J., Vialle, J., Broyer, M. (1993), *Chem. Phys. Lett.* 205, 13.
Balian, R, Werthammer, N. (1963), *Phys. Rev.* 131, 1533.
Baltensperger, W. (1958), *Physica* 248 (Suppl.), 153.
Barabanov, A., Maksimov, L., Uimin, G. (1988), *JETP Lett.* 47, 622.
Barabanov, A., Berezovsky, V. (1994), *JETP* 79, 627.
Barabanov, A., Hayn, R., Kovalev, A., Urazaev, O., Belemuk, A. (2001), *JETP* 92, 677.
Bardeen, J. (1954), *Phys. Rev.* 94, 554.
Bardeen, J. (1961), *Phys. Rev. Lett.* 6, 57.
Bardeen, J., Cooper, L., Schrieffer, J. (1957), *Phys. Rev.* 108, 1175.
Barisic, S., Barisic, I., Friedel, J. (1987), *Europhys. Lett.* 3, 1231.
Barišić, N., Chan, M., Li, Y., Yu, G., Zhao, X., Dressel, M., Smontara, A., Greven, M. (2013), *PNAS* 110, 12235.
Barone, A., Paterno, G. (1982), *Physics and Applications of the Josephson Junction Effect*, Wiley, NY.
Baroni, S., de Gironcoli, S., Dal Corso, A., Gianozzi, P. (2001), *Rev. Mod. Phys.* 73, 515.
Bar'yakhtar, V., Krivoruchko, V., Yablonskii, D. (1984), *Green's Functions in Magnetism Theory*, Naukova Dumka, Kiev.
Barzykin, V., Gor'kov, L. (2009), *Phys. Rev. B* 79, 134, 510.
Baskaran, G. (2003), *Phys. Rev. Lett.* 91, 097003.
Bauer, E., Hilscher, G., Michor, H., Paul, C., Scheidt, E., Gribanov, A., Seropegin, Y., Noel, H., Sigrist, M., Rogl, P. (2004), *Phys. Rev. Lett.* 92, 027003.
Becca, F., Sorella, S. (2017), *Quantum Monte Carlo Approaches for Correlated Systems.* Cambridge University Press, Cambridge.
Bechgaard, K., Jacobsen, C., Mortenson, K., Pedersen, H., Thorup, N. (1979), *Sol. State Comm.* 33, 1119.
Becker, T., Streng, C., Luo, Y., Moshnyaga, V., Damashke, B., Shannon, N., Samwer, K. (2002), *Phys. Rev. Lett.* 89, 237203.
Bednorz, G., Mueller, K. (1986), *Z. Phys B* 64, 189.
Bednorz, G., Mueller, K. (1987), *Ang. Chem.* 100, 5.
Bednorz, G., Mueller, K., Takashide, M. (1987a), *Science* 236, 73.
Bednorz, G., Mueller, K., Takashide, M. (1987b), *Europhys. Lett.* 3, 379.
Belinicher, V., Chernyshev, A., Shubin, V. (1996), *Phys. Rev. B* 53, 335.
Bell Burnell, J. (2007), *Science*, 318, 579.
Belova, L. (2000), *J. Supercond. Nov. Magn.* 13, 305.
Belyaev, S. (1959), *Mat. Phys. Medd. Dan. Selsk.* 31, 131.
Berestetskii, V., Lifshitz, E., Pitaevskii, L. (1971), *Relativistic Quantum Theory*, Pergamon Press, Oxford.
Berezinskii, V. (1972), *JETP* 34, 610.
Bergemann, C., Tyler, A., Mackenzie, A., Cooper, J., Julian, S., Farrell, D. (1998), *Phys. Rev. B* 57, 14387.
Bergmann, G., Rainer, D. (1973), *Z. Phys.*, 263, 59.
Bergomi, L., Jollicoeur, T. (1994), *C.-R. Acad. Sci.* 318, II, 283.
Berk, N., Schrieffer, J. (1966), *Phys. Rev. Lett.* 17, 433.

Bermoh, S. (1960), *Tech. Rept. 1, University of Illinois, Urbana*, National Science Foundation Grant NSF GP 1100.

Bernstein, N., Hellberg, C., Johannes, M., Mazin, I., Mehl, M. (2015), *Phys. Rev. B* 91, 060511.

Bersuker, I. (2006), *The Jahn–Teller Effect*, Cambridge University Press, Cambridge.

Bertsch, G. (2013), in *Fifty Years of Nuclear BCS*, Broglia, R., Zelevinsky, V., Eds., World Scientific, Singapore.

Bi, T., Zarifi, N., Terpstra, T., Zurek, E. (2019), in *Reference Module in Chemistry, Molecular Sciences and Chemical Engineering*, Reedijk, J., Ed., Elsevier, NY.

Bianconi, A. (1994a), *Sol. State Comm.* 89, 933.

Bianconi, A. (1994b), *Sol. State Comm.* 91, 1.

Bianconi, A., Jarlborg, T. (2015a), *Europhys. Lett.* 112, 37001.

Bianconi, A., Jarlborg, T. (2015b), *Nov. Superconduct. Mat.* 1, 15.

Bianconi, A., Missori, M. (1994), *J. Phys. I* 4, 361.

Bickers, N. E., Scalapino, D. J, Scalettar, R. T. (1987), *Int. J. Mod. PhysB* 1, 687.

Bickers, N. E., Scalapino, D. J., White, S. (1989), *Phys. Rev. Lett.* 62, 961.

Bill, A., Kresin, V. Z., Wolf, S. (1998), *Phys. Rev. B* 57, 10814.

Bill, A., Morawitz, H., Kresin, V. Z. (2003), *Phys. Rev. B* 68, 144519.

Billinge, S., Proffen, T., Petkov, V., Sarrao, J., Kycia, S. (2000), *Phys. Rev. B* 62, 1203.

Binnig, G., Baratoff, A., Hoenig, H., Bednorz, G. (1980), *Phys. Rev. Lett.* 45, 1352.

Binnig, G., Rohrer, H. (1982), *Helv. Phys. Acta* 55, 726.

Bistritzer, R., MacDonald, A. (2011), *PNAS* 108, 12233.

Blanter, Y., Kaganov, M., Pantsulaja, A., Varlamov, A. (1994), *Phys. Rep*, 245, 161.

Blasé, X., Bustarret, E., Chapelier, C., Klein, T., Mareenat, C. (2009), *Nat. Mat.*, 8, 375.

Blatter, G., Feigel'man, M., Geshkenbein, V., Larkin, A., Vinokur, V. (1994), *Rev. Mod. Phys.* 66, 1180.

Bloch, F. (1929a), *Zs. Phys.* 57, 545.

Bloch, F. (1929b), *Zs. Phys.* 61, 206.

Blonder, G., Tinkham, M., Klapwijk, T. (1982), *Phys. Rev. B* 25, 4515.

Boel, L., Kovtus, J., Anderson, O. (2004), *Phys. Rev. Lett.* 93, 237004.

Boeri, L., Dolgov, O. V., Golubov, A. A. (2008), *Phys. Rev. Lett.* 101, 026403.

Bogoluybov, N. (1958), *JETP* 7, 41.

Bogoluybov, N., Tolmachev, N., Shirkov, D. (1959), *A New Method in the Theory of Superconductivity*, Consultant Bureau, NY.

Bohm, D., Pines, D. (1953), *Phys. Rev.* 92, 609.

Bohr, A., Mottelson, B. (1974), *Nuclear Structure*, Vol. 2, Benjamin, NY.

Bohr, A., Mottelson, B., Pines, D. (1958), *Phys. Rev.* 110, 936.

Bok, J., Bouvier, J. (2012), *J. Supercond. Nov. Magn.* 25, 657.

Bonca, J., Prelovsek, P., Sega, I. (1989), *Phys. Rev. B* 39, 7074.

Bonča, J., Trugman, S. (2001), *Phys. Rev. B* 64, 094507.

Bonesteel, N., Wilkins, J. (1991), *Phys. Rev. Lett.* 66, 2684.

Boninsegni, M., Manousakis, E. (1993), *Phys. Rev. B* 47, 11897.

Bono, D., Bakharev, O., Schepf, A., Hartig, J., Schockel, H., de Jongh, L. (2007), *Z. Anorg. Allg. Chem.* 633, 2173.

Booth, C., Bridges, F., Kwei, G., Lawrence, J., Cornelius, A., Neumeier, J. (1998), *Phys. Rev. Lett.* 80, 853.

Booth, C., Bridges, F., Snyder, G., Geballe, T. (1996), *Phys. Rev. B* 54, 15606.

Boren, (1933), *Ark. Kem. Min. Geol.* 11A, 2.

Borggreen, J., Crowdhurry, P., Kebaili, N., Lundsberg-Nielsen, L., Lutzenkirchen, K., Nielsen, M., Pedersen, J., Rasmussen, H. (1993), *Phys. Rev. B* 48, 17507.

Born, M., Huang, K. (1954), *Dynamic Theory of Crystal Lattices*, Oxford University Press, NY.

Born, M., Oppenheimer, R. (1927), *Ann. Phys.* 84, 457.

Bornemann, H., Morris, D., Liv, H. (1991), *Physica C* 182, 132.

Borzi, R., Grigera, S., Farrell, J., Perry, R., Lister, S., Lee, S., Tennant, D., Maeno, Y., Mackenzie, A. (2007), *Science* 315, 214.

Bouquet, F., Fisher, R., Phillips, N., Hinks, D., Jorgensen, J. (2001), *Phys. Rev. Lett.* 87, 047001.

Bozin, E., Kwei, G., Takagi, H., Billinge, S. (2000), *Phys. Rev. Lett.* 84, 5856.

Bozovic, I., Eckstein, J., Klausmeierbrown, M., Virshup, G. (1992), *J. Supercond.* 5, 19.

Bozovic, I., Logvenov, G., Verhoeven, M., Caputo, P., Goldobin, E., Beasley, M. (2004), *Phys. Rev. Lett.* 93, 157002.

Braganca, H., Sakai, Sh., Aguiar, M., Civelli, M. (2018), *Phys. Rev. Lett.* 120, 067002.

Braginski, A. (2019), *J. Supercond. Nov. Magn.* 32, 23.

Braunecker, B., Japaridze, G. I., Klinovaja, J., Loss, D. (2010), *Phys. Rev. B* 82, 045127.

Braunecker, B., Simon, P. (2013), *Phys. Rev. Lett.* 111, 147202.

Bravyi, A., Kitaev, S. (2001), *Quant. Comp. Comp.* 2, 43.

Bredl, C., Steglich, F., Schotte, K. (1978), *Phys B* 29, 327.

Brenig, W. (1995), *Phys. Rep.* 251, 153.

Brinkmann, M., Bach, H., Westerholt, K. (1996), *Phys. Rev. B* 54, 6680.

Brockhouse, B., Arase, T., Caglioti, K., Rao, K., Woods, A. (1962), *Phys. Rev.* 128, 1099.

Broglia, R., Zelevinskii, V., Eds. (2013), *50 Years of Nuclear BCS*, World Scientific, Singapore.

Brovman, Y., Kagan, Y. (1967), *JETP* 25, 365.

Bulaevskii, L., Kuzii, V., Sobyanin, A. (1977), *JETP Lett.* 25, 290.

Bulayevskii, L., Nagaev, E., Khomskii, D. (1968), *JETP* 27, 836.

Bulut, N. (2002), *Adv. Phys*, 51, 1587.

Bulut, N., Scalapino, D. J. (1996), *Phys. Rev. B* 54, 14971.

Bussmann-Holder, A., Keller, H. (2008), *J. Phys.: Conf. Ser.* 108, 012019.

Bussman-Holder, A., Mueller, K. (2012), in *100 Years of Superconductivity*, Rogalla, H., Kes, P., Eds., Taylor & Francis, Boca Raton.

Buzdin, A., Bulaevskii, L., Panyukov, S. (1982), *JETP Lett.* 35, 178.

Bychkov, Y., Gor'kov, L., Dzyaloshinskii, I. (1966), *JETP* 23, 489.

Bychkov, Y., Rashba, E. (1984), *JETP Lett.* 39, 78.

Canfield, P., Gammel, P., Bishop, D. (1998), *Phys. Today* 51, 40.

Cao, B., Neal, C., Starace, A., Ovchinnikov, Y., Kresin, V. Z., Jarrold, M. (2008), *J. Supercond. Nov. Magn.* 21, 163.

Cao, C., Hirschfeld, P., Cheng, H. P. (2008), *Phys. Rev. B* 77, 220506.

Capone, M., Kotliar, G. (2006), *Phys. Rev. B* 74, 054513.

Carbotte, J. (1990), *Rev. Mod. Phys.* 62, 1027.

Carbotte, J., Greeson, M., Perez-Gonzales (1991), *Phys. Rev. Lett.* 66, 1789.

Caretta, B., Lascialfari, A., Rigamonti, A., Rosso, A., Varlamov, A. (2000), *Phys. Rev. B* 61, 12420.

Carrington, A., Bonalde, I., Prozorov, R., Giannetta, R., Kini, A., Schlueter, J., Wang, H., Geiser, U, Williams, J. (1999), *Phys. Rev. Lett.* 83, 4172.

Castro Neto, A., Guines, F., Peres, N., Neroselos, K., Geim, A. (2009), *Rev. Mod. Phys.* 81, 109.

Cawthorne, A., Barbara, P., Shitov, S., Lobb, C., Weisenfeld, K., Zangwill, A. (1999), *Phys. Rev. B* 60, 7575.

Cen, C., Thiel, S., Mannhart, J., Lery, J. (2009), *Science*, 323, 1026.

Chadwik, J. (1932), *Nature* 129, 312.
Chahara, K., Ohno, M., Kasai, M., Kosono, Y. (1993), *Appl. Phys. Lett.* 63, 1990.
Chandrasekhar, B. (1962), *Appl. Phys. Lett.* 1, 7.
Chao, K., Spalek, J., Oles, A. (1977), *J. Phys.: Cond. Mat.* 10, 271.
Chatterjee, U., Ai, D., Zhao, J., Rosenkranz, S., Kaminski, A., Raffy, H., Li, Z., Kadowaki, K., Randeria, M., Norman, M., Campusano, J. (2011), *PNAS* 108, 9346.
Chechersky, V., Nath, A., Isaak, I., Franck, K., Ghosh, K., Ju, H., Greene, R. (1999), *Phys. Rev. B* 59, 497.
Chen, K., Cybart, S., Dynes, R. (2005), *IEEE Trans. Appl. Supercond.* 15, 149.
Cheng, P., Shen, B., Hu, J., Wen, H. (2010), *Phys. Rev. B* 81, 174529.
Cherkez, V., Mallet, P., Le Quang, T., Magaud, L., Veuillen, J. (2018), *Phys. Rev. B* 98, 195441.
Chevrier, J., Suck, J., Capponi, J., Perroux, M. (1988), *Phys. Rev. Lett.* 61, 554.
Cho, D., Bastiaans, K., Chatzopoulos, D., Gu, G., Allan, M. (2019), *Nature* 571, 541.
Cho, K., Smith, B., Coniglio, W., Winter, L, Agosta, C., Schlueter, J. (2009), *Phys. Rev. B* 79, 220507.
Choi, H., Romdy, D., Sum, H., Cohen, M., Louie, S. (2002), *Nature*, 418, 758.
Chosal, A., Randeria, M., Trivedi, N. (1998), *Phys. Rev. Lett.* 81, 3940.
Chosal, A., Randeria, M., Trivedi, N. (2001), *Phys. Rev. B* 65, 014501.
Choudhury, N., Walter, E., Kolesnikov, A., Loong, C.-K. (2008), *Phys. Rev. B* 77, 134111.
Choy, T., Edge, J., Akhmerov, A., Beenakker, C. (2011), *Phys. Rev. B* 84, 195442.
Chu, C. (2002), *Phys. Scripta T* 102, 40.
Chu, C. (2017), in *High* T_c *Cooper Oxide Superconductors and Related Novel Materials, Bussmann-Holder*, Keller, H., Bianeoni, A., Eds., Springer, Cham.
Chu, C., Deng, L., Gooch, M., Hujan, S., Lv, B., Wu, Z. (2019), *J. Supercond. Nov. Magn.* 32, 7.
Chu, C., Xue, Y., Wang, Y., Hellman, A., Lorenz, B., Meng, R., Cmaidalka, J., Dezaneti, L. (2000), *J. Supercond. Nov. Magn.* 13, 679.
Chu, J., Kuo, H., Analytis, J., Fisher, I. (2012), *Science* 337, 710.
Chubukov, A., Efremov, D., Eremin, I. (2008), *Phys. Rev. B* 78, 134512.
Chubukov, A., Eremin, I., Efremov, D. (2016), *Phys. Rev. B* 93, 174516.
Chubukov, A., Maslov, D., Yudson, V. (2014), *Phys. Rev. B* 89, 155126.
Civelli, M., Capone, M., Kancharla, S., Parcollet, O., Kotliar, G. (2005), *Phys. Rev. Lett.* 95, 106402.
Clar, E. (1964), *Polycyclic Hydrocarbons*, Academic Press, NY.
Clark, K., Hassanien, A., Khan, S., Braun, K., Tanaka, H., Hia, S. (2010), *Nat. Nano*, 41, 2061.
Clarke, J. (1969), *Proc. Roy. Soc. A* 308, 447.
Clogston, A. (1962), *Phys. Rev. Lett.* 9, 266.
Cobert, D., Schoolwock, U., von Delft (2004), *Eur. Phys. J. B* 38, 501.
Coey, J., Viret, M., von Molnar, S. (1999), *Adv. Phys.* 48, 167.
Cohen, M. (1964), *Phys. Rev.* 134, A511.
Cohen, M. (1969), in *Superconductivity*, Vol. 1, Parks, R., Ed., Marcel Dekker, NY.
Cohen, M., Falicov, L, Phillips, J. (1962), *Phys. Rev. Lett.* 8, 316.
Coleman, L., Cohen, M., Sandman, D., Yamagishi, F., Garito, A., Heeger, A. (1973), *Sol. State Comm.* 12, 1125.
Cooper, L. (1956), *Phys. Rev.* 104, 1189.
Cooper, L. (1961), *Phys. Rev. Lett.* 6, 689.
Covaci, L., Marsiglio, F. (2006), *Phys. Rev. B* 73, 014503.
Crawford, M., Farneth, W., McCarronn, E., Harlow, R., Moudden, A. (1990), *Science* 250, 1390.
Crawford, M., Kunchur, M., Farneth, W., McCarron, III, E., Poon, S. (1990), *Phys. Rev. B* 41, 282.

Cribier, D., Jacrot, B., Madhov, R., Farnoux, B. (1964), *Phys. Lett.* 9, 106.
Croitoru, M., Buzdin, A. (2013), *J. Phys.: Cond. Mat.* 25, 125702.
Croitoru, M., Shanenko, A., Kaun, C., Peeters, F. (2011), *Phys. Rev. B* 83, 214509.
Cuk, T., Baumberger, F., Lu, D., Ingle, N., Zhou, X., Eisaki, H., Kaneko, N., Hussain, Z., Devereaux, T., Nagaosa, N., Shen, Z. (2004), *Phys. Rev. Lett.* 93, 117003.
Cyberey, M., Fahhaki, T., Lu, J., Kerr, A., Weikle, R., Lichtenberger, A. (2019), *IEEE Trans on Appl. Super.* 29, 1100306.
Cyr-Choinière, O., Grissonnanche, G., Badoux, S., Day, J., Bonn, D., Hardy, W., Liang, R., Doiron-Leyraud, N., Taillefer, L. (2015), *Phys. Rev. B* 92, 224502.
Dagotto, E. (1991), *Int. J. Mod. Phys. B* 5, 907.
Dagotto, E. (1994), *Rev. Mod. Phys.* 66, 763.
Dagotto, E., Hotta, T., Moreo, A. (2001), *Phys. Rep.* 344, 1.
Dagotto, E., Ortolani, F., Scalapino, D. (1992), *Phys. Rev. B* 46, 3183.
Dagotto, E., Riera, J, Young, A. (1990), *Phys. Rev. B* 42, 2347.
Dagotto, E., Riera, J. (1993), *Phys. Rev. Lett.* 70, 682.
Dagotto, E., Riera, J., Scalapino, D. (1992), *Phys. Rev. B* 45, 5744.
Damascelli, A. (2004), *Physica Script. T* 209, 61.
Damascelli, A., Hussain, Z., Shen, Z. (2003), *Rev. Mod. Phys.* 75, 473.
Daou, R., Chang, J., LeBoeuf, D., Cyr-Choinière, O., Laliberté, F., Doiron-Leyraud, N., Ramshaw, B., Liang, R., Bonn, D., Hardy, W., Taillefer, L. (2010), *Nature* 463, 519.
Darhmaoui, H., Jung, J. (1998), *Phys. Rev. B* 57, 8009.
Das Sarma, S., Adam, S., Hwang, E., Rossi, E. (2011), *Rev. Mod. Phys.* 83, 407.
Das Sarma, S., Nayak, C., Tewari, S. (2006), *Phys. Rev. B* 73, 220502.
Daybell, M., Steyert, W. (1968), *Rev. Mod. Phys.* 40, 380.
de Gennes, P. (1960), *Phys. Rev.* 118, 141.
de Gennes, P. (1964), *Phys. Cond. Matter* 3, 79.
de Gennes, P. (1964), *Rev. Mod. Phys.* 36, 225.
de Gennes, P. (1966), *Superconductivity in Metals and Alloys*, Benjamin, NY.
de Heer, W. (1993), *Rev. Mod. Phys.* 65, 611.
De Vault, D., Chance, B. (1966), *Biophys. J.* 6, 825.
Deaver, B., Fairbank, W. (1961), *Phys. Rev. Lett.* 7, 43.
Delft, D., Kes, P. (2010), *Phys. Today* 46, 38.
Deng, L., Hujan, S., Wu, S., Zhao, K., Lv, B., Gooch, M., Yuan, H., Chu, C. (2018), *Quan. Stud. Math. Found.* 5, 103.
Deng, L., Lv, B., Wu, Z., Xue, Y., Zhang, W., Li, F., Wang, L., Ma, X., Xue, Q., Chu, C. (2014), *Phys. Rev. B* 90, 214513.
Deng, M., Vaitiekėnas, S., Hansen, E., Danon, J., Leijnse, M., Flensberg, K., Nygård, J., Krogstrup, P., Marcus, C. (2016), *Science* 354, 1557.
Desgreniers, S. (2020), *Nature* 255, 626.
Deutscher, G. (1987), in *Chance and Matter*, Souletie, J. Vannimenus, J., Stora, R., Eds., Elsevier, Amsterdam.
Deutscher, G. (2005a), *New Superconductors: From Granular to High T_c*, World Scientific, Singapore.
Deutscher, G. (2005b), *Rev. Mod. Phys.* 77, 109.
Deutscher, G. (2012), *J. Supercond. Nov. Magn.* 25, 581.
Deutscher, G., de Gennes, P. (1969), *in Superconductivity*, Parks, R., Ed., Marcel Derker, NY.

Deutscher, G., Fenichel, H., Gershenson, M., Grunbaum, E., Ovadyahu, Z. (1973), *J. Low Temp. Phys.* 10, 231.

Devoret, M., Schoelkopf, R. (2013), *Science* 339, 1169.

Dewar, M. (1969), *The Molecular Orbital Theory of Organic Chemistry*, McGraw-Hill, NY.

Dias, R., Silrera, I. (2017), *Science* 355, 715.

Ding, H., Norman, M., Campusano, J., Randeria, M., Bellman, A., Yokoya, T., Takahashi, T., Mochiki, T., Kadowaki, K. (1996), *Phys. Rev. B* 54, 9678.

Doll, R., Nabauer, M. (1961), *Phys. Rev. Lett.* 7, 51.

Doniach, S., Engelsberg, S. (1966), *Phys. Rev. Lett.* 17, 750.

Drozdov, A., Eremets, M., Troyan, I., Ksenofontov, V., Shylin, S. (2015), *Nature* 525, 73.

Drozdov, A., Kong, P., Minkov, V., Besedin, S., Kuzovnikov, M., Mozaffari, S., Balicas, L., Balakirev, F., Craf, D., Prakapenka, V., Greenberg, E., Knyazev, D., Tkacz, M., Eremets, M. (2019), *Nature* 69, 528.

Duan, D., Liu, Y., Tian, F., Li, D., Huang, X., Zhao, Z., Yu, H., Liu, B., Tian, W., Cui, T. (2014), *Sci. Rep.* 4, 6968.

Duan, D., Yu, H., Xie, H., Cui, T. (2019), *J. Supercond. Nov. Magn.* 32, 53.

Dubitsky, G., Blank, V., Baga, S., Semonova, E., Kulbashinskii, V., Krecketov, A., Kutin, V. (2005), *JETP* 81, 260.

Duffee, S., Irawan, T., Bieletzki, M., Richter, T., Sieben, B., Yin, C., von Issendorff, B., Moseler, M., Hövel, H. (2007), *Eur. Phys. J. D* 45, 401.

Duprat, R., Bourbonnais, C. (2001), *Eur. Phys. J. B* 21, 219.

Durajski, A., Szczesniak, R., Li, Y. (2015), *Physica C* 515, 1.

Dynes, R. (1972), *Sol. State Comm.* 10, 615.

Dynes, R., Garno, J. (1981), *Phys. Rev. Lett.* 46, 137.

Dynes, R., Narayanamurti, V. (1973), *Sol. State Comm.* 12, 341.

Dynes, R., Narayanamurti, V., Chin, M. (1971), *Phys. Rev. Lett.* 26, 181.

Dynes, R., Sharifi, P., Valles, J. (1992), in *Lattice Effects in High* T_c *Superconductors*, Bar-Yam, Y., Egami, T., Mustre-Leon, J., Bishop, A., Eds., World Scientific, Singapore.

Dyson, F. (1970), *Physics Today* 23, 204.

Dzebisashvilli, D., Komarov, K. (2018), *Eur. Phys. J. B* 91, 278.

Dzero, M., Gor'kov, L., Kresin, V. Z. (1999), *Sol. State Comm.* 112, 707.

Dzero, M., Gor'kov, L., Kresin, V. Z. (2000), *Eur. Phys. J. B* 14, 459.

Dzyaloshinskii, I., Katz, E. (1968), *JETP* 55, 338.

Dzyaloshinskii, I., Larkin, A. (1972), *JETP* 34, 422.

Eagles, D. (1969), *Phys. Rev.* 186, 456.

Edegger, B., Muthukumar, V., Gros. C. (2007), *Adv. Phys.* 56, 927.

Eder, R., Ohta, Y. (1994), *Phys. Rev. B* 50, 10043.

Eder, R., von der Brink, J., Sawatsky, G. (1996), *Phys. Rev. B* 54, 732.

Efremov, D., Korshunov, M., Dolgov, O., Golubov, A., Hirschfeld, P. (2011), *Phys. Rev. B* 84, 180512(R).

Eilenberger, G. (1966), *Z. Phys.* 190, 142.

Eilenberger, G. (1967), *Phys. Rev.* 153, 584.

Eilenberger, G. (1968), *Z. Phys.* 214, 195.

Einaga, M., Sakata, M., Ishikawa, T., Shimizi, K., Eremets, M., Drozdov, A., Tpoyan, I., Hirao, N., Ohishi, Y. (2016), *Nat. Phys.* 12, 835.

Einstein, A. (1922), *Naturwiss* 10, 105.

Eisenmerger, W. (1969), in *Tunneling Phenomena in Solids*, Burstein, E., Lundqvist, S., Eds., Plenum, NY.
Eisenmerger, W., Dayem, A. (1967), *Phys. Rev. Lett.* 18, 125.
Ekardt, W. (1984), *Phys. Rev. B* 29, 1558.
Ekimov, E., Sidorov, V., Bauer, E., Melnik, N., Curro, W., Thompson, J., Stishor, S. (2004), *Nature* 428, 542.
Eliashberg, G. (1960), *JETP* 11, 696.
Eliashberg, G. (1961), *JETP* 12, 1000.
Eliashberg, G. (1963), *JETP* 16, 780.
Elsinger, H., Wosnitza, J., Wanka, S., Hagel, J., Schweitzer, D., Strunz, W. (2000), *Phys. Rev. Lett.* 84, 6098.
Emery, V. (1987), *Phys. Rev. Lett.* 58, 2794.
Emery, V. (1987), *Phys. Rev. Lett.* 58, 3759.
Emery, N., Herold, C., d'Astuto, M., Carcia, V., Bellin, C., Mareche, J., Lagrange, P., Loupias, G. (2005), *Phys. Rev. Lett.* 95, 078003.
Emery, V., Reiter, G. (1988), *Phys. Rev. B.* 38, 4547.
Eremets, M. (1996), *High Pressure Experimental Methods*, Oxford Science Publ., Oxford.
Eremets, M., Drozdov, A. (2016), *Phys. Usp.* 186, 1257.
Eremets, M., Drozdov, A., Kong, P., Wang, H. (2017), arXiv:1708.05217.
Eremets, M., Troyan, I., Drozdov, A. (2016), arXiv:1601.04479v1.
Eremets, M., Troyan, I., Medvedev, S., Tse, J., Yao, Y. (2008), *Science* 319, 1506.
Errea, I., Calandra, M., Pickard, C., Nelson, J., Needs, R., Li, Y., Liu, H., Zhang, Y., Ma, Y., Mauri, F. (2015), *Phys Rev. Lett.* 114, 157004.
Eskildsen, M., Kugler, M., Tanaka, S., Jun, J., Kazakov, S., Kar-pinski, J., Fisher, O. (2002), *Phys. Rev. Lett.* 89, 187003.
Essmann, V., Trauble, H. (1967), *Phys. Lett.* 24A, 526.
Fang, M., Pham, H., Qian, B., Liu, T., Vehstedt, E., Liu, Y., Spinu, L., Mao, Z. (2008), *Phys. Rev. B* 78, 224503.
Fath, M., Freisem, S., Menovsky, A., Tomioka, Y., Aarts, J., Mydosh, J. (1999), *Science* 285, 1540.
Feldman, B., Randeria, M., Li, J., Jeon, S., Xie, Y., Wang, Z., Drozdov, I., Bernevig, B., Yazdani, A. (2017), *Nat. Phys.* 13, 286.
Felner, I., Asaf, U., Levy, Y., Millo, O. (1997), *Phys. Rev. B* 55, 3374.
Felner, I., Tsindlekht, M., Gorodetsky, G., Rosenberg, F. (2000), *J. Supercond. Nov. Magn.* 13, 659.
Feng, X., Zhang, J., Gao, G., Liu, H., Wang, H. (2015), *RSV Adv.* 5, 59292.
Fernandes, R., Haraldsen, J., Wolfle, P., Balatsky, A. (2013), *Phys. Rev. B* 87, 014510.
Fernandes, R., Chubukov, A., Schmalian, J. (2014), *Nat. Phys.* 10, 97.
Ferraris, J., Cowan, D. O., Walatka, W., Perlstein, J. H. (1973), *J. Am. Chem. Soc.* 95, 948.
Ferrell, R. (1964), *Phys. Rev. Lett.* 13, 330.
Ferrell, R., Fulde, P. (1964), *Phys. Rev.* 135, A550.
Ferrero, M., Cornaglia, P., De Leo, L., Parcollet, O., Kotliar, G., Georges, A. (2009), *Phys. Rev. B* 80, 064501.
Fert, A. (2007), *Nobel Lecture.*
Fetter, A. (1974), *Ann. Phys.* 88, 1.
Feynman, R., Cohen, M. (1956), *Phys. Rev.* 102, 1189.
Feynman, R., Leighton, R., Sands, M. (1963), *The Feynman Lectures on Physics*, Vol. 3, Addison, Reading.

Flores-Livas, J., Amsler, M., Heil, C., Sanna, A., Boeri, L., Profeta, G., Wolverton, C., Goedecker, S., Gross, E. (2016), *Phys. Rev. B* 93, 020508.

Flores-Livas, J., Sanna, A., Gross, E. (2016), *Euro. Phys. J. B* 89, 63.

Flouquet, J. (2019), *J. Phys. Soc. Jpn* 88, 043702.

Fong, H., Bourges, P., Sidis, Y., Regnault, L., Ivanov, A., Gu, G., Koshizuka, N., Keimer, B. (1999), *Nature* 398, 588.

Fortunato, L. (2009), *Europhys. News* 40, 25.

Fradkin, E., Kivelson, S. A., Tranquada, J. M. (2015), *Rev. Mod. Phys.* 87, 457.

Francavilla, T., Skelton, E., Pande, C., Wolf, S., Hein, R. (1987), *Jpn J. Appl. Phys.* 26, 951.

Franck, J. (1994), in *Physical Properties of High Temperature Superconductors*, Ginsberg, D., Ed., World Scientific, Singapore.

Franck, J., Harker, S., Brewer, J. (1993), *Phys. Rev. Lett.* 71, 283.

Franck, J., Isaac, I., Chen, W., Chrzanowski, J., Irwin, J. (1998), *Phys. Rev. B* 58, 5189.

Franck, J., Jung, J., Mohamed, A., Gydax, S., Sproule, G. (1991), *Phys. Rev. B* 44, 5318.

Fratini, M., Poccia, N., Ricci, A., Campi, G., Burghammer, M., Aeppli, G., Bianconi, A. (2010), *Nature*, 466, 841.

Frauendorf, S., Guet, C. (2001), *Ann. Rev. Nucl. Part. Sci.* 51, 219.

Frenkel, D., Hanke, W. (1990), *Phys. Rev. B* 42, 6711.

Friedel, J. (1992), *J. Phys.* 2, 959.

Friedel, J. (1999), *Philos. Mag.* 79, 1251.

Fröhlich, H. (1950), *Phys. Rev.* 79, 845.

Fröhlich, H. (1968), *J. Phys. C* 1, 544.

Fu, L., Kane, C. (2008), *Phys. Rev. Lett.* 100, 096407.

Fujita, K., Kim, C., Lee, I., Lee, J., Hamidian, M., Firmo, I., Mukhopadhyay, S., Eisaki, H., Uchida, S., Lawler, M., Kim, E., Davis, J. (2014), *Science* 344, 612.

Fulde, P. (1991), *Electron Correlations in Molecules and Solids*, Springer-Verlag, Berlin.

Fulde, P., Ferrell, R. (1964), *Phys. Rev.* 135, A550.

Furness, J., Zhang, Y., Lane, C., Buda, I., Barbiellini, B., Markiewicz, R., Bansil, A., Sun, J. (2018), *Comm. Phys.* 1, 11.

Gaididei, Yu., Loktev, V. (1988), *Phys. Stat. Sol. b* 147, 307.

Galitski, V., Migdal, A. (1958), *JETP* 7, 96.

Gamble, F., McConnell, H. (1968), *Phys. Lett.* 26A, 162.

Ganzuly, B. (1973), *Z. Phys.* 265, 433.

Gao, G., Oganov, Bergara, A., Martinez-Canales, M., Cui, T., Iitaka, T., Ma, Y., Zou, G. (2008), *Phys. Rev. Lett.* 101, 107002.

Gao, L., Xue, Y., Chen, F., Xiong, Q., Meng, R., Ramirez, D., Chu, C., Eggert, J., Mao, H. (1994), *Phys. Rev. B* 50, 4260.

Garcia-Garcia, A., Urbina, Y., Yuzbahyan, E., Richter, K., Altshuler, B. (2011), *Phys. Rev. B* 83, 014510.

Garland, J. (1963), *Phys. Rev. Lett.* 11, 111.

Gastiasoro, M., Chubukov, A., Fernandes, R. (2019), *Phys. Rev. B* 99, 094524.

Gaudefroy, L., Daugas, J., Haas, M., Grevy, S., Stodel, C., Thomas, J., Perrot, L., Girod, M., Rosse, B., Angelique, J., Balabanski, D., Fiori, E., Force, C., Georgiev, G., Kameda, D., Kumar, V., Lozeva, R., Matea, I., Meot, V., Morel, P., Nara Singh, B., Nowacki, F., Simpson, G. (2009), *Phys. Rev. Lett.* 102, 092501.

Gavaler, J., Janocko, M., Jones, C. (1974), *J. Appl. Phys.* 45, 3009.

Gavrichkov, V, Ovchinnikov, S. (2008), *Phys. Sol. State* 50, 1081.

Gavrichkov, V., Ovchinnikov, S., Borisov, A., Goryachev, E. (2000), *JETP* 91, 3693.
Gavrichkov, V., Ovchinnikov, S., Nekrasov, I., Kokorina, E., Pchelkina, Z. (2007), *Phys. Sol. State* 49, 2052.
Gebhard, F.(1997), *The Mott–Hubbard Metal Insulator Transition: Models and Methods*, Springer, Berlin.
Gedik, N., Yang, D., Logvenov, G., Bozovic, I., Zewait, A. (2007), *Science*, 316, 425.
Geerk, J., Xi, X., Linker, G. (1988), *Z. Phys. B* 73, 329.
Gegenwart, P., Si, Q., Steglich, F. (2008), *Nat. Phys.* 4, 186.
Geilikman, B. (1966), *Phys. Usp.* 8, 2032.
Geilikman, B. (1971), *J. Low Temp. Phys.* 4, 189.
Geilikman, B. (1973), *Phys. Usp.* 16, 141.
Geilikman, B. (1958), *JETP* 7, 721.
Geilikman, B. (1965), *JETP* 48, 1963.
Geilikman, B. (1975), *Phys . Usp.* 18, 190.
Geilikman, B., Kresin, V. Z. (1966), *Sol. State* 7, 8659.
Geilikman, B., Kresin, V. Z. (1968a), *Semicond.* 2, 639.
Geilikman, B., Kresin, V. Z. (1968b), *Sol. State* 9, 2453.
Geilikman, B., Kresin, V. Z. (1972), *Phys. Lett.* 40A, 123.
Geilikman, B., Kresin, V. Z., Masharov, N. (1975), *J. Low Temp. Phys.* 18, 241.
Geilikman, B., Zaitsev, R., Kresin, V. Z. (1967a), *Sol. State* 9, 642.
Geilikman, B., Zaitsev, R., Kresin, V. Z. (1967b), *Fiz. Met. Metallov.* 21, 796.
Geilikman, B., Zaitsev, R., Kresin, V. Z. (1968), *Fiz. Met. Metallov.* 25, 171.
Georges, A, Kotliar, G. (1992), *Phys. Rev. B* 45, 6479.
Georges, A., Kotliar, G., Krauth, W., Rozenberg, M. (1996), *Rev. Mod. Phys.* 68, 13.
Geshkenbein, V., Larkin, A. (1986), *JETP Lett.* 43, 395.
Ghigo, G., Torsello, D., Ummarino, G., Gozzelino, L., Tanatar, M., Prozorov, R., Canfield, P. (2018), *Phys. Rev. Lett.* 121, 107001.
Giaever, I. (1960), *Phys. Rev. Lett.* 5, 464.
Giaever, I., Mergerle, K. (1961), *Phys. Rev.* 122, 1101.
Gilabert, A. (1977), *Ann. Phys.* 2, 203.
Gilabert, A., Romangan, J., Guyon, E. (1971), *Sol. State Comm.* 9, 1295.
Gill, W., Greene, R., Street, G., Little, W. (1975), *Phys. Rev. Lett.* 35, 1732.
Ginsberg, D., Hebel, L. (1969), in *Superconductivity*, Parks, R., Ed., Marcel Dekker, NY.
Ginzburg, V. (1965), *JETP* 20, 1549.
Ginzburg, V. (1957), *JETP* 4, 153.
Ginzburg, V. (1960), *Sov. Sol. State Phys.* 2, 61.
Ginzburg, V. (1970), *Phys. Usp.* 13, 335.
Ginzburg, V., Landau, L. (1950), *Zh. Eksp. Teor. Fiz.* 20, 1064 [English translation in Landau, L. D. (1965), *Collected Papers*, Pergamon Press, Oxford].
Ginzburg, V., Kirzhnits, D., Eds. (1982), *High Temperature Superconductivity*, Consultant Bureau, NY.
Giustino, F., Cohen, M., Louie, S. (2008), *Nature* 452, 975.
Glover, III, R., Tinkham, M. (1957), *Phys. Rev.* 108, 243.
Glover, R. (1967), *Phys. Rev. Lett. A* 25, 542.
Goedecker, S. (2004), *J. Chem. Phys.* 120, 9911.
Goeppert-Mayer, M. (1949), *Phys. Rev.* 75, 1969.
Goldman, J. (1947), *JETP* 17, 681.

Goldstein, R., Bialek, W. (1983), *Phys. Rev. B* 27, 7431.
Golubov, A., Dolgov, O., Boris, A., Charnukha, A., Sun, D., Lin, C., Shevchun, A., Korobenko, A., Trunin, M., Zverev, V. (2011), *JETP Lett.* 94, 333.
Golubov, A., Mazin, I. (1997), *Phys. Rev. B* 55, 15146.
Gomes, K., Pasupathy, A., Pushp, A., Ono, S., Ando, Y., Yazdani, A. (2007), *Nature*, 447, 569.
Gomjayi, A., Rozenberg, M., Chitra, R. (2007), *Phys. Rev. B* 76, 195108.
Goncharov, A., Lobanov, S., Kruglov, I., Zhao, X., Chen, X., Oganov, A., Konopkova, Z., Prakapenka, V. (2016), *Phys. Rev. B* 93, 174105.
Goncharov, A., Lobanov, S., Prakapenka, V., Greenberg, E. (2017), *Phys. Rev. B* 95, 14011.
Goncharov, A., Struzhkin, V., Mao, H., Hemley, R. (1999), *Phys. Rev. Lett.* 83, 1998.
Gor'kov, L. (1958), *JETP* 7, 505.
Gor'kov, L. (1959), *JETP* 9, 1364.
Gor'kov, L. (1960), *JETP* 10, 998.
Gor'kov, L. (2012), *Phys. Rev. B* 86, 060501.
Gor'kov, L. (2016a), *Phys. Rev. B* 93, 054517.
Gor'kov, L. (2016b), *Phys. Rev. B* 93, 060507(R).
Gor'kov, L. (2017), *J. Supercond. Nov. Magn.* 30, 845.
Gor'kov, L., Gruner, G., Eds. (1989), *Charge Density Waves in Solids*, North-Holland, Amsterdam.
Gor'kov, L., Kresin, V. Z. (1998), *JETP Lett.* 67, 985.
Gor'kov, L., Kresin, V. Z. (1999), *J. Supercond. Nov. Magn.* 12, 243.
Gor'kov, L., Kresin, V. Z. (2001), *Appl. Phys. Lett.* 78, 3657.
Gor'kov, L., Kresin, V. Z. (2002), *Physica C* 367, 103.
Gor'kov, L., Kresin, V. Z. (2004), *Phys. Rep.* 400, 149.
Gor'kov, L., Kresin, V. Z. (2016), *Nat. Sci. Rep.* 6, 25608.
Gor'kov, L., Kresin, V. Z. (2017), arXiv:1705.03165.
Gor'kov, L., Kresin, V. Z. (2018), *Rev. Mod. Phys.*, 90, 011001.
Gor'kov, L., Melik-Barkhudarov, T. (1963), *JETP* 18, 1031.
Gor'kov, L., Rashba, E. (2001), *Phys. Rev. Lett.* 87, 037004.
Gor'kov, L., Sokol, A. (1987), *JETP Lett.* 46, 420.
Gordon, J., Marcenat, C., Franck, J., Isaac, I., Zhang, G., Lortz, R., Meingast, C., Bouquet, F., Fisher, R., Phillips, N. (2001), *Phys. Rev. B* 65, 024441.
Gorter, C., Casimir, H. (1934), *Phys. Zs.* 35, 963.
Grafe, H., Paar, D., Lang, G., Curro, N., Behr, G., Werner, J., Hamann-Borrero, J., Hess, C., Leps, N., Klingeler, R., Buechner, B. (2008), *Phys. Rev. Lett.* 101, 047003.
Graser, S., Maier, T., Hirschfeld, P., Scalapino, D. (2009), *New J. Phys.* 11, 025016.
Gray, K., Hawley, M., Moog, E. (1987), in *Novel Superconductivity*, Wolf, S., Kresin, V. Z., Eds., Plenum, NY.
Gray, K., Kim, D. (1983), *Phys. Rev. Lett.* 70, 1693.
Greene, R., Street, G., Suter, L. (1975), *Phys. Rev. Lett.* 34, 577.
Grigorov, L., Chernavskii, D. (1972), *Biofizika* 17, 95.
Grimvall, G. (1969), *Phys. Kondens. Mat.* 9, 283.
Grimvall, G. (1981), *The Electron–Phonon Interaction in Metals*, North-Holland, Amsterdam.
Grinenko, V., Efremov, D., Drechsler, S., Aswartham, S., Gruner, D., Roslova, M., Morozov, I., Nenkov, K., Wurmehl, S., Wolter, A., Holzapfel, B., Buchner, B. (2014), *Phys. Rev. B* 89, 060504.
Grinenko, V., Kikoin, K., Drechsler, S., Fuchs, G., Nenkov, K., Wurmehl, S., Hammerath, F., Gueron, S., Pothier, H., Birge, N., Esteve, D., Devoret, M. (1996), *Phys. Rev. Lett.* 97, 3025.

Grismes, C., Shapiro, S. (1968), *Phys. Rev.* 169, 397.
Gros, C., Joyn, R, Rice, T. (1987), *Phys. Rev. B* 36, 8190.
Guertler, S., Wang, Q., Zhang, F. (2009), *Phys. Rev. B* 79, 144526.
Gull, E., Ferrero, M., Parcollet, O., Georges, A., Millis, A. J. (2010), *Phys. Rev. B* 82, 155101.
Gull, E., Millis, A. (2015), *Phys. Rev. B* 91, 085116.
Gull, E., Parcollet, O., Millis, A. (2013), *Phys. Rev. Lett.* 110, 216405.
Gull, E., Parcollet, O., Werner, P., Millis, A. (2009), *Phys. Rev. B* 80, 245102.
Gull, E., Werner, P., Wang, X., Troyer, M., Millis, A. (2008), *Europhys. Lett.* 84, 37009.
Gumeniuk, R., Schnelle, W., Rosner, H., Nicklas, M., Leithe-Jasper, A., Grin, Y. (2008), *Phys. Rev. Lett.* 100, 017002.
Gunnarsson, O., Andersen, O., Jepsen, O., Zaanen, J. (1989), *Phys. Rev. B* 39, 1708.
Gunnarsson, O., Schafer, T., LeBlanc, J., Gull, E., Merino, J., Sangiovanni, G., Rohringer, G., Toschi, A. (2015), *Phys. Rev. Lett.* 114, 236402.
Guo, Y., Zhang, Y., Bao, X., Han, T., Tang, Z., Zhang, L., Zhu, W., Wang, E., Niu, Q., Qiu, Z., Jia, J., Zhao, Z., Xue, Q. (2004), *Science* 306, 2015.
Gurevich, V., Larkin, A., Firsov, Y. (1962), *Sol. State* 4, 131.
Gutfreund, H. (2016), *J. Supercond. Novel Magn.* 29, 9.
Gutfreund, H., Little, W. (1969), *Phys. Rev.* 183, 68.
Gutfreund, H., Little, W. (1979), *Highly Conducting One-Dimensional Solids*,Devresee, J., Evrard, R., den Doren, V., Eds., Plenum, NY.
Gutzwiller, M. (1963a), *Phys. Rev.* 137, A1726.
Gutzwiller, M. (1963b), *Phys. Rev. Lett.* 10, 159.
Gutzwiller, M. (1964), *Phys. Rev.* 134, A923.
Haddon, R., Hebard, A., Rosseinsky, M., Murphy, D., Duclos, S, Lyons, K., Miller, B., Rosamilia, J., Fleming, R., Kortan, A., Glarum, S., Makhija, A., Muller, A., Elick, R., Zahurak, S., Tycko, R., Dabbagh, G., Thiel, F. (1991), *Nature* 350, 321.
Hagel, J., Kelemen, M., Fisher, G., Pilawa, B., Wosnitza, J., Dormann, E., H. Lohneysen, A. Schnepf, H. Schnockel, U. Neisel, J. Beck, J. (2002), *J. Low Temp. Phys.* 129, 133.
Haldane, F. (1981), *J. Phys. C* 14, 2585.
Halder, A., Kresin, V. Z. (2015), *Phys. Rev. B* 92, 214506.
Halder, A., Liang, A., Kresin, V. Z. (2015), *Nano Lett.* 15, 1410.
Halperin, B., Hohenberg, P. (1969), *Phys. Rev.* 168, 898.
Hamo, A., Benyamini, A., Shammass, L., Khivrich, I., Waissmann, J., Kaasbjerg, K., Oreg, Y., Von Oppen, F., Ilani, S. (2016), *Nature* 535, 395.
Hannay, N., Geballe, T., Mattias, B., Andres, K., Schmidt, F., MacNair, D. (1965), *Phys. Rev. Lett.* 14, 225.
Hao, X., Moodera, J., Meservey, R. (1991), *Phys. Rev. Lett.* 67, 1342.
Hardy, G., Hulm, J. (1953), *Phys. Rev.* 89, 884.
Harland, M., Katsnelson, M., Lichtenstein, A. (2016), *Phys. Rev. B* 94, 125133.
Harrison, W. (1961), *Phys. Rev.* 123, 85.
Hartnoll, S., Hofman, D., Metlitski, M., Sachdev, S. (2011), *Phys. Rev. B* 84, 125115.
Hasan, M., Kane, C. (2010), *Rev. Mod. Phys.* 82, 3045.
Hasegawa, Y., Poilblanc, D. (1989), *Phys. Rev. B* 40, 9035.
Hashimoto, M., Vishik, I., He, R., Devereaux, T., Shen, Z. (2014), *Nat. Phys.* 10, 483.
Haskel, D., Stern, E., Hinks, D., Mitchell, D., Jorgenson, J. (1997), *Phys. Rev. B* 56, 521.
Haule, K., Kotliar, G. (2007), *Phys. Rev. B* 76, 104509.
Hawley, M., Gray, K., Terris, B., Wang, H., Carlson, K., William, J. (1986), *Phys. Rev. Lett.* 57, 629.

He, S., He, J., Zhang, W., Zhao, L., Liu, D., Liu, X., Mov, D., Ov, Y., Wang, Q., Li, Z., Wang, L., Peng, Y., Liu, Y., Chen, C., Yu, L., Liu, G., Dong, X., Zhang, J., Chen, C., Xu, Z., Chen, X., Ma, X., Xue, Q., Zhou, X. (2013), *Nat. Mat.* 12, 605.

Hebard, A. (1992), *Phys. Tod.* 45, 26.

Hebard, A., Rosseinsky, M., Haddon, R., Murphy, D., Glarum, S., Palstra, T., Ramirez, A., Kortan, A. (1991), *Nature* 350, 600.

Heffner, R., Le, P., Hundley, M., Neumeier, J., Luke, G., Kojima, K., Nachumi, B., Uemura, Y., MacLaughlin, D., Cheong, S. (1996), *Phys. Rev. Lett.* 77, 1869.

Heid, R., Bohnen, K. (2005), *Phys. Rev. B* 72, 134527.

Heikkila, T., Volovik, G. (2016), in *Basic Physics of Functionalized Graphite*, Esquinazi, P. D., Ed., Springer, Cham.

Helfand, E., Werthamer, N. (1964), *Phys. Rev. Lett.* 13, 686.

Helfand, E., Werthamer, N. (1966), *Phys. Rev.* 147, 288.

Hepting, M., Chaix, L., Huang, E., Fumagalli, R., Peng, Y., Moritz, B.Kummer, K., Brookes, N., Lee, W., Hashimoto, M., Sarkar, T., He, J., Rotundu, C., Lee, Y., Greene, R., Braicovich, L., Ghiringhelli, G., Shen, Z., Devereaux, T., Lee, W. (2018), *Nature* 563, 374.

Herring, C. (1960), *J. Appl. Phys.* 31 (Suppl.), 3S.

Hettler, M., Tahvildar-Zadeh, A., Jarrell, M., Pruschke, T., Krishnamurthy, H. (1998), *Phys. Rev. B* 58, R7475.

Hicks, C., Lippman, T., Huber, M., Ren, Z., Yang, J., Zhao, Z., Moler, K. (2009), *J. Phys. Soc. Jpn* 78, 013708.

Higasi, K., Baba, H., Rembaum, A. (1965), *Quantum Organic Chemistry*, Interscience, NY.

Hill, D., Wheeler, J. (1953), *Phys. Rev.* 89, 1102.

Hirsch, J. (1985), *Phys. Rev. Lett.* 54, 1317.

Hirsch, J., Tang, S. (1989), *Phys. Rev. Lett.* 62, 591.

Hirschfeld, P., Korshunov, M., Mazin, I. (2011), *Rep. Prog. Phys.* 74, 124508.

Hirschfeld, P., Atkinson, W. (2002), *J. Low Temp. Phys.* 126, 881.

Hock, C., Schmidt, M., Issendorff, B. V. (2011), *Phys. Rev.* 128, 113401.

Hock, K., Nickisch, H., Thomas, H. (1983), *Helv. Phys. Acta* 56, 237.

Hofer, F., Schmidt, F., Grogger, W., Kothleitner, G. (2016), *IOP Conf. Ser.: Mat. Sci. Eng.* 109, 012007.

Hohenberg, P. (1967), *Phys. Rev.* 158, 383.

Holcomb, M., Perry, C., Coleman, J., Little, W. (1996), *Phys. Rev. B* 53, 6734.

Holczer, K., Klein, O., Huang, S., Kaner, R., Fu, K., Whetten, R., Diederich, D. (1991), *Science* 252, 1154.

Holstein, T. (1959), *Ann. Phys.* 8, 343.

Holstein, T., Primakoff, H. (1940), *Phys. Rev.* 58, 1098.

Honerkamp, C., Fu, H., Lee, D. (2007), *Phys. Rev. B* 75, 014503.

Hor, Y., Williams, A., Checkelsky, J., Roushan, P., Seo, J., Xu, Q., Zandbergen, H., Yazdani, A., Ong, N., Cava, R. (2010), *Phys. Rev. Lett.* 104, 057001.

Horsch, P., Stephan, W. (1993), in *Electronic Properties of High-T_c Superconductors*, Kuzmany, H., Mekring, M., Fink, J., Eds., Kirchberg, Tirol.

Horvatić, M., Ségransan, P., Berthier, C., Berthier, Y., Butaud, P., Henry, J. Y., Couach, M., Chaminade, J. P. (1989), *Phys. Rev. B* 39, 7332(R).

Hoshi, Y., Kuroda, T., Okada, M., Moriya, R., Masubuchi, S., Watanabe, K., Taniguchi, T., Kitaura, R., Machida, T. (2017), *Phys. Rev. B.* 95, 241403.

Hosoi, S., Matsuura, K., Ishida, K., Wang, H., Mizukami, Y., Watashige, T., Kasahara, S., Matsuda, Y., Shibauchi, T. (2016), *PNAS* 113, 8139.
Hosono, H. (2012), in *100 Years of Superconductivity*, Rogalla, H., Kes, P., Eds., CRC Press, Boca Raton.
Hotjtiak, G., Zandstra, P. (1960), *Mol. Phys.* 3, 371.
Hsu, F., Luo, J., Yeh, K., Chen, T., Huang, T., Wu, P., Lee, Y., Huang, Y., Chu, Y., Yan, D., Wu, M. (2008), *Natl. Acad. Sci USA* 105, 14262.
Hu, C. (1972), *Phys. Rev. B* 6, 1756.
Hu, R., Bozin, E., Warren, J., Petrovic, C. (2009), *Phys. Rev. B* 80, 214514.
Huang, X., Wang, X., Duan, D., Sundqvist, B., Li, X., Huang, Y., Li, F., Zhou, Q., Liu, B., Cui, T. (2016), arXiv:1610.02630.
Hubbard, J. (1963), *Proc. Roy. Soc. A* 276, 238.
Hubbard, J. (1964), *Proc. Roy. Soc. A* 281, 401.
Hubbard, J. (1965), *Proc. Roy. Soc. A* 285, 542.
Huckel, E. (1931), *Physik* 70, 204.
Huebener, R., Luebbig, H. (2008), *A Focus of Discoveries*, World Scientific, Singapore.
Huy, N., Gasparini, A., de Nijs, D., Huang, Y., Klaasse, J., Gortenmulder, T., de Visser, A., Hamann, A., Gorlach, T., von Löhneysen, H. (2007), *Phys. Rev. Lett.* 99, 067006.
Hybertsen, M., Stechel, E., Foulkes, W., Schluter, M. (1992), *Phys. Rev. B* 45, 10032.
Iavarone, M., Karapetrov, G., Koshelev, A., Kwok, W., Crabtree, G., Hinks, D., Kang, W., Choi, E., Kim, H. J., Kim, H., Lee, S. (2002), *Phys. Rev. Lett.* 89, 187002.
Ichimura, K., Takami, M., Nemura, K. (2008), *J. Phys. Soc. Jpn* 77, 114107.
Igushi, I., Yamagreshi, T., Sugimoto, A. (2001), *Nature* 412, 420.
Ihm, J., Cohen, M. L., Tuan, S. (1981), *Phys. Rev. B* 23, 3258.
Ishida, K., Hosoi, S., Teramoto, Y., Usui, T., Mizukami, Y., Itaka, K., Matsuda, Y., Watanabe, T., Shibauchi, T. (2020), *J. Phys. Soc. Jpn* 89, 064707.
Ishida, K., Ozaki, D., Kamatsuka, T., Tou, H., Kyogaku, M., Kitaoka, Y., Tateiwa, N., Sato, N. K., Aso, N., Geibel, C., Steglich, F. (2002), *Phys. Rev. Lett.* 89, 037002.
Ishiguro, T., Ito, H. (1996), in *From High-Temperature Superconductivity to Micro Miniature Refrigeration*, Cabreva, B., Gutfreund, H., Kresin, V. Z., Eds., Plenum, NY.
Ishiguro, T., Yamagaji, K., Saito, G. (2001), *Organic Superconductivity*, Springer, Berlin.
Ishihara, S., Nagaosa, N. (2004), *Phys. Rev. B* 69, 144520.
Issendorff, B. (2011), in *Handbook of Nanophysics*, Sattler, K., Ed., CRC Press, Boca Raton.
Ito, H., Ishiguro, T., Komatsu, T., Matsukawa, N., Saito, G., Anzai, H. (1994), *J. Supercond.* 5, 667.
Ivanov, D. (2001), *Phys. Rev. Lett.* 86, 268.
Izumi, M., Manako, T., Konishi, Y., Kawasaki, M., Tokura, Y. (2000), *Phys. Rev. B* 61, 18187.
Izyumov, Yu., Proshin, Yu., Khusainov, M. (2000), *JETP Lett.* 71, 138.
Izyumov, Yu., Proshin, Yu., Khusainov, M. (2002), *Phys. Usp.* 45, 109.
Izyumov, Yu. (1995), *Phys. Usp.* 38, 385.
Jaclic, J., Prelovsek, P. (2000), *Adv. Phys.* 49, 1.
Jahn, H., Teller, E. (1937), *Proc. Roy. Soc. A* 161, 220.
Jaime, M., Lin, P., Chun, S., Salamon, M., Dorsey, P., Rubinstein, M. (1999), *Phys. Rev. B* 60, 1028.
Jaklevic, R., Lambe, J., Silver, A., Mercereau, J. (1964), *Phys. Rev. Lett.* 12, 159.
Jarrell, M. (1992), *Phys. Rev. Lett.* 69, 168.
Jayaram, B., Chen, H., Callaway, J. (1995), *Phys. Rev. B* 52, 3742.
Jedrak, J., Spalek, J. (2011), *Phys. Rev. B* 83, 104512.
Jerome, D. (2012), *J. Supercond. Nov. Magn.* 25, 633.

Jerome, D., Grenzet, F. (1987), in *Novel Superconductivity*, Wolf, S., Kresin, V. Z., Eds., Plenum, NY.

Jerome, D., Mazaud, A., Ribault, M., Bechgaard, K. (1980), *J. Phys. Lett.(Paris)* 41, L95.

Ji, S., Zhang, T., Fu, Y., Chen, X., Ma, X., Li, J., Duan, W., Jia, J., Xue, Q. (2008), *Phys. Rev. Lett.* 100, 226801.

Jin, B., Ketterson, J. (1989), *Adv. Phys.* 38, 189.

Jin, S., Tiefel, T., McCormack, M., Fastnacht, R., Ramesh, R., Chen, L. (1994), *Science* 264, 413.

Jin, X., Meng, X., He, Z., Ma, Y., Liu, B., Cui, T., Zou, G., Mao, H. (2010), *PNAS* 107, 9969.

Jollicoeur, T. (1998), in *Pair Correlations in Many-Fermion Systems*, Kresin, V. Z., Ed., Plenum, NY.Jones, M., Marsh, R. (1954), *JACSAT* 76, 1434.

Jonner, G., Van Santen, J. (1950), *Physica* 16, 337.

Jorgensen, J., Veal, B., Paulikas, A., Nowicki, L., Crabtree, G., Claus, H., Kwok, W. (1990), *Phys. Rev. B* 41, 1863.

Josephson, B. (1962), *Phys. Lett.* 1, 251.

Josephson, B. (1965), *Adv. Phys.* 14, 419.

Jung, J., Boyce, B., Yan, H., Abdelhadi, M., Skinta, J., Lemberger, T. (2000), *J. Supercond.* 13, 753.

Junod, A. (1982), in *Proceedings of the 4th Conference on the Superconductivity of α- and f- Band Metals*, Buckel, W., Weber, W., Eds., Kernforsdukngszentrum, Karlsruhe.

Kaczmarczyk, J., Bünemann, J., Spaek, J. (2014), *New J. Phys.* 16, 073018.

Kaczmarczyk, J., Spaek, J, Schickling, T., Bünemann, J. (2013), *Phys. Rev. B* 88, 115127.

Kadanov, L. (1966), *Physics* 2, 263.

Kadin, A. (1999), *Superconducting Circuits*, Wiley Interscience, NY.

Kadowaki, K., Campuzano, J. (2008), *Phys. Rev. Lett.* 101, 137002.

Kamerlingh-Onnes, H. (1911), *Commun. Phys. Lab. Univ. Leiden* 124C.

Kamihara, Y., Watanabe, M., Hirano, M., Kawamura, R., Yanagi, H., Kamiya, T., Hosono, H. (2008), *J. Am. Chem. Soc.* 130, 3296.

Kaminski, A. (2013), *Phys. Rev. Lett.* 111, 157003.

Kaminski, A., Kondo, T., Takeuchi, T., Gu, G. (2014), *Phil. Mag.* 95, 5.

Kampf, A. (1994), *Phys. Rep.* 249, 221.

Kampf, A., Schrieffer, J. (1990), *Physica B* 163, 267.

Kampf, A., Schrieffer, J. (1990), *Phys. Rev. B* 41, 11663.

Kampf, A., Schrieffer, J. (1990), *Phys. Rev. B* 42, 6399.

Kanamori, J. (1961), *J. Appl. Phys.* 31 (Suppl.), 145.

Kane E. (1969), in *Tunneling Phenomena in Solids*, Burstein, E., Lindqvist, S., Eds., Plenum, NY.

Kanigel, A., Chatterjee, U., Randeria, M., Norman, M., Koren, G., Kadowaki, K., Campusano, J. (2008), *Phys. Rev. Lett.* 101, 137002.

Kanigel, A., Norman, M., Randeria, M., Chatterjee, U., Souma, S., Kaminski, A., Fretwell, H., Rosenkranz, S., Shi, M., Sato, T., Takahashi, T., Li, Z., Raffy, H., Kadowaki, K., Hinks, D., Ozyuzer, L., Campuzano, J. (2006), *Nat. Phys.* 2, 447.

Kaplan, M., Vekhter, B. (1995), *Cooperative Phenomena in Jahn–Teller Crystals*, Plenum, NY.

Karakosov, A., Maksimov, E., Mashkov, S. (1976), *JETP* 41, 971.

Karkin, A., Werner, J., Behr, G., Goshchitskii, B. (2009), *Phys. Rev. B* 80, 174512.

Kasahara, S., Watashige, T., Hanaguri, T., Kohsaka, Y., Yamashita, T., Shimoyama, Y., Mizukami, Y., Endo, R., Ikeda, H., Aoyama, K., Terashima, T., Uji, S., Wolf, T., von Löhneysen, H., Shibauchi, T., Matsuda, Y. (2014), *PNAS* 111, 16309.

Kasumov, A., Kociak, M., Gueron, S., Revlet, B., Volkov, V., Klimov, D., Bouchiat, H. (2001), *Science*, 291, 280.

Katakuse, I., Ichihara, T., Fujita, Y., Matsuo, T., Sakurai, T., Matsuda, H. (1986), *Int. J. Mass Spect. Ion Proc.* 69, 109.

Kawano, H., Kajimoto, R., Yoshizawa, H., Tomioka, Y., Kuwahara, H., Tokura, Y. (1997), *Phys. Rev. Lett.* 78, 4253.

Keller, H. (1989), *IBM J. Res. Dev.* 33, 314.

Keller, H. (2003), *Physica B* 326, 283.

Keller, H. (2005), in *Structure and Bonding*, Mueller, K., Bussmann-Holder, A., Eds., Springer, Berlin.

Kent, P., Jarrel, M., Maier, T., Pruschke, T. (2005), *Phys. Rev. B* 72, 060411.

Khasanov, R., Eshchenko, D., Luetkens, H., Morensoni, E., Prokscha, T., Suter, A., Garifianov, N., Mali, M., Roos, J., Conder, K., Keller, H. (2004), *Phys. Rev. Lett.* 92, 057602.

Khasanov, R., Shengelaya, A., Conder, K., Morenzoni, E., Savic, I., Karpinski, J., Keller, H. (2006), *Phys. Rev. B* 74, 064504.

Khasanov, R., Shengelaya, A., Morenzoni, E., Angst, M., Conder, K., Savic, I., Lampakis, D., Liarokapis, E., Tatsi, A., Keller, H. (2003), *Phys. Rev. B* 68, 220506.

Khusainov, M. (1991), *JETP Lett.* 53, 579.

Khusainov, M. (1996), *JETP* 82, 278.

Kihlstrom, K., Hovda, P., Kresin, V. Z., Wolf, S. (1988), *Phys. Rev. B* 38, 4588.

Kim, E., Castro-Neto, A. (2008), *Europhys. Lett.* 84, 57007.

Kim, H., Tanatar, M., Liu, Y., Sims, Z., Zhang, C., Dai, P., Lograsso, T., Prozorov, R. (2014), *Phys. Rev. B* 89, 174519.

Kitaev, A. (2001), *Phys. Usp.* 44, 131.

Kitatani, M., Schafer, T., Aoki, H., Held, K. (2019), *Phys. Rev. B* 99, 041115(R).

Kitatani, M., Tsuji, N., Aoki, H. (2015), *Phys. Rev. B* 92, 085104.

Kivelson, S., Fradkin, E., Emery, V. (1998), *Nature* 393, 550.

Kjaergaard, M., Wölms, K., Flensberg, K. (2012), *Phys. Rev. B* 85, 020503.

Klehe, A., Gangopadhyay, A., Diederichs, J., Schilling, J. (1993), *Physica C* 213, 266.

Klein, B., Cohen, R.(1992), *Phys. Rev. B* 45, 12405.

Kleiner, R., Mueller, P. (1994), *Phys. Rev. B* 49, 1327.

Kleiner, R., Steinmeyer, F., Kunkel, G., Muller, P. (1992), *Phys. Rev. Lett.* 68, 2394.

Klinovaja, J., Stano, P., Yazdani, A.Loss, D. (2013), *Phys. Rev. Lett.* 111, 186805.

Knight, W. (1987), in *Novel Superconductivity*, Wolf, S., Kresin, V., Z, Eds., Plenum, NY.

Knight, W., Clemenger, K., de Heer, W., Saunders, W., Chou, M., Cohen, M. (1984), *Phys. Rev. Lett.* 52, 2141.

Kobayashi, H., Tomota, H., Udagawa, T., Naito, T., Kobayashi, A. (1995), *Synth. Met.* 70, 867.

Kohno, M. (2012), *Phys. Rev. Lett.* 108, 076401.

Kohno, M.(2014), *Phys. Rev. B* 90, 035111.

Kondo, J. (1964), *Prog. Theor. Phys.* 32, 37.

Kondo, T., Palczewski, A., Hamaya, Y., Takeuchi, T., Wen, J., Xu, Z., Gu, G., Kondo, T., Malaeb, W., Ishida, Y., Sasagawa, T., Sakamoto, H., Takeuchi, T., Tohyama, T., Shin, S. (2015), *Nat. Comm.* 6, 7699.

Kong, P., Minkov, V., Kuzovnikov, M., Besedin, S., Drozdov, A., Mozaffari, S., Balicas, L., Balakirev, F., Prakapenka, V., Greenberg, E., Knyazev, D., Eremets, M. (2019), arXiv:1909.10482.

Kopaev, Y. (1982), in *High-Temperature Superconductivity*, Ginzburg, V., Kirzhnits, D., Eds., Consultant Bureau, New York.

Kordyuk, A. (2012), *Low Temp. Phys.* 38, 888.

Kordyuk, A. (2015), *Low Temp. Phys.* 41, 319.

Korshunov, M., Efremov, D., Golubov, A., Dolgov, O. (2014), *Phys. Rev. B* 90, 134517.
Korshunov, M., Gavrichkov, V., Ovchinnikov, S., Nekrasov, I., Pchelkina, Z., Anisimov, V. (2005), *Phys. Rev. B* 72, 165104.
Korshunov, M., Gavrichkov, V., Ovchinnikov, S., Pchelkina, Z., Nekrasov, I., Korotin, M., Anisimov, V. (2004), *JETP* 99, 559.
Korshunov, M., Ovchinnikov, S. (2007), *Eur. Phys. J. B* 57, 271.
Korshunov, M., Ovchinnikov, S., Shneyder, E., Gavrichkov, V., Orlov, Yu., Nekrasov, I., Pchelkina, Z. (2012), *Mod. Phys. Lett. B* 26, 24, 1230016.
Korshunov, M., Togushova, Yu., Dolgov, O. (2016), *Phys. Usp.* 59, 1211.
Kosterlitz, J., Thoulles, D. (1972), *J. Phys. C* 5, L124.
Kosugi, T., Miyake, T., Ishibashi, S., Arita, R., Aoki, H. (2009), *J. Phys. Soc. Jpn* 78, 113704.
Kotliar, G., Liu, L. (1988), *Phys. Rev. B* 38, 5142.
Kotliar, G., Savrasov, S., Haule, K., Oudovenko, V. S., Parcollet, O., Marianetti, C. (2006), *Rev. Mod. Phys.* 78, 865.
Kotliar, G., Savrasov, S., Palsson, J., Biroli, J. (2001), *Phys. Rev. Lett.* 87, 186401.
Kouznetsov, K., Sun, A., Chen, B., Katz, A., Bahcall, S., Clarke, J., Dynes, R., Gajewski, D., Han, S., Maple, M., Giapintzakis, J., Kim, J., Ginsberg, D. (1997), *Phys. Rev. Lett.* 79, 3050.
Kozlov, A., Maksimov, L. (1965), *JETP* 21, 790.
Krasnopolin, Y., Khaikin, M. (1973), *JETP* 37, 883.
Kresin, V. Z. (1967), *Phys. Lett. A* 24, 749.
Kresin, V. Z. (1972), *JETP* 34, 527.
Kresin, V. Z. (1972), *Sol. State* 13, 2468.
Kresin, V. Z. (1973), *J. Low Temp. Phys.* 11, 519.
Kresin, V. Z. (1974), *Phys. Lett. A* 49, 117.
Kresin, V. Z. (1982), *Phys. Rev. B* 25, 147.
Kresin, V. Z. (1984a), *J. Low Temp. Phys.* 57, 549.
Kresin, V. Z. (1984b), *Phys. Rev. B* 30, 450.
Kresin, V. Z. (1986), *Phys. Rev. B* 34, 7587.
Kresin, V. Z. (1987a), *Phys. Lett. A* 122, 434.
Kresin, V. Z. (1987b), *Phys. Rev. B* 35, 8716.
Kresin, V. Z. (1987c), *Sol. State Comm.* 63, 725.
Kresin, V. Z. (1998), in *Pair Correlation in Many-Fermion Systems*, Kresin, V. Z., Ed., Plenum, NY.
Kresin, V. Z. (2008), *J. Chem. Phys.* 128, 094706.
Kresin, V. Z. (2010), *J. Supercond. Nov. Magn.* 23, 179.
Kresin, V. Z. (2018), *J. Supercond. Nov. Magn.* 31, 3391.
Kresin, V. Z., Bill, A., Wolf, S., Ovchinnikov, Y. (1997), *Phys. Rev. B* 56, 107.
Kresin, V. Z., Friedel, J. (2011), *Europhys. Lett.* 93, 13002.
Kresin, V. Z., Gurfreund, H., Little, W. (1984), *Sol. State Comm.* 51, 339.
Kresin, V. Z., Gurfreund, H., Little, W. (1984b), *Sol. State Comm.* 51, 12.
Kresin, V. Z., Lester, W., Jr. (1984), *Chem. Phys.* 90, 935.
Kresin, V. Z., Litovchenko, V., Panasenko, A. (1975), *J. Chem. Phys.* 63, 3613.
Kresin, V. Z., Little, W., Eds. (1990), *Organic Superconductivity*, Plenum, NY.
Kresin, V. Z., Morawitz, H. (1990), *Phys. Lett.* 145, 368.
Kresin, V. Z., Morawitz, H. (1991), *Phys. Rev. B* 1341, 2691.
Kresin, V. Z., Morawitz, H., Wolf, S. (2014), *Superconducting State. Mechanisms and Properties.* Oxford University Press, Oxford.
Kresin, V. Z., Ovchinnikov, Y. (2006), *Phys. Rev. B* 74, 024514.

Kresin, V. Z., Ovchinnikov, Y. (2020), *Ann. Phys.* 417, 168141.
Kresin, V. Z., Ovchinnikov, Y., Wolf, S. (2003), *Appl. Phys. Lett.* 83, 722.
Kresin, V. Z., Ovchinnikov, Y., Wolf, S. (2004), *Physica C* 408, 434.
Kresin, V. Z., Ovchinnikov, Y., Wolf, S. (2006), *Physics Reports* 431, 231.
Kresin, V. Z., Parchomenko, A. (1975), *Sol. State* 16, 2180.
Kresin, V. Z., Tavger, B. (1966), *JETP* 23, 1124.
Kresin, V. Z., Wolf, S. (1990), *Physica C* 169, 476.
Kresin, V. Z., Wolf, S. (1992), *Phys. Rev. B* 46, 6458.
Kresin, V. Z., Wolf, S. (1994), *Phys. Rev. B* 49, 3652.
Kresin, V. Z., Wolf, S. (1995), in *Anharmonic Properties of High T_c Cuprates*, Mihailovich, D., Ruani, G., Kaldis, E., Mueller, K., Eds., World Scientific, Singapore.
Kresin, V. Z., Wolf, S. (1995), *Phys. Rev. B* 51, 1229.
Kresin, V. Z., Wolf, S. (1998), *J. Appl. Phys.* 63, 7357.
Kresin, V. Z., Wolf, S. (2009), *Rev. Mod. Phys.* 81, 481.
Kresin, V. Z., Wolf, S. (2012), *J. Supercond. Nov. Magn.* 25, 581.
Kresin, V. Z., Wolf, S., Ovchinnikov, Y. (1996), *Phys. Rev. B* 53, 118311.
Kresin, V. Z., Zaitsev, G. (1978), *JETP* 47, 983.
Kresin, V. Z. (1992), *Phys. Rep.* 220, 1.
Kresin, V. Z., Knight, W. (1998), in *Pair Correlations in Many-Fermion Systems*, Kresin, V. Z., Ed., Plenum, NY.
Krinitsyn, A., Nikolaev, S., Ovchinnikov, S. (2014), *J. Supercond. Nov. Magn.* 27, 955.
Krivenko, S., Avella, A., Mancini, F., Plakida, N. (2005), *Physica B*, 359, 666.
Krusin-Elbaum, L., Greene, R., Holtzberg, F., Malozemoff, A., Yeshurun, Y. (1989), *Phys. Rev. Lett.* 62, 217.
Kuboki, K., Fukuyama, H. (1989), *J. Phys. Soc. Jpn* 58, 376.
Kubozono, Y., Mitamura, H., Lee, X., He, X., Yamanari, Y., Takahashi, Y., Suzuki, Y., Kaji, Y., Eguchi, R., Akaike, K., Kambe, T., Okamoto, H., Fujiwara, A., Kato, T., Kosugi, T., Aoki, H. (2011), *Phys. Chem. Chem. Phys.* 13, 16476.
Kuchinskii, E., Nekrasov, I., Sadovskii, M. (2010), *JETP Lett.* 91, 518.
Kuchinskii, E., Sadovskii, M. (2008), *JETP Lett.* 88, 192.
Kudo, K., Miyoshi, Y., Sasaki, T., Kobayashi, N. (2005), *Physica C* 426, 251.
Kugel, K., Khomskii, D. (1973), *JETP* 37, 725.
Kugel, K., Khomskii, D. (1982), *Phys. Usp.* 25, 231.
Kulik, I., Yanson, I. (1972), *The Josephson Effect in Superconductive Tunneling Structures*, Israel Program for Scientific Translations, Jerusalem.
Kuroki, K., Usui, H., Onari, S., Arita, R., Aoki, H. (2009), *Phys. Rev. B* 79, 224511.
Kusco, C., Zhai, Z., Hakim, N., Markiewicz, R., Sridhar, S., Colson, D., Viallet-Guillen, V., Nefyadov, Y., Trunin, M., Kolesnikov, N., Maignan, A., Daignere, A.Erb, A. (2002), *Phys. Rev. B* 65, 132501.
Kuzmin, V., Nikolaev, S., Ovchinnikov, S. (2014), *Phys. Rev. B* 90, 245104.
Kuzmin, V., Nikolaev, S., Ovchinnikov, S. (2019), *J. Supercond. Nov. Magn.* 32, 1909.
Kyung, B., Kancharla, S., Senechal, D., Tremblay, A, Civelly, M., Kotliar, G. (2006), *Phys. Rev. B* 73, 165114.
Kyung, B., Landry, J., Tremblay, A.-M. (2003), *Phys. Rev. B* 68, 174502.
Labbe, J., Baristic, S., Friedel, J. (1967), *Phys. Rev. Lett.* 19, 1039.
Labbe, J., Friedel, J. (1966), *J. Phys.* 27, 153.
Lanczos, C. (1950), *J. Res. Natl. Bur. Stand.* 45, 255.

Landau, L. (1933), *Phys. Z. Sowj.* 3, 664.
Landau, L. (1965), in *Collected Papers of L. D. Landau*, Ter Haar, D., Ed., Pergamon Press, London.
Landau, L. (1937), *Zh. Eksp. Teor. Fiz.* 7, 19.
Landau, L., Plugge, S., Sela, E., Altland, A., Albrecht, S., Egger R. (2016), *Phys. Rev. Lett.* 116, 050501.
Landau, L., Lifshitz, E. (1963), *Electrodynamics of Continuous Media*, Pergamon Press, NY.
Landau, L., Lifshitz, E. (1975), *The Classical Theory of Fields*, Pergamon Press, Oxford.
Landau, L., Lifshitz, E. (1977), *Quantum Mechanics*, Pergamon Press, Oxford.
Landau, L., Lifshitz, E. (1980), *Statistical Physics*, Vol. 1, Elsevier, Amsterdam.
Landau, L., Lifshitz, E. (1994), *The Classical Theory of Fields*, Pergamon Press, NY.
Landau, L., Lifshitz, E. (2000), *Classical Theory of Field*, Elsevier, Oxford.
Landau, L., Lifshitz, E., Pitaevsky, L. (2004), *Electrodynamics of Continuous Media*, Elsevier, Amsterdam.
Landau, L., Pekar, S. (1946), *JETP* 16, 341.
Lang, G., Grafe, H., Holzapfel, B., van den Brink, J., Büchner, B., Schultz, L. (2011), *Phys. Rev. B* 84, 134516.
Lang, K., Madhavan, V., Hoffman, J., Hudson, E., Eisaki, Pan, S., H., Ushida, S., Davis, J. (2002), *Nature* 415, 412.
Langenberg, D., Scalapino, D., Taylor, B. (1966), *Proc. IEEE* 54, 560.
Lanzara, A., Bogdanov, P., Zhou, X., Kellar, S., Feng, D., Lu, E., Yoshida, T., Eisaki, H., Fujimori, A., Kishio, K., Shimoyama, J., Noda, T., Uchida, S., Hussain, Z., Shen, Z. (2001), *Nature* 412, 510.
Lanzara, A., Zhao, G., Saini, N., Bianconi, A., Conder, K., Keller, H., Mueller, K. (1999), *J. Phys.: Cond. Mat.* 11, L541.
Larkin, A. (1960), *JETP* 37, 186.
Larkin, A., Ovchinnikov, Y. (1964), *JETP* 47, 1137.
Larkin, A., Ovchinnikov, Y. (1969), *JETP* 28, 1200.
Larkin, A., Ovchinnikov, Y. (1983), *Phys. Rev. B* 28, 6281.
Larkin, A., Ovchinnikov, Y. (1986), in *Non-Equilibrium Superconductivity*, Landenberg, D., Larkin, A., Eds., Elsevier, Amsterdam.
Larkin, A., Ovchinnikov, Y. (2001), *JETP* 92, 519.
Larkin, A., Pikin, S. (1969), *JETP* 29, 891.
Lascialfari, A., Mishonov, T., Rigamonti, A,, Zucca, I., Berg, G., Loser, W., Drechsler, S. (2003), *Eur. Phys. J. B* 35, 326.
Lascialfari, A., Rigamonti, A., Romano, L., Tedesco, P., Varlamov, A., Embriaco, D. (2002), *Phys. Rev. B* 65, 1445232.
Laughlin, R., Lonzarich, G., Monthoux, P., Pines, D. (2001), *Adv. Phys.* 50, 361.
Lawler, M., Fujita, K., Lee, J., Schmidt, A., Kohsaka, Y., Kim, C., Eisaki, H., Uchida, S., Davis, J., Sethna, J., Kim, E. (2010), *Nature* 466, 347.
Lawrence, W., Doniach, S. (1971), in *Proceedings of the 12th International Conference on Low Temperature Physics*, Kanda, B., Ed., Academic Press of Japan, Kyoto.
Lazar, L.Westerholt, K., Zabel, H., Tagirov, L., Goryunov, Yu., Garif'yanov, N., Garifullin, I. (2000), *Phys. Rev. B* 61, 3711.
Lebegue, S. (2007), *Phys. Rev. B* 75, 035110.
Lee, J., Schmitt, F., Moore, R., Johnston, S., Cui, Y., Li, W., Yi, M., Liu, Z., Hashimoto, M., Zhang, Y., Lu, D., Devereaux, T., Lee, D., Shen, Z. (2014), *Nature* 515, 245.

Lee, J., Fujita, K., McElroy, K., Slezak, J. A., Wang, M., Aiura, Y., Bando, H., Ishikado, M., Mazui, T., Zhu, J., Balatsky, A., Eisaki, H., Uchida, S., Davis, J. (2006), *Nature* 442, 546.
Lee, P., Nagaosa, N., Wen, X.-G. (2006), *Rev. Mod. Phys.* 78, 17.
Leggett, A. (1975), *Rev. Mod. Phys.* 47, 331.
Leggett, A. (1980), *J. Phys. Coll. C* 41, 7.
Lehmann, H. (1954), *Nuo. Cim.* 11, 342.
Lerme, J., Dugourd, P., Hudgins, R., Jarrold, M. (1999), *Chem. Phys. Lett.* 304, 19.
Levanyuk, A. (1959), *JETP* 9, 571.
Levi, Y., Millo, O., Sharoni, A., Tsabba, Y., Leitus, G., Reich, S. (2000), *Europhys. Lett.* 51, 564.
Levy, F., Sheikin, I., Grenier, B., Huxley, A. (2005), *Science* 309, 1343.
Li, H., Zhou, X., Parham, S., Reber, T., Berger, H., Arnold, G., Dessau, D. (2018), *Nat. Comm.* 9, 26.
Li, J., Wang, Y. (2009), *Europhys. Lett.* 88, 17009.
Li, S., de la Cruz, C., Huang, Q., Chen, Y., Lynn, J., Hu, J., Huang, Y., Hsu, F., Yeh, K., Wu, M., Dai, P. (2009), *Phys. Rev. B* 79, 054503.
Li, Y., Balédent, V., Barišić, N., Cho, Y. C., Sidis, Y., Yu, G., Zhao, X., Bourges, P., Greven, M. (2011), *Phys. Rev. B* 84, 224508.
Li, Y., Hao, J., Liu, H., Li, Y., Ma., Y. (2014), *J. Chem. Phys.* 140, 174712.
Li, Y., Hao, J., Liu, H., Tse, J., Wang, Y., Ma., Y. (2015), *Sci. Rep.* 5, 9948.
Li, Z., Ooe, Y., Wang, X., Liu, Q., Jin, C., Ichioka, M., Zheng, G. (2010), *J. Phys. Soc. Jpn* 79, 083702.
Licciardello, S., Buhot, J., Lu, J., Ayres, J., Kasahara, S., Matsuda, Y., Shibauchi, T., Hussey, N. E. (2019), *Nature* 567, 213.
Lichtenstein, A., Katsnelson, M. (2000), *Phys. Rev. B* 62, R9283.
Lieb, E., Wu, F. (1968), *Phys. Rev. Lett.* 20, 1445.
Lifshitz, E., Pitaevskii, L. (1980), *Statistical Physics*, Part 2, Elsevier, Amsterdam.
Lifshitz, E., Pitaevskii, L. (1999), *Physical Kinetics*, Elsevier, Oxford.
Lifshitz, E., Pitaevskii, L. (2002), *Statistical Physics*, Part 2, Elsevier, Oxford.
Lifshitz, I. (1960), *JETP* 11, 1130.
Lindenfeld, Z., Eisenberg, E., Lifshitz, R. (2011), *Phys. Rev. B* 84, 064532.
Little, W. (1964), *Phys. Rev.* 134, A 1416.
Little, W. (1965), *Scientific American*, 212, 21.
Little, W. (1983), *J. Phys. Coll. C* 3, 819.
Little, W. (2016), *J. Supercond. Nov. Magn.* 29, 3.
Little, W., Collins, K., Holcomb, M. (1999), *J. Supercond. Nov. Magn.* 12, 89.
Little, W., Holcomb, M., Ghiringhelli, G., Braicovich, L., Dallera, C., Piazzalunga, A., Tagliaferri, A., Brookes, N. (2007), *Physica C* 460, 40.
Little, W., Parks, R. (1964), *Phys. Rev. A* 133, 97.
Liu, A., Mazin, I., Kortus, J. (2001), *Phys. Rev. Lett.* 87, 087005.
Liu, D., Zhang, W., Mou, D., He, J., Ou, Y., Wang, Q., Li, Z., Wang, L., Zhao, L., He, S., Peng, Y., Liu, X., Chen, C., Yu, Li., Liu, G., Dong, X., Zhang, J., Chen, C., Xu, Z., Hu, J., Chen, X., Ma, X., Xue, Q., Zhou, X. (2012), *Nat. Comm.* 3, 931.
Liu, H., Naumov, I., Geballe, Z., Somayazulu, M., Tse, J., Hemley, R. (2018), *Phys. Rev. B* 98, 100102.
Liu, H., Naumov, I., Hoffmann, R., Ashcroft, N., Hemley, R. (2017), *PNAS* 114, 6990.
Liu, Q., Yang, H., Qin, X., Yu, Y., Yang, L., Li, F., Yu, R., Jin, C., Uchida, S. (2006), *Phys. Rev. B* 74, 100506.

Loeser, A., Shen, Z., Dessau, D., Marshall, D., Park, C., Fournier, P., Kapitulnik, A. (1996), *Science* 273, 325.
Lofland, S., Bhagat, S., Ghosh, K., Greene, R., Karabashev, S., Shulyatev, D., Arsenov, A., Mukovski, Y. (1997), *Phys. Rev. B* 56, 13705.
Löhneysen, H. v., Pietrus, T., Portisch, G., Schlager, H. G., Schröder, A., Sieck, M., Trappmann, T. (1994), *Phys. Rev. Lett.* 72, 3262.
Loktev, V. M. (1996), *Sov. J. Low Temp. Phys.* 22, 3.
London, F. (1937), *J. Phys. (Paris)* 8, 397.
London, F., London, H. (1935), *Proc. Roy. Soc. A* 149, 71.
Loram, J., Mirza, K., Wade, J., Cooper, J., Liang, W. (1994), *Physica C* 235, 134.
Loret, B., Auvray, N., Gallais, Y., Cazayous, M., Forget, A., Colson, D., Jullien, M.-H., Paul, I., Civelli, M., Sacuto, A, (2019), *Nat. Phys.* 15, 771.
Louca, D., Egami, T. (1997a), *J. Appl. Phys.* 81, 5484.
Louca, D., Egami, T. (1997b), *Physica B* 241, 842.
Louca, D., Egami, T. (1999a), *Phys. Rev. B* 59, 6193.
Louca, D., Egami, T. (1999b), *J. Supercond.* 12, 23.
Louie, S., Cohen, M. (1977), *Sol. State Comm.* 10, 615.
Loybeyre, P., Occelli, F., Dumas, P. (2020), *Nature*, 577, 631.
Lu, Y., Wang, Z. (2013), *Phys. Rev. Lett.* 110, 096403.
Lubashevsky, Y., Lahoud, E., Chashka, K., Podolsky, D., Kanigel, A. (2012), *Nat. Phys.* 8, 309.
Lubashevsky, Y., Garg, A., Sassa, Y., Shi, M., Kanigel, A. (2011), *Phys. Rev. Lett.* 106, 047002.
Luders, G. (1967), *Z. Naturforsch.* 22, 845.
Luke, G., Fudamoto, Y., Kojima, K., Larkin, M., Merrin, J., Nachumi, B., Uemura, Y., Maeno, Y., Mao, Z., Mori, Y., Nakamura, H., Sigrist, M. (1998), *Nature* 394, 558.
Lutchyn, R., Sau, J., Das Sarma, S. (2010), *Phys. Rev. Lett.* 105, 077001.
Luttinger, J. (1968), *Math. Phys.* 4, 1154.
Ma, M., Yuan, D., Wu, Y., Zhou, H., Dong, X., Zhou, F. (2014), *Supercond. Sci. Technol.* 27, 12200.
MacDonald, A., Tsoi, M. (2011), *Phil. Trans. R. Soc. A* 369, 3098.
Mackenzie, A., Maeno, Y. (2003), *Rev. Mod. Phys.* 75, 657.
Mackenzie, A., Julian, S., Lonzarich, G., Carrington, A., Hughes, S., Liu, R., Sinclair, D. (1993), *Phys. Rev. Lett.* 71, 1238.
Mackenzie, A., Haselwimmer, R., Tyler, A., Lonzarich, G., Mori, Y., Nishizaki, S., Maeno, Y. (1998), *Phys. Rev. Lett.* 80, 161.
Maeda, H., Tanaka, Y., Fukutomi, M., Asano, T. (1988), *Jpn J. Appl. Phys.* 27, L209.
Maekawa, S., Tohyama, T., Barnes, S., Ishihara, S., Koshibae, W., Khaliullin, G. (2004), *Physics of Transition Metal Oxides*, Springer, Berlin.
Maeno, Y., Hashimoto, H., Yoshida, K., Nishizaki, S., Fujita, T., Bednorz, J., Lichtenberg, F. (1994), *Nature* 372, 532.
Magneli, A. (1949), *Ark. Kem.* 1, 223.
Mahan, G. (1993), *Many-Particle Physics*, Plenum, NY.
Mahendiran, R., Ibarra, M., Maignan, A., Millange, F., Arulraj, A., Mahesh, R., Raveau, B., Rao, C. (1999), *Phys. Rev. Lett.* 82, 2191.
Maheshwari, P., Raghavendra Reddy, V., Awana, V. (2018), *J. Supercond. Nov. Magn.* 31, 1659.
Maier, T., Jarrell, M., Pruschke, T., Hettler, M. H. (2005a), *Rev. Mod. Phys.* 74, 1027.
Maier, T., Jarrell, M., Pruschke, T., Keller, J. (2000), *Eur. Phys. J. B* 13, 613.
Maier, T., Jarrell, M., Schulthess, T., Kent, P., White, J. (2005b), *Phys. Rev. Lett.* 95, 237001.

Makarov, I., Gavrichkov, V., Shneyder, E., Nekrasov, I., Slobodchikov, A., Ovchinnikov, S., Bianconi, A. (2019), *J. Supercond. Nov. Magn.* 32, 1927.

Makarov, I., Ovchinnikov, S. (2015), *JETP* 121, 457.

Makarov, I., Shneyder, E., Kozlov, P., Ovchinnikov, S. (2015), *Phys. Rev. B* 92, 155143.

Maki, K. (1964), *Physica (New York)* 1, 21.

Maki, K. (1968), *Prog. Theor. Phys.* 39, 897.

Maki, K. (1969), *in Superconductivity*, Parks, R., Ed., Marcel Dekker, NY.

Mancini, F., Avella, A. (2004), *Adv. Phys.* 53, 537.

Mannella, N., Rosenhahn, A., Booth, C., Marchesini, S., Mun, B., Yang, S., Ibrahim, K., Tomioka, Y., Fadley, C. (2004), *Phys. Rev. Lett.* 92, 166401.

Mannhart, J. (2018), *J. Supercond. Nov. Magn.* 31, 1649.

Mannhart, J. (2020), *J. Supercond. Nov. Magn.* 33, 249.

Mannhart, J., Bednorz, J., Mueller, K., Schlom, D. (1991), *Z. Phys. B* 83, 307.

Mannhart, J., Braak, D. (2019), *J. Supercond. Nov. Magn.* 32, 17.

Mannhart, J., Chaudhari, P. (2001), *Phys. Tod.* 53, 48.

Mao, H., Hemley, R. (1994), *Rev. Mod. Phys.* 66, 671.

Mao, Z., Maeno, Y., Fukazawa, H. (2000), *Mat. Res. Bull.* 35, 1813.

Mao, Z., Mori, Y., Maeno, Y. (1999), *Phys. Rev. B.* 60, 610.

Maple, M. (1973), in *Magnetism*, 5th ed., Rado, G. T., Suhl, H., Eds., Academic Press, NY.

Maradudin, A., Montroll, E., Weiss, G, Ipatova, I. (1971), *Theory of Lattice Dynamics in the Harmonic Approximation*, Academic Press, NY.

Marshall, D., Dessau, D., Loeser, A., Park, C, Matsuura, A., Eckstein, J., Bozovic, I., Fournier, P., Kapitulnik, A., Spicer, W., Shen, Z. (1996), *Phys. Rev. Lett.* 76, 4841.

Martin, I., Morpurgo, A. (2012), *Phys. Rev. B* 85, 144505.

Masaki, S., Kotegawa, H., Hara, Y., Tou, H., Murata, K., Mizuguchi, Y., Takano, Y. (2009), *J. Phys. Soc. Jpn* 78, 063704.

Massa, W. (2004), *Crystal Structure Determination*, Springer, Berlin.

Matano, K., Li, Z., Sun, G., Sun, D., Lin, C., Ichioka, M., Zheng, G. (2009), *Europhys. Lett.*, 87, 27012.

Matano, K., Ren, Z., Dong, X., Sun, L., Zhao, Z., Zheng, G. (2008), *Europhys. Lett.* 83, 57001.

Mathur, N., Grosche, F., Julian, S., Walker, I., Freye, D., Haselwimmer, R., Lonzarich, G. (1998), *Nature* 394, 39.

Matsubara, T. (1955), *Prog. Theor. Phys.* 14, 351.

Matsukawa, H., Fukuyama, H. (1989), *J. Phys. Soc. Jpn* 58, 2845.

Matsuyama, K., Gweon, G. (2013), *Phys. Rev. Lett.* 111, 246401.

Mattheiss, L. (1987), *Phys. Rev. Lett.* 58, 1028.

Mattias, B., Geballe, T., Geller, S., Corenzwit, E. (1954), *Phys. Rev.* 95, 1435.

Mattis, D. C., Bardeen, J. (1958), *Phys. Rev.* 111, 412.

Matveev, K., Larkin, A. (1997), *Phys. Rev. Lett.* 78, 3749.

Mauri, F. (2016), *Nature* 532, 81.

Maxwell, E. (1950), *Phys. Rev.* 78, 477.

Mazin, I., Singh, D., Johannes, M., Du, M. (2008), *Phys. Rev. Lett.* 101, 057003.

Mazin, I. (1989), *Phys. Usp.* 32, 469.

McClure, J. (1957), *Phys. Rev.* 108, 612.

McMahan, A., Martin, R., Satpathy, S. (1989), *Phys. Rev. B* 38, 6650.

McMahon, J., Morales, M., Pierleoni, C., Ceperley, D. (2012), *Rev. Mod. Phys.* 84, 1607.

McMillan, W. (1968a), *Phys. Rev.* 167, 331.

McMillan, W. (1968b), *Phys. Rev.* 175, 537.
McMillan, W., Rowell, J. (1965), *Phys. Rev. Lett.* 14, 108.
McMillan, W., Rowell, J. (1969), *in Superconductivity*, Vol. 1, Parks, R., Ed., Marcel Dekker, NY.
Meirwes-Broer, K., Ed. (2000), *Metal Clusters at Surfaces*, Springer, Berlin.
Meissner, H. (1960), *Phys. Rev.* 117, 672.
Meissner, W., Hom, R. (1932), *Z. Physik* 74, 715.
Meissner, W., Ochsenfeld, R. (1933), *Naturwissenschafen* 21, 787.
Meissner, W., Voight, B. (1930), *Ann. Phys.* 399, 761.
Merchant, L., Ostrick, J., Barber, B., Jr, Dynes, R. (2001), *Phys. Rev. B* 63, 134508.
Mermin, N., Wagner, H. (1966), *Phys. Rev. Lett.* 17, 1133.
Mesot, J., Furrer, A. (1997), *J. Supercond.* 10, 623.
Metzner, W., Vollhardt, D. (1988), *Phys. Rev. B.* 37, 7382.
Metzner, W., Vollhardt, D. (1989), *Phys. Rev. Lett.* 62, 324.
Miao, L., Liu, S., Xu, Y., Kotta, E., Kang, C., Ran, S., Paglione, J., Kotliar, G., Butch, N, Denlinger, J., Wray, L. (2020), *Phys. Rev. Lett.* 124, 076401.
Michel, C., Raveau, B. (1984), *Rev. Chem. Min.* 21, 407.
Michon, B., Girod, C., Badoux, S., Kačmarčík, J., Ma, Q., Dragomir, M., Dabkowska, H., Gaulin, B., Zhou, J., Pyon, S., Takayama, T., Takagi, H., Verret, S., Doiron-Leyraud, N., Marcenat, C., Taillefer, L., Klein, T. (2019), *Nature* 567, 218.
Migdal, A. (1958), *JETP* 7, 996.
Migdal, A. (1959), *Nucl. Phys.* 13, 655.
Migdal, A. (1960a), *JETP* 10, 176.
Migdal, A. (1960b), *JETP* 37, 176.
Migdal, A. (1967), *Theory of Finite Fermi Systems and Application to Atomic Nuclei*, Interscience, NY.
Migdal, A. (2000), *Qualitative Methods in Quantum Theory*, Pergamon Press, NY.
Mihailovich, D., Kabanov, V., Mueller, K. (2002), *Europhys. Lett.* 57, 254.
Mikelsons, K., Khatami, E., Galanakis, D., Macridin, A., Moreno, J., Jarrell, M. (2009), *Phys. Rev. B* 80, 140505(R).
Mikheenko, P. (2019), *J. Supercond. Nov. Magn.* 32, 1121.
Millis, A., Littlewood, P., Shraiman, B. (1995), *Phys. Rev. Lett.* 74, 5144.
Millis, A., Shraiman, B., Mueller, R. (1996a), *Phys. Rev. B* 54, 5389.
Millis, A., Shraiman, B., Mueller, R. (1996b), *Phys. Rev. Lett.* 77, 175.
Millis, A., Norman, M. (2007), *Phys. Rev. B* 76, 220503(R).
Mineev, V. (1983), *Phys. Usp.* 26, 160.
Mineev, V. (2017), *Phys. Usp.* 60, 121.
Mineev, V., Samokhin, K. (1998), *Vvedenie v Teoriyu Neobychnoi Sverkhprovodimosti*, MFTI, Moscow [English translation: Mineev, V., Samokhin, K. (1999), *Introduction to Theory of Unconventional Superconductivity*, Gordon and Breach Sci. Publ., Amsterdam].
Mishchenko, A. (2009), *Phys. Usp.*. 52, 1193.
Mishchenko, A., Nagaosa, N. (2004), *Phys. Rev. Lett.* 93, 036402.
Mitsuhashi, R., Suzuki, Y., Yamanari, Y., Mitamura, H., Kambe, T., Ikeda, N., Okamoto, H., Fujiwara, A., Yamaji, M., Kawasaki, N., Maniwa, Y., Kubozono, Y., 2010, *Nature*, 464, 76.
Miyake, K., Schmitt-Rink, S, Varma, C. (1986), *Phys. Rev. B* 34, 3544.
Moeser, J., Steglich, F., von Minnigerode, G. (1974), *J. Low Temp. Phys.*15, 91.
Monthoux, P., Pines, D., Lonzarich, G. (2007), *Nature* 450, 1177.
Moodera, J., Hao, X., Gibson, G., Meservey, R. (1988), *Phys. Rev. Lett.* 61, 637.

Moore, C., Stanescu, T., Tewari, S. (2018), *Phys. Rev. B* 97, 165302.
Morawitz, H., Bozovic, I., Kresin, V. Z., Rietveld, G., van der Marel, D. (1993), *Z. Phys. B* 90, 277.
Morel, P., Anderson, P. (1962), *Phys. Rev.* 125, 1263.
Morell, E., Correa, J., Vargas, P., Pachco, M., Barticevic, Z. (2010), *Phys. Rev. B* 82, 121407.
Moreo, A. (1993), *Phys. Rev. B* 48, 3380.
Moritomo, Y., Akimoto, T., Fujishiro, H., Nakamura, A. (2001), *Phys. Rev. B* 64, 064404.
Morse, R., Bohm, H. (1957), *Phys. Rev.* 108, 1094.
Moskalenko, V. (1959), *Fiz. Met. Metallov.* 8, 503.
Mota, A., Visani, P., Pollini, A (1989), *J. Low Temp. Phys.* 76, 465.
Mott, N. (1949), *Proc. Phys. Soc. A* 62, 416.
Mott, N. (1968), *Rev. Mod. Phys.* 40, 677.
Mott, N. (1974), *Metal–Insulator Transitions*, Taylor & Francis, London.
Mourachkine, A. (2004), *Room-Temperature Superconductivity*, Cambridge International Science Publishing, Cambridge.
Mourik, V., Zuo, K., Frolov, S., Plissard, S., Bakkers, E., Kouwenhoven. L. (2012), *Science* 336, 1003.
Mueller, K. (1990), *Z. Phys. B* 80, 193.
Mueller, K. (2007), *J. Phys.: Cond. Mat.* 19, 251002.
Mueller, K. (2012), *J. Supercond. Nov. Magn.* 25, 2101.
Mueller, K., Burkard, H. (1979), *Phys. Rev. B* 19, 3593.
Mueller, K., Shengelaya, A. (2013), *J. Supercond. Nov. Magn.* 26, 486.
Mühge, Th., Westerholt, K., Zabel, H., Garif'yanov, N., Goryunov, Yu., Garifullin, I., Khaliullin, G. (1997), *Phys. Rev. B* 55, 8945.
Muller, J. (1980), *Rep. Prog. Phys.* 43, 641.
Muller, P. (1994), *Physica C* 35-40, 289.
Murakami, H., Ohbuchi, S., Aoki, R. (1994), *J. Phys. Soc. Jpn* 63, 2653.
Muranaka, T., Akimutsi, J. (2012), in *100 Years of Superconductivity*, Rogalla, H., Kes, P., Eds., CRC Press, Boca Raton.
Murayama, H., Sato, Y., Kurihara, R., Kasahara, S., Mizukami, Y., Kasahara, Y., Uchiyama, H., Yamamoto, A., Moon, E.-G., Cai, J., Freyermuth, J., Greven, M., Shibauchi, T., Matsuda, Y. (2019), *Nat. Comm.* 10, 1038.
Nadj-Perge, S., Drozdov, I., Li, J., Chen, H., Jeon, S., Seo, J., MacDonald, A., Bernevig, B., Yazdani, A. (2014), *Science* 346, 602.
Nagaev, E. (1996), *Phys. Usp.* 39, 781.
Nagamatsu, J., Nakagawa, N., Muranaka, T., Zenitani, Y., Akimutsu, J. (2001), *Nature* 410, 63.
Nakai, Y., Iye, T., Kitagawa, S., Ishida, K., Kasahara, S., Shibauchi, T., Matsuda, Y., Terashima, T. (2010), *Phys. Rev. B* 81, 020503.
Nakamura, K., Arita, R., Ikeda, H. (2011), *Phys. Rev. B* 83, 144512.
Narlikar, A. (2014), *Superconductors*, Oxford Press, Oxford.
Nayak, C., Simon, S., Stern, A., Freedman, M., Das Sarma, S. (2008), *Rev. Mod. Phys.* 80, 1083.
Nedorezov, S. (1967), *JETP* 24, 578.
Neil, C., Boeri. L. (2015), *Phys. RevB* 92, 06058.
Nicol, J., Shapiro, S., Smith, P. (1960), *Phys. Rev. Lett.* 5, 461.
Niedermayer, C., Bernard, C., Blasius, T., Golnik, A., Moodenbaugh, A., Budnick, J. (1998), *Phys. Rev. Lett* 80, 3843.
Nikolaev, S., Ovchinnikov, S. (2010), *JETP* 111, 634.

Ning, F., Ahilan, K., Imai, T., Sefat, A. S., Jin, R., McGuire, M., Sales, B., Mandrus, D. (2009), *J. Phys. Soc. Jpn* 78, 013711.
Niu, Q., Thouless, D., Wu, Y. (1985), *Phys. Rev. B* 31, 3372.
Norman, M., Ding, H., Randeria, M., Campuzano, J., Yokoya, T., Takeuchi, T., Takahashi, T., Mochiku, T., Kadowaki, K., Guptasarma, P., Hinks, D. (1998), *Nature* 392, 157.
Norman, M. (2014), in *Novel Superfluids*, Vol. 2, Bennemann, K., Ketterson, J., Eds., Oxford University Press, Oxford.
Novoselov, K., Geim, A., Morozov, S., Yiang, D., Zhang, Y., Dubonos, S., Grigorieva, I., Firsov, A. (2004), *Science*, 306, 666.
Nowack, A., Poppe, U., Weger, M., Schweitzer, D., Schwenk, H. (1987), *Z. Phys. B* 68, 41.
Nowack, A., Weger, M., Schweitzer, D., Keller, H. (1986), *Sol. State Comm.* 60, 199.
Nozieres, P. (1974), *J. Low Temp. Phys.* 17, 31.
Nozieres, P. (1995), in *Bose–Einstein Condensation*, Griffin, A., Snoke, D., Stringari, S., Eds., Cambridge University Press, Cambridge.
Nozieres, P., Schmitt-Rink, S. (1985), *J. Low Temp. Phys.* 59, 195.
O'Malley, T. (1967), *Phys. Rev.* 152, 98.
Ogata, M., Shiba, H. (1990), *Phys. Rev. B* 41, 2326.
Ohta, Y., Tsutsui, K., Koshibae, W., Shimozato, T., Maekawa, S. (1992), *Phys. Rev. B* 46, 14022.
Ohtamo, A., Hwang, H. (2004), *Nature* 427, 423.
Ohtamo, A., Miller, D., Grazul, J., Hwang, H. (2002), *Nature* 419, 378.
Okada, H., Takahashi, H., Mizuguchi, Y., Takano, Y., Takahashi, H. (2009), *J. Phys. Soc. Jpn* 78, 083709.
Okazaki, R., Shibauchi, T., Shi, H. J., Haga, Y., Matsuda, T. D., Yamamoto, E., Onuki, Y., Ikeda, H., Matsuda, Y. (2011), *Science* 331, 439.
Okimoto, Y., Tokura, Y. (2000), *J. Supercond. Nov. Magn.*, 13, 271.
Okuda, T., Asamitsu, A., Tomioka, Y., Kimura, T., Taguchi, Y., Tokura, Y. (1998), *Phys. Rev. Lett.* 81, 3203.
Onsager, L. (1949), *Nuo. Cim.* 9 (Suppl.), 6249.
Oreg, Y., Rafael, G., von Oppen, F. (2010), *Phys. Rev. Lett.* 105, 177002.
Osofsky, M., Soulen, Jr, R., Wolf, S., Broto, J., Rakoto, H., Ousset, J., Coffe, G., Ashkenazy, S., Pari, P., Bozovic, I., Eckstein, J., Virshup, G. (1993), *Phys. Rev. Lett.* 71, 2315.
Otaki, T., Yahagi, Y., Matsueda, H. (2015), *J. Phys. Soc. Jpn* 86, 084709.
Otsuki, J., Hafermann, H., Lichtenstein, A. (2014), *Phys. Rev. B* 90, 235132.
Ott, H., Rudigier, H., Fisk, Z., Smith, J. (1983), *Phys. Rev. Lett.* 50, 1595.
Ovchinnikov, S. (1994), *Phys. Rev. B* 49, 9891.
Ovchinnikov, S. (1995), *Phys. Sol. State* 37, 2007.
Ovchinnikov, S. (1997), *Phys. Usp.* 40, 993.
Ovchinnikov, S. (2003a), *Acta Phys. Pol. B* 34, 431.
Ovchinnikov, S. (2003b), *Phys. Usp.* 46, 21.
Ovchinnikov, S, Nikolaev, S. (2011), *JETP Lett.* 93, 517.
Ovchinnikov, S., Orlov, Yu. S., Nekrasov, I., Pchelkina, Z. (2011), *JETP* 112, 140.
Ovchinnikov, S., Sandalov, I. (1989), *Physica C* 161, 617.
Ovchinnikov, S., Shneyder, E. (2003), *Cent. Eur. J. Phys.* 3, 421.
Ovchinnikov, S., Shneyder, E. (2007), *J. Magn. Magn. Mat.* 310, 93.
Ovchinnikov, S., Shneyder, E. (2010), *J. Supercond. Nov. Magn.* 23, 733.
Ovchinnikov, S., Shneyder, E., Kordyuk, A. (2014), *Phys. Rev. B* 90, 220505.
Ovchinnikov, S., Shneyder, E., Korshunov, M. (2011b), *J. Phys.: Cond. Mat.* 23, 045701.

Ovchinnikov, S., Val'kov, V. (2004), *Hubbard Operators in the Theory of Strongly Correlated Electrons*, Imperial College Press, London.

Ovchinnikov, S., Petrakovsky, O. (1991), *J. Supercond.* 4, 437.

Ovchinnikov, Y., Kresin, V. Z. (1996), *Phys. Rev. B* 54, 1251.

Ovchinnikov, Y., Kresin, V. Z. (1998), *Phys. Rev. B* 58, 12416.

Ovchinnikov, Y., Kresin, V. Z. (2000), *Eur. Phys. J. B* 14, 203.

Ovchinnikov, Y., Kresin, V. Z. (2002), *Phys. Rev. B* 65, 214507.

Ovchinnikov, Y., Kresin, V. Z. (2005), *Eur. Phys. J. B* 45, 5.

Ovchinnikov, Y., Kresin, V. Z. (2010), *Phys. Rev. B* 81, 214505.

Ovchinnikov, Y., Kresin, V. Z. (2012), *Phys. Rev. B* 85, 064518.

Ovchinnikov, Y., Kresin, V. Z. (2013), *Phys. Rev. B* 88, 214504.

Ovchinnikov, Y., Wolf, S., Kresin, V. Z. (1999), *Phys. Rev. B* 60, 4329.

Ovchinnikov, Y., Wolf, S., Kresin, V. Z. (2001), *Phys. Rev. B* 63, 069524.

Overend, N.Howson, M., Lawrie, I. (1994), *Phys. Rev. Lett.* 72, 3238.

Overhauser, A. (1962), *Phys. Rev.* 128, 1437.

Owen, C., Scalapino, D. (1971), *Physica* 55, 691.

Palstra, T., Zhou, O., Iwasa, Y., Sulewski, P., Fleming, R., Zegarski, B. (1995), *Sol. State Comm.* 93, 327.

Pan, H., Cole, W., Sau, J., Das Sarma, S. (2020), *Phys. Rev. B* 101, 024506.

Papaconstantopoulos, D., Klein, B. Mehl, M., Pickett, W. (2015), *Phys. Rev. B* 91, 184511.

Paramekanti, A., Randeria, M., Trivedi, N. (2001), *Phys. Rev. Lett.* 87, 217002.

Paramekanti, A., Randeria, M., Trivedi, N. (2004), *Phys. Rev. B* 70, 054504.

Park, T., Tokiwa, Y., Bauer, E., Ronning, F., Movshovich, R., Sarrao, J, Thompson, J. (2008), *Physica B* 403, 943.

Park, W., Nepijuko, S., Fanelsa, A., Kisker, E. (1994), *Sol. State Comm.* 91, 655.

Parker, C., Pushp, A., Pasupathy, A., Gomes, K., Wen, J ., Xu, Z., Ono, S., Gu, G., Yazdani, A. (2010), *Phys. Rev. Lett.* 104, 117001.

Parkin, S. (1995), *Ann. Rev. Mat. Sci.* 25, 357.

Parmenter, H. (1968), *Phys. Rev.* 166, 392.

Parolla, A., Sorella, S. (1990), *Phys. Rev. Lett.* 64, 1831.

Patashinskii, A., Pokrovsky, V. (1966), *JETP* 23, 1292.

Pawlak, R., Kisiel, M., Klinovaja, J., Meier, T., Kawai, S., Glatzel, T., Loss, D., Meyer, E. (2016), *npj Quantum Inf.* 2, 1.

Pellarin, M., Baguenard, B., Bordas, C., Broyer, M., Lerme, J., Vialle, J. (1993), *Phys. Rev. B* 48, 17645.

Pena, V., Gredig, T., Santamaria, J., Schuller, I. (2006), *Phys. Rev. Lett.* 97, 177005.

Penc, K., Mila, F., Shia, H. (1995), *Phys. Rev. Lett.* 75, 894.

Pereivo, J., Petroric, A., Panagopoulos, C. (2011), *Phys. Expr.* 1, 208.

Pfleiderer, C. (2009), *Rev. Mod. Phys.* 81, 1551.

Phillips, J. (1963), *Phys. Rev. Lett.* 10, 96.

Pickett, W. (1989), *Rev. Mod. Phys.* 61, 433.

Pickett, W. (1991), *J. Supercond.* 4, 397.

Pickett, W., Singh, S. (1996), *Phys. Rev. B* 53, 1146.

Piekarz, P., Konior, J., Jefferson, J. (1999), *Phys. Rev. B* 59, 14697.

Pineiro, A., Pardo, V., Baldomir, D., Rodriguez, A., Cortes-Gil, R., Gomes, A., Aries, J. (2012), *J. Phys.: Cond. Mat.* 24, 275503.

Pippard, A. (1953), *Proc. Roy. Soc. A* 216, 547.

Pitaevskii, L. (1960), *JETP* 10, 1267.

Plakida, N. (2001), *JETP Lett.* 74, 36.

Plakida, N., Oudovenko, V. (2007), *JETP* 104, 230.

Plakida, N., Oudovenko, V. (2013), *Eur. Phys. J. B* 86, 115.

Plakida, N., Oudovenko, V. (2014), *JETP* 119, 554.

Plakida, N., Oudovenko, V., Horsch, P., Liechtenstein, A. (1997), *Phys. Rev. B* 55, R11997.

Plevert, L. (2011), *Pierre-Gilles de Gennes: A Life in Science*, World Scientific, Singapore.

Pogorelov, Yu., Santos, M., Loktev, V. (2011), *Low Temp. Phys.* 37, 633.

Poilblanc, D., Riera, J, Dagotto, E. (1994), *Phys. Rev. B* 49, 12381.

Pokkia, N., Fratini, M., Ricci, A., Campi, G., Barba, L., Vittertini-Orgeas, A., Bianconi, G., Aeppli, G., Bianconi, A. (2011), *Nat. Mat.* 10, 733.

Pokrovsky, V. (1961), *JETP* 13, 447.

Polyakov, A. (1969), *JETP* 28, 533.

Pompa, M., Bianconi, A., Castellano, A., Della Londa, S., Flank, A., Lagarde, P., Udron, D. (1991), *Physica C*, 184, 51.

Ponomarev, Y., Tsokur, E., Sudakova, M. Tchesnokov, S., Shabalin, S., Lorenz, M., Hein, M., Muller, G., Piel, H., Aminov, B. (1999), *Sol. State Comm.* 111, 513.

Potthoff, M. (2003), *Eur. Phys. J. B* 32, 429.

Potthoff, M. (2011), in *Theoretical Methods for Strongly Correlated Systems*, Avella, A., Mancini, F., Eds., Springer-Verlag, Berlin.

Prelovsek, P. (1988), *Phys. Lett. A* 126, 287.

Preosti, G., Muzikar, P. (1996), *Phys. Rev. B* 54, 3489.

Preuss, R., Muramatsu, A., Von der Linden, W., Deiterich, P., Assaas, F., Hanke, W. (1994), *Phys. Rev. Lett.* 73, 732.

Proshin, Yu., Khusainov, M. (1997), *JETP Lett.* 66, 562.

Puchkov, A., Basov, D., Timusk, T. (1996), *J. Phys.: Cond. Mat.* 8, 10049.

Pullman, B., Pullman, A. (1963), *Quantum Biochemistry*, Interscience, NY.

Radovic, Z., Ledvij, M., Dobrosavljević-Grujić, L., Buzdin, A., Clem, J. (1991), *Phys. Rev. B* 44, 759.

Raghu, S., Kivelson, S., Scalapino, D. (2010), *Phys. Rev. B* 81, 224505.

Raimondi, R., Jefferson, J., Feiner, L. (1996), *Phys. Rev. B* 53, 8774.

Ramirez, A., Kortan, A., Rosseinsky, M., Duclos, S., Mujsce, A., Haddon, R., Murphy, D., Makhija, A., Zahurak, S., Lyons, K. (1992), *Phys. Rev. Lett.* 68, 1058.

Ran, S., Eckberg, C., Ding, Q.-P., Furukawa, Y., Metz, T., Saha, S., Liu, I., Zic, M., Kim, H., Paglione, J., Butch, N. (2019), *Science* 365, 684.

Randeira, M., Varlamov, A. (1994), *Phys. Rev. B* 50, 10401.

Randeria, M., Campuzano, J. (1998), in *Proceedings of the International School of Physics "Enrico Fermi" on Conventional and High-Temperature Superconductors*, Iadonisi, G., Schrieffer, J. R., Chiafalo, M. L., Eds., IOS Press, Amsterdam.

Randeria, M., Sensarma, R., Trivedi, N. (2011), in *Theoretical Methods for Strongly Correlated Systems*, Avella, A., Mancini, F., Eds., Springer-Verlag, Berlin.

Rashba, E. (1960), *Sov. Phys. Sol. State* 2, 1109.

Rashba, E. (1982), in *Excitons*, Rashba, E., Sturge, M., Eds., North-Holland, Amsterdam.

Raveau, B., Michel, C., Hervieu, M., Groult, D. (1991), *Crystal Chemistry of High-*T_c *Superconducting Cooper Oxides*, Springer, Berlin.

Read, N., Green, D. (2000), *Phys. Rev. B* 61, 10267.

Reber, T., Plumb, N., Sun, Z., Cao, Y., Wang, Q., McElroy, K., Iwasawa, H., Arita, M., Wen, J., Xu, Z., Gu, G., Yoshida, Y., Eisaki, H., Aiura, Y., Dessau, D. (2012), *Nat. Phys.*, 8, 606.
Reber, T., Zhou, X., Plumb, N., Parham, S., Waugh, J., Cao, Y., Sun, Z., Li, H., Wang, Q., Wen, J., Xu, Z., Gu, G., Yoshida, Y., Eisaki, H., Arnold, G., Dessau, D. (2019), *Nat. Comm.* 10, 5737.
Reeves, M., Ditmars, D., Wolf, S., Vanderah, T., Kresin, V. Z. (1993), *Phys. Rev. B* 47, 6065.
Reich, S., Tsabba, Y. (1999), *Eur. Phys. J. B* 9, 1.
Reittu, H. (1995), *Am. J. Phys.* 63, 940.
Renker, B., Compf, F., Ewert, D., Adelmann, P., Schmidt, H., Gering, E., Hinks, H. (1989), *Z. Phys. B* 77, 65.
Renker, B., Compf, F., Gering, E., Nucker, N., Ewert, D., Reichardt, W., Rietschel, H. (1987), *Z. Phys. B* 67, 15.
Renner, Ch., Revaz, B., Genoud, J., Kadowaki, K., Fisher, O. (1998), *Phys. Rev. Lett.* 80, 149.
Reynolds, C., Serin, B., Wright, W., Nesbitt, L. (1950), *Phys. Rev.78*, 487.
Reyren, N., Thiel, S., Caviglia, A., Kourkoutis, L., Hammerl, G., Richter, C., Schneider, C., Kopp, T., Ruetschi, A., Yaccard, D., Gabay, M., Muller, D., Triscone, J., Mannhart, J. (2007), *Science*, 317, 1196.
Riblet, G., Winzer, K. (1971), *Sol. State Comm.* 9, 1663.
Rice, T. (1965), *Phys. Rev.* 140, A889.
Rice, T., Sigrist, M. (1995), *J. Phys.: Cond. Mat.* 7, L643.
Riera, J, Young, A. (1989), *Phys. Rev. B* 39, 9697.
Ring, P., Schuck, P. (1980), *The Nuclear Many-Body Problem*, Springer, NY.
Rohringer, G., Hafermann, H., Toschi, A., Katanin, A., Antipov, A., Katsnelson, M., Lichtenstein, A., Rubtsov, A., Held, K. (2018), *Rev. Mod. Phys.* 90, 025003.
Romberg, H., Alexander, M., Nucker, N., Adelmann, P., Fink, J. (1990), *Phys. Rev. B* 42, 8768.
Ronning, F., Helm, T., Shirer, K., Bachmann, M., Balicas, L., Chan, M., Ramshaw, B., McDonald, R., Balakirev, F., Jaime, M., Bauer, E., Moll, P. (2017), *Nature* 548, 313.
Rossat-Mignod, J., Regnault, L., Vettier, C., Bourges, P., Burlet, P., Bossy, J., Henry, J., Lapertot, G. (1991), *Physica C* 185, 86.
Rosseinsky, M., Ramirez, A., Glarum, S., Murphy, D., Haddon, R., Hebard, A., Palstra, T., Kortan, A., Zahurak, S., Makhija, A. (1991), *Phys. Rev. Lett.* 66, 2830.
Rowell, J. (1969), in *Tunneling Phenomena in Solids*, Burnstein, E., Lindqvist, S., Eds., Plenum, NY.
Rozhkov, A., Sboychakor, A., Rachmanov, A., Nori, F. (2016), *Phys. Rep.*, 648, 1.
Ruby, M., Pientka, F., Peng, Y., von Oppen, F., Heinrich, B., Franke, K. J (2015), *Phys. Rev. Lett.* 115, 197204.
Ruckenstein, A., Hirschfeld, P., Appel, J. (1987), *Phys. Rev. B* 36, 857.
Ruppel, M., Rademann, K. (1992), *Chem. Phys. Lett.* 197, 280.
Ruvalds, J. (1981), *Adv. Phys.* 30, 677.
Sabo, J. (1969), *Phys. Rev.* 131, 1325.
Sacramento, P. (2003), *J. Phys.: Cond. Mat.* 15, 6285.
Sadovskii, M. (1997), *Phys. Rep.* 282, 225.
Sadovskii, M. (2008), *Phys. Usp.* 51, 1201.
Sadovskii, M. (2016), *Phys. Usp.* 59, 947.
Saint-James, D., Sarma, G., Thomas, E. (1969), *Type II Superconductivity*, Pergamon Press, Oxford.
Sakai, S., Motome, Y., Imada, M. (2009), *Phys. Rev. Lett.* 102, 056404.
Sakai, S., Sangiovanni, G., Civelli, M., Motome, Y., Held, K., Imada, M. (2012), *Phys. Rev. B* 85, 035102.
Salem, L. (1966), *The Molecular Orbital Theory of Conjugated Systems*, Benjamin, NY.

Sales, B., Sefat, A., McGuire, M., Jin, R., Mandrus, D., Mozharivskyj, Y. (2009), *Phys. Rev. B* 79, 094521.
Sano, W., Korensune, T., Tadano, T., Akashi, R., Arita, R. (2016), *Phys. Rev. B* 93, 094525.
Santiago, R., de Menezes, O. (1989), *Sol. State Comm.* 70, 835.
Sarazin, F., Sarajois, H., Mittig, W., Nowacki, F., Orv, N., Ren, Z., Roussel-Chomaz, P., Auger, G., Balboredin, D., Belozyorov, A., Borcea, C., Caurier, E., Dlouhy, Z., Gilibert, A., Lalleman, A., Lewitowicz, M., Lukyanov, S., de Oliveira, F., Penionzhkevich, Y., Ridikas, D., Sakurai, H., Tarasov, O., de Vismes, A. (2000), *Phys. Rev. Lett.* 84, 5062.
Sarma, G. (1963), *J. Phys. Chem. Sol.* 24, 1029.
Sarrao, J., Morales, L., Thompson, J., Scott, B., Stewart, G., Wastin, F., Rebizant, J., Boulet, P., Colineau, E., Lander, G. (2002), *Nature* 420, 297.
Sato, M., Ando, Y. (2017), *Rep. Prog. Phys.* 80, 7.
Sato, M., Fujimoto, S. (2009), *Phys. Rev. B* 79, 094504.
Sato, M., Takahashi, Y., Fujimoto, S. (2010), *Phys. Rev. B* 82, 134521.
Sato, Y., Kasahara, S., Murayama, H., Kasahara, Y., Moon, E.-G., Nishizaki, T., Loew, T., Porras, J., Keimer, B., Shibauchi, T., Matsuda, Y. (2017), *Nat. Phys.* 13, 1074.
Sau, J., Lutchyn, R., Tewari, S., Das Sarma, S. (2010), *Phys. Rev. Lett.* 104, 040502.
Sau, J. D., Tewari, S. (2012), *Phys. Rev. B* 86, 104509.
Savini, G., Ferrari, A., Guistino, F. (2010), *Phys. Rev. Lett.* 105, 037002.
Sawatzky, G., Elfimov, I., van den Brink, J., Zaanen, J. (2009), *Europhys. Lett.*, 86, 17006.
Saxena, S., Agarwal, P., Ahilan, K., Grosche, F., Haselwimmer, R., Steiner, M., Pugh, E., Walker, I., Julian, S., Monthoux, P., Lonzarich, G., Huxley, A., Sheikin, I., Braithwaite, D., Flouquet, J. (2000), *Nature* 406, 587.
Scalapino, D. (1969), in *Superconductivity*, Parks, R., Ed., Marcel Dekker, NY.
Scalapino, D. (1987), *Phys. Rev. B* 35, 6694.
Scalapino, D. (1991), *Physica C* 185, 104.
Scalapino, D. (1995), *Phys. Rep.* 250, 329.
Scalapino, D. (2012), *Rev. Mod. Phys.* 84, 1383.
Scalapino, D., Loh, E., Hirsch, J. (1986), *Phys. Rev. B* 34, 8190.
Scalapino, D., Schrieffer, J., Wilkins, J. (1966), *Phys. Rev.* 148, 263.
Scalapino, D., White, S., Zhang, S. (1992), *Phys. Rev. Lett.* 68, 2830.
Schafroth, M. (1955), *Phys. Rev.* 100, 463.
Schenck, A. (1985), *Muon Spin Rotation Spectroscopy: Principles and Applications in Solid State Physics*, Adam Higler, Bristol.
Schenkel, T., Lo, C., Weis, C., Schuh, A., Persaud, A., Bokor, J. (2009), *Nucl. Instr. Meth. Phys. Res. B* 267, 2563.
Scher, H., Zallen, R. (1970), *J. Chem. Phys.* 53, 3759.
Schiffer, P., Ramirez, A., Bao, W., Cheong, S. (1995), *Phys. Rev. Lett.* 75, 3336.
Schilling, A., Cantoni, M., Guo, J., Ott, H. (1993), *Nature*, 363, 56.
Schilling, J., Klotz, S. (1992), in *Physical Properties of High-Temperature Superconductors*, Vol. 3, Ginzberg, D., Ed., World Scientific, Singapore.
Schirber, J., Northrup, C. (1974), *Phys. Rev. B* 10, 3818.
Schlom, D. (2015), *Appl. Mat.* 6, 062403.
Schlueter, J., Kini, A., Ward, B., Geiser, U., Wang, H., Montasham, J., Winter, R., Gard, G. (2001), *Physica C* 351, 261.
Schluter, M., Hybertsen, M., Christensen, N. (1988), *Physica C* 153, 1217.
Schluter, M., Lannoo, M., Needles, M., Baraff, G., Tomanek, D. (1992), *Phys. Rev. Lett.* 68, 526.

Schmalian, J., Pines, D., Stojkovich, B. (1998), *Phys. Rev. Lett.* 80, 3839.
Schmiedeshoff, G., Detwiler, J., Beyermann, W., Lacerda, A., Canfield, P., Smith, J. (2001), *Phys. Rev. B* 63, 134519.
Schneider, R., Geerk, J., Reitschel, H. (1987), *Europhys. Lett.* 4, 845.
Schrieffer, J., Wen, X., Zhang, S. (1989), *Phys. Rev. B* 39, 11663.
Schriver, K., Persson, J., Honea, E., Whetten, R. (1990), *Phys. Rev. Lett.* 64, 2359.
Sedrakyan, T., Chubukov, A. (2010), *Phys. Rev. B* 81, 174536.
Sénéchal, D., Pérez, D., Pioro-Ladrière, M. (2000), *Phys. Rev. Lett.* 84, 522.
Senechal, D., Tremblay, A. (2004), *Phys. Rev. Lett.* 92, 126401.
Senga, Y., Kontani, H. (2008), *J. Phys. Soc. Jpn* 77, 113710.
Shanenko, A., Croitoru, M., Peeters, F. (2006), *Europhys. Lett.* 76, 498.
Shapiro, S. (1963), *Phys. Rev. Lett.* 11, 80.
Sharma, M., Rani, P., Sang, L., Wang, X., Awana, V. (2020), *J. Supercond. Nov. Magn.* 33, 565.
Sharma, R., Xiong, G., Kwon, C., Ramesh, R., Greene, R., Venkatesan, T. (1996), *Phys. Rev. B* 54, 10014.
Shaw, W., Swihart, J. (1968), *Phys. Rev. Lett.* 20, 1000.
Shen, K., Ronning, F., Lu, D., Baumberger, F., Ingle, N., Lee, W., Meevasana, W., Kohsaka, Y., Azuma, M., Takano, M., Takagi, H., Shen, Z. (2005), *Science* 307, 901.
Shen, K., Ronning, F., Lu, D., Lee, W., Ingle, N., Meevasana, W., Baumberger, F., Damascelli, A., Armitage, N., Miller, L., Kohsaka, Y., Azuma, M., Takano, M., Takagi, H., Shen, Z. (2004), *Phys. Rev. Lett.* 93, 267002.
Shen, K., Ronning, F., Meevasana, W., Lu, D., Ingle, N., Baumberger, F., Lee, W., Miller, L., Kohsaka, Y., Azuma, M., Takano, M., Takagi, H., Shen, Z. (2007), *Phys. Rev. B*75, 075115.
Shen, Z., Dessau, D. (1995), *Phys. Rep.* 253, 1.
Sheng, Z., Hermann, A. (1988), *Nature* 332, 138.
Shengelaya, A., Conder, K., Mueller, K. (2020), *J. Supercond. Nov. Magn.* 33, 306.
Shengelaya, A., Mueller, K. (2019), *J. Supercond. Nov. Magn.* 32, 3.
Sherman, A. (2006), *Phys. Rev. B* 73, 155105.
Sherman, A. (2018), *J. Phys.: Cond. Mat.* 30, 195601.
Shestakov, V., Korshunov, M., Dolgov, O. (2018a), *Symmetry* 10, 323.
Shestakov, V., Korshunov, M., Togushova, Y., Efremov, D., Dolgov, O. (2018b), *Supercond. Sci. Technol.* 31, 034001.
Shiina, Y., Shimada, D., Mottate, A., Ohyagi, Y., Tsuda, N. (1995), *J. Phys. Soc. Jpn* 64, 2577.
Shim, H., Chaudhari, P.Logvenov, G., Bozovic, I. (2008), *Phys. Rev. Lett.* 101, 247004.
Shimada, D., Tsuda, N., Paltzer, U., de Wette, F. (1998), *Physica C* 298, 195.
Shimahara, H., Takada, S. (1991), *J. Phys. Soc. Jpn* 60, 2394.
Shklovskii, B., Efros, A. (1984), *Electronic Properties of Doped Semiconductors*, Springer, Berlin.
Shmidt, V. (1997), in *The Physics of Superconductors*, Muller, P., Ustinov, A., Eds., Springer-Verlag, Berlin.
Shneyder, E., Makarov, I., Zotova, M., Ovchinnikov, S. (2018), *JETP* 126, 683.
Shneyder, E., Ovchinnikov, S. (2006), *JETP Lett.* 83, 394.
Shneyder, E., Ovchinnikov, S. (2009), *JETP* 109, 1017.
Shubnikov, L., Khotkevich, V., Shepelev, Y., Riabinin, Y. (1937), *Zh. Eksp. Teor. Fiz.* 7, 221.
Shukhareva, I. (1963), *JETP* 16, 828.
Sigai, S., Shamoto, S., Sato, M., Takagi, H., Ushida, S. (1990), *Sol. State Comm.* 76, 371.
Sigmund, E., Mueller, K., Eds. (1994), *Phase Separation in Cuprate Superconductors*, Springer, Berlin.

Simon, R., Chaikin, P. (1981), *Phys. Rev. B* 23, 4463.
Simon, R., Dalrymple, B., Van Vechten, D., Fuller, W., Wolf, S. (1987), *Phys. Rev. B* 36, 1962.
Singh, D., Du, M. (2008), *Phys. Rev. Lett.* 100, 237003.
Singleton, J., Mielne, C. (2002), *Contemp. Phys.* 43, 63.
Skalski, S., Betbeder-Matibet, O., Weiss, P. (1964), *Phys. Rev.* 136, A1500.
Slater, J. (1951), *Phys. Rev.* 82, 538.
Slonczewski, J., Weiss, P. (1958), *Phys. Rev.* 109, 272.
Sluchanko, N., Glushkov, V., Demishev, S., Samarin, N., Savchenko, A., Singleton, J., Hayes, W., Brazhkin, V., Gippius, A., Shulgin, A. (1995), *Phys. Rev. B* 51, 1112.
Snider, E., Dasenbrock-Gammon, N., McBride, R., Debessal, M., Vindana, H., Cantasamy, K., Lawler, K.Salamat, A., Dias, R. (2020), *Nature* 586, 373.
Solymar, L. (1972), *Superconducting Tunneling and Applications*, Clapman, London.
Solyom, J. (1979), *Adv. Phys.* 28, 201.
Somayazulu, M., Muhtar, A., Mishra, A., Geballe, Z., Baldini, M., Meng, Y., Struzhkin, V., Hemley, R. (2019), *Phys. Rev. Lett.* 122, 027001.
Sommerfeld, A., Bethe, H. (1933), *Electronen Theorie der Metalle*, Springer-Verlag, Berlin.
Song, H., Duan, D., Cui, T., Kresin, V. Z. (2020a), *Phys. Rev. B* 102, 014510.
Song, H., Zhang, Z., Cui, T., Pickard, C., Kresin, V. Z., Duan, D. (2020b), arXiv:2010.12225.
Sonier, I., Ilton, M., Pacradouni, V., Kaiser, C., Sabok-Sayr, S., Ando, Y., Komiya, S., Hardy, W., Bonn, D., Liang, R., Atkinson, W. (2008), *Phys. Rev. Lett.* 101, 117001.
Sonier, J., Brewer, J., Kiefl, R., Miller, R., Morris, G., Stronach, G., Gardner, J., Dunsiger, S., Bonn, D., Hardy, W., Liang, R., Heffner, R. (2001), *Science* 292, 1692.
Sordi, G., Haule, K., Tremblay, A. (2011), *Phys. Rev. B* 84, 075161.
Sordi, G., Semon, P., Haule, K., Tremblay, A. (2012), *Phys. Rev. Lett.* 108, 216401.
Sorella, S., Martins, G., Becca, F, Gazza, C., Capriotti, L., Parola, A., Dagotto, E. (2002), *Phys. Rev. Lett.* 88, 117002.
Souma, S., Mashida, Y., Sato, T., Takahashi, T., Matsui, M., Wang, S., Ding, H., Kaminski, A., Campusano, J., Sasaki, S., Kadowaki, K. (2003), *Nature* 423, 65.
Spalek, J. (2015), *Phil. Mag.* 95, 661.
Sparn, G., Thompson, J., Whetten, R., Huang, S., Kaner, R., Diederich, F., Grüner, G., Holczer, K. (1992), *Phys. Rev. Lett.* 68, 1228.
Spille, H., Rauchschwalbe, U., Steglich, F. (1983), *Helv. Phys. Acta* 56, 165.
Stanescu, T., Kotliar, G. (2006), *Phys. Rev. B* 74, 125110.
Stechel, E., Jennison, D. (1988), *Phys. Rev. B* 38, 4632.
Steglich, F., Aarts, J., Bredl, C., Lieke, W., Meschede, D., Franz, W., Schäfer, H. (1979), *Phys. Rev. Lett.* 43, 1892.
Steglich, F., Armbrüster, H. (1974), *Sol. State Comm.* 14, 903.
Steglich, F., Wirth, S. (2016), *Rep. Prog. Phys.* 79, 084502.
Steward, G. (2015), *Physica C* 514, 28.
Stoner, E. (1938), *Proc. Roy. Soc. A* 165, 372.
Strongin, M. R., Kammerer, O., Paskin, A. (1965), *Phys. Rev. Lett.* 14, 949.
Stuffer, D., Aharony, A. (1992), *Introduction to Percolation Theory*, Taylor & Francis, London
Subedi, A., Zhang, L., Singh, D., Du, M. (2008), *Phys. Rev. B* 78, 134514.
Suh, B., Borsa, F., Torgenson, D., Cho, B., Canfield, P., Johnston, D., Rhee, J., Harmon, B. (1996), *Phys. Rev. B* 54, 15341.
Suhl, H., Mattias, B., Walker, L. (1958), *Phys. Rev. Lett*, 3, 552.
Sun, J., Ruzsinszky, A., Perdew, J. (2015), *Phys. Rev. Lett.* 115, 036402.

Sun, X., Ono, S.Abe, Y., Komiya, S., Segawa, K., Ando, Y. (2006), *Phys. Rev. Lett.* 96, 017008.

Sun, Y., Kittaka, S., Sakakibara, T., Machida, K., Wang, J., Wen, J., Xing, X., Shi, Z., Tamegai, T. (2019), *Phys. Rev. Lett.* 123, 027002.

Sun, Y., Lv, J., Xie, Y., Liu, H., Ma, Y. (2019), *Phys. Rev. Lett.* 123, 097001.

Tabis, W., Yu, B., Bialo, I., Bluschke, M., Kolodziej, T., Kozlowski, A., Blackburn, E., Sen, K., Forgan, E., Zimmermann, M., Tang, Y., Weschke, E., Vignolle, B., Hepting, M., Gretarsson, H., Sutarto, R., He, F., Le Tacon, M., Barišić, N., Yu, G., Greven, M. (2017), *Phys. Rev. B* 96, 134510.

Tagirov, L. (1998), *Physica C* 307, 145.

Takada, K., Sakurai, H., Takayama-Muromachi, E., Izumi, F., Dilanian, R., Sasaki, T. (2003), *Nature* 422, 53.

Takada, Y. (1977), *J. Phys. Soc. Jpn* 45, 786.

Takagi, H., Uchida, S., Kitazawa, K., Tanaka, S. (1987), *Jpn J. Appl. Phys.* 26, L123.

Takano, Y. (2009), *J. Phys.: Cond. Mat.* 25, 253201.

Takigawa, M., Hammel, P., Heffner, R., Fisk, Z., Smith, J., Schwarz, R. (1989), *Phys. Rev. B* 39, 300.

Tanigaki, K., Ebbesen, T., Saito, S., Mizuki, J., Tsai, J., Kubo, Y., Kuroshima, S. (1991), *Nature* 352, 222.

Tao, H., Lu, F., Wolf, E. (1997), *Physica C* 282, 1507.

Tarascon, J., Wang, E., Kivelson, S., Bagley, B., Hull, G., Ramesh, R. (1990), *Phys. Rev. B* 42, 218.

Tavger, B. (1986), *Phys. Lett. A* 116, 123.

Tavger, B., Zaitsev, B. (1956), *JETP* 30, 564.

Tedrow, P., Tkaczyk, J., Kumar, A. (1986), *Phys. Rev. Lett.* 56, 1746.

Tedrow, P., Meservey, R. (1994), *Phys. Rep.* 238, 173.

Temprano, D., Mesot, J., Janssen, S., Conder, K., Furrer, A., Mutka, H., Mueller, K. (2000), *Phys. Rev. Lett.* 84, 1990.

Teo, J., Fu, L., Kane, C. (2008), *Phys. Rev. B* 78, 045426.

Tersoff, J., Hamann, D. (1985), *Phys. Rev. B* 31, 85.

Testardi, L. (1975), *Rev. Mod. Phys.* 47, 637.

Testardi, L., Wernick, J., Royer, W. (1974), *Sol. State Comm.* 15, 1.

Tewari, S., Gumber, P. (1990), *Phys. Rev.* 41, 2619.

Thiel, L., Rohner, D., Ganzhorn, M., Appel, P., Neu, E., Muller, B., Kleiner, R., Koelle, D., Maletinsky, P. (2016), *Nat. Nanotech.* 11, 677.

Thiel, S., Hammerl, G., Schmehl, A., Schneider, C., Mannhart, J., 2006, *Science* 313, 1942.

Thompson, R. (1970), *Phys. Rev. B* 1, 327.

Thurston, T., Birgenau, R., Kastner, M., Preyer, N., Shirane, G., Fujii, Y., Yamada, K., Endoh, Y., Kakurai, K., Matsuda, M., Hidaka, Y., Murakami, T. (1989), *Phys. Rev. B* 40, 4585.

Tilley, D., Tilley, J. (1986), *Superfluidity and Superconductivity*, Adam Hiller, Bristol.

Ting, S., Pernambuco-Wise, P., Crow, J., Manousakis, E., Weaver, J. (1992), *Phys. Rev. B* 46, 11772.

Tinkham, M. (1996), *Introduction to Superconductivity*, McGraw-Hill, NY.

Tinkham, M., Hergenrother, J., Lu, J. (1995), *Phys. Rev. B* 51, 12649.

Tokura, Y. (2003), *Phys. Tod.*, 56, 50.

Tokura, Y., Tomioka, Y. (1999), *Jour. Magn. Magn. Mat.* 200, 1.

Tomic, S., Jérome, D., Mailly, D., Ribault, M., Bechgaard, K. (1983), *J. Phys.* 44, C3: 1075.

Tomonaga, S. (1950), *Prog. Theor. Phys.* 5, 544.

Tou, H., Kitaoka, Y., Ishida, K., Asayama, K., Kimura, N., Ōnuki, Y., Yamamoto, E., Haga, Y., Maezawa, K. (1998), *Phys. Rev. Lett.* 80, 3129.

Tou, H., Maniwa, Y., Koiwasaki, T., Yamanaka, S. (2001a), *Phys. Rev. B* 63, 020508.
Tou, H., Maniwa, Y., Koiwasaki, T., Yamanaka, S. (2001b), *Phys. Rev. Lett.* 86, 5775.
Tou, H., Maniwa, Y., Yamanaka, S. (2003), *Phys. Rev. B* 67, 100509.
Toyli, D., Weis, C., Fuchs, G., Schenkel, T., Awschalon, D. (2010), *Nano Lett.* 10, 3168.
Toyozawa, Y., Shinozuka, Y. (1983), *J. Phys. Soc. Jpn* 52, 1446.
Tranquada, J. (2014), *Physica B* 460, 1016.
Tranquada, J., Axe, J., Ichikawa, N., Moodenbaugh, A., Nakamura, Y., Uchida, S. (1997), *Phys. Rev. Lett.* 78, 338.
Tranquada, J., Sternlieb, B., Axe, J., Nakamura, Y., Uchida, S. (1995), *Nature* 375, 561.
Troyan, I., Gavriliuk, A., Ruffer, R., Chumakov, A., Mironovich, A., Lyubutin, I., Perekalin, D., Drozdov, A., Eremets, M. (2016), *Science* 351, 1303.
Tsuda, N., Shimada, D., Miyakawa, N. (2007), in *New Research on Superconductivity*, Martins, B., Ed., Nova Science, NY.
Tsuda, S., Yokoya, T., Kiss, T., Takano, Y., Togano, K., Kiko, H., Ihara, H., Shin, S. (2001), *Phys. Rev. Lett.* 87, 177006.
Tsuda, S., Yokoya, T., Takano, Y., Kito, H., Matsushita, A., Yin, F., Itoh, J., Harima, H., Shin, S. (2003), *Phys. Rev. Lett.* 91, 127001.
Tsuei, C., Kirtley, J. (2000), *Rev. Mod. Phys.* 72, 969.
Tsuei, C., Kirtley, J., Chi, C., Yu-Jahnes, L., Gupta, A., Shaw, T., Sun, J., Ketchen, M. (1994), *Phys. Rev. Lett.* 73, 593.
Tyablikov, S., Tolmachev, V. (1958), *JETP* 7, 867.
Urushibara, A., Moritomo, Y., Arima, T., Asamitsu, A., Kido, G., Tokura, Y. (1995), *Phys. Rev. B* 51, 14103.
Usadel, K. (1970), *Phys. Rev. Lett.* 25, 507.
Vaks, V., Larkin, A. (1966), *JETP* 22, 678.
Val'kov, V., Dzebisashvili, D. (2005), *JETP* 100, 608.
Val'kov, V., Dzebisashvili, D., Korovuskin, M., Barabanov, A. (2017), *JETP* 125, 810.
Val'kov, V., Mitskan, V., Shustin, M. (2019a), *JETP* 129, 426.
Val'kov, V., Mitskan, V., Zlotnikov, A., Shustin, M., Aksenov, S. (2019b), *JETP Lett.* 110, 140.
Val'kov, V., Ovchinnikov, S. (2004), *Hubbard Operators in the Theory of Strongly Correlated Electrons*, Imperial College Press, London.
Val'kov, V., Val'kova, T., Dzebisashvili, D., Ovchinnikov, S. (2002), *JETP Lett.* 75, 378.
Val'kov, V., Zlotnikov, A. (2012), *JETP Lett.* 95, 350.
Val'kov, V., Zlotnikov, A. (2013), *J. Supercond. Nov. Magn.* 26, 2885.
Val'kov, V., Zlotnikov, A. (2019), *JETP Lett.* 109, 736.
Val'kov, V., Zlotnikov, A., Shustin, M. (2018), *J. Magn. Magn. Mat.* 459, 112.
Val'kov, V., Kagan, M., Aksenov, S. (2019c), *J. Phys.: Cond. Mat.* 31, 225301.
Van Delft, D., Kes, P. (2010), *Phys. Today*, 38.
Van Duzer, T., Turner, C. (1999), *Principles of Superconducting Devices and Circuits*, Prentice Hall, NY.
Van Harlingen, D. (2012), in *100 Years of Superconductivity*, Rogalla, H., Kes, P., Eds., Taylor & Francis, Boca Raton.
Van Hove, L. (1953), *Phys. Rev.* 89, 1189.
Van Vleck, J. (1932), *The Theory of Electric and Magnetic Susceptibilities*, Oxford University Press, Oxford.
Varelogiannis, G. (2002), *Phys. Rev. Lett.* 88, 117005.

Varlamov, A., Larkin, A. (2005), *Theory of Fluctuations in Superconductors*. Clarendon Press, Oxford.
Varma, C., Littlewood, P., Schmitt-Rink, S., Abrahams, E., Ruckenstein, A. (1989), *Phys. Rev. Lett.* 63, 1996.
Varma, C., Simons, A. (1983), *Phys. Rev. Lett.* 51, 138.
Varma, C., Zaanen, J., Raghavachari (1991), *Science* 254, 989.
Varma, C.Schmitt-Rink, S., Abrahams, E. (1987), *Sol. State Comm.* 62, 681.
Varma, C., Zhu, L. (2006), *Phys. Rev. Lett.* 96, 036405.
Velicky, B., Kirkpatrick, S., Ehrenreich, H. (1968), *Phys. Rev.* 175, 747.
Venkatesan, T., Wu, X., Inam, A. (1988), *Appl. Phys Lett.* 53, 1431.
Vignale, G., Singwi, K. (1989), *Phys. Rev. B* 39, 2956.
Vijay, S., Hsieh, T., Fu, L. (2015), *Phys. Rev. X* 5, 041038.
Vinetskii, V. (1961), *JETP* 13, 1023.
Viret, D., Ranno, L., Coey, J. (1997), *Phys. Rev. B* 55, 8067.
Vishik, I. (2018), *Rep. Prog. Phys.* 81, 062501.
Visscher, P., Falicov, L. (1971), *Phys. Rev. B* 3, 2541.
Vladimir, M., Moskalenko, V. (1990), *Theor. Math. Phys.* 82, 301.
Vladimirov, A., Ihle, D., Plakida, N. (2009), *Phys. Rev. B* 80, 104425.
Vollhardt, D., Byczuk, K., Kollar, M. (2011), in *Theoretical Methods for Strongly Correlated Systems*, Avella, A., Mancini, F., Eds., Springer-Verlag, Berlin.
Volovik, G. (2018), *JETP Lett.* 107, 516.
Volovik, G. (2009), *The Universe in a Helium Droplet*, Oxford University Press, Oxford.
Volovik, G. (1997), *JETP Lett.* 66, 522.
von Delft, D., Ralph, D. (2001), *Phys. Rep.* 345, 61.
von Delft, J., Braun, F. (1999), arXiv:cond-mat/99/1058v1.
von Helmholt, R., Wecker, J., Holzareff, B., Schultz, L., Samuer, K., Fert, A. (2008), *Phys. Usp.* 178, 1336.
von Issendorf, B. (2010), in *Handbook of Nanophysics: Clusters and Fullerenes*, Sadder, K., Ed., Taylor & Francis, Oxford.
von Issendorff, B., Cheshnovsky, O. (2005), *Ann. Rev. Phys. Chem.* 56, 549.
Von Szchepanski, K., Horsch, P., Stephan, W., Ziegler, M. (1990), *Phys. Rev. B* 41, 2017.
Vonsovskii, S., Izyumov, Y. (1963), *Phys. Usp.* 5, 547.
Vucicevic, J., Ayral, T., Parcollet, O. (2017), *Phys. Rev. B* 96, 104504.
Wade, J., Loram, J., Mirza, K., Cooper, J., Tallon, J. (1994), *J. Supercond.* 7, 261.
Waldram, J. (1996), *Superconductivity of Metals and Cuprates*, Inst. of Phys. Pull., Bristol.
Wallace, P. (1947), *Phys. Rev.* 71, 622.
Walmsley, P., Putzke, C., Malone, L., Guillamón, I., Vignolles, D., Proust, C., Badoux, S., Coldea, A. I., Watson, M., Kasahara, S., Mizukami, Y., Shibauchi, T., Matsuda, Y., Carrington, A. (2013), *Phys. Rev. Lett.* 110, 257002.
Walstedt, R., Warren, W. (1990), *Science* 248, 1082.
Wang, H., Tse, J., Tanaka, K., Litaka, T., Ma, Y. (2012), *PNAS* 109, 6463.
Wang, Q., Li, Z., Zhang, W., Zhang, Z., Zhang, J., Li, W., Ding, H., Ov, Y., Deng, P., Chang, K., Wen, J., Song, C., He, K., Jia, J., Ji, S., Wang, Y., Wang, L., Chen, X., Ma, X., Xue, Q. (2012), *Chin. Phys. Lett.* 29, 037402.
Wang, Y., Chubukov, A. (2013), *Phys. Rev. B* 88, 024516.
Wang, Y., Chubukov, A. (2014), *Phys. Rev. B* 90, 035149.
Wang, Y., Li, L., Naughton, M., Gu, G., Uchida, S., Ong, N. (2005), *Phys. Rev. Lett.* 95, 247002.

Wang, Y., Ma, Y. (2014), *J. Chem. Phys.* 140, 040901.
Wang, Y., Plackowski, T., Junod, A. (2001), *Physica C* 355, 179.
Warren, W., Walstedt, R., Brennert, G., Cava, R., Tycko, R., Bell, F., Dabbagh, G. (1989), *Phys. Rev. Lett.* 62, 1193.
Watson, M., Kim, T., Haghighirad, A., Davies, N., McCollam, A., Narayanan, A., Blake, S., Chen, Y., Ghannadzadeh, S., Schofield, A., Hoesch, M., Meingast, C., Wolf, T., Coldea, A. (2015), *Phys. Rev. Lett.* 91, 155106.
Weber, W. (1988), *Z. Phys. B* 70, 323.
Weht, R., Filippelli, A., Pickett, W. (1999), *Europhys. Lett.* 48, 320.
Wen, X., Wilczek, F., Zee, A. (1989), *Phys. Rev. B* 39, 11413.
Westwanski, B., Pawlikovski, A. (1973), *Phys. Lett. A* 43, 201.
Weyeneth, S. (2017), in *High-T_c Cooper Oxide Superconductors and Related Novel Materials*, Bussmann-Holder, A., Keller, H., Bianconi, A., Eds., Springer, Berlin.
Weyeneth, S., Muller, K. (2011), *J. Supercond. Nov. Magn.* 24, 1235.
Whangbo, M., Williams, J., Schultz, A., Emge, T., Beno, M. (1987), *J. Am. Chem. Soc.* 109, 90.
Wheatley, J. (1975), *Rev. Mod. Phys.* 47, 415.
White, B., Thompson, J., Maple, M. (2015), *Physica C* 514, 246.
Wigner, E., Huntington, H. (1935), *J. Chem. Phys.* 3, 764.
Wilson, K. (1971), *Phys. Rev. B* 4, 3174.
Wilson, K. (1983), *Rev. Mod. Phys.* 55. 583.
Wolf, E. (2012), *Principles of Electron Tunneling Spectroscopy*, Oxford University Press, NY.
Wolf, E., Noer, R., Arnold, G. (1980), *J. Low Temp. Phys.* 40, 419.
Wolf, S., Kennedy, J., Nisenoff, M. (1977), *J. Vac. Sci.* 13, 875.
Wolf, S., Kresin, V. Z. (2012), *J. Supercond. Nov. Magn.* 25, 165.
Wolf, S., Lowrey, W. (1977), *Phys. Rev. Lett.* 39, 1038.
Wriggle, G., Astruc Hoffman, M., von Issendorff, B. (2002), *Phys. Rev. A* 65, 063201.
Wu, G., Xie, Y., Chen, H., Zhong, M., Liu, R., Shi, B., Li, Q., Wang, X., Wu, T., Yan, Y., Ying, J., Chen, X. (2009), *J. Phys.: Cond. Mat.* 21, 142203.
Wu, J., Bollinger, A., He, X., Božović, I. (2017), *Nature* 547, 432.
Wu, J., Bollinger, A., He, X., Božović, I. (2018), *J. Supercond. Nov. Magn.* 32, 1623.
Wu, M., Ashburn, J., Torng, P., Hor, P., Meng, R., Gao, L., Huang, Z., Wang, Y., Chu, C. (1987), *Phys. Rev. Lett.* 58, 908.
Xiang, H., Li, Z., Yang, J., Hou, J., Zhu, Q. (2004), *Phys. Rev. B* 70, 212.
Xiang, Y., Wang., F., Wang, D., Wang, Q., Lee, D. (2012), *Phys. Rev. B* 86, 134508.
Xiao, G., Cieplak, M., Xiao, J., Chien, C. (1990), *Phys. Rev. B* 42, 8752(R).
Xie, H., Yao, Y., Feng, X., Duan, D., Song, H., Zhang, Z., Jiang, S., Redfern, S., Kresin, V. Z., Pickard, C., Cui, T. (2020), *Phys. Rev. Lett.* 125, 217001.
Yakovlev, D., Pethick, C. (2004), *Ann. Rev. Astron. Astrophys.* 42, 169.
Yamanaka, S., Hotehama, K., Kawaji (1998), *Nature* 392, 580.
Yan, H., Jung, J., Darhmaoui, H., Ren, Z., Wang, J., Kwok, W. (2000), *Phys. Rev. B* 61, 11711.
Yanson, I. (1974), *JETP* 39, 506.
Yanson, I., Svistunov, V., Dmitrenko (1965), *JETP* 21, 650.
Yao, H., Lee, D., Kivelson, S. (2011), *Phys. Rev. B* 84, 012507.
Yeh, K., Huang, T., Huang, Y., Chen, T., Hsu, F., Wu, P., Lee, Y., Chu, Y., Chen, C., Luo, J. (2008), *Europhys. Lett.* 84, 37002.
Ying, T., Chen, X., Wang, G., Jin, S., Zhou, T., Lai, X., Zhang, H., Wang, W. (2012), *Sci. Rep.* 2, 426.

Yip, S. (1990), *Phys. Rev. B* 41, 2612.
Yoshida, T., Zhou, X., Lu, D., Komiya, S., Ando, Y., Eisaki, H., Kakeshita, T., Uchida, S., Hussain, Z., Shen, Z., Fujimori, A. (2007), *J. Phys.: Cond. Mat.* 19, 125209.
Yu, G., Lee, C., Heeger, A., Herron, N., McCarron, E., Cong, L., Spalding, G., Nordman, C., Goldman, A. (1992), *Phys. Rev. B* 45, 4964.
Zaanen, J, Sawatzky, G. (1990), *J. Sol. State Chem.* 88, 8.
Zaanen, J, Sawatzky, G., Allen, J. (1985), *Phys. Rev. Lett.* 55, 418.
Zaanen, J. (2000), *Nature* 404, 714.
Zaanen, J., Oles, A. (1988), *Phys. Rev. B*37, 9423.
Zaitsev, R. (1975), *JETP* 68, 207.
Zaitsev, R. (1976), *JETP* 43, 574.
Zavaritskii, N., Itskevich, E., Voronovskii, A. (1971), *JETP* 33, 762.
Zech, D., Conder, K., Keller, H., Kaldis, E., Mueller, K. (1996), *Physica B* 219, 136.
Zech, D., Keller, H., Conder, K., Kaldis, E., Liarokapis, E., Poulakis, N., Mueller, K. (1994), *Nature* 371, 681.
Zegrodnik, M., Spalek, J. (2018), *Phys. Rev. B* 98, 155144.
Zelikovic, I., Xu, Z., Wen, J., Gu, G., Markiewicz, R., Hoffman, J. (2012), *Science*, 337, 320.
Zener, C. (1951a), *Phys. Rev.* 81, 440.
Zener, C. (1951b), *Phys. Rev.* 82, 403.
Zhang, F., Gros, C., Rice, T., Shiba, H. (1988), *Supercond. Sci. Technol.* 1, 36.
Zhang, F., Rice, T. (1988), *Phys. Rev. B* 37, 3759.
Zhang, H., Liu, C., Gazibegovic, S., Xu, D., Logan, J., Wang, G., van Loo, N., Bommer, J., de Moor, M., Car, D., Veld, R., van Veldhoven, P., Koelling, S., Verheijen, M., Pendharkar, M., Pennachio, D., Shojaei, B., Lee, J., Palmstrm, C., Bakkers, E., Das Sarma, S., Kouwenhoven, L. (2018), *Nature* 556, 74.
Zhang, X., Oh, Y. Liu, Y., Yan, L., Kim, K., Greene, R., Takeuchi, I. (2009), *Phys. Rev. Lett.* 102, 147002.
Zhao, G., Conder, K., Keller, H., Mueller, K. (1996), *Nature*, 381, 676.
Zhao, G., Hunt, M., Keller, H., Mueller, K. (1997), *Nature*, 385, 236.
Zhao, K., Lu, B., Deng, L., Hujan, S., Xue, Y., Chu, C. (2016), *PNAS* 113, 12968.
Zhou, P., Chen, L., Liu, Y., Sochnikov, I., Bollinger, A., Han, M., Zhu, Y., He, X., Bozovic, I., Natelson, D. (2019), *Nature* 572, 496.
Zhou, P., Liyang, C., Yue, L., Sochnikov, I., Bollinger, A., Han, M., Zhu, Y., He, X., Božović, I., Natelson, D. (2019), *Nature* 572, 496.
Zhuang, Q., Jin. X., Cui, T., Ma, Y., Lu, Q., Li, Y., Zhang, H., Meng, X., Bao, K. (2017), *Inorg. Chem.* 56, 3901.
Ziman, J. (1960), *Electrons and Phonons*, Clarendon Press, Oxford.
Ziman, J. (1964), *Principles of the Theory of Solids*, Cambridge University Press, Cambridge.
Zolotavin, P., Guyot-Slonnest, P. (2010), *ACS Nano* 4, 5599.
Zubarev, N. (1960), *Phys. Usp.* 3, 320.

Index